Philipp Spitta
JOHANN SEBASTIAN BACH

VOLUMES II & III

PHILIPP SPITTA

JOHANN SEBASTIAN BACH

HIS WORK AND INFLUENCE ON THE MUSIC OF GERMANY, 1685-1750.

TRANSLATED FROM THE GERMAN BY

CLARA BELL

AND

J. A. FULLER-MAITLAND.

IN THREE VOLUMES
VOL. II

Lees-McRae College Library
Banner Elk, N. C.

LONDON
NOVELLO & CO., LTD.

NEW YORK
DOVER PUBLICATIONS, INC.

*Copyright 1951 by
Dover Publications, Inc.
180 Varick St., New York 14, N. Y.*

*This edition has been produced
under special arrangement with
Novello & Company, Ltd.,
London.*

Printed and bound in the United States of America

CONTENTS.

BOOK IV.
CÖTHEN, 1717-1723.

I.—ARRIVAL AT CÖTHEN. DEATH OF BACH'S FIRST WIFE AND JOURNEY TO HAMBURG IN 1720. REINKEN. BACH AND HANDEL AS ORGAN-PLAYERS... 1

II.—BACH'S CLAVIER MUSIC. TOCCATAS. HIS REFORM IN FINGERING. ADJUSTMENT OF PITCH. INVENTION OF THE PIANOFORTE. BACH AS A TEACHER. THE CLAVIER-BÜCHLEIN. INVENTIONEN UND SINFONIEN 30

III.—BACH AS A VIOLINIST. THE SUITE AND THE SONATA. WORKS FOR VIOLIN, VIOLONCELLO, FLUTE, &C. 68

IV.—BACH'S SECOND MARRIAGE. CHANGE OF POSTS. THE FRENCH SUITES. THE WOHLTEMPERIRTE CLAVIER ... 146

BOOK V.
LEIPZIG, 1723-1734.

I.—BACH'S APPOINTMENT AND INSTALLATION AS CANTOR AT LEIPZIG 181

II.—THE THOMASSCHULE. DUTIES OF THE CANTOR. STATE OF MUSIC IN LEIPZIG... 189

III.—BACH'S OFFICIAL DUTIES AS CANTOR, HIS DISPUTE WITH THE TOWN COUNCIL, AND ENDEAVOURS TO IMPROVE THE CONDITION OF THE MUSIC. HIS LETTER TO ERDMANN. GESNER'S APPOINTMENT AS RECTOR OF THE SCHOOL ... 213

IV.—THE PLAN AND ARRANGEMENT OF THE CHURCH SERVICES IN LEIPZIG. MUSIC USED IN IT; THE ORGANS AND BACH'S TREATMENT OF ACCOMPANIMENTS. DIFFICULTIES OF PITCH AND TUNE 263

V.—KUHNAU. THE CHURCH CANTATA. TEXTS BY NEUMEISTER AND PICANDER. COMPARISON OF THEIR MERITS. BACH'S CHURCH CANTATAS. THE "MAGNIFICAT"... 332

VI.—BACH'S CANTATAS (CONTINUED) 437

VII.—PASSION MUSIC BEFORE BACH. THE ST. JOHN PASSION. THE ST. MATTHEW PASSION 477

VIII.—BACH'S COMPOSITIONS FOR CHRISTMAS, EASTER, AND ASCENSION 570

IX.—BACH'S MOTETTS 594

X.—"OCCASIONAL" COMPOSITIONS 612

APPENDIX (A, TO VOL. II.) 649

BOOK IV.
CÖTHEN, 1717—1723.

I.

ARRIVAL AT CÖTHEN.—DEATH OF BACH'S FIRST WIFE AND JOURNEY TO HAMBURG IN 1720.—REINKEN.—BACH AND HANDEL AS ORGAN-PLAYERS.

PRINCE LEOPOLD of Anhalt-Cöthen was born November 2, 1694; and, at the time when he invited Bach to his court, was at the end of his twenty-third or the beginning of his twenty-fourth year. He had entered into possession of his little dominions on the last day of the year 1715; and a few weeks later a marriage was solemnised between his sister and Prince Ernst August of Sax-Weimar, in the royal castle of Nienburg, on the Saale. This was the dower-house of his mother, Gisela Agnes, an active, energetic, and prudent woman, who had governed during Leopold's minority, and had given a careful education to the boy whom death had deprived of his father when he was ten years old. Leopold had been for some time at the "Ritter Akademie" at Berlin, to which at that time many young princes were attracted by its celebrity. Then, in October, 1710, he had set out on the usual grand tour—first to Holland and England, and then through Germany to Italy; and in the spring of 1713 he had returned, by Vienna, to Cöthen. His conspicuous musical tastes and talents had been promoted and cultivated, particularly during his stay in Italy. In Venice he was assiduous in visiting the opera-house. In Rome he attracted to himself the German composer Johann David Heinichen, and under his guidance he became familiar with the promised land of music.[1] The

---

[1] Gerber, N. L., II., col. 615, according to Hiller, Wöchentliche Nachrichten, I., p. 213. But Heinichen must have been in Rome first and afterwards in Venice. From a very slight sketch of his travels we gather that Prince Leopold was in Rome from March 2 till June 6, 1712, and then turned towards Vienna by Florence. Heinichen must have accompanied him part of the way, and have remained behind in Venice, where he composed two operas in 1713. The manuscript of the diary of his journey is in the Castle Library at Cöthen.

famous organ in Santa Maria Maggiore, at Trient—where Handel had carried away his audience by his masterly playing a few years previously[2]—may have been played upon before the music-loving Prince on Sundays, during the service. He also showed an intelligent taste for pictures, admired Michael Angelo's "Moses," and had a number of masterpieces copied for him in the galleries of Rome. He had, indeed, a frank and independent nature, alive to every impression; his information was general and sound; and, at a later period, he laid the foundation of the library of the castle at Cöthen.[3]

His open countenance—with a high brow and large clear eyes, and its setting, contrary to the fashion of the time, of long, naturally waving hair—has a most winning expression of youthful freshness, and an unmistakable trace of his artistic bent. Of this Prince's deeds as a ruler there is little to be told; but that little corresponds to the promise of the face. The court was of the Reformed Church, as also was a large part of the population. His predecessor, Emanuel Leberecht, had granted to the Lutherans the free and public exercise of their religious observances, probably under the influence of his wife, who was of that confession. A Lutheran church had been built in 1699; and, in 1711, Gisela Agnes founded a Lutheran home and school for women and girls. One of Leopold's first enactments was not merely to confirm, but to increase, the liberty granted by his father, " because it was the greatest blessing when the subjects in a country were protected in their freedom of conscience." The results were visible in the happy and flourishing condition of the little capital and of the whole province.

The court was held on a small and modest scale; it had never possessed a theatre, and the Reformed services did not encourage music. Bach had nothing to do with the organ service in either of the three churches in the town. Christian Ernst Rolle was Organist in the Lutheran church,

---

[2] Chrysander, Händel, I., p. 229.

[3] Joh. Christoph Krausen, Fortsetzung der Bertramischen Geschichte des Hauses in Fürstenthums Anhalt, Part II., p. 672 (Halle, 1782). Stenzel, Handbuch der Anhaltischen Geschichte, p. 279 (Dessau, 1820).

and Joh. Jakob Müller held that post in the principal Reformed church until 1731.[4] He probably also undertook to serve the castle organ. It was the same here, no doubt, as at Arnstadt. Its small dimensions and compass would scarcely fit it for any use but that of playing chorales; and this amply sufficed for the requirements of the reformed service. The two manuals had together ten stops; the pedal had three.[5] Though, even in cases where Bach gives at full length the titles he bore at that time, he never calls himself "Hoforganist," it need not be inferred that he never played on this little instrument.

The strength of music there, however, lay in chamber music, and in this it is evident that the Prince himself took part. If we may judge from an inventory of the instruments in his private possession, he played not only the violin, but the viol-di-gamba and the clavier[6]; and he was also a very good bass singer. Bach himself said of him later, that he had not merely loved music, but had understood it. Under whom he studied is not known. Bach's predecessor as capellmeister was Augustin Reinhard Stricker, the same who in 1708, as chamber musician at Berlin, had composed the festival music for the marriage of the King with the Princess Sophie Louise of Mecklenburg.[7] At about this time Leopold must have been at the "Ritter Akademie" there, and the hypothesis is probably not unfounded that the connection he then formed with Stricker may have led to the composer's obtaining the

---

[4] Walther, *sub voce* "Rolle"; and the Cathedral Registers at Cöthen.

[5] I conclude from this that the present very dilapidated organ is the one which existed in Bach's time. From an inscription on the bellows, which have been renewed not long since, it would seem that they were constructed in 1733. It does not therefore follow that the organ itself is no older, for the bellows are often the part that first needs mending. It may very likely have been built at the time when the wing of the castle was finished in which the chapel stands. But, even if it were not so, no organ there could have been any larger than the present one, for there is not space for it. It would seem easier to believe that the chapel had no organ until 1733.

[6] This inventory, which his brother and successor, August Ludwig, entered under April 20, 1733, is in the ducal archives of Cöthen. See also Gerber, op. cit.

[7] Walther, Lexicon.

post of capellmeister at Cöthen. Afterwards in Italy, and later still by a renewal of his intercourse with his capellmeister, he sought to cultivate still farther his taste in music, and Stricker was in his service as early as 1714. However, to judge from all that we can learn concerning his labours as a composer, Stricker was more devoted to vocal than to instrumental music, and this no doubt is the reason of his having soon left Cöthen, for that place had but small vocal resources. Mattheson has preserved the memory of two young singers—two *Demoiselles de Monjou*—from Cöthen, who sang at Berlin in July, 1722, before the Queen of Prussia, and then retired to their native town. "The younger," we are told, "had a fine clear voice and great perfection in music." It is said they both went to Hamburg, and were there engaged in the opera.[8] There must also have been among the cantors and teachers in the town some good bass to be found, and probably also a tenor, but of any regular vocal band or trained chorus like that at Weimar we find not the smallest trace. If there had been, Bach would not have failed to avail himself of it in composing his birthday serenade for the Prince.

Among the members of the Prince's band we find the names of Johann Ludwig Rese, Martin Friedrich Marcus, Johann Friedrich Torlée, Bernhard Linike, the "Premier Kammer musicus," Josephus Spiess, and the "Viol-di-gambist," Christ. Ferdinand Abel. This cannot have constituted the whole of the band; still those here named were no doubt the most important members; at the same time the only one, even of these, who became more widely known, was Abel. He, like Sebastian Bach's brother Joh. Jakob, had in his youth followed Charles XII. into the field; he was already employed in Cöthen when Bach was invited thither and was still living there in 1737. Of his two gifted sons, Leopold August and Karl Friedrich, both born at Cöthen, the second, as is well known, attained European celebrity.[9]

---

[8] Mattheson, Crit. Mus., Vol. I., Part III., p. 85.

[9] From a document in the archives of Cöthen, preserved at Zerbst, entitled "Protocoll über die Fürstl Capell- und Trompeter-Gagen von 1717-18." Also Gerber, Lexicon, I., cols. 3 and 4.

A pupil of Bach's at this time was Johann Schneider, born near Coburg; he played the organ, clavier, and violin, and entered the band at Weimar as violinist in 1726; in 1730 he went to Leipzig as Organist to the Church of St. Nikolaus.[10] He must no doubt have been employed in the band. Bach himself, as "Capellmeister and director of the Prince's chamber music," as he describes himself with his own hand, received a salary of 400 thalers a year: a good round sum at that period. And the terms of his appointment prove the high estimation in which the Prince held him, for it was dated, and the salary paid, from August 1, 1717, though Bach cannot have entered his service before the end of November. This, with a few other meagre notices, is all that is known to us concerning his official position in Cöthen. Time has effaced or overgrown almost every trace of his labours, as the grass has overgrown the castle-yard which the master must so often have crossed; and his name has died out among the people of the place almost as completely as the sounds with which he once roused the echoes of the now empty and deserted halls.

It must not be supposed, indeed, that even at that time his efforts resulted in much outward display. They were quite private and unpretending, as were his surroundings, and barely extended beyond the limits of the castle concert-room. It was only by his journeys that Bach kept up any connection with the outer and wider world; in his place of residence he had nothing to do with public life. Nevertheless, it was here that he passed the happiest years of his life; here, for a time, he felt so far content that he was resigned to end his days in the peaceful little town. This seems quite incomprehensible so long as we conceive of Bach's artistic side as directed exclusively to sacred music; then, indeed, his residence in Cöthen, where he was debarred from any occupation in the church, must appear to be lost time, and his own satisfaction in it as mere self-deception.

---

[10] Walther's article in the Lexicon shows again how little he interested himself in the biography of his great contemporary Bach. He does not even know that Bach was still in Cöthen in 1720, or does not think it worth mentioning.

But it all becomes natural and intelligible when we do not lose sight of the fact that instrumental music—that is to say, music for music's sake—was the aim and essence of his being: a fact I have endeavoured to insist on from the first. It must have been with a feeling of rapture that for once he found himself thrown back exclusively into this his native element, to drink from it fresh strength for new struggles towards the high ideal that remained hidden from his fellow-men. An essential feature of German art becomes more conspicuous in Bach at this period of his life than at any other: that meditative spirit which is never happy till it dwells within narrow bounds—the joy of occupation and the pleasure of a quiet and homelike circle of a few appreciative friends, whose sympathetic glance responds to the deepest feelings of the heart. From this German characteristic the quartet took its rise; and its very embodiment was the delicious chamber music of Sebastian Bach, which took form chiefly in Cöthen, and in the first rank of it was the Wohltemperirte Clavier.[11] The musical performances that now took place in the castle, when and how often we know not, were of an intimate and thoughtful character, and always undertaken with a genuine zeal for art; the gifted young Prince threw himself into it, heart and soul; all the more so since he was as yet unmarried. He soon became aware of the treasure he had found in Bach, and showed it in the frankest manner. He could not bear to part with him—he took him on his travels, and loved him as a friend; and after his early death Bach always cherished his memory.

An act of homage, dating probably from the first year of his residence in Cöthen, was a serenade for the Prince's birthday. Having to rely on the modest musical resources of the place, he employed in it only one soprano and one bass, with an accompaniment of two flutes and one bassoon, besides the quartet of strings and harpsichord.[12] The writer

---

[11] The Well-tempered Clavier—*i.e.*, preludes and fugues to be played on a clavier tuned according to the system of "equal" temperament, by which system all keys are equally in tune, whereas in the unequal temperament, formerly in use in tuning claviers, many of the keys could not be employed. The work is better known in England by the title Forty-eight Preludes and Fugues.

[12] The autograph is in the Royal Library at Berlin.

of the congratulatory words is not named; if this was out of modesty he had ample reason. At a later period, Bach himself is known to have written certain texts for music, and, knowing this, we cannot altogether avoid suspecting that he may have written these words. Of course some other dabbler in verse is quite as likely to have been the criminal; for they are wretched, be the writer who he may. But the music covers every deficiency; in it we find a perfect reflection of Bach's spirit at this period. It fits the tone of festive feeling in a merely general manner, and within the limits of this idea disports itself freely, developing all that charm of novel invention and elaborate artistic structure of which Bach availed himself with such fascinating grace in his chamber music. There are seven numbers in all: the soprano begins with a recitative and aria in D major; the bass follows with an aria in B minor, and all the music for this voice is pitched very high, showing that it was written for a particular singer. He then goes on to the graceful and dignified minuet in G major, the soprano continues it in D major, and they presently combine in A major, the bass singing the melody, as leading in the dance. Then comes a duet in recitative, again an air for soprano and one for bass, in D major and A major respectively, and finally the closing piece, in two parts and in the leading key, inscribed *Chorus*, by which it is intended to distinguish the crowning finale, for the way in which the parts are treated prohibits all notion of a multiplicity of performers. A happy and self-contented spirit smiles from it throughout. In later years the composer thought it a pity to leave this noble music wedded to its text, and made use of it for a Whitsuntide cantata, as he also did of the music written for the birthday of the Duke of Weissenfels.[13]

On May 9, 1718, the Prince set out to take the baths at Carlsbad, which was at that time a favourite resort of all the high personages of Germany, and took with him Bach and six members of his band. Again, in 1720, Bach had to

---

[13] "Erhöhtes Fleisch und Blut": this autograph also is in the Royal Library at Berlin.

accompany the Prince to Carlsbad;[14] an old tradition still survives of the way in which Bach was wont to occupy his more or less involuntary leisure on these journeys—to this I will return presently. He received another mark of favour in the autumn of the same year, when Maria Barbara gave birth to their seventh child—a boy—on November 15, and on the 17th, the Prince stood godfather to the infant, with his younger brother August Ludwig, his sister Elenore Wilhelmine—who had married into the house of Weimar—with Privy Councillor Von Zanthier, and the wife of Von Nostiz, Steward of the Household.[15] From this it is very evident in what high favour Bach must have stood at court. The child held at the font with so many honours did not survive his first year; he was buried September 28, 1719. A pair of twins had already died in February and March, 1713, soon after their birth; but four children lived to grow up as witnesses of a calm and happy family life. The firstborn was a daughter, Katharina Dorothea, born December 27, 1708; she remained unmarried. On November 22, 1710, followed Wilhelm Friedemann, his father's remarkable and gifted favourite; then Karl Philipp Emanuel (March 8, 1714), who was the most distinguished of the family, though he was not, perhaps, the most talented. Finally, Johann Gottfried Bernhard, born May 11, 1715.[16] We shall have occasion to speak again of all these sons.

---

[14] The dates of these journeys are derived from the orders for special prayers on both occasions issued by the Chancellor of the Duchy, in the archives at Cöthen. It seems certain that during the time from 1718 to 1733 the Prince was at Carlsbad only on these two occasions, since this agrees with an old chronicle of Carlsbad, as I am obligingly informed by Dr. Hlawacek, of that town. The musicians were paid in advance, on May 6, their salaries for the month of June.

[15] Parish Register of the cathedral church of St. James: "1718, the 17th of November, the Prince's Capellmeister, Herr Johann Sebastian Bach, and his wedded wife, Maria Barbara, had a son baptised in the castle chapel, born on the 15th ult., named Leopold Augustus." The names of the sponsors follow.

[16] The Parish Registers give the names of all their sponsors, many of whom were distinguished by birth or office. Philipp Emanuel, in the Genealogy, gives his birthday as March 14; but I have adhered to the date in the Register, though I must confess it is hardly likely that he should be mistaken as to his own birthday.

As has been said, Bach did not give up his own journeys in pursuit of art, even in Cöthen; indeed, his personal need for them was perhaps greater there than in Weimar. Only a few weeks after quitting that town he accepted an invitation to the university town of Leipzig, in order to test the large new organ completed in the Church of St. Paul there, on November 4, 1716. The examination took place December 16, 1717, and was highly favourable to the builder, Johann Scheibe; Bach was greatly satisfied, not only with the quality and construction of the separate portions, but also with the general arrangement, which he declared to be among the completest in Germany. He conducted the examination by himself; only two competent witnesses accompanied him.[17]

In the autumn of 1719 he made another journey, which took him to Halle; this town, no doubt, was not the only goal of the excursion, but we hear of his being there from a circumstance connected with him. Handel had arrived in Germany in the spring, from England, to find singers for the newly founded operatic academy in London; on his return journey he remained for a short time with his family at Halle, and Bach sought him out, but was so unlucky as to find that Handel had that very day set out for England. Another attempt made by Bach, ten years later, to make a personal acquaintance with the only one of his contemporaries who was in any way his equal was just as unsuccessful. Inferences, unfavourable to Handel, have been drawn from these incidents, but there is no sufficient reason for supposing that he would have repelled Bach's courteous advances. We nowhere find any indication that he intentionally took himself out of Bach's reach by leaving Halle on the day of Bach's arrival there; while, on the other hand, it is difficult to overlook the fact that Bach, in this first attempt at a meeting, merely availed himself of an opportunity. Otherwise, as Handel had been in Germany since the previous

---

[17] This incident is recorded by Christoph Ernst Siculs in his Anderer Beylage zu dem Leipziger Jahr-Buche (Leipzig, 1718), p. 198; and it is to A. Dörffels that the praise is due of having first brought it to light. See Musikal. Wochenblatt (Leipzig: E. W. Fritzsch); Annual Series, I., p. 335.

March, he might have arranged a meeting somewhere or other.[18]

On the second occasion, in June, 1729, Bach, who was prevented by illness from travelling himself, sent his eldest son from Leipzig to Halle, with an invitation to Handel, who was staying there on his way from Italy. Handel regretted his inability to accept it, and it seems most probable that the time he had left was in fact too short. It may, however, be confidently denied that it is in any way very regrettable in the interests of art that these two men should thus have failed to meet; it would have been interesting, no doubt; and a desire to hear them compete is said to have been very prevalent among the lovers of music in Leipzig.[19] But the whole object of their meeting would have merged in this, and would have ended, very certainly, without any decision being arrived at on the vexed question as to which of the two should bear off the palm, seeing how totally different their natures were. Any intercourse leading to reciprocal incitement, such as can only be developed from long acquaintance and contact, was out of the question, from the dissimilarity of their outward circumstances. On the other hand, the judgment goes against Handel, without any bias being given by our appreciation of Bach's artistic greatness; for in 1719, when Handel spent eight months in Germany, he certainly might have found time for making a visit which might have originated with him more properly than with Bach, who was occupied with his official duties. Added to this, he resided in Dresden and Halle, places where Bach's importance as an artist was fresh in the minds of living witnesses, and he must there have heard the most splendid reports of the great composer who was working in his immediate neighbourhood. No facts have come to light that prove him to have taken any interest in Bach's works; Bach, on the contrary, not only purposed more than once to make Handel's personal acquaintance, but bore emphatic witness to the value he

---

[18] I entirely agree in Chrysander's views as to this incident. Händel, II., p. 18, note.
[19] Forkel, p. 47.

attributed to his works. Handel's music to Brock's text for a "Passion Music" is still extant in a manuscript of sixty leaves, of which the twenty-three first (exclusive of the last two staves) were copied by Bach's own hand, and the remainder by his wife. The parts of a very meritorious Concerto Grosso in seven movements, by Handel, also exist in Bach's handwriting.[20] The same is the case with a solo cantata by Handel, of which Bach even seems to have possessed the original autograph, for this and Bach's parts are now in the hands of the same owners.[21] It is with particular satisfaction that I am able to point out these indications of a magnanimous artist, free from all envy or prejudice. A few words will yet remain to be said as to their reciprocal relations as organists.

On May 27 of the following year Prince Leopold again went to Carlsbad; he must have returned in July. When Bach entered his home, full of the happy prospect of seeing his family, he was met with the overwhelming news that on the 7th of the month his wife had been buried. He had left her in good health and spirits; now a sudden death had snatched her away in the bloom of life, not yet thirty-six years old, without any news of it having reached her distant husband, who, indeed, had probably begun his return journey. When his son, Philipp Emanuel, thirty-three years after, compiled the notice of his father for the Necrology, though he treated many family events with the brevity of a chronicler, his mother's death and the circumstances connected with it dwelt so vividly in his memory that he

---

[20] Both these MSS. are in the Royal Library at Berlin. The name of the author is wanting to the last, of which Dr. Rust has arranged a score. Dr. Chrysander informs me that the work is unquestionably Handel's, since the motives of the concerto reappear in later works by him. It has indeed struck me that certain passages of the third number, a fugue, have a remarkable resemblance to the double canon treatment of the final chorus of the "Messiah." In the fifth movement, on the other hand, there are passages which recur almost precisely in the B flat minor Prelude in Part I. of the Wohltemperirte Clavier, bars 20-22.

[21] Messrs. Breitkopf and Härtel, of Leipzig. The cantata is entitled "Armida Abbandonata"; the parts written out by Bach at his Leipzig period are those for the first and second violins and continuo. Chrysander pronounces the autograph to be Handel's, and he is an unimpeachable authority.

reported it in detail. But we do not need his evidence to believe in Sebastian's grief; it is easy to guess the feelings that must have tortured his strong, deep nature as he stood by the grave of the wife who had been his companion through the years of his youthful endeavour, and of his first success, only to be snatched from him when fortune was at its height.

We know too little of Maria Barbara Bach to attempt to sketch her character. But, remembering the intelligent nature of her father, and the happy, naïve temper of her second son—who appears to have greatly resembled his mother, while the father fancied he saw himself reproduced in the eldest—we have grounds for picturing to ourselves a calm and kindly nature, with enough musical gift to sympathise keenly with her husband's labours; a wife who enabled him to enjoy in his home that which was one of his deepest needs, the family life of an honourable and worthy citizen.

His terrible loss did not crush Bach's energy; he bore it manfully. A journey to Hamburg, which he had planned for the autumn, was not given up; still there is reason to suppose that he put it off for several weeks. The cantata, "Wer sich selbst erhöhet der soll erniedrigt werden,"[22] stands as evidence of this. The text is taken from a cycle of poems which was printed in 1720, by order of the government secretary, Johann Friedrich Helbig, at Eisenach, for the use of the band there.[23] In Cöthen itself, as no church music was performed there, no such texts were procurable. When Bach desired to compose a cantata he had to look elsewhere for the poetic materials; and, again, the impulse to such a work could only arise from journeys which took him to places

---

[22] B.-G., X., No. 47.
[23] "Auffmunterung | Zur | Andacht, | oder: | Musicalische | Texte, | über | Die gewöhnlichen Sonn- und | Fest- Tags Evangelien durchs | gantze Jahr, | Gott zu Ehren | auffgeführet | Von | Der Hoch-Fürstl. Capelle | zu Eisenach. | [Incitements to Devotion, or Musical Texts on the Gospels in use for Sundays and Holydays thoughout the whole year. Performed in God's honour by the Prince's band at Eisenach.] Daselbst gedruckt und zu finden bey Johann | Adolph *Boëtio*, 1720. | " A copy is in the library of the Count of Stolberg, at Wernigerode. There is a notice of Helbig in Mattheson, Ehrenpforte, under "Melchior Hofmann," p. 118.

where church music was cherished, and into the society of famous church composers. Thus he selected the text for the Seventeenth Sunday after Trinity out of this very indifferent poetry—which, however, came home to him as being that of a fellow-countryman—intending to visit Hamburg and to have his cantata performed there. It also seems highly probable that he set it to music during his excursion to Carlsbad; crushed, however, immediately after by the blow he had suffered, he was unable to carry out this project, and did not reach Hamburg till November.[24] Whether the cantata was then and there performed it is impossible to know; perhaps it was once, irrespective of Divine service. From beginning to end it is the expression of the most complete structural power, and it surpasses his earlier works more particularly by the scope and extent of the grand introductory chorus on the final words of the gospel for the day: "He that exalteth himself shall be abased, and he that humbleth himself shall be exalted." No chorale is introduced, but the whole, in agreement with the text, is made into a double fugue; still its second theme undergoes no independent development.

The great stride forward marked by this piece, when compared with Bach's earlier works for instruments and voices, is conspicuous, not merely in the bold, broad grasp of the parts and their stately, free movement, even where they are most intricate—not only in the grandeur of the proportions, but above all in the fact that the master was no longer satisfied with allowing the instruments to take part in the fugal treatment or to work out a particular motive, but, on the contrary, gave them a theme of their own, thus building up his palace of sound out of the material of three distinct ideas. Still, as the instrumental theme appears in homophonic harmony, the structure necessarily became something else than that of a triple fugue. When we hear the beginning of the movement, in G minor, common time, *allegro*, it is difficult to believe that a choral work can grow out of it; it is like the beginning of an Italian concerto, with its

---

[24] See Appendix A, No. 1.

stringed instruments, oboes and organ. First comes a *tutti* theme, to which is added the contrast of busy passages; the dominant is led up to in the regular way, and the same development is worked out upon it; then, at bar 45, we return to the leading key. But there the tenor takes us by surprise with the entrance of the eight-bar theme, which rises and falls through an octave; after which the counter-subject, nine bars long, comes in, descends an octave and a half, and then shoots up again with a swift, strong flight. Meanwhile the instruments continue to work out their *tutti* theme, *piano*, and in the manner of an episode, but at last the main current sweeps them up also, and bears them on. After a cadence on B flat there comes a shorter episodic interlude, derived from the alternation of the chorus and the instruments; then, again, a grand fugal movement, as at first; another interlude; again a fugal passage; and, to form a *coda*, as it were, the whole chorus takes up the instrumental subject of the beginning; the whole forty-five bars are gone through with the entire body of sound, so that the picture is grandly rounded off, and it closes in dignified magnificence. It is a composition born of the most supreme command over all the forms of music, great and small, and which at the same time affords the most perfect solution conceivable of the problem as to the amalgamation of instrumental and vocal music. After this introductory movement no increase of mere effect could lie within Bach's view; he was content in this, as in all similar cases, to let the cantata flow directly into the symbolical and significant form of the simple chorale. Intermediate between the two are two arias, connected by a recitative. The first, of which the moralising words were most intractable to poetic treatment, is a most ingenious trio for soprano, continuo and organ obbligato or solo violin, and no less a masterpiece in its way. The second is still finer in polyphonic richness, and at the same time glows with noble poetic feeling.

Telemann, who had always been capellmeister at Eisenach, and still was so, also composed music to this text.[25] The

---

[25] An ancient MS. in the Gotthold Library at Königsberg (Prussia), No. 250,862 of the catalogue.

Bible words themselves made the commencement with a double fugue obvious; it is a singular coincidence that he should have chosen the same key as Bach. In everything else the gulf between them, which was visible in their earlier works, yawns more widely than ever. Telemann works out his fugue of thirty-eight bars, common time (Bach's has two hundred and twenty-eight bars), shortly and dryly, and without warming to it at all; the instruments only strengthen the voice parts. Of the rest of the text he only composed the second aria and set the chorale; the air is not even cast in the Italian form, and has the simplest possible accompaniment.

Bach subsequently troubled himself no farther about Helbig's texts, which he had used merely for want of better, with one exception, and that again must have been because he had no choice; for it is evident that in hurriedly arranging the cantata for the third Sunday in Advent of that same year, " Das ist je gewisslich wahr "—" This is certainly the truth "—he worked up earlier compositions in the first chorus (G major, common time); it is impossible to doubt that it is founded on what was originally a duet, particularly if we compare it with the almost identical style of the opening chorus in the second arrangement of the Whitsuntide cantata, " Wer mich liebet "—" Whoso loves me."[26] In bars 52 and 53 of the first aria the bad adaptation to the text betrays him. We cannot expect to find any great merit in such hasty work, though it contains much that is pleasing —nay, beautiful.[27]

Johann Reinken was still living in Hamburg, and, in spite of his ninety-seven years, still officiated as Organist in the Church of St. Katharine with much energy and vigour. He was held in higher respect than any of his fellow officials in the city, not only by reason of his great age, but also on account of his distinction as an artist, of which I have already spoken in detail. To Bach, who as a youth had derived benefit in the most direct way from Reinken's art, it

---

[26] B.-G., XVIII., No. 74.
[27] I only know this cantata from the copy in the Royal Library at Berlin. The final chorale is absent; according to the text it should be " Christe, du Lamm Gottes."

must have been very delightful to present himself to the veteran master as a perfectly accomplished musician. What we learn of Reinken's character is not, on the whole, favourable; he was not only conscious of his own merit, but vain, and jealous of other artists. His predecessor in office had been Heinrich Scheidemann, so fine an organist that any one might have been regarded as somewhat rash who ventured to succeed him; so, at least, a great Dutch musician said, when he learnt that Reinken was to take his place. Reinken, hearing of this, sent him his arrangement of the chorale, "An Wasserflüssen Babylon," with this note: That in this he might see the image of the rash man. The Dutchman, finding that through this certainly very remarkable work he had made the acquaintance of a man who was his superior, came to Hamburg, heard Reinken, and when he met him kissed his hands in admiration.[28]

Mattheson, who in the matter of vanity could do something himself, indignantly observes that Reinken, in the title-page to his *Hortus Musicus*, styles himself, *Organi Hamburgensis ad Divæ Catharinæ Directorum celebratissimum*,[29] and in general he finds nothing good to say of him; but this was Reinken's own fault, for he would never forgive Mattheson for the fact that at one time it had been proposed to give him his (Reinken's) place.[30] Mattheson took his revenge by all sorts of petty hits in the "Beschütztes Orchestre"; and, in the brief obituary notice which he bestowed on him,[31] he even says: "As regards his social character, one and another of the clergy have at times been known to say that he was a constant admirer of the fair sex, and much addicted to the wine-cellar of the Council." Still, he cannot avoid admitting that he kept his organ at all times in beautiful order and in good tune, and that he

---

[28] Walther, Lexicon, *sub voce* "Scheidemann."
[29] Ehrenpforte, p. 293. But Mattheson, consciously or unconsciously, does not here speak the truth. Reinken, on the title-page of the *Hortus Musicus*, writes, "*Organi Hamburgensis ad D. Cathar. celebratissimi Director.*" Hence it is the organ, and not the organist, that he says is famous.
[30] Adlung, Anl. zur Mus. Gel., p. 183. Reinken's deputy and successor was Johann Heinrich Uthmöller (1720-1752).
[31] Crit. Mus., I., p. 255.

could play it in so exceptional and pure a style "that, in his time, he had no equal in the matters he was practised in"; adding, maliciously, that he was always talking of his organ, for it was really very fine in tone.

These observations are a necessary introduction to our comprehension, in all its significance, of Reinken's attitude towards Bach. At an appointed hour the magistrate of the town, and many other important personages, met .in St. Katharine's Church to hear the stranger perform. He played for more than two hours, to the admiration of every one; but his greatest triumph was won by an improvisation on "An Wasserflüssen Babylon," which he carried on for nearly half an hour in the broad, motett-like manner of the northern masters, with which we are already familiar. Reinken came up to him, having listened attentively throughout, and said: "I thought this art was dead; but I perceive that it still lives in you."[82] Irrespective of the high recognition it conveys, there is more in this than mere self-conceit; but, in fact, Bach's mental culture had long since advanced far beyond that stage of organ chorale treatment. Still it is an evidence of his extraordinary mastery over the whole realm of form in music, that he could deliberately revert to it so promptly and so perfectly. He has not left us any written version of what he played, then and there, on the spur of the moment; but it has been already remarked, when speaking of the organ chorale with double pedal, on this same melody, that it may have had some connection with the Hamburg journey. It is quite conceivable that he should have worked it up previously, and have laid it before Reinken as a specimen of imitative art in both subject and treatment; and then, still further to accommodate himself to Reinken's comprehension, have gone on following out his imagination in the way to which "the worthy organists of Hamburg were formerly accustomed in Sunday vespers." Possibly he added something to it, as he did, twenty-seven years later, to a theme set him by Friedrich the Great. At any rate, Reinken was so well satisfied that

---

[82] Mizler, Nekrolog., p. 165.

he invited Bach to visit him, and treated him with distinguished attention. Two years later he departed this life, November 24, 1722, and was buried, by his own wish, in the Church of St. Katharine, at Lübeck, where his kindred spirit, Buxtehude, had already been resting for fifteen years. Reinken had seen, in its full and glorious bloom what Buxtehude had only noted in its bud—the genius of the man who was destined to reach the summit to which they had so successfully opened the way.

Bach was perfectly happy with the organ of St. Katharine's (Hamburg), with its four manuals and pedal. It is interesting to learn that he was greatly in favour of good reeds, and these he found in abundance.[33] The organ also possessed a posaune, and a "principal" (*i.e.*, diapason) of thirty-two feet which spoke clearly and quickly down to C, and Bach subsequently asserted that he had never heard another "principal" of such a size which had this merit. The instrument was not new; it dated at least from the sixteenth century, and had been renovated, in 1670, by the organ-builder Besser, of Brunswick.[34] It preserved a remnant of older taste in a mixture of ten ranks. But this was not the only organ for which Hamburg was famous. The instrument in the Church of St. James was still more powerful as regards the number of stops, and had likewise four manuals and pedals; it was built, between 1688 and 1693, by the Hamburg organ-builder Arp Schnitker,[35] who had given proof of his remarkable skill in other churches in the city. Among this crowd of fine instruments Bach's affectionate longing for his own special province of music revived all the more strongly because, quite unexpectedly, a prospect opened before him of finding a suitable position in Hamburg. Heinrich Friese, the Organist

---

[33] Adlung, Mus. Mech., I.; p. 66, note. The specification is given in Niedt. Mus. Handl., II., p. 176.

[34] H. Schmahl, Nachrichten über die Orgel der St. Catharinen-Kirche in Hamburg. Hamburg: Grüning, 1869, pp. 4-8.

[35] H. Schmahl, Bericht über die Orgel der St. Jacobi-Kirche. Hamburg, 1866. Gerber, N. L., IV., col. 106. This organ had sixty sounding stops, and was restored in 1865-66.

of St. James's, had died, September 12, 1720; so short a time, therefore, before Bach had arrived there that it is probable that he had not heard of it till then. It is, at any rate, certain that the aim and end of his journey was not to seek this appointment, since he had prepared for the expedition in the summer by composing a cantata. However, as it had so happened, he offered himself. Not a little tempting must have been the circumstance that Erdmann Neumeister was chief preacher at this church; he could hardly have pictured to himself a more promising perspective for composing for the organ as well as cantatas, and for the practice of every branch of his art, than that which opened on him here.

Seven other candidates came forward besides himself, mostly unknown names; but a son of the excellent Vincentius Lübeck, and Wiedeburg, Capellmeister to the Count of Gera, were among the number. On November 21 the elders of the church, among whom was Neumeister, resolved on holding the examination on the 28th, and to select as experts, to aid in their decision, Joachim Gerstenbüttel, the cantor of the church, Reinken, and two other organists of the town, named Kniller and Preuss. Bach could not wait so long; his Prince required his return by November 23. Three of the other candidates, including Wiedeburg and Lübeck, had already retired, so that there were but four to submit to the tests, which consisted in the performance of two chorales—"O lux beata Trinitas," and "Helft mir Gott's Güte preisen"—with an extemporised fugue on a given theme. The election was not to be held till December 19. Bach had promised to announce, by letter from Cöthen, whether he chose to accept or decline the appointment; and that it should have come to this, without his being required to pass any examination, proves that he had been regarded as particularly eligible, and distinguished above the other candidates. Unfortunately, nothing is known as to the contents of his answer. So much as this alone is certain—he did not decline the post; the letter was publicly read to the committee, and then, by a majority of votes, they elected—Johann Joachim Heitmann. What he had ever done in his art is less well known than the fact that, on

January 6, 1721, he paid over to the treasury of the Church of St. James "the promised sum of four thousand marks current," in acknowledgment of having been elected. A transaction in view of this, by the church committee, had already been recorded with astonishing frankness, on November 21. They had come to the conclusion: "That, no doubt, many reasons might be found why the sale of the organist's appointment should not be made a custom, because it appertained to the service of God; therefore the choice should be free, and the capability of the candidate be considered rather than the money. But if, after the election, the elected person, of his free will, desired to show his gratitude, this should be favourably looked upon by the church."

Neumeister was extremely indignant at this proceeding, which he had not been able to prevent; he probably would rather have brought Bach into his church than any one. After the election, he would not wait till the nominee came, but left the room in a rage. What further happened, and what the public opinion was of the choice made, we will let Mattheson tell, as he was intimately acquainted with all the details. "I remember," says he in 1728—" and, no doubt, many other people still remember likewise—that some years ago a great musician, who since then has, as he deserves, obtained an important appointment as cantor, appeared as organist in a certain town of some size, boldly performed on the largest and finest instruments, and attracted universal admiration by his skill. At the same time, among other inferior players, there offered himself the son of a well-to-do artisan, who could prelude with thalers better than he could with his fingers, and the office fell to him, as may easily be guessed, although almost every one was angry about it. It was nigh upon Christmastide, and an eloquent preacher, who had not consented to this simony, expounded very beautifully the gospel concerning the angelic music at the birth of Christ, which, very naturally, gave him the opportunity of expressing his opinions as to the recent event as regarded the rejected artist, and of ending his discourse with this noteworthy *epiphonema*. He believed quite certainly that if one of the angels of Bethlehem came from heaven, who played divinely, and desired to be organist to St. James's church, if he had no money he would have nothing to do but to fly away again."[36]

The homage to Bach's merits which Mattheson was obliged to pay—and, in this passage, does pay—was somewhat bitter to him, if we are not deceived by appearances. The only place in which he speaks of him with warm admiration is in the "Beschütztes Orchestre," written four years earlier.[37] If we collect out of all his numerous writings the few paragraphs which refer to Bach, we come to the conclusion that, though he never under-estimates him, he judges him narrowly; his heart is always cold towards him; and we feel as though Bach must have been to him one of those distant and puzzling natures whom we are compelled by our intellects to admire, but who have no hold on our feelings. It is not safe to assume that Bach ignored and so offended a man who was no doubt eminent, though he had more reason for ignoring him than had Handel, who, from 1703 to 1706, had lived in constant intercourse with Mattheson, and who never again sought his acquaintance, though he often was in Germany afterwards, and passed through Hamburg. Mattheson had politely requested Bach to furnish him with the facts of his life for the "Ehrenpforte." He already enjoyed a considerable reputation as a writer on musical subjects; and Bach must have seen that this request was a compliment, although he never acceded to it. But their natures were too dissimilar. It is perfectly evident that Bach did not think much of the Hamburg composer's music: he could copy out the works of Keiser and Telemann, but it is not known that he ever did the same with Mattheson's, though he formed the third of the trio; and as a busy, practical musician, he had neither time nor inclination to form an estimate of his literary work.

---

[36] Mattheson, Der Musicalische Patriot (Hamburg, 1728), p. 316. Herr Schmahl, the present Organist of St. James's, has been so obliging as to write out for me all that refers to the matter in question in the archives of the church.
[37] See ante, Vol. I., p. 393.

Now Mattheson was very vain; he thought that artists ought to flock to him to prove their devotion, to crave his counsel and instruction; and such as did he mentioned at great length in his books: "In August, 1720, an organist came from Bremen and had himself taught composition by Mattheson, for which he paid highly"; "My Lord Carteret arrived in Hamburg (November 8) from his embassy to Sweden, and took such pleasure in our Mattheson's music, that he once sat for two whole hours listening to him, without stirring from the spot; and at last, in presence of the whole assembly, he pronounced this judgment: 'Handel certainly plays the clavier finely and skilfully, but he does not make it sing with so much taste and expression.'" So the man writes of himself.[38] It sounds almost as if he were trying to indemnify himself when we consider that this compliment was paid him by the ambassador at the very time when Bach happened to be in Hamburg. If there had been anything to accrue to his fame in his meeting with Bach, he certainly would not have forgotten it in his autobiography; but he does not say a syllable about it.

The carping criticism which he published a few years after of the cantata "Ich hatte viel Bekümmerniss," and which is fully discussed in its place, can only be accounted for by wounded vanity. And when he praised Bach subsequently it was always with an invidious distinction as to his execution or his clever settings.[39] We learn that he already knew a good deal of Bach's work in the year 1716, both of chamber pieces and church music. On the present occasion he had no doubt heard and seen much that was new—perhaps the criticised cantata and new organ pieces. It is singular, and again only explicable by his feelings towards Bach, that, several years after, he betrays some such knowledge, and, contrary to his usual boasting of information, tries to

---

[38] Ehrenpforte, p. 206.

[39] "The famous Bach, whom I have already mentioned honourably, and here mention again especially for manual skill," &c., Volk. Capellm., p. 412; "The artistic Bach, who was particularly happy in this department" (the construction of fugues on a given theme), &c.; ibid., p. 369.

conceal it. In the second edition of the "General Bass-Schule"[40] he states that, in a test performance on the organ, the candidates had given to them the following theme—

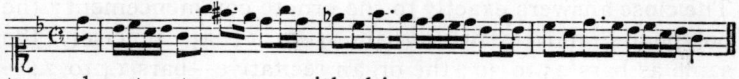

to work out extempore, with this—

to be treated as the counter-subject. Now this is the theme of one of Bach's most magnificent organ fugues, and the counterpoint is also of his invention. Mattheson does not mention this, and only observes, in a note, that he knows very well with whom the idea originated, and who formerly had worked it out with great skill. He chose it because it was wiser in such cases to take something familiar, so that the examinee might get through with a better grace. Nevertheless, we are grateful to him for this notice. It tells us that this fugue of Bach's was known to a wide circle as early as in 1725. It leads us to suppose that the composer may have taken it with him in 1720, and to imagine the sensation made by its appearance in the circle of organists there—for the organist examined by Mattheson must have been a native of Hamburg, or of the neighbourhood. Finally, it tells us that the fugue in its present state must be a later remodelling, for the theme is now wonderfully improved by two small alterations. The prelude belonging to this work is strong evidence of its having been originally composed on purpose for the Hamburg journey. It is conspicuously different from the thematic treatment of the later Weimar period, reverting, indeed, to the imaginative style of the northern masters.[41]

Here, too, Bach seems to have wished to meet the Hamburg organists on their own most peculiar ground. Bursting torrents of ornament, imitative episodes, organ recitatives,

---

[40] Called also Exemplarische Organisten-Probe, Hamburg, 1731, p. 34.
[41] Prelude and fugue, B.-G., XV., p. 177. P. S. V., C. 2 (241), No. 4. In this a variant is also given, which, however, seems to me to be due to accident. No manuscript has as yet come to light containing the oldest form of the fugue.

the boldest modulations, and broad, resonant progressions of chords—all are here in apparent disorder. And yet the mature genius of Bach presides over it and informs it all. The close answers exactly to the ornate commencement; the polyphonic movement in bars 9 to 13 are precisely the same as bars 25 to 30; the organ recitative—bars 14 to 24— are balanced by the free harmonies of bars 31 to 40. Even in the modulations, which almost beat Buxtehude in audacity, a plan is clearly traceable; they rise from *a*—bar 14— by degrees through B minor, C minor, up to *d*, then to E flat minor, of which the bass seizes the dominant, and thence proceeds upwards by six chromatic steps; to correspond to this, the harmonic body subsequently descends from *d*—bar 31—through C minor, B flat minor, A flat minor, after which the six chromatic steps up from B begin in the pedal. It is necessary to keep the outline of the prelude resolutely in mind at first, in order not to be confused by the swift runs, and deafened by the heavy masses of sound; but presently we become accustomed to it, and not only recognise the purpose and method of the work, but feel it too; while in similar pieces by Buxtehude, and particularly in the interludes to his fugues in several parts, there is rarely any plan at all. We are also struck by the wide difference between this and earlier works by Bach, in which Buxtehude's influence could be seen; its remarkable peculiarities can only be accounted for by referring it to some special incident in Bach's history; and it seems most obvious to suppose that this was the Hamburg journey.

The most beautiful contrast to this is offered by the grand, calm modulations and strict four-part treatment of the fugue, which is a long one, and which, on the other hand, certainly bears some relationship to Reinken; hence the hypothesis that it was composed in 1720 is further confirmed. The theme, particularly, has an unmistakable resemblance with the fifth sonata of Reinken's *Hortus Musicus*. It begins thus :—

We may unhesitatingly view this resemblance as an

intentional allusion, and therefore a certain homage to
Reinken. A musician of the last century spoke of the
G minor fugue as "the very best pedal piece by Herr Johann
Sebastian Bach." I modify this verdict only so far as to say
that no other fugue appears to stand above it. It is in view
of such a production as this that we are justified in making
what may seem an exaggerated assertion—namely, that there
never was a fugue written by any other composer that could
compare with one of Sebastian Bach's.[42] We have shown
how highly Buxtehude's works of the same kind are to be
esteemed, and it cannot be disputed that in a few clavier
fugues Handel proved himself Bach's equal; but this soaring
imagination, this lavish and inexhaustible variety of form—
again, this crystal lucidity and modest naturalism, this lofty
gravity and deep contentment which strikes awe into the
hearer, and at the same time makes him shout with joy—
all this is so unique in its combined effect, that every notion
of a comparison with others appears preposterous. Once, and
only once, has anything been produced in the whole realm
of instrumental music which can be set by the side of these
most perfect organ fugues by Bach—namely, Beethoven's
symphonies.

The mention of Mattheson brings us once more to a
comparison and contrast of Bach and Handel—this time,
however, not as men, but as organists. That Bach had no
equal in Germany in playing the organ was soon an admitted
fact; friends and foes alike here bowed to the irresistible
force of an unheard-of power of execution, and could hardly
comprehend how he could twist his fingers and his feet so
strangely and so nimbly, and spread them out to make the
widest leaps without hitting a single false note, or displacing
his body with violent swaying.[43] But from England, on the
other hand, Handel's growing fame had reached Germany,
not only as a composer of opera and oratorio, but as an
unapproachable organ-player. So far as England was concerned, that was not saying too much, but other foreigners

---

[42] Forkel, p. 33.
[43] Scheibe, Kritischer Musikus. New ed., 1745, pp. 839, 875.

who had heard him there brought the same news; and as he was a German, the comparison with Bach was obvious, while Bach's cantatas, Passion Music, and masses were scarcely appreciated in the contemporary world as compared with Handel's music. The attempt made by his Leipzig friends, in 1729, to bring about a meeting of the two players miscarried, so opinions and assertions could spread unchecked. Some came from England full of Handel's praises, but saying nevertheless that there was but one Bach in the world, and that no one could compare with him; others, on the other hand, were of opinion that Handel played more touchingly and gracefully, Bach with more art and inspiration, and it was always the one then playing who at the moment seemed the greatest.[44]

In one thing all were agreed: that if there was any one who could depose Bach, it could be none but Handel; as, however, the names of those who formed this judgment have remained unknown, and we are no longer able to determine how far they were competent, it may be considered a happy accident that Mattheson heard both the masters, and has recorded his opinion.[45] Soon after the transactions of 1720, he writes that among the younger composers he had met with no one who displayed such skill in double fugues as Handel, whether in setting them or in extemporising, as he had heard him do, with great admiration, a hundred times.[46] A very laudatory general opinion of Bach has been already quoted; and in a remark written later they are set in direct comparison, as follows: " Particularly, no one can easily surpass Handel in organ-playing, unless it were Bach, of Leipzig; for which reason these two are mentioned first, out of their alphabetical order. I have heard them in the prime of their powers, and

---

[44] Scheibe, ibid., pp. 843, 875, note 15.
[45] Peter Kellner also heard them both play (F. W. Marpurg, Historisch-Kritische Beyträge, I., p. 444. Berlin, 1754), but his verdict remains unknown. Ph. Emanuel Bach, on the other hand, records his judgment, but only from hearsay and suppositions, as to Handel's playing. See his letter to Eschenburg, in Nohl, Musikerbriefe, ed. 2, p. XLIX. Leipzig: Duncker und Humblot, 1873.
[46] Crit. Mus., I., p. 326.

have often competed with the former, both in Hamburg and in Lübeck."[47] It is beyond a doubt that Mattheson was quite competent to pronounce judgment in such a case; he was a musician of incontestably sound training. But I regard it as equally beyond a doubt that in this instance his information is wholly worthless. Mattheson's recollection of Handel's organ-playing dated from the days of their youth, when they were much together—days which, as he grew older, he recalled with peculiar pleasure. The experience is universal that favourable judgments cherished in youth are apt to persist, in spite of our progressive development, even when the subject of our interest is never again within reach for the verification of the opinion; and this was the case here. Mattheson had never heard Handel play since 1706.[48] Even if he had, his decision might have remained the same, because Handel's proclivities as an artist were far more sympathetic than Bach's to Mattheson, who had grown up under the influence of the opera—more particularly of Keiser's opera—and who, while still young, had become indifferent to organ music.[49] And this sympathy did not cease to exist, in spite of Handel's distant behaviour; still, it is an error to assert that after 1720 Mattheson showed a warm interest in Bach.[50] I have already stated that this was not the case, and a collation of the passages from Mattheson's writings relating to Handel and Bach reveals his attitude very clearly. Finally, it is of some importance to note that vanity would prompt him to set Handel's importance as an organist as high as possible, for had he not competed with him in Hamburg and Lübeck? The notable mode of expression used in the sentence quoted— not free from partisanship, but only wavering — also had its origin in the want of lucidity and the indecision of the writer, whose inclination and judgment balanced on

---

[47] Vollk. Capellmeister (Hamburg), 1739, p. 479.
[48] Chrysander has shown that Handel was already in Italy in 1707. Händel, I., p. 139.
[49] This is very conspicuous in the "neu eröffnetes Orchestre."
[50] Compare Chrysander, Händel, III., pp. 211-213; his views throughout are opposed to mine.

opposite sides. All attempts to explain it away are vain; for this purpose he is useless.

We may, however, accept his statements about Handel as a player and composer of double fugues, for there is at any rate something characteristic in it; but this brings us back to deciding upon internal grounds, which is, in fact, what we must do with the whole question. It must all rest on this: to which of the two musicians organ music was of the deepest vital significance. Handel, too, had derived his first training from a German organist, and had been one himself, for a while, in his youth; but he turned towards other aims, ending at last by using the organ as a musical means, one among others in the general mass of instruments he employed, but merely as a support or to introduce external embellishments. Bach started from the organ, and remained faithful to it to the last day of his life. All his productions in other departments—or, at any rate, all his sacred compositions—are merely an expansion and development of his organ music; this was to him the basis of all creation, the vivifying soul of every form he wrought out. Consequently in this he, of the two composers, must have been capable of the greatest work—the greatest, not merely in technical completeness, but also in the perfect adaptation of its purport to the instrument. When once we are clear as to this, the accounts handed down to us are equally clear, and leave no doubt in our minds that Handel's organ-playing was not, properly speaking, characterised by *style* in the highest sense—was not that which is, as it were, conceived and born of the nature of the instrument. It was more touching and graceful than Bach's; but the proper function of the organ is neither to touch nor to flatter the ear. Handel adapted to the organ ideas drawn from the stores of his vast musical wealth, which included all the art of his time, just as he did to any other instrument. In this way he evolved an exoteric meaning, intelligible to all, and hence the popular effect. To him the organ was an instrument for the concert-room, not for the church. It corresponds to this conception that we have no compositions by Handel for the organ alone, while it was precisely by these that Bach's fame was to a great extent kept up till this century; but we have by Handel a considerable number of organ concertos with instrumental accompaniment, and adapted with brilliant effect to chamber music.

His fondness for the double fugue—an older, simpler and not very rich form, of which, however, the materials are easier to grasp, and which is therefore more generally intelligible, can also be referred to his exceptional attitude towards the organ ; and so no less may the improvisatory manner which was peculiar to his playing and to his clavier compositions, which came close to the limits of organ music; while the organ—which, both in character and application, is essentially a church instrument—must be handled with the utmost collectedness of mind and an absolute suppression of the mood of the moment. It is in the highest degree probable that Handel—whose technical skill was certainly supreme—with his grand flow of ideas, and his skill in availing himself of every quality of an instrument, produced unheard-of effects in his improvisations on the organ. But even the more fervid and captivating of these effects must have been very different from Bach's sublimer style. I must at least contravene what has been asserted by an otherwise thoughtful judge[51]—namely, that he was surpassed on this one point—taking it for granted that improvisation is to be criticised by its intrinsic musical worth, and not merely by its transient and immediate effect. At a time when so much importance was attached to extempore music, which indeed, as an exercise in thorough-bass, was part of the musical curriculum everywhere, it would have been most strange if the man whose whole being as an artist was wrapped up in the organ, and who had exhausted its powers in every direction, had not risen to a corresponding height in this point also. The express testimony of his sons and pupils as to his "admirable and learned manner of fanciful playing"—*i.e.*, improvising—as to the "novelty, singularity, expressiveness and beauty of his inspirations at the moment, and their perfect rendering," stands in evidence. "When he

---

[51] Chrysander, p. 213.

sat down to the organ, irrespective of Divine service, as he was often requested to do by strangers, he would choose some theme, and play it in every form of organ composition in such a way that the matter remained the same, even when he had played uninterruptedly for two hours or more. First he would use the theme as introductory, and for a fugue with full organ. Then he would show his skill in varying the stops, in a trio, a quartet or what not, still on the same theme. Then would follow a chorale, and with its melody the first theme would again appear in three or four different parts, and in the most various and intricate development. Finally, the close would consist of a fugue for full organ, in which either a new arrangement of the original theme was predominant, or it was combined with one or two other subjects, according to its character." [52]

So far as concerns the other aspects of organ music the author of the Necrology might with justice appeal to Bach's existing compositions, which call into requisition the highest technical means in order to express the profoundest ideal meaning and "which he himself, as is well known, performed to the utmost perfection," and so confirm his statement that "Bach was the greatest organ-player that has as yet been known."

II.

BACH'S CLAVIER MUSIC.—TOCCATAS.—HIS REFORM IN FINGERING.—ADJUSTMENT OF PITCH.—INVENTION OF THE PIANOFORTE.—BACH AS A TEACHER.—THE CLAVIER-BÜCHLEIN.—INVENTIONEN UND SINFONIEN.

WE must now investigate more closely the field in which Bach had been especially invited to labour at Cöthen. At

---

[52] Kirnberger, Die wahren Grundsätze zum Gebrauch der Harmonie, p. 53, note. Berlin und Königsberg, 1773. Mizler, p. 171. Forkel, p. 22. Forkel observes that the method of organ improvisation attributed to Bach is precisely that form of organ music which Reinken had supposed to have died out; and we must assume that it was so, inasmuch as northern masters depended much upon the stops, used independent themes in contrast to the lines of the chorale, were fond of dissecting and remodelling the ideas of the fugue, and of extending and enlarging upon it generally.

that time the harpsichord was the instrument nearest to the organ; its soulless tone—which could only acquire a certain amount of expressiveness by the use of several keyboards—indicated the necessity for infusing an intrinsic animation by means of polyphony and rich harmonic treatment, of a steady and thoroughly progressive melodic development; and, in addition to these—since it was defective in duration of sound—of increased rapidity of action. Henceforth he cultivated both these instruments with equal devotion, and endeavoured to extend the province of each in its style by reciprocal borrowing. Just as, on one hand, he adapted to the harpsichord the tied and *legato* mode of playing which the organ imperatively demands, so, on the other hand, he transferred to the organ so much of the florid execution of the clavier style as could be engrafted on its nature. Hence, though the organ, as was due to its superior importance, always had the precedence, his art was developed quite equally on both instruments; and, in the very year which saw the end of his official work as organist, he was required to stand a triumphant comparison with one of the great French clavier-masters. Hitherto no particular attention has been directed to the clavier compositions of the Weimar period, with the express purpose of not confusing our general purview, which has been cast in other directions. We will now hastily sketch, in broad outlines, what has been neglected so far, and thus directly lead on to the consideration of the whole department of his clavier and other chamber music in Cöthen.

When speaking of the cantata "Nach dir, Herr, verlanget mich" (Vol. I., p. 433), it was said that the fugal thema of the first chorus had undergone further development in a toccata for the clavier in F sharp minor. This theme is, no doubt, a favourite subject of Bach's, and recurs frequently; nevertheless, the identity—alike in the whole and in the details, in the feeling and in the expression—is so complete that we may regard the piece as a remodelling of the chorus quite as certainly as we detect in the beautiful organ fugue in A major the further development of the subjects of the overture to "Tritt auf die Glaubensbahn." That the chorus is not the later and the toccata the earlier work is proved by the greater musical completeness of the latter piece, and, in the second place, by its superiority to the three clavier toccatas previously mentioned, in D minor, G minor and E minor (Vol. I., p. 433), from which, indeed, it differs widely as to form. Like them, however, it is not unique in its way, for Bach, following out his old principles, wrought out at least two such pieces, thus giving us a right to regard them as constituting a new species of toccata.[53]

The essential improvement that characterises it consists in the introduction of a slow subject thoroughly worked out as an organic element, and in reducing the two fugues formerly included to one—if not in the strict sense of the word, at any rate so far as the thematic material is concerned. The ornate portions at the beginning remain, and have also appropriated a certain space in the middle. The slow subject follows immediately on the introductory runs; it is worked out with great skill and feeling on the themes—

and :—

A half-close prepares us for the fugue, which in one place runs to sixty-one bars, and in the other only to forty-seven. The process is only apparently different in the two works; in the toccata in F sharp minor an episode of one bar serves for the development of a free interlude, which is, it cannot be denied—like the second part of the clavier prelude in A minor—somewhat digressive and fatiguing, in spite of several modifications of the motive. In the C minor toccata the composer is content with a few bars full of brilliant passages, and then the fugue begins again; but now, by the addition of a second subject, it becomes a double fugue, while in the F sharp minor toccata he returns to the theme of the *adagio*, and constructs on that a quite

---

[53] B.-G., III., pp. 311, 322. P. S. I., C. 4, Nos. 4 and 5; Vol. 210, pp. 10, 20. See Appendix A, No. 2.

new fugue, distinct even in time. Why he should have done this here is clear by a reference to his model—the cantata—for there also the theme at first appears broad in style and full of longing, and then, after a highly varied interlude, it returns *agitato* and with intricate elaboration.

These two toccatas are superior to the former set, not merely by the greater concentration of their form, but also by the solidity and significance of their subject-matter; the E minor toccata alone is worthy to be compared with them in its peculiar dreamy and longing expression. The rather tame interlude of the F sharp minor toccata renders it somewhat inferior to its fellow, although, from its prevalent imaginative character, it does not seriously disturb the flow of the piece. For, when the *adagio* comes in with its deep accents, after the introductory passages, which seem to have met fortuitously, as it were, it is as though spirits innumerable were let loose—whispering, laughing, dancing up and down—teasing or catching each other—gliding calmly and smoothly on a translucent stream—wreathed together into strange and shadowy forms; then suddenly the phantoms have vanished, and the hours of existence are passing as in every-day life, when the former turmoil begins afresh—only now the memory of a deep grief pierces through it unceasingly.

It is otherwise with the second toccata. After a stormy beginning, the *adagio* sinks into grave meditation, from which the fugue springs forth with a most original repetition of the first phrase of the theme; this indeed is conspicuous throughout the arrangement, and serves to determine its general character. It is a proud and handsome youth, swimming on the full tide of life, and never weary of the delightful consciousness of strength. Compare with this, again, the closing fugue of the E minor toccata, and marvel at the master's inexhaustible creative wealth.

By the side of these two toccatas stands a three-part fugue in A minor, which is prefaced by a short *arpeggiato* introduction.[54] This is the longest clavier fugue that Bach

---

[54] B.-G., III., p. 334. P. S. I., C. 4, No. 2; Vol. 207, p. 36. Andreas Bach has a manuscript copy.

left complete; it consists of a hundred and ninety-eight bars, 3-4 time; besides this, it moves in uninterrupted semiquavers, so that it may be called another "moto perpetuo," and placed by the side of Weber's well-known movement.

It is hard to decide which most to admire—the unbridled and incessant flow of fancy or the firm structure which connects the whole; or, again, the executive skill and endurance that it presupposes. The theme, consisting of six bars, appears only ten times; more than two-thirds of the composition are worked out episodically, from the material it affords, and the nearer we get to the end, the less do we hear the theme in a regular form—only three times, in fact, in the last hundred bars. The mighty rush of the initial portion gradually swells to a raving storm, which almost takes the hearer's breath away; but, of course, without an *accelerando*, only by an increment of internal effects. When we now learn that Bach was accustomed to take the *tempo* of his compositions very fast, a degree of execution is suggested compared to which the hardest tasks of any other composer are as child's play. Bach owed his own attainment of it not merely to his iron perseverance, but also to the formative force of genius, which taught him to find the means of giving an adequate clothing to the world of ideas that seethed within him.

At the beginning of the seventeenth century the musical world was very indifferent as to the mode of using and playing on keyed instruments. A musician who was conspicuous for this way of thinking, Michael Prætorius, despised all who even spoke of them in real earnest, declaring that if a musical note were produced clearly and agreeably to the ear, it was a matter of indifference how this was done, even if it had to be played with the nose.[55] At a later date, the advantage—indeed, the absolute necessity—of a regular scheme of fingering was better understood; but it was not till the beginning of the eighteenth century that a rational and methodical practice began to prevail. Up to that time the thumb was almost excluded from use, and the exercise

---

[55] Prætorius, Syntagma Musicum.

of the little finger very lame, to say the least. The reason for this lay in the conspicuous difference in length between these and the three middle fingers, which seemed to disable them from equal efficiency. Still, as it was necessary to attend to the slurring of the notes one into another, particularly on the organ, the middle fingers were slipped over or under each other; the thumb simply hung down. It is, indeed, beyond a doubt that the school of Sweelinck and his compeers—that is to say, all the northern composers, more or less—who did so much to improve rapidity of execution, also did much in the regulation of the principles of fingering.[56] At the same time, even these never used the thumb, excepting under necessity; for, when Sebastian Bach told his son Philipp Emanuel Bach that he, as a lad, had heard great men play who could only make up their minds to use this despised finger for very wide stretches, we cannot understand him as meaning any but the northern masters, and Böhm, who was so closely allied to them. To Bach himself the unnatural conditions of such a limitation were soon obvious; he began to turn the thumb to the same account as the other fingers, and he must at once have perceived that the whole art of playing had thereby undergone a complete revolution. While the useless hanging of the thumb had resulted in an outstretched position for the other fingers, the use of it, being so much shorter, naturally necessitated a curved position for the others. This curving at once excluded all rigidity; the fingers remained in an easy, elastic attitude, ready for extension or contraction at any moment, and they could now hit the keys rapidly and accurately as they hovered close over them. Thus, by diligent practice, the greatest possible equality of touch, strength and rapidity was acquired in both hands, and each was made quite independent of the other.[57]

---

[56] There have been hitherto hardly any direct authorities that are copious on this subject. C. F. Becker, in his work Hausmusik in Deutschland, p. 60, has given a few examples of fingering for scales and passages, dating from the seventeenth century, and Hilgenfeldt has repeated them (p. 173); and a complete and very valuable MS., of 1698, with clavier pieces, many of them very precisely fingered, is in my possession.

[57] Mizler, p. 171.

Practical insight and a talent for composition combined to discover the surest and quickest road to these ends; every finger must be made equally available for every purpose; and Bach learned to perform trills and other embellishments with the third and little finger just as evenly and roundly as with the others. Nay, he even found it quite easy, meanwhile, to play the melody lower down with the same hand. The natural tendency of the thumb to bend towards the hollow of the hand made it of admirable use in passing it under the other fingers, or them over it. The scales—those most important of all the sequences of notes—were newly fingered by Bach; he established the rule that the thumb of the right hand must fall immediately after the two semitones of the scale in going up, and before them in coming down, and *vice versâ* in the left hand.[58] To release the note, the tips of the fingers were not so much lifted as withdrawn; this was necessary to give equality to the playing, because the passing of one of the middle fingers over the little finger or the thumb could only be effected by drawing back the latter; and it also contributed to a *cantabile* effect, as well as to clearness in executing rapid passages on the clavichord. The result of all this was that Bach played with a scarcely perceptible movement of his hands; his fingers hardly seemed to touch the keys, and yet everything came out with perfect clearness, and a pearly roundness and purity.[59] His body, too, remained, perfectly quiescent, even during the most difficult pedal passages on the organ or harpsichord; his pedal *technique* was as smooth and unforced as his fingering.[60] His peculiar fertility of resource enabled him to overcome incidental difficulties; in keyboards placed one above the other he preferred short keys, so as to be more easily able to move from one to another, and he liked the upper row to be somewhat shallower than the lower, because

---

[58] Kirnberger, Grundsätze, &c., p. 4, note 2. Compare Ph. Em. Bach, p. 18.
[59] Quantz, Versuch einer Anweisung die Flöte Traversiere zu spielen, third edition, 1789, p. 232. Forkel, p. 12. Compare Ph. Em. Bach, p. 13. See App. A., No. 3.
[60] Scheibe, p. 840.

he could slip down imperceptibly, without any change of finger.[61]

It was not Bach alone, however, of the musicians of that period who had hit upon the more extensive use of the thumb; the whole art of organ and clavier music, now so rapidly developing, cried out for the introduction of more ample methods of rendering. In France, François Couperin (1668-1733), Organist of the church of St. Gervais, opened the way to a more rational rule of fingering by his work, "L'Art de Toucher le Clavecin" (Paris, 1717). Johann Gottfried Walther, Bach's contemporary—and, at one time, his colleague in Weimar—has left us a few organ chorales, with marked fingering, in which the thumb is variously employed.[62] Heinichen, who has been already mentioned, invariably requires the employment of all five fingers for the performance of his directions for playing from figured bass.[63] Handel also brought the thumbs into constant play, as necessarily follows from the bent position of his fingers, as described by eye-witnesses,[64] by which they fell upon the notes almost by themselves.

Still the new method was not methodically worked out either by Couperin or by Walther. In the scales Couperin certainly prescribes starting with the thumb on the first note, but not the turning under of the thumb in the progress upwards; he is very ready to employ the thumb in changes on the same note, and also in extensions, where he freely allows it to strike the black keys, but hardly ever in such a way as that it is passed under the middle finger, or the middle finger over it. Two solitary cases occur among the vast selection of examples and test-pieces in "L'Art de Toucher le Clavecin," in which the method is different; one of these is for the left hand, and it is remarkable that the left-hand thumb

---

[61] Adlung, Mus. Mech., II., p. 24.
[62] See the Königsberg autograph. There are arrangements of "Allein Gott in der Höh," "Wir glauben all an einen Gott," "Wo soll ich fliehen hin" (verse 3).
[63] J. D. Heinichen, Der General-Bass in der Composition, p. 522. Dresden, 1728.
[64] Chrysander, Händel, III., p. 218; from Burney.

seems to have been brought into frequent use at a much earlier date, and here the passing over of the middle finger is several times indicated. The other has the following fingering [65]—

a decisive piece of evidence as to the want of practice in the use of the thumb; by turning the thumb according to Bach's rule—that is to say, on the c''—the passage runs of itself. In the three organ chorales Walther only twice crosses the middle finger over the thumb, and then in the left hand; in other cases only the first finger. Of Handel's method we know nothing exact; however, Mattheson supplies this deficiency to a certain extent, since, as has been said already, he thought he could compete with him in clavier-playing, and he, in the most crucial instance—namely, in the scales—does not know the method of turning the thumb under, but in ascending he puts the middle finger over the third in the old manner, in the right hand, and, in descending, the middle finger over the first.[66] Philipp Emanuel Bach, himself one of the most remarkable, if not the most remarkable, of the clavier-players of the middle of the eighteenth century, has laid down his views on the method of teaching the clavier in an admirable and very thorough work.[67] In the lesson on fingering, section 7, he speaks of its extension and improvement by his father, "so that now everything that is possible can be easily performed"; and he there explains that his wish is to base his teaching on the progressive development of his father's method.

It has been universally assumed that Emanuel Bach's method is the same as his father's, not merely in fingering, but in the other elements of instruction, although there is no statement in the book which satisfactorily proves it. How-

---

[65] L'Art de Toucher le Clavecin, p. 66, lowest stave. From the second book of Pièces de Clavecin; in the new edition by J. Brahms, p. 121 (Denkmäler der Tonkunst, IV.).

[66] Kleine Generalbassschule, p. 72. Hamburg, 1735.

[67] Versuch über die wahre Art Clavier zu Spielen.

ever, two small pieces with the fingering marked throughout in Sebastian Bach's own hand have come down to us, and a comparison of these with the rules laid down by his son proves that they differ widely. Philipp Emanuel prohibits the passing of the middle finger over the first; Sebastian prescribes it in the fifth bar of the first piece and in bars 22 and 23 of the second, agreeing with Couperin, as is shown by the example given above. Emanuel does not allow the third finger to cross over the little finger; Sebastian requires it of the left hand in bars 38 and 39 of the second piece. The practice of crossing under Emanuel limits to the thumb; Sebastian makes the little finger pass under the third in bars 34 and 35 of the same piece. Of the crossing of the little finger over the thumb, which Emanuel also forbids, there is, as it happens, no example in Sebastian's little pieces, but we find it in one of Walther's chorales. Yet more: although Sebastian's rule for using the thumb after the semitone intervals of the scale is most distinctly authenticated,[68] he himself has not observed it at the beginning of the first of these pieces, but has fingered it in the old manner, and though in the third bar the left hand advances, it is true, by a turn over the thumb it is only with the first finger, in the old fashion. Thus, though his fingering is distinguished from that of his predecessors and contemporaries by the regular use of the thumb, it differs from his son's method by certain peculiarities, some of which are retained from the older method of playing, while others were naturally derived from it; the origin of Sebastian Bach's method is thus tolerably clear. It took into due consideration all the combinations which the use of the thumb now rendered possible, but without abandoning the technical accomplishment which the earlier method had afforded; still, we may be permitted to suppose that Bach, who always followed the path pointed out by Nature, avoided, as far as possible, passing a smaller finger over a larger: for instance, the first or the third over the middle finger.

This combination of methods gave him such an unlimited

---

[68] By Mizler, as well as by Kirnberger; he also was a pupil of Bach's, and in the Mus. Bib., II., p. 115, he thus fingers certain scale passages.

command of means that it is easy to understand how it was that difficulties had ceased to exist for him. And, as though he had been destined in every respect to stand alone and at the summit of his art, he remained the only master of clavier-playing who acquired such stupendous technical facility. All who came before him, and all who succeeded him, worked with a much smaller supply of means; he stood on an eminence commanding two realms, and ruled that which lay before him as well as that he had left behind. His son even, who represents the actual starting-point of modern clavier-playing, greatly simplified his father's rules of fingering. He limited the crossing of the longest fingers to that of the middle finger over the third, and cultivated a more extensive use of the thumb. He did not, indeed, require such a wealth of resource for his far easier and more homophonic style of composition; and in art all that is superfluous is faulty. Then, with the introduction of the modern pianoforte, the door was finally closed on the old method of fingering, because the mechanism of hammers demands an elastic tap on the key from above, and prohibits the oblique blow which is given by crossing the middle fingers. Thus, even in these days, when we boast of a sovereign command of all the resources of clavier *technique*, Sebastian Bach's own mode of playing can only be restored to the extent to which it was carried out by his son, and is even now indispensable to enable us to perform Bach's compositions. It would be lost to us as a whole, even if we could be conversant with all its details; but the abnormal difficulty of his compositions is in great part grounded on this; for all that modern skill has gained on one side it has necessarily lost on another, from the very nature of the instrument. For this reason it cannot be denied that the modern *technique* is, after all, not superior to Bach's, or at any rate that he overcame many difficulties far more easily. What holds good for the clavier does so still more for the organ; indeed, Bach stamped the character of the organ on the clavier, without, however, detracting from its intrinsic value. But here, where there were no hindrances arising from the construction of the instrument, a further development of *technique* on the lines he had laid down would not be impossible ; and, so soon as this instrument attained somewhat more importance in our artistic life, the attempt was immediately made.

The utmost improvement of finger practice was indispensable to Bach, if for this reason only : that he was accustomed to play on claviers of equal temperament, and could therefore avail himself indifferently of all the twenty-four keys. The idea of establishing the equal temperament by a regular distribution of the ditonic comma—that is to say, by an adjustment of the difference resulting from twelve fifths, as compared with the twelve degrees included in the scale of an octave—had been already thought of at the beginning of the seventeenth century, and soon found universal acceptance. Many of the musicians already so frequently mentioned had made theoretical use of it—as Andreas Werkmeister (1644-1706) and Johann Georg Neidhardt (died 1740)—still the discovery can hardly have led to any practical application, since the differences of pitch were so minute as to be finally distinguishable by the ear alone. The methods of tempering which were evolved from the theory were at first singular enough. About the year 1739, the three following general rules still obtained: (1) The octave, the minor sixth and third must be absolutely pure ; (2) the major sixth and the fourth were to be somewhat enlarged ; (3) the fifth and major third were to be somewhat diminished.[69] How far this may have sounded well or ill may be approximately estimated when we remember that the determination of only the octave, fourth and fifth has any foundation in the nature of things ; all the intervals which have a more complicated relation to the keynote, like the major and minor sixths and the minor third, will bear, as is well known, greater deviations from their pure relation ; and how the major thirds can have been diminished is quite inconceivable, since the sum of three major thirds, even when purely tuned, does not equal an octave, which is what equal temperament demands.

---

[69] Mattheson, Vollkommener Capellmeister, p. 55.

It is very satisfactory to know that in this, too, Bach was in advance of his time, and had already made himself master of the method of tuning which is now universally followed. It is expressly stated that he took all the major thirds a little sharp—that is to say, slightly augmented—which is indispensable for the equalisation of the diesis.[70] But as it is impossible that he should have tuned from nothing but major thirds, he must have proceeded as we still do at the present day—that is to say, by four successive fifths, each slightly flattened, so that the last note forms a major third with the key-note, and, with the aid of the first fifth, a common chord on it. Of the various artifices which are used to facilitate the application of this method, he must have known, at any rate, that which consists in testing the deviation of the fifth by striking it, together with its octave, the fourth below the key-note, and taking the fifth up again from thence.[71] That he evolved all this by his own study and reflection, and not from reading theoretical treatises, would be very certain, even if we had not the testimony of his contemporaries;[72] and he carried out his method with such rapidity and certainty that it never took him more than a quarter of an hour to tune a harpsichord or a clavichord.[73]

---

[70] F. W. Marpurg, Versuch über die Musikalische Temperatur, p. 213. Berlin, 1776. "Herr Kirnberger himself has often told me and others how, during the time when he enjoyed the instruction of the celebrated Joh. Sebastian Bach, he would intrust to him the tuning of his clavier, and expressly enjoined him to make all the major thirds sharp."

[71] Emanuel Bach speaks of this, as well as of the tuning of the fifth downwards and the testing of the third, in section 14 of the introduction to the Versuch über die wahre Art, &c.; and he could scarcely have had any inclination to deviate from his father's practice in the matter of temperament.

[72] Mattheson was a sworn enemy to certain folks who insisted on making music a branch of mathematical science, and in this he was one with Bach. In Mizler's autobiography in the Ehrenpforte he adds, on p. 231, à propos to Mizler's intercourse with Bach, this observation: "Bach very certainly would no more have brought forward this mathematical basis of composition than the present writer [Mattheson]; that I will warrant." And what he says as to composition naturally holds good for the other branches of art: "Our late friend Bach never entered into deep theoretical considerations about music, and was all the more efficient in performance." Necrology, p. 173.

[73] Forkel, p. 17.

We soon shall see the splendid use in creative work which he made of the newly opened realm of tone, now for the first time accessible. But he did not allow himself to be led away into excursive modulations; this was quite opposed to his style. It was only under special circumstances that he now and then showed how keenly alive his ear was to the inner connection of keys, and how admirably he could avail himself of enharmonic transitions when he chose. An instance occurs in the prelude, previously mentioned, to the Hamburg fugue in G minor; another in the Chromatic Fantasia, as it is called, to which we shall return later.

I cannot decide, in view of the slender evidence, whether a piece called "Das kleine harmonische Labyrinth"—"The Little Labyrinth of Harmony"—is by Bach or not; it consists of an entrance full of enharmonic wandering, leading up, as to a central goal, to a little fugue, which is worked out, and then has its exit through similar mazy paths, returning to daylight in the key of C major. It was Heinichen—Prince Leopold's companion in Italy—who first clearly displayed, and practically applied, the sequence and connection of the twenty-four keys. We know of no similar attempt from Bach's hand; the soaring independence of his genius was averse to every merely mechanical device.[74]

We have designated the tone of the harpsichord as soulless, and, so far, similar to that of the organ. At the same time it cannot have remained unnoticed by a delicate ear that it responded more kindly under the hands of one player than those of another; hence it is not altogether unreasonable to speak of a subjective mode of treatment, even of the harpsichord. The possibility of such a treatment depends partly on the indefinable peculiarities of touch, and then on the yet more indescribable art of calling forth in the hearer those responsive emotions which are indispensable to the appreciation of the artistic idea; a power which clavier music demands with peculiar insistence. We have reason

---

[74] Heinichen (Gen. Bass, p. 837, Dresden 1728); and even before this in the title work published in 1711. The test piece, pp. 885-895, is wrongly assigned to Bach in a MS. in the Royal Library at Berlin (press-mark p. 295). The "kleine harmonische Labyrinth" is also to be found there.

to suppose that Bach must have possessed this power. That he had a peculiar charm of touch is a matter of course, with his new modes of playing; and that his playing, even on the harpsichord, which he always furnished with quills himself, must have been, in a certain way, inspired. His son Philipp Emanuel points out that the only way to attain this is the diligent cultivation of the clavichord,[75] and this was precisely Sebastian's favourite instrument. Even though it had not much strength, the tone was wonderfully capable of light and shade, and comparatively persistent. It was possible to play *cantabile* on it, and this *cantabile* style was regarded by Bach as the foundation of all clavier-playing. In view of these indisputable facts, the opinion we find expressed here and there, that, in the performances of his clavier pieces, Bach gave no light and shade of expression, and that the introduction of such a rendering is a presumptuous modern innovation, must fall to the ground as an unfounded hypothesis.

Nowhere can we see more strikingly than here how a great genius can contain within itself the aim and end of a long process of historical development, and foresee it across the lapse even of centuries. The ideal instrument which floated in the mind of Bach for the performance of his *Inventions* and *Sinfonias*, of his suites and clavier fugues, was not altogether the clavichord; the ideas brought down by him from the sublime heights of the organ were too ponderous, and weighed too heavily on its delicate frame. But it was not the organ either. From the organ, no doubt, emanated that craving for more abundant alternations of feeling which sought its satisfaction in chamber-music, just as the endeavour after definition of feeling gave rise to the main idea of the church cantata as proceeding from the organ chorale. Its solemn, calm and rapt solitude blossomed out into blooming beauty and the living speech of man. The harpsichord could not here satisfy him; no

---

[75] Ph. Em. Bach, op. cit., section 17: "By constant playing on the harpsichord we get into the way of playing with one sort of tone, and the different varieties of tone which can be brought out even by an ordinarily good clavichord-player are entirely lost."

instrument but one which should combine the volume of tone of the organ with the expressive quality of the clavichord, in due proportion, could be capable of reproducing the image which dwelt in the master's imagination when he composed for the clavier. Every one sees at once that the modern pianoforte is in fact just such an instrument. Nothing can be more perverse than to wish to have the old clavichord restored in order to play Bach's clavier pieces— or even the harpsichord, which, indeed, was of the very smallest importance in Bach's musical practice; this might do for Kuhnau, for Couperin, and Marchand; Bach's grander creations demand a flowing robe of sound, an inspired mien and expressive motions.

If, in recent times, more and more attention has been paid to Bach's clavier works, one reason for this, among others, and by no means the least, is that we have felt that at last the means were not altogether inadequate to the purpose. Of course this is not said with reference to executive embellishment; but, indeed, the danger in this direction is not imminent; the fabric of these compositions is so compact, the progression of the parts so melodious throughout, that any arbitrary insertion of heterogeneous details is all but impossible, after a little studious attention to their organic structure. Where a phrase is intended to be conspicuous, the composer has taken care that it shall become so of itself. The echo-like contrasts of *forte* and *piano*, which are indicated by the character of the harpsichord, with its several keyboards, are almost always marked, and where they are not they are easily recognisable—they invariably refer each to a complete phrase. Whatever more than this depends on the performer will infallibly be clear to him when once he has accustomed himself to follow, in his own mind, the vocal phrasing, so to speak, of the separate parts, and their symphonic combined effect. He will then breathe life into the emotions they embody, in due proportion as they swell and fall, and give more or less fulness of tone in sympathy with their agreement or antagonism to the fundamental harmony at the moment. Then will he become aware of that melody which pervades the inner parts of the harmonic progression of every piece by Bach. He will ride on its wings, whether it roars with the force of a storm, or, again, whispers like the breath of May—intangible, invisible, and yet all-pervading. Bach's clavier compositions are a heritage into which it has been left to this generation to enter in the fullest extent—an inestimable legacy to a period when the spring of musical inspiration no longer flows with its former abundance—an immovable rock in the midst of the troubled waters of passionate aberrations, and a solemn warning to all who still have ears to hear never to neglect the dignity of art.

The master lived to see the early youth of the pianoforte, and aided it by severe criticism. Gottfried Silbermann, of Freiburg, somewhere between 1740 and 1750 constructed two claviers with hammer action, probably after the invention of Cristofori, the Florentine. Bach played on one of these, praised the tone highly, and found fault only with the heavy touch and the feebleness of the upper notes. Deeply as Silbermann felt this criticism, he nevertheless was willing to bow to it; he worked for years at the improvement of his hammer action, and at last earned Bach's unqualified praise.[76] It is not likely that Bach ever became himself the possessor of such an instrument, for, if he had, his pupil Agricola, through whom we hear of the affair, would have mentioned it. And the reason is very clear: the hammer mechanism did not accommodate itself readily enough to all the appliances of Bach's method of fingering. Still, his satisfaction at Silbermann's instrument shows very clearly whither his clavier music tended.

To remedy at least one main defect in the harpsichord—namely, its brief resonance—in the year 1740 (or thereabout) he devised a "Lauten-clavicymbel" (Lute-harpsichord), which was constructed by the organ-builder Zacharias Hildebrand, under his direction; the greater duration of tone was produced by gut strings, of which it had two to each note, and these were supplemented by a set of metal strings giving a four-foot tone. When the ringing tone of

---

[76] Adlung, Mus. Mech., II., p. 139.

these was checked by a damper of cloth the instrument sounded much like a real lute, while without this it had more of the gloomy character of the theorbo. In size it was shorter than the ordinary harpsichord.[77] The thorough comprehension of the construction of instruments which Bach here displays, and the experience he had already proved in the department of organ-building, together with his skill in tuning and his perfection of ingenuity in fingering, are the outcome of his technical talent. His superb nature stands firm on the true foundations of all art—an inexhaustible depth of imagination; while his thorough technical knowledge includes even the humblest mechanical means of casting the precious material in the noblest forms. We need only remember the talents of Joh. Michael and Joh. Nikolaus Bach respectively to verify once more the statement that in Sebastian all the capacity of his family converged.[78] An admirable musical connoisseur of the last century exclaims that "the immortal Joh. Seb. Bach combined all the great and different talents of a hundred other musicians."[79]

Nor was it only that all the ways and means of artistic production and utterance were at his command; he was besides a distinguished teacher of music. Of all the great German composers, Bach is the only one round whom are grouped any great number of disciples—men, too, who do not owe their chief glory to their master. Irrespective of his sons, Ziegler, Agricola, Altnikol, Ernst Bach, Homilius, Kirnberger, Goldberg, Müthel, Kittel, Transchel, Vogler and, above all, Joh. Ludwig Krebs were musicians of undoubted merit, and some of them of great eminence. Though no one of them opened out new paths in composition, the reason of this lay partly in the isolated supremacy

---

[77] Adlung, Mus. Mech., II., p. 139.

[78] Bitter, in his book on J. Seb. Bach, I. (p. 141), states that the composer constructed a musical clock for the castle at Cöthen, which still exists in the Castle of Nienburg, on the Saale. Herr Albert, the minister there, was good enough to examine this clock at my request; it bears on a disc in the interior the words "*Johann Zacharias Fischer Fecit. a. Halle.*"

[79] Marpurg, Loc. Cit., p. 234.

of their master himself; it was hard to make any approach to that, and creative power is a thing that can neither be imparted nor acquired. The strong point of Bach's pupils lay in their executive art, to which industry and good guidance are the chief aids. That Bach could so well cultivate these is due in the first place to the moral worth of his character, which prompted him to place his own acquirements with self-denying liberality at the service of his fellow-men. "Dem höchsten Gott allein zu Ehren, Dem Nächsten draus sich zu belehren," was what he wrote (see Vol. I., p. 598) on the title-page of the precious "Little Organ Book," and he acted up to his motto. It both commands our reverence and quickens our heart to see this man, whose Titanic imagination could at one moment lift its hand to grasp a sublime ideal, sitting down, the next hour, among his scholars, the sons of organists and cantors of the most modest pretensions, explaining patiently the mechanical use of the fingers, generously helping a blunderer by writing out a special exercise, and urging them on to higher aims with all the earnestness of a teacher, by performing the examples he had set them. Thus he began in Mühlhausen, and thus he continued forty years later, when declining into old age. The native Bach spirit, the great German ideal, penetrated him throughout in all its depth and modesty.

Besides this, even his teaching promoted his own progress. No doubt one reason why most of the great masters have proved more or less unfitted for teaching is to be sought in their lack of patience in explaining clearly to others the things they have drunk in, as it were, instinctively; but there is, more certainly, another—namely, that they are all merely carrying forward and completing a process already begun, and that therefore they fail in that living experience which gives an interest even in the simplest elements. In the province of the organ this was Bach's attitude also. But in clavier music he had not only entered into the inheritance of his predecessors; he had brought to it so much from the stock of organ music that it acquired a perfectly new aspect, and in the same way he had so completely transformed its vehicle of expression—namely, the *technique* of fingering—by his intelligent and ingenious novelty of method that it was radically different. Here he felt himself the creator of the art from its most elementary principles; here he had tried and tested the best methods of training by his own indefatigable labours; and from the nature of things that method of instruction took its rise in and from the clavier. It was one more instance of the truth long since uttered by Socrates: that every one can be eloquent on a subject he understands. And how could this eloquence fail of convincing effect, when his pupils saw to what results Bach's methods had brought him? how in him the most exact knowledge and the utmost executive power were combined? Then a third aspect of his gift for teaching lay in this: that he could, when needful, interrupt himself, set aside the clear logic of the intellect in favour of the flight of genius, and by a perfect revelation of his own powers show his pupils the goal which, under his guidance, they had begun to approach. Thus he refreshed and invigorated their courage; and though on one hand he required the severest application, he at the same time had hours in store for them which, by their own admission, were among the happiest in their lives.[80]

We have some information, too, as to his course of instruction. In the first instance he only gave exercises in touch, in fingering and in the equal and independent action of every finger of both hands. To this he kept the pupil for at least a month, but would sweeten the bitter dose by giving him graceful little pieces, in each of which some special technical difficulty was dealt with. Even embellishments and *maniers*, as they were called, had to be practised persistently in both hands from the very first. When a certain proficiency had been attained in these elements he went on at once to the root of the matter in difficult compositions, by preference in his own. Before the pupil began to study one, he played it to him, thus rousing his zeal and a desire not to fail of a

---

[80] So says Heinrich Gerber, Lexicon, I. col. 492.

happy result.[81] He set the highest value on industry, and set himself up as an example to them in this alone. " I have to be diligent," he would say, " and any one who is equally so will get on equally well." He never seemed to be aware of his wonderful gifts.

A happy circumstance has also enabled us to overhear, as it were, a whole practical course of teaching by Bach, so far at least as it is worked out by purely mechanical exercises rounded off into complete musical pieces. When his eldest son was nine years old, finding he had great musical gifts, his father began to cultivate them. On January 22, 1720, he projected the "Clavier-Büchlein" (Little Clavier-Book) for Wilhelm Friedemann Bach,[82] in which, beginning with the simplest elements, he introduced, by degrees, compositions of progressive difficulty, and here and there even let the boy himself write some. On the first page the keys and the principal ornaments are explained.[83] Then follows the little piece before mentioned as having the fingering marked. It is called *Applicatio*, and headed with the pious words, *In nomine Jesu*. Here scale passages and ornaments are combined with a special view (as is shown in bars 2, 6 and 8) to the practice of the shake with the third and little fingers of the right hand. The second piece (a *preamble* of eighteen bars in C major) is for the practice of embellishments in the left hand, with a perfectly equal semiquaver movement for the right in precise alternation with the left.[84] Then comes —and this is highly significant in its bearing on Bach's attitude towards clavier music—the three-part chorale "Wer nur den lieben Gott lässt walten," lavishly embellished in both hands. It is, in an improved form, the same subject as he had some years previously arranged for church purposes by the addition of a prelude, interlude and postlude.[85] Since that time he had given up the use of such elaborate

---

[81] " My late father had a happy way of putting his pupils to the proof. With him his scholars had to go at once to his by no means easy pieces." Ph. Em. Bach, I., p. 10.—Forkel, pp. 38 and 45.

[82] See ante, Vol. I., note 21, p. 12. The size is a small oblong quarto.

[83] The little diagram arranged for this purpose is to be found, B.-G. III., p. 14.

[84] P. S. I., C. 9, No. 16, I. (Vol. 200, p. 3).

[85] See ante, Vol. I., p. 313, note 135.

accompaniments to congregational singing, and had arrived at the conclusion that they were only serviceable for purposes of practice on the clavier; and to perform this smoothly and roundly demands a skill far beyond that of a beginner.

The fourth lesson consists of a somewhat longer prelude in D minor, a calm movement in quavers closing with a cadence in semiquavers for both hands alternately.[86] Bars 9 and 13 each have a slurred passage lasting into the next bar. But as a true *legato* could only be produced on the clavichord by increased pressure, involving added intensity of tone, the slur indicates at the same time a shading in the force of tone; this is all the clearer because it does not cover a complete phrase, but is lost in the following bar; thus the passage began *forte*, then *diminuendo*, down to *piano*—a practical hint as to Bach's attention to expressive execution.

The fifth place is occupied by another three-part chorale, "Jesu meine Freude," coloured and ornamented like the former one; but it is not written out to the end.[87]

Two easy allemandes follow as a pleasing change, both in G minor; but the second of these is also only a fragment. Then follow three preludes, in F major and G minor—this one has the fingering marked—and F major again.[88] The two first again aim at rapidity, and a smooth execution of semiquavers and quavers, to which it is evident that Bach gave much attention; but the third is already of that category of more difficult pieces to which the pupil was ere long introduced. The polyphony which governs all the parts, and which is so essentially Bach's, combined as it is with an equally characteristic variety in the musical ideas, presupposes a by no means contemptible independence and rapidity of finger. The polyphony, however, is restricted to three parts, and these are used at first with caution; but, in spite of this, they demand smooth handling and some stretching and grasping power in the hands.

---

[86] P. S. I., C. 9, No. 16, V. (Vol. 200, p. 7).

[87] The fragment occurs in P. S. V., C. 5 (Vol 244), after the variants.

[88] P. S. I., C. 9, No. 16, VIII., XI., IX. (Vol. 200, pp. 9, 10, 11). More exactly the two first are called *preambles;* however, there is no perceptible difference between this and a prelude.

Corresponding to the three preludes are an equal number of minuets—in G major, G minor, G major[89]—in which the study of polyphonic treatment is carried farther; in the third, too, a rhythmical figure is introduced, of the greatest utility in practising clearness of touch.

This is, as it were, a stage reached; the next exercise represents a higher level of study. It consists of eleven preludes, which recur later in a more or less altered shape in the Wohltemperirte Clavier. The order in which they stand shows that their purpose was the attainment, in due sequence, first, of increased rapidity, and then of a sustained and equal execution, going on to a cantabile and polyphonic style of playing. The keys follow thus: C major, C minor, D minor, D major, E minor (this is for left-hand practice only), then E major, F major, C sharp major, C sharp minor, E sharp minor, F minor. The preludes are not all finished to the end, but we shall of course consider them again with reference to the relation they bear to the pieces collectively of the Wohltemperirte Clavier, as well as with regard to their intrinsic merit. After them we come, for the first time, on a composition not by Bach—an allemande in C major, by J. C. Richter.[90] The courante which follows it may be by the same composer.

Then, among a number of trifles and fragments, we may distinguish a prelude in D major and a three-part fugue in C major.[91] In the fugue, among other technical aims, it is easy to perceive a special adaptation to the exercise of the third and little fingers of the right hand; however, for this prelude there is a general demand on the true Bach mode of playing. What had before been an end is now merely a means; the student is one step nearer to perfection. This is confirmed by the rest of the little work, which is filled almost exclusively by the Inventionen und Sinfonien, the first of the three great master-works for the clavier which owe their existence to the Cöthen period. There

---

[89] P. S. I., C. 13, No. 11, I., II., III. (Vol. 216, pp. 30 and 31).

[90] Probably the same who was afterwards Court Organist in Dresden, Joh. Christoph Richter. See Gerber, N. L. III., col. 855. Nothing further is known as to any acquaintance between him and Bach.

[91] P. S. I., C. 9, No. 16, IV., and No. 9 (Vol. 200, pp. 6 and 24).

remain only two little suites, of which one in three parts (A major) is not, it is true, in Sebastian Bach's handwriting; it may nevertheless be of his composition; the other, in four parts (G minor), is by G. H. Stölzel, Capellmeister of Gotha. Bach amused himself by adding to its minuet a trio, which is as delightful as it is learned.[92] All the original compositions here mentioned not only perfectly fulfil their instructive purpose, but are masterly productions when viewed as works of art—a varied and fragrant wreath, in which roses, lilies and perfumed stocks have their place, as well as wilder growths—each in their degree, but each with its own peculiar charm. The forms are at first quite simple, but with the advance of technical acquirement they become gradually broader till the fugue is reached. We are not now speaking of the peculiar structure of the Inventionen und Sinfonien; that of the latter approaches very nearly to the four-part construction of the fine prelude in D major.

There are still many more pieces written by Bach especially for technical practice, though most of them were no doubt dispersed and lost among his pupils. Very admirable is a little prelude in C minor, which runs whispering on in harp-like tones from one set of harmonies to another, and yet lets the mystical romanticism of Bach's genius pierce through it all.[93] We find even fugues, both with and without preludes, which, like the former one, probably served as pieces for testing the progress of the pupil, three-part fugues of really enjoyable perfection as to form and purport, and conceived with that concentration of structure which does not allow of a single superfluous note, nor say a word too much—the distinguishing mark, in short, of all the fugues written at Cöthen or later. The preludes are as artistic as they are profound, particularly that grave and melancholy one in D minor, in which we might fancy we had found an

---

[92] P. S. I., C. 9, No. 16, X. (Vol. 200, p. 81).

[93] P. S. I., C. 9, No. 16, III. (Vol. 200, p. 4). Also for others, see II., VI., VII. and XII.

organ piece, if Bach had not with his own hand added to it a fugue unmistakably written for the clavier.[94]

Our study of Bach's qualities of *technique* and as a teacher has led us back again to his work as a composer. His nature is a grand homogeneous whole ; all his various characteristics reacted on each other—interpenetrated each other—to compose an indivisible unity. Just as every exercise he wrote is a true work of art, so, on the other hand, every independent composition is full of technical instruction. He never wrote a clavier piece which did not serve as a healthy gymnastic for the fingers ; but, on the other hand, he never composed anything which fulfilled no other end than that of an exercise. It is precisely in one of his profoundest masterpieces that he addresses himself directly to "the young who desire to learn." Their advancement and culture in the comprehension of art were to him objects of the warmest interest, and inspired him to creative effort. The way in which, by patient waiting and diligence, he gradually educated his pupils to be his public is a brilliant example for every artist who cherishes the natural desire to make a way for his ideas. But he was far from all affectation —from all delight in unintelligent admiration. Instrumental music, more than any other art, demands a certain understanding, and claims a higher degree of musical culture in those who would deserve her favours ; her votaries must be specially trained in her service, or she turns her blessings to a curse—a mere futile and demoralising means of luxury. This Bach knew very well; even his zeal as a teacher was at bottom merely an emanation from that true art which gives dignity to humanity, and makes no distinction between the good and the beautiful.

---

[94] Four fugues with two preludes are published in P. S. I., C. 9, Nos. 4, 5, 8 (Vol. 200, 11, Vol. 212, p. 3). See Griepenkerl's preface to the volume. In it there are also two other fugues, in D minor and A minor, Nos. 12 and 6 (Vol. 212, p. 5 ; Vol. 200, p. 33), which, however, do not fit with the two other preludes. No data as their origin and purpose are forthcoming, but their internal character indicates beyond a doubt that they were not written later than the others. They are by no means inferior: the A minor fugue has a strongly marked organ type.

When, at the beginning of 1723, he revised the Inventionen und Sinfonien, and wrought them into the form of an independent volume, he gave it the following title: "An honest guide by which the lovers of the clavier, but particularly those who desire to learn, are shown a plain way, not only (firstly) to learn to play neatly in two parts, but also, in further progress (secondly), to play correctly and well in three obbligato parts; and, at the same time, not only to acquire good ideas, but also to work them out themselves, and, finally, to acquire a *cantabile* style of playing, and, at the same time, to gain a strong predilection for and foretaste of composition."[95] Here, once more, we have the whole confession of faith of the musical instructor: "A Guide": here the instructional purpose is most clearly indicated—"an honest guide"—for true art can be served by no hollow mockery. "The lovers of the clavier"—that is, the clavichord, the foundation of all Bach's teaching, on which alone a *cantabile* or flowing and expressive mode of execution is indeed possible—"particularly those who desire to learn": the persevering youth, whose sympathetic intelligence must be won, for the future is theirs. Pieces, first in two and then in three parts *obbligato*, are given, and the development of polyphonic playing is the highest goal; in these, again, purity, accuracy and grace are required. The musical idea contained in the piece was intended to ripen the imagination of the learner and encourage him to produce both improvised pieces (*inventiones*) and more artistic works, duly arranged and worked out (*composition*). Finally, in the carrying out the ideas, he was to study the organism of a grand composition.

How far it was from Bach's views that a clavier pupil need only be trained in mere finger-work—in *riding* the

---

[95] " Auffrichtige Anleitung, Wormit denen Liebhabern des *Clavires*, besonders aber denen Lehrbegierigen, eine deutliche Art gezeiget wird, nicht alleine (1) mit 2 Stimmen reine spielen zu lernen, sondern auch bey weiteren *progressen* (2) mit dreyen *obligaten Partien* richtig und wohl zu verfahren, anbey auch zugleich gute *inventiones* nicht alleine zu bekommen, sondern auch selbige wohl durchzuführen, am allermeisten aber eine *cantable* Art im Spielen zu erlangen, und darneben einen starcken Vorschmack von der *Composition* zu überkommen." B.-G., III., Preface.

clavier, as he used to call it—how thoroughly, on the contrary, he guided the player at the same time through the intricacies of construction and the feeling of the piece he was playing, is here made very plain; nay, that he knew, too, how to incite him to living and original production by arousing his own formative faculty. The plan of the programme seems at first sight somewhat confused; still it is not very difficult to disentangle the different ideas that seem to cross each other. He wished, in the first place, to produce an exercise-book for the clavier-player; but, with the mechanical practice, he proposed also to cultivate the pupil's artistic powers generally, both on the side of impromptu invention—which was so essential for the use of the figured bass, then thought very important—and on that of serious composition. Having formerly been a first-class scholar in St. Michael's School, at Lüneburg, he had not so far forgotten the terminology of rhetoric as not to know that *collocatio* (order) and *elocutio* (expression) are indispensable to *inventio* (invention); and thus, immediately after his observations on good inventions, we find order or arrangement discussed, and a *cantabile* handling; otherwise, certain other sections might have seemed more nearly connected with it. The ancient rules of rhetoric come in again in another place, when he teaches that in two-part pieces purity of execution is essential, but in three-part pieces correct and finished playing—not meaning, of course, that purity is less requisite in three parts, or correctness and finish in two. It is perfectly clear that these words stand for the *emendatum* (correct), *perspicuum* (pure—*i.e.*, clean and neat) and *ornatum* (finished—*i.e.*, winning or graceful) of the old rhetoricians, the three chief requisites of a good image or statement.[96]

It is extremely interesting to observe that, in spite of his musical occupations at Lüneburg, Bach cannot have been a

---

[96] Compare Mattheson, Grosse General-Bass Schule, p. 8 (Hamburg, 1731). " For there were then already wise folks who were not satisfied that the figured bass should be carried out *correctly* (*recht*)—that is to say, without mistakes—but demanded that it should also be *good*—that is, artistic and elegant. Hence it is not without reason that we contrast right (or correct) and good (or beautiful)."

very bad Latin scholar, since twenty years after he still had a present memory of these matters, and could apply his knowledge so aptly and correctly. But what is more important still is the direct parallel he institutes between music and human speech. This he could not possibly have done if he had not felt that the art of music was a perfectly evolved language of emotion—that the progression of each part in his polyphonic pieces was like the utterance of a distinct personage—that the composer was indeed in some sort a dramatic poet. It would seem that he often applied this comparison himself in order to disclose to his pupils the inner life and purport of his music.[97] What a performance it must have been that was inspired by this idea needs to be no further enlarged upon. And it is now quite clear what the association of ideas must have been that led him to call the two-part pieces *inventions*, when the name *preamble*, which he had applied to them in Friedemann's book, did not satisfy him. Indeed, the true prelude style is certainly not recognisable in these strict and simple compositions, with the exception, perhaps, of two of them; still the name *inventions* is not particularly happy either; the pieces are too far from mere inventions, too carefully worked out, and, in contrast with the *sinfonias* which follow them (this was the new name so happily bestowed on what had been first called "fantasias," and are now more generally known as "inventions in three parts."), can at most be accepted as pictures more lightly projected and more directly invented.[98]

If we now consider this work—which more than any other displays an instructive aim—from the side of its artistic

---

[97] Since Forkel, p. 24, says quite the same thing as we have here derived from our analysis of Bach's words, I have no doubt that his statement is founded on a direct communication from Friedemann or Ph. Em. Bach. Birnbaum also states it distinctly in Scheibe, Critischer Musikus, p. 997.

[98] It is not probable that Bach was the first to use the name "invention" for a piece of music. See App. A, No. 7. In Breitkopf's list for Easter, 1763, we find, on page 73: "Bach, Joh. Seb., Capellm. und Musik-Director zu Leipzig, XXII. *Inventiones* vors Clavier: Leipzig, fol. *a*, 1 thl., 12 gr." This established the interesting fact that a printed edition of the Inventionen existed so early as 1763. Only the number is puzzling; it is perhaps a misprint for XXX., which would include the sinfonias.

value, it is a striking illustration of the fact that Bach's fertility and inspiration grew in direct proportion as he more distinctly formulated his educational purpose. In extent alone is it inferior to the two parts of the Wohltemperirte Clavier and the Kunst der Fuge, in its more modest dimensions and the limitations imposed by the fewer means employed, but certainly in no other respect. Nay, in one way it is superior to them and to all Bach's later clavier music—namely, in its perfect novelty of form. The master had good reason to seek for a suitable name for these pieces, for there was nothing like them in all the clavier music of the time. It is not merely the treatment of the polyphony, which pursues its two or three parts without an instant's interruption, and nevertheless reveals the harmony throughout with absolute distinctness and fulness, never diminishing in interest by monotony of changes, never wearying us by repetition: more than all this is the whole development of each tone-picture—the sovereign independence with which all the forms of music are applied—the canon, the fugue, free imitation, double and triple counterpoint, episodic working-out, inversions of the theme—all combining and following each other in pieces of very moderate extent, without anywhere obtruding themselves on our notice; these are what render the Inventionen und Sinfonien unique in the whole body of clavier music. A slight leaning towards the Italian music of the time is certainly discernible, and somewhat more decidedly in the sinfonias than in the inventions. Still these lovely blossoms have sprung principally from Bach's own organ and clavier pieces: a quintessence, as it were, of all he had accomplished. And yet we can perceive the efforts he has made to ripen his work, for, besides the fifteen inventions in two parts and the sinfonias in three, there are among the works he has left more pieces in the same style, which prove that he only gave his more earnest labour to what seemed to him best or most suitable for the work out of the abundance he could produce.

He seems to have struggled longest after the ideal form of the inventions. A two-part fugue in C minor is, as it were, the butterfly half-escaped from the chrysalis; it is, properly speaking, a fugue only to the end of the sixth bar, and afterwards more and more of an "invention" in its freedom of theme and episode. Another aspect of the process of evolution is discoverable in three small pieces in D minor, E major and E minor, of which the first especially shows already in a high degree that bewitching play of inversion and double counterpoint which is equally characteristic of the inventions and the sinfonias. But they all three have that verse or song form in two divisions, which Bach with one exception excluded from the collection that forms the book, because it disturbed the flow of polyphonic development. He has perfectly attained his aim in another piece in C minor, only it is difficult to decide whether it should be designated *fantasia* or *invention*.[99] If the rest could be regarded merely as studies this is a *paralipomenon*, which, from its completeness of form, is worthy to be called either. The case is the same with two two-part and two three-part compositions which, however, he thought worthy to grace another work; these are the preludes in C sharp major, F sharp major and A major in the first part of the Wohltemperirte Clavier, and that in B flat minor in the second. It is certainly doubtful whether the last piece can have been written so soon as this; still, there are in the second part several pieces which can be proved to be of early date. In the case of the other three their early origin is certain, since the date of the first part of the Wohltemperirte Clavier is well ascertained. That Bach should here have given the name of prelude to what he elsewhere calls a sinfonia shows, again, how unique was the style here unfolded. We have already seen three names applied to one and the same kind of piece. Another three-part piece must, on the contrary, be regarded as a study—we might say, indeed, is a study—for the first sinfonia in C major, and it is in the same key. This also is entitled a prelude.[100]

---

[99] P. S. I., C. 7, No. 2; No. 1, III., V., VI. (all in Vol. 201); C. 9, No. 10 (Vol. 212, p. 2).

[100] P. S. V., C. 8, No. 7 (Vol. 247). Here it is placed among organ works, and appears indeed to have been used for that instrument. Its connection with the sinfonia will be apparent to any careful examiner.

Bach, it is evident, was no less doubtful as to the order of the thirty pieces than as to their designation. This is interesting to observe, because we can detect that it was always ultimately decided on instructive grounds. The work exists in three distinct autographs. In Friedemann Bach's Little Clavier Book the inventions are separated from the sinfonias, but the principle of the arrangement is the same in each, since, so far as the number and keys of the pieces allow, they proceed first upwards and then downwards.[101] A second autograph copy, which also seems to have been written in Cöthen, gives the pieces in the same order; but the sinfonia in the same key is placed immediately after each invention. The third copy, on the contrary, is arranged on the principle of using the keys only in the ascending order of the scale; here all the inventions are given first, and then the sinfonias in the same order.[102] At the present day we should rather attempt to group such a collection of clavier pieces with reference to a pleasing contrast in their various characters, but such an idea seems never to have occurred to Bach; indeed, there was no need for it, for each is so different from every other that, in whatever order they may be played, the effect of contrast will necessarily be produced.

The scheme of the inventions is, for the most part, that they are divided into three sections, and have a remote resemblance to the form of the Italian aria. The first part is generally obviously disjoined from the rest by a decided cadence on the dominant or supermediant (*i.e.*, the relative major, if the piece be in the minor); it comes in again in a more or less shortened form at the close. The sixth invention alone is in the two-part "song-form," with repeats; but here, too, at the end of the second section,

---

[101] Thus: C major, D minor, E minor, F major, G major, A minor, B minor, B flat major, A major, G minor, F minor, E major, E sharp major, D major, C minor. But there are only twelve bars written of the D major sinfonia, and that in C minor is wholly wanting.

[102] The second autograph is a little book in oblong quarto in the Royal Library at Berlin. The writing is not quite that of the Leipzig period, but sharper and more pointed; still, it is essentially different from Bach's writing when he was at Weimar. The B.-G. edition is founded on the third autograph.

the first is practically introduced again, constituting a regular sonata movement in miniature. The first and seventh inventions are in three sections, but without being in cyclic form. Even in this contracted sphere an astonishing variety of conception is displayed.

The first invention grows from this germ—

and is treated in imitation; at the third bar there begins a moderately long episode on the inverted subject, which, in the course of the piece, is set in opposition to the subject in direct motion, giving rise to playful alternations of each. Of all fifteen this one has the most reserved and dispassionate character, and even the theme is somewhat conventional, only revealing its importance by degrees.

The second, in C minor, is quite different. A passionate and eager phrase comes rushing in, followed by its exact facsimile and companion, and the two figures pursue one another, first one and then the other taking the lead. It is a canon on the octave; at first the upper being followed by the lower part, at the distance of two bars, until bar 10 is reached, when they change places; the lower part leads, now followed by the upper one, bar for bar, in exact imitation; then, after a short digression, the first order of parts is resumed until the last bar.

No. 3, in D major, is of a merry character, and consists of free imitation; it is followed and eclipsed by the gloom of No. 4, which is treated now in direct, now in inverse motion, in D minor.

No. 5, E flat major, which sets off at once in two parts, is in double counterpoint on the octave throughout, modulating through B flat major, C minor, and F minor, and then, by means of episodical extensions, back to the original key. It is a piece full of grace and dignity.

The one in E major (No. 6) is full of roguish fun, and also begins in two parts; a prominent part is taken by the double counterpoint, and the formation of episodes on the chief theme.

No. 7, in E minor, shows an affinity in form with No. 1; but its expression is different; it is suppliant and mournful; still, in spite of its disturbed character, has extraordinary melodic beauty.

In contrast to this, the next one, in F major (No. 8), is full of a happy and innocent contentment. It begins in canon, and after the twelfth bar it becomes freer, and most lovely little episodes are developed.

The companion piece in the corresponding minor key (No. 9, F minor), which is similar in form to No. 5, but richer in episodic formations, again is full of impassioned strains of sadness, which rise to a great intensity of effect in bars 21 to 26.

No. 10, in G major, begins like a fugue, but without any of the fetters of that form; it flits to and fro—now in imitation, and now in episodic extension. There is a piece of quite wanton fun at the *reprise*, when the upper part takes upon itself the double duty of a theme and response (*dux* and *comes*).

The piece which follows has a character of tormented restlessness. A chromatic counter-subject of two bars long, attached from the outset to the chief theme, evolves, by means of inversion in the fourth bar, an episode of the most painful and insistent kind, which reappears in alternation with the original counterpoint at bar 14. The phrases consist of six bars, but at the last two recurrences it is only five bars and a half long, and these two are merged directly in one another without any cadences to give a moment's rest (see bars 12, 13, and 18); each beat of the bar is more restless than the last.

A feeling of honest German fun is given by No. 12, in A major, which corresponds in form to Nos. 5 and 9.

The two next inventions, in A minor and B flat major, both have somewhat the character of preludes, because the subjects and the workings-out alike move almost exclusively in harmonic passages; the second betrays a close relationship to the prelude of the B flat partita in the first portion of the Clavierübung. The three-fold division is, however, preserved in this case. In the B flat invention the first subject comes in in canon at its repetition (bar 16, in the middle), and in this place it has a bold, soaring character, while in the former it was dreamy and rather melancholy.

The last invention comes in gravely, yet not without a certain dignified grace; in its fugal working-out it alternates with episodical interludes, founded on the counterpoint to the theme. It is remarkable that the theme does not come in alone, but is supported by short notes in harmony in the bass. In no other of the inventions is this the case; but it always happens so in the sinfonias (or inventions in three parts), and I believe that these have had a reactive influence on this invention. The form of the Sinfonias in their barest outlines is founded on that of the Italian instrumental trio, as settled by Corelli and diligently cultivated by Albinoni, Vivaldi, and many others; it had also become widely known throughout Germany.

We have already seen how Bach could make the Italian forms serve his purpose. He was the more likely to go on assimilating these forms since, in Cöthen, chamber-music demanded his chief energies. It was a favourite mode of construction in the fugal movements of these trios—which were generally written for two violins, string bass, and figured bass—that the theme should not be given out quite alone, but supported by a figured bass, played by a fourth performer—some particular accompanist—with proper harmonies. Subsequently the figured bass was drawn into the fabric of the fugue, and the accompanying harmonies had to follow and support the other parts, so as to fill up any gaps left by the instruments in the harmonies. It is plain that Bach got this manner of treatment from the trio form. But that this influence was purely external is shown by the fact that absolutely nothing remains of the supporting bass but a slight remnant in those few bars at the beginning; and even this appears, not as the groundwork of a set of accompanying chords, but as a free and independent part, and it soon establishes its full and individual right to a share in the polyphonic working-out. For this reason a further comparison would be out of place; the Bach clavier trios (or inventions in three parts) are so thoroughly original that we cannot but doubt whether he thought at all of the Italian instrumental trio, and whether he did not rather think of his own works of that kind, to which we shall presently draw the reader's attention. The style which pervades these last has at least a general similarity to the sinfonias, although the treatment is much broader and bolder. The original source of both sets of works, however, is the organ. The polyphony is chiefly fugal, and but seldom (Nos. 2, 5, 15) canonic, although there is no actual fugue or canon in the set. It is difficult to say anything more on the general characteristics of the form of the sinfonias. With the greatest freedom, and yet with a marvellous order and arrangement, every device of thematic and episodic polyphony is employed; each piece is a microcosm of art—a vessel of richly cut crystal filled with the purest and most precious essence. The effect of these is heightened if the (two-part) invention corresponding in number to each is played before it. For it cannot be doubted that the composer conceived each pair at the same time; in Nos. 15, 12, and 6 the themes agree, although not perhaps note for note, yet in their chief features. No less do the emotional characteristics of each correspond, and even in respect of form certain connections can be seen.

The one in C major has the same brilliantly polished, reserved, and dispassionate nature as the invention in the same key, and the theme is treated in the same masterly manner, whether in its direct or inverted form:—

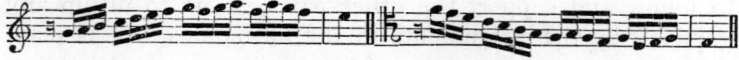

The C minor invention, with its feverish restlessness, is followed by a sinfonia full of the deepest yearning, which, however, is interrupted throughout by quicker passages; the imitations are in canon form, as in the invention, but are thrown into the background in the second half by episodic developments of the loveliest kind.

The one in D major exhibits the same cheerful character in three parts that its forerunner did in two, and adds to it some tender prattling. We might call this sinfonia, above all, a golden fruit in a silver shell. What a charming theme is this—

which, when it is taken up by the second part in A major, is met by this as second subject, which pleasantly accompanies it—

while both parts are surmounted with this lovely and expressive phrase:—

And now it proceeds in triple and double counterpoint with delicious animation, and between whiles little episodes peep out roguishly and vanish again; there is something to be said of nearly every bar. By this means of performing each pair together the player will be able to appreciate in every number the inner connection between the invention and its companion sinfonia; how that which was foreshadowed there is brought out with firmer strokes; what was there abrupt and stiff becomes gentler in contour; what was trivial becomes deepened; what was restless and capricious becomes calm and firm; and sometimes, too, how anxious complaint is intensified into the deepest woe and the most acute suffering.

Particular mention must be made of the caressing sweetness—almost in the style of Mozart—of the sinfonia in E flat, coming after the haughty grace of the invention—a piece which is also distinguished above all the others in its form, the upper parts being in free canon, while the bass repeats the same figure in each bar.

Then comes the touching lament of the E minor sinfonia, which, however full of character, is yet in contrast to the pathos of the invention, and of an organic beauty such as Bach alone could create.

Allied to its invention, both in feeling and in contrast, is the sinfonia in G minor (No. 7), but in this a lovely melody, with broad and sustained notes, is continuously heard above the lower parts in a way that could hardly be deemed possible in a piece of such polyphonic character; it also has the character of an aria.

How splendid is the intensifying of emotion in coming from the invention to the sinfonia in A minor, the theme of which, with its working-out by passages in thirds and sixths, bears a distinct resemblance to the beautiful organ fugue in A major!

The theme of the B minor sinfonia is strictly evolved from that of the invention—at first in canon, but then treated in a more episodical way, as in No. 2; but, in the meantime, impetuous passages in demi-semiquavers rush up and down, overtake, and cross each other in contrary motion, which, by the way, must have been a difficult task for the fingers on the clavichord, which never possessed two manuals.

Finally, the sinfonia in F minor (No. 9) is positively steeped in anguish and pain; but to compensate for this abnormal emotion the form is as strict and concentrated as possible; or, to speak more accurately, the feeling first acquires its intensity by means of the form, and the form achieves its astounding concentration by means of the feeling, so that the two factors are indissolubly heightened in effect, in and through one another. The piece consists of three themes in triple counterpoint, neither of them inferior to the others in force of expression; and, though they are externally in contrast to each other, they nevertheless all reflect the same particular emotion:—

At first only the first and second themes appear together, but then all three come in together nine different times in four permutations, and various though nearly related keys. As a relief to these there are five interludes, of which the first is in free form, but the others are built episodically upon the first theme; direct and inverse motion and augmentation unite to give a most complex effect. Of the workings-out of the themes, sometimes two follow closely on one another, without an interlude between them, sometimes one stands alone; but this is always in accordance with a fixed plan, as the interludes correspond closely with each other. The following scheme may serve to exhibit the wonderful arrangement of the phrases; the Arabic numerals stand for the number of times that the workings-out occur consecutively, and the Roman for the different interludes—

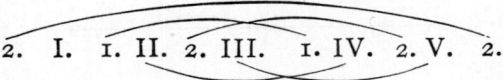

(the curved strokes indicate the connection between the different parts of the piece). The first interlude alone (bars 5 and 6) stands by itself and without any corresponding part afterwards; it releases the ear for a moment from the strain caused by the first bars, and lets it become acquainted with the themes themselves. In the development of the piece the greatest possible daring is shown in the way of bold leaps of intervals, discords resolved by skips, and false relations. But we must not suppose that this originated in a forced and artificial correctness. Bach shows in the sinfonias how he can combine the most elaborate art with the most exquisite loveliness of effect. It was no caprice of pedantry that gave rise to the sinfonia in F minor; on the contrary, it bears the impress of true imaginative work. This will be the ultimate feeling about the piece if, instead of being repelled by the disjointed impression produced perhaps on most people on a first acquaintance with it, we pay due attention to the course of the separate parts, and in playing the piece try and give to each its full effect as a living individual. Then perhaps it will strike the sensitive hearer with awe that such a deep abyss of woe could open in the human breast but he will enjoy the comforting thought that the moral force of the will can by perfection of form triumphantly bridge over even such depths as these. Kirnberger, Sebastian Bach's theoretical pupil, regarded the F minor sinfonia as an experiment which was bold even to obscurity, and quoted it as a proof of Bach's having infringed the rule which forbids the unprepared entrance of the fourth in the bass—the so-called chord of the six-four.[103] The passages where Bach took this liberty (bars 4, 14, 19, 27, and 32) sound indeed strange, and at first unsatisfactory; they are justified by his general view of the nature of part-writing, which, according to him, took its rise no longer from the polyphonic system, but the harmonic. More will be said on this subject in another place.

III.

BACH AS A VIOLINIST.—THE SUITE AND THE SONATA.—WORKS FOR VIOLIN, VIOLONCELLO, FLUTE, ETC.

BACH'S first musical impressions arose from his hearing his father's violin-playing. His own first public post was that of violinist in Weimar. He afterwards held this position in the Duke's band for nine years, and in course of time was promoted to be concertmeister. In his later years, too, he did not neglect his string-playing, and in instrumental pieces in several parts he preferred to play the viola, since he enjoyed, as it were, surveying the harmonies on both sides from the middle position; besides, good viola-players, and such as satisfied his requirements, were seldom to be met with.[104] It is not indeed necessary for a concertmeister to be an extraordinary performer—a thorough musician with moderate technical qualities, if they are genuine and sound,

---

[103] Kunst des Reinen Satzes, II., 2, p. 39, ff. All the six possible permutations of the three themes are given here, the second and sixth of which are not employed by Bach in the sinfonia.

[104] Forkel, p. 45. Quantz, Versuch einer Anweisung, &c., p. 207.

will often be much more useful in this place[105]—and considering that no contemporary, not even his son Philipp Emanuel, mentions Bach's violin-playing, and that he devoted his chief energies to the organ and clavier, we shall hardly be wrong in supposing that he was not possessed of any extraordinary facility on the violin. But it is not intended to give the impression of his having been an insignificant player. He was not the only one of the great musicians of Germany in whom the defects possibly arising from an insufficient technical method have been made up for by the individuality and magnificence of their creative genius. Thus, the pianoforte playing of C. M. von Weber lacked much in neatness and equality, and nevertheless it had in it a soaring flight and a magic charm which enraptured those who heard him. Nay, even Handel's violin-playing, although he laid less stress upon it after his residence in Hamburg, was sufficiently full of fire and importance to induce great performers to come to learn of him.[106] Bach's familiarity with stringed instruments extended so far that he even undertook to make changes in their shape and build to suit his purposes; while in Cöthen he invented an instrument, something between a tenor and a violoncello, which was held like a violin, and had five strings tuned to the notes C, G, d, a, and e'; he called it *viola pomposa*, and wrote a suite for it; he also had it used in Leipzig for the easier performance of his difficult and rapid basses.[107] But from a consideration of his compositions for strings, and especially those for violins alone, it follows that his knowledge of this branch of art must have been enormous. Granted that he may not have been able to execute these himself quite perfectly—and yet he must also have been a good violoncellist, for he wrote

---

[105] Quantz, p. 179: "But there is no absolute necessity that he (*i.e.*, the leader of the music) should possess the skill to execute passages of peculiar difficulty; this can well be left to those who try to distinguish themselves only by playing to please, of whom there are plenty to be met with."

[106] Chrysander, Händel, I., p. 228.

[107] Compare on this App. A, No. 4. A composition for flute and *viola pomposa* without bass, by Telemann, is to be found in his Getreuen Musikmeister (Hamburg, 1728, pp. 77 and 84).

similar solo compositions for that instrument—at all events, only one who had the most thorough experience of the capabilities of the instrument and their utmost limits could produce such works. And such experience is not to be attained by theoretical speculation, but only by practical exercise.

It is easy to see, from the individuality of Bach's violin compositions, from the number of parts employed, from certain types of figures, and from the interweaving of one or even two more instruments *obbligato*, that their peculiar style did not, at all events, take its rise solely in the nature of the instrument. The overpowering influence of the organ style, which relentlessly overmastered all that came within its reach, is even here too evident to be overlooked. With special regard to the employment of double stopping, it must be added that Corelli had already raised it to an important place in the art by his violin sonatas with harpsichord accompaniment, and had even attempted to employ fugal treatment, as far as it could be conveniently adapted to the instrument; and that the Germans, who at the end of the seventeenth century were in other respects far inferior to the Italian violinists in execution and inventive faculty, had cultivated this very branch of *technique*—viz., playing in more than one part at a time—with especial energy, which is very significant, as showing how it was their nature to strive after harmonic richness much more than after clearness of melody. Nikolaus Bruhns, Buxtehude's talented pupil, who was mentioned before as an eminent violinist, attained such a proficiency in double stopping that it seemed as if three or four violins were being played together; and then he would sometimes sit down in front of the organ with his violin, and with his feet add a pedal-part to the full harmonies he elicited from the strings.[108] In the case of that native of Celle, Nikolaus Strungk—of whom we have spoken before (Vol. I., p. 201), and to whom Corelli, after hearing him play, was forced to cry in amazement, "I am called *Arcangelo*, but you must be called *Arcidiavolo*"

---

[108] Mattheson, Ehrenpforte, p. 26.

—the chief feature of his performance was probably the playing in several parts, since he, as well as Bruhns, was an organ and clavier player.[109] The secretary of the elector of Mainz, Johann Jakob Walther (born 1650), who was also a violinist, gives no little attention to this particular branch of *technique* in his "*Hortulus Chelicus*," published in 1694, and especially alludes to it in the title.[110] So that in adopting this form Bach was aiding and furthering a tendency which was particularly German; but he wedded to it all that had been acquired by the Italian feeling for form, and improved this by means of his incomparably greater power of construction.

He wrote a book containing six compositions in several movements, without any accompaniment, for the violin, and a similar one for the violoncello (or the viola pomposa). I do not know whether he had any predecessor in the isolated treatment of a stringed instrument, but I should be inclined to doubt it, because the Italians, who were the general exemplars in matters of this kind, in defiance of all art, put a cantabile and one-part style of playing in the foremost place, whereby the music must have lost half the intended effect through being deprived of the supporting harmonies.[111] All that can be said with certainty as to the date of these compositions is that it cannot be later

---

[109] Gerber, Lex. II., col. 604. Strungk tuned his violin in such a way as to facilitate the performance of passages in harmony.

[110] "*Hortulus Chelicus*. Das ist Wohl-gepflanzter *Violini*scher Lust-Garten Darin—auch durch Berührung zuweilen zwey, drey, vier Seithen, auff der Violin die lieblichiste Harmonie erwiesen wird."—" Garden of the Lyre. That is, the well-stocked pleasure-garden of violin practice, in which is shown how to produce the loveliest harmony on the violin by occasionally touching two, three, or four strings."

[111] A remark of Mattheson's in Critica Musica may be here quoted, I. (1722), p. 224, i.: "I was lately shown a *Suonata per Violino solo del Sigr: M. M.*, which, to say nothing of the key being in F minor, demands such long fingers that I know no one who could easily execute the passages ('*praestanda praesti*ren'). And yet I cannot blame such a work if its object is intended for showing his own exceptional advantage in the way of long fingers, or else for an exercise, rather than for everybody's execution for them to boast of." The violin solo sonatas by Telemann and Pisendel were certainly all composed later than Bach's.

than the Cöthen period. The six violin solos consist of three sonatas and three suites; and if at the present day we are accustomed to speak and write of Bach's six violin sonatas, it is an inaccuracy for which Bach is not to blame.[112] The difference of the two generally is clearly definable, since the suite consists principally of dance-forms, which are mostly introduced by a prelude.

The suite-form, by which a new laurel branch was added to the immortal crown of Bach's fame—for he it was who brought it to its highest perfection—stretches back its roots into the sixteenth century. Its development I believe to be easily discernible in general, although in the details much is obscure.[113] It was in dance-music that the song-tunes from which they took their rise were first transferred to the imitating instruments, and then were independently enlarged and extended, the song-form being retained. It followed naturally, from this, that people wanted to hear such dance-music on other festive occasions, so that, as its popularity increased, the composers turned their attention to this kind of composition. Wandering musicians carried the most popular of these from place to place and from country to country. About the year 1600, the Italian *paduanas* and *gagliardas*, or *romanescas*, became very widely known; and how charmingly they lent themselves to instrumental treatment is seen in the five-part pieces in this form which Johann Moller (the court organist at Darmstadt) published in the years 1610 and 1611. Besides these, much attention was given to the forms of the *volta*, the *passamezzo*, the *balletts*, and the *intradas*, which last were called "Aufzüge," or "processions," by the German composers, and indicated a particular kind of solemn music which preceded a more intricate dance. The ring-dance (*branle*) and *courante* came from France, unless, indeed, the last was originally Italian. The only

---

[112] P. S. III., C. 4 (vol. 228).—B. G. xxvii., 1., *Vide* App. A, No. 4.
[113] I must own that my opinions are founded on very incomplete materials. Any one who knows the state of musical history with regard to the seventeenth century will pardon this.

THE HISTORY OF THE "SUITE." 73

German dance which figures here is the *allemande*, showing, as it would seem by its name, that there were no different varieties of dance in Germany. But to make up for this the Germans showed their individuality in the working-up of the foreign forms; thus, in 1604, Johann Ghro, of Dresden, published thirty paduanas and gaillards, and announced in the preface that they were " set in the German manner."[114] No general name could be given to such collections of dances, seeing that they were not yet arranged according to any comprehensive principle. The only arrangement was that the paduana was followed by the gagliarda, because of the contrast between their rhythms (the first being in common time and the second in triple time).

At this stage came the Thirty Years' War, which, although it brought the most fearful misery upon Germany, nevertheless appears to have forwarded the development of the *suite* in that country. The idea of choosing out from among the dance-forms of civilised Europe the most original and adaptable, and uniting them in an artistic whole, received a certain impulse in the unhappy state of Europe, which had driven Italians, Spaniards, Frenchmen, Swedes, Danes, and Poles to jostle one another in the busy ferment of warfare for a series of years. When affairs became settled again, efforts were plainly made to arrive at a higher form of art. To this end it was, before all, needful for the clavier composers to step in and preserve the adaptable musical essence of the dance-tunes by transferring them from the province of the unruly guilds of German town musicians to the quieter, purer atmosphere of domestic music. All evidence goes to prove that the invention of the clavier suite must be sought for in the school of Sweelinck. That it was German is plainly seen from the order of the pieces, which had by this time become firmly settled, and in which the allemande held the first place, followed by the courante;

---

[114] Compare Carl Israël, Die Musikalischen Schätze der Gymnasialbibliothek und der Peterskirche zu Frankfurt a. M. (Frankfurt. a. M., 1872, p. 41).

for the finale two new dance-forms were employed — the Spanish *sarabande* and the English *gigue*, either both together or singly. The Germans continued to work up foreign materials "in a German style," and to associate them with their own forms, as they had done at the beginning of the century. At the same time dance-music naturally did not decrease among the town pipers, but was carefully cherished by them, though more or less disjoined from its original practical purpose. It was very likely that for "table-music" or other festive occasions several contrasted dance-tunes would be played one after another. Whether from this circumstance a sort of customary and regular order arose—as we saw in the case of the juxtaposition of paduanas and gagliardas—is at present uncertain. At all events, the town pipers had a common name for such collections of dances, which name was appropriated by the composers of the clavier suites; its general meaning was that of a complete whole consisting of many parts, and thence it came to be applied to clavier variations. This name is *partie*, or, in Italian, *partita*.[115]

In the sets of variations, as in the sets of dances called by this name, the same key was generally adhered to in all the sections, which shows that their origin was that of mere external juxtaposition. The form invented by the German clavier masters was now adopted in the Italian chamber sonatas of Corelli and his followers. But the different technical requirements of the violin, and the nature of the Italians, who gave the greatest attention to melodic beauty, threatened to obliterate the characteristics of the separate types until they could no longer be recognised. Even the German nature, with its predilections for harmonic elaboration, could not entirely counteract this. Then the French, who loved strongly marked rhythms, adopted these sets of dances. French orchestral music had long been familiar

---

[115] The Lustige Cotala (compare Vol. I., p. 20, note 36) says, at p. 181: "One asked us if we had by us any *sonatas*, or any other things set for *instrumenta*. I said, Yes, and opened my portfolio and took out several pieces and parthies." On Kuhnau's *parties* see p. 237, note 79; on the *partie* as a variation see Vol. I., p. 127.

to the court bands and guilds of town pipers,[116] and its influence had extended to clavier music as well. It was no less a man than Pachelbel who was first infected by it, and who transferred the French overture to the clavier (see Vol. I., p. 124). But this was not enough; the French must lay their hands even on the clavier dances. But the order of the parts was already so firmly fixed that they could not venture to alter it. The component sections even among them were still the *allemande,* the *courante,* the *sarabande,* and the *gigue;* but they introduced them by an overture, and added dances of their own at the end, such as the *gavotte,* the *minuet,* the *rigaudon,* the *passepied,* the *bourrée,* and the *chaconne,* which was properly Italian, inserting it before the gigue, or else in substitution for it; but in all these they had recourse to the most pronounced rhythm. And as this is the most important element in any dance, it was only natural that they should give these compositions the names by which they became known. The form returned to Germany under the name of "suite," there to attain its fullest perfection under Sebastian Bach, who had been preceded by George Böhm, and whose contemporary Handel treated it in a few important compositions. Bach ultimately rejected the French titles; he restored the name *partie* in one of his chief clavier works, as well as in the three so-called suites for violin solo. The suite—the oldest form of instrumental music in many movements—is a German production, in the perfecting of which all the then important nations of Europe took a more or less active part.

It is more difficult to define the limits of the sonata, the history of which is contemporary with that of the suite. It does not entirely dispense with dance forms, but never consists of them alone. What was understood by the sonatas of Giov. Gabrieli at the beginning of the seventeenth century—how this form influenced Sebastian Bach's

---

[116] Compare Vol. I., p. 201. The oboe, or the "French Schalmey," was quite a common instrument among the town pipers in the last decade of the seventeenth century, as appears from "*Battalus, der vorwitzge Musikant.*" Freyburg, 1691, pp. 63, 64. Compare, also, the catalogue on Vol. I., p. 169.

cantatas, how sometimes it keeps to its original unity of movement, and again sometimes is extended to two movements—has been shown before (see Vol. I., p. 124). When, in the second half of the century, chamber music and solo violin-playing made such gigantic strides in Italy, Corelli adopted the form in two movements, and by freely combining two such pairs of movements, made up a whole constituting a three-part *sonata da chiesa* (church sonata), which he transferred from chamber music back to sacred music, and in its new form it was accompanied on the organ. When it was not intended for church performance dances might be inserted; and this was done sometimes in the manner of a suite, with an allemande at the beginning, and more frequently by concluding with a gigue. The chief principle, then, of the sonata consisted in the alternation of slow, broadly treated movements with quick and generally fugal ones; they must also contrast with each other in rhythm, and, if dance-forms were introduced, they had to be adapted to this rule. As in the suite, so also here, the normal number of movements is four. But, inasmuch as the second slow movement was by preference in another key, the sonata resembled the concerto, of which the influence was also felt in the construction of the several movements, especially the last. Thus the Gabrieli sonata assisted in forming a new type of art without being absorbed into it; indeed, even in the latter half of the seventeenth century, the secular sonata for full instrumental band was still retained and kept up by the town pipers, who used to play it as a beautiful piece of music in the manner of a motett[117] at their performances of "table music" and on other suitable occasions.[118] These two types are thus quite distinct from Corelli's church and chamber sonatas. While in course of time the suite

---

[117] So runs the well-known definition in M. Prætorius. Syntagma Musicum, III., 2.

[118] The "Lustige Cotala" (p. 44) says, in an account of a "table music" at a wedding: "We then played a *sonata*, in which there was a *fuga;* he himself played the viola." Battalus (Loc. Cit., p. 63) says: "Then the musicians began to make music. They played a sonata with two trumpets, two hautbois, and a bassoon, which sounded very well."

was left entirely to the clavier, the reverse was the case with the chamber sonata, in so far that, having been properly a violin composition, it remained so for a time; however, as we have seen (Vol. I., p. 237), it was transferred to the clavier by Kuhnau. No direct step is perceptible from this stage to the modern sonata form; but the polyphonic nature of the allegro movement, which no longer appealed to the spirit of the time, had to be replaced by another kind of treatment. It was another Italian—Domenico Scarlatti—who detected this; he wrote clavier sonatas, of which each movement was in song form, homophonic, and decorated with new and tasteful passages. The three-movement form of the concerto was already adopted, and this opened the way by which the modern sonata could reach its final perfection, coming down by way of Philipp Emanuel Bach and Haydn to Beethoven.

Sebastian Bach's three sonatas for violin solo display the form in its strictest and purest development. All are in four movements. But, inasmuch as the second slow movement is in another though nearly allied key, while the rest keep to the original key, the fundamental scheme is still in three movements; the first adagio unites with the following allegro to form one section, and in the majority of cases leads directly into it by means of a cadenza on the chord of the dominant. The difference of the modern sonata consists only in the style of the several movements: in other respects the conditions are similar. In both, the first part is followed by a second in thorough contrast to it, and the last movement is an attempt to resolve into itself the different meanings of the other two; thus, by a psychological process, as it were, it serves as a bond between them. In both, the greatest musical importance is given to the first allegro movement, while the finale is of lighter calibre as regards both substance and form. The introductory adagio is not considered indispensably necessary in the later sonata; but in the most important branch of the genus, the orchestral symphony, it is almost always retained, although in a shortened form. Still, even here the introductory adagio is in its scheme very different from the middle one, retaining throughout the nature of a prelude; while the second adagio enters as a piece of music in strict form. This fundamental rule is indeed not always adhered to in the master's other works in sonata-form; it is enough that it appears generally in such a way as to leave no doubt as to this method of treatment being intentional.

In spite of their being written for an instrument which, in comparison with the organ and clavier, and considering the direction in which the composer's chief power and importance lay, is confined within the narrowest limits, these sonatas have something very great about them. By the extension of the chords produced by double stopping, and the skilful employment of the open strings, an almost inconceivable fulness of tone is produced; the sharply defined rhythms, the bold and often almost violent execution made necessary by the polyphonic treatment, and especially the fire and force of the fugal allegro movements, give to the sonatas more perhaps than to any other of Bach's instrumental works a certain demoniacal character. The type of the first movement is settled by Corelli in his violin sonatas (Op. 5); it is broad and melodious, but a free and fantastic character is imparted to it by the introduction of many ornamental figures of various types. This character is rendered even more conspicuous in Bach by the form which his polyphonic treatment takes; because, for practical reasons the progression of the subsidiary parts frequently can only be indicated, and has to be filled up by the hearer. In the beautiful and impassioned introductory adagio of the first sonata, in G minor, the melody first appears in the middle part; the upper part meanwhile is progressing in single notes and phrases, and seems to vanish away; it is then lightly touched in the course of the melody, and so brought to sight again; but it is there all along for him who can hear it. From bar 14, where the melodic phrases of the opening are repeated in C minor, the upper part plays the principal *rôle;* the middle part is not on that account inactive, but often displays remarkable independence. The same method is of course pursued with the lower part; the melody has often to be interrupted for a moment in order to play a short bass note, and often the bass is vaguely heard through the ornamental figures. It is quite an exception when the treatment is in more than three parts, allowing of course for the single four-part chords occasionally thrown in for the sake of fulness. In the case of fugues, it is a matter of course that the counterpoint can only be very simple—mere chords often having to suffice for the accompaniment of the theme—and in spite of the more animated time, much can only be indicated. Passages of runs and arpeggios in one part, are introduced in Corelli's manner to rest and prepare the mind for the polyphony that is to follow. For the rest we may be sure that the master of the fugue himself would be as careful as possible to satisfy the strictest requirements; we find not only free fugatos, but genuine and thoroughly worked-out fugues, displaying the most marvellous wealth of combination and invention. At present the best known fugue is the one in the first sonata; Mattheson, in two of his writings, draws attention to that in the second sonata (A minor), as being a model in its kind; which circumstance is of importance when we remember what his feeling was towards Bach. He says:[119] "The length of the theme in a fugue is, in some measure, left to the taste of the composer, but, as a general rule, it may be said that the earlier and the closer the response follows the theme the better will the fugue sound. Frequently the most excellent working-out is found in a fugue on the fewest notes. Who would ever think that these eight short notes—

could be fruitful enough to give rise naturally to a counterpoint of more than a whole sheet of music without any considerable extension? And yet this has been done, as is plainly

---

[119] Kern melodischer Wissenschaft, Hamburg, 1737, p. 147.—Vollkomm.-Capellmeister, p. 369. Mattheson's citations are incorrect in both places, but chiefly so in the first, where he writes the theme in 3-4 time; the sharp before d'' is lacking in the second piece.

to be seen, by the great Bach, of Leipzig, who was particularly happy in this kind of composition; and more than that, it is treated directly and in inversion."

But the fugue in the third sonata in C minor must be allowed to surpass these two in grandeur and importance; the only obstacle which prevents its attaining a wide-spread popularity is its enormous difficulty. It will presently be shown that this difficulty may probably be explained by the history of its composition. The third movement of the G minor sonata consists of a charmingly conceived *Siciliano* in B flat major, with marvellous polyphonic working-out, but the tender character of this dance-form is injured by the strength and harshness of tone necessarily resulting from the employment of several parts; this is one of the cases in which the alien character of the style is very prominent. The corresponding movement of the A minor sonata is in C major and in song-form, with two sections; a broad and expressive melody comes in supported by short, staccato notes in the lowest part, while the middle part takes a small share in the development of the melody. The sonata in C major has, in this place, a largo in F major of quite as expressive a character, which is not separated from the other movements by any pause. In all three the treatment of the last movement is identical. The form is in two sections, and in only one musical part; it flies along in almost incessant semiquavers; the type is exactly that of the last movement of a concerto, which has been before described (Vol. I., p. 409).

Bach's standpoint at the time of his writing these sonatas is plainly shown by the circumstance that all three reappear either in parts or in their entirety, in the form of clavier or organ pieces. The whole of the middle one is arranged as a clavier sonata, and transposed for this from A minor to D minor.[120] Although it does not exist in Bach's handwriting, the wonderful genius displayed in the arrangement leaves no room for doubting that it is from the composer's own hand. In its clavier form it is so much richer in treatment that at times the original appears

---

[120] P. S. I., C. 3, No. 3 (Vol. 213, p. 24).

a mere sketch beside it; the natural way in which the polyphonic richness is brought out shows what was the proper birthplace of Bach's violin compositions of this kind. At the same time it is quite certain that the sonata was originally written for the violin; this is shown not only by many details, but also by the selection of D minor as the key, whereby a great deal is brought into depths of pitch which we are not accustomed to in Bach; this transposition was necessary in order to avoid too great extension in the compass of the parts. The fugue of the G minor sonata exists in an arrangement for the organ; that it, too, was first intended for the violin is shown by the nature of the theme.[121] Its connection with the original is here not so close, and in two places there is an extension of about one bar; the arrangement must have been made very soon after its composition, for a copy exists which was made in the year 1725. A more complicated method has been taken with regard to the C major sonata. A clavier arrangement of the first movement was made by Bach, and its lower pitch (G major) shows again that the violin form is the older.[122] In this, more than any other place, it is clear that the composer's imagination clung to the clavier style, even in the original conception. Here there appears no melody with fanciful figures suitable to the violin, but that sort of soft progression of slowly changing harmonies which owes its origin neither to the nature of the violin nor to any Italian influence; indeed its source is very evident. Even with the most perfect performance the intention of the composer can never be realised on the violin; the execution of chords of three or four notes has inevitably a violent and harsh effect, which contradicts the character of the movement. When played on the clavier in that enriched form which the composer himself gave it, it is discovered to be one of the most marvellous productions of Bach's genius; one of those preludes which is pervaded by a single rhythm throughout, and in which the harmonies softly melt

---

[121] B.-G. XV., p. 149, transposed into D minor.—P. S. V., C. 3, No. 4 (Vol. 242). Compare Appendix A., No. 4.

[122] P. S. I., C. 3, Appendix pp. 1 and 2 (not in English edition of Peters).

into one another like cloud shapes, while from beneath their magic veil comes a long-drawn and yearning melody. All that the heart feels, and that the tongue vainly endeavours to utter, is here revealed at once, and yet remains remote and unapproachable. No human being since has ever created such tones!

Nothing remains of the other sonata movements arranged for the clavier. Did any such arrangement exist? As to the fugue this question may safely be answered in the negative. I take it to be rather a transcription of an organ piece. Its theme consists of the first line of the chorale tune " Komm, heiliger Geist, Herre Gott," an unheard-of procedure in a violin sonata. The contrapuntal artifice is so complicated for a solo violin that it demands impossibilities from the player, and skilled players have assured me that at times the method of writing is as contrary to what is playable as if the composer had never set eyes upon a violin at all.

Special attention must be drawn to the fact that Mattheson, in his " grosse Generalbasschule," gives a description of an organ fugue on the same theme, which agrees almost entirely with Bach's treatment. He gives the theme as follows—

and remarks "(1) that it is the beginning of a chorale; (2) that in the response not the least artifice is attempted; (3) that the fugal counter-subject might be chromatic, and that the fugue might therefore be treated in augmentation since it is too simple without such treatment; (4) that the chief subject allows of being turned both ways; (5) that direct and inverse motion could be united and harmonised together; (6) also that there are many other neat combinations which might be made into the subject and the response by bringing them into closer contact," &c. He then (p. 38) gives his own views as to the right way of treating the work.[123]

---

[123] The theme is also quoted in the Vollkomm.-Capellmeister, p. 363.

Now in Bach's violin fugue there occurs at the very beginning the chromatic countersubject which he requires; and we here find that complicated stretto which he speaks of under No. 6, with the entrances now after the first, and now after the fourth note of the theme (compare bars 93 ff and 109 ff); here, too, is the inversion (bars 201 ff), and of course there is plentiful employment of double counterpoint. Those of Mattheson's precepts which Bach fails to comply with are either unnecessary for the free and irrepressible swing of the fugue (such as the inversion of the theme, retaining the exact semitones), or else tasteless (such as the combination of the direct and inverse motions of the theme); such a combination could only be pleasing if one part entered after the other. But, in fact, Bach worked in a richer material than Mattheson could elaborate. Thus he adds to the chromatic countersubject a second countersubject—

(compare for example bars 135-136 and the episodical extension in the bars that follow; also bars 293-294 and bars 107-108); possibly these are included in Mattheson's "&c." Now it certainly cannot be said that the whole treatment as devised by Mattheson, and actually executed by Bach, was altogether an obvious one—only the introduction of the chromatic countersubject was not unusual, but occurs several times in works of this period [124]—so that Mattheson must have been familiar with Bach's fugue, though certainly not in its present form. It would be very easy to point to the Hamburg journey as the time when Mattheson became acquainted with it, but Bach would hardly have taken his violin solos with him, if indeed they were written at this time, but rather his organ pieces only, and possibly some vocal compositions. Mattheson for the first time betrays his acquaintance with the violin sonatas in the year 1737, but he had already set the fugue theme quoted above

---

[124] For example, in a fugue by Pachelbel on a theme nearly similar to this one in Commer. Musica Sacra I., p. 156, and also in Sebastian Bach's Organ Canzone.

as an exercise at a trial of organists on October 8, 1727. It may be remembered that in the same passage of the "grosse Generalbasschule," the theme and countersubject of Bach's great G minor fugue had been quoted, which the author had once employed for a similar purpose (see p. 23). Apparently there was a chorale fugue by Bach on the hymn "Komm, heiliger Geist, Herre Gott," which he had performed in Hamburg in 1720, and which he then made free use of for the violin fugue. Little as it profits a man to deck himself in borrowed plumes, yet it was not an unheard-of trait in Mattheson, since he could not prevail upon himself to name Bach as the composer of the G minor fugue. Many of his remarks sound as if he wanted to justify himself in his own eyes—for instance, when he calls the fugue theme "easy," and insists that it is borrowed from a chorale, and when he suggests the introduction of the chromatic countersubject with the remark that without it the fugue would be too simple. And yet here it is, in its place.

Among Bach's chief works I place the three sets of compositions for the clavier known by the name of the French Suites, the English Suites, and the six Partitas. We may go farther, and extend the observation, with some inconsiderable limitations, to the three other violin solos, and all the compositions for solo violoncello, with which we shall have more to do subsequently. But we must remark in passing that in them Bach only adheres quite slightly to the Italian form, but keeps very closely to that of the clavier suite as it had been developed first by the Germans, then by the French, and lastly by himself. There is a greater adaptation of style in these than in the sonatas. What was gained from the French was the careful marking of the rhythm, which was almost unrecognisable among the Italians, although it is the most essential feature of dance-music. Corelli's Sarabandes are often nothing more than slow Sicilianos, and sometimes are divested of every characteristic, even the three-time. His Gavottes lack the important feature of beginning with the second half of the bar of common time, and once he even begins with a short note before the bar. In the Allemandes no rhythmic type whatever can be recognised; a dignified movement and a polyphonic style seem to be considered sufficient; the dignity is often sacrificed in an Allegro or Presto. The old courante, according to its etymology, indicated among the Italians a piece full of flying, running passages, but the Germans and French gave it a grave, sustained, and impassioned character. But the Italian type had shown itself too prolific in art, and too firmly settled, and in particular had exercised too important an influence upon the last movement of the concerto (see Vol. I., p. 409), to be easily driven from its position; thus there existed side by side two utterly different types. It would be well to distinguish once for all between the *corrente* and the *courante*.

From the attempt to smooth away all peculiarities of rhythm, there grew up among the Italians an inclination to confuse the characteristic divisions of the suite with those of the sonata and the concerto. The French, on the other hand, not only gave important assistance to the development of the suite form, by giving definiteness and prominence to the contrasts of rhythm, but they also made an important advance by the suppleness and pliability of their passages, and by the elegance and richness of their adornments. Since this was willingly acknowledged by every one, they wished to be taken as the universal models of suite compositions. But, happily, the musical world was not so blinded by this as to set aside the service, at least as important to this form of music, rendered by the Germans, and Mattheson says, quite justly and with happy preciseness: " The French indeed wrote, or rather pretended to write, Courantes and Allemandes for the clavier, and indeed they made very free with these particular forms; but he who will compare without prejudice their bald, thin, and empty jingling with a well-written, *nerveuse*, German courante with its distributed polyphony, will see how little truth there is in their pretensions."[125] He might have said yet more; for the French not only did nothing towards the combination of the dances to a com-

---

[125] Neu eröffnetes Orchestre, S. 187.

plete whole, but rather checked the growth of the best type of arrangement. Marchand, in a suite in D minor, introduces after a prelude, first an allemande, two courantes, a sarabande, and a gigue, and then a chaconne in four "couplets," a gavotte and a minuet. The true idea of concluding with a gigue is either misunderstood or ignored. Couperin's sets of pieces are scarcely to be called suites at all; the second set of his *Pièces de Clavecin* (D minor) contains allemande, two courantes, sarabande; then a free interlude in D major; a gavotte, minuet, *les Canaries* (a kind of gigue) with a variation, a passepied and trio, a rigaudon and trio—eleven independent pieces varying between major and minor—a rondeau, and then for the close a piece in the style of a gigue but very freely written. The fifth set (A major) consists of an allemande, two courantes, a sarabande, a gigue, and six rondos, intermingled with pieces in free style. In spite of this he never entirely quits the ground of the suite, for he keeps to the same key throughout, even when he does not begin with the usual pieces. But it is clear that he never felt the necessity of welding together the various constituent parts to one perfect whole of many members. Why he failed is seen in the titles that Couperin bestowed on his clavier pieces, in which he either followed or introduced a custom that was general, though I have previously found it only in the case of Gaspard de Roux. He tries to make them represent definite personifications, or a connected series of actions, or even of public events. Thus, titles such as these occur: The Sublime, The Majestic, Industry, Shyness, Gloom, Danger; and still more individual names in the rondos and free pieces, such as The Florentine Lady, The Sailor of Provence, the Gossip, Nanette, Manon, Mimi; and in one case a piece called "The Fair Pilgrims" is in three sections, of which the first represents the pilgrimage (treated of course in a thoroughly French and frivolous manner), the second a petition for alms, and the third gratitude for it.[126] Another piece in three sections is called *les Bacchanales*, and

---

[126] Couperin's Werke, edited by J. Brahms. Vol. I. Bergedof, near Hamburg, 1871, p. 55. f., from whence the examples quoted above are taken.

is divided into *Enjouements bachiques, Tendresses bachiques,* and *Fureurs bachiques.* Pictures from nature are of rare occurrence, and it is always the idea of motion that is represented, as clouds, bees, or a floating veil. The chief point of interest is nearly always outside the pieces, as it were, and music is a mere accessory; briefly it is a refined kind of ballet music, and is like the orchestral dance tunes in Lully's operas transcribed for the clavier. This corresponds, as we saw in another case (Vol. I., p. 246), to the theatrical nature of the French, but restrains and destroys the activity of free musical genius. We must, however, consider that every day the French either saw or performed in the theatres the other orchestral types of dances, if not actually the allemande, and so got to connect them in their minds with certain definite ideas and representations. In Germany the case was different; in the courts it is true that they aped the French ballet music, but the people were, happily, not affected by this, and so could enter into the purely musical value of the dance-form without any disturbing connections with the stage.

It had become usual among the German suite-composers before Bach to work out the courante on the lines of the allemande. This custom exactly corresponds with the fugue form of two or three sections, so much in favour with the northern organ masters, in which the same theme is worked out in a variety of ways. This has been fully gone into in speaking of Buxtehude's organ works, and the analogy with the suite form was then pointed out. (See Vol. I., pp. 264 and 275.) The contrasts of rhythm between the allemande and the courante, and between the courante and the gigue, are precisely the same as those between the three sections of Buxtehude's great E minor fugue. It is clear that, with the exception of Froberger, it must have been chiefly the northern masters who improved and enlarged the suite in the second half of the seventeenth century, for they have stamped it with a lasting impression of their own specific individuality. All through Reinken's *Hortus Musicus* the courantes are nothing more than modifications of the allemandes which precede them, and the sarabandes, and even the gigues, show a plain connection with the allemandes. Walther says of the allemande: "In a musical partie (*i.e.*, suite) it is, as it were, the proposition, from which flow forth the other movements, for instance the courante, the sarabande, and the gigue, like the constituent parts of it (*partes*)."[127] Handel, too, followed this method in all essential particulars. The suites of the second and third collections of his clavier works especially show a connection between the allemande and the courante; it is found, too, in the first collection, and in particular in the E minor suite, where it goes farther and is carried on into the gigue.[128]

From these circumstances, as well as from productions like Buxtehude's suites on the chorale "Auf meinen lieben Gott,"[129] it is evident that the Germans first sought for the unity by which the different sections might be combined, in the use and treatment of variations. They must soon have become aware, however, that by this means the characteristic types of the dance-forms were too severely weighted, and so for the most part they contented themselves with treating the courante alone as a variation. But when once the rule was given up this custom might easily be abrogated. Sebastian Bach saw that unity could be attained by a scheme of internal treatment alone, since the four fundamental types had been arranged in so happy an order that each contrasted with, and made up for, the deficiencies of the others. He, therefore, adhered steadily to these four types, and the few exceptions he allowed himself only prove the rule. It is not difficult to recognise even in the suite that far-reaching principle in art: that of the triple form. The allemande and the courante are closely connected even when their subject-matter is not the same. The allemande has always a medium character, being neither fast nor slow, neither solemn nor impassioned; it is, as Mattheson says, "the picture of a contented and satisfied mind, delighting in order and repose."[130] It is always in common time: it con-

---

[127] Lexicon, p. 28.
[128] German Handel Society's Edition, Vol. II. Peters, Vol. IV., A and B.
[129] See my edition of Buxtehude's Organ Works, Vol. II., sect. ii., No. 33.
[130] Vollkomm.-Capellmeister, p. 232, s. 128

sists of two sections, tolerably equal as to length, of from eight to sixteen bars each on an average, and has this peculiarity, that it begins with either one or three short notes before the bar (Böhm in one case begins with seven, Bach with four, semiquavers). The harmonies are broad, and by preference in broken chords, and the upper part has various figures. This character is not decided enough to produce the effect of contrast; but it gains in intensity in the courante which follows it, which, even when not treated in the Italian manner, gives the effect of animation by means of its triple rhythm. Besides the notes before the bar at the beginning and the similar length of the sections—which it has in common with the allemande—its typical characteristics consist in certain disturbing syncopations of accent produced by the mixture of triple and double rhythm; for the 6-4 time runs into the 3-2 time, and *vice versâ* (according to rule at the end of each section). According to Mattheson the courante expresses "Hope," but this is saying too much, for a definite emotion of that kind is not to be attained by instrumental music; still the view has some foundation.[131] Thus the allemande prepares the way for the courante, and both form one whole, just as the introductory adagio and the fugue do in the sonata.

Then the sarabande fills the same place in the suite as the second adagio in the sonata, or the slow movement in the modern form of the sonata. Its movement is quiet and solemn, suggesting Spanish haughtiness, and its tone is grave and calm. It is in triple time, and begins, as a rule, with a whole bar. The accent falls by preference on the second beat, which is so prolonged as to include either half or the whole of the last beat. Its length was originally limited to two sections of eight bars each. This number was seldom exceeded in the first section, even in later times, but the second section was extended to twelve, sixteen, or even more bars, and sometimes even a third section is added.

Lastly, the concluding gigue corresponds entirely to the

---

[131] Vollkomm.-Capellmeister, p. 231, s. 123.

last movement of the sonata and concerto—in place of which it was frequently employed—its quick running and capering form, which is inconsistent with thoughtful intensity, forms a vivid contrast to the allemande and the courante as well as the sarabande. The more grave impressions produced by the movements that have gone before are gathered up into a cheerful and animated form, and the hearer goes away with a sensation of pleasant excitement. The rhythm of the gigue is chosen from the most animated kinds of triple time—12-8 (or common time in triplets), 6-8, and 3-8 times are of the most frequent occurrence, but 6-4, 9-8, 9-16, 12-16, and 24-16 are also found. It is of course in two sections, and its length, which cannot be well compared by numbering the bars because of the various *tempi*, is proportional to the rest of the dances. The Italian and the German modes of treatment have not, it is true, resulted in a complete division into two different types—as in the case of the courante—but yet its structure has become modified. In the former treatment it is essentially homophonous, accompanied in chords by a figured bass and other instruments, but in the latter it is developed polyphonally even to a genuine fugue. This is fresh evidence that the northern masters had a hand in the formation of the Suite. Just as in their organ fugues in several movements, the last was in 12-8 or 6-8 time, so here they write the concluding piece of the suite in 12-8 or 6-8 time. They were the first to follow the method of thematic working out with any great powers of invention, and it is to them that we owe the plan of the second section of the fugal gigue—which has been a type and model since the end of the seventeenth century—namely, the treatment of the theme of the first part in inversion in the second.[132] It is evident that by this means, without detracting from the cheerfulness of the concluding movement, a balance was struck between that and the gravity and importance of the other movements, and the suite form was consolidated, and made worthier to receive and utilise material from wherever it might come.

---

[132] Thus it is, for instance, throughout Reinken's *Hortus Musicus*.

In Bach's clavier compositions the fugal gigue with the inversion in the second part is the only form employed, whereas Handel nearly always treats his in the Italian manner; where he does not, he supports the entry of the theme with harmonies, and in only one instance—in the F minor suite in the first collection of his clavier works—does he make use of an inversion in the second section. The French did nothing worthy of mention towards the development of the gigue.

The form was now complete in itself, and when new numbers were introduced—as they were sure to be from the multiplicity of unemployed and piquant French dance types—a suitable place was found for them between the sarabande and the gigue. Since the first section of the suite was composed of the allemande and the courante together, something might be inserted before the gigue without disturbing the balance; nay, the greater the importance given to the gigue by the use of polyphony, the more would the need be felt of some light, short, bright intermezzo, in contrast to the measured gravity of the allemande, the passionate eagerness of the courante, and the calm dignity of the sarabande. So it became customary to insert one, or even two or three such pieces, according to circumstances; the forms of the *gavotte*, the *passepied*, and the *bourrée* were found ready to the composer's hands, and they ultimately gave rise to the scherzo and minuet of the modern symphony.[133] Whether the impulse to do this came first from the French or the Germans must be specially inquired into.

In the suites of Dieupart and Grigny, which Bach copied out for himself, and which seem to date from the year 1700, a gavotte and a minuet are found inserted between each sarabande and gigue (Vol. I., p. 202). And in Germany, Johann Krieger published, in the year 1697, six *Parties* " consisting

---

[133] These inserted pieces are well called "intermezzi" by G. Nottebohm, who has written a series of well-considered articles on the nature of the suite in the "Wiener Monatsschrift für Theater und Musik"; Vol. for 1855, pp. 408-412, 457-461; Vol. for 1857, pp. 288-292, 341-345, 391-396. He has also pointed out the reciprocal internal relations of the other movements.

of Allemandes, Courantes, Sarabandes, *Doubles* (*i.e.*, Variations on a dance-tune), and Gigues, besides interspersed Bourrées, Minuets, and Gavottes." In every case the French were misled by their theatrical proclivities. When once the admission of such an intermezzo was decided upon, it could be more freely used, and just as in Beethoven's later and latest works the scherzo often comes before the adagio, so in Bach the sarabande is several times preceded by a gavotte and a passepied, or the like.

If we now compare the form of the suite with that of the sonata in respect of their general value, we find that the comparison is not, as we should expect, so greatly in favour of the latter, but that they must be considered as of equal value. In the sonata, the inner connection is so close that an element of contrast has to be brought in by the introduction of a movement in another key; and the very existence of this form depends on the adequate treatment of this contrast. The sonata proceeds with the inexorable precision of a causal nexus; its very essence is emotion or *Pathos*. The suite has no internal self-contradiction to overcome; it presents, on the level ground of one unchanging key, a concordant and reasonably differing variety; its spirit is that of repose or *Ethos*. The love for the sonata form, which increased from Bach's time onwards, corresponds to the love for subjective and impassioned expression, and to the decided leaning towards poetry, which appears more prominently from this time forth in German instrumental music; while in the suite a simpler and more purely musical view of the art is taken. Accordingly the materials of the sonata were invented by individual composers, while those of the suites had their rise in the natural forces of nationalities. The suite, in spite of the multiplicity of its movements, is simple when compared to the sonata; it is a single stone cut with many facets, and the sonata is a ring composed of many stones. Thus the movements of the suite could never give rise to such expansions as those of the sonata; a development corresponding to that of the sonata into the symphony was quite impossible. But the introduction of the minuet or scherzo in the same key as the first and last movements, shows that even the sonata, when the number of movements was increased beyond three, could not transgress the law of the suite, since it was that of all instrumental music. The relations of the movements in both forms are based on the catholic and inherent laws of art. But the more purely musical the character of the piece, the more freely can the question of the propriety of different kinds of forms and of details be decided by feeling. So that if it is difficult to prove the necessary connection of the movements in a sonata, in each particular case, the difficulty is much greater in the case of the suite. Nevertheless, the demands of art are always valid, and the reason why the diligent study of the masterpieces of this class has so great an influence in the formation of musical taste is because it leads, as scarcely any other means can, to the appreciation of the finer and more delicate degrees of proportion and feeling between the sections and the whole.

It still remains for us to glance at the suites for violin and violoncello[134] in detail. The three *Parties* correspond nearly in character to the three sonatas, with which they were united in one work by the composer. He seems to employ the contrast between the two forms as a structural plan, for each sonata is followed by a suite. All three are remarkably irregular in their formation. In the B minor suite each movement has a variation which follows it like its shadow. It is very probable that the addition of variations in the suite was an after effect, resulting from the attempt to work up the dance-forms that followed the allemande as variations on it. At all events this method occurs at a very early period; for instance, in an excellent suite in F♯ minor by Christian Ritter (Kammerorganist at Dresden from 1683-1688,[135] subsequently Capellmeister in Sweden), in which the sarabande is followed by two

---

[134] B.-G., XXVII., 1.
[135] Fürstenan, Zur Geschichte der Musik am Hofe zu Dresden, I., pp. 267 and 299.

variations; in a *Partie* by Johann Ernst Pestel (b. 1659), where the treatment is the same; and in the violin suites in Walther's *Hortus chelicus* (for instance, Nos. 20 and 23), where each dance is followed by a variation, as in Bach. When employed in moderation there was nothing to complain of in a method which impressed the import of a piece more plainly on the hearer, and gave it in certain ways a more emphatic resonance; only the fundamental relations of the parts must not be disturbed by it, and the number of permissible variations must never exceed two.[136] Bach in almost every case contented himself with one. Since in the B minor suite he wished to avail himself of variations in all the sections, he could hardly use the gigue as the concluding movement, as it is ill adapted for variations. In its stead he chose the Bourrée, a dance-form of light, pleasant, and somewhat reckless character (in common time, beginning with the last beat of the bar, moderately quick, and smooth in its style), which, however, has here an air of uncouth jollity, only coming back to its proper character in the *Double*. For the rest, it is wonderful how sharply defined, in spite of the limited means, are the individualities of the types; the most difficult task was in the case of the allemande, which combines richness of harmony and polyphony with varied figures in the upper part. The courante in the French and German style is contrasted with one in the Italian style which follows it, as a variation, rushing by in a wild and irresistible manner. After this, the sarabande comes in heavily and proudly in three and four part harmony.

The second suite, in D minor, has the customary four movements. In the quick time of the gigue no fugal style can be expected of course from the single instrument; it is throughout in one part, but produces the effect of harmonic

---

[136] Joh. Jak. Walther begins his *Scherzo da Violino solo* which appeared in 1676 with a regular suite in four movements, in which the allemande has no fewer than six variations, while the courante has only one, and the sarabande and gigue none at all. Since, however, the courante, the sarabande, and the gigue are all formed on the same subject as the allemande, it is strictly nothing but a continuous series of variations.

fulness by the way in which the passages are written. This is followed by a chaconne. It is longer than all the rest of the suite put together, and must not be considered as the last movement of it, but as an appended piece; the suite proper concludes with the gigue. The French were fond of introducing chaconnes, but in a somewhat different form from that now known to us. They were accustomed to treat both the chaconne and the passacaglio in clavier music with a much greater freedom. Either no ground theme at all was taken, but a number of phrases of four bars long and in the same rhythm in 3-4 time were put together, in which case the artifice consisted in making them grow more animated and louder (as is done in a chaconne in Muffat's *Apparatus musico-organisticus*); or a subject of four bars with a *reprise* was taken and repeated without alteration after each of a number of independent phrases, or *couplets*. Couperin and Marchand usually followed this method, and Muffat, in the work just mentioned gives a passacaglio constructed on this scheme. The form was closely allied to that of the rondo, and even the essential triple rhythm was not always adhered to by Couperin; the only characteristic that is retained is the somewhat grave and solemn style. Bach so far adopted this rondo form that in several cases, and with great effect, especially in the middle and at the end he returns to the eight bars of the opening and introduces new ideas between the repetition of them, but in general he remains true to the fundamental working-out of the chief theme in the old and thoughtful way. In all cases his manner of treatment corresponds exactly with the definition given above of the passacaglio (Vol. I., p. 279); a free handling of the theme was necessary inasmuch as it had to be played solo on a violin. An analysis of the whole will not be unacceptable, since the notes of the themes are often dispersed through different octaves in the whirl of the figurations, so that their connection is not always easy to recognise. The first and principal theme is as follows—

it is once gone through, and then comes (bar 17):—

In the manner of a rondo, but in a new dress, the first theme returns once; and the second, which soon is wrapped up in smooth semiquaver figures, recurs twice. The third, from bar 49 onwards, comes in, but never in a simple form; without the ornaments it would be approximately in this shape—

in which it must be noticed that the skips of thirds are afterwards enlarged into tenths or inverted into sixths. This is gone through four times, then, at bar 81, the second returns, still with new figures, and resulting in its second section in a new modification of the first;[137] this in its turn prepares for a fourth subject, which comes in with bar 97—

and is carried on to bar 121; then, to conclude, all four themes come in, combined with marvellous genius, the third being in this form—

alternating during four bars with rushing demi-semiquavers and semiquavers; the first then recurs in a broad and heavy style, as at the beginning; and lastly the second and fourth together for four bars, so that the former lies in the

---

[137] The meaning of this passage (bars 89—97) is not doubtful, but we do not get a clear idea of the first theme. It is, however, quite in accordance with rule that after so long a silence it should reappear once more. Mendelssohn and Schumann were of the same opinion, as appears from their arrangements.

upper part, and the latter in the form it took in bar 113, in the lowest part. In gavottes, minuets, bourrées, and in chaconnes, contrasting trios were in great favour; such an one now makes its appearance in D major with a modification of the third theme, which must be reckoned as a fifth subject on account of its independent treatment:—

In bars 133—209 it becomes larger and freer, and at last is varied once only, the ground rhythm being retained; then the minor mode recurs, and all five themes are gone through again in it: the third until bar 229, the fifth (in the form adopted in bar 161) combined with the second until bar 237, the fourth until bar 241, and again the third until 249; and at last this production, so prodigious of its kind, is crowned by the first theme in its original form. The hearer must regard this chaconne as some phenomenon of the elements, which transports and enraptures him with its indescribable majesty, and at the same time bewilders and confuses him. The overpowering wealth of forms pouring from a few and scarcely noticeable sources displays not only the most perfect knowledge of the *technique* of the violin, but also the most absolute mastery over an imagination the like of which no composer was ever endowed with. Consider that all this was written for a single violin! And what scenes this small instrument opens to our view! From the grave majesty of the opening, through the anxious restlessness of the second theme to the demi-semiquavers which rush up and down like very demons, and which are veiled by the weird form of the third subject—from those tremulous arpeggios that hang almost motionless, like veiling clouds above a gloomy ravine, till a strong wind drives and rolls them together and scourges them down among the tree tops, which groan and toss as they whirl their leaves into the air—to the devotional beauty of the movement in D major where the evening sun sets in the peaceful valley. The spirit of the master urges the instrument to incredible utterance; at the end of the major section it sounds like an organ, and sometimes a whole band of violins might seem to be playing. This chaconne is a triumph of spirit over matter such as even he never repeated in a more brilliant manner. There have been many attempts in later days to melt down the precious material for other instruments. Little as this is to be blamed on æsthetic considerations—for Bach himself led the way with his own additional arrangements[138]—yet it is certain that it needs a master hand to do it with success, and it was no contemptible task for two of the greatest musicians of modern times, Mendelssohn and Schumann, to make an adequate pianoforte accompaniment to the chaconne. The wonderful result shows how profound and fruitful is the original theme. And yet Schumann, who is known to have arranged accompaniments for all six violin solos in this way, not only intensified the general musical import, but also shed a clearer light on the chaconne form by following it out phrase for phrase in the most exact way. The fear that by this means an incoherent effect might be produced is as unfounded as it would be were it a whole suite; for it is the principle of the suite which animates the organism of this chaconne. In both there are movements and groups of movements of different characters in juxtaposition which must be all in the same key; in spite of all changes of emotion and all their passionate character, one ruling feature is evident to every one, the undisturbed unity of repose. And so the union of the chaconne with the suite had at last a still deeper issue; the amalgamation of two equally complete forms to a more perfect whole, so as to give the greatest possible importance and value to the idea which permeates them both.

At the beginning of the third partie, in E major, there stands a wonderfully fresh prelude moving in incessant semiquavers, now in runs, and now in arpeggios. It was not unusual to begin the clavier suites with a prelude. That

---

[138] Thereby confuting his pupil Kirnberger's remarkable assertion that no other part could be added to the violin and violoncello solo without harmonic faults. (Kunst des reinen Satzes, I., p. 176.)

there is here a transference of style is proved by the
composer himself, for he subsequently arranged the movement for organ obbligato and orchestra (transposing it into
D major), and used it as an instrumental introduction to a
cantata written for the election of senators in 1731.[139] The
prelude is followed neither by an allemande, a courante, nor
a gigue; all these forms are lacking in this suite. Bach has
for once given the reins to his love for contrast as he has
done nowhere else, excepting in his suites for orchestra,
where he had historical precedent for it. Thus, there comes
next a *loure* in 6-4 time, moderato—a kind of gigue, but slower
and graver.[140] Then comes a gavotte in rondo form, with its
rollicking merriment, a genuine piece of fun in the style
of the older Bachs. Two minuets about fill the place
of the sarabande, the first fine and solemn, the second
tender and delightful—a charming little pair. Between these
and the concluding gigue is inserted a bourrée. This last
partie has perhaps the same meaning with regard to the
whole collection of the six solos as the gigue has in the
single suite; its bright cheerfulness almost takes away the
impression produced by the solemn greatness of the others;
but the connection of the different emotions is brought
about by the concluding allegro of the C major sonata.

In the six compositions for violoncello alone[141] a general
character may also be perceived, which is distinct from that
of the works for the violin in proportion to the difference of
the instruments in readiness of expression. The passionate
and penetrating energy, the inner fire and warmth which
often grew to be painful in its intensity, is here softened
down to a quieter beauty and a generally serene grandeur,
as was to be expected from the deeper pitch and the fuller
tone of the instrument. In the same ratio (four to two) in
which the minor keys preponderated in the other case, do

---

[139] B.-G. V., 1, No. 29. The whole suite is also extant in an arrangement for
clavier (Royal Library in Berlin), and the autograph is still existing.

[140] Mattheson (Vollkomm.-Capellmeister, p. 228, § 102) says that the loures
had a proud and inflated style; but Bach's loures, at all events, are the very
opposite to this.

[141] P., S. IV., C. 1 (vol. 238a). B.-G. XXVII., 1.

the major keys preponderate here; while there one-half consisted of sonatas, here there are only suites; and while there all the suites differed in form from one another, here they all agree entirely. Each begins with a grand prelude, boldly constructed out of broad arpeggios and weighty passages, and in the fifth suite Bach introduced in their place a complete overture in the French style, in which the adagio, with its long pedal points on C and G, has an imposing and glowing character. Then follow, according to rule, the allemande, courante and sarabande, and before the concluding gigue in each case there are two intermezzos which consist in the first two suites (in G major and D minor) of minuets, in the third and fourth (in C major and E flat major) of bourrées, and in the last two (C minor and D major) of gavottes. The uniformity of design in all the suites shows, too, that the last suite is conceived of as one whole with all the rest, and hence we may include it without further remark among the violoncello solos, although it was written for the *viola pomposa* invented by Bach. The great extent of tone opened up by this instrument may have been one reason for the remarkable and quite unique beauty of the work, and it is to be most deeply deplored that, with this *viola* has vanished the possibility of ever hearing this suite which was destined for it in its original form.[142] Since Bach himself devised the instrument, he must have played it himself, and this suite upon it: we can the more easily imagine this, since we are told he was a skilful player on the tenor. I, not being a proficient, cannot judge of the technical difficulties of the work, as compared with the violin solos, but they seem to be very considerable. At all events, for the violoncello he possessed a friend in the gamba-player Abel, who could be at hand to give his advice on technical points, and for whom the suites were probably written. Their value is much more than this, however; the decisive character of the dance-forms places them almost

---

[142] In the Peters' edition, superintended by Fr. Grützmacher, and unfortunately much disfigured by many arbitrary additions, it is arranged for violoncello and transposed into D major, whereby much is of course lost.

above the violin suites, and they show just as much inexhaustible fulness of invention. In a single case—in the C major suite—the courante is evolved from the allemande; this is the exception, before alluded to, to the general statement made on page 85, as to the rule of contrast in the movements. The majestic structure of the C minor courante, which is built on a subject rising gradually from the depths at intervals of a bar, and in the second part sinking down again in a scale passage of equal length, should be noticed as a remarkable point.

The way in which Bach treated the violin and violoncello as solitary instruments was of course entirely altered as soon as, by the introduction of another and a supporting instrument, the duty of elucidating the harmonies no longer fell to their share; for indeed, although treated with the most masterly skill, they could never be entirely free from a certain feeling of constraint. The most usual combinations were those of one, two, or even three stringed instruments with the clavier, the first called a solo and the last trio, which was not quite a consistent name, inasmuch as in the trio the string bass when it was added only strengthened the bass of the harpsichord, while in the solo an accompanying harpsichord bass is taken for granted. The task of the accompanist was of only secondary importance; he had only to put in the background before which the other parts were to move, and so his part was not written out in full, but the harmonies indicated by numerals over the bass part were sufficient, and had to be turned into a complete fabric of harmony without gaps or mistakes, on the spur of the moment. Bach followed this custom, although not without modifying it to suit his own views. An inserted part in mere progression of chords, and without intrinsic importance, was little in accordance with his artistic soul, which always strove after organic unity. The *Basso continuo*, from the beginning of the seventeenth century, when an Italian, Ludovico Viadana, first employed it in vocal works for one or more voices, until Bach's period, had exercised an universal influence in all branches of the art—since there was scarcely one worth mentioning that is altogether without it; with its assistance all the component parts of the music could be grasped with freedom and certainty, a result which, without its aid, could scarcely have been obtained at all, and certainly not in so wonderfully short a time. But the chief object was now either to free itself entirely from this support, or else to cause it to strike root and become alive, so that the branches might embrace and grow together into one organism. The whole development of art was directed towards the latter course. Bach transferred the polyphonic style of the clavier into this sphere of art also, and thus his mode of writing for instruments supported by the harpsichord though less strange than that of the solo compositions just mentioned, is not exclusively formed on the inherent nature of those instruments, but upon the character of Bach's polyphony, which had already attained its full growth and took its rise from the organ.

It is of importance to make as clear as possible Bach's fundamental principles with regard to the performance of accompaniments, for this, the highest attainment in the art of that time, has now quite died out, and yet an essential part of the possibility of making Bach's works accessible in our time rests on its due reawakening. Before all else—we are speaking now only of chamber music—two different cases must be clearly distinguished. Disregarding the custom of the time, Bach in his most important works treats the clavier as an obbligato instrument. These works are almost entirely in strict trio form, really in three-part writing, in which the clavier takes two parts, and a violin, viol da gamba or flute the third; there is one largo in quartet form in which the clavier takes three parts. The background of harmony is almost entirely dispensed with. Only when, at the opening of a movement, or of a new working out in a movement, the theme is first given out by the chief instrument over a supporting bass, full chords must be struck in order to give especial distinctness and emphasis. Bach, who was generally most particular about writing out his works with the figured bass, and indeed had the more reason to be so the more his practice departed, as is here the case, from custom—clears up all doubts in this respect by his own handwriting. Besides there are a few scattered passages, in which the part for the chief instrument is written over a simple harpsichord bass and nothing more, where light supplementary chords should be inserted; this is indicated by figures or else by a written direction and may be regarded as an exception. In general, however, any completion of the harmony would not only be superfluous by reason of the wonderfully animated and perfected three-part writing, but would also be impossible to insert without ruining the beauty of the outlines. Whenever a part is added for fulness, not counting one separate full chord as such, it is done in a perfectly organic way, and is indicated not only in the part for the clavier, but in that for the violin. Whenever a movement is not in three parts only, from the first note to the last, it is to be ascribed to the freedom of chamber-music style, and to the dry, ineffective tone of the harpsichord. This was Bach's ideal; we see it plainly in the six great organ sonatas for two manuals and pedals, which remain to be noticed later on, and which agree in form with most of the chamber trios; we have already seen it in the three-part clavier sinfonias which we spoke of as unique in their kind, and the accompanied violin sonata is also idealised by Bach in such a manner as to admit of no comparison with any, even the best, of his contemporaries.[143]

But compositions are not lacking in which, according to universal custom, the harpsichord has to accompany from a figured bass. Even at the end of the seventeenth century a three-part accompaniment was often considered sufficient, but in the following period the four-part accompaniment became universal, and, if desired, the parts might be doubled to help out the poverty of the harpsichord.[144] That Bach also was fond of accompanying with full parts we know for certain from several of his scholars. This of course means not a continuous treatment in four or more parts, but an accompaniment varied in fulness accord-

---

[143] See Appendix A, No. 5.
[144] Heinichen Der Generalbass in der Composition. Dresden, 1728, pp. 131 and 132.

ing to the circumstances, since it is the part of a good accompanist to accommodate himself at every moment to the form and expression of the particular work.[145] Johann Christian Kittel, one of the last of Bach's scholars (b. 1732), gives an interesting account of how he used to go to the rehearsals of a cantata under the master's direction in Leipzig. "One of his most proficient pupils had to accompany on the harpsichord. It may be imagined that he could not venture on playing too meagre an accompaniment from the figured bass. Notwithstanding, he had always to be prepared to find Bach's hands and fingers suddenly coming in under his own, and without troubling him any farther, the accompaniment completed with masses of harmony, which amazed him even more than the unexpected proximity of his strict master."[146] Here the talent for improvisations, which Bach possessed in so remarkable a degree (see p. 27), found its right place. But it had most opportunity for its display in a solo. "Whoever," says his Leipzig friend, Mizler, "wants to hear true delicacy in figured bass playing, and what is called really good accompanying, need only trouble himself to hear our Capellmeister Bach, for he accompanies a given figured bass in such a manner for a solo, that one would think it was a concerto, and that the melody he is making with his right hand had been composed before."[147] It must not, however, be understood from this, as will be shown in another place, that Bach always accompanied in polyphony *ex tempore*. In a minuet in a sonata in C major for the flute,[148] Bach's own accompaniment is extant, fully written out, and really makes an independent piece of itself. It is, in accordance with the tender character of the minuet, chiefly in three

---

[145] Quantz, Versuch einer Auweisung, &c., p. 223 : "The general rule as to a figured bass is to play always in four parts. If you want to accompany really well, however, it often has a better effect not to keep too strictly to this."

[146] Johann Christian Kittel. Der angehende praktische Organist. Section 3. Erfurt, 1808, p. 33.

[147] Musikalische Bibliothek, Pt. IV. Leipzig, 1738, p. 48. This is confirmed to Heinichen. Op. cit. p. 547 f.

[148] P., S. III. C. 6, No. 4. (vol. 235).

parts. The upper part goes smoothly on in graceful quavers, always emphasising the chief progressions of the melody, and now soaring above it and now going below with great freedom. We must imagine that Bach, when accompanying, often gave the reins to his talent for improvisation, and adorned the accompaniment in a wonderfully charming way, with freely inverted counter-melodies. We cannot but lament that this charm was utterly and irrecoverably lost when the master died; and yet, if he had considered this kind of accompaniment essential to the full effect, he would assuredly have fixed it in all his works by an obbligato clavier part. But he allowed them to be spread among the people by his pupils with only a figured bass, so that he must have supposed that he had indicated all that a discerning player would need, and we may hope that an accompaniment of quite simple form would not be contrary to his intentions.

The fuller the writing, the less room was there left for free improvisation. It must be plain from the art with which Bach treated three-part writing, and the pleasure he took in it, that nothing was left to be desired in the way of fulness, even in the trio with figured bass. And, by a happy circumstance, an irrefragable testimony offers itself to this point. A trio for two flutes and harpsichord was afterwards changed by the composer into a sonata for the viol da gamba with obbligato accompaniment for the clavier.[149] The autograph of both is in existence. In the first shape the bass part is carefully figured, but in the second there is not the least sign of any figuring. This shows that in the first case the accompaniment cannot possibly have been very independent, and that its aim was not so much to produce fulness of harmony as to amalgamate the differing qualities of tone. If the harpsichord was entrusted with the bass alone, a certain medium must be interposed between its dry short tone and the liquid fulness of that of the flute or, in other cases, the flexible and pathetic tones of the violin; this was not necessary when the harpsichord took

---

[149] B.-G. IX., pp. 175 ff, and 260 ff.

parts and thus came into the same register as the other instrument. The accompaniment in four parts is arranged by one of the master's best pupils, Johann Philipp Kirnberger, who did the same thing in a trio-sonata of Bach's, and declared plainly that it was by Bach's desire.[150] This style of accompaniment always follows the progression of the parts that are written down, doubling them with the addition of a fourth part, or it repeats their harmonies in another position, but in accordance with the rule that the hands should not be too widely separated in accompanying;[151] there is nowhere a trace of any arbitrary or independent additions. Bach himself may often have proceeded differently. He sometimes would exercise his harmonic genius, when a trio was put before him, in extemporising a real fourth part in addition to the other three; and what he did with the works of others he may well have done with his own.

But these were probably the effects of a happy fancy, or of a joyful sense of power, just as he would play off a complete trio or quartet from nothing but the mere bass part, or read an unknown work from the separate parts placed side by side.[152] As a rule the simple supporting four-part accompaniment remained in vogue. The character of the harpsichord tone prevented the outlines of the principal parts from being entangled or obliterated. To this purpose the modern pianoforte is much less adapted, and demands double care and discretion.

Now in concertos and orchestral suites the accompaniment throughout follows the harmonic changes of the parts which

---

[150] Namely, to the one in the "Musikalischen Opfer" for flute, violin, and clavier, in P., S. III., C. 8, No. 3 (vol. 219). The accompaniment to the third movement is also quoted by Kirnberger in his " Grundsätzen des Generalbasses," who says of it: " Lastly, as an overwhelming testimony for the necessity of knowledge as to the different kinds of figured basses, I have added (fig. LI.) an example from a trio by John Sebastian Bach, which although it is only a trio must notwithstanding be accompanied in four parts ; and this may serve to confute the common opinion, as also may the case of trios, sonatas, &c., for a ' concerto' part with bass; likewise cantatas that are only accompanied on the clavier should not be accompanied in four parts."

[151] Quantz. op. cit., p. 233.

[152] Forkel, pp. 16 and 17.

are written out, so that it is of necessity simple. Passing and non-essential notes are generally omitted, and the accompaniment is played in the middle range of notes, so that the accompanist's only task is to represent just the germ from which the harmony springs. In fugal passages it used to be the custom in time past, when counterpoint was very much simpler, to indicate whatever part had the theme in written notes above the figured bass, and there still exist two fugues for clavier by Bach himself, which, with the help of figuring, are written on a single stave, but then they are of a very simple construction as compared with his others.[153]

In other cases in fugues he used to express all that he wanted in the accompaniment by means of figures, with such wonderful clearness that even at the present day any musician moderately skilled in the rules of playing from figured bass could, without much trouble, produce a good and flowing accompaniment from it; and at that time an accompanist who was accustomed to Bach's style of writing could easily perform the task without a mistake. His mode of writing was indeed, in many points, exceptional, and adapted to his own style; he required his pupils to learn to read it aright, just as they would the ordinary notation. His system was such as to exclude all doubt as to the proper harmonies to be added in the case of those sustained basses which so often occur in his works; he always wrote out the figures representing the desired chord, reckoning upwards from the first bass-note of the group, whether it were a dissonance or not; the harmony was sustained until the next figure, or until it was clear that it

---

[153] P., S. I., C. 4, Nos. 7 and 8 (apparently not in English edition). In the royal library at Berlin there is a book with the title, *Praeludia et Fugen | del signor | Johann Sebastian | Bach | Possessor | A. W. Langloz | Anno 1763.* | It contains 62 preludes and fugues, in every case on a single stave with figuring. There is no single fugue theme which can be recognised as like anything of Bach's elsewhere, and the composition is so poor that I do not believe it to be by him. Possibly they were pieces for practising figured bass playing, collected by a pupil of Bach's, and transcribed by the said Langloz. As to this old manner of writing out fugues in the figured bass part, compare Niedt. Musikal-Handleitung, I., Hamburg, 1710, sheet *E*.

had to be resolved into a triad.[154] In the case of Bach's concertos for clavier and orchestra it is worth remarking that the figured bass accompaniment is to be played on a second harpsichord, so that the first may come out prominently as the solo instrument.[155] In the vocal chamber music, in the ritornels of the airs sung to the harpsichord alone, a special demand is made on the accompanist, for a whole phrase has to be treated melodically and perfectly finished off. In a general way the material can be borrowed from the vocal melody which follows or precedes; but sometimes this melody lies in the bass part itself, and in such cases the task is to produce in the upper parts a correct and flowing counterpoint.

It follows from all this that Bach, in the obbligato treatment of the clavier in free accompaniment, left out nothing whatever except in a few quite distinct cases; that in performing an accompaniment from figured bass he delighted in indulging his talent for improvisation and playing in this way against a solo instrument or voice, except in the case of trios or pieces of rich texture; and that wherever he wrote a mere figured bass a correct accompaniment in four parts was all that he required. In his writing in three or more parts, his harmonies do not stand in need of any amplification, and the harpsichord is only brought in for the sake of blending the differences in qualities of tone. But its importance is in no way lessened by this. Although almost withdrawn from sight, it exercises a powerful influence in settling the artistic form, since the solo instruments are assimilated to it and not contrasted with it. It is the hidden root by which nourishment is supplied to the tree. This root, however, derived its nourishment from the organ. This accounts for the somewhat *foreign* effect which strikes us even in this chamber music, for a peculiar quality, which was common to the organ and the harpsichord, is lacking in the piano, our

---

[154] Kirnberger, op. cit., p. 87.
[155] W. Rust was the first to remark this, in B.-G. IX., p. xvii. Compare B.-G. VII., p. xv.

modern substitute; a certain amount of habit, however, overcomes this feeling. The place of the harpsichord, thus influenced by the organ, in chamber music is exactly the same as that of the organ itself in church music, and all that was said about the place of that instrument in accompanying applies here equally well and unconditionally.

We have already shown, in an earlier part of the work, that the style of Bach's church music, with all its individualities, resulted from organ music. In order to keep this point always in view, it is necessary to bear in mind continually the power which this instrument possesses of governing and uniting different heterogeneous elements, and the fact that by its aid alone could true church music be produced. An important, though obvious, testimony to this is that although, under Italian influence, an attempt was made to introduce harpsichords into the churches, yet Bach invariably used the organ for accompanying from figured bass. Remove the organ, and the soul has gone: only a machine remains. Distinct evidence that this was Bach's view is provided by five movements from his church cantatas, arranged by himself for the organ alone, three of which are in three parts, and two in four parts; in the cantatas, where they are set for voices and instruments over a figured bass, these have an accompaniment in figured bass, while, as organ pieces, they are quite independent of it.[156] Of course the nature of the accompaniment was changed to suit the character of the instrument, the strict four-parts were more strongly insisted on, and the full chords used on the harpsichord were excluded, because the effect could be produced by stops. But that the organ

---

[156] The organ chorales "Ach bleib bei uns, Herr Jesu Christ," "Meine Seele erhebet den Herren," "Wachet auf, ruft uns die Stimme," (B.-G. xxv., 2, Part II. P., S. V., C. 6 (Vol. 245), No. 2, and C. 7 (Vol. 246), Nos. 42 and 57, were published with three others, arranged by the composer himself, by G. Schübler at Zell; they are taken from the cantatas " Bleib bei uns, denn es Abend werden " (B.-G., I., No. 6. P. Vol. 1015), " Meine Seele erhebet den Herren " (B.-G., I., No. 10. P. Vol. 1278), and " Wachet auf, ruft uns die Stimme " (Winterfeld, Evang. Kischeng, III. Supplement p. 172).

took so important a part as to be entrusted with the special and essential part of every piece, and that, consequently, all Bach's church music, in the form in which it has come down to us, consists of mere sketches, is by no means proved. It was enough for the organ to define the general expression; as an instrument participating in the effect, it was quite in a secondary position, the chief parts alone being written down in notes.[157]

Of the sonatas for violin with clavier obbligato six were again united by Bach into a whole set. The year of their composition cannot be ascertained with any certainty, but there is a very credible tradition to the effect that they were written at Cöthen.[158] So, probably, were three sonatas for viol da gamba and clavier, and three for flute and clavier. A comparison of these three collective works shows with wonderful clearness how much attention Bach paid to the nature of the instruments; for although he did not directly form these works from the idea of a violin, a gamba, or a flute, yet, putting aside for a moment the general style which all these works have in common, the character of each of these instruments really is reflected in a clear and distinct manner in the compositions designed for them. The violin sonatas are throughout pervaded with that feeling of manly vigour which, although capable of the most various shades of expression, is the true characteristic of the violin. To this feature, which they have in common, must be added agreement in form; this, with the single exception of the last sonata, is the four-part structure with which we are already acquainted. The description of the several forms given us in the solo sonatas is here insufficient; it was not for nothing that Bach transferred the violin sonata to a sphere especially his own. In parts he enlarged the structure of the separate movements to such bold proportions that he seems to bridge over a whole century, and approaches nearly to the fully perfected forms of the Beethoven sonata.

---

[157] See Appendix A, No. 6.
[158] Forkel, p. 57, asserts it quite decisively, so that it must have emanated from Bach's son.

The chief advance is the employment of the Italian aria-form and the genius with which it is united with the fugal style of chamber music; by this means the triple form is seen as distinctly or even more distinctly than in the Beethoven sonata form, and the proportion of the sections to each other is the same, so that the third repeats the first, and the second works up for the most part the material thus supplied; the only difference consists in the fact that the modern sonata is built on the song or dance form in two sections (Lied-form), while the older is developed from the fugue, and accordingly in the former homophony, and in the latter polyphony with its auxiliaries predominates. It has already been said that the contrasting relations of the movements of the older sonata, if the first adagio is regarded as an introductory movement, are not very different from those of the modern sonata. In these, and for the most part in Bach, the first allegro dispenses with the strict and typical organism in three sections; in the last allegro he was accustomed to employ the dance-form in two sections, which was in general use at the time, combining the fugal form with it in no less remarkable a way. That he did not go on from this point to the perfect Beethoven form in three sections was because the development of his style was too much fettered by the form in two sections. In the extended and amplified form, afterwards employed by Philipp Emanuel Bach, and generally, too, in the clavier sonatas of Haydn and Mozart, it had been long known to him, as is shown by the Invention in E major; the solo movement for harpsichord in E minor from the last violin sonata, is also a perfect model of this type. As a rule no new instrumental forms were created after Bach's time; those which occur are only modifications of those existing before, and worked out by Bach; and all the varieties which they took in the following century put together do not nearly amount to the number of the forms which he alone brought to perfection.

The six violin sonatas are in B minor, A major, E major, C minor, F minor, and G major,[159] and in spite of their

---

[159] B.-G., IX., pp. 69-172, P., S. III., C. 5 (Vols. 232 and 233).

general unity of design show a marvellous variety. The first opens with an adagio in 6-4 time, of which both the melody and the harmony are equally broad and beautiful. Notwithstanding its introductory character, its form is complete, for, in the first place, a distinct bass subject and a fluctuating quaver figure are retained through the whole piece, and, secondly, the phrase which appears in the dominant in bars 13 to 20 recurs in the tonic in bars 24 to 31, after which in the last passages reference is made to the beginning of the movement, so that the two chief divisions have two subdivisions each, which correspond in an inverted order :—

(A B̑ b̲ a̲).

This is followed by a bold fugal allegro in aria form with three sections ; in the second section (bars 41-101) Bach displays his wonderful power—which he got from the northern school—of episodical development, which is applied to the theme that was strictly worked out in the first section ; the third is an unaltered repetition of the first. At the beginning of these fugal sections the theme is never brought in without a supporting bass ; this license came from the Italian chamber music style, and first occurs in Bach in the clavier sinfonias. Now comes the second, the real adagio, which is here, however, an andante in D major, a piece of wondrous beauty, wrought as if with wreaths of flowers, and an organism as perfect in construction as any even of Beethoven's adagios ; attention should be paid to the fine artistic feeling with which the tender and expressive subsidiary theme appears in the sub-dominant after the return of the chief subject (bars 22 ff.). The finale is a movement in two sections with repeats, in fugal form, but of such a kind that the theme is always brought in with two-part counterpoint ; its character is martial and defiant ; observe, besides the splendid theme, the sudden change to the dominant at the close of the first section, and the bold introduction of the chord of the sixth of C major just before the end.

The second sonata, now the best known of all the six,

begins with a movement in 6-8 time, very tenderly developed from a theme of one bar in length, and afterwards (from the eighth bar) combined with a whispering subject in semiquavers. In marked contrast is the splendid *Allegro assai* in 3-4 time which follows. The form is the same as that of the first movement of the B minor sonata, the last section being an exact repetition of the first, while the second is different in structure; for in bars 30—33 a new subject is introduced which alternates with the chief theme, so that a working-out in the style of the concerto before described is the result; finally the new subject generates broad violin arpeggios, while the harpsichord works out the chief theme episodically, supported by a splendid pedal point of nineteen bars, and then the third section is brought in. As to the well known canon in F sharp minor, with its deep thoughtfulness and melodic beauty, it need only be said that its two chief sections correspond to one another in a reversed form, like those of the first movement of the B minor sonata: the first period, consisting of four phrases, leads into C sharp minor and begins anew from that key, but by the insertion of a middle section and by repeating the second phrase it returns to F sharp minor; for the end the expressive and melancholy notes of the opening are heard again like the echo of a vanished past, and the way is prepared by a half close for the last Presto. This is in two sections, and fugal, but the second section prefers to work out a theme of its own, and never takes up the first theme until the end, when it is brought in in playful strettos.

In the first movement of the third sonata the violin wanders freely and melodiously over a subject in the accompaniment, worked out in the usual way; the second agrees in form with the corresponding movement of the A major sonata, except that the repetition of the third division is abridged. The third movement is an adagio in C sharp minor full of the most touching expression. It is a chaconne of which the bass subject is repeated fifteen times, and besides this Bach has worked out an independent theme in the upper parts; it is in the same form as the canon in F sharp minor, but of grander dimensions. The last movement is in three sections instead of two, and the second is, as before, in concerto form, and each part is repeated.

At the outset of the fourth sonata, instead of a largo, we meet with a Siciliano full of grief and lamentation, the beginning of which is almost identical with that of the celebrated air in B minor from the Matthew-Passion, "Erbarme dich, mein Gott"—" Have mercy upon me, O God." An unusually bold and important allegro snatches us from this melancholy mood; it is the richest and broadest movement of the kind in the whole set of sonatas. This great wealth of ideas resolves itself into four sections, a comprehensive epilogue (bars 89—109) being added on after the third (bars 55—89), which does not consist of mere repetition, but is treated with great freedom, and affords more opportunity than the second section (bars 34—55) for interesting developments of episodes. The adagio, in E flat major, goes by in a lovely restful way, calm and gentle as a summer evening; the violin has the melody, accompanied by simple triplets, and now and again breaks off to listen, as it were, to the echo of its tones, and the parts are first united in a full stream of emotion quite at the end. An allegro in two sections, full of Bach's delight in his work, forms the final movement; in this also the second section has a characteristic fugal treatment.

The fifth sonata is introduced by a largo, the only one in which four real parts are employed. Besides being especially distinguished by this fact, it is also one of the most powerful pieces in the collection, and of Bach's chamber music altogether. The three-part clavier part is so independent that almost the whole of it might be played alone; the violin has now passages of broken arpeggios, and now an apparently unending flow of broadly treated melody in 3-2 time. And the violin is far from bringing a foreign element into the organism; it rather raises it to a higher level; it does not cripple the effects but gives them a loftier and more general tendency—as if carrying out in music some eternal law of nature. Its hundred and eight bars fall naturally into four sharply defined sections, the first of which closes in the relative major (bar 37); the second, extended by imitation and by episodes, gets into C minor (bar 59); the third refers back at this point to the first section, still in this key, but only by way of reminding us of it; it then diverges again, and only gets back to the original key at bar 88; in the fourth section there is a real repetition, which combines the beginning and end of the first section together in a concentrated form. The theme on which the clavier part is exclusively formed—

occurs again, hardly altered, in an eight-part motett, "Komm, Jesu, komm, gieb Trost mir Müden, Das Ziel ist nah, die Kraft ist klein,"[160] and the sentiment of the two works is very similar. The movement is imbued with a desire, not agitated, but of inexpressible intensity, for redemption and peace, and spreads its wings at last with such a mighty span that it seems as though it would throw off every earthly tie. Some passages (such as bar 90 ff) sank deep into the sensitive mind of Schumann, there to put forth new blossoms—like the andante of his quartet for piano and strings. The allegro movements follow in inverted order, the one in two sections being the second movement, and that in three the fourth; between them there is an adagio in C minor, formed on the same plan as those dreamy clavier preludes whose development is only harmonic; the violin has two-part harmony in slow quaver movement, which derives a greater fulness from arpeggio semiquavers, alternating between both hands in the accompaniment.[161]

The last sonata as has been said, differs essentially in its general plan. First of all there are five movements, three slow movements being as it were enclosed by two in quick time, the first allegro being repeated at the end. There is no second example of such an extension of form in Bach's

---

[160] Motetten, von Johann Sebastian Bach, Leipzig. Breitkopf und Härtel, No. 4.

[161] Bach subsequently replaced these by groups of demi-semiquavers; the earliest form is given in B.-G. IX., p. 250 f.

works, and it is doubly surprising in a composition which was designed to be united into one work with five others, all agreeing among themselves in form. It can hardly have been on artistic grounds that such an irregularity is found in this particular place. I am inclined rather to ascribe it to some personal motive arising from some incident in his life, for the investigation of which no means are, however, at hand. For the rest the form is ruled by the highest artistic intelligence; the third movement, judging by its importance, is to be regarded as the germ of the organism; it is *Cantabile, ma un poco Adagio,* and around it the two other adagio movements in E minor and B minor, and outside them again the allegros complete the harmony of the grouping. They are written with a creative power all his own, and with evident inspiration. In the first movement, which is in three sections, passages of demi-semiquavers glide busily and unceasingly up and down, enticing or mocking cries are heard here and there; it is as though we were looking at a merry, busy throng of people. The grave second movement, *largo,* as well as the fourth, *adagio,* with its passionate longing, are wisely made quite short, so as not to crowd too closely upon the heart of the whole work. This is a fully developed and extended piece in 6-8 time and in three sections, remarkable for a singularly bridal feeling: it is marked by a sweet fragrance and a breath of lovely yearning such as are seldom found in Bach. The lengthy superscription—which Bach was wont to disdain as a rule—is remarkable; and there is developed in the two upper parts a kind of loving intercourse, a dialogue as from mouth to mouth, carried on above a bass which has nothing to do but to support the harmony. All these are quite at variance with the style of Bach's trios in other places; and what is just as unique is that all three parts do not conclude at the same time: the clavier melody ceases twelve bars before the end, while the violin repeats the whole of the opening phrase of the melody, supported by the bass. For justification of the epithet "bridal" it may serve to refer to certain arias in Bach's wedding music, particularly the one in A major which was afterwards rearranged for the Whitsuntide cantata,

"O ewiges Feuer, o Ursprung der Leibe,"[162] and also to the aria in G major from the cantata, "Dem Gerechten muss das Licht immer weider aufgehen."[163] It need hardly be said that I have no idea of drawing conclusions from the creation of this adagio as to events in Bach's life which may have given rise to it. But circumstances of a personal and intimate kind certainly influenced him; this is clear from the fact that in later life, dissatisfied with the whole form of the sonata, he altered its shape twice, the last time quite in the decline of life, and that neither of these alterations pleased him; and it is well known that any especially subjective productions become more and more difficult to alter as life goes on. I, at least, cannot see that these alterations—in the first two dance-forms designed for a clavier Partita are inserted, while in the second, against all precedent, a sonata movement in two sections for harpsichord alone is made the third movement, and followed by two entirely new concluding movements—have in any way improved the form, beautiful as the three last mentioned movements are in themselves. The central point, the G major adagio, is omitted in both cases, perhaps because its personal element no longer pleased the mighty spirit of its creator as time went on; but the whole structure was endangered by its removal.[164]

The viol da gamba was an instrument with five strings or even more, somewhat like Bach's *viola pomposa* in compass, the lowest string giving D and the highest *a;* but it differed essentially from that in being tuned by fourths and thirds, and also in being held between the knees, like the violoncello. It afforded a great variety in the production of tone, but its fundamental character was tender and expressive rather than full and vigorous. Thus

---

[162] B.-G. VII., p. 146 ff. P., vol. 129.

[163] B.-G. XIII., 1., p. 34 ff.

[164] As to the relations of the different attempted recensions, the last of which is given in the Bach Society's edition, compare the careful discussion by W. Rust in the preface to Vol. IX., p. XX f. The differences in the first recension are there given in an appendix, p. 252 ff.

Bach could rearrange a trio originally written for two flutes and bass, for viol da gamba, with harpsichord obbligato, without destroying its dominant character.[165] This sonata in four movements in G major is the loveliest, the purest idyl conceivable. In the romantic andante alone (in E minor) is there a gentle yet awful whispering and fluttering, as of leaves softly moved at night, and a ghostlike murmur runs over the still depths (represented in a marvellous manner by an *e* sustained for four bars on the viol da gamba). With this exception all is happy, bright with sunshine in a blue sky. In the last movement, a fugue of that mingled strength and grace that is so typical of Bach, there are introduced, between the separate groups in the working-out, light and charmingly worked episodes on the Corelli model, after each of which the unexpected and yet natural entry of the theme has a delightful effect. This sonata was not united with the two others into a collective work by the composer, who, as it appears, did not intend to do so, since the very carefully written autographs of two of them still exist.

The second sonata (in D major) is somewhat inferior to the others in merit and, moreover, is not free from a certain stiffness in the first allegro.[166]

The third, on the contrary (in G minor), is a work of the highest beauty and the most striking originality.[167] It has only three movements, like a concerto, and the concerto form has had a very important share in the construction of the allegros. The first allegro begins, indeed, in the manner of a sonata, but the long and prolific theme attains a freer development. No fugal working-out in the dominant follows, but a repetition, more richly ornamented, in the tonic, and then episodical work until the end of the first section (bar 25). Part of the principal subject fugally treated and answered, serves to bring in the second sec-

---

[165] B.-G. IX., p. 175 ff (in its older form at p. 260 ff). P., S. IV., C. 2, No. 1 (vol. 239).
[166] B.-G. IX., p. 189 ff. P., S. IV. C. 2 (vol. 239), No. 2.
[167] B.-G. IX., p. 203 ff. P., S. IV. C. 2 (vol. 239), No. 3.

tion, and soon is followed by a new tributary of half a bar in length:—

We must notice, too, the passage of four bars which comes in at bar 53—

this, with the chief subject, constitutes the whole material from which the movement is developed, quite in the style of a concerto. After this there is no need of a third section: the movement goes its restless way, unceasingly renewing itself from within. If we are forced to marvel at the absolutely inexhaustible wealth of fancy in almost every new work of Bach's, this especially shows how Bach's style, in spite of its polyphonic nature, was capable of assuming a characteristic picturesqueness of the most marked kind. Here we have a composition in Magyar style: a rushing as of wild and fiery steeds across an open space; the impetuous tributary themes sound like strokes of a whip; now the figures fall confusedly into the discord of the diminished seventh, resolved by means of a bright shake in the upper part; and now they unite in the chief subject in heavy unison—an effect so seldom found in this master—beneath its tread the very earth groans. The irresistible swing which keeps up the movement and action by new and unexpected impulses is almost the same as that which we admire in Weber's overtures. How much Bach himself was carried away by it is seen both in the frequent unisons in bar 64, where the chief theme suddenly appears in three parts on the clavier, the harmony thus becoming fourfold; and then the magnificent end (from bar 95 onwards), where the whole array of notes rushes tumultuously from one diminished seventh to another. In this movement no use is made of the tender character of the viol da gamba, but only of its wide compass and its flexibility. A lovely adagio in B flat major (3-2 time) satisfies our desire for melody with a devotional and earnest strain, of which the beginning is a clear foreshadowing of Beethoven. In the last allegro there seems to be an inexhaustible fund of the loveliest melodies: the most extraordinary number of subjects is produced out of materials already given out, in the manner which we now call the art of thematic working. We have all along used the expression "episodical formation" to distinguish the imitative working-out of an unaltered theme from the alterations incident upon "thematic development." Except at the beginning the concerto form governs the whole. The theme—

is twice worked out in all the parts, and closes in B flat major. From the first bar is evolved a softly pulsating figure for the harpsichord bass, above which the viol da gamba gives out a new and expressive melody, while the right hand has broken chords in semiquavers, and then changes place with the viol da gamba in F major. After the chief theme has been once again brought in, this accompanying semiquaver figure is episodically extended, and a new and no less charming melody appears (bars 37—55). Then comes a thematic and episodical working-out of the chief theme, and the first tributary theme is brought in, in D minor; with this is contrasted a third subject, and these two engage in a pleasing contention (bars 69—79). In the cadence of this phrase there comes in a fourth working-out of the chief theme in C minor, which leads back into G minor; and at bar 90 a fourth tributary is brought in on the viol da gamba, accompanied in the same manner as was the first; reappearing subsequently, after a fifth working-out of the chief theme on the clavier, at the end of the movement. Thus on the stem of the theme one flower replaces another in a way which is marvellous in itself, and not only for the time when it was written. Even in Beethoven's day, when, in accordance with the altered style of instrumental music, episodical work was much more employed than thematic, it would be hard to find anything of this kind more masterly or richer in invention. Bach held as absolute a sway over the art of episodical treatment as over that of thematic treatment; and while his predecessors often preferred the former method, both alike found favour with him, and were used to complete and raise each other.

The Sonatas for the flute are influenced by the form of the concerto both in outline and in details, and in some points correspond exactly with it. They are all in three movements, and the E flat major sonata is a concerto from the first to the last bar.[168] In the first movement the form is somewhat timidly handled; this sonata may have been one of the first attempts to construct a trio on this entirely new plan. The middle movement—a Siciliano—and the final allegro are quite perfect, the soft and pleasing nature of the whole expression agreeing admirably with the character of the flute. This was the emotional character that lent the works of Philipp Emanuel Bach and his successors their peculiar stamp. Joseph Haydn's clavier sonatas had their root in this expression, and it remained in force to Mozart's time; indeed it is the distinguishing characteristic of the period. The height and depth, the sublimity and strength of Sebastian Bach's music were beyond the apprehension of the next generation; still they could drink of the living fount in such a degree as was suited to their capacities and needs. The connection between Bach and Haydn is not indeed self-evident, but it nevertheless exists, and is proved by other works besides the E flat major sonata; the same feeling equally pervades a sonata in G minor, which in its present form is intended for violin and harpsichord, but which was certainly meant by the composer for the flute; Bach wrote it at the same time as the E flat major sonata, and the construction is identical even in the smallest particulars.[169] And among his later

---

[168] B.-G. IX., p. 22 ff. P., S. III. C. 6 (vols. 234—235), No. 2.

[169] B.-G. IX., p. 274 ff. That the sonata cannot be considered spurious as long as the authenticity of the E flat major sonata is undoubted is shown by W. Rust on p. xxv. of the same volume. Besides this, in the adagio there is an unmistakable resemblance to the largo of the concerto for two violins.

works the resemblance to the works of Haydn—I am speaking only of his clavier music—is very prominent in the great organ prelude with which the third part of the Clavierübung begins,[170] a sufficient evidence that this element of feeling was deeply rooted in Bach's inmost nature.

In the flute sonata in A major,[171] the first movement, which unfortunately is incomplete, is quite on the lines of a concerto, not, of course, that the clavier and flute have each a theme belonging to themselves alone, which they contrast with one another; here, as in the E flat sonata, it is only the general musical principle that Bach has assimilated. The fresh and important finale, this time the crowning point of the work, is in three sections; of these the middle one falls into two groups, in F sharp minor (bars 53—118) and E major (bars 118—209), in each of which the chief theme is combined with a new subject in a masterly way: a more ingenious and striking way of coming back to the theme than that employed in bars 160—166 could hardly be imagined.

By far the best of the three, however, is the sonata in B minor; the magnificent freedom and beauty of its form, its depth and overpowering intensity of expression, raise it to the position of the best sonata for the flute that has ever existed. There is none of equal merit in the works of any great master of a later time; and so perfectly does it correspond to the character of the instrument, which indeed is soft and pleasing, but of only moderate capabilities of expression as compared to the violin and instruments of that class, that it reaches the highest level of Bach's style, with its calm surface and its depths of passionate intensity ever craving for expression. The first movement is in three sections and of the broadest proportions; but the composer has here departed entirely from his customary method of treating the first and middle sections. Neither a fugal entrance with episodical or concerto-like working-out, nor a concerto form from the beginning, could fit an imaginative

---

[170] B.-G. III., p. 173 ff. P., S. V., C. 3 (vol. 242), No. 1.
[171] B.-G. IX., p. 245 ff and 32 ff. P., S. III., C. 6 (vols. 234—235), No. 3.

work which was to flow on in unchecked expression of a single and deeply felt emotion like a grand elegy. So the master begins by forming one section from two subjects smoothly worked out side by side. They are not themes, but two melodies of imperishable beauty: the first is carried on for twenty bars in the most lovely way, supported by a soft rocking accompaniment; the second closes in the same key and then goes into D major. The process of development consists of this whole section being repeated, first in F sharp minor and then again in B minor for the close; but between the two last groups is brought in a passage formed of parts of the first and second melodies worked together (bars 61—77) so as to emphasise the return to the principal key. No one who has not seen and heard the work can form any idea of the genius with which Bach varies the theme. The development is carried on bar by bar, but is constantly altered in some way—ornamented more richly or made longer, especially by means of beautiful imitations in canon, which are, as it were, imperceptibly generated from it by a natural power; a special charm is given by reversing the order of several single phrases. This incomparably beautiful piece is closed by a wonderful little coda formed of phrases from the first melody (bars 111—117). The fact that the form is built upon the Italian aria betrays itself by one feature at the opening; the manner in which the melody begins in a kind of tentative way, breaking off after two bars to begin again at the fourth bar, is the same as that so frequently employed by Bach in the sacred arias,[172] The second movement in D major, *largo e dolce*, is simply in two sections with repeats, and is in all respects worthy of what goes before it; in particular the painful yet sweet expression in the last bar but one, where the flute rises in long syncopated notes on the chord of the diminished seventh, must appeal to every heart. The presto begins with a passionate and beautiful three-part fugue, but soon a pause comes on

---

[172] It has been already noticed (Vol. I., p. 26) that the beginning of this melody is identical with a fugal theme from one of Bernhard Bach's orchestral suites.

the dominant and, in accordance with the requirements of the concerto form, a gigue in Italian style (12-16 time) is introduced, which is quite new and yet familiar, since it is evolved most beautifully in Buxtehude's manner from the fugal theme.

All of Bach's independent chamber trios with clavier obbligato which exist have now been enumerated, except one, which is in neither sonata nor concerto form, but has a certain resemblance to that of the suites. This composition for violin and clavier in A major, is not a true suite in the strictest sense, and it is only designated as a "Trio." For, although it adheres to unity of key, and consists chiefly of dances, it is abnormal both in the number and order of the dances. It comprises seven sections worked out at some length, of which the last is an independent allegro in common time, while the first is a free fantasia.[173] The work stands by itself among Bach's works, an exceptional production such as a master who has perfect command over all forms might allow himself; because it has a certain affinity with the independent style of the orchestral suite. The attempt was crowned with success; the pieces with their masterly forms, though they nowhere attain grandeur, are models of graceful and delicate workmanship; fresh, bright music, the full expression of a vigorous and healthy mind.

Solo sonatas with harpsichord accompaniment were less to Bach's taste, as they were not as yet a thoroughly developed form of art. Only four of the kind are known, besides one fugue by itself. One in E minor is for violin and harpsichord; a prelude of running or *arpeggiato* semiquavers is followed by a lovely adagio, an allemande, and a gigue in the Italian style, on the old Corelli pattern. The great fugue in G minor is intended for the same instruments; such compositions may have served as preparations for the fugues in the sonatas for violin solo.[174] The other three sonatas are for flute and clavier. The one in C major has rather an old-fashioned aspect, the fourth (and last)

---

[173] B.-G. IX., p. 43 ff. P., S. III., C. 7 (vols. 234—236), No. 1.
[174] P., S. III., C. 7 (vols. 234—236), Nos. 2 and 3.

movement consisting of a couple of minuets, to the first of which Bach has written out the accompaniment in full. The others, in E minor and major respectively, are in regular form, but their allegros are for the most part in two sections, and of course no such rich display of variety is possible as in the sonata with harpsichord obbligato. Still, they are full of interest and beauty.[175]

There are also very few examples of the trio for two instruments and figured bass. It has already been said that the sonata for viol da gamba in G major originated in a trio of this kind for flutes. A sonata in the same key for flute, violin, and bass is a gem of polished and concise form, full of delightful beauty.[176] Another sonata for two violins and bass, in C major, is not quite equal to this in merit; a gigue serves as the last movement, otherwise the forms are regular.[177]

The principle of concerto-like form which plays so important a part in Bach's creations had hitherto been applied by him only to works which were not actually concertos in external construction. In fact, he only availed himself of the principle to turn it to account for his own purposes (Comp. Vol. I., p. 408 *et seq.*); he can hardly have written any true concertos before the Cöthen period. In order to understand this point historically and fully, we must remember a licence of which the composers of that time availed themselves to a considerable extent in writing concertos. According to rule a tutti subject and a solo subject were placed side by side, and the solo instrument and the tutti instruments vied with each other in producing the greatest amount of material from their respective subjects. The principal key and those nearest to it were the fields

---

[175] P., S. III., C. 6 (vols. 234—236), Nos. 4, 5, and 6.

[176] B.-G. IX., p. 221 ff. P., S. III., C. 8 (vol. 237), No. 2. The autograph parts, which were written in the Leipzig period, are in the possession of Herr J. Rietz, Capellmeister at Dresden.

[177] B.-G. IX., p. 231 ff. P., S. III., C. 8 (vol. 237), No. 1. The resolution of the figured bass given in this and other pieces in the edition of Peters by Fr. Hermann is clever and very good musically, but it might be wished that he had kept more closely to the original figuring.

on which these contests were alternately displayed; when the disputants returned to their original position the combat was over. According to the quality of tone of the contrasting instruments the one theme was heavy and firm, and the other light and pliable. But there were also cases in which one chief subject was considered sufficient. Then it was given out by the tutti and taken up and worked out by the solo instrument. When strictly carried out this plan gave the work a rather poor effect, but when the composer possessed the power of inventing and devising episodes, he might take a phrase of the tutti subject, and by making new matter out of it for the solo instrument, give the form a particular charm. The feeling of dramatic contest between two individualities was, however, much weakened by this method; the form more strictly belonged to the realm of pure music. But it was just this which chiefly interested Bach: the purely musical duality, its contrasts, its combinative fertility, and the impulse given to episodic development by its antagonisms. Thus, even in the concerto-like sonatas for the flute, he made the themes alternate between the clavier and the flute on purely musical grounds. And thus it happens, too, in his concertos, that the tutti passage comprises all the material for the solo subjects. The effectiveness of this departure from the rule of formation depends on the way in which the instruments are treated. This is especially the case with the violin concertos. Here, where the solo violin is set against the string band completed by the harpsichord, the contrast of the two bodies of sound is of course natural and obvious. The class of work had a great interest for Bach, as will be easily understood, after his thorough study of the structure of Vivaldi's concertos. We possess three concertos in their original shape, and three only in a later remodelled form for clavier with instrumental accompaniment; out of the three original ones two have been treated in the same way.[178] These rearrangements were made in Leipzig, to judge from the nature of the

---

[178] P., S. III, C. 1, 2, and 3 (vol. 229, 230, &c.). Compare the dissertation by W. Rust, in B.-G. XVII., p. xiii ff, and B.-G. XXI., 1; p. xiii, f.

autographs; we have no direct evidence that the originals
are of the Cöthen time, but we conclude this to be the case
from a series of other instrumental concertos to which these,
with their far simpler construction, form the natural stepping-
stones; it is also probable from the official post held by Bach
at Cöthen. Notwithstanding the lack of sterling compositions for violin with orchestra, these concertos have hitherto
not achieved the wide popularity which is due to their high
musical worth; the reason is partly to be found in the comparative neglect of the simple and generally intelligible
cantilena style—since the animated harpsichord style, which
had taken possession of the concerto-form, prefers passages
and figurations; and a second reason is, that the form has
become strange to us. We can, of course, get over these
peculiarities, and particularly the second, for the older concerto form is more comprehensible than the new, which
has more or less become merged in the modern sonata
form. In fact, the charm of the episodical working-out is
not less in Bach than in the best concerto composers of
Beethoven's time. In this respect the first movement of
the E major concerto is especially remarkable, with the
working-out of the subject—

which Bach cast in the three-section form that we have seen
so much of in the violin sonatas with harpsichord. In the
second movement we have one of those free adaptations of
well-known forms which Bach alone knew how to treat. It
is a chaconne, such as had been already employed in the E
major violin sonata; but the bass theme not only wanders
freely through different keys, but is also extended and cut up
into portions of a bar long; it often ceases altogether, but
then a few notes revive the conviction that, in spite of all,
it is the central point on which the whole piece turns.
The middle movement of the A minor concerto has what is
seldom found with such definiteness in the adagios, a heavy
tutti subject contrasted with a light figure for the solo instrument; the organism is built on the interchange of these,
without becoming a strict violin *cantilena*. The D minor

concerto[179] is without doubt the finest of the set, and is held in due esteem by the musical world of the present day. Two solo violins are here employed, but it is not in any way a double concerto, for the two violins play not so much against one another, as both together against the whole band. Each is treated with the independence that is a matter of course in Bach's style. In the middle movement, a very pearl of noble and expressive melody, the orchestra is used only as an accompaniment, as was usual in the adagios of concertos.

The free and purely musical concerto form, however, achieved its perfect and untrammelled development in a collection of six concertos which was completed in March, 1721. The occasion was a very special one. Several years before Bach had met, possibly at Carlsbad, a Prussian Prince who was a lover of art, and who delighted in his playing; he had desired Bach to send him some compositions for his private band.[180] This was Christian Ludwig, Margraf of Brandenburg (b. May 14, 1677), the youngest son of the great Elector by his second wife. A sister of his was second wife of Duke Ernest Ludwig of Saxe Meiningen, with whose court Bach had been connected apparently ever since he had been to Weïmar. The Margraf, who was at this time provost of the cathedral at Halberstadt and unmarried, lived alternately at Berlin and on his estates at Malchow; he was especially devoted to music, over and above the ordinary aristocratic amateur dabbling in science and art, and he spent a great part of his income on music.[181] In the spring of 1721 he was living in Berlin, and thither Bach must have sent these six concertos which he had finished as the execution of his honourable commission on the 24th of March. The French dedication in which he mentions the

---

[179] P., vol. 231.

[180] Bach himself specifies the space of time which had elapsed since this—"*une couple d'années.*" If we take this quite literally, as is hardly necessary, we get the year 1719, but no journey of Duke Leopold to Carlsbad is known of in that year.

[181] Amounting sometimes to 48,945 thalers, but this was not always sufficient for him.

occasion which gave rise to the composition may have been written by some courtier at Cöthen. He himself was evidently not skilled enough in French to trust to his own knowledge in such a case, and the mistaken fashion of the time, when nothing but French was spoken or written at the German Courts, was in vogue here. How the offering was received by the Margraf is not known. His band was not lacking in members capable of executing these difficult works in a fit manner; we know the name of one of his private musicians, Emmerling, and that he was distinguished as a composer and performer on the clavier and viol da gamba.[181] After the Margraf's death, which took place at Malchow, September 3, 1734, Bach's precious manuscript experienced the risk of being carelessly sold off among a lot of other instrumental concertos at a ridiculously low price, but a happy fate has preserved it to us. These works exhibit the highest point of development that the older form of the concerto could attain.[182]

Bach calls them *Concerts avec plusieurs instruments*. According to the custom of the time, this means the so-called *Concerti Grossi*, in which, instead of one single instrument, several (generally three) play against the tutti. But to this category belong only the second, fourth, and fifth concertos; the common feature which unites them to a single unity is rather the concerto-like form which is here developed to the greatest musical freedom. Bach had for a long time been on the track of this ideal. The reader will remember the great instrumental introductions

---

[181] Walther, Lexicon.

[182] The few facts I have been able to give concerning Margraf Christian Ludwig are the results of my researches in the royal domestic archives at Berlin. The musical property, which was considerable, was catalogued and valued. Compared with concertos by Vivaldi, Venturini, Valentini, Brescianello, &c., Bach's work was not thought worthy the honour of a special mention by name, and so it must have been in one of the two following lots, " 77 concertos by different masters, and for various instruments, at 4 ggr. (altogether) 12 thlr. 20 ggr."; and " 100 concertos by different masters for various instruments. No 3, 3 16 thlr." As to the subsequent fate of the autograph, see B.-G. XIX., Preface. These concertos are also published with a facsimile of the dedication in P. S., VI., Nos. 1-6 (Vol. 261-266).

to the Weimar cantatas "Uns ist ein Kind geboren," "Gleichwie der Regen," and "Der Himmel lacht" (Vol. I., pp. 487, 492, and 541). But it is not only the separate movements, but the whole form of many parts, that he set in so definite a manner on its ideal musical basis. Throughout his concertos the disposition into three movements is employed, which had indeed been elevated into a canon of art by Vivaldi's delicate instinct; but the *Concerto grosso* was not always confined to this: ere long four movements or more were introduced, giving a resemblance to the sonata, and even dance-forms were intermingled. But the three-movement form was amply sufficient for the materials which were to be displayed in the concerto: for the grave and exciting strife between the bold and active solo instruments and the strong and mighty tutti; for the broad *cantilena*, with its ingenious and beautiful ornamentation; and for the joyful triumphant bravura close which carries all before it. For this reason the three-movement form has remained in general use for instrumental concertos until the present day. The orchestration of Bach's concertos is very strong, and in particular the wind instruments appear in greater force, though they were already employed in the chamber concerto. Such an application of them as we find here had, it is true, never been dreamed of by any one before; like the stringed instruments they were altogether brought under subjection to that polyphony of Bach's by which everything was quickened and compelled to his will. Let us now consider the concertos separately.

First concerto, in F major. Instruments: the string quartet, strengthened in the bass by the *Violone grosso* (the double-bass), and in the first violin part by the *Violino piccolo* (a bright-toned and smaller violin tuned a fourth higher), two horns, two oboes, bassoon, and harpsichord of course as the accompaniment (*Basso continuo*). The usual relations between solo and tutti are disregarded in this concerto, and there are no special subjects for each respectively. The material for the first movement is given out by all the instruments together in bars 1-13. They then divide into three groups—horns, wood, and strings—and work out this material in the style of a concerto. The first bar—

is now raised to the rank of a tutti subject, and is used to mark the beginning of each new section; the rest of the phrase serves for the contrasting solo subject. When distributed on the instruments, this antagonism is no longer prominent, the working-out obeys the laws of free writing, but so that the concerto style is preserved between the three groups, which are united together at the climaxes to a magnificent body of sound in ten parts. The divisions are very clear and intelligible, as they are in all well written movements in the concerto style; they are as follows, the "exposition," or first few bars being reckoned in: A, bars 1-13 (F major); B, 13-27 (F major); C, 27-43 (D minor); D, 43-52 (C major); E, 52-57 (G minor); F, 57-72 (F major); G, 72-84 (F major). The reciprocal relations of the divisions are of especial interest; the first two in the principal key return at the close in inverted order, enclosing the others in the middle, so that they correspond as follows:

A B C D E F G

Thus the form is cyclic, in the same way as that of the violin sonata in G major, and the cantata "Gottes Zeit" (see Vol. I., p. 456), except that in the latter case it extends over the whole work. Exactly the same disposition is found again in the third movement, where bars 1-17 correspond to the closing bars 108-124, and bars 17-40 to bars 84-108, while the working-out is enclosed in the middle. And again these two divisions of similar form and fresh and exuberant character enclose, in the adagio (D minor, 3-4 time), the true kernel of the whole. The adagio is one of the most impassioned songs of woe ever written. The melody gives expression to a piercing grief, often rising to a shrill cry; the oboe begins it in an apparently objectless way on the dominant, and then the high violin and the gloomy bass take it up one after another, after which it is carried on in close canon on the oboe and violin; while below, the quavers of the accompanying instruments keep on in a calm and mournful manner. The finale, as bold and full of genius as the opening, breaks in upon this movement, as that of Beethoven's Eroica Symphony does upon its funeral march; the unsatisfied cries of woe suddenly cease: only a soft sob is heard in the empty air. Appended to the concerto are a Minuet diversified by a Polacca and two trios. These are fine music and a work of genius, but have nothing to do with the true concerto. Dance-forms were much in favour, as has been said, even in orchestral concerto, although they were in entire opposition to the ideal of the form. This is the only instance of Bach having made a concession to the taste of the time; and as the dances can be separated from the rest of the work, if desired, they hardly impair its beauty.

Second concerto, in F major. Instruments: trumpet, flute, oboe, violin, and the string band as tutti. It is thus a true *Concerto grosso*, excepting that the *concertino—i.e.*, the group of solo instruments which is contrasted with the tutti—here consists of four, all of high register: namely, one string and three wind; so that a departure is made in every way from the custom which decrees that the concertino shall consist of two violins and a violoncello. The plan of the first movement is a model of clearness and simplicity, but an indescribable wealth of episodical invention and the most delicate combination sparkles and gushes forth from all sides. The andante (in D minor) consists of a quartet of flute, oboe, violin, and violoncello, with harpsichord; the finale, *allegro assai*, is a fugue in the concertino parts supported by the bass and accompanied by the tutti in a modest and masterly manner. On account of its crystal, clear, and transparent organism, this concerto is a greater favourite than the closer fabric of the first; the feeling, moreover, is throughout of a kind easily entered into. The marvellously beautiful andante is only soft and tenderly simple, while the first and last movements rush and riot with all the freshness and vigour of youth. Truly, even if Bach could not avail himself of the full colours used by later musicians, yet his instrumental music is steeped in the true spirit of German romance.

This first movement: how it goes past like a troop of youthful knights with gleaming eyes and waving crests! One begins a joyful song which echoes through the tree-tops in the forest; a second and a third take it up, and their comrades chime in in chorus; now the song loses itself in the distance: it gets fainter and fainter: anon it is heard for an instant, and then wafted away by the wind and drowned by the fluttering of the leaves.

> "Fainter and fainter still, upon the air
> The music dies away—but where? ah, where?"

And this is evolved from the simple concerto form!

Third concerto, in G major. Instruments: three violins, three violas, three violoncellos, violin and harpsichord. The first movement is similar in its development to that of the first concerto, but is superior, from the art and charm with which it is treated. The violins, violas, and violoncellos play in three groups, sometimes treated polyphonally among themselves, sometimes not, and frequently combined in unison. What is made out of these subjects—

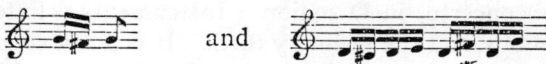

is astounding—in fact, the whole movement is built on them. It is throughout instinct with life and genius. One passage (from bar 78 onwards) is as fine as anything in the whole realm of German instrumental music; the chief subject is given out in the second violin part, the first violin then starts an entirely new subject which next appears on the second violin, drawing in more and more instruments, and is at last taken up by the third violin and the third viola, and given out weightily on their G string; this is the signal for a flood of sound to be set free from all sides, in the swirl of which all polyphony is drowned for several bars. There is no adagio in regular form. Two long-held chords alone release the imagination for a moment, and then begins the concluding movement, a true concerto finale in 12-8 time.

Fourth concerto, in G major. Instruments: violin, two flutes, and the strings as tutti. It is a *Concerto grosso* in the manner of No. 2.[183] The first movement, allegro in 3-8 time, is of a very pleasing character. The material is given out in bars 1-83, for the most part by the concertino, the tutti only interrupting it now and then. Here again we meet with the "cyclic" arrangement, showing that this still was the master's ideal of form. The "exposition" (A) is followed in bars 83-157 by a working-out (B) going into the relative minors; then there is a further working-out (C) up to bar 235, after which (B) returns with some alterations and extensions until bar 345, when (A) is brought back for the close. The adagio in E minor, which is entirely taken up with alternations between the tutti and the concertino, is a beautiful and grave piece, in a mournful measure, like music for a funeral procession. The last movement consists of a fugue—*presto* and in common time—which is grand in every respect. It is 244 bars long, and for animation, for importance of subjects, for wealth of invention, for easy mastery over the most complicated technical points, for brilliancy and grace it is in the very first rank of Bach's works of this kind.

Fifth concerto, in D major. Instruments: flute, violin, harpsichord and the ordinary tutti. It is not a strict clavier concerto with accompaniment, but the clavier combines with the violin and flute to form a contrasting group with the tutti; in this a second harpsichord for accompanying only was probably introduced, in accordance with Bach's usual practice, even in concertos for the clavier only. In this way this work belongs properly to the class of *Concerti grossi*, or at least it is derived from them. But that the clavier must have taken the lion's share in this combination is obvious from its character, and this is more easy to understand when we consider the great subjective importance of the place held by the harpsichord in this class of Bach's chamber compositions. Two subjects for tutti and solo respectively are given out in perfectly developed form, and engage in the most

---

[183] W. Rust, in the B.-G. edition, is wrong in calling it a violin concerto. The word *ripieni* in the title applies only to the violins, since there are no *flauti ripieni*. Besides this, the intention is clear from the work itself. Dehn, in the Peters edition, gives it the right title.

charming alternations. One fragment of the tutti in particular—

is worked out with the loveliest combinations. In the middle there is evolved a new subject in F sharp minor of a quiet and calm character—

it sets out in its course over the gently moving waves of harmony, and loses itself as on an immeasurable ocean, guided only by a rhythm on one note, until the wind swells the sails and brings us to the wished-for destination (bars 71-101). Before the last tutti there is a great clavier solo, which demands, as does also the other clavier part, a finger dexterity which no one except Bach could have possessed at that time. A lovely and tender *affettuoso* in B minor stands for the middle movement. The general character of the concerto is not so much deep or grand as cheerful, delicate, and refined. The last movement is of the same character. It is in the form which was noticed first in the violin sonatas with harpsichord obbligato—for instance, in the second movement of the sonatas in A major. The structure is in three sections, after the pattern of the Italian aria; the first, which is completely repeated for the third, is fugal, and the second introduces a subsidiary theme and combines it with the chief subject. This subsidiary, however, is here derived from the chief theme and is of exceptional melodic charm; in the harmonic treatment a false relation which constantly recurs and quickly vanishes again is very remarkable.

Sixth concerto, in B flat major. Instruments: two violas, two viol da gambas, violoncello, and violin, with harpsichord. There are two subjects for tutti and solo respectively (bars 1-17 and 17-25), but only in idea, not specified by particular instruments. That for tutti consists of a canon for the two violas at the distance of a quaver,[184] while the other instru-

---

[184] Kirnberger, in his "Kunst der reinen Satzes," II. 2, p. 57, quotes it as a model.

ments have a simple harmonic accompaniment of quavers, so that the result is a movement similar to the church sonatas of Gabrieli and Bach. In the solo subject this phrase—

is taken up in all parts in an animated manner. The whole movement has a strangely mysterious character, such as Bach alone could give it, and doubly strange when we consider the original object of a concerto. The theme of the adagio (E flat major, 3-2 time) is a lovely melody, given out by the two violas alone over the basses. For a long time they keep the theme to themselves, treating it fugally, until at last it is taken up by the basses with beautiful effect. The final close is, curiously enough, in G minor. This movement is unusually noble and grand in character. The last movement, a concerto finale in 12-8 time, is powerful, without abandoning the fundamental character of the first movement, and it requires very good viola-players. While retaining the general character of the Italian gigue, it is in three sections, and yet altogether concerto-like in treatment.

From a production of the highest genius and mastery in art, as these six Brandenburg concertos must be called, one glance may not unfitly be directed to corresponding works by Bach's contemporaries. The *Concerto grosso* rapidly came into favour at the beginning of the eighteenth century, and the best musicians tried experiments in it. But they only followed the pattern of Vivaldi's style to a limited extent, and there was another style, as has been said, which originated with Corelli's sonatas. These composers retained the four-movement form in the recognised order: an adagio, a fugue in the same key, an adagio in an allied key, and a finale; but they did not forget that the form admitted of more movements, and did not exclude dance types. But at the same time this imitation and dependence on a previous style exercised a powerful influence on the shaping of the separate movements. The dialogue between solo and tutti remained

only an external alteration of different bodies of sound on
the same material, and was scarcely more than a contrast,
phrase by phrase, between strong and weak tones ; in the
clavier and organ music this was represented by the different
manuals—in the fugal movements of the French overtures by the contrast of the whole orchestra with the trio
of oboes and bassoon. Thus in their essential nature these
works were not concertos at all, but orchestral sonatas.
Telemann liked this form, but did not devote himself exclusively to it ; the man who did the greatest things with it was
Handel. Handel's *Concerti grossi* cannot be compared with
Bach's, since they have scarcely anything in common but the
name. It might have been said that there was nothing in
common, if he had wholly avoided the form of Vivaldi's concertos in them. Where he uses it, however, he is always the
great artist ; but the fact is plain that his genius was unsuited
to this class of composition.[185] In the broad adagios, in the
fugues, and in the simple dances of the Corelli sonata he found
the impulses which most certainly set in motion the purely
musical side of his nature. In accordance with his aim,
which was to produce something brilliant and showy, he
gave this form larger proportions and filled it with meaning.
A precisely similiar case is that of Handel's organ concertos,
the mention of which is suggested by the fifth of the Brandenburg concertos. In these, too, the form and order of the
movements are influenced in the most striking way by the
sonata. But with him the organ is only a more powerful
clavier, and of true organ style there is as good as none. But
here, more than in the *Concerti grossi*, we meet with the real
form of the concerto movements, because it was suggested
so plainly by the independent and complete nature of the
clavier or the organ.[186] As far as the form had been developed
by the Italians, especially by Vivaldi, so far Handel wielded
it with mastery; but he in no degree furthered its true growth.

---

[185] For instance, in the C major concerto (Handel Society's edition, XXI.,
p. 63), which, excepting in the last gavotte, has quite the Vivaldi form ; and in the
second movement of the great G major concerto (same ed., XXX., No. 1).

[186] Handel Society, XXVIII. Compare particularly concertos 1, 2, 4, and 6
from Op. 4 ; from Op. 7, concertos 3 and 6.

It is a significant fact of his musical nature that no single instrumental form of the many that were being developed at the time received any furtherance of growth from him. He appropriated what had been done in this way up to the time of his own work, and his incomparable wealth of ideas enabled him easily to surpass even the important works of other composers. When he lighted upon anything of this kind which was, comparatively speaking, formed, he was successful in producing instrumental works of lasting worth. Irrespective of the undeveloped condition of the Corelli sonatas as regards the arrangement and connection of the movements, Handel's *Concerti grossi*, so far as they depend upon those, are important enough to take a lasting place of honour in German instrumental music; and we do not wish to see the time when works like the concertos in E minor, A minor, and G minor shall have lost their effect,[187] for they contain at least separate sections of a solid and concise kind. But for the form of the concerto the Italians had scarcely done more than prepare the skeleton; the best was yet to be done, and chiefly by means of the art of treating episodes. Like the Italians, Handel possessed but little of this art, and that explains the unsatisfied feeling left more or less by all his movements in concerto style. Nothing is developed: all is ready made from the beginning, and only awaits the moment when it shall be displayed.[188] Other German composers, such as Telemann, or the Kammer-musicus at Dresden, Dismas Zelenka, produced works of this class which were more conformable to rules, though not so rich in ideas. These, however, after all, are too inferior in talent to Bach to be allowed to share in the fame of having brought the concerto form to its highest point of development.

The Brandenburg concertos form a class by themselves in German orchestral music, since they must be reckoned as such. As among mountains the highest points seem close to one another, and the ravine that lies between them, and that will take many toilsome hours to traverse, is almost

---

[187] Handel Society, XXX., No. 3, 4, 6.
[188] Compare the intelligent and clever dissertation on Handel as an instmental composer in Chrysander's Handel, III., p. 174 ff

indistinguishable at a distance, so these seem nearly allied to the modern symphony, and yet no direct way lies between them. They stand upon another and a much narrower foundation, upon which only a gigantic creative power could raise such a structure. The orchestral music proper of the period was not the *Concerto grosso*, then hardly invented, but the orchestral suite. This form, together with the clavier suite which had sprung from its root in the seventeenth century, practically reached its completion and end in the first half of the eighteenth century. So rational a unity as that presented by the clavier suite was out of the question in the orchestral suite, on account of the surroundings among which it grew. Whether there was ever any half-established custom with regard to the arrangement of the dances is for the present uncertain, but it is clear that the most eminent employers of this form recognised no such rule, and that the separate component parts are always grouped quite in an arbitrary way. But this lack of definite form was only the reverse side of an important advantage—namely, that the orchestral suite had sprung directly from the life of the German people. The freshly flowing fount of popular melody of the older centuries sprang forth from the ruins of the Thirty Years' War and divided into two streams : the sacred song or hymn, which was soon appropriated to the use of the organ, and the instrumental dance, which fell to the care of the town pipers. It matters not that so many other nations, especially the French, should have contributed some of their tunes and style. On the contrary, by this means the German spirit was kindled into that activity so peculiarly its own, which absorbs and amalgamates foreign elements to be part of its own strength ; this, as has been said before, was an advantage won directly from the turmoil of war. The French at all events contributed largely to the more delicate bringing out of rhythm in German dance music, and not only that, but we owe to them the first orchestral form of a secular character—namely, the so-called French overture. But they had hardly any share in the artistic development and elevation of this overture and of the dance types. Nor did they attempt to form an artistic whole from these elements. Even the Italians were far superior to them in these respects: there are overtures by Antonio Lotti, in the French style, such as no true Frenchman could have written in so excellent a way, to say nothing of Handel with his Italian education.

But the Germans formed a suitable set of dance pieces into a purely musical collective whole, prefixing to them a French overture. This is evident from the remarkable circumstance that the name "Suite" was not used for the analogous orchestral form, as it surely would have been had the French done as much for it as for the dance series for the clavier. But no general and collective name exists for it. With that modesty which is so characteristic of the true German musician, and which confines itself to the matter in hand, careless about outward appearance, the composers either indicated the separate component parts in the title of such a work, or were content to abbreviate it in this way, "Ouverture, &c.," then giving a list of the instruments employed. But the separate dances contained in any particular set were called by the town pipers "Parties" ("Partheyen") and we only do justice to the Germans in calling the class of music henceforth by the German name of "Orchester-partien."

If any one was destined to bring forth something extraordinary in this class, it was assuredly Sebastian Bach. In order to prove this it is sufficient simply to look back to his ancestors. His father, uncle, and grandfather had followed exclusively the calling of town pipers. How could this tendency of German art-life fail to find its completion in the musician who was to comprise in himself all the abilities which his family had developed for a hundred years? If the number of his "Orchester-partien" is not large—since the whole form was not earnest or prolific enough for this to be the case, and the clavier suite had absorbed a good part of his creative impulse in this direction—yet their very existence goes to show how thoroughly national was Bach's individuality. Let no one, as we have said before,[189] think

---

[189] See Vol. I., p. 162.

lightly of the importance of the place held in music by the town pipers of the seventeenth century. Granted that there were no little roughness and disorderliness among them; the *Volkssänger* (people's singers) of the sixteenth century were also rough and disorderly in their way. Nevertheless their songs handed down a genuine element of the old German spirit, and became a model of its expression; and so it was with the instrumental dance of the later period. Add to this that the Bach family made the most strenuous efforts to keep as clear as possible from the vulgarity of their companions. The great composer truly had no reason to be ashamed of coming into this part of the inheritance of his ancestors. And in fact he had taken possession of it with joy, with full intent to apply all the wealth of his powers to this ideal of national art. His four *Orchester-partien* are altogether works of a master hand, and in this respect of equal excellence. The keys are C major, B minor, and (twice) D major.[190] They all begin with a French overture worked out at some length. First, there is a *grave*, which is repeated and followed by a fugue leading back into the *grave*, and also repeated. The typical character, consisting of the contrast between broad beauty and eager impetus, is plainly recognisable, but is marvellously refined, so that we hear no opera music, but the most delicate chamber music, especially in the overture in B minor.

After that, in the C major partie, comes a courante, a gavotte, a *Forlane* (a Venetian dance in 6-4 time, resembling a gigue), a minuet, a bourrée, and a passepied. All the pieces except the courante and the forlane are double, so as to bring out in each the favourite contrast between the strong and the tender, and make each complete in itself. The name "trio," now universally known, came from this custom, for the tender subject was played by only three instruments, or was in only three parts; but soon the number of parts was not restricted, although the general character of the music remained the same. Here only the

---

[190] Three of these are published in P. S., VI., Nos. 7, 8, 9 (Vols. 267-269). See Appendix A, 9.

bourrée and the passepied have trios in the strictest sense;
the last-named dance is repeated in a different setting for
its own trio, in a wonderfully ingenious manner, all the
violins and tenors playing the melody in the inner part,
while the oboes have a rocking motion above it in quavers.
The trio of the gavotte is properly only in three parts, but
the united violins and tenors give out at intervals, and
without finishing it, a soft passage in the style of a fanfare—
a fancy which Bach had introduced before on the horns, in
the first movement of the first Brandenburg concerto. The
trio of the minuet, on the other hand, is played by the
strings alone; it goes by with an elastic step, and a sweet
and caressing character.

In the B minor partie the overture is followed by a rondo,
a sarabande, a bourrée, a polonaise, a minuet, and a little
piece in free style in 2-4 time, entitled "Tändelei"
(*Badinerie*); the bourrée has a trio, and the polonaise a
variation. This partie, in which there is only a flute in
addition to the string quartet, has throughout a distinguished
and debonair character peculiar to itself; it thus stands
in a certain contrast to the other, without ever entirely
casting off the popular feeling. The rondo form, which
we meet with here for the first time in Bach's works, seems
to have been imported from France; in it a short phrase,
generally of eight bars, alternates with an arbitrary number of
somewhat longer interludes. The rondo in question is freely
constructed on this plan, but the chief theme is heard even in
the interludes; it is a real pearl of musical invention, and
steeped in Bach's peculiar melancholy. In the sarabande
the ear is occupied with following an interesting canon
between the upper part and the bass; the first bourrée has a
delightful burlesque working-out of a basso ostinato—

the variation on the lovely polonaise keeps the melody in the
bass from beginning to end, while the flute has a figure above
it, supported by chords on the harpsichord. With this
should be compared the beautiful and effective polonaise in
G major, from Handel's E minor concerto, the whole of

which throws a very clear light on the different character of
the two masters, even in this class of composition.[191] The
Badinerie at the close represents indeed no clearly defined
dance type, but completely retains the form of two sections.
The introduction of such pieces was taken, as its name
implies, from the French. Even real dances were given
names *à la Couperin;* thus once Bernhard Bach calls a
bourrée "*les plaisirs*," and another time a piece of like form
"*la joye;*" but even this composer introduced pieces that
entirely departed from the dance form. On the other hand,
I know an orchesterpartie by Telemann, in which all the
pieces are in dance rhythm, but no single one has any
name. Thus, as we see, the greatest liberty prevailed. A
general title for such free dance forms was "*Air*," which
was not especially used for simple or *cantabile* pieces.[192]

One of the parties in D major concludes in the same way
as that in B minor. The finale is here called *Réjouissance*
and has a bold and vigorous motion in triple time. The
other numbers are, after the overture: Bourrée 1 and 2,
Gavotte, Minuet 1 and 2. The component parts of the other
D major partie are: Air, Gavotte 1 and 2, Bourrée and
Gigue in the Italian style. Besides having the same key
they are both more strongly orchestrated; for, besides the
string quartet, three trumpets, three and two oboes respectively, and drums are employed. The last-mentioned
partie is a favourite in our time and is often performed, but
the others are no less worthy. It is to be hoped that in time
all the orchestral works of Bach will take their proper
place in our public performances, as soon as the material
hindrances are removed which have hitherto stood in the
way of the performance of a great part of them. Before all
else the restoration of the old trumpet, so rich in animation,
compass, and expression, is indispensable. The instrument
which is now substituted for it can either not perform what
is required of it at all, or else by its piercing tones distorts

---

[191] Handel Society, XXX., p. 40.

[192] Comp. Vol. I., p. 576. Dismas Zelenka also gives complicated and various forms under this name.

the delicate proportions of Bach's outlines in such a way that only a caricature is the result.

It was pointed out earlier in this work[193] that Bernhard Bach, Sebastian's cousin, had done remarkable things as a composer of orchester-partien. In this composer, too, the influences of the family of town pipers from which he sprung are very clearly to be seen. He has a right to be considered as the foremost in this branch of art after Sebastian Bach. Ludwig Bach, of Meiningen, is only known to posterity by a single partie, but in this there are traces of that old style which are all the more remarkable when we consider his leaning towards softness and Italian charm of sound. At all events, all the orchester-partien by other composers with which I am acquainted are far inferior to the productions of the Bachs. Handel, so far as I know, made no attempt in this class of composition.

We have called Handel's the more universal talent as contrasted with Bach's; and justly, so far as his relation to the culture of nations and his effect upon it are concerned. He was educated in Germany, travelled in Italy, studied French music, and lived in England. He succeeded, as no other of our great masters have ever done, in setting in vibration those cords of the human heart, which are independent alike of nationality and of time, and more or less the same all over the world. But if we look at the musical material presented in the whole body of his work we see that he left a considerable part of the elements with which the musical atmosphere of the time was filled entirely unused. It was not he, but Bach, who was universal in amalgamating all the musical forms of the national culture of the time. The course of our investigations justifies us in saying that there was no single musical form existing all through the seventeenth century or in the beginning of the eighteenth that was not brought to lasting perfection either by Bach alone or by Bach and Handel together. At the close of the narrative of the Weimar period I drew attention to the vast wealth of forms worked upon by Bach. Add to these the Chamber

---

[193] See Vol. I., p. 26.

Sonata, the Suite, and the "Orchester-partie" with the French overture, and we have before us all that Germany, Italy, and France can offer us in the sphere of pure music. If the verdict that Handel was broader and Bach more profound is to remain in force, it must not be understood to mean that Bach restricted himself to one or to a few branches of art. The very essence of music is depth, and the more this is the case, the richer will its outcome be. To Handel the poetic aspect of the art was the chief object, and this, by means of sung words, is universally intelligible; Bach devoted himself to what was purely musical. Without question many a true German characteristic found noble and worthy expression in Handel also—for instance, his predisposition to devote himself to what was foreign in order to absorb it into his own personality, while purifying and completing it; consider, too, his fearlessness, his perseverance, his right-mindedness, and high morality! On these grounds he is and he remains German; but in his whole nature rather than in a specially musical way, for he neglected the most characteristic German art of his time—namely, that of the organ, with the chorale as its central point. It is the very fact that in Bach this was the true focus in which every ray of light was concentrated, from thence to radiate with new effects, which renders him in the most emphatic sense a national musician. The activity which permeated all the art elements of the time was not due to his personality alone, but to that music which was at the time the fullest and purest expression of the German nature, and of which he was merely the most famous representative. On this foundation he constructed the church cantata in Weimar, and from it he evolved there—and with still more energy at Cöthen—every musical form which is now universally accepted, and imbued it with nobler substance. But this was not enough. These newly created musical entities showed their vitality by twining round each other, sending out shoots hither and thither, and meeting again, from opposite poles as it were, to unite once more and become the parents of newer and greater forms. Bach's development, when we once recognise the motive power of it, grows and blossoms like a flower; it is as though we saw into nature's mighty workshop:—

> Each to the whole its influence gives,
> Each in the other works and lives!
> Like heavenly angels upward, downward soaring,
> And fragrant odours from their vials pouring,
> All joy, all bliss abounding,
> The earth with heavenly life surrounding,
> And all the Eternal's praise resounding!

Yea, verily! for it is not that Bach's creations reproduce —more purely perhaps than those of any other German master—the inmost soul of music; no, it is that his very being, his moral essence, and the breath of his life were music music in that deepest sense in which it is conceived in Goethe's "Faust" as a reverberation of the sempiternal harmony of the Macrocosm. That effect which is produced alike by an absorbed contemplation of nature and by the enjoyment of any truly great music—the strengthening of our moral tone combined with the purest pleasure—is also to be found in the apparently simple and undisturbed current of the great man's life. Hitherto it has lain dormant: would that it might wake to bring gladness and exaltation to his country!

IV.

BACH'S SECOND MARRIAGE.—CHANGE OF POST.—THE FRENCH SUITES.—THE WOHLTEMPERIRTE CLAVIER.

KNOWING the views of life that prevailed in the Bach family, it is hardly necessary to say that Sebastian did not live long in the state of widowhood to which he had been brought by the death of his first wife. His father, under similar circumstances and at a far more advanced age, had remarried at the end of only seven months. Though the son could not, like him, console himself at the end of so short a time for such a bitter loss, he nevertheless was making preparations for a second marriage at the end of the year 1721. He had long been known among the ducal band at Weissenfels; in 1714 he had chosen one of its members— a chamber musician—to be godfather to his son Philipp

Emanuel. In the youngest daughter of the Court trumpeter, Johann Caspar Wülken, he found the woman who could reconstitute the household which had been so suddenly broken up. Anna Magdalena Wülken was at that time twenty-one years old; the wedding took place on the 3rd of December, 1721, in Bach's house. This was in obedience to the commands of Prince Leopold, who felt a personal interest in the important step his friend was taking; all the more so because, just eight days later, his own marriage was to be solemnised with Frederike Henriette, a princess of Anhalt-Bernburg, aged nineteen.[194]

His young wife was to the master a source of deep and permanent happiness. She was extremely musical and took a part in her husband's labours, which extended far beyond a mere enjoyment of them. She was endowed with a fine soprano voice, and assisted in the performance of Sebastian's compositions—not, it is true, in public, but all the more zealously in the family circle—and she was the centre of the little domestic band which Bach was beginning to gather round him, formed of his nearest relations. He writes feelingly of this on October 28, 1730, to his friend Georg Erdmann : " They are one and all [his children] born musicians, and I can assure you that I can already form a concert, both vocal and instrumental, of my own family, particularly as my present wife sings a very clear soprano and my eldest daughter joins in bravely." Anna Magdalena was skilful, too, with the pen, and not seldom, when her household work was done, she would help her too busy husband in copying his own or other music. In this way she assisted with her beautiful handwriting in copying out the solo sonatas for violin and violoncello, and a manuscript copy of Handel's music to Brocke's text on the Passion is in great part executed by her. Her notation is rather less

---

[194] Gerber, Lexicon, I., col. 76. Register of the Cathedral of Cöthen. It would seem that Anna Magdalena was not born at Weissenfels, since the register does not mention her. Even before her marriage she was Court singer at Cöthen and betrothed to Sebastian Bach as early as September, 1721. As such she stood sponsor with him to a child of Christian Halen, cellar clerk to the Prince. Baptismal register of the Cathedral of Cöthen.

light than Sebastian's and differs from it in the form of the
C clef, the naturals, and the sharps, and in a few other trifles,
but it is very flowing and free, without a trace of feminine
ineptitude, as also is the alphabetical hand, which also differs
from her husband's in certain particulars; still the whole
effect both of notes and letters is often so like Bach's as to be
difficult to distinguish. But she was not satisfied with this:
she was her husband's diligent pupil in clavier-playing, and
even in playing from figured bass.

Two music-books, kept in common by the husband and
wife, display very touchingly their intimate and tender
relations.[195] They are full of the most miscellaneous
matters; the older of the two is in small oblong quarto
and is modestly bound in dull green boards with back
and corners of brown leather. On the inside of the board
is written, not very regularly, in Gothic letters, " Clavier
Büchlein | vor | Anna Magdalena Bachin | ANNO 1722."
Then follows the letter B on a fresh line as if something was to have been added; and then, in Bach's
hand:

"*Anti Calvinismus* und ⎫
Christen Schule *item*   ⎬ von *D.* Pfeifern."[196]
Anti Melancholicus   ⎭

Thus, this little book must have been begun immediately
after their marriage. The words written under the title
are a playful, but perfectly serious indication of the purpose
of the work, which was to oppose that dry Calvinistic
doctrine, inimical to all art, which reigned supreme in
Cöthen, and to counteract all the sorrows and bitter
experiences of life—the " School of the Christian "—all
gloomy thoughts and dismal moods. How could the
fountain of music, either in the church or in home-life,
be better described? *Dominus* Pfeiffer was a theological

---

[195] Both in the Royal Library at Berlin.

[196] More correctly " Ante Calvinismus," either a slip of the pen or perhaps
only indecipherable. Dr. F. L. Hollmann, in Lübeck, soon after the appearance of this first volume, suggested to me that it was the Leipzig professor
Dr. August Pfeiffer who was probably meant by these words, and the
accuracy of this view was subsequently confirmed by the list of Bach's Theological Library. See Vol. III., Appendix B., VI.

writer of the seventeenth century. Bach had his works—
Evangelische Christen Schule, Anti-Calvinismus, and *Anti-Melancholicus*—on his own bookshelves, and the Clavierbüchlein was to be, to a certain extent, their musical reflection. Its contents were for the most part worked out in the French suites, to which we shall come presently. Besides these it contains an ornate chorale in three parts, "Jesus, mein Zuversicht,"[197] a fragment of a fantasia for the organ—perhaps Anna Magdalena wished learn to play the organ too—an air with the beginning of some variations on it, and a minuet.[198]

The second and larger book has a green binding stamped with gold, and gilt edges, and has a more imposing appearance; it is tied with a band of brown silk fastened to the upper cover. In the middle of the cover there is stamped in gold—

A. M. B.
1725.

It belongs, therefore, to the Leipzig period, and must have been a gift from her husband. Besides two clavier partitas (those in A minor and E minor of the first part of the "Clavierübung"), two of the French suites, the C major prelude of the Wohltemperirte Clavier, and the air for the Goldberg variations published in Part IV. of the Clavierübung, it chiefly contains little pieces written out by Anna Magdalena herself — polonaises, minuets, marches, and such like—which are not indeed all by Sebastian himself; for one minuet (page 70) bears the express statement "*fait par Mons. Böhm.*" However, we come upon various vocal pieces : first, the beautiful hymn by Paul Gerhard "Gieb dich zufrieden und sei stille in dem Gotte deines Lebens"—"Be still, my soul, and rest contented in the hand of God thy Maker." It must have been a favourite with Bach, for it is to be found three times in succession, and with two quite new melodies in F major and E minor (or G minor). With regard to the last, Bach is stated to be the

---

[197] P. S., V., Appendix No. 2.
[198] P. S., I., C. 13. No. 11, I.

composer of it, and a special importance is very justly attached to this melody, for it is one of the most impressive sacred airs in existence, and any one who hears it under conditions worthy of it, in Bach's own four-part setting, will carry away an impression which he will not forget so long as he lives.[199] Towards the end of the book Bach has written another beautiful composition of his own on the hymn by B. Crasselius " Dir, Dir Jehovah will ich singen;"[200] before and after this are the hymns " Schaffs mit mir, Gott, nach deinem Willen" and "Wie wohl ist mir o Freund der Seelen."[201]

Besides these compositions, which stand halfway between the congregational hymn and more secular music, there are a few true arias written for Anna Magdalena's voice. The first place must be awarded to the lovely piece "Schlummert ein, ihr matten Augen, fallet sanft und selig zu," with the recitative belonging to it, taken out of the sacred cantata " Ich habe genug, ich habe den Heiland," and transposed to suit the singer from E flat major to G major.[202] A second and more song-like aria in E flat major, "Gedenke doch, mein Geist, zurücke ans Grab und an den Glockenschlag"— " Consider, oh, my soul, remember the grave and ponder on the end "—is a warning to prepare for death; this likewise is a composition by Sebastian in Anna Magdalena's handwriting. This is followed by the chorale " O Ewigkeit, du Donnerwort "—" Eternity, oh word of might "—not, it is true, in the same key, but evidently connected with the former in the mind of the transcriber. A third aria, similar to these, in F minor, "Warum betrübst du dich und beugest dich zur Erden, mein sehr geplagter Geist "—" Wherefore art thou so sad, and why so crushed and broken, oh, much tormented soul "—treats of submission to the will of God. The interest she evidently took in these compositions shows

---

[199] Published by L. Erk, Johann Sebastian Bach's mehrstimmige Choralgesänge und geistliche Arien. I., 43, 44; II., 208. Leipzig: Peters.

[200] L. Erk, ibid., I., 19 and 20.

[201] L. Erk, ibid., I., 111, has given the air of the first of these. Both were well-known hymns, and are to be found in Schmelli's collection.

[202] B.-G., XX., 1, No. 82. The accompaniment is not written down, since Bach would have transposed it at sight from the score of the cantata.

how near the young wife's sympathies must have been to the grand world of ideas, with its " dim religious light," in which her husband had his being.

Two other songs are of a more familiar character. The " Edifying reflections of a Smoker" show us Bach in the comfortable attitude of a citizen and *house-father*, as the Germans say; still, even here, his reflections take a sober hue:—

> So oft ich meine Tabakspfeife,
> Mit gutem Knaster angefüllt,
> Zur Lust und Zeitvertreib ergreife,
> So giebt sie mir ein Trauerbild,
> Und füget diese Lehre bei,
> Dass ich derselben ähnlich sei.

> Whenever in an hour of leisure,
> With Knaster good my pipe I fill,
> And sit and smoke for rest or pleasure,
> Sad pictures rise without my will.
> Watching the clouds of smoke float by,
> I think how like this pipe am I.

This comparison of the fragile clay pipe and its fleeting fire, so soon burnt out, with the brevity of human existence, is carried through five stanzas. The song occurs twice, once in D minor and then transposed for a soprano into G minor; Anna Magdalena desired to sing it and has transcribed it herself. The second song having a da capo form is still more a true aria. The text—

> Bist du bei mir, geh ich mit Freuden
> Zum Sterben und zu meiner Ruh.
> Ach wie vergnügt wär so meine Ende,
> Es drückten deine schönen Hände
> Mir die getreuen Augen zu.

> Be thou but near, and I, contented
> Will go to Death, which is my rest.
> How sweet were then that deep reposing
> If thy soft hand mine eyes were closing
> On thee, their dearest and their best!—

is evidently supposed to be addressed by a husband to a beloved wife, and has a peculiarly delicate and tender sentiment bordering on hyperbole and still perfectly true in feeling. Bach has given it a setting full of fervour and of purity (E flat major, 3-4 time). This also is intended for a soprano, and Anna Magdalena herself has it written out; but a few resolutions of discords have been added subsequently by her husband, if I am not mistaken. This receptivity and sympathy with the moods of a man's mind marks a tender and childlike devotion in the wife.[203] The musical portion of the book extends to the chorale "O Ewigkeit, du Donnerwort," on page 121, where the paging ceases. Then, after a blank page, come some wedding verses; of course they can only refer to Anna Magdalena. That they should find their place here after the lapse of several years is a striking proof of a happy married life:—

> Ihr Diener, werthe Jungfer Braut,
> Viel Glücks zur heutgen Freude!
> Wer sie in ihrem Kränzchen schaut
> Und schönen Hochzeit-Kleide,
> Dem lacht das Herz vor lauter Lust
> Bei ihrem Wohlergehen;
> Was Wunder, wenn mir Mund und Brust
> Vor Freuden übergehen.

> Your servant, sweetest maiden bride:
> Joy be with you this morning!
> To see you in your flowery crown
> And wedding-day adorning
> Would fill with joy the sternest soul.
> What wonder, as I meet you,
> That my fond heart and loving lips
> O'erflow with song to greet you?

On the other side of the leaf we come upon rules for figured bass playing, which are continued over four pages. The first and smaller portion, in which the major and minor scales and major and minor triads are explained, Anna Magdalena wrote out from a sketch or precis by Sebastian; all that follows, and which contains serious instructions for playing from a figured bass, has been inserted by Bach's

---

[203] This aria begins on p. 75 and goes on to p. 78, the copyist probably turned over two leaves by mistake. On the vacant pages 76 and 77 the air for the Goldberg variations was subsequently written. With regard to the spuriousness of the song attributed to Bach, which is certainly to be found in this book, see App. A, No. 10.

own hand, and in a note at the end he says that the sequel must be taught by word of mouth. I shall take a future opportunity of returning to these rules for thorough bass.

In the course of twenty-eight years of married life, Anna Magdalena brought him thirteen children, six sons and seven daughters: thus, by his two wives, Bach had in all twenty children. A portrait of her in oils, twenty-five inches high by twenty-three wide and painted by Cristofori, was afterwards in the possession of her stepson Philipp Emanuel.[204] In their rank of life it was an unusual distinction to have a portrait taken, and she must have had it done by Sebastian's desire: a fresh proof of the affection and high estimation on which the married life of this pair of artists was founded—a model to all.

When, in 1707, Bach was married for the first time, he had had the agreeable surprise of a legacy from his uncle, Tobias Lämmerhirt, of Erfurt, then lately deceased.[205] It was a strange coincidence that, a few months before his second marriage, that uncle's widow also died without surviving heirs, and by her will part of her fortune fell to him. Sebastian had been on excellent terms with his aunt and had made her godmother to one of his elder children. He now had the opportunity of proving that his regard for her endured even beyond the grave, for a lawsuit was immediately begun as to the property she had left. Tobias Lämmerhirt, not long before his death, had made a will to the effect that in the event of his decease legacies of various amounts were to be paid first to the children of his brothers and sisters, and to his godparents and half-brothers and sisters. The remainder was to go to his wife as residuary legatee, but with this proviso: that, if she remained a widow, at her death half the fortune was to revert to his nearest relations. The widow paid the legacies, and on October 8, 1720, made a will on her own part, in which she treated her husband's fortune as being her own inheritance and property, from which, at her death, she alienated a whole catalogue of legacies,

---

[204] Gerber, Lexicon, II., App., p. 60; it is now unfortunately lost.
[205] See ante, Vol. I., p. 339.

and then divided the remainder among ten legatees, five of whom, in agreement with her husband's will, were his nearest relations and five her own. The will was read September 26, 1721; and at first the distribution was agreed to, the legacies were deducted, and the residue divided into ten equal parts. It was not till after this that the idea occurred to some of Tobias Lämmerhirt's relations that his will might be interpreted to their greater advantage. They demanded for their share, first, half of the whole fortune left by Tobias Lämmerhirt, and they calculated that it should amount to 5,507 thl., 6 gr. Out of the other half, the legacies left by the widow should then be paid and the residue divided into ten portions. A petition to this effect was filed January 24, 1722, in the names of the five relations who preferred the claim—namely, Joh. Christoph Bach, of Ohrdruf, Joh. Jakob Bach, Joh. Sebastian Bach, Maria Salome Wiegand—born Bach—and Anna Christine Zimmermann—born Lämmerhirt—the daughter of a brother of Tobias Lämmerhirt. But practically it emanated only from the two last-named petitioners, who, to give the claim more weight, had taken for granted the consent of the brothers Bach in this proceeding.

The carelessness with which they had gone to work may be understood when we remember that Christoph Bach had been already dead ever since February 22, 1721; at the same time they had given the attorney who had drawn up the document such insufficient information that he allowed Jakob Bach, of Stockholm, to answer for the consent of his brother Sebastian, of Cöthen; though it might also be inferred from this that they, knowing his magnanimous nature, simply dared not mention the transaction to him at all. In point of fact, it was only through a third person that it came to Sebastian's knowledge. He at once sent the following letter to the Council of Erfurt:—

MOST NOBLE, PRUDENT AND VERY LEARNED, MOST JUDICIOUS GENTLEMEN, AND, MORE PARTICULARLY, MOST WORSHIPFUL PATRONI,

It is already known to your worships how that I and my brother, Joh. Jacob Bach (at present in the service of the King of Sweden), are co-heirs under the will of the late Lemmerhirt. Because whereas I learn by hearsay that the other co-heirs are minded to bring a lawsuit with regard to this will, and whereas I and my absent brother have not been served with a notice, since I am not minded to dispute the Lemmerhirt will by law, but am quite satisfied with what is thereby given and allotted to me and my brother, I desire by this letter to renounce on my own part, and *sub cautione rati nomine* on my brother's part, all part in any such lawsuit, and to have it ratified by the usual form of protest. I have, therefore, esteemed it necessary humbly to lay this letter before your Worships, with due submission, and to beg you favourably to receive this my protest and renunciation, and that of my brother, and to restore whatever is to come to us out of the money, both that which has already been paid in and that which yet remains, for which great favour I beg humbly to thank you, and remain

Your Worships'
Most devoted servant,
JOH. SEB. BACH,
Capellmeister to the Prince of Anhalt-Cöthen.

CÖTHEN, *March* 15, Ao. 1722.

(Addressed to) The most noble, prudent, learned, and judicious Gentlemen, the Provosts, Burgomasters, *Syndic*, and other members of the Council, more particularly to my most gracious masters and patrons in Erffurth.[206]

After this emphatic declaration, no steps could be taken to initiate the proceedings, and no documents exist which have any bearing on the matter. To hinder the unfilial conduct of his relations, Sebastian came forward at once in the name of his brother Jakob, whose opinion he was sure would be identical with his own. Joh. Jakob Bach, after quitting his quiet home, in 1704,[207] had been a brave follower of Charles XII. of Sweden through all his wild campaigns, had taken part in the battle of Pultawa, and had followed his royal leader as far as Bender, in Turkey. There he had remained on duty till 1713, and then obtained leave to retire in peace to Stockholm as Court Musician there. But first he had gone from Bender to Constantinople, and had there studied playing the flute under Pierre Gabriel Buffardin (afterwards Chamber Musician at Dresden and teacher of the famous Quantz), who happened to be there in

---

[206] This letter, and the documents on which the petition was based, are in the town archives of Erfurt, Part IV., No. 116. I owe my knowledge of this letter to Herr Ludwig Meinardus, of Hamburg.

[207] See Vol. I., p. 235.

the suite of the French Ambassador, and who afterwards related the circumstances to Sebastian Bach.[208] Whether he then passed through Germany to Sweden and took that opportunity of visiting his relations in Thuringia we have no means of knowing. It can be proved that he received his salary from the privy purse of the Court of Stockholm from 1713 till 1721 inclusive. He must have died in 1722, hardly more than forty years old, and probably much broken by the terrible fatigues of the Russian campaign. Thus he probably never even heard of the circumstances under which Sebastian had answered for him, while Sebastian had to mourn the loss of this the last of his brothers, not long after the death of Joh. Christoph, who had formerly been his teacher, and of another highly esteemed relative.[209]

Thus, between joy and sorrow, more than four years were passed at Cöthen; but that which lay at the foundation of Bach's happiness there remained unchanged. The eager and intelligent interest the Prince took in his art had enabled him entirely to forget how narrow was the musical circle within which he moved there, its exclusive limitation to chamber-music, and the absence of all development in the direction of sacred composition. Since, however, it was for this that Bach must have felt himself especially fitted, nothing was needed but some external impulse to make him aware that his genius would not permit him to set up his tent for the rest of his life in this spot, however delightful he might feel it. This impetus was given by the Prince's marriage. His wife had no love of music, and she absorbed her husband's whole attention, all the more because she was delicate and needed every care. The Prince's interest in music seemed falling off, and it now suddenly became clear to Bach that it was no part of his work in life to make his transcendent gifts subservient to one single dilettante prince.

---

[208] According to the Genealogy and Fürstenau, II., p. 95.

[209] Johann Jakob left no children; I cannot even find out whether he ever married. In the private accounts of the Court of Sweden he figures as "Johann Jakob Back." The salary, for various obvious reasons, was not then paid with great punctuality, so that "Back" almost every year had to prefer a petition to the crown for arrears of pay. This occurred for the last time in 1723, and the payment was probably made to his relations in Germany.

In the letter to Erdmann, before quoted,[210] he states this in plain terms. "From my youth up," he writes, "my fate has been known to you until the last change which took me to Cöthen. There lives a gracious Prince who both loves and understands music, and with him I purposed to spend the closing term of my life. However, as it fell out, the above-mentioned *Serenissimus* married a Princess of Berenburg, and as then it began to appear as though the said Prince's musical inclination was growing somewhat lukewarm, and at the same time the new Princess seemed to despise my art, it was the will of God that I should be called to be *Director Musices* here, and Cantor in the Thomas Schule. Still, at first it did not perfectly suit me to become Cantor from having been Capellmeister, for which reason my resolution was delayed for a quarter of a year; however, this position was described to me as so favourable that at last, particularly as my sons seemed to incline to study here, I ventured in the name of the Highest and betook myself to Leipzig, passed my examination, and then undertook the move to Leipzig."

It is clear, however, that the temporary cooling of the Prince's interest in music was in fact only the external impetus to a step for which the necessity lay in the general conditions of Bach's artistic nature; and we see this in the fact that his decision remained unaltered, although that "music-hating" personage, the Princess Friederike Henriette, died so early as April 4, 1723, and it was not till May that he pledged himself in Leipzig to take the place of Cantor to the school of St. Thomas. Meanwhile the obsequies of the deceased Princess took place in Cöthen without any musical adjuncts.[211] The Prince married for the second time, June 2, 1725, Charlotte Frederike Wilhelmine, a Princess of Nassau Siegen.

Though Bach had to quit the spot where his patron resided,

---

[210] See ante, p. 144.

[211] The funeral sermon preached on the occasion was published in folio, with all the poems in praise and lamentation of the deceased Princess, in 1724. Among them there is no funeral cantata or text for music of any kind to which Bach could have composed. The castle library at Cöthen possesses a copy of these memorials, which is graced with an engraved portrait of the Princess.

he continued to be his honorary Capellmeister.[212] In this capacity he composed for November 30, 1726, in honour of the first birthday of the second Princess after her marriage, a congratulatory cantata for which the Leipzig "occasional" poet Christian Friedrich Henrici—or Picander, as he was wont to style himself—composed the words.[213] It begins with a chorus, "Steigt freudig in die Luft zu den erhabnen Höhen"—"Rejoice and soar aloft to distant heights ethereal" (D major, 3-4 time); this is followed by four recitatives, alternating with three graceful airs, of which the second is much the most interesting, not unintentionally perhaps, since it is written for a bass voice, and Prince Leopold himself was a good bass singer. The finale consists of a cheerful homophonic chorus in a gavotte rhythm, and little recitative subjects are introduced; its beginning, it may be observed incidentally, is almost identical with the theme of Beethoven's Choral Fantasia. This pleasing, though not very important, work was afterwards adapted, with the text somewhat altered, to another birthday ode, and finally it was remodelled into a cantata for the first Sunday in Advent, where the recitatives are eliminated and chorale arrangements inserted in their place.[214]

Not long after, this beloved patron ended his short life, November 19, 1728,[215] and Bach had to contribute to the funeral solemnity. This he did by composing a grand mourning ode (Trauer Musik) which he himself conducted at Cöthen in 1729, probably early in the year. The musical performers he took with him from Leipzig (he had most likely done the same for the birthday cantata); in Cöthen itself nothing of the kind could have been got up.

---

[212] So says the genealogy. This connection must still have existed in the year 1735.

[213] They are printed in "Picander's | Ernst-Schertzhafft | und | Satyrische | Gedichte | Mit Kupffern. | Leipzig, | in *Commission* zu haben bey Boetio. | Anno 1727," p. 14.

[214] It is given in this form, B.-G., VII., No. 36. See the preface and appendix to that volume.

[215] Not November 17, as is stated in J. Ch. Krause's History of the House of Anhalt.

The text was again by Picander.[216] It consists of four parts, and is intended for a double choir. The music was still in existence in 1819; it then vanished leaving no trace, perhaps for ever, and we have nothing to indemnify us for this loss but the enthusiastic praise of its last possessor;[217] there can be no doubt that the master would have put forth his whole strength in it. Thus death broke the tie which distance could not sever.

It must have been with a heavy and sorrowful heart that Bach moved to Cöthen;[218] but what Cöthen could give him was now a thing of the past. More than five years almost exclusively devoted to instrumental chamber music had invigorated his genius from the purest and freshest fountain of musical art, and he now could aim directly at that sublime goal which he was born to reach.

He had turned the time to good account. We have tried to glance over the vast mass of chamber compositions which were written—some certainly and some probably—in Cöthen. Still there are wanting to the complete picture the two works which, with the Inventions and Sinfonias, represent the highest summit of his clavier compositions at that period; these are the French Suites and the "Wohltemperirte Clavier."

The French Suites are, as has been said, contained for the most part in Anna Magdalena's first book, and almost fill it.[219] The name "French" was given to them later, without the master's concurrence, on account of the meagre form of their component sections, which, even in external dimensions, adhere as closely as possible to the dance type on which they are founded. In this respect they offer a conspicuous contrast to the broad symphonic forms of the

---

[216] " Picander's | bis anhero herausgegebene | Ernst-Scherzhafte | und | Satyrische | Gedichte, | auf das neue übersehen, | und in einer bessern Wahl und Ordnung | an das Licht gestellet." | Vierte Auflage." | Part I., p. 328. Leipzig, 1748.

[217] This was Forkel, who died 1818. He mentions it on p. 36.

[218] Mizler, Nekrology, p. 166.

[219] P. S., I., C. 7 (Vol. 202). B.-G., XIII., 2, pp. 89-127. See Appendix A, No. 11.

later partitas and the "English" Suites, as they are called. Beyond this there is no idea of imitating or carrying out any specially French characteristics; none such are to be discerned anywhere in Bach, nor could they be possible except in his very earliest work.[220] It would be more natural to detect a certain affinity with the Suites of Georg Böhm, who, no doubt, for his part, was strongly influenced by French art; but this affinity even is only one of feeling. The arrangement of the French Suites is always that which has already been fully described—Allemande, Courante, Saraband, Giga, are the essential sections; between the last two pieces intermezzi are introduced. Not one of them has a prelude; there would seem to have been one originally to the fourth, which was afterwards cut out for the sake of uniformity.[221] The whole work does not give the impression of being a collection made or determined by accident; on the contrary, it is arranged with artistic intelligence—a whole cast in one mould. As in the Inventions and the Sinfonias, we here too find a considerable number of " Paralipomena " which prove the care with which the master selected the best. No less than three complete suites exist besides these, and identical with them as to the character of the details and the whole arrangement. They are in A minor, E flat major, and E minor, and are so admirable that only something of very superior beauty could have a right to displace them.[222]

It was careful consideration which gave the first place in the French Suites to three in minor keys (D minor, C minor, B minor) and the last to three in major keys

---

[220] Compare Vol I., pp. 202 and 210.

[221] It is to be found in the Royal Library at Berlin, in a MS. copy, press mark P. 289.

[222] P. S., I., C. 3 (Vol. 214), Nos. 6, 7, and 8. Besides these there are a Prelude, Saraband, and Giga in F minor (P. S., I., C. 9, Vol. 212, No. 17); and in MS. Allemande and Giga in C minor; Prelude, Fugue, Saraband, and Giga in C minor. This last work seems to hesitate between the clavier and the violin, and perhaps, as it lies before us now, it is only an arrangement of a violin piece. Ph. Em. Bach is our authority for its genuineness. In other manuscripts in the Royal Library at Berlin are to be found the beginnings of the subjects in the thematic catalogue, p. 84, No. 2. The two first-named pieces are in the same Library, but in a more modern MS.

(E flat major, G major, E major). But even the minor Suites are of a pensive and elegiac character rather than profound or grave, and the giga at the close gives a sense of vigorous and elastic reaction. The giga of the D minor Suite is exceptional in form; it is in common time, and strides on heavily and steadily almost like the *grave* movement of a French overture. A most delightful feeling pervades the three last Suites—a happy mood of joy and blessing; a sentiment of content that the world is so fair and that men may rejoice in its beauty; a radiance as of spring sunshine and an atmosphere as of the scent of violets prevail throughout. Truly, indeed, *Antimelancholicus*. The separate numbers are each more beautiful than the other in their indescribable and constantly varying charm. It would be in vain to try to say anything of each in particular. The forms are of the very simplest. Schumann once observed [223] there were some things in the world of which nothing can be said— for instance, of the C minor Symphony with fugue by Mozart, and a few things by Beethoven: if we add of many things by Bach, particularly of the French Suites, this still remains quite within Schumann's meaning. And that this very work had quite captivated a spirit so nearly akin to that of Bach himself he unintentionally proved by the resemblance which exists in one of his string quartets to the gavotte in the E major Suite.[224]

Many independent examples have already shown us the transcendent mastery which Bach had achieved in the fugue, chiefly on the organ, but also on the clavier. It must have occurred to him often to collect a number of compositions of this type, and arrange them in a single work. He accomplished this in the year 1722, and gave the work the following title: "Das wohl temperirte Clavier oder *Praeludia* und *Fugen* durch alle *Tone* und *Semitonia* so wohl *tertiam majorem* oder *Ut Re Mi* anlangend, als auch *tertiam minorem* oder *Re Mi Fa* betreffend. Zum Nutzen und Gebrauch der Lehrbegierigen *Musical*ischen Jugend als auch derer in

---

[223] Gesammelte Schriften, I., p. 198 (of 1st Ed.).
[224] The *Quasi Trio* in the finale to the Quartet in A major, Op. 41, No. 3.

diesem *Studio* schon *habil* seyenden besondern *Z*eit Vertreib aufgesetzet und verfertiget von *Johann Sebastian Bach p. t.* Hochfürstl. Anhalt. Cöthenischen *Capell*-Meistern und *Directore* derer Cammer-*Musiquen. Anno* 1722." ("The well-tempered Clavier, or preludes and fugues in all the tones and semitones, both with the major third or 'Ut, Re, Mi,' and with the minor third, or 'Re, Mi, Fa.' For the use and practice of young musicians who desire to learn, as well as for those who are already skilled in this study, by way of amusement; made and composed by Johann Sebastian Bach, Capellmeister to the Grand Duke of Anhalt-Cöthen and Director of his chamber-music. In the year 1722.") Thus the instructive purpose is here distinctly set forth, and it was this certainly which prompted the principle on which the order of the collection was based—a course, namely, through all the twenty-four keys, major and minor, a few of which at that time were never used at all; so that it was Bach who, by his new principles of fingering and his method of tuning the clavier, first made them accessible.[225] Even in this the instructional aim stands out in all its simplicity, since Bach has not arranged the twenty-four keys according to the rule of their relationship, as Heinichen had laid it down ten years previously in his Musikalische Zirkel, but in simple chromatic order. And this direct simplicity is equally characteristic of the separate compositions: they are bereft of all superficial embellishment; the severest solidity and chastest treatment, purposeful to the very last note, is the stamp of them all.

By far the larger number of them, at any rate, were written by Bach during the Cöthen period; probably all under the same impulse, and quickly one after the other. A trustworthy tradition informs us that this was in a place and under circumstances when he was deprived of all musical occupation—nay, even of any instrument whatever; he strove to preserve himself against depression and tedium by such

---

[225] Heinichen, in his Generalbasslehre, p. 511, § 17, says, "Nowadays we play but rarely B major and A flat major, and pieces are never set in F sharp major or C sharp major." This was published in 1728.

an exercise as this.[226] This very possibly occurred during that journey on which he had to accompany the Prince. Still this work is not cast in a single mould like the French Suites or the Inventions and Sinfonias; in the first place, some of the fugues, though but a few, bear clear traces of an earlier origin; also they are not all of them duly thought out and connected with their preludes. The older handling most conspicuously betrays itself in the fugue in A minor, first by the pedal note which enters at the end—a license which we have already noticed in different places, and which was altogether abjured by the composer in riper years. Besides, from this pedal note it is clear that the fugue was originally written for the harpischord. Thus, it is in opposition to the intention of the whole collection, which is properly intended to be performed on the clavichord. When we consider the position given to the clavichord by Bach, this is almost self-evident in a work like the Wohltemperirte Clavier: it is shown, however, by bars 15 and 16 of the E flat minor fugue, where each time the upper part is not continued up from c''' flat to d''' flat, because this last note was lacking in most clavichords; and, moreover, by bar 30 of the A minor fugue, where on account of the limited compass the regular imitation in the right hand is altered.[227] In like manner the bass does not go below C, except in the case of some unimportant doubling of octaves, though the harpischord had a larger compass both ways, which Bach employed without hesitation.[228] But, moreover, the A minor fugue is an evident imitation of one of Buxtehude's organ fugues in the same key.[229] This deserves special description.

First there comes a working-out *in motu recto* till bar 14,

---

[226] Gerber, Lexicon, I., col. 80. The tradition is trustworthy for this reason—that the lexicographer would have heard it from his father Heinrich Gerber, who was Bach's pupil in Leipzig soon after 1722.

[227] In the second part of the Wohltemperirte Clavier no d''' flat occurs with the single exception of the A flat major prelude.

[228] Compare in the overture in the partite in D major (Clavierübung, Part I., No. 4), bars 68–70 and 90–91 of the fugue, and for the three-stroke octave many passages in the Goldberg variations.

[229] Comp. Vol. I., p. 276.

then *in motu contrario* till bar 27 ; then a stretto *in motu recto* till bar 48, stretto *in motu contrario* till bar 64 ; at this point there follows a stretto in two parts *in motu recto*, and in two others *in motu contrario*, then another such stretto between the alto and bass parts in F major, and finally from bar 76 the theme is brought in in inverted motion, the response a note higher in direct motion, and a stretto in direct motion from bar 80, and then a close on a pedal point. The playing about between direct and inverted motion is exactly the same in Buxtehude's fugue, only that he extended the structure, in his own way, by change of rhythm and episodical treatment, adding a coda rich in ornamental passages. Bach retains externally a greater concentration, but his whole plan appears to have been thought out collectively and coolly rather than conceived directly in the imagination. The fugue is somewhat scholastic, is lacking in emotional development, and has no climaxes. One chief reason is that the theme is not adapted for such extended strettos, which move chiefly in intervals of sixths and thirds, and so sound only like harmonies supporting the theme, while they display much harmonic and polyphonic development ; the rhythm is also tedious. The inversion of the theme is even less happy. The characteristic skip of the seventh from F to G sharp is inverted, and in the inversion seems not like a necessary sequence of the melody, but like an unmelodic arbitrary transposition of the theme into the higher octaves, since the ear demands at each repetition of the passage to be carried up a semitone higher. The tonality is also uncertain, the theme wandering from minor to major and from major to minor. One glance at the fugue in B flat minor in the second part of the Wohltemperirte Clavier, which is quite similar in scheme, is all that is necessary in order to see how the mature master dealt with materials of this kind. We can hardly be wrong in assigning the A minor fugue to the years 1707 or 1708. The G sharp minor fugue, too, is more or less clearly recognisable as a work of his youth. The theme strikes us as somewhat stiff, when contrasted with the incomparable elasticity of other subjects by Bach ; the counterpoint in chords, which is of frequent occurrence here, is well enough in Buxtehude, Buttstedt,

THE FORTY-EIGHT PRELUDES AND FUGUES. 165

and other composers of the older generation, but we do not expect it in Bach, any more than we do the repetition of one and the same phrase in a higher octave. In its manner of treatment certain similarities with a fugue in A minor before mentioned [230] are apparent, which is of the same date as this, but much more graceful and charming.[231]

With reference to the Preludes, Robert Schumann—who in certain respects was the most competent judge of Bach's work in recent times—was of opinion that many of them had no original connection with the Fugues.[232] In fact we already know that Bach cultivated the prelude as an independent form;[233] and it can moreover be proved, not only that all the preludes of both parts of the Wohltemperirte Clavier had been collected into an independent whole by Bach himself, even without the fugues—which is certain from the state of an autograph copy to be fully described elsewhere—but also that several of those belonging to the first part were originally conceived of as independent compositions. For instance, in Friedemann Bach's Little Clavier Book, which was begun in 1720, we find in an isolated form the eleven preludes in C major, C minor, D minor, D major, E minor, E major, F major, C sharp major, C sharp minor, E flat minor, and F minor. There is not, on the face of it, the smallest ground for assuming that these were less independent pieces than the other preludes in this volume, but their distinct origin is all the more surely proved by the fact that several of them are used in the Wohltemperirte Clavier as subjects of a more extensive elaboration. This can be demonstrated as regards the preludes in C major, C minor,

---

[230] Compare Vol. I., p. 432.

[231] The opinion that several youthful works are contained in the first part of the Wohltemperirte Clavier is given by Forkel (p. 55), who, I believe, derived much general information from Bach's sons. I cannot approve of his judgment in particular points: especially he is in evident error when he takes the fugues in C major and F minor for early works. Those in F major, G major, and G minor seem to me not to belong to the most important in the collection, but I can find no indication of their being of a different date from the most important of the set.

[232] Gesammelte Schriften, II., p. 102.

[233] See Vol. I., p. 432.

D minor and E minor.[234] Nor is it difficult to perceive that the feeling frequently does not altogether harmonise with that of the fugue, particularly in the case of the C major prelude; and the insignificant A minor prelude is not in its place as leading to the fugue that follows, a stately piece of workmanship, attired in all the panoply of its race.

Nevertheless the Wohltemperirte Clavier, as a whole, remains a masterpiece among Bach's instrumental works. All of it that does not stand on the very highest eminence is important enough to hold its place worthily; otherwise the master who criticised himself so severely and so constantly would certainly have cast it out; he would have been in no difficulty to find a substitute. That he himself set a high value on the work is proved by the three copies extant in his own handwriting (possibly, indeed, a fourth)—an unusual number for a work of such extent. However, he hardly can have thought of publishing it, though Mattheson challenged "the famous Herr Bach, of Leipzig, who is a great Master of Fugue," in print, to do something of the kind.[235] This profoundly conceived and original music could have no success with the ordinary class of clavier-playing amateurs, and Mattheson described the organists of the time as ignorant folks, ready enough to take lucrative places, but who would do nothing and learn nothing "but what they might pick up by chance." Bach used the work as material for the practice and improvement of his advanced pupils,[236] and at a later period composed, as a fellow work to this, twenty-four more preludes and fugues, which we shall discuss in their proper place. They are usually included in the work now under consideration under the general title of the "Well-tempered Clavier," though this name was originally given by Bach to the older series only.[237]

---

[234] The C major prelude is given in the supplement (No. 5) to this work in the form in which it exists in Friedemann Bach's Little Book. See also App. A, No. 12.

[235] Vollkommener Capellmeister, p. 441, § 66.

[236] Gerber, Lexicon, I., col. 492.

[237] Of the different editions I will here name only that of the B.-G., XIV., edited by Franz Kroll. In the introduction to it there is a very careful enumeration of the various MSS. and printed editions. With reference to an autograph copy, hitherto unknown, see App. A, No. 13.

THE FORTY-EIGHT PRELUDES AND FUGUES. 167

In considering the general æsthetic aspect of this work, what is most striking is the wonderful variety in the character of the twenty-four fugues, each of which is entirely different from all the rest. This is equally true even of the least important; and the endeavour after variety was probably the reason why Bach selected characteristic pieces from among the works of his earlier time. The preludes are no less various, though most of them are kept in one and the same form—that, namely, to which Bach was accustomed to adhere in his independent preludes; the whole subject is worked out from an animated phrase that sometimes becomes definite enough to be called a theme, but often is only distinct in rhythm, or wanders on dreamily from one harmony to another. A model of this form of composition, of which we have already pointed out several examples, is the famous C major prelude, a piece of indescribable fascination, in which a grand and beatific melody seems to float past, like the song of an angel heard in the silence of night through the murmur of trees, groves, and waters. The fugue belonging to it is worked up, and not without good reason, to the highest pitch of finish and intricacy; it was to hold its high position with fitting dignity. A marvellous art is displayed in the various strettos on the fifth, octave, third, seventh, and fourth, which are brought in by turns at the third, fifth, and seventh quavers, for the most part in double counterpoint; in the direct and inverted motions of the counter-subject, and its treatment with counterpoint at the twelfth.[238] It is no light task even for the player. The theme of the fugue begins on the second quaver of the bar, and it must be noted that the fervid and culminating force which is characteristic of Bach is here strongly marked; for it is not till nearly a bar later that we feel the strongest accent, though all that has gone before has tended towards it with peculiar yearning. It is an internal *crescendo*, to which, in playing it, the master would also have given as much expression as possible. By far the greater number of his clavier fugues are constructed

---

[238] Compare Kirnberger, Kunst des reinen Satzes, II., 2, p. 192 f.

in the same or a similar way. Of the forty-eight numbers of the two parts of the Wohltemperirte Clavier, eighteen begin after the first quaver (or semiquaver, as it may be), seven after the first crotchet, and three after the first crotchet and a half. Indeed, in most of Bach's other clavier fugues—for instance, in those in E minor, F sharp minor, and C minor—the same is observable. In the whole two parts of the Wohltemperirte Clavier only fourteen themes begin with the bar, and only six at the half bar.

In the organ fugues the conditions are somewhat different: here the entrance of the theme with the bars predominates; still, instances to the contrary are to be found, particularly in early works, where they are not unfrequent; and here it is all the more perceptible, because the organ is incapable of accent, and therefore the feeling of the true rhythmical value can only be given gradually, and by other means. Bach had to submit to the natural conditions of the instrument, and subsequently restricted his use of these modes of utterance, full of internal emotion and unrest, to the more sympathetic clavier; but on this instrument he still further developed this rhythmical extension, of which the F sharp minor fugue in the second part of the Wohltemperirte Clavier is an admirable example.[239] The C minor prelude has the same general plan, still, irrespective of the key, it is both sadder and more intricate; the motive does not consist merely in a broken chord, but has besides something of a melodic character—nay, towards the close a vehement passion betrays itself. Even its indescribably graceful and charming fugue, which has something peculiarly piquant in the bold use of false relations in the harmony, is not devoid of pensive passages.[240]

We have already studied the C sharp major prelude as the finished sketch for the invention in two parts; this with the fifth, ninth, and twelfth are in one category. This, however, refers only to its first and shortest form, which was subsequently extended to nearly forty bars. It may be remembered that Bach had originally called the Inventions

---

[239] See Vol. I., p. 253.
[240] Marpurg gives an analysis of it. Kritische Briefe, I., p. 218.

"Preambles." In this, as in those, both hands are employed alternately in working out a complete melodic phrase, and an extremely graceful composition is evolved, sporting gaily up and down. The fine fugue, which carries out and intensifies the happy and vigorous sentiment of this prelude, is based on a bold theme which only the mind of a genius could have devised.

The C sharp minor prelude is more akin to a type of work which we have already had occasion to notice in the compositions of the later Weimar period; it is founded on a real theme worked out in imitation.[241] The triple fugue in five parts, which follows it, suits admirably with this noble and deeply pathetic movement; it is one of the grandest creations in the whole realm of clavier music. The main theme, consisting of four notes, massive as if hewn out of granite, is associated, after the thirty-fifth bar, with another in a smooth flow in quavers, and finally, in bar 49, with a more energetic and insistant one; and then, for sixteen bars more, it expands into a composition of such vast breadth and sublimity, of such stupendous—almost overwhelming—harmonic power, that Bach himself has created but few to equal it. It is as though we were drifting rapidly over a wide ocean; wave rises over wave crested with foam, as far as the eye can reach, and the brooding heavens bend solemnly over the mighty scene—the surging forces of nature and helpless, devoted humanity.

The preludes in D major and D minor are wrought out on ornate motives in semiquavers, and they are almost exclusively in two parts or homophonic. The first is graceful and playful, the second restless and yearning. The D major fugue has a very distinctive character; it seems to march in defiantly, and then stride on proudly with a somewhat rigid dignity. A not inconsiderable space is taken up by some highly interesting episodical figures, which are rendered necessary by the peculiar structure and the brevity of the theme; after bar 17 it never appears again, and it is precisely here that the composition attains its

---

[241] See Vol. I., p. 590.

greatest brilliancy from the contrasts suggested by the theme: sudden bursts and pathetic grandeur developed side by side. The D minor fugue is remarkable for its artistic inversions and strettos, and the extraordinary economy it displays in the use of musical material; the expression is bitter and capricious, as the composer's humour could be at times.[242]

A very peculiar composition lies before us in the E flat major prelude. It is broadly, artistically, and firmly constructed, in four parts on two themes; but they are first carried through independently, one after the other, in a free and, as it were, explanatory manner—the *agitato* theme first, to bar 10, then the calmer one in crotchets and minims as far as bar 25. The strong contrast between the two phrases reminds us at once of the toccata form as Bach was wont to begin it; indeed, we are already acquainted with a similar instance of explanatory treatment.[243] What was attempted rather than achieved in the last movement of the D minor toccata is here carried out with perfect mastery in every respect. The feeling is most noble, deep, and purposeful. All this, of course, tells to the disadvantage of the fugue that follows it, which, notwithstanding its grace and sweetness, is too light when compared with the prelude, which, properly speaking, ought only to lead up to it and prepare us for it. The two pieces cannot possibly have been originally designed at the same time; Bach must have wished to make use of the lovely toccata movement in this work, and as it was too ponderous as a true prelude, he inverted the order for once, and intentionally supplemented it with a very short fugue in three parts, and not above half as long.

The prelude in E flat minor is one of those which bears the clearest stamp of genius. From this germ—

which is applied in various ways—first in the right hand and

---

[242] This fugue is exhaustively analysed by S. W. Dehn. Analysen dreier Fugen aus Joh. Seb. Bach Wohltemperirten Clavier und einer Doppel-fuge A. M. Bononcini's, pp. 1—7. Leipzig: C. F. Peters, 1858.

[243] See Vol. I., p. 437.

then in the left, now dismembered, now lost in figurations, while this rhythm ♩♩♩♩ follows it all through in ponderous chords—is developed a piece unique among Bach's works. The triumph achieved here by episodical art is all the greater because we are quite unconscious of it under the spell-bound feeling which envelopes us, heavy and oppressive as a sultry stormy evening, when not a breath is stirring and lurid lightning flickers along the horizon. From bar 29 the sentiment is sad as death, and the change to the major at the close is awful. The three-part fugue suits this to perfection: a real *ricercar* again, and the only piece of the first part in which he has availed himself of the enlargement of the theme (bar 62 ff). Art is here raised to such a pitch that, after all the strettos and inversions that we are already familiar with have been applied, from the sixty-second bar the theme is carried on in augmentation as well as in the ordinary form, both at once, and in direct as well as inverted motion; and from bar 77 in all the three parts. This closely compacted fabric of parts leaves an impression of nervous excitation, of anxious and passionate seeking, and at the same time we still hear the passages of the upper part in bars 15 and 16, and 48 to 52: the same form as we find again in the contrapuntal violin part to the chorale "Ich ruf zu dir, Herr Jesu Christ, ich bitt, erhör mein Klagen" which forms the finale of the cantata "Barmherziges Herze der ewigen Liebe"[244]—"All-merciful heart of the love everlasting." This fugue made so deep an impression on Ludwig Krebs that he attempted to compose an imitation of it.[245]

We have in E major a bright and charming prelude worked out on a motive of six quavers, and the fugue is still more delightful, with its theme that sets out with an audacious leap and then proceeds so deliberately. To do anything like full justice to Bach's incredible flow of invention we must study other fugues of this period, in which it is conspicuous in the endless variety of the themes; they are so many pictures which once known and understood can never be forgotten.

---

[244] See Vol. I., p. 546.
[245] In A minor. In MS. in my possession.

The E minor prelude, as we now have it, is the working-out of a little piece which was written for Friedemann Bach, and, as it would seem, to exercise the left hand; semiquavers roll up and down, while the right hand strikes short chords. With that mastery which was his alone, Bach has devised an independent melody for the bass. It is evident that the ideal form he had in his mind was the adagio movement of the Italian Violin Sonata, of which he has left us such admirable specimens in his own Sonatas. From bar 23 the bass motive is subjected (in a quicker tempo) to a farther elaboration in both hands, and it rises by degrees from two parts to four. This most original prelude is followed by an equally remarkable fugue, the only one in two parts in the whole work. A liberty quite unheard of for a two-part fugue is the use of unison which occurs twice, bars 19 and 38: a license which we do not expect to meet with in Bach's work, and least of all here. There are, however, a few places from which we learn that the master did not scorn even these means when he required a particular effect. We have already noticed one, in the cantata "Bereitet die Wege," where, in the first recitative, the voices and instrumental bass twice concur in ascending and descending passages in unison, as an illustration of the union of the Christian with his Redeemer.[246] Another example is in the wild first movement of the Sonata in G minor for viol di gamba; a third occurs in the little G minor "Preamble" in Friedemann Bach's Little Book, the fifth bar from the end,[247] and a fourth in the *Burlesca* of the A minor partita in the first part of the Clavier Uebung, bar 16 of the second section.[248] And in the E minor fugue under discussion the object is unmistakably a peculiar effect. On both occasions the parts do not coalesce in their natural course, but one breaks in assertively and wilfully on the quiet flow of the other. This character of wilful caprice, indeed of pugnacity, is stamped on the whole fugue, and is still farther confirmed by the

---

[246] See ante, p. 558. Note 277.
[247] P. S., 1, C. 9, No. 16, XI.
[248] B.-G., III., p. 78.—P. S., 1, C. 5, III.

pertinacious assertiveness of the semiquavers that force their way through the maze of sounds.

The prelude and fugue in F major are pleasing and sweet, but have no conspicuous peculiarities of form. The following pair, on the contrary, in F minor, are deep and passionate; the prelude is fine, founded almost throughout on this motive—

and the theme alone—

sufficiently proves that the Fugue is worked out on the broadest lines.

The prelude in F sharp major again is a two-part composition, and, with its happy sportive fugue, it forms an indescribably delightful whole. That in F sharp minor starts from a motive in rolling semiquavers one bar in length, which is then developed with the most wonderful imaginativeness; the form is crisp and round, the sentiment sad and weird. This truly Bach-like counterpoint—

is used in contrast to the long-winded fugal theme in 3-4 time, and in the progress of the fugue it increases in intensity, particularly from bar 35 onwards, by coming out in doubled thirds and sixths. This is the kind of counterpoint Kirnberger means when he says:[249] "When the *cantus floridus* (where more than one note is opposed to one) is the composer's intention, Bach adopts at once a definite phrase to which he adheres throughout the piece." In this wide sense the statement is certainly not accurate; on the contrary, it is precisely in the invention of constantly new counterpoint that Bach is so great and inexhaustible. Kirnberger, however, as may be seen by

---

[249] Gedanken über die verschiedenen Lehrarten in der Composition, p. 8. Berlin, 1782.

the context, had something quite different in his mind—namely, the skill with which Bach was wont to work out his counterpoint from the first counter-subject, for it was by this means that he attained in great measure that admirable homogeneity and characteristic purpose which give to each of his fugues a distinct individuality, while most of his predecessors and contemporaries were satisfied with using contrapuntal treatment in a perfectly arbitrary manner, just as they would have worked out an episode in an organ chorale.

The gay and jovial prelude in G major is followed by a very fresh and merry fugue full of positive audacity, particularly in the inversion of the theme. The violin adagio seems again to have suggested the form of the G minor prelude—a melody composed of long-held notes and varied figures lies above a series of interesting harmonies; presently they change parts, the bass takes up the melody for a time and then joins the upper part. The sentiment is grave and deep, and it continues the same in the fugue, which is marked by great moderation.

The lively prelude in A flat major owes its origin entirely to this motive:—

In the fugue, besides the brevity of the theme, we are struck by its hardly moving out of the principal key, while the melody is insignificant; hence its progress is worked out very quietly and inconspicuously.

The G sharp minor prelude is a really inspired composition of the most subtle construction. I have already spoken of its fugue.

The A major prelude is of the same type as the three-part sinfonias, and worthy to stand side by side with those glorious works of art. The theme of its fugue is a grand invention, which with its first note seems to knock at a door and then, after a pause of three quavers, to walk quietly in; presently greater vigour is introduced by the counterpoint in semiquavers. The merits of the A minor fugue and its relations to its prelude have been already discussed.

In B flat major we have a fiery prelude in demisemiquavers now rocking softly and now storming up and down, followed by a fugue of a soothing and peculiarly sweet character, reminding us in many ways of the beautiful D major sinfonia. The unusual equality of the phrases contributes to give it its character.

The B flat minor prelude is of a deeply melancholy cast of beauty; Bach works it out with consummate genius from this germ:—

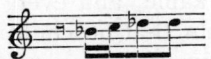

In bars 20 to 22 the resemblance is very remarkable to the fifth movement of Handel's *Concerto grosso* in F minor, of which Bach copied out the parts.[250] A grand fugue follows it, remarkable for its massive theme, mighty harmonies, and skilful strettos. The two last preludes and fugues once more vividly illustrate at the close the spirit of contrast which prevails throughout the twenty-four pairs of pieces which constitute the work. The B major prelude starts up before us from this motive—

in the most perfect order and freedom; its bright, fresh feeling revives the soul, and it flows on for nineteen bars, polished and smooth down to the most insignificant detail, a perfect gem of chamber-music.

The B minor prelude on the other hand—the only one of the whole first set which is in two sections with a repeat—is a duet in imitation above a bass in unflagging quavers, and is equally masterly even in the minutest details, but still apparently too compact and self-contained for a prelude. Apparently—and so long as the fugue is left out of the question. While the fugue which belongs to the B major

---

[250] See ante p. 11.

prelude goes on its way contentedly, *debonnair*, and without pause, this one—

proceeds slowly, sighing, saddened, and pain stricken; its feeling is akin to that of the F minor sinfonia, only here the suffering is so intensified as to be almost unendurable.[251] And we must beware of regarding the piercing bitterness of the effect in this fugue as a mere result of contrapuntal skill. From this point of view indeed it is in no way remarkable, and even if it were, Bach has proved again and again that he could preserve a sweet and pleasing character even with the greatest intricacy of construction. No, it was his purpose to produce a picture of human misery, to give it full utterance here, in his favourite key, and at the close of this glorious work in which all his deepest sympathies with human feeling had found expression. For to live is to suffer. This is the idea—persistent as an organ point—which asserts itself through all the manifold, motley, and endless variety in this work, gradually built up by the master's unresting industry, and which asserts itself once more in its closing chords.

There is another reflection which again forces itself upon us as we close the Wohltemperirte Clavier. How little can a composer who finished one of his most important instrumental works, conceived and produced as a grand whole, with such a crown of thorns, have counted on the sympathy of the great music-loving public! Still, that which the god prompted his deep heart to utter, that he spoke without reservation or calculation; he appealed only to a restricted circle of docile pupils and intelligent friends. But their sympathy, on which he could no doubt fully rely, did not

---

[251] Kirnberger has analysed its harmonic structure, Die wahren Grundsätze zum Gebrauch der Harmonie, p. 55. I avail myself of this opportunity of alluding to Carl van Bruyk's Technische und aesthetische Analysen des Wohltemperirten Clavier. Leipzig: Breitkopf und Härtel, 1867. Though I cannot agree with all the views expressed in it, the work contains many charming observations and is written with a real inspiration of love for the subject.

betray him into pouring out his feelings in capricious imagery; he must always refine and purify them to be the soul of the severest possible forms. It is impossible to speak with too high praise of this supreme artistic morality.

It is very difficult to say anything that will convey a general idea of the character of these fugues. Their forms are so complete within their narrow limits that what distinguishes them could only be made clear by a thorough technical analysis of each, or of most of them, and that, it is evident, is impossible here. And—in spite of their strong individuality—their character as a whole is even more inaccessible to verbal description than that of the other instrumental works, by reason of the lofty idealisation which the feeling they express derives from the severity of the form which expresses it. There is a legend which tells us of a city of marvels that lies sunk beneath the sea; the sound of bells comes up from the depths, and when the surface is calm, houses and streets are visible through the clear water, with all the stir and turmoil of busy, eager human life—but it is infinitely far down, and every attempt to clutch the vision only troubles the waters and distorts the picture. We feel the same thing as we listen to this music. All that stirred the soul of the composer—love and hatred, joy and sorrow, with their fortuitous and transient impulses—lie deep below the surface: faintly, remotely, we hear their echoes, and as we gaze through the crystal flood of sound we see the living soul within, and perceive that it suffered or was gay even like ourselves, only what it was that stirred it we may not see. But each of us can recognise with kindred feelings the experience of his own life: every one of all the human hearts which, for the last century and a half, has duly studied and absorbed this work; and this it is which has made it, to our own day, a perennial source of joy and of spiritual refreshment and strength. Indeed, what has already been said of Bach's clavier works in general is especially true of this—that he wrote them for an ideal instrument, which it was left to our own time to realise. A movement so pervaded with profound melancholy as the C sharp minor prelude and its fugue, through which the spirit of God seems to rush with sublime terrors, could find no adequate interpretation on the clavichord. So that it is to us in fact that all the glories which filled the master's fancy have been first revealed; we hear more clearly the sweet bells from the deep, and stand more nearly face to face with the forms that people it. But the work will long survive our generation: it will stand as long as the foundations of the art endure on which Bach built. It finds a fitting place at the end of this section, for it reflects at parting the whole of the Cöthen period of Bach's life, with its peace and contemplation, its deep and solemn self-collectedness.

BOOK V.

LEIPZIG, 1723—1734.

BOOK V.

LEIPZIG, 1723—1734.

I.

BACH'S APPOINTMENT AND INSTALLATION AS CANTOR AT LEIPZIG.

THE post of Cantor to the town-school of St. Thomas at Leipzig was not a brilliant one; but those who were familiar with its conditions knew that it had certain valuable advantages. Kuhnau died June 5, 1722, and a month later the Council had had a choice among six candidates. For the most part these were men who from their own knowledge were aware of what the Cantor of the Thomasschule[1] had to expect. Fasch, Rolle, and Telemann were conspicuous among them. Johann Friedrich Fasch, Capellmeister to the Prince of Anhalt-Zerbst, had been at the Thomasschule from 1701 to 1707, and had enjoyed Kuhnau's instruction. As a student of law he got up a musical society among the Leipzig students, and with them provided a part of the church music in the University church of St. Paul, in 1710. After this he led a life of vicissitude, and he had only been a few weeks in his place at Zerbst when one of his Leipzig patrons, the Hofrath Lange, suggested to him that he should become a candidate for the vacant place of Cantor.[2]

Christian Friedrich Rolle we have already met with in the course of Bach's history; it was with him and Kuhnau that he had tested the organ of the Church of the Blessed Virgin at Halle in 1716; at that time Rolle had already

---

[1] So called in German; the designation is retained here for convenience and brevity.
[2] Records of the University of Leipzig; and Gerber, n. d., II., col. 9.

been for a year at Quedlinburg; he was now musical director in Magdeburg.[3] Telemann, finally, who had gone from Eisenach to Frankfort am Main, and from thence, in 1721, to Hamburg as musical director and Cantor, had formerly first shown his great musical powers at Leipzig; he had gone thither in 1701, intending to study jurisprudence and to suppress all his musical proclivities. But his talent was discovered, and he was immediately engaged to produce a composition once a fortnight for the Church of St. Thomas, where Kuhnau had lately been made Cantor. He also formed a *Collegium Musicum*, a Musical Union of students, which rapidly rose to importance, and during the first ten years of the eighteenth century was a power in the musical life of Leipzig. He soon found occupation as a dramatic composer, and wrote a number of operas for the Leipzig Theatre, for some of which he also wrote the text, and even appeared in them himself. When at last he obtained the place of Organist to the New Church (August 8, 1704) the Council hastened to instal him (August 18). "He was a very good composer—he was to give his services at the Thomasschule on occasion—he was not only to play on the organ, but to direct all the music—but he must refrain from theatres and give up acting."[4] In the same year he was invited to Sorau, as Capellmeister, and thenceforth Germany rang with his fame. Hence, when the point under discussion was the selection of a new Cantor for the Thomasschule, he was preferred above all the other candidates, and some days after, when he had passed the customary tests, his appointment was definitely settled by the Council. The only difficulty raised was the obligation under which the Cantor lay to teach some other branch of knowledge in the school; to this Telemann would not consent. The Council, however, declared their willingness to make other arrangements in this respect, and prepared to

---

[3] See Vol. I., p. 520. In the documents preserved in the archives at Magdeburg he is called Johann Christian Rolle, but there is no doubt as to their identity.

[4] Acts of the Leipzig Council. Telemann's Autobiography in Mattheson's Grosse General Basschule, p. 173. Ehrenpf., p. 238.

instal the famous musician in his office, when he returned to Hamburg and wrote from thence that he could not accept it.

The Council, much provoked, proceeded to a new election. Meanwhile, Georg Friedrich Kauffmann, of Merseburg,[5] and after him Christoph Graupner, Capellmeister of Darmstadt, had come forward as candidates, and the town was represented by Georg Balthasar Schott, the highly-esteemed Organist of the New Church. The decision fell on Graupner, for whom Kauffmann voluntarily made way. Even Graupner might regard himself as an old Leipziger, so far as that he owed most of his musical and general culture to a nine years' discipline at the Thomasschule. From being a prize scholar under Kuhnau, in clavier playing and composition, he had become a master who, as a composer for the clavier, may rank as one of the best of his time.[6] In his application for the post of Cantor he had been strongly recommended by his old friend Heinichen, the Capellmeister at Dresden. Graupner came to Leipzig, and seems to have passed his tests and presented his testimonials; but, when all had proceeded so far, the Landgrave of Hesse Darmstadt refused to part with him. As the transaction had been conducted privately, Graupner was able to retire more honourably than Telemann.

Besides Graupner, and as it would seem rather later—at any rate, not till the end of the year 1722—Bach came forward[7] to offer himself. It is not probable that he should only now have heard for the first time of the vacant post; his late appearance must have had other causes. He was, in fact, in a critical position. Prince Leopold's failing interest in music, his own anxiety for the higher education of his sons, the feeling that in the service of the Court only one side of his artistic genius could thrive and labour—all this made a further residence in Cöthen seem undesirable. On the other hand, he no doubt did not undervalue the comfort-

---

[5] See Vol. I., p. 118.
[6] Mattheson, Ehrenpf., p. 410.
[7] Documents of the Leipzig Council. The appointment, dated Dec. 21, says that several had become candidates—namely, Capellmeister Graupner, of Darmstadt, and Bach, of Cöthen.

able and honourable position, free from all petty anxiety, which the Prince's favour secured to him. In Leipzig a wider circle of labour awaited him; he would be standing midway in a broader and fuller stream of public life, but "from a Capellmeister to become a Cantor" was not at all to the mind of a man who was both proud and famous. Even after he had actually put out his hand to gather the inheritance of Kuhnau, for fully three months he doubted whether he should not do better to withdraw. But certain persons whose counsel he asked urged it so strongly that at last he took the decisive step. He went to Leipzig at the beginning of February, 1723; on the 7th, being the Sunday called *Estomihi* (Sunday next before Lent), he performed as his test piece the cantata on "Jesus nahm zu sich die Zwölfe"—"Jesus called the Twelve unto Him."[8]

His appointment did not immediately follow; the Council were still in treaty with Graupner, who, three weeks previously, had passed his tests, and besides him Kauffmann and Schott were still candidates. However, when Graupner had retired, no long consideration was needed to discern the worthiest of the three remaining competitors. Bach had been acquainted with Kuhnau; he knew Leipzig and Leipzig knew him. He had already been invited thither in 1717 to inspect the great organ in St. Paul's Church, and the Council knew that it was strengthening itself by such a selection. They reflected that he was a distinguished clavier-player, a man for whose sake even Telemann might be forgotten, the equal of Graupner, and one who was famous enough to attract even the students to take part in his musical performances, which in the then state of affairs was highly desirable. Besides this, Bach seemed to be willing to fulfil the Cantor's duties in every branch, and in this he was distinguished from the other candidates; he was willing even to undertake the general instruction required of him. This consisted in giving five Latin lessons weekly to the third and fourth classes; in these the course included written exercises,

---

[8] So says a note on a copy of this cantata, which, though not an autograph, was revised and completed by Bach. It is in the Royal Library at Berlin. See also ante, p. 157.

grammar, the *Colloquia Corderii*,[9] and an explanation of Luther's Latin Catechism.[10] At first Bach would seem to have resisted the demand that he should be Latin teacher as well,[11] or else, from the refusal of all the other candidates it was taken for granted in his case. However, when the Council met for final decision on the 22nd of April, Burgomaster Lange was in a position to state that Bach had expressly pledged himself both to hold his official classes and to give private lessons in the Latin tongue. He cannot have been ignorant that in Telemann's case a dispensation from these duties had been contemplated, and this relief would undoubtedly have been at once granted to him also, since the gentlemen of the Council declared of their own accord that if he could not accomplish all the instruction required in Latin no objection would be raised to paying a deputy to do it for him. Bach, however, felt equal to performing his own duties, and no doubt regarded it also as a point of honour to be in no respect behind his predecessors; after such a man as Kuhnau this was saying something, but we have already had occasion to observe that Bach from his schooldays had been a sound Latin scholar. Still it must no doubt have seemed to him a strange experience to stand in front of the third class of boys with the Latin Grammar in his hand, a church cantata, perhaps—who knows?—running in his head. Beyond instructing his own children, perhaps, such teaching had never been any part of his duties. Indeed, he soon felt the task a burthen, and paid his colleague, Magister Pezold, the sum of fifty thalers per annum to relieve him of the greater part of his teaching; after this he held the class only when Pezold was ill or otherwise prevented, and then he would dictate an exercise to the boys for them to *elaborate* (construre and parse).[12]

---

[9] *Maturini Corderii Colloquia Scholastica, pietati, literarum doctrinis, decoro puerili, omni muneri, ac sermoni præcipue scholastico, utiliter concinnata.* 3rd ed. Leipzig, 1595.

[10] Acts of the Leipzig Council concerning the "Schuel zu S. Thomas."

[11] "All three"—namely Bach, Hauffmann and Schott—"will not be able to teach (Latin) as well." Document dated April 9.

[12] This condition of affairs was reported to the Consistory by Superintendent Deyling, Feb. 24, 1724.

About a fortnight after the transaction above mentioned, Bach, who had appeared in person before the Council, received an official intimation that he was considered the best of the candidates, and had been unanimously elected; the office was therefore conferred upon him on the same conditions as those on which his predecessors had held it. He then had to sign a contract deed which had been prepared for Telemann the year before (and which, twenty-seven years after was used again for his successor); this contained the customary stipulations, as to leading a respectable and sober life, to fidelity and diligence in the performance of his official duties, and to due and proper respect and obedience to the worshipful Council; it pledged him, among other conditions, not to make the church music too long nor too operatic, to instruct the boys not only in singing but—for the avoidance of expense—in instrumental music also, to treat them with humanity; not to send any incapable singers to join the chorus of the New Church, which was exempt from his supervision, not to make any journeys without permission from the Burgomaster, nor to accept any office in the University without the consent of the Council.[13] And even after all this the appointment was not an accomplished fact. Its confirmation was needed by the Consistory of Leipzig, a superior municipal body, composed partly of ecclesiastics and partly of laymen.[14] When the Council desired to appoint to any post in the town churches or schools, the candidate had to present himself before the Consistory, which then put him through a sort of examination on its own account, with the object of ascertaining the religious principles of the examinee. If the result was satisfactory his appointment was forthwith confirmed by the Consistory. Bach was presented, on May 8, by Deyling, Superintendent and Consistorial Assessor, and his examiner was the Consistorial Assessor Dr. Schmid. The two assessors then

---

[13] See App. B, VI.

[14] Sicul gives a list of the names of the members of the Consistory for 1724, with the dates of their election; Leipziger Jahrbuch, Vol. iii, p. 358. It had at that time six Assessors—the Doctors Wagner, Lange, Schmid, Packbusch, Deyling, and Mascov. The director, since 1709, was Dr. Johann Franz Born.

testified that Herr Johann Sebastian Bach had answered the questions put to him in such wise that he might be permitted to assume the post of Cantor in the Thomasschule.[15] On the 13th he was confirmed in his appointment by the Consistory; he had to subscribe the *concordia* formula,[16] and be sworn.

On Monday, May 31, his formal installation at last took place.[17] At nine in the morning two deputies from the Council—namely, one Lehmann, who was at that time superintendent of the school, and who held the civic office of " Baumeister,"[18] and Menser, the chief town clerk—proceeded to the Thomasschule, where they were received at the door by the Rector (or warden) Joh. Heinrich Ernesti, and conducted to the hall appointed for the examination of the deed. Here they were met by the licentiate Weisse, preacher at the Church of St. Thomas, who appeared as the representative of Superintendent Deyling, and as the ambassador from the Consistory. The six other masters of the school now joined them, with their new colleague—namely, Licentiate Christian Ludovici, the sub-warden (Conrector), Magister Carl Friedrich Pezold, Master of the third class; Christoph Schmied, of the fourth; Johann Döhnert, of the fifth; Johann Breunigke, of the sixth; and Christian Ditmann, of the seventh.[19]

They took their seats, the pastor and the two reporters of the Council in one row, and opposite to them the school

---

[15] "Du. Jo. Sebastianus Bach ad quaestiones a me propositas ita respondit, ut eundem ad officium Cantoratus in Schola Thomana admitti posse censeam.
" D. Jo. Schmidius.
" Consentit. D. Salomon Deyling."
Act of the Leipzig Consistory.

[16] The "concordia formula" is an abridgement of the contents of the Concordienbuch, a kind of religious statute book, in which are embodied the tenets and doctrines of the Reformed Church.

[17] The official documents date it on June 1. Deyling, in a letter written a month later, says May 31; and this is certainly right, since it fell in 1723 on a Monday, and Bach would begin his school work after conducting the services of the previous day.

[18] This answered rather to the Roman Aedile, and does not mean an architect; Baumeister Lehmann was a lawyer. See Sicul, Leipziger Jahrbuch, Vol. IV., p. 764.

[19] From E.E. Hochw. Raths | der Stadt Leipzig | Ordnung | Der Schule | Zu S. THOMAE | Gedruckt bey Immanuel Tietzen, 1723, p. 11.

officials according to their rank. The choir first sang a piece of music at the door, and then all the scholars came in. The town clerk made a speech bearing on the installation, and the pastor then pronounced the fact of installation, adding the customary admonitions and injunctions. Bach replied in a few words, he was congratulated on his new appointment, and the ceremony concluded with another musical performance.

It was plainly shown on this occasion that in the Consistory the Council had a thorn in the flesh, for the superior court fettered its liberty and independence in various ways. Up to this time it had never been the custom for the Consistory to interfere in this direct manner with the installation of a school official. The deputies of the Council declared then and there that it was an infringement of their rights, that the superintendent and pastor when present at such a ceremony had no more share in it than to congratulate the new officer. It was owing only to the moderation of Weisse's conduct that matters did not come to an outbreak between him and the irate councilmen in the presence of the whole assembly. The Council immediately proceeded to draw up a formal protest, however, and the Consistory appealed against it to the regulations of the canon law of the electorate of Saxony.[20]

Bach had an official residence in the left wing of the school buildings; this had probably been the Cantor's dwelling from time immemorial, for Kuhnau, at any rate, had inhabited it before him.[21] The building at that time had only two storeys, and was much too small for its purpose. At the beginning of 1731 it was added to, and an additional storey was built, and meanwhile Bach had a temporary residence assigned to him, from the spring of 1731 till the New Year of 1732, probably, in the house of Dr. Cristoph Dondorff, who from 1730

---

[20] See Appendix B, III., of the German for the entire document.
[21] Das jetzt lebende und florirende Leipzig; Leipzig, bey Joh. *Theodori Boetii* seel. Kindern, 1723. "Joh. Sebastian Bach, *Director Music*, und *Cantor*, am Thomas Kirchofe auf der Thomas-Schule." We know that Kuhnau lived here, from a note as to his interment in the Leipzig Register.

had owned the Mill of St. Thomas,[22] and who was a friend of the Bach family.[23] St. Thomas' Mill stood outside the city walls, which ran round the back of the schoolhouse, where the Schlobach estate now lies, on the farther side of the Promenade. It may be mentioned, as characteristic of the conditions of life at that time, that the rent paid by the Council for this house, which Bach occupied for nearly a year, was sixty thalers. His residence, as Cantor, was meanwhile somewhat altered; a room on the first floor was lost in consequence of the rebuilding, and Bach had another instead on the third floor. It would seem that the dwelling was made on the whole more commodious. Bach never quitted it again till his death; and after him it was, with very little alteration, the residence of all the Cantors of the school down to the time of Moritz Hauptmann. The view of the open place near the church, to which the school buildings turn their front, and the houses which enclose two sides of it, must be much the same now as they were then; only the great stone fountain, which at that time graced the middle of the quadrangle, has now disappeared.[24]

II.

THE THOMASSCHULE—DUTIES OF THE CANTOR—STATE OF MUSIC IN LEIPZIG.

LEIPZIG, at the time of which we are speaking, had three public schools: those of St. Thomas and St. Nicholas, and the Orphanage. The first of these was by far the oldest, dating from the thirteenth century as a foundation school under the Augustine monks (or Austin Friars). It had not become a town school until four years after the introduction of the reformed doctrines into Leipzig, when,

---

[22] According to the Ward-book of the town, preserved there in the Lower Court of Justice.

[23] He was godfather to a son of Bach's, born in 1732.

[24] A view of this quadrangle, engraved on copper, is prefixed to the school regulations printed in 1723. But the Thomasschule, in the autumn of 1877, quitted this venerable and memorable home for a new building outside the town.

in the year 1543 the Council took possession of the Monastery of St. Thomas and all its dependencies. The Monastery had had an *Alumneum,* or foundation school, in which a number of boys were maintained for the proper performance of the choral portions of the Liturgy and other parts of Divine worship. As a Protestant establishment the school was soon considerably extended. At first it had four classes and the same number of masters. When the St. Nicholas school (founded in 1511) could no longer contain the pupils that resorted to it, the Council decided that, at any rate till further orders, little boys should be admitted to the Thomasschule. Their instruction was at first carried on by monitors called *Locates,* but under-masters were afterwards substituted for these. The school now consisted of seven classes, of which the three lowest, particularly, were for a long time very much crowded. The foundation school was kept up, and the number of scholarships had been gradually increased by a succession of endowments. For a time there were thirty-two, these increased to fifty-four, and at last, by the munificence of Privy Councillor Born, to fifty-five.[25]

The principal aim and end of the multiplication of the scholarships was the cultivation of church music. Formerly most of the Leipzig churches, and among them those of St. Thomas and St. Nicholas, had been under the management of the Augustine choir brethren; the Reformation clung to a close alliance between Church and school, and the foundation scholars of the Thomasschule were the means most obviously at hand for the musical requirements of the Protestant service. It is well known how urgently Luther has insisted on the use of music, and how he relied upon it greatly for securing the extension of the reformed doctrines.

Thus the Cantor of such an institution was a personage of much importance, doubly so since he also was required to take a share in the general course of instruction. This, in fact, was sufficiently recognised by the position he took

---

[25] Various documents referring to the school are preserved among the Archives of Leipzig. Stallbaum, Die Thomasschule zu Leipzig. Leipzig, 1839. See, too, his work, Ueber den innern Zusammenhang musikalischer Bildung, &c. Leipzig: Fritzsche, 1842.

among his colleagues. He ranked third in order,[26] and, while the other teachers had to give four hours of lessons daily, he, like the Rector, had only to give three. This moderate requirement which, as time went on, was even farther reduced, was the reason why at last eight masters were needed for seven classes. The hours of work were from seven to ten in the morning, and from twelve to three in the afternoon. Before the issue of the school regulations of 1634, the Cantor had daily to give a lesson in Latin grammar from seven to eight—Luther's Latin Catechism was used only on Saturday—in music from twelve to one; and from one to two Latin syntax with the third class. In accordance with those regulations, the hours of singing lessons were somewhat increased; those of the Latin lessons were considerably diminished, quite irrespective of the usual division into three lessons daily. Instruction in singing was now given on Monday, Tuesday, and Wednesday, at nine and at twelve; on Friday at twelve only. It comprised all the classes at once; that is to say, the four upper and original classes, to which alone the foundation boys belonged. On Thursday, at seven in the morning, the Cantor had to take the boys to church, and then was free for the rest of the day; on Saturday, at the same hour, he had to expound the Latin catechism to the third and fourth classes; on other days he had to give one Latin lesson to the third class. This plan of lessons was still kept up with remarkable regularity till the time of Bach's arrival, and for a time was carried on by him with undiminished regularity, excepting that he went to early church with the boys on Friday morning, and so had Thursday entirely free. The Cantor gave instructions in singing only to the four upper classes, his few lessons in

---

[26] In the Act of election one of the members of the Council designates the Cantor as *Collega Quartus*. This is an inaccurate statement, and can only have referred to the instruction in Latin (or what not) in which the Cantor did in fact rank below the third master; but in order of rank he came next to the *Conrector*, and he next to the *Rector*. (See Ordnung der Schule zu S. *Thomas*, 1723.) But when this mentions, besides the Rector, eight other masters, this certainly does not refer to Kuhnau's time. From the Acts of 1717 we learn that there were only seven masters besides the Rector, just as in Bach's time; there was no Quartus above the two Bachelors.

Latin to the third alone, the *Tertius*, as he was called, being their master in other things. He, with the Rector, the Conrector, and the Cantor, formed the circle of the four upper masters (*superiores*), who held themselves aloof from the others, the *Baccalaureus funerum*, *Baccalaureus nosocomii* and the first and second *collaborators* (under masters), or, as they were called after 1723, the *Quartus*, *Quintus*, *Sextus*, and *Septimus*. The four upper masters, including the Cantor, were also required to inspect the foundation boys, and took this duty in weekly rotation. They had them to live entirely with them, and to comply with the regulations of the schoolhouse, which required them to rise at five in the morning (at six in winter); to dine at ten; sup at five in the afternoon; and go to bed at eight.[27]

These were the Cantor's duties in the school itself. With regard to the public, further duties arose from the position he took as director of certain church choirs which were formed of the foundation scholars. The two most important churches of the town were those of St. Thomas and St. Nicholas. But in 1699 the church of the Franciscan friars had been repaired and restored to use, under the name of the New Church, and after Telemann's appointment in 1704 had had music of its own. After this, in 1711, the services, which for a while had entirely ceased in St. Peter's church, had also been revived, so that from that time the scholars of the Thomasschule had had to supply the music every Sunday in four churches, and on high festivals in the church of the Hospital of St. John.[28] Thus they were divided into four choirs. The beginners and weaker singers were assigned to St. Peter's, where only chorales were sung; this choir probably also served the church of St. John, as its festival and that of St. Peter would not

---

[27] Acts of the Council relating to the School of St. Thomas. Under Gesner's wardenship these arrangements were somewhat modified. See Gesetze der Schule zu S. Thomae, 1733.

[28] For this they received a special gratuity, at first consisting of food and cakes, but afterwards of 13 thlrs. 3 gr., a year. Accounts of the Hospital are preserved in the Town Hall at Leipzig.

interfere with each other.[29] The rest of the singers were pretty equally distributed; still the service in the New Church was comparatively the easiest, since there the scholars had only to sing motetts and chorales under the direction of the choir Prefect; while on holy days, and during the great Fair times other church music of a concerted character was performed, and not by the boys.[30] Since Telemann's time the director of this had always been the organist then in office, and the function of the Cantor of the school extended no further than the selection of the hymns, and perhaps of the motetts which were to be sung.[31]

He had nothing to do with the churches of St. Peter and St. John, but the music in St. Thomas' and St. Nicholas' was under his direction. On ordinary Sundays a cantata and a motett were performed in only one of the churches, each in turn; the first choir sang the cantata under the leading of the Cantor. But on the first two days of each of the great festivals, and at the New Year, Epiphany, Ascension Day, and Trinity Sunday, and on the festival of the Annunciation, concerted music was performed twice a day, and in both churches at once, the plan being that the first choir sang at St. Thomas' in the afternoon the same cantata that it had performed in the morning at St. Nicholas', and on the next holy day following sang at St. Thomas' in the morning and St. Nicholas' in the afternoon: the second choir taking the reverse order. This second choir sang under the conduct of its Prefect.[32] The rehearsals of the Sunday music took place in the church regularly on Saturday after two o'clock vespers, and lasted till four o'clock.

The direction and performance of music for wedding festivals and funeral processions were also regarded as part of the Cantor's official church duties, as being in direct

---

[29] Bach observes in a tabulated list of the four choirs which he drew up: "And this last choir must also serve the *Petri Kirche*." Documents of the "Schuel zu St. Thomas," Vol. IV.

[30] See Appendix B, VII. Sicul, *Neo-Annalium Lipsinesium Continuatio*, II., 2nd Ed., 1719, p. 508.

[31] See Appendix B, VII. [32] Sicul, ibid., 568.

connection with the divine services. If the funeral procession was of a grand and solemn character, the whole school—or at least the larger half of it, *i.e.*, the three first classes and the fifth—accompanied the body, and were accustomed first to sing a motett at the door where the deceased lay. While the bier was being carried to the churchyard, the boys marching in front of it sang a simple chorale, and only performed part-music on specially grand occasions. The Cantor had at all times to accompany the funeral train, and to decide on what should be sung, and as a rule he led the motett himself. The fact that the Cantor not unfrequently escaped this task, and left his musical duties to the choir Prefect, gives us a clear insight into various official utterances; for instance, Bach was specially enjoined in the deed he signed at all times to accompany his scholars in funeral processions as often as possible. As regarded wedding services, the use of music on such occasions was also very various, depending on the position and wishes of the persons chiefly concerned. In all cases the arrangement of the music lay with the Cantor, even when he himself took no part in it and it consisted merely of chorale singing. He had a representative in the Prefect of each choir, who could relieve him in many ways, not merely of the labour of leading, but also of the rehearsals. This custom prevailed so early as in the seventeenth century. After the school regulations of 1634 had instituted two hours of singing lesson for each of the first three days of the week, the Rector, Conrector, and Tertius petitioned the Council that the Cantor should be relieved of the singing lesson at twelve o'clock, and give a Latin lesson instead, since this singing lesson was very inconvenient to him by reason of his dinner hour, and he therefore but seldom presided at it; nor indeed was it needful that he should, as the boys could and did sing without him. The old order of instruction was, however, adhered to, and, as we have seen, it still continued in Bach's time. But so, to be sure, did the lax practice of the Cantor, which, in fact, gave rise to a complaint on the part of the Rector, Joh. Aug. Ernesti, that Bach held but one hour's singing class, whereas he ought to have held two, and that

THE SCHOOL PROCESSIONS.

consequently the boys had not enough practice in music. But it was in the nature of things that an opportunity for the development of a certain independence should be afforded to the Prefects, for, irrespective of the fact that a school choir was not unfrequently required at weddings and festivities by personages connected with the school to sing during the banquet, the scholars also had to perform their perambulations at fixed seasons of the year with processional singing, and in both cases they had to rely upon themselves for the conduct of the singing.

The processions took place at Michaelmas and the New Year, and on St. Martin's and St. Gregory's Days (the 11th of November and 12th of March, N.S.).[33] On these occasions likewise the boys who could sing were divided into four choirs, each with a Prefect of its own, and apparently each of these had one of the four quarters of the city assigned to it as the scene of its performances. For instance, in 1718 the four choirs were thus distributed; the first included three basses, three tenors, two altos, and three trebles; the second, two voices for each part; the third, two basses, two tenors, two altos, and three trebles; the fourth, two basses, three tenors, two altos, and three trebles, and each choir had besides one or two torch-bearers. The Cantor's duty was restricted to selecting and composing the choirs, to determining generally what should be sung during the perambulations, and to superintending from time to time the rehearsals held for the purpose; all else was the Prefects' affair. The Prefects of the first two choirs especially held an important position.[34] They had to lead the motetts on Sundays and

---

[33] According to the report of the school visitation held in the year 1717. In the school regulations for 1723 a payment is mentioned for music performed in the summer ("Music-Gelde so im Sommer colligiret wird.") When and how these summer processions took place I am not able to say.

[34] It must be supposed that the four church choirs were somewhat differently constituted to those which were selected for the processional singing. These, indeed, cannot at all times have been equally strong. At the New Year only thirty-two foundation scholars sang, eight in each choir, an arrangement which had evidently survived from the time when there had been only thirty-two scholarships. This also explains the case when in the acts and school regulations here and there mention is made of the eight *concentores*.

festivals, and start the hymns in church; the Prefect of the second had to conduct the cantata on those festivals when the Cantor was not in the church, while the Prefect of the first could distinguish himself in his duties of leading the vocal music at wedding feasts and similar occasions, in taking his choir on its Michaelmas perambulation, and in representing the Cantor as conductor of the cantata, when he was prevented attending.[35]

If to all this we add that the Cantor was director of the music in the two other town churches, and required to inspect their organs and to superintend the town musicians—both singers and players—who had to bear a part in the church music, all his official duties have been enumerated. It cannot be said that they were oppressively heavy. Besides five lessons in Latin—from which, as we have seen, Bach soon released himself—he had to give seven singing lessons weekly; but of these he commonly left the afternoon practice to the Prefects. There was no lack of holidays at the Thomasschule. At each of the Fair times—*i.e.*, at Easter, Michaelmas, and the New Year—there was a week of whole holidays, and a second week when the afternoons were free. In the dog-days there were four weeks of half-holidays. Morning lessons were pretermitted on Saints' days, on the occasion of funeral orations in the university church, and on the quarterly academical speech days. A whole holiday was given in honour of the fête or name days of the four upper masters. Eight days were to be given up to the rehearsals for the processional singing at the New Year, St. Martin's, and St. Gregory's, but so arranged that morning lessons should be attended on Monday, Tuesday, Wednesday, and Saturday, and that "no one should sleep through them"—so runs the school regulation. So in point of fact not less than four weeks were given up to preparing for the New

---

[35] All these details are derived from original documents, of which a considerable number are preserved in the Archives of the ancient town of Leipzig. It has not been thought necessary to trouble the English reader with the press-marks and references in detail. The duty of starting the hymn properly belonged to the Cantor, but common custom had deputed it to the Prefects and the practice continued, like many others, in spite of the rules laid down by the school regulations of 1723.

Year's singing. In 1733 the Rector made an attempt to restrict the preparation for St. Gregory's Day to six or eight afternoons, for St. Martin's Day to four, and for the New Year to twelve or at most fourteen. While these rehearsals were going on all the Cantor's afternoon lessons were omitted. Even his church duties did not continue the whole year through. All through Lent no concerted church-music was performed excepting on the festival of the Annunciation, and the same was the case on the three last Sundays in Advent.[36] Excepting on these occasions the Cantor had to conduct one cantata every Sunday. It was only on high festivals that he was very much occupied, particularly at the time of the three holy seasons, when he had to provide two sacred pieces for two holy days in each, and to give two performances of each on both days, but the second Prefect relieved him of one.

As has been already shown, the Cantor had a residence given him free of rent; the rest of his income amounted to about 700 thlrs.,[37] though, from its nature, it could not be exactly estimated. The fixed salary as paid by the Town Council was only 100 gülden (= 87 thalrs. 12 ggr.), and 13 thlrs. 3 ggr. in money for wood and lights. Besides, the Cantor received—at any rate, in Bach's time—1 thlr. 16 ggr. on the Berger foundation and a similar sum on that of Frau Berger and that of Adlershelm, with 5 ggr. yearly from the Meyer foundation; he also had a variable sum (3 thlrs. and 18 ggr., 2 thlrs. 1 ggr., 10 ggr. 6 pf., and so on) forming a share in certain bequests to the school; and, finally, in kind, 16 bushels of corn, 2 cords of firelogs, and from the Church revenues two measures of wine at Easter, Whitsuntide and Christmas. Everything beyond this came from incidental fees. These, of course, were derived chiefly from the school money. Twice a week eight of the scholars went

---

[36] Leipziger Kirchen Staat, Das ist Deutlicher Unterricht vom Gottes Dienst in Leipzig, wie es bey solchem so wohl an hohen und andern Festen, als auch an denen Sonntagen ingleichen die gantze Woche über gehalten wird," &c. Leipzig, 1710, p. 32. (This work is to be found at Halle in the Ponickau Library.)

[37] As we learn from Bach's statement in his letter to Erdmann.

round the town with boxes to collect small donations from a certain number of benefactors to the school—the "*runners' money*," as it was called. Out of this 6 pfennigs were deducted weekly as school payment for each scholar, and this was divided among the four upper masters.[38] The very small sum is accounted for by the principle observed in the Thomasschule of bringing up by preference the children of parents without means. Out of the money collected at the Michaelmas and the New Year perambulations, after one thaler was deducted for the Rector, the Cantor received one-eleventh, one-eleventh more was taken for the Conrector, and sixteen thirty-thirds for the singers; the Cantor then took one quarter of the residue. The money collected in the summer was divided in like manner. Out of what was obtained on St. Gregory's Day the Rector first had one-tenth to give an entertainment to the four upper masters, and out of the residue the Cantor took one-third. Funeral money was another source of income; if the whole school accompanied the procession, and if a motett was sung outside the house of mourning, the Cantor received 1 thaler 15 ggr.; without a motett he had 15 ggr.; for the larger half of the school he took 1 thlr., for the smaller half 4 ggr., for a quarter of the school 6 pf. The Cantor received 2 thlrs. for a wedding service. An income which consisted mainly of fees had of course its unsatisfactory side. It could never be calculated on with certainty beforehand, and was dependent on all sorts of accidents—nay, literally on wind and weather; for, as Bach writes to Erdmann, "when the air of Leipzig is wholesome there are fewer funerals," and consequently a perceptible diminution in the Cantor's receipts. On this theory the comfort of the Cantor would naturally increase with the mortality of his neighbours. Many, indeed, strove to deprive the Cantor of his dues by evading the prescriptions

---

[38] This was the custom under the wardenship of Rector Joh. Heinrich Ernesti even after the school regulations of 1723 had decided that school money should only be deducted for the foundation scholars at the rate of 12 pfennigs each weekly. The state of affairs generally above described as existing at the time of Bach's election is not in accordance with the rules laid down by the Council. They by no means corresponded on every point.

of the law and breaking through ancient usages. Kuhnau had had to complain to the Council that many distinguished couples chose to be united without any music and singing, or even away in the country; that in those Sunday and weekday hours when formerly only solemn and profitable weddings had ever been permitted, now everybody was allowed to be married; that many deceased persons even were quietly buried with only the smaller half of the school, and without music because they were ashamed of making this public, so that new compositions were hardly ever ordered for such occasions, even of him. However, it is very evident—in spite of many differences of opinion, both now and formerly—that there was a strong sense of the undignified attitude of an institution which allowed an important school and church official to derive his means of subsistence in groschen and pfennigs, which moreover were partly obtained by the agency of begging scholars. Rector Gesner was of opinion that " in the increasing conceit of youth it was much to be desired, since it would cut off the roots of many evil temptations, that the payment of the teachers should no longer be subtracted from the runners' money." Joh. Hein. Ernesti took the distribution of the funeral money into his own hands merely to avoid the complaints, vexation, and dissatisfaction which at all times had arisen on account of this money; but in doing so he found much to put up with, and was presently traduced before the Council and the whole town as a most iniquitous man. But, in spite of everything, so much as this remains certain—the income of the Cantor allowed a man such as Bach, even with his numerous family, to live comfortably in the fashion of a simple artisan. We have evidence of this in his well-managed finances and the well furnished and fitted house he left behind him at his death.

With regard to its official conditions and labours, the post of Cantor to the Thomasschule may also be considered to have been a satisfactory one. Merely glancing at the surface of things, some dark shadows are certainly to be seen in the bright picture. From the beginning of the eighteenth century the school had been falling into frightful decay.

Part of the blame was due to the organisation. In compliance with the conditions of the foundations round which it had grown up, it was to be on one hand a nursery and academy for church music, and on the other a *schola pauperum*. A thorough and uniform intellectual training, with strict and incessant supervision of the scholars, had become almost impossible in the course of the many and various employments which arose from their musical vocation. And yet, among children of the humbler classes, particularly those of that generation, these were doubly needful; but for a long time men really fit for their position had been wanting among the masters. When Bach became Cantor he found in the Rector of the school an old man who had held his office for nearly forty years. Johann Heinrich Ernesti, born in 1652, was the son of a village minister of Saxony, and had studied theology and philosophy in Leipzig; in 1680 he was appointed Sunday preacher at the church of St. Nicholas and Conrector at the Thomasschule; in 1684 he became Rector there, and in 1691 *Professor Poeseos* in the University of Leipzig.[39] Though a learned man enough, he does not seem to have been fitted to be at the head of a public school of this kind; he could keep neither masters nor scholars well in hand. The college of masters lived in disunion and jealousy, and but meagrely discharged their duties; the scholars fell into undisciplined and slovenly habits. All the year round the school was a centre of disease, which took the deeper root because the accommodation was to the last degree limited—but for this, to be sure, the Rector was not to blame.

Before the extension of the school buildings in 1731, the second and third classes on one side, and the fifth, sixth, and seventh classes on the other, were all held at once, and in the same hall, by their respective masters.[40] The result was that the number of scholars was rapidly diminishing. No doubt there were always plenty of applicants for scholarships, since these provided a maintenance almost

---

[39] Sicul, Leipziger Jahrbuch, Vol. IV., p. 920, and Neue Zeitungen von Gelehrten Sachen. Leipzig, 1729, p. 791.
[40] See Appendix B, VII.

free of cost and, besides this, a not inconsiderable income. Boys were attracted even from remote places to the foundation of the Thomasschule just as they were also to the richly endowed convent of St. Michael, at Lüneburg. But beyond this the better classes began to keep aloof from a place of such evil repute. This is very clearly proved by the class lists of the lowest classes. In Ernesti's early time they had often numbered one hundred and twenty scholars; in 1717 they show altogether no more than fifty-three. Those who did not send their children to the school of St. Nicholas preferred to send them to one of the many private schools, or kept a tutor—which was not too extravagant, for the price of private lessons in Leipzig at that time was not more than from twelve to eighteen pfennigs. The lowest classes of the Thomasschule were frequented only by boys of the very worst character, who wanted to make a profit of singing at funerals, and had to be kept by the masters from running barefoot after them, and who were not at all above begging about the town.[41] The Council could at length no longer shut their eyes to this wretched state of things, and determined on a visitation of the school and a revision of its statutes. For a while, however, it was content with good intentions; the Superintendent of the school appointed by the Council, Dr. Baudiss, warned them in 1709 that the school was in the direst need of the long-planned visitation, both as regarded the masters and the scholars. It was evident that the authorities were afraid to meddle, and let matters go on in their wild way.

At last, in 1717, the reform was begun apparently in real earnest; the visitation was carried out, and each master was required to write a report on the condition of the school, and to state his wishes as to any improvements. Here much that was far from pleasant came to light. The venerable Ernesti himself was obliged to confess: "On this opportunity I cannot conceal to what an extent a very sad state of things has hitherto obtained in the *Classes inferiores*, which, indeed, have almost ceased to exist. Also I cannot do otherwise

---

[41] According to a sketch penned probably by Gesner.

than write, with deep regret, of the condition at this time of the *Superiores*, and particularly of the *Chorus musicus*, that there is more of evil to be guarded against than of good to be hoped for." Then a few more years slipped away; in 1721 a project was drawn up for a new school code, and in 1722 it was made public. In point of fact, little or no improvement was to be gained from it: Ernesti withstood to the utmost every kind of change. So far as it bore upon the distribution of the school and choir funds and other money matters, he read it as an insult to his old age, and an abridgement of his emoluments; and in this he was seconded by the Cantor. It then seemed only reasonable not to hurt the old man's feelings, so the old state of things was allowed to continue, or rather, to grow worse and worse, till his death, October 16, 1729. On November 11 of the same year all the foundation scholars were cited to appear before the Council and earnestly admonished as to the unseemly irregularity of their lives and insubordination to the masters. On August 30, 1730, the Superintendent of the school, Dr. Stiglitz, the Counsellor of Appeal, reported a disagreement among the "Herren Praeceptoren," and that, because all and each did not duly fulfil his official duties, a falling-off in discipline, and disorders in the lives and conduct of the scholars were only too rife. The school, he said, was "fast going to ruin, and had almost run wild." Soon after this the number of scholars in the lower classes was so small that it was proposed to close them altogether.

Of course all these circumstances reacted on the character of the choirs. We could not, in any case, speak very favourably of their efforts in the seventeenth and eighteenth centuries. The bold statement with which J. A. Hiller opens his Anweisung zum Musikalisch richtigen Gesange (1774) — "Every one sings, and the greater number sing badly"—was an even more pointed truth fifty or a hundred years previously. At that time there was nowhere in Germany any true art of singing, much less could it have found a refuge in the school choirs. No available material for a good choir could have been found among uneducated boys' and immature men's voices, even if more favourable conditions of general culture and higher views of art had prevailed than those which, in the prospectus of studies, considered that, next to the glory of God, the first aim and end of the singing classes was the promotion of the scholars' digestion.[42] The universal and secular institution of professional singing had, no doubt, done much to cherish a feeling for music, and to preserve the connection between the people and that branch of art; but there is no doubt, too, that the custom had done at least as much to hinder the development of singing as an art. Those long, slow perambulations, almost always in the most inclement seasons, during which the scholars either sang from house to house for hours in the cold foggy air, or else scampered in breathless haste up and down long flights of stairs, so as to sing before the door of each separate inhabitant, were absolute ruin to the voices. Kuhnau spoke from an experience of many years when he stated, in 1717, that the best singers, and particularly the trebles, if they were not taken proper care of in all these funerals, weddings, and perambulations, lost their voices long before they had reached a moderate proficiency in the art of using them.[43] Now we must remember, too, that these singers were untrained youths who wasted all the money they earned by singing in prohibited pleasures, and were often enfeebled and miserable from disease. It cannot have been a very pleasing task to work with such materials.

There was another thing which, before 1710, had brought the choir of the Thomasschule completely to ruin. Leipzig was in dangerous proximity to Dresden and Weissenfels, two courts much addicted to opera music. In the prevailing influx of foreigners it seemed to be a timely and promising undertaking when, in 1693, Nikolaus Strungk opened an

---

[42] The *hora Cantoris* was always from twelve to one, thus immediately after dinner (see Ungewitter, Die Entwickelung des Gesangunterrichtes in den Gymnasien seit die Reformationszeit, Königsberg, in Pr. 1872, p. 11). This, to be sure, was not the case in the Thomasschule, since the dinner hour was ten. But even so refined a mind as Gesner's could go so far as to propose that the dinner hour should be eleven, in order that the singing lesson might follow it, as this was the healthiest form of exercise after eating.

[43] See Appendix B, VII.

opera-house of his own in the Brühl, in which at Fair time "certain operas were performed." The Leipzig opera it is true had not existed any longer than that at Hamburg, and had certainly always stood far behind it, because it was only open at certain short seasons of the year. It was closed in 1729, and the opera-house pulled down. Still, it had lasted long enough to produce a marked influence for some few decades on the musical life of the place. In the licence for its opening the Elector expressed a hope that the Leipzig opera might contribute to the advancement of art, and at the same time prove a sort of preliminary school for the Dresden opera, meaning with regard to the instrumentalists, since he could not use the German vocalists among his Italians.[44] In the first place, it is certain that the opera for a long period ruined the native musical tendencies of Leipzig. The man who, perhaps unconsciously, dealt them the first decisive blow was Telemann; and it is a singular coincidence that, after Kuhnau's death, the Council exerted itself greatly to place him at the head of an institution he had done so much to damage. It has been mentioned that Telemann, while a student in Leipzig, developed great talents as a poet, composer, and director of opera. In 1704 he was appointed to the post of organist and director of the music in the New Church. The church choir was usually composed of scholars from St. Thomas'; and, as the Cantor of the school had been completely ignored in the question, Kuhnau took offence, and not without reason.

The direct connection between opera and sacred music which thus took form in the person of Telemann at once exerted its baleful influence. Formerly the St. Thomas' choir had derived a by no means contemptible amount of support from students with musical tastes and good voices, some of whom had belonged to the school and still clung to its traditions. Even when the opera was opened, and the students who could sing joined it for amusement and profit,

---

[44] Geschichte des Theaters in Leipzig. Von dessen ersten Spuren bis auf die neueste Zeit. Leipzig, 1818.

the custom survived among them of joining the choir in the Sunday and festival performances. But since one of themselves had written the operas for them, had formed a musical society among them, and was now directing the church music, they attached themselves to him, and left Kuhnau in the lurch.[45] The performances in the New Church found a rapidly increasing popularity; not only was a lively and operatic style of music to be heard there, but a fresh and excellent method of execution. The musical union which had originated these performances soon assumed important dimensions, and for twenty years or more it was the most important musical institution of Leipzig. Its directors were the organists of the New Church, and so it naturally followed that they were always closely connected. Among Telemann's successors, Melchior Hoffmann (1705 to 1715) seems to have presided over it at its most brilliant period. It often numbered as many as sixty members, who met twice a week, on Wednesdays and Fridays, from eight till ten in the evening, for general practice. Their performances, which only took place on grand occasions, or at Fair time, were always regarded as public events. The union kept up its connection with the opera through its directors, who also trod in Telemann's footsteps in composing for the Leipzig stage. The circle was in truth a jolly one; during the day they made music in pleasant society, and at night serenaded in the streets; besides, at the regular practisings, which were held in a coffee-house in the market place, and from which listeners were not excluded, there was a general cheerfulness which was in strong contrast to the school practisings. The hope which the Elector had expressed as to the Leipzig opera was to a great extent fulfilled by the Musical Union of the town. On various occasions when the ruling heads of Saxony and other provinces came to Leipzig they had to perform before them, and their best members found engagements in the bands of the Elector and other princes. Thus Pisendel and Blochwitz went to Dresden, Böhm to

---

[45] See Appendix B, VII.

Darmstadt, and the singers Bendler and Petzhold to Wolfenbüttel and Hamburgh. Others, who had previously been opera singers, when they came to Leipzig, joined the Union. But, on the whole, vocal music was less well represented than instrumental, as was very usual at that time.

For the performances in the New Church the vocal parts were allotted to single voices; Stölzel, who belonged to the Union between 1707 and 1710, has handed down to us the names of the four singers at that time. The bass was Langmasius, afterwards Kammerrath at Eisenach; the tenor was Helbig, afterwards secretary to the Government at Eisenach, and a writer of cantata texts;[46] the treble was Markgraf, who, at a later period was Conrector in Augsburg; and, as alto, Stölzel thinks he remembers a certain Krone, who died at Weimar as private musician to the Duke. Hoffmann himself was an excellent musician, who endeavoured to extend the circle of musical knowledge; in 1710, he is said to have made a tour of two years' duration in the interests of art, and to have visited England; in the meantime his place was taken in the musical union by Pisendel, a famous violinist.[47] Hoffmann's successor was Johann Gottfried Vogler, "a lively composer and good violinist," as Telemann says. A certain "liveliness" appears to have characterised his life as well as his music; he ran into debt, and in 1719, at the time of the Michaelmas fair, he secretly vanished from the town.[48] The affair attracted much notice, and he seems to have been caught and brought back, for he received his salary up to the first quarter of the next year, inclusive; but that for the second quarter was withheld because he had made away with some instruments belonging to the church, and had not yet restored them. At the third quarter his place was taken by Georg Balthasar Schott, who is already known to us as a competitor with Bach for the post of Cantor to the Thomasschule.

---

[46] See p. 12.
[47] Mattheson, Grosse Gen. Basschule, p. 173. Ehrenpf., p. 117. Sicul, op. cit., p. 414. Gerber, L., I., col. 656.
[48] "*Continuation* derer Leipzigischen Jahrbücher von *Anno*, 1714, bis. 1728." MS. in fol. in the town Library at Leipzig; Pressmark, Vol. 18.

So long as the musical union of students and the opera gave the tone to the music of Leipzig, the Cantor fell on evil days. The lack of sympathisers and assistants was all the more keenly felt, because it was always most conspicuous on festivals and at Fair times, when he was in greatest need of reinforcement, and wanted to appear in a favourable light before strangers; for it was difficult to see what he could do with a wild mob of dirty schoolboys who had shouted themselves hoarse in the streets, and a few very mediocre town musicians. Pieces of a high class could no longer be sung at all; if he ever attempted them the performance was such a miserable result that he could only feel himself shamed before the audience. Formerly the Town Council had lent some assistance to the formation of the choir. In the time of Johann Schelle they had always four or five more foundation scholars to be maintained in the school than the revenues allowed, and as there were ample means for the purpose, this bounty benefited the music. But when Schelle died, and his wife was allowed to provide the food for the scholars, the Council, to assist her, removed these supernumeraries, and she had fewer to provide for, with the same money. Kuhnau was indefatigable in his exertions to restore the former state of things, or, at any rate, some substitute for it; he represented that an increase of the musical resources of the school had never been more necessary, and the Superintendent supported his statement, but without any result worth mentioning. He could not even obtain that two trebles should be assigned to sing church music only and released from all other vocal duties. Since they then could have no share in the funeral and "runners'" money, some compensation would be indispensable, and the Council would not vote an equivalent. They granted a permission to release two boys each year from the New Year's perambulations, which were the most injurious, and they were to have in compensation four gülden a year for the two.[49] This was the end of it. Obviously the interest in supporting the

---

[49] Accounts of the churches of St. Thomas and St. Nicholas.

music of St. Thomas and St. Nicholas was waning when so much better could be heard in the New Church.

After Vogler had withdrawn, Kuhnau made a last attempt to recover the lost ground. He represented that the tendencies of the "*Operists*" in the New Church were destroying all feeling for true church music among the citizens; that the organ was belaboured first by one and then by another pair of "unwashed hands," since the director of the music either could not play or, after the manner of the "Operists," was constantly away. It would be better that a really permanent organist should be appointed, and that the direction of the music should be given to the Cantor of St. Thomas'. Then every Sunday music could be properly performed in the three churches alternately, just as now was done in two; and when, on festivals, cantatas were to be given in all three churches, as heretofore, the Cantor could very well give these also, since he would then have at his command a large number of students, and could distribute them among the churches according to his judgment. The students must have a gratuity in order that they might be ready and willing in the matter, and a student might also be appointed as organist in the New Church. If, however, the Council would not consent to all this, some means must at least be thought of to attach those young men who quitted the school for the University to the choir of St. Thomas.[50] But again Kuhnau wrote in vain. Schott was appointed organist and director of the music, and all went on in the old way. Kuhnau once more made a timid attempt at opposition in 1722, when he made difficulties as to lending the scholars for the Passion Music in the New Church, but he was ruthlessly called to order by the Council.

To put a climax to the confusion the St. Thomas' boys caught the opera fever. This was, indeed, not to be wondered at when they had the temptation perpetually before their eyes. As soon as they had attained a certain proficiency in singing and music in the school, they pined to escape from its narrow bounds and find themselves at liberty, dreamed of

---

[50] See Appendix B, VII.

artists' laurels and ceased to be of any use where they were. If they then could make acquaintance with any operatic *impresario*, the most advanced among them could obtain engagements; they straightway demanded to be released from their school indentures, and if this were not granted they ran away, returning with some opera troupe at Fair time, and exciting the envy of their former schoolmates by their appearance on the boards and at the New Church. A treble named Pechuel had gone, by permission of the Council, from time to time to Weissenfels to perform in the opera; in the course of time he wished to extend these excursions and to take a leading part at Naumburg. The Council forbade this, and then the young genius broke his bonds and ran away; two years later a bass named Pezold[51] followed his example. A respectable citizen of Leipzig had aided and abetted them both, and it is not hard to see on which side the sympathy of the public was. When Kuhnau died the music of the Thomasschule was at the lowest possible ebb, and the twelve months of interregnum which followed before another Cantor was appointed certainly did nothing to mend matters.

It is impossible that Bach should not have been exactly informed of all these circumstances, as he was intimately acquainted with Kuhnau, and had been in Leipzig several times. As has been said, he doubted for a long time whether he should do wisely in becoming Kuhnau's successor, and in more than one respect these doubts were well founded. Did Leipzig then offer him some special temptation as a musician, some opening and incitement to profitable effort in his art? It is difficult to answer this question in the affirmative. There were no remarkable musicians there at the time; the only one, besides Kuhnau, who had done anything important in his own branch of art was Daniel Vetter,[52] who had died two years before,

---

[51] Probably the same who afterwards distinguished himself in the Musical Union.

[52] "Herr Vetter, our able Organist here." See the report on the Organist of St. Paul's church in Die andere Beylage zu dem Leipziger Jahrbuche, 1718, p. 198.

as Organist to the church of St. Nicholas. This absence
of fellow artists would certainly have been no great grief to
Bach, a man of such vigorous productivity and such
emphatic independence, if only he had had at his command
some means and resources worthy of him, to enable him
to do himself justice. Leipzig was a populous town, a
centre of resort, and stirred by various interests; but a focus
of art, such as Dresden, Vienna, Munich, or even Hamburg,
it certainly was not. It was, no doubt, to some extent
musical, but that was common to all Germany. No special
effort on the part of the better class of citizens to develop
anything truly artistic is at that time discernable; it was not
till Bach was growing old, and it was too late, so far as he
was concerned, that a different spirit began to show itself.
It was only among the student class that any love and
taste for music were manifested. Kuhnau might, indeed,
have succeeded in attaching the Academic youth to himself
if he had been less hesitating and less conservative; even
side by side with Telemann's flourishing musical union it
might have been accomplished, if only he had known how to
grasp the matter; but he did not. He saw Fasch—who for
many years was his pupil—founding a second Musical Union
among the students, which established itself as securely in
the University Church as Telemann's had done in the New
Church.[53] The new organ, which was finished in the
autumn of 1716, was entrusted to a very diligent musician
in the person of Gottlieb Görner. It cannot now be
ascertained whether the Musical Union directed by Görner
was that founded by Fasch, or whether, indeed, that society
survived to his time; we only know for certain that Görner
was at the head of such a *Collegium musicum* when Bach was
appointed Cantor.[54]

Görner, who was born in 1697, had been made organist of
St. Nikolaus after Vetter's death. When, in the autumn of
1729, old Christian Gräbner died, he succeeded him as
organist to St. Thomas'. He was thus, to a certain extent,

---

[53] See the document by Bach, given presently in the account of his quarrel with the University.

[54] Das jetzt lebende und florirende Leipzig, 1723, 8, p. 59.

under Bach's direction; but it did not at all meet Görner's views to give way to the greater man; on the contrary, he boldly put himself forward as his rival. When, in the winter of 1727-8, there was a public mourning, he asked permission to continue, notwithstanding, his musical gatherings. For—said he—in his Union the students coming from the school brought to perfection any skill they might have acquired there, and were enabled to make themselves heard at the performances at the Fair times, and thus found their way to places as cantors and organists.[55] Thus when the scholars left Bach's hands they received the finishing polish from Görner! His audacious pretensions are all the more singular, because he seems to have been in fact but a very mediocre musician. A contemporary Leipzig musician, Johann Adolf Scheibe, in the year 1737, recorded a very bitter opinion of him, which may, no doubt, have been influenced by personal feeling, but cannot, on the whole, be very far from the truth: "He has been engaged in music for many years, and it might be supposed that experience would have brought him into the right way; but nothing can be more disorderly than his music. The real meaning of the different modes of writing according to their distribution (of parts) is wholly unknown to him. Rules are things he must daily dispense with, for he knows them not. He can never set a pure line (of music), and the grossest blunders grace—or disgrace—every bar. In a word he can depict disorder in his music to perfection." Then, as to his character: "He is so completely possessed by conceit and rudeness that through the first he does not know himself, and through the second asserts his pre-eminence among a large number of his equals." On a subsequent occasion Scheibe still further embitters this verdict and adds: "Nor would he be even what he is if a certain man had not done everything for him. And the result he has shown is that on a certain occasion when he could and ought to have proved himself grateful, he was anything rather than grateful,

---

[55] "Ephoralarchiv," at Leipzig: "Trauer Feiern beim Absterben der Sächsischen Fürsten," Vol. I.

but repaid the kindness that had been done him by a piece of treacherous spite."[56] Who is meant by this benefactor does not appear; but it throws a significant light on the state of affairs in Leipzig that such a man as Görner should have played his part by the side of Bach for a whole generation. And at the University Church he had planted his foot so firmly that Bach could not succeed in getting him removed in spite of his own powerful name and influence.

We will now contemplate Bach's position in all its aspects. This time, more than ever before, he had taken a step into the unknown; he had made the venture—to use his own words—in the name of the Most High. The craving and need of his artist soul to live once more under circumstances where there was work worth doing to be done for music seemed likely to find some satisfaction in Leipzig. The downward step from Capellmeister to Cantor—for so it was deemed at that time—was made easier to him by the high position the Cantor of St. Thomas' held among musicians. Seth Calvisius, Hermann Schein, Tobias Michael, Sebastian Knüpffer, Johann Schelle, and Johann Kuhnau, who had held the appointment in succession during one hundred and twenty-five years, had all been distinguished—some of them highly distinguished—practical musicians and learned men. To continue the series was an honour, and Bach felt it as such. Besides, the post gave him some tangible advantages, and it would seem that it was this which turned the balance. The place was endowed and the duties were light; at the same time this did not mean that for their complete fulfilment a man of merit and of mettle was not required; but there was no overwhelming load of official work; Bach would have time enough for his own occupations. Finally, he would now be enabled to give his sons a superior education without too great a pecuniary sacrifice. How near his heart this matter lay is shown by a little circumstance. On December 22, 1723, when he had been about six months in Leipzig,

---

[56] Johann Adolf Scheibe, Critischer Musikus, New Ed., Leipzig, 1745, p. 60. Görner's name is not mentioned, nor that of the place; but there is ample evidence that he and Leipzig are meant. The passage was written in 1737.

he applied to the university to have his son, Wilhelm Friedemann, then thirteeen years old, entered on the register as a future student (academic citizen), although it was not till April 5, 1729, that he actually became a member of the University.[57] Such an early nomination was not unheard-of; it even sometimes happened that a matriculation at the University was a christening gift from a godfather. Bach appears to have given it to his favourite son as a Christmas-box. If against these advantages he could not but weigh the dark side of the appointment, he no doubt hoped that the great fame he enjoyed and the influence of his strong individuality would bring the choir into better condition, and that by degrees he might get the management of all the musical concerns and undertakings of Leipzig into his own hands. However great the talents of his predecessors, in celebrity he beyond a question stood far above them; the name of Bach was famous far and near—the great player, who came from a Prince's court, and was the friend of Princes. Indeed, he not only continued to hold the honorary post of Capell-meister at Cöthen, but in the very year of his removal to Leipzig the same honour was conferred on him by the Court of Weissenfels.[58]

III.

BACH'S OFFICIAL DUTIES AS CANTOR, HIS DISPUTE WITH THE TOWN COUNCIL, AND ENDEAVOURS TO IMPROVE THE CONDITION OF THE MUSIC.—HIS LETTER TO ERDMANN.—GESNER'S APPOINTMENT AS RECTOR OF THE SCHOOL.

THE direction of the music in the University Church was not inseparable from the office of Cantor to St. Thomas'; it was, however, customary, and had been from time immemorial allowed by the Council, that he should have the charge of it. So long as the University Church was

---

[57] "Bach, Wilhelm Friedemann, *Vinario-Thuringensis*," under the heading of *Depositi, nondum inscripti*, of December 22, 1723.

[58] Walther, Lexicon, p. 64. Among the deeds of appointment of the Weissenfels court from 1712 to 1745, which are preserved among the State Archives at Dresden, all those belonging to 1723 are wanting, including, of course, that of Bach.

opened only on the three great festivals, on the festival of the Reformation, and for the quarterly speeches, no severe tax was thus laid on the Cantor. But since 1710 a regular Sunday service had been performed there, and consequently the post of the director of the music became very important. Kuhnau had been able to secure it for himself, though, in the first instance, Fasch had attempted to establish there a musical union that should be independent of the Cantor. By great efforts and sacrifices Kuhnau had succeeded in exploding this scheme; he declared himself fully prepared to fulfil these new duties without any emolument, and this was a consideration [59] to which the University was keenly alive. After Kuhnau's death Görner for a time took his place as director of the music at the University Church. Bach's accurate comprehension of the position is shown by the fact that he made it his first business to get this function out of Görner's hands. Unless he could create for himself a strong following among the University students, there could be no prospect of his moulding the musical affairs of the town according to his own views. His appointment was confirmed to him on the 13th of May. The first church cantata he composed as Cantor of St. Thomas' he conducted on the 30th of May, the first Sunday after Trinity, before he took his place in the school, thus, as it were, inaugurating his musical labours as Cantor. He had already begun his duties as Musical Director to the University a fortnight earlier, on Whit Sunday—at any rate, he had supplied them with a composition of his own.[60] But Görner knew very well what the upshot must be, and had determined to save as much for himself as possible. On the four great festivals and the quarterly speeches he knew he must retire into the background; the Cantor's claims were too strongly supported by ancient rights of custom; but it was different with regard to the regular Sundays and the other church holidays. It appears that Görner had set it very clearly before the University that, with all his duties in the churches of St.

---

[59] See Appendix B, VII.
[60] See Bach's detailed statement as to his quarrel with the University, towards the end of it.

Thomas and St. Nicholas, the Cantor could not always provide with due punctuality for the performance of the service in St. Paul's, which began when the others were only about half over. At any rate, Görner still continued to officiate in the New service, as it was called, as Musical Director to the University.

In the third place there were the extraordinary University high days to be considered. With regard to these Bach took his stand on the ground that they had been customary long before the arrangement was made for the New service, and that the Cantor had always presided over them; that, consequently, they constituted part of his duties, and he proceeded to act on this basis. On Monday, August 4, 1723, the birthday of Duke Friedrich II., of Sax-Gotha—a prince who had specially distinguished himself in promoting the cause of learning and of the church—was solemnly kept; a Bachelor of Philosophy, named Georg Grosch,[61] delivered a discourse, *De meritis Serenissimi Friderici in rem litterariam et veram pietatem;* which was followed by a Latin ode composed and conducted by Bach—"an admirable piece of music," says the Leipzig chronicler; ."so that this solemnity was concluded to everybody's satisfaction by about eleven o'clock in the morning."[62] Heinrich Nikolaus Gerber, who in 1724 went to the Leipzig University, at a later date told his son that he had heard many concerts at that time under Bach's direction.[63] As any music but church music is out of the question, and as Bach did not direct any Union of his own, this can only mean academic performances.[64]

Meanwhile Bach's vigorous self-assertion from the very first, by no means settled the matter at once. Görner evidently was a favourite with those who gave the cue

---

[61] Grosch was a native of Gotha, and in 1724 became the Prince's tutor: he subsequently held various livings in that province. (Archives of Gotha.)
[62] Vogel; Continuation Derer Leipzigischen Jahrbücher.
[63] Gerber, L., I., col. 491.
[64] In the *ACTA LIPSIENSIUM ACADEMICA*, Leipzig, 1723, p. 514, Bach is called "*Cantor* and *Collegii Musici Director.*" Possibly *Collegii* is a slip of the pen or a misprint for *chori;* at any rate, it is quite certain that the St. Thomas' choir is what is meant.

to the University, and he also received an *honorarium* out of the fund set aside for the Cantor's services. At last, Bach, who knew very well how to reckon, and was particularly precise in money matters, thought this beyond a joke. After swallowing the affront for two years with as good a grace as he might, he resolved that, even if he could not take the whole direction out of the hands of Görner, who was tough to deal with, he could at any rate secure his income. He, therefore, addressed a petition to the King-Elector at Dresden, by which he thought he should obtain full security for his interests, all the more as he was in favour at court:—

    Most Serene,
        Most Potent King and Elector,
            Most Gracious Sovereign,

May your Royal Majesty and Most Serene Highness graciously permit me to represent, with the humblest submission, with regard to the Directorship of the Music for the Old and New services of the church in the Worshipful University of Leipzig, that, together with the salary and usual fees, they had always been associated and joined with the place of Cantor at St. Thomas', even during the lifetime of my predecessor; that after his death, and while the post was vacant, they were given to the Organist of St. Nicholas, Görner; and that, on my assuming my office, the direction of the so-called Old service was restored to me again, but the payment was withheld and assigned, with the direction of the New service, to the above-mentioned organist of St. Nicholas; and, although I have sued duly to the Worshipful University, and made application that the former regulations may be restored, I have nevertheless not been able to obtain anything more than that I should have half of the salary, which formerly amounted to twelve gülden.

Nevertheless and notwithstanding, Most Gracious King and Elector, the Worshipful University expressly required and assumed that I should appoint and direct the music for the Old service, and I have hitherto fulfilled this function; and the salary which has been given to the director of the New service did not formerly belong to it but properly to the Old services; and at the same time the New were connected with the Old; and, if I were not to dispute the right of directing the New service with the organist of St. Nicholas, still the retention of the salary which formerly and at all times—nay, even before the new cultus was instituted—belonged to the Cantor, is extremely painful and prejudicial to me: and church patrons are not wont to dispose otherwise of what it assigned and fixed as the regular payment of a church servant, either withholding it altogether, or reducing it, while I have already for more than two years been forced

to fulfil my duties concerning the above-mentioned Old service for nothing. Now, if my humble suit and petition may find favour with your Royal Majesty and Most Serene Highness, you will graciously communicate it to the Worshipful University, to the end that they may restore the former state of things, and assign to me, with the direction of the Old service that also of the New, and more particularly the full salary of the Old service and the enjoyment of the fees accruing from both. And for such Royal and gracious favour,
I shall ever remain,
Your Royal Majesty's and Serene Highness's
Most humble and obedient

LEIPZIG, *Sept.* 14, 1725.   JOHANN SEBASTIAN BACH.

Addressed to the Most Serene and Most Potent Prince and Lord— The Lord Friedrich Augustus, King of Poland (Here follow all his Titles) my Most Gracious King, Elector, and Sovereign.[65]

Bach was not deceived in his presumption. On the 17th of September a requisition was forwarded from the Ministry at Dresden to the University to relieve the petitioner or to adduce their reasons to the contrary. The expedition brought to bear on the matter was so great that there seems not even to have been time to read Bach's petition; in the references to his statement of grievances there are inaccuracies by which the circumstances are placed in a false light.[66]

The University attempted to justify itself in every particular, and caused Bach to be informed that they had forwarded their version of the case. Bach, however, had reason to suppose that in this statement the affair had not been truly represented, and to forefend an unfavourable decision he wrote a second time to the king:—

THE MOST SERENE,
    MOST POTENT KING AND ELECTOR,
        MOST GRACIOUS SOVEREIGN,

After that your Royal Majesty had most graciously been pleased to issue your orders in the matter of the request preferred by me on the one part, and by the University of this town on the other part

The said University submitted the required very humble report, and duly notified me of its departure; and I, on the other hand, for my further need, deem it necessary to observe that if my most humble petition may find favour with your Royal Majesty and Serene Highness you will communicate to me a copy of the said report, and be

---

[65] This letter is in the Archives at Dresden, and was obligingly copied for me by Herr Moritz Fürstenau.

[66] This and the following documents are in the Archives at Leipzig.

graciously pleased to wait, and defer your Sovereign determination till I again have made the necessary representations; and I will not fail to hasten with them as much as possible, and for the whole of my life remain with the deepest submission,

Your Royal Majesty's and Serene Highness's
Most obedient and humble

LEIPZIG, *Nov.* 3, 1725.   JOHANN SEBASTIAN BACH.
[Addressed as before.][67]

This petition was granted. He then preferred a thorough refutation of the justification drawn up by the University. The document is extremely interesting, for it displays, as no other does, Bach's keen and business-like intelligence and incisive mode of expressing himself:—

MOST SERENE AND MOST POTENT
KING AND ELECTOR,
MOST GRACIOUS SOVEREIGN,

I beg to acknowledge, with the humblest thanks, your Royal Majesty's and Serene Highness's favour in graciously condescending to allow to be communicated to me the copy of that document in which the University of this place objected to my accusation brought against it, as concerning the direction of the music in the Old and New services in the Church of St. Paul, and the salary belonging to the former of the two, which has hitherto been withheld. Although I am of opinion that the University will at once indemnify me, as is proper, and grant my well-founded suit without further formality, it must yet be examined how they make various excuses, and give themselves the trouble to make themselves out innocent, that is to say :—

(I.) That I had stated without reason that the direction of the music for the Old and New services was necessarily connected with the office of Cantor to St. Thomas'—nay, that the University in giving the above-mentioned Direction was *in libertate naturali*, whereby, however, they cannot contest with me the direction of the Old divine service, nor can they deny that on account of the fulfilment of those duties, they paid me an *honorarium*. Moreover,

(II.) That my representations that I had hitherto had to do my duty for nothing surprised them all the more, since it was clear from the *Rationes Rectorales*, that at all the Quarterly *Orationes*, and at the three great Festivals, as well as the *Festum Reformationis Lutheri* a special and profitable *honorarium* of 13 thlrs. 10 gr. was paid me, and that I have hitherto received it. Also

(III.) That I have hitherto not generally presided in person at the Quarterly *Orationes*, but, as the register shows, have allowed the Prefects to direct the singing of the motetts. Likewise

---

[67] This letter is autograph throughout. The seal seems to be the rose with a crown. See Vol. I., p. 39, note 76.

(IV.) That, in consequence of his Sunday and holy day duties, the Cantor of St. Thomas' is quite unable to undertake at the same time the direction of the music in the University Church without prejudice to it, and confusion; since he would also have, at very nearly the same hour, to direct the music in the churches of St. Thomas and St. Nicholas. Especially

(V.) It is very expressly stated that a new *honorarium*, of 12 fl. was newly granted to my predecessor on account of the direction of the music for the new service. Moreover

(VI.) That so many difficulties were made on the part of the Council with reference to the St. Thomas' scholars and the town musicians that the University availed itself of the services of its students, and was forced to consider of the selection of another individual who might in his own person preside unhindered over the direction of the music, and who could better maintain that good understanding with those students who refused to assist the Cantor without additional payment. To which it was added

(VII.) That during the long vacancy of the office after the death of the former Cantor, the University had given over the direction of the music for the New service to Johann Gottlieb Görner, and assigned to him the new salary devoted to it of 12 fl., so that this salary had nothing to do with the former direction of the Old service, but was a new institution.

But, Most Gracious King, Elector, and Sovereign, these objections brought forward by the University are not founded on fact, and are quite easy to refute. For, in the first place—

(I.) As to the connection of the New Service with the Old, I did not say that the connection was a necessary one, only that the direction of the latter had formerly been combined with that of the former, and it was not for me to inquire as to the power and liberty of joining or separating them; that can be settled in the proper place: On the contrary, I admit that the direction of the Old service, according to previous custom (as set forth) in the very humble report would be granted and vouchsafed to me. But, if this were so, the Direction of the music at the solemn *Acts*, ceremonies of the universities, of conferring Doctors' degrees, and others, which take place in St. Paul's church, with the fees accruing from them, ought not to be withheld from me, because all this, at any rate as regards the music, was in direct connection with the Old service, according to custom before the New service was instituted. In the next place

(II.) It surprises me not a little how the University can refer to a profitable *honorarium* of 13 thlrs. 10 gr., which I ought to have received from them, and deny and contradict that I have hitherto performed the work for nothing, since the *honorarium* is something apart from the salary, which is 12 fl., and this gratuity does not include the salary; how too, my complaint can have been regarded as concerning not the *honorarium* but the ordinary salary of 12 fl. attached to the Old service, which, however, has hitherto been withheld from me—nay, since it can be proved from the *Rationes Rectorales* put forward by the University itself, that this honorarium, which ought to have amounted to 13 thlrs. 10 gr., has not once been paid in full, but that each quarter the two beadles, as they could depose on oath, have paid to me instead of the 20 gr. 6 pf., as set down in the *Rationes Rectorales*, no more than 16 gr. 6 pf.; and at the three high festivals, as also on the *Festum Reformationis Lutheri*, each time instead of 2 thlrs. 12 gr. no more than one thaler. Thus, instead of 13 thlrs. 12 gr. altogether, only 6 thlrs. 18 grs. in the year; also my predecessors, Schelle and Kuhnau (witness the attestations of their widows *sub lit.* A and B), received a no larger sum for the quarterly and festival music, and consequently never gave receipts for any larger sum, and yet in the extract from this *Rationes Rectorales* a much higher *quantum* is set down. That

(III.) I have frequently not attended the quarterly *Orationes*, and that the register of October 25, 1725, proves this, is of no importance; for, from the month and the date, it appears that the entry was made in the register after I had previously complained against the University, while before that time nothing had been registered against me; thus my absence did not happen more than once or twice, and then, indeed, of *impedimenta legitima*, since I was travelling on necessity, and, in particular, several times had business in Dresden; moreover, the Prefects are appointed under the Cantor to the before-mentioned quarterly music, so that my predecessors, Schelle and Kuhnau, never conducted these in person, but the singing of the motetts was arranged and directed by the Prefects.

(IV.) Neither can it have any foundation when the University urges that the attendances for the music in the two churches are not compatible for one person; for certainly the instance which might be given of Görner, the Organist of St. Nicholas' Church in this town, is far more striking, since it was even less compatible for him to direct the music in both churches in his own person, because the Organist had not only in the same way to be at one and the same time at the Church of St. Nicholas and in that of St. Paul, to attend to the music before and after the sermon, but also had to play the organ even to the very last hymn, while on the other hand the Cantor, after having performed his music, can go out, and need not stay for the hymns at the close of divine service; and the late Kuhnau in his time did both quite well, without prejudice and confusion, and in the church, where no formal music is ordered, common music can be directed perfectly well by the *Vicarii* and *Præfecti*.

(V.) As to what particularly regards the 12 fl. under discussion, the University can never again assert with reason that they began giving it to my predecessor as an independent gratuity on account of the direction of the music in the New service. The state of the case is rather that the

12 fl. having been from time immemorial the salary for the arrangement of the music in the Old service, my predecessor, in order to avoid other consequences disadvantageous to himself which might be feared from the division of the director's duties, directed the music in the New service for nothing, and never demanded a penny for it, and thus never before enjoyed the said new gratuity of 12 fl. Nay, not only by Kuhnau but also by Schelle, and even before that, before anybody had ever thought of the New services, a receipt was always given for these 12 fl. And, as the widows of Schelle and Kuhnau, in their attestations, *sub lit.* A and B, distinctly state, the 12 fl. were always the regular salary for the arranging of the music in the Old service. The University cannot escape making public the above-mentioned receipts. Therefore

(VI.) The salary connected with the music of the Old service cannot be tampered with, notwithstanding that the ordering of the New service was not well received by the students, and they would not assist the Cantor for nothing. Now, while this can be neither proved nor gainsaid, and it is well known that students who are lovers of music are always ready and willing to assist, I, for my part, have never had any unpleasantness with the students; they are wont to assist me in both vocal and instrumental music without hesitation, and to this hour gratis and without payment. Moreover

(VII.) If the *directorium* of the music in the New service at that time and as far as regards Görner himself was to remain *in statu quo;* if, besides, no one had any doubt that a new salary could be granted on account of such new arrangements; then the salary of 12 fl. hitherto assigned to him was in no respect a new institution, nor assigned to the direction as anything new, but this was withdrawn from the *directorium* of the music in the Old service and not received by Görner until subsequently, during the vacancy of the post of Cantor at St. Thomas'; and when Görner had the new direction assigned to him it was granted to that new direction.

All the foregoing had even before been proved, nay, it is all a matter of notoriety to those who hitherto have had to do with the music in both churches, and by their deposition it could be still further confirmed and made public. In fact, I feel compelled here and now to adduce this particular circumstance: that two years since, when I took occasion to speak of the direction to the then *Rector Magnificus Junius* and he wanted to demonstrate the contrary to me out of a written account-book, which probably was a *Liber Rationum Rectoralium*, it must needs happen that on the page he opened appeared written the account, and my eyes fell on the plain words, " To Schelle, *pro Directorio Musices*, 12 fl., *Salarium;* " and this entry was then and there shown and pointed out by me to the *Rector Magnificus Junius*.

Finally, the University have already granted and offered me the half of the payment of these 12 fl. through D. *Ludovicus*, who during last summer administered the rectoral affairs, and they certainly would not have done this had they not been convinced that the matter rests on sound foundations. Hence this alteration seems to me all the harder, when they choose to ignore all salary whatever, and to deprive me of it altogether. Afterwards, too, I expressly mentioned this offer, in my very humble memorial, but the University, in their counter-statement, pass over this point, and have answered nothing to it. Thus, in fact, by their silence the ground of my pretension and the justice of my case are established afresh and, as they themselves have been convinced, are tacitly acknowledged.

Since the University, according to their own confession, offer me, for the *Quartal-Orationes*, 3 thlrs. 10 gr. per annum, and for the three high festivals and the Reformation Festival, a peculiar *honorarium* of 10 thlrs. per annum, thus making in all 13 thlrs, 10 gr. by reason of the custom already referred to; and as I, from the time when I entered into my duties under the University at Whitsuntide, 1723, until the end of 1725, which makes $2\frac{3}{4}$ years, ought altogether to have received 36 thlrs. 18 grs. 6 pf., and have not received so much, but only 11 thlrs., in payment for many festival performances, and 7 thlrs. 13 grs. 6 pf. for eleven *Quartal-Orationes*, in all 18 thlrs. 13 gr. 6 pf.; I thus have to require 18 thlrs. 5 gr., the regular salary of 12 fl. for $2\frac{3}{4}$ years—*i.e.*, 33 fl. remaining owing. They, the University, since they are willing to agree as to the salary, and since they have already offered to give me the half of it, cannot *eo ipso* regard my request as unjust and unfounded, but must admit it; also, since they in their humble report pass this over in silence, and thus once more tacitly admit the facts to me, and moreover, have not been able to adduce the smallest thing of any importance—when my most submissive prayer is presented to your Royal Majesty and Electoral Serenity, be graciously pleased immediately to command the University that they not only acquiesce in the previous order of things, and henceforth confer upon me the full payment, consisting of 12 fl., for the old service, together with the fees of the *Promotiones Doctorales* and other solemn occasions formerly attached to it, but also that they shall hand over to me the arrears of *honorarium* amounting to 18 thlrs. 5 gr. and the regular salary already owing, amounting to 33 fl. and moreover, allow me all expenses incurred by me in this business, or else, inasmuch as the University may not be convinced by what has hitherto been adduced, that it shall be made to publish the receipts given by Schelle and Kuhnau, both as regards the special *honorarium* as well as the regular salary. This great favour I will recognise with humblest thanks all my life, and remain,

Your Royal Majesty's and Electoral Serenity's

      Most humble and obedient,

LEIPZIG, *Dec.* 31, *Anno.* 1725. JOHANN SEBASTIAN BACH.

[Addressed with the full title as given above.][68]

---

[68] Without seal. The transcript of this document, preserved among the deeds of the University, is only signed by Bach.

This was followed up by a document dated from Dresden Jan. 21, 1726, not very definite in tone, still apparently deciding in Bach's favour on the whole; the presentation of this document to the University did not take place, strange to say, before May 23. Whether during these four months attempts were made to bring about a friendly compromise can only be conjecture; even as regards the settlement of the money question, which became more and more prominent, we can come to no more definite conclusion. From a comparison of various intimations, culled here and there during the following years, we are led to infer that Görner remained at the head of the "New service." In the solemnities of the University, sometimes one and sometimes the other of the rivals seems to have been called in, but more frequently Bach. On August 3, 1725, he had composed, to order, a "*Dramma per musica*" in honour of the name day of Professor August Freidrich Müller ("Der zufriedengestellte Aeolus"—"Aeolus satisfied"), and after this, so soon as December 11 of the following year, he wrote another cantata, in honour of the promotion of *Magister* Gottlieb Korte to be Professor extraordinary; again, on May 12, 1727, for the birthday of King Friedrich August, who happened just then to be in Leipzig, another *Drama musicum*, which was performed under his direction by the prizemen of the University; then the music for the mourning celebration held in the University Church, October 17, of the same year, in memory of Queen Christiane Eberhardine, who died September 5. Görner, on the other hand, was commissioned to compose the Latin Ode which was sung in the University Church on that same Royal birthday. For the two-hundreth anniversary of the introduction of evangelical doctrine into Saxony, which was celebrated in the University on August 25, 1739, Görner also composed the music to a Latin Ode, of which the first portion was performed before and the second after the sermon.[69] On the first of these occasions we find him styled quite plainly *Director Chori Musici*

---

[69] Gretschel, Kirchliche Zustände Leipzigs vor und währund der Reformation, p. 293.

Academici, of the New service in the *Pauline*.[70] A report of the year 1736, in fact, names him alone as the Musical Director of the Academy, and adds that on grand occasions solemn music was performed by the students and other musicians, under his direction.[71] Still, this need not refer to academical performances, since the Musical Union was wont to hold independent festival concerts; thus, for instance, in that very year, 1736, Görner's Union gave a cantata in honour of the King's birthday with words by Joh. Joachim Schwabe. And it is very precisely pointed out, in the year 1728, that St. Paul's church had a special musical director, Herr Joh. Gottlieb Görner for ordinary Sunday and holy day music, but that on other festivals and for the quarterly speeches the Cantor of St. Thomas' filled his place, as from time immemorial.[72]

Finally, Bach may very well have been content with the issue of his efforts. He had, at any rate, gained an established position among the music-loving youth of the University; and this was still farther secured when, in 1729, Schott went to Gotha as Cantor, and the direction of the famous old Musical Union founded by Telemann, fell into his hands. So far as the general condition of things at that time can be said to have allowed it, a good and favourable time for public musical performances would seem now to have dawned upon him. He performed regularly once a week with his Union; in the summer season from four to six on Wednesday afternoons, in the Zimmermann Garden, in Windmühlengasse (Wind Mill Street); in winter from eight to ten on Friday evenings, in the Zimmermann Coffee-house, in the Katharinenstrasse—the corner house of the Böttcher Gässchen, now No. 7. During Fair times they played twice a week, on Wednesdays and Fridays. Under his direction the Union distinguished itself by several festival concerts. On December 8, 1733, he produced a *Dramma per Musica* for the Queen's birthday, "Tönet ihr Pauken, erschallet

---

[70] Sicul, (Christoph Ernst) *ANNALIVM | LIPSIENSIVM | MAXIME ACADEMICORVM | SECTIO* XXIX. | &c., Leipzig, 1728. 8.
[71] Das jetzt lebende und florirende Leipzig, p. 32.
[72] *ANTONII WEIZII* Verbessertes Leipzig, 1728, p 12.

Trompeten"—"Sound ye drums, peal forth ye trumpets"; in January, 1734, another work of the same character, "Blast Lärmen, ihr Feinde, verstärket die Macht," for the coronation festival of August III., both, of course, of his own composition; and the old Weimar Cantata "Was mir behagt ist nur die muntre Jagd" had to serve again for the King's birthday, with different words.[73] And what was more, the Union ceased to perform in the New Church, and so was at liberty to do so in Bach's church music. The organist who had been appointed in Schott's stead, Carl Gotthelf Gerlach, was a protégé of Bach's, and had obtained the place by his recommendation,[74] so he had to submit when his patron deprived him of the assistance of the Union; but at the same time—whether from a sense of propriety, or from their old preference for the musical performances in the New Church —the Council supported him with, relatively speaking, abundant means for forming a small choir of his own for the needs of the service there. In later years Gerlach was promoted to be the director of Telemann's Musical Union, Bach himself retiring from the post. It is not known precisely when this took place, it is only certain that it was after 1736. However, the good old days never returned for the New Church; the Union seems even to have stood more completely aloof from it, as it lost its first importance and was passed from hand to hand. The focus of musical art in Leipzig had centred elsewhere by the year 1740.[75]

Bach of course took the management of the church music at St. Thomas' and St. Nicholas' vigorously in hand from the first. It is highly characteristic of him that he should have regarded his position as Cantor less as being that of a teacher in a public school—which is what it certainly was, first and foremost—than as a civic and official Conductorship, with the additional duty of giving certain lessons. His predecessors in office had been simply entitled Cantors; if the words *Director Musices* were added, they referred only to the

---

[73] See Vol. I., p. 567.
[74] Gerlach had four rivals, but he was "praised by Herr Bach."
[75] See Appendix A., No. 14.

University church.[76] Bach subscribes and describes himself almost always, and from the first, as *Director Musices*, or *Chori Musici* and *Cantor*, or even *Director Musices* alone;[77] and it is the exception, in speaking of singing rehearsals in the school itself, when he calls himself only *Cantor*. Even his pupils give him the title.[78] In a Leipzig address book of 1723[79] he is thus designated; he evidently chose to assert his position as being an essentially musical and independent official, and would do so with all the more determination in proportion as the petty officials persisted in giving him the simple title of Cantor. This is very characteristic of the determined temper which supported him through many conflicts, for his outward conduct bespoke the inner man. The protestant church music had always depended on the school, and had become what it was through its instrumentality. It certainly was not mere caprice or arrogance which led Bach to regard his connection with the school as a secondary consideration. Bach's music is no doubt true church music, characterised by a style of its own; but it is impossible not to perceive that it also contains the germ of an independent branch of concert music, and in the course of Bach's own labours this is now and then very prominent. He was conscious of this peculiarity in his art, and his determined insistance on his position as musical director, and not as a school and church employé, clearly proves this.

The Town Council had received him with due respect, but in order to raise the standard of church music as he, on his

---

[76] Johann Schelle "was in 1677 Cantor at the Thomasschule, with which the University entrusted to him the *Directorium Chori Musici* in St. Paul's Church." MS. addition in G. M. Telemann's copy of Mattheson's Ehrenpforte, in the Royal Library at Berlin. In the Leipzig register the entry of Kuhnau's death styles him "*Director Musices* bey der Löbl. Universitaet und *Cantor* bey der Schulen zu *St. Thomæ*."

[77] As in his memorial as to "well appointed church music" of the year 1730, although this almost exclusively concerns the St. Thomas' scholars.

[78] Heinrich Nickolaus Gerber always wrote in his copies of the French and English Suites "*Joh. Seb. Bach*. H. (ochfürstlich) A. (nhalt) C. (öthenischer) *Capell*meister, *Dir.* (*ector*) *Ch.* (*ori*) *M.* (*usici*) *L.* (*ipsiensis*) *et* (sometimes "*auch*") *Cant.* (*or*) *S.* (*ancti*) *T.* (*homæ*) *S.* (*cholæ*) *Lips.* (*iensis*)."

[79] Das jetzt lebende und florirende Leipzig, 1723, p. 78.

part, deemed fitting, he had dipped more deeply into their money-bags than they approved. Bach's immediate superior in the church services was the Superintendent of the Diocese of Leipzig, at that time Dr. Salomo Deyling. He took pleasure and interest in the district under his official jurisdiction, and enjoyed a well-deserved esteem far beyond its limits. He was a man of extensive learning, strong character, and undoubted administrative capacity. He was born, the son of poor parents, in 1677, at Weida in Voigtland, and by indefatigable energy had worked up to becoming a student at Wittenberg; in 1703 he qualified in the philosophical faculty, and two years later became Archdeacon of Plauen; by 1708 he was Superintendent at Pegau, and in 1716 General Superintendent at Eisleben. Meanwhile he had obtained the degree of licentiate in theology, and in 1710 had been made *Dr. Theol.* This was followed by his call to Leipzig as minister of the Church of St. Nicholas and Superintendent of the Diocese, in 1720. He entered on these offices in 1721, and at the same time was made Professor extraordinary of the University, which subsequently resulted in his being Professor in ordinary;[80] he was also Assessor to the Consistory. His abundant labours in all these relations ended only with his death, which took place in 1755.[81] The most interesting question to us, is what attitude he took up as regards church music. He had ample opportunity for expressing his views on the subject, for the number of his published writings is considerable. They treat of philosophy, philology, mathematics, and antiquities, but principally of theological matters, alike exegetical, dogmatic, historical, and practical. The greater part of his Latin dissertations is contained in his *Observationes Sacræ*, which were published in three parts, each containing fifty

---

[80] "Professor extraordinary," is used to denote those who had no fixed post in the university; in this case he became regular professor.

[81] Sicul. Leip. Jahr. Gesch., 1721, p. 227.—Adelung Fortsetzung zu Jöchers Gelehrten Lexicon, Leipzig, 1787, V. II., p. 684. A portrait in oil of Deyling hangs in the choir of St. Thomas' Church, and a copperplate from it is to be found in Geographischer Schau-platz Aller vier Theile der Welt, von Christian Ehrhardt Hoffmann, V. II. (Town Library, Leipzig.)

essays, at Leipzig, 1708, 1711, and 1715; in these he displays great learning and a strictly conservative, high-Lutheran tendency.

Of all the 150 dissertations, one only deals with things musical. It is entitled *Hymni a Christianis decantandi* (V. III., No. XLIV., p. 336 to 346). But even this is almost entirely of a theological and antiquarian character. What the Greeks understood by a hymn, how many kinds of song the Jews distinguished, on what occasions the Greeks used singing, how far the early Christians may have imitated them in this and how far not; all this and much more is amply and learnedly discussed, with a final reference to two passages in the Epistles to the Ephesians and to the Colossians. It is not till the last paragraph that he mentions that among the heathen certain persons were appointed as superintendents to lead public singing on the occasions of great festivals and he closes the treatise with this practical application:[82]

"Therefore, since profane men, strangers to the true worship of God, used to institute societies of singers of hymns and pæans for the purpose of singing to the praise and glory of false gods publicly in temples and other places of resort, and even used to sing to their praise in their very feasts, how much more ought Christians to be 'singers of hymns and pæans' to the true God? How much more does it beseem them to celebrate God's goodness and glory in 'psalms and hymns and spiritual songs,' both in their feasts and in their temples, both in public and private?"

This is all Deyling could find to say in his *Observationes Sacræ* to indicate his attitude towards church music, but it is enough. The parallel drawn between the hymn and pæan singers on one hand, and the choirs in protestant churches on the other, is an obvious one, and when he

---

[82] "Cum igitur profanus hominum coetus, et a vero Dei cultu peralienus, in commentitiorum numinum honorem, et ad laudes eorum decantandas, publica quondam Ὑμνῳδῶν et Παιανιστῶν in Templis, aliisque conventibus instituerint Collegia, ac inter ipsas epulas consueverint ὑμνολογεῖν; quanto magis Christianos decet esse Hymnologos et Paeanistas veri Dei? Quanto hos magis decit ψαλμοῖς καὶ ὕμνοις, καὶ ᾠδαῖς πνευματικαῖς Dei beneficia et laudes in conviviis, ac Templis, publice privatimque celebrare?"

speaks of convivial singing he evidently has in his mind the prevailing custom of having the choir of boys to sing at festivities, particularly at wedding banquets. We must remember, too, that this was written at a time when the question of part-music being introduced into the church service was under eager dispute, and especially as to whether, and how far, it was suitable for independent choirs. Thus Deyling considered church music, as represented by Bach, to be desirable. From the meagreness of his treatment of the whole subject it may be doubted whether he took any intelligent or sympathetic interest in music. However, all that was needed was that he should give Bach full liberty to act. All that has been said as to a re-organisation of the services in which Bach and Deyling co-operated rests on unfounded assumptions and baseless opinions.[83] A confusion between the words of the church music and the hymns sung by the congregation could not occur if the Cantor did but understand his duty, since the congregational hymns to be sung on holy days were fixed once for all, and the choice of the hymns for ordinary Sundays was one of the traditional duties of the Cantor. Any lack of connection between the cantata and the Epistle and Gospel for the day was hardly possible either, because the texts to which the cantatas were composed for each Sunday and holy day were usually written to suit their ecclesiastical significance and the contents of the portions of Scripture appointed for the day. Any alteration in the musical arrangements of the form of worship would have been an infringement of Bach's official rights, since, on his induction, the Council had expressly enjoined him not to allow any innovations in the services; they were, and remained throughout Bach's

---

[83] Rochlitz, Sebastian Bach's Grosse Passionsmusik nach dem Evangelisten Johannes, Vol. IV., 3rd ed., Leipzig, 1868, p. 271. Part of it is printed as a preface to the B.-G., IV., p. 13—15. It will be seen from ch. IV. of this section that the picture given by Rochlitz of the services before and during Bach's time is in part perfectly inaccurate. Rochlitz says that at the beginning of each week Bach sent several, usually three, texts of cantatas suitable for the following Sunday to the Superintendent, who made a selection. I can find nothing to confirm this statement, but it is not impossible that Rochlitz, who himself was a foundation scholar at St. Thomas', followed a credible tradition.

life, the same as in Kuhnau's time. They afforded abundant opportunity for the use of music; indeed, Bach did not even usually avail himself of them to the full. Certainly he had to submit the text he chose for his composition to the censorship of the Superintendent. It can hardly be said that they co-operated in the matter; it was rather a check on the composer's liberty, which he much resented, for Bach always evaded every kind of surveillance, and liked to feel himself perfectly independent in his own sphere of work.

A circumstance which occurred—at a later period, it is true, but which may find a fitting place here—serves to prove this statement. On Good Friday, 1739, Bach had announced the programme of a "Passion Music," by the plan, then customary, of sending round printed copies of the text. The Council must needs once more assert its superiority at a wrong time; one of their subordinates was commissioned to announce, verbally, to the Cantor that the Passion Music must be postponed till a regular permission had been granted by the superior authorities. Bach, however, was refractory. He had proceeded on this occasion just as he had done on all others, and, as regarded the text, there was nothing reprehensible in that, since the work had already been performed several times. And besides, he did not care whether the performance took place or not, for he had nothing but the trouble of it and no profits. He would explain to the "Herr Superintendent that the Council had forbidden it." The Council and Consistory were often at loggerheads, and the fray resulting from this must have been an additional incentive to Bach to act, as far as possible, on his own account.[84]

The congregational hymns were subject to the censorship of the church authorities, as well as the texts of the Cantatas. A certain series of hymns was sanctioned once for all; within these limits the Cantor was free to choose, but he might not go beyond them. It is possible that Bach

---

[84] As a parallel case we find that on April 4, 1722, Kuhnau had received a reproof from the Burgomasters, because he had asked permission of the Consistory and not of the Council with regard to a "Passion Music" in St. Thomas'.

may at some time have made the attempt, and that it was reported to the Consistory. On February 16, 1730, a warning reached the Superintendent that he should take heed lest hymns not hitherto in use should be sung without the concurrence of the authorities, as had recently been done.[85] When we see what a subtle and profound feeling Bach shows in the selection of those chorales, for instance, which are introduced as Madrigal texts into the Passion according to St. Matthew, it is quite credible that he should be very ready to avail himself of the Cantor's right to select the congregational hymns, in order to produce the greatest and most symbolical variety of effect in the different musical portions of the church services. And even if he had not been a determined man who would not yield an inch in the department which belonged to him, it would still be intelligible that he should allow no one to interfere in the arrangement of the hymns. However, an attempt to do so was made ere long. *Magister* Gaudlitz, the sub-dean of St. Nicholas, began in 1727 to select the hymns for the Vesper sermons given by him—at first with the knowledge of the Superintendent and the Cantor's consent. After he had done this about a year, our master would no longer submit to his interference; he chose to ignore the sub-deacon's decision, and made the choir sing the hymns he himself selected. Gaudlitz reported him to the Consistory, who, somewhat over-hastily, sent a notice to the Cantor, through the Superintendent, to the effect that for the future he was to have the hymns sung which the preacher had selected. Bach now thought it time to appeal to the Council, and he wrote as follows:—

Magnifici,
MOST NOBLY-BORN, MOST NOBLE, POWERFUL, HIGH, AND LEARNED AND MOST WISE,
MOST HONOURABLE LORDS AND PATRONS,

Will your *Magnifici*, well-born and noble lordships condescend to remember how I was admonished by your *Magnifici*, well-born and noble lordships on the occasion of my being called to the Cantorate of the School of St. Thomas in this place, of which I was always to

---

[85] I have not been able to discover the document which is given by Bitter, II., p. 86, and must therefore refer to him.

perform the traditional usages in the public divine service, duly in all respects, and not to introduce any innovations; and how, under the same contract, you were pleased to assure me of your high protection? Among these usages and customs was also the right of ordering the hymns before and after the sermons, which right was left entirely to me and to my predecessors in the Cantorate; provided that the hymns chosen be in conformity with the gospels and the use of the Dresden hymn-book regulated by these, and as may seem suitable according to time and circumstances; and certainly, as the worthy *Ministerium* can well attest, no contradiction to this has ever arisen.[86] But, to the contrary of this, the *Sub-diaconus* of St. Nicholas Church, Herr *Magister* Gottlieb Gaudlitz, has attempted to introduce an innovation, and instead of the hymns hitherto ordered in accordance with church customs, has ordered other hymns; and when I scrupled to yield to this because of serious consequences which might result, he brought an accusation against me before the worshipful *Consistorium* and obtained an injunction against me, by the contents of which I, for the future, am to let those hymns be sung which shall be commanded by the preachers. But it seemed to me not proper, without the knowledge of your *Magnifici*, well-born, and noble lordships, the patrons of the churches in this place, to carry this into effect; and all the less so because hitherto the arrangement of the hymns by the Cantor had for so long a time remained undisturbed, the afore-mentioned Herr *Magister* Gaudlitz having himself allowed in the document presented to the most worshipful *Consistorium*, of which a copy is subjoined, that when once or twice he had been allowed to do it, my consent as Cantor had been required. In addition to which, when the hymns which had to be sung as part of the church music were of inordinate length, the service would be prolonged, and thus all kinds of irregularities would have to be provided for, putting aside the fact that not one of the officiating clergy, with the exception of Herr *Magister* Gaudlitz, as *Sub-diaconus*, seeks to introduce this innovation. Thus, I esteem it necessary most submissively to bring before your *Magnifici*, well-born, and noble lordships the humble prayer that you will most graciously protect me in the use and ordering of these hymns, as has hitherto been usual. And with life-long devotion,
    I remain,
        Your *Magnifici*, well-born, and noble lordships'
                Most obedient,
                      JOHANN SEBASTIAN BACH.[87]
Leipzig, September 20, 1728.

---

[86] "As I appoint the hymns for all three churches": Kuhnau, Memorial of December 4, 1704, see Appendix B., VII.
[87] Council deeds "Schuel zu *St. Thomas.* Vol. IV., Stift, VIII., B., 2." Fol. 410. Only the signature of Bach's memorial is autograph.

Thus the Council once more found itself in conflict with the Consistory, but how the affair terminated among them is not known.

Under the circumstances, all that Bach proposed and hoped to do for the improvement of the music in the principal churches in the direction which he had always pursued in art, involved nothing less than the training of singers and players alike to a higher pitch of executive power, and the education of their feeling for art by keeping them engaged on fine and important works. And indeed this was inevitably the case, since he himself composed for them as much as possible; it was quite in accordance with his own wish, and, as we shall see, he at once began to develop this side of his genius, and during a long series of years was incessantly productive. With regard to the improvement and extension of his choir, he had made a great step quite at the beginning of his official life by establishing a connection with the students. He knew very well that without them little could be done, still he was wholly dependent on their free-will for whatever support they contributed to his Sunday performances; the foundation scholars, who were bound to sing, still formed the main body of the choir. They were not required to sing only; since by his deed of installation Bach was pledged to instruct them diligently, not only in vocal but also in instrumental music. As things then stood this was highly necessary, for the performance of the instrumental accompaniments by the corps of musicians kept up by the Town Council was neither efficient nor strong enough by itself to fulfil his requirements. It consisted of only seven members—four "Stadtpfeiffer," and three "Kunstgeiger," who, with a single exception, belonged to the class of "doubtful characters."[88] There was indeed no lack of instrumentalists, and of good ones, in Leipzig,[89] but their assistance would

---

[88] During the first twelve years of Bach's official life their names were— Gottfried Reiche, Christian Rother, Joh. Cornelius Gentzmar, Joh. Caspar Gleditsch, "Stadtpfeiffer." The "Kunstgeiger" were Heinrich Christian Bezer, Christian Ernst Meyer, Joh. Gottfried Kornagel. Accounts of the churches in the Archives of the Foundation Library, Leipzig.

[89] "*Dunkeln Ehrenmänner.*" A party of five musicians performed, under Schott, in the New Church. See the accounts from 1725 to 1729.

cost money, while the scholars had to fiddle for nothing. We must picture to ourselves Bach's plan of instruction in clavier, violin, and organ-playing, as consisting very much in attaching to himself any scholar he found qualified, and endeavouring to urge him forward by his example and by opportune instruction. Many young men, whose names we shall meet with again, sought instruction at the Thomasschule, less in general knowledge than in music; and these became Bach's pupils in the strictest sense of the word, both in playing and in composition, and quitted the school, not as good general scholars, but as accomplished artists.

However, the singing choir was the first musical object which the scholars had to keep in view. What kind of training it was that Bach put them through to this end is a very interesting question, but one not easy to answer. A number of testimonials in Bach's own hand have come down to us—reports written after the examinations of the singers. They refer for the most part to such youths as had applied for admission to the foundation of the Thomasschule. In the summer of 1729 a certain Gottlieb Michael Wünzer was candidate for a scholarship. Ernesti, the Rector, gave him a testimonial for Latin, and under this Bach wrote:—

> The above-named Wünzer has a rather weak voice, and still but a poor method, still he may—if he is diligent in private practice—be available in time.
>
> JOH: SEBAST: BACH.
> Leipzig, *Jun* 3, 1729. Cantor.[90]

On another occasion he writes:—

> This present Erdmann Gottwald Pezold of Auerbach, *ætatis* 14, has a fine voice and tolerable proficiency. Witness my own hand,[91]
>
> JOH: SEB: BACH.

---

[90] Obig benandter Wünzer hat eine etwas schwache Stimme und noch Wenige *profectus* dörffte aber wohl (so ein *privat exercitium* fleissig getrieben würde) mit der zeit zu gebrauchen seyn.

[91] Vorzeiger dieses Erdmann Gottwald Pezold von Auerbach, *ætatis* 14 Jahr, hat eine feine Stimme und ziemliche *Profectus*. So hiermit eigenhändig attestiret wird

von
JOH: SEB: BACH.

And again:—

This present Johann Christoph Schmied of Bendeleben, in Thuringia, *ætatis* 19, has a fine tenor voice and reads well at sight.[92]

            JOH: SEB: BACH.
Or:—   *Director Musices.*

Carolus Heinrich Scharff, *ætatis* 14, has a tolerable alto voice and a moderate proficiency in music.[93]

              J. S. BACH.
               Cantor.

A number of similar testimonials will presently be given on another occasion, since they do not contribute any new information as to the present question. Here, however, I may adduce another piece of evidence of a later date, though, it is true, it does not deal with the entrance of any member of the Thomasschule. In 1740 a new office of *Collaborator* in the Thomasschule was to be instituted. Since the new master was to be charged with the duty of grounding the boys in the elements of music, the candidates were sent to Bach to be examined. He reported the result of this examination in these words:—

By command of their Excellencies the Vice-Chancellors, the three competing persons have been with me and I found them as follows:—

(I.) Herr *Magister* Röder renounced the examination, having altered his determination and taken the place of Steward in a noble family in Merseburg.

(II.) Herr *Magister* Irmler has a fine method of singing, but he is somewhat wanting in accuracy of ear.

(III.) Herr Wildenhayn plays a little on the clavier, but in singing he is not skilled, by his own confession.

LEIPZIG, *Jan.* 18, 1740.   JOH: SEB: BACH.[94]

---

[92] Vorzeiger Dieses Johann Christoph Schmied von Bendeleben aus Thüringen, *ætatis* 19 Jahr, hat eine feine Tenor Stimme und singt vom Blat fertig.

[93] Carolus Henrich Scharff, *ætatis* 14 Jahr, hat eine ziemliche Alt Stimme, und mittelmässige *Profectus in Musicis.*

[94] Irmler got the place.

Auf Ihro *Excellence* des Herrn *Vice-Cancellarii* hohe *Ordre* sind die drey *competenten* bey mir gewesen, und habe Sie folgender massen befunden:

(I.) Der Herr *M.* Röder hat die *probe depreciret,* weiln er seine *resolution* geändert und eine Hoffmeister Stelle bey einer Adelichen *Familie* in Merseburg angetreten.

(II.) Der Herr *M.* Irmler hat eine gar feine Singart; nur fehlet es ihm in etwas am *judicio aurium.*

(III.) Der Herr Wildenhayn spielet etwas auf dem Clavier, aber zum Singen ist er eigenem Geständniss nach, nicht geschickt.

LEIPZIG, 18 *Januar,* 1740.   JOH: SEB: BACH.

Brief and general as these documents are, we learn from them to what aspect of vocal art Bach principally directed his attention. The word *profectus*, which was much used at that time in the sense of proficiency or productive power, may certainly include everything that can be required of a singer; but it here bears a more limited meaning, if we remember that the first consideration was a serviceable choir singer, and then weigh what is said of one and another of the candidates. Bach required of his singers, in the first place, accuracy of pitch and time, a pure intonation, fertility of resource, and also, if possible, a pleasing quality of voice— " er hat eine feine Stimme." In all this it is the musician that speaks, and not the singing master; there is no allusion even to the cultivation, utterance, equality of register, or any other vocal technicality. It would be absurd to suppose that Bach was insufficiently acquainted with all these matters; besides, it must be remembered that his second wife was a well-trained and accomplished singer. It by no means follows from his silence that he left them out of consideration in the education of his singers, but we may venture to assert that he thought it no part of his task to make accomplished vocalists of the scholars who could sing in at all the same way as he trained Krebs, Ernst Bach, his own sons, and others, to be distinguished players and composers. There was not time for this in seven hours of singing a week, especially when forty pupils took part in them, as was the case when the choirs were complete. He would have had to take the best of them under his private tuition, and before he could do this he must have acquired a more thorough knowledge of the art of singing than he actually possessed, since we know nothing of his having had any practical experience of it since the days of his boyhood.

Bach was, before and above everything, a composer for the organ, as the whole course of his development shows us; all his writing for other instruments has something about it that suggests the organ, and his vocal compositions might be designated as the last and utmost embodiment of the true Bach organ style. Nor does the admission of this

view necessarily imply any reproach; not even if we insist
on regarding the subject exclusively from the most purely
musical and artistic point of view. As long as instrumental
music has existed it has been on terms of mutual interchange with vocal music, and they have borrowed reciprocally.
But to Bach, as we have already seen, the organ was imbued
with a peculiar, more essentially ideal, character, which gives
a higher justification to much that might otherwise seem
faulty in style in his vocal music. At any rate, he required
of the singers something different in itself, as adapted to the
views he held of church music, from what was expected by
the musical world he lived in. He required less, in so far
that he attached less importance to all that characterises
skilled singing as such, all that must be made as prominent
as possible where the human voice is the principal element
in the piece; but, on the other hand, he required more, inasmuch as in his compositions he often put before the singers
technical difficulties in the music which could not have been
suggested by an imagination working with the single idea of
writing for the voice. It is self-evident that in such music
as this the distinction between the demands on a chorus
singer and soloist is far less than in Handel's oratorios. A
scholar who was thoroughly trained in singing Bach's choral
music might soon be qualified to perform an aria, since its
effect would depend far less on him alone than is the case
with the singer of an opera or oratorio air. Certainly the
feeling of Bach's arias could not have found that flow of
passionate utterance when sung by youths and boys which
they are capable of when the finest singers regard their due
rendering as one of their noblest tasks. But we may rest
convinced that Bach himself would not have endured a mere
mechanical reading of them. Johann Friedrich Agricola,
Bach's pupil from 1738 to 1741,[95] says it is indispensable
that a singer should learn elocution, or at least acquire, by
the verbal instructions of a good speaker or by accurate
observation of his mode of delivery, what sort of sound of
the voice is requisite for the due expression of certain

---

[95] Gerber, L. I., col. 17. Burney's Diary. Rolle, Neue Wahrnehmngen, &c. Berlin, 1784, p. 93.

emotions or figures of speech, and that he should also diligently practise himself in reading or declaiming, according to these rules, emotional passages from the works of good orators and poets.[96] And we know that Bach himself was fond of illustrating the proper method of musical performance by the rules of rhetoric. "He so perfectly understood the resemblance which the performance of a musical piece has in common with rhetorical art that he was listened to with the utmost satisfaction and pleasure when he discoursed of the similarity and agreement between them; but we also wonder at the skilful use he made of this in his works." So writes his friend *Magister* Birnbaum.[97]

Though these were the principles which Bach regarded as the foundation of his vocal training, it by no means follows that he should have been successful in practically carrying them out. For this, two things were necessary: a real talent for teaching youth on his part, and musical talent on that of his pupils. It seems to be a contradiction to what was said in a former chapter,[98] as to Bach's great gift for teaching, when the younger Ernesti (who was Rector of the Thomasschule after 1734) states that he could maintain no discipline among the choir boys, and that after Bach's death it was stated in the Council that "Herr Bach was indeed great as a musician, but not as a schoolmaster." But it is one thing to guide one single teachable and reverently disposed pupil, and quite another to quell an unintelligent and unruly mob of boys. Bach was peculiarly fitted for the former task by his gifted, sympathetic, and essentially wise and humane nature; but in the latter case his artist's irritability hindered him, and this was all the boys could see, who were incapable of understanding his greatness. In this respect the man in Leipzig in no whit differed from the youth at Arnstadt; and we shall presently see that the musical qualifications of the scholars received but a meagre share of his attention. Bach frequently visited Dresden from Leipzig, heard there the

---

[96] Tosi, Anleitung zur Singkunst. Mit Erläuterungen und Zusätzen von Joh. Friedrich Agricola. Berlin, 1757, p. 139.

[97] See Scheibe, Critischer Musikus. Leipzig, 1475, p. 997. Birnbaum's statements refer to the year 1739. See ante, p. 56. [98] Ante, p. 47.

beautiful performances of the Italian singers, and the admirable playing of the Court Band, and was himself a much admired personage both among his fellow artists and in court circles. It was but human nature that under these circumstances he should often fulfil his proper duties with his foundation boys and town musicians in a grudging spirit; and work done without any heart in it is but rarely successful.

The fact thus indicated did not become publicly evident, however, till after the lapse of a few years. At first the charm of novelty, the natural desire to justify the expectations formed of him by those who were most intimately concerned, and, above all, the delight of being able at last to perform and compose church music to his heart's content, would easily have outweighed many disagreeables and disadvantages; at any rate, we have no information which stands in the way of this assumption. The first traces of a misunderstanding became visible in 1729. At Easter in that year nine foundation boys had finished their studies and quitted the school. They had been useful musicians and among them indeed there was one of distinguished talent, Wilhelm Friedemann, Bach's eldest son. On this occasion it came to light that the Council, who were still as negligent in all that concerned the school choir as they had been in Kuhnau's time, had for a long time ceased to pay the requisite attention to the question whether the new foundation scholars who were admitted had any musical gifts. The choir in consequence had fallen into such a wretched condition that some very decisive steps had to be taken, if the music were to be carried on at all in the way that had become traditional. Nor was it only Bach who represented in the strongest terms that the vacant places must be filled up by boys of musical qualifications; even the old Rector Ernesti requested it, and Dr. Stiglitz, the Inspector of schools, who had to transmit their demands to the Council, supported them by a petition, couched in the most emphatic terms, dated May 18. He also forwarded, as supplementary, a report drawn up and written by Bach as to the new candidates and their capabilities, as well as on the indispensable constituents of the different church choirs. This is as follows:—

The boys, who under the present vacancies in the school of St. Thomas, desire to be received into it as *Alumni* are the following:—
I. Such as have musical qualifications, and, firstly, *trebles*.
  1. Christoph Friedrich Meissner, of Weissenfels, *ætatis* 13, has a good voice and a fine method.
  2. Johann Tobias Krebs, of Buttstädt, *ætatis* 13, has a good strong voice and a fine method.
  3. Samuel Kittler, of Bellgern, *ætatis* 13, has a tolerably strong voice and a very pretty method.
  4. Johann Heinrich Hillmeyer, of Gehrings Walde, *ætatis* 13, has a strong voice and a fine method.
  5. Johann August Landvoigt, of Gaschwitz, *ætatis* 13, has a passable voice; his method is tolerable.
  6. Johann Andreas Köpping, of Grossboden, *ætatis* 14; his voice is tolerably strong and his method moderate.
  7. Johann Gottlob Krause, of Grossdeuben, *ætatis* 14; has rather a weak voice and very mediocre method.
  8. Johann Georg Leg, of Leipzig, *ætatis* 13; his voice is rather weak and method indifferent.

Alti.

  9. Johann Gottfried Neucke, of Grima, *ætatis* 14, has a strong voice and tolerably fine method.
  10. Gottfried Christoph Hoffmann, of Nebra, *ætatis* 16, has a fairly good alto voice, but his method is rather faulty.
II. Those who did not offer themselves as musicians:—
  1. Johann Tobias Dieze.
  2. Gottlob Michael Wintzer.[99]
  3. Johann David Bauer.
  4. The son of Johanna Margaretha Pfeil.
  5. Gottlob Ernst *Hausius*.
  6. Friedrich Wilhelm, the son of Wilhelm Ludwig.
  7. Johann Gottlieb Zeymer.
  8. Johann Gottfried Berger.
  9. Johann Gottfried Eschner.
  10. *Salomon* Gottfried Greülich.
  11. Michael Heinrich Kittler, of Prettin.

JOH: SEBAST: BACH,
Direct: Musices,
u. *Cantor* at S. Thomae.

Then, as supplementary to the former list, we have:—
  Gottwald Pezold, of Aurich, *ætatis* 14, has a fine voice and tolerable method.[100]

---

[99] Evidently the same who, on the 3rd of June, obtained a somewhat warmer testimonial.

[100] No doubt the same to whom Bach had already given a separate testimonial.

Johann Christoph Schmid, of Bendeleben, *ætatis* 19, has a tolerably strong tenor voice and sings very prettily.

And, finally, we have the following supplement :—

In the Church of St. Nikolaus, there are belonging to the first choir:	Of St. Thomas, to the second choir:	In the New Church, to the third choir:
3 Trebles.	3 Trebles.	3 Trebles.
3 Alti.	3 Alti.	3 Alti.
3 Tenors.	3 Tenors.	3 Tenors.
3 Basses.	3 Basses.	3 Basses.

To the fourth choir:
2 Soprani.
2 Alti.
2 Tenors.
2 Basses.

And this last choir must serve the Church of St. Peter as well.

Nevertheless, this representation was only half attended to by the Council. On May 24, it granted, it is true, five scholarships to the musical boys named as Meissner, Krebs, Kittler, Hillmeyer, and Neucke; on the other hand three were given to those who "did not offer themselves as musicians" (Dieze, Zeymer, and Berger), and the last scholarship was given to a candidate named Feller, who is not mentioned by Bach, and who, therefore, had evidently not even come before him. Soon after this there must again have been a vacancy, for on June 3 Wintzer was admitted. Pezold and Schmid, who had both been recommended by Bach, were passed over; Krause, who is probably the lad named by Bach as No. 7 of list 1, was not admitted till October of the following year.

In the Holy week of 1729, Bach had for the first time conducted a performance of his Passion Music according to St. Matthew. From this we see that this work had made so little impression on the gentlemen of the Town Council of Leipzig, that they did not even accede to the request of the composer so far as to choose the nine musical scholars out of the candidates for the scholarships. After so signal a proof of their ignorance of his value, it seems almost surprising that Bach should not at once have sought to resign his office; and it certainly is not strange that for a time he was thoroughly disgusted with it. It happened just at the same time that he undertook the direction of the Musical Union formed by Telemann, and might there hope to find rather more love of music and intelligent sympathy. It may be that this prospect may have inspired him with courage for the future. But so far as his functions as Cantor were concerned, the Musical Union was a source of fresh vexation. Kuhnau had, before this, vainly striven to obtain that a sufficient stipend might be allotted to prove an inducement to the students to co-operate regularly and in considerable numbers with the choir of St. Thomas. As we shall see from a memorial drawn up by Bach (given below), the only result up to this time had been that a few donations, and those very insufficient, had been made to the students who had assisted; and in Bach's time these had become more and more meagre, till at length they entirely ceased. The Council no doubt thought them now unnecessary, since from Bach's position at the head of the Union the students would join the choir without payment. Their conduct really bears the aspect of dishonesty, and yet it was simply the result of mental narrowness; for, when the church music became obviously worse, they were very indignant with the Cantor.

On October 16, 1729, the Rector, Joh. Heinrich Ernesti, died;[101] the place remained vacant for many months, and then, on June 8, 1730, the elders (*Väter*—fathers) of the town agreed in inviting Johann Matthias Gesner to fill it, since the choice of a native of the town would have " caused jealousy." One member of the Council expressed a wish that they might "fare better in this appointment than in that of the Cantor." The neglect of his duties which the Council thought it had observed of late had already given occasion to his receiving various warnings and admonitions, but the gentlemen must have discovered to their amazement that they did not produce the expected effect. The offended and defiant artist roundly refused to give any explanation, and this must have happened several times, for one of

---

[101] Neue Zeitungen von Gelehrten Sachen. Leipzig, 1729, p. 791.

the Council declared point blank that the Cantor was
"incorrigible." This occurred at a meeting of the Council
on August 2, 1730, on the discussion of a question as to the
restoration of the school buildings. As has been said, Bach,
at his installation, had received permission to employ a
substitute, as far as might be necessary, in giving general
instruction. This substitute had hitherto been found in the
person of the *Tertius*, afterwards Conrector, *Magister* Pezold,
but he had failed to give satisfaction in this, and indeed in
his work generally, and a proposal was therefore made to
employ as Bach's substitute, instead of Pezold, a younger
master, *Magister* Abraham Krügel. It would have been the
most obvious course to restore to Bach the duty of giving
the lessons; but the strong feeling against him now came
to light. He had not conducted himself as he should—for
instance, he had sent a member of the choir into the country
without any previous intimation to the municipal authorities;
he had gone on a journey himself without asking leave; he
did nothing; he did not attend to the singing classes, not to
mention other accusations. The meeting proposed to put
him down to one of the lowest classes, where he could either
give the elementary instruction himself or, as they would
not withdraw the permission originally given, put a deputy
to do so for him. But this motion was not passed, and it
was resolved instead "to sequestrate the Cantor's income,"
besides addressing to him an admonition and appealing to his
conscience. The accusation that the Cantor "did nothing"
is really startling when we learn that, within the few weeks
previous, on the occasion of the Jubilee of the Augsburg
Confession, the 25th, 26th, and 27th of June, Bach had produced and conducted three grand cantatas; that, besides
such a monumental work as the Passion according to St.
Matthew, during the seven years that he had been Cantor
he had composed a series of cantatas which, to any other
musician, would have represented the labours of half a lifetime. But this sort of work counted for nothing with the
Council; what they required was that the Cantor should
hold his classes regularly, and should not neglect his
duties as an instructor in so independent a fashion.

There can be no doubt that Bach had taken things easily as regarded the singing lessons, and it was a neglect of duty which we shall not attempt to screen or defend; still, there were so many and such important circumstances to excuse it that they might almost constitute a justification. It has been already said that the afternoon practising had always been left entirely to the management of the Prefect since the time of Tobias Michael. Thus, when Bach in the same way only gave one singing lesson in the morning of Monday, Tuesday, and Wednesday—and it would seem that he was certainly accustomed to do so much as this—he was only following the established custom.[102] Up to this time hardly any one had found anything to say against this; nay, Michael's colleagues had even opined that the Cantor's presence at the afternoon singing was quite unnecessary. Even if we go farther, and suppose that Bach had been irregular in giving the three morning lessons, we must remember above all the inefficient state of the choir, and that the Council did not even think it necessary to do what lay in their power to improve it and meet the difficulty which Bach experienced in dealing in the right way with this half-wild troop of boys. Nor must we forget that the state of affairs in the Thomasschule was constantly most unsatisfactory, and that the picture given of it above continued to be true during the first seven years of Bach's official life. The careful re-organisation of the institution had been a merely superficial labour; regulations can have no value if no one is concerned to carry them out, and this was the case here. Bach must soon have perceived that in this state of entire demoralisation, this war of each against all carried on by the body of teachers, and

---

[102] From an act of the Council referring to the school. At the end of this document we find certain "remarks" as to the regulations of the school from the pen of the younger Ernesti, in which he speaks of a proportionally small preparation for the processional singing at the three seasons being limited to certain afternoons. Finally he says, "It must be considered whether these practisings could not be transferred to the hours after the lesson (*Lection*), from three till six o'clock. In this way the singing lesson which the Cantor has to give from twelve till one would not be lost, but, on the contrary, more strictly kept up, for at present he gives only one hour's lesson, whereas he ought to give two, and thus the boys do not have enough practice in music." This was written about 1736.

the confusion which resulted from the constant disputes as to their authority between the Council and the Consistory—in which he was especially involved—the only safe course was to look neither to the right hand nor to the left, but to act as independently as possible. To this course his natural disposition inclined him, and he no doubt indulged it all the more since, in point of fact, the post of school Cantor was a singularly independent position. The Rector was little accustomed to trouble himself about the singing lessons, of which Bach's duties as a teacher principally consisted. Since it was one of the chief tasks of the scholars to perform the church services, the Cantor was, in their eyes, almost as important a personage as the Rector himself. In the time of the elder Ernesti, candidates offered themselves to Bach in the first instance quite as often as to the Rector himself; that is to say, they applied to him in person and were examined in music, and he then procured their admission.[103] No doubt that pleased the old Rector best which gave him the least possible trouble, and it is plain that Bach always remained on excellent terms with him, since he asked his wife and daughter to stand sponsors to the two children born to him in 1724 and 1728. Such a position, which really forced him into independence, might easily lead to his occasionally kicking at those limitations which necessarily existed, particularly when they were fixed by an authority before which Bach, as an artist, could not bow, however great his respect for it in other ways. Added to all this, as a factor of special significance, we must count the deep mortification which must have filled the master's mind when he found the best he could do so absolutely unintelligible to his superiors in office.

In short, he was to be called to order. Documents exist which show that the mere order to withhold his emoluments was not the end of the matter. His actual salary and the fees could hardly be interfered with; but there were endowments, holiday money, and other revenues to be distributed,

---

[103] On May 18, 1729, Dr. Stiglitz writes to the Council that the new applicants for the scholarships had either addressed themselves in writing, and with the recommendation of some witness, to the Rector or Conrector, or else had been introduced merely by the nomination of " Cantor Bach."

and these afforded opportunities for aggrieving those who were out of favour. The fees that had accrued to the Rector during the time that elapsed between Ernesti's death and Gesner's appointment were to be divided into three equal parts by order of the Council—for the widow, the Conrector, and the *Tertius*, since the two last had to fill the place of the Rector in the school. As against this, Dr. Stiglitz, the Superintendent of the school, represented in a statement made September 23, 1730, that the Conrector had had much more trouble than the *Tertius*, since he had had to take the Rector's place in other matters besides the lessons, and that the Cantor had also had to take his turn in the duties every three weeks instead of every four. That, therefore, it was fit and proper that more should be given to the Conrector than to the *Tertius*, and that something should also be allotted to the Cantor. The sum in question amounted to 271 thlrs. 7 ggr. 3 pf., and the Council decided as follows on November 6, 1730: the widow was to have forty-one thlrs., the Conrector 130 thlrs. 7 ggr. 3 pf., and the *Tertius* 100 thlrs.; Bach getting nothing.

A similar case had already occurred during the course of the year, before the noble determination to curtail the Cantor's finances had been formally recorded, though an ill feeling already prevailed against him in the Council. A certain citizen, named Philippi, had bequeathed to the Thomasschule a sum of money of which the interest, amounting to twenty thlrs. yearly, was to be subdivided in such wise that twelve thlrs. were given to twelve poor scholars, one thlr. each to the Rector, Conrector, *Tertius*, and first bachelor, sixteen ggr. to each of the collaborators, and two thlrs. to the scholar who made the memorial speech of the year. Under this distribution there was a residue of sixteen ggr., and the question arose as to whether it should not be given to the Cantor or to the other bachelor, neither of whom were mentioned. The Council decided, December 3, 1729, in favour of the other bachelor. Bach was the only member of the college sent empty away.

When, twenty-four years previously, Bach had had to answer before the Consistory of Arnstadt for a dereliction of duty, similar to that he was now accused of, he avoided any farther verbal discussion by promising to clear himself in writing.[104] He now at first would not be drawn into any discussion whatever with the Leipzig Council, but he subsequently was compelled to change his views, and he drew up a memorial, which we may regard partly as a formal declaration in writing and farther development of what he had said by word of mouth in the previous year to the school Superintendent, and partly as an outline of that officer's report to the Council:—

A short, but indispensable sketch of what constitutes well-appointed church music, with a few impartial reflections on its present state of decay.

For well-appointed church music, vocalists and instrumentalists are requisite. In this town the vocalists consist of the foundation scholars of St. Thomas, and these are of four classes: trebles, altos, tenors, and basses.

If the choir are to perform church pieces properly and as is fitting, the vocalists must again be divided into two classes: *concertists* and *ripienists* (soloists and choristers). The concertists are usually four, but sometimes five, six, seven, up to eight, according to the requirements of the case and if the music is *per choros*.

The ripienists must be at least eight, two to each part.

The instrumentalists are of various sorts—as violinists, oboeists, flautists, trumpeters, and drummers. N.B.—Among the violinists are reckoned those who play the viola, violoncello, and violin.

The number of the foundation scholars of St. Thomas is fifty-five. These fifty-five are divided into four choirs, for the four churches in which they partly perform music (musiciren),[105] partly sing motetts, and partly chorales. In three of the churches, *i.e.*—St. Thomas, St. Nicholas, and the New Church—all the scholars must understand music, and the residue go to St. Peter's; those, that is to say, who do not understand music, but can only sing a chorale at need.

To each of those musical choirs there must belong, at least, three trebles, three alti, three tenors, and as many basses, so that if one is unable to sing—which often happens, and particularly at this time of year, as can be proved by the recipes sent from the school of medicine to the dispensary—a motett may be sung with, at least, two voices to each part.[106] (N.B.—How much better it would be if the *Coetus* were

---

[104] See Vol. I., p. 326.

[105] The word "Musiciren" in the common parlance of Bach's time, when applied to church music, never meant anything but the performance of part music with obbligato instrumental accompaniments.

[106] The Scholars were very fond of declaring they were ill, often for a month, or even a quarter, or a half-year; they threw the medicine out of the window and revelled in the strengthening food allowed in case of sickness. Niclas, in Eyring's *Biographia Academica Gottingensis*, Vol. III., p. 52.

so arranged that four singers could be available for each part, each choir thus consisting of sixteen persons.)

From this it appears that the number of those who must understand music is thirty-five persons.

The instrumental music consists of the following parts :—

Two or even three	Violino 1º.
Two or three	Violino 2do.
Two	Viola 1º.
Two	Viola 2do.
Two	Violoncello.
One	Double bass.
Two or three, according to need ...	Oboes.
One or two	Bassoons.
Three	Trumpets.
One	Drum.

In all eighteen persons, at least, for the instruments. N.B.—Added to this since church music is also written for flutes (*i.e.*—they are either *à bec* or *Traversieri*, held sideways), at least two persons are needed for that; altogether, then, twenty instrumentalists. The number of persons appointed for church music is eight, four town pipers, three town violinists, and one assistant. Diffidence, however, forbids my speaking truly of their quality and musical knowledge; however, it ought to be considered that they are partly inefficient and partly not in such good practice as they should be. This is the list of them:—

Herr Reiche[107]	plays	first trumpet.
Herr Genssmar	,,	second trumpet.
Vacant	,,	third trumpet.
,,	,,	drum.
Herr Rother	,,	first violin.
Herr Beyer	,,	second violin.
Vacant	,,	viola.
,,	,,	violoncello.
,,	,,	double bass.
Herr Gleditsch	,,	first oboe.
Herr Kornagel	,,	second oboe.
Vacant	,,	third oboe or taille.
The assistant	,,	the bassoon.

---

[107] Reiche was the only prominent musician of the whole of this worthy society, but when Bach wrote the above he was already 64 years old. A volume of 24 *Quatricina* for a cornet and 3 trumpets, published by him in 1696, exists in the Royal Library at Berlin. He died in 1734, unmarried, in the Stadtpfeiffer Gasschen, whither he was carried home struck by apoplexy and lay there from Oct. 6 till he died (Register at Leipzig), and his funeral was followed by the larger half of the school. Gerber, Lexicon II., col. 258, says he was born at Weissenfels, Feb. 5, 1667; according to the register he was 68 years of age.

Thus the most important instruments for supporting the parts, and the most indispensable in themselves, are wanting, to wit:

        Two first violins,   Two violoncellos,
        Two second violins,   One double bass,
        Two to play the viola, Two flutes.

The deficiency here shown has hitherto had to be made good partly by the University students, but chiefly by the scholars. The students used to be very willing to do this, in the hope that in time they might derive some advantage from it, or perhaps receive a stipend or honorarium (as was formerly customary). But as this has not been the result, but, on the contrary, the little chance perquisites[108] which formerly fell to the *chorus musicus* have been in succession altogether withdrawn, the readiness of the students has likewise disappeared, for who will labour in vain, or give his service for nothing? Moreover, it must be remembered that, as the second violins generally, and the viola, violoncello, and double bass at all times, have been played by students (for lack of more efficient performers), it is easy to estimate what has thus been lost to the vocal choir; this refers only to Sunday music. But if I come to speak of the music for festivals, when music must be provided for both the principal churches at the same time, the lack of necessary performers will at once be still more striking, since then I have to give up such scholars as can play this or that instrument, and I am obliged to do altogether without their assistance.

Besides this it must not pass unnoticed that through the admissions hitherto granted to so many boys unskilled and ignorant in music the music has necessarily dwindled and fallen into decay. For it is easy to understand that a boy who knows nothing about music, who cannot even sing a second, can have no natural musical gifts, and consequently can never be of any use in music. And even those who bring some elementary knowledge to school with them, still are not of use so soon as is requisite and desirable. For time will not allow of their being duly trained for a year, or till they are skilled enough to be of use, but as soon as they are admitted they are placed in the choirs; and they ought to be at least sure of time and tune to be of any use in the service. Now, as every year some of those who have done something in music leave the school and their places are filled by others, many of whom are not immediately available, and most of them never of any use at all, it is easy to see that the choirs must by degrees diminish, and it is indeed notorious that the gentlemen, my predecessors Schell and Kuhnau, were obliged to have recourse to the assistance of

---

[108] These perquisites, "*Beneficia*," must have been derived from small savings, the residue of endowments, &c. No special funds were applied to the purpose or they must have been mentioned in the school and church accounts. In Kuhnau's time 50 gülden a year = 43 thlr. 18 ggr., were paid for church music out of the revenues of the church of St. Nicholas; Bach cannot refer to this sum for it was always regularly paid.

the students when they desired to perform complete and well-sounding music; which they were so far warranted in doing that several vocalists, a bass, a tenor, and an alto, as also instrumentalists, particularly two contra-bassists were favoured with salaries from a certain noble and learned councillor, and thereby were induced to strengthen the church music. Thus, since the present *status musices* is quite different to what it used to be formerly—the art being much advanced and taste marvellously changed, so that the old fashioned kind of music no longer sounds well in our ears, and competent assistance is thus rendered all the more necessary—such performers ought to be selected and appointed as may be able to satisfy the present musical taste and to undertake the new kinds of music, and at the same time be qualified to give satisfaction to the composer by their rendering of his work; and yet the few perquisites have been altogether withheld from the choir, though they ought to have been increased rather than diminished. And this, moreover, is passing strange, since it is expected of German musicians that they should be capable of performing extempore every kind of music, whether Italian or French, English or Polish, like some of those *virtuosi* before whom it may be placed and who have studied it for a long time beforehand, or even know it almost by heart, and who besides have such high salaries that their pains and diligence are well rewarded; but all this is not duly considered, and these (before spoken of) are left to take care of themselves, so that in the need of working for their bread many can never think of attaining proficiency, much less of distinguishing themselves. To give one instance of this statement we need only go to Dresden and see how the musicians are paid there by the king; it necessarily follows that all care as to maintenance is taken from these musicians; they are relieved of anxiety, and as, moreover, each person has to play but one instrument, it must be admirable and delightful to hear.

The conclusion is easy to arrive at: that in ceasing to receive the perquisites I am deprived of the power of getting the music into a better condition. Finally, I find myself compelled to depend on the number of the present scholars, to teach each one the methods of music, and then to leave it to more mature consideration whether under these circumstances the music can be carried on any longer, and whether its many and increasing deficiencies may be remedied.

It is, however, desirable to divide the whole into three classes. The really efficient boys are the following:—

1. Pezold, Lange, Stoll; *Prefecti* Frick, Krause, Kittler, Pohlreüter, Stein, Burckhard, Siegler, Nitzer, Reichhard, Krebs *major* and *minor*, Schöneman, Heder, and Dietel.

The motett singers, who must farther improve themselves in order to be efficient in the course of time for part-singing,[109] are the following:—

---

[109] Bach uses the word part-singing (figural music) in a restricted sense, which does not include the motett.

THE MATTER FINALLY DROPPED. 251

2. Jänigke, Ludewig *major* and *minor*, Meissner, Neücke *major* and *minor*, Hillmeyer, Steidel, Hesse, Haupt, Suppius, Segnitz, Thieme, Keller, Röder, Ossan, Berger, Lösch, Hauptmann, and Sachse.

Those of the last class are not musicians at all, and are named:

3. Bauer, Gross, Eberhard, Braune, Seyman, Tietze,[110] Hebenstreit, Wintzer, Osser, Leppert, Haussius, Feller, Crell, Zeymer, Guffer, Eichel, and Zwicker.

Total: seventeen available, twenty not yet available, and seventeen useless.

Leipzig, August 23, 1730.   Joh: Seb: Bach.
Director Musices.

Certainly this document is not couched in very dutiful, much less very submissive, language. We need only compare it with Kuhnau's memorial which contains representations to the same effect, to understand how new and strange such a mode of address must have seemed to the Town Council, and it may be easily supposed that Bach's demeanour was not likely to dispose them to take steps to remedy the abuses he had pointed out. They simply let the memorial fall through with a certain expression of opinion. After the sitting of August 2, the Burgomasters called Bach before them to communicate to him the admonition that has been mentioned, and at the same time asked him whether he would be disposed to give the general lessons again in the place of *Magister* Pezold. Whether this interview took place before or after August 25, the date of Bach's memorial, at any rate the subject of it also came under discussion. When, on August 25, Dr. Born again presided at a meeting of the Council to which he reported his conversation with Bach, it is most probable that he already had Bach's document in his hands, or, at any rate, he knew perfectly well that it was on its way, and what its contents were. What he proposed, however, was only as follows : That the Cantor had shown very little inclination to resume the school work in question, and it was therefore advisable to entrust the teaching to *Magister* Krugel. The Council agreed unanimously and the whole affair was dropped.

Nor was Bach's application reconsidered later, or at any rate not in a sense favourable to him. This is quite evident

---

[110] Identical, no doubt, with Joh. Tobias Dieze named above.

from his letter to Erdmann; and the account books of the Council place it beyond a doubt that, up to the year 1746, no real addition was made to the funds devoted to keeping up the music in the churches of St. Thomas and St. Nicholas. But if any one would fain try to infer from these unsatisfactory and discreditable proceedings that the town authorities felt a certain longing to make their mere material superiority felt by the genius that was striving to escape into light and air, that they intentionally oppressed him and so tried indirectly to hinder him in the full development of his artistic faculties, it would be altogether an error of judgment. The Council were for the time very wroth with Bach, and, in this frame of mind, allowed themselves to be carried away and to pass an odious measure—carried out, however, for only a short time, and they kept the purse strings tight whenever they thought outlay unnecessary; but they never deliberately hindered the progress of the music under Bach's direction. Indeed, when the authorities closed the year's accounts they could prove by figures that considerably more had been voted and paid for musical requirements in the last few years than in the previous years. A complete restoration of the great organ in St. Thomas', costing 390 thalers, had been effected in 1720 and 1721; in 1724 and 1725 a thorough renovation of the "very dilapidated and injured" organ of St. Nicholas was effected at a cost of 600 thalers. In 1725 forty thalers were again expended in repairing the organ of St. Thomas, and two years later fifteen thalers more, and in 1735, when Bach's relations with the Council were at the highest strain, a sum of fifty thalers was voted for making the *Rückpositiv* of the same organ an independent and separable instrument. In 1729, two new and "fine" violins, a "similar" viola and a violoncello were bought for church use at a cost of thirty-six thalers, and in the same year Bach was enabled to purchase the *Florilegium Portense* by Bodenschatz, at the price of twelve thalers,[111] for the use of the St. Thomas scholars

---

[111] In 1736 another copy of this collection of motetts was acquired for the special use of the church of St. Nicholas; this cost ten thalers, and another in the following year for St. Thomas' at the price of eight thalers. It would thus seem that one copy was insufficient.

in church music. This outlay to us seems but small, even at the higher value of money at that time. But it was not so when compared with the usual average of expenditure in matters musical, though, on the other hand, it was not large in proportion to what such a musician as Bach was justified in expecting. The fault of the Council lay in their ignorance of what a genius and power they possessed in him, and that double and treble the opportunities and materials they afforded would not have sufficed for his full and free development.

After this last experiment it may well be supposed that Bach was deeply embittered against his superiors in authority, and he seriously considered the question of quitting Leipzig. If at this moment any advantageous prospect had offered itself he very certainly would have followed it up, and his official existence as Cantor of St. Thomas would have closed at the end of his seven years' tenure, with small credit to the town of Leipzig. But this was not the case, otherwise it would not have occurred to Bach to apply to Erdmann, the friend of his youth, who had meanwhile been appointed the agent in Dantzig for the Emperor of Russia, to procure him an appointment there. What could Bach want to be at in Dantzig? It will be seen that it was but a clutch in the empty air. However, we owe to this idea the most interesting letter that exists in the master's hand:—

EXCELLENT AND RESPECTED SIR,

Your Excellency will forgive an old and faithful servant for taking the liberty of troubling you with this letter. Nearly four years have now elapsed since your Excellency did me the pleasure of kindly answering my last sent to you;[112] though, as I remember, you were graciously pleased to desire that I should give you some news of my vicissitudes in life, and I hereby proceed to obey you. From my youth up my history has been well-known to you, until the change which led me to Cöthen as Capellmeister. There lived there a gracious Prince, who both loved and understood music, and I thought there to spend my life and end my days. As it turned out, however, his Serene Highness married a Princess of Berenburg, and then it appeared as though the musical dispositions of the said Prince had grown somewhat lukewarm,

---

[112] Erdmann had come from Dantzig in 1725 to Sax Gotha, his native place, to arrange some matters of business; but, as we gather from this letter, he had not seen Bach on this occasion.

while at the same time the new Princess served as an amusement to him, and it pleased God that I should be called to be Director Musices and Cantor to the Thomasschule in this place. At first it did not altogether please me to become a Cantor from having been a Capellmeister, and for this reason I deferred my decision for a quarter of a year; however, the position was described to me in such favourable terms that finally (and especially as my sons seemed inclined to study here) I ventured upon it, in the Name of the Most High; I came to Leipzig, passed my examination, and then made the move. And here, by God's pleasure, I remain to this day. But now, since I find (I.) that this appointment is by no means so advantageous as it was described to me; (II.) that many fees incidental to it are now stopped; (III.) that the town is very dear to live in; (IV.) and that the authorities are very strange folks, with small love for music, so that I live under almost constant vexation, jealousy, and persecution, I feel compelled to seek, with God's assistance, my fortune elsewhere. If your Excellency should know of, or be able to find, a suitable appointment in your town for your old and faithful servant, I humbly crave you to give me the benefit of your favourable recommendation. Nothing shall be wanting on my part to give satisfaction (and justify) your favourable recommendation and intercession, and to use my best diligence. My present position secures me about 700 thlrs., and when there are rather more deaths than usual the fees increase in proportion; but it is a healthy air, so it happens, on the contrary—as in the past year—that I lost above 100 thalers of the usual funeral fees. In Thuringia I can do more with 400 thalers than here with twice as many hundred, by reason of the excessive cost of living. I must now make some small mention of my domestic circumstances. I am now married for the second time, and my first wife died in Cöthen. Of my first marriage, three sons and a daughter are living, which your Excellency saw in Weimar, as you may be graciously pleased to remember. Of my second marriage, one son and two daughters are living. My eldest is *Studiosus Juris*, the other two are one in the first and the other in the second class, and my eldest daughter is still unmarried. The children of my second marriage are still little, the eldest, a boy, being six years old. They are all born musicians, and I can assure you that I can already form a concert, both vocal and instrumental, of my own family, particularly as my present wife sings a very clear soprano, and my eldest daughter joins in bravely. I should almost overstep the bounds of politeness by troubling your Excellency any further, so I hasten to conclude with most devoted respects, and remain your Excellency's life long and most obedient and humble servant,

LEIPZIG, *October* 28, 1730.[113]   JOH: SEBAST: BACH.

---

[113] The history of this letter is given in the Preface; the address is wanting. Since quoting certain passages of this letter on p. 147, a friend has put me in possession of a photograph copy of the whole document.

This letter had not the hoped for result; Bach remained in Leipzig, and, as we may safely conclude, not altogether unwillingly in the end. If the conditions of his life had been really unendurable, a man of such energetic character would never have rested till he had set himself free; and, enjoying such fame as he did, it certainly could have been a matter of no great difficulty. Though Bach dwells with some insistance on the high price of the necessaries of life in Leipzig, and complains of the insufficiency and uncertainty of his income—and the eloquence with which he alludes, in his memorial to the Council, to the financial advantages of the musicians in Dresden is highly significant—some immediate pressure of circumstances and his general sense of discomfort evidently made him take too black a view. We may counterbalance these complaints merely by the facts that at Bach's death his private concerns were left in good order, that his household was established on a comfortable footing, and that he even left a small sum of money. When Kuhnau even could say that, with his fixed salary, a musician like himself—who was always receiving visits from his fellow artists, who often had to treat the students who sang with the choir, and had besides a large household to keep up—could not make much show in the world, how much more must this have been the case with Bach, whose family was so much larger and whose house no musician on his travels ever passed by? That he should, under these circumstances, have laid anything by, speaks volumes.

Then he speaks of suffering from the jealousy of his fellow artists, we think at once of Görner. But where could such a man as Bach have gone without finding others jealous of him? The musical resources of the place were no doubt meagre, but this he must have known from the first, and in some respects they had actually increased and improved. And, finally, if the attitude of the authorities was not very encouraging—nay, at times, oppressive and offensive—Bach himself must have seen that the matters at issue were not of so crucial a nature as to lead to the question "To be, or not to be?" So by degrees the black clouds rolled over and the sky cleared again, and an occurrence which took place in that same year must have had a considerable influence over his affairs, an event which may be said to have led to the happiest period of Bach's life in Leipzig.

In September, Johann Matthias Gesner, the newly elected Rector of the Thomasschule, came to take up his appointment. Gesner was born in 1691, at Roth, near Nuremberg, he had studied at Jena, and in 1715 he became Conrecter of the Academy at Weimar.[114] Here he remained till 1729, and for seven years had been Principal of the Ducal Library. He then was placed at the head of the Academy at Anspach, but at the end of a year he gave up his appointment in order to go to Leipzig. Here he only remained till 1734; he had pledged himself not to hold a professorship in the University at the same time as his office as Rector (or Warden) of the school. But as this had always been the custom with the former Rectors, Gesner's position was much injured by this condition of remaining outside the pale of the University, and besides this, his distinguished abilities qualified him for the highest academical honours. In 1734 he therefore obeyed a call to the University of Göttingen, where he died, in 1761, after many years of brilliant and successful labours.[115] All that Gesner did for classical learning in Germany—how he revived the study of Greek, and was the first to contemplate the works of the ancients from a higher standpoint as to their purport and form, giving them life and value in the mental training of his pupils—is familiar to every student of philology. But even more important to the purpose of the present work was the talent he immediately displayed as a master of schoolboys. In him the Council had found the very man they needed, if the regulations drawn up for the resuscitation of the institution were to be anything but a dead letter. Under Gesner's mastership a new period dawned on the fallen fortunes of the school. Together with a vast store of practical learning he possessed in an eminent

---

[114] Vol. I., p. 390.

[115] *J. Aug. Ernesti* narratio de Jo. Matthia Gesnero ad Davidem Ruhnkenium. Addita opusculis oratoriis, 1762, p. 306. H. Sauppe, Johann Matthias Gesner, Weimar, 1856.

degree the power of governing; resolute firmness was combined in his character with humanity and gentleness; in his conduct to the Council he was uniformly polite, but decided. It was inevitable that he should soon win their high esteem and complete confidence; indeed, he was from the first treated by the authorities with a distinction which proves that they were not so absolutely devoid of all sense of intellectual superiority as we might perhaps infer from Bach's experience. We may here quote a fact recorded by his successor, Johann August Ernesti. Gesner, whose health was feeble, and who during his residence in Leipzig had two severe illnesses, at first had a residence assigned to him at some distance from the school, which was being rebuilt; and to relieve him of the inconvenience of the daily walk to the schools he was always carried thither in a chair, and back again to his house when lessons were over, at the cost of the Council. He was also so far relieved of the exercise of his functions as to be released from the duty of inspecting the school, which the Rector usually was required to do in weekly rotation with the three upper masters, and which was now undertaken by the *Quartus, Magister* Winkler.

Gesner introduced a better mutual understanding among his colleagues, and set them an admirable example in the fulfilment of his duties. He secured the affection of his scholars by his new and intelligent methods of instruction, by his unwearying interest in their progress and welfare, and by the determination with which he enforced discipline and morality. He carried out the new regulations with exactitude, endeavouring to amplify and emphasise certain provisions and to modify the existing condition of things in harmony with their purport. Thus, for instance, under the existing system the Latin prayers, morning and evening, were replaced by German prayers; Gesner, who attached the greatest importance to the practice of speaking Latin, proposed that the Latin services should be restored, because otherwise "rudeness—*i.e.*, want of culture—and ignorance would once more get the upper hand." Again, in 1733, by his suggestion, rules in German, and in conformity with the new school regulations, were printed for the scholars.[116] One, who was a pupil at the Thomasschule under Gesner, describes his person and proceedings in the following pleasing manner:—"In discipline he guided himself very precisely by the laws of the school, at that time just revised; he was cautious in punishing, and in order to avoid undue severity would let a few days pass after the delinquency was committed. Then in the evening, after prayers, and when the motett was sung, he would come among the scholars, call up the criminal, point out with impressive gravity the impropriety and sinfulness of his fault, and then pronounce, besides the admonition, his verdict as to the punishment. This way of delivering judgment had a wonderful effect; all the more because he was universally respected. Every week all the scholars, even the outside day-boys, had to give in their diaries, and when they were returned to them they found ample evidence that nothing had escaped him. Ernesti also did this at first, but only, as before, with the foundation boys, and it was soon altogether given up, as Ernesti commonly forgot to give the diaries back. Gesner was in other respects very affable and affectionate in his intercourse with the boys, and would look in upon them even during the singing lessons, with which the Rectors did not usually trouble themselves, and would listen with pleasure to the practising of a piece of church music. If he found any boy at work in his room at anything which, though not part of his school task, was useful in itself—for instance, if he were writing from a copy—he did not fall upon him with a storm of indignation, but, if he saw a real talent for caligraphy, would recommend him to further study and practice, because, said he, the state had need of every variety of talent and skill. And to all he would preach, when opportunity offered, 'Always do something that is of some definite

---

[116] E. E. Hochweisen Raths der Stadt Leipzig Gesetze der Schule zu S. THOMAE. Leipzig, druckts Bernhard Christoph Breitkopf, 1733, pp. 4, 39.

use, and which you can turn to account in your calling in life.' "[117]

Bach and Gesner had already known each other in Weimar; the acquaintance was now renewed and soon grew to be a hearty friendship between the two colleagues. Our authority has just told us how much interest Gesner took in Bach's musical efforts, how much he enjoyed his performances, even visiting him in lesson hours. He exerted himself also in other ways to give the music in the school a helping hand so far as lay in his power, and when, in 1732, Bach wished to acquire a MS. collection of motetts and responses for the choir of St. Thomas, he himself applied to the Council for the necessary sum, which was readily voted. He had a strong feeling for music and could do full justice to Bach's greatness; years after he had not forgotten the overwhelming impression made on him by that grand musician. We find eloquent witness to this in the note which he makes in his edition, published in 1738, of the *Institutiones Oratoriæ* of Marcus Fabius Quinctilianus to a passage (I., 12, 3) where Quinctilian is speaking of the capacity possessed by man of comprehending and doing several things at once; adducing as an example a player on the lyre, who can at the same time utter both words and tones and besides play on the instrument and mark time with his foot. To which Gesner remarks, " All these, my Fabius, you would deem very trivial could you but rise from the dead and see Bach (whom I mention because not long ago he was my colleague in the Thomasschule at Leipzig); how he with both hands, and using all his fingers, plays on a key board which seems to consist of many lyres in one, and may be called the instrument of instruments, of which the innumerable pipes are made to sound by means of bellows; and how he, going one way with his hands, and another way with the

---

[117] Historia Scholarum Lipsiensium collecta a Joh. Frid. Köhlero, pastore Tauchensi, 1776 seqq., p. 160. MS. preserved in the Royal Public Library at Berlin. The author adds, " These remarks are from the words of a scholar of Gesner, who spoke often, and always with enthusiasm, of his master's great qualities."

utmost celerity with his feet, elicits by his unaided skill many of the most various passages, which, however, uniting produce as it were hosts of harmonious sounds; I say, could you only see him, how he achieves what a number of your lyre-players and six hundred flute-players could never achieve, not as one who may sing to the lyre, and so perform his part, but by presiding over thirty or forty performers all at once, recalling this one by a nod, another by a stamp of the foot, another with a warning finger, keeping time and tune; and while high tones are given out by some, deep tones by others, and notes between them by others, this one man, standing alone in the midst of the loud sounds, having the hardest task of all, can discern at every moment if any one goes astray, and can keep all the musicians in order, restore any waverer to certainty and prevent him from going wrong; rhythm is in his every limb, he takes in all the harmonies by his subtle ear, as it were uttering all the different parts through the medium of his own mouth. Great admirer as I am of antiquity in other respects, I yet deem this Bach of mine, and whosoever there may chance to be that resembles him, to comprise in himself many Orpheuses and twenty Arions."[118]

With such feelings of admiration and liking as Gesner had for Bach he must have made it his concern to lighten, as far as possible, his colleague's sense of discomfort, and also to bring him into pleasanter relations with the Council. And we can see that he actually attempted both, from two memorials which he addressed to the Council and which—as the petitioner was Gesner—were no doubt successful. Although Bach had been for the most part represented by a deputy in giving the Latin lessons, he was not altogether released from them, and at any rate had to hold himself in readiness in case of need. Gesner now proposed that the Council should entrust the Cantor with the general supervision of the school during week days, which he had hitherto undertaken only on Fridays, since it was a function which most naturally allied

---

[118] The first person to draw attention to this passage in Gesner's commentary was Constantin Bellermann in the *Parnassus Musarum*, p. 41 (see Vol. I., p. 801); it was subsequently pointed out again by Joh. Adam Hiller.

itself with his other duties. In exchange for this, and for a
few hours more of singing lessons, he was to be exonerated
altogether from general teaching of any kind. This suggestion was probably made by Gesner on the occasion of *Magister*
Pezold's death, May 30, 1731, since on the arrival of a new
functionary it was easy to modify the position of affairs; in
the beginning of 1732 Joh. August Ernesti was appointed
Conrector. While the place was vacant the Conrector's fees
amounted to a sum of 120 gülden 10 ggr. 5 pf., of which
Pezold's heirs received forty gülden, and Gesner begged the
Council to divide the remainder equally among the Rector,
Cantor, *Tertius*, and *Quartus*. Bach had not indeed ever given
lessons in the place of the Conrector, but the Inspector's work
had of course come round in more frequent rotation. Hence
Gesner wrote: "The duties of teaching, it is true, have
occasioned no trouble to Herr Bach. Still he hopes, this
time, to have his equal share in the division, because on the
last occasion (when Ernesti died) he was altogether passed
over." So here we may suppose the great "sequestration"
question was brought to a final issue.

Bach, however, was not so easy to conciliate. The same
obstinacy which supported him in the pursuit of his aims
characterised any antipathy he had once conceived. When,
in 1733, he dedicated to the King and Elector the two first
sections of his B minor Mass he frankly says, in the letter
which accompanied it, that his object in this dedication was
to obtain some court appointment. In his present position
he had no doubt suffered from one and another unmerited
insult and occasionally from a diminution in his fees;[119] all
this he would then be relieved from. But although the
appointment he desired was not for the present accorded
to him, no farther serious differences arose between Bach
and the Council. They had learnt to know each other
and henceforth sought to accommodate matters. Bach's
position in Leipzig could never be more favourable than it

---

[119] "Accidentien," which commonly means fees, in this letter had no doubt a
more general meaning, including all moneys accruing from exceptional sources.
The fees for funerals and weddings, and the Cantor's share of the money collected in the processional singing, were secured to him by law.

was now. He had the command of the most famous Musical Society in the town, of which he could also avail himself for the church services, and some excellent pupils, as Johann Ludwig Krebs, the son of his old Weimar pupil Tobias Krebs; and his own three eldest sons—who already deserved to be considered important, or at any rate highly promising artists—were of efficient use in his work. The organist of the New Church was a musician devoted to his service, and at St. Nicholas' he procured the appointment, in 1730, of a former pupil, Johann Schneider, who before that had been Kammermusicus in Weimar.[120]

It is a sufficient proof of Bach's consequence and influence that two of his former pupils should be among the candidates for the vacant post of organist. Johann Caspar Vogler, an admirable musician and court organist at Weimar, competed with Schneider, but the decision arrived at was that Vogler misled the people by playing too fast, so Schneider was preferred. He was at the same time a good violin player and could be of use to Bach in other ways on various occasions. He held his post for years after Bach's death.[121] And now, last and best of all, he had the advantage of Gesner's friendship, and the support of his authority, which was favourable to music. That this was in fact a source of happiness to him we may take as a matter of course, though there is no direct evidence of it. An altered version of the cantata composed in 1726, for the birthday of the Princess of Anhalt-Cöthen, may probably be attributed to Gesner, though no proof can be adduced; but we may regard as evidence of their hearty co-operation the solemn ceremonial with which the enlarged and improved school buildings were opened on June 5, 1732. In the Latin speech made by Gesner on this occasion he did not fail to mention, with a few words of warm approval, the care given to music in the institution, and Bach conducted a

---

[120] Walther gives 1729 as the date of Schneider's appointment. But this is an error; the place was no doubt vacant in 1729 but it was not given to Schneider before August 1, 1730.

[121] In 1766 Schneider applied for an assistant which was granted, as we learn from documents concerning the organists of the church of St. Nicholas.

cantata of which the words were written by his colleague Winkler. The music to these verses—and very bad they are—is not known to exist.[122]

IV.

THE PLAN AND ARRANGEMENT OF THE CHURCH SERVICES IN LEIPZIG.—MUSIC USED IN IT; THE ORGANS AND BACH'S TREATMENT OF ACCOMPANIMENTS.—DIFFICULTIES OF PITCH AND TUNE.

THE arrangement of the Lutheran service throughout Electoral Saxony was regulated by an act issued by Duke Heinrich, in 1540. This decree aimed not merely at establishing a uniform order of divine worship throughout the Duchy of Saxony and its dependencies at that time, but also at laying down a line of limitation within which different parishes and congregations might regulate their respective services according to their needs and wishes. This was in accordance with Luther's own views, as expressly stated in his treatise on the German Mass and the ordering of divine service (Wittenberg, 1526), to the effect that each one should, in all Christian freedom and in accordance with his own pleasure, use only such customs of those he laid down "how, when, where, and so long as circumstances suited and permitted." And this was actually continued even after the church ordinances of the Elector August, in 1580, had expressly introduced a greater uniformity in the divine services in the different towns and villages of his dominions. Thus there were certain usages in the church services at Leipzig which gave them a quite peculiar character. The Lutheran form of worship was a modification of the Roman Catholic Mass. In many towns and districts this was more rapidly and completely abandoned;

---

[122] They are given in App. B., x., 4, of the German edition of this work. In the invitation circular, sent out the day before, Gesner expresses a wish that the scholars may be "*juvenes ad humanitatem, quæ litteris censetur et musice, probe instituti atque exculti.*"

in Leipzig, for a long time—indeed even during Bach's residence there—it retained a close resemblance to it. This was visible partly in the external ceremonial and usage, as, for instance, in the use of a little bell at the consecration of the bread and wine at the Lord's Supper, in the retention of ceremonial robes for the officiating ministers and surplices for the singing boys, and partly in the continued use of the original form of certain portions of the Catholic ritual, and, as connected with this, of a more extended use of Latin. These traditions prevailed very extensively even in the more accessory details of religious life. The bequests which, even so late as Bach's time, were not unfrequently made to the school foundation for the public performance of certain chorales on the anniversary of the testator's death, plainly indicate a Roman origin.[123] Latin hymns and responses were sung even in the processional singing, as well as German hymns and songs, and there was no lack of timid souls who would gladly have seen the great resemblance of the Protestant services at Leipzig with that of the Roman church radically altered. Indeed, in 1702, the Council thought it ought to take the matter in hand, and on February 13 addressed a petition to the King-Elector, to request that devout and approved hymns, prayers, and texts in German might be for the future introduced in all the churches of Saxony. At the time of the Reformation matters had been otherwise. The authorities had been anxious not to cause a collision between the clergy and the simpler folks and new converts by any too rapid and conspicuous changes in the church ceremonial; they had hoped, indeed, to attract greater numbers to the Lutheran evangelical church by this moderation, but the pressing danger now was that many persons, not seeing any very great difference in the services, might be misled into a wrong interpretation of the Lutheran doctrines.

---

[123] To give a single instance, from November 19, 1736, the chorale " O Jesu Christ mein's Lebens Licht," was to be sung every year in the New Church in memory of Dame Anna Elisabeth Seeber, who bequeathed, in consideration, fifteen thalers a year for distribution. See the account book of the New Church for that year.

"Spiritual songs" in German were far better adapted to rouse feelings of devotion than the old Latin responses and versicles, which were, for the most part, not understood by the people. However, the petition was not adequately supported. The townspeople of Leipzig knew full well that these forms of divine service were made interesting by their peculiarity, and even the clerical body were of opinion that the Latin hymns still pleased some people, and particularly strangers. It had, therefore, but small results; only, with regard to the school perambulations, it was determined—but not before 1711—that instead of the response "Sint lumbi vestri circumcincti" (Luke xiii. 35), certain German hymns should be sung referring to the Day of Judgment—namely, "Es ist gewisslich an der Zeit," "Wachet auf, ruft uns die Stimme," and "O Ewigkeit du Donnerwort."

An exact comprehension of the order of divine service in the churches of St. Nicholas and St. Thomas is of the greatest importance to an estimate of Bach's church music, for it is only perfectly intelligible when regarded in all its bearings and relations to the nature and scheme of these services. We have a note in Bach's own hand of the order of the service on the first Sunday in Advent when he was in Leipzig, in 1714.[124] As it is interesting on this account it may find a place here, although it will be seen from what follows that it is not altogether exact and complete:—

The order of divine service in Leipzig on the first Sunday in Advent, in the morning: 1. The Prelude. 2. Motett. 3. Prelude to the *Kyrie*, which is accompanied throughout. 4. Intoning at the altar. 5. The Epistle read. 6. The Litany sung. 7. Prelude to the chorale. 8. The Gospel read. 9. The Prelude to the principal music. 10. The Creed sung. 11. The sermon. 12. After the sermon a few verses of a hymn are commonly sung. 13. *Verba institutionis*. 14. Preludes and chorales alternately until the celebration of the Communion.

The services were different on ordinary Sundays, on Holy Days, and in Passion week. We will begin with ordinary Sundays. The series of services, during which the houses

---

[124] See Vol. I., p. 519.

were closed all day and all public transactions were prohibited, began at 5.30 in the morning with matins in the church of St. Nicholas. There was a special choral institute attached to this church, under its own Cantor, besides the choir from St. Thomas', and this was supported by a municipal stipend. It consisted of students and was governed by strict regulations of its own.[125] At the same time the Cantor of St. Thomas exercised a certain supervision, since the musical property was entrusted to his care.[126] The choir sung the Psalm "*Venite, exultemus Domino*," then a Psalm, a *Responsorium;* afterwards one of them read the Gospel for the day from the desk in Latin, followed by another who read it in German. Then came the *Te Deum*, started by the organist, and played and sung verse by verse antiphonally between him and the choir. The service ended with "*Da pacem*," or some other versicle adapted to the season, and last of all "*Benedicamus Domino.*"

Morning service began precisely at seven in both churches. An organ prelude introduced a motett suited to the Gospel for the day, and usually sung in Latin. In Lent, or in times of general mourning, when the organ was not played, the motett was omitted, and the service began with the Song of Zacharias "*Benedictus Dominus*" (Luke i. 68), and "*Vivo ego.*"[127] It was not till then that the Introit was performed, with which, according to Duke Heinrich's *Diary*, the service was said to begin; after that the Kyrie. This was sung alternately in Latin and German in the two churches; when it was in German the version beginning "Kyrie Gott Vater in Ewigkeit" was used. Sometimes the Kyrie was

---

[125] These rules were drawn up in 1767, when the institute probably came into existence. They were re-written in 1678 by Gottfried Vopelius, well known as the compiler of the New Leipzig Hymn Book of 1682, and Cantor at the time to St. Nicholas. The MS. exists in the library of the church. The Cantor in Bach's time was *Magister* Johann Hieronymus Homilius.

[126] In 1767, when the hymn for Palm Sunday, "*Gloria, laus et honor tibi,*" was to be sung, Doles, the Cantor, confessed that he had lost the music.

[127] The disuse of the organ in Lent was so far limited in the year 1780, that from that time the Communion hymns at least might be sung with an organ accompaniment; see the accounts of the Church of St. Nicholas, from Candlemas to Crucis, 1780.

in a concerted form, probably during the fasts—that is to say, on the first Sunday in Advent, which agrees with Bach's statement quoted above; since on the last three Sundays in Advent, during Lent, and on the Ember days no concerted music was used. Another day on which the Kyrie was given in a concerted form will presently be mentioned in enumerating the festivals. Next the Lord's Prayer was said by the minister, kneeling at the altar, and the sacramental cup was placed on the altar by the sacristan; one of the deacons intoned the "*Gloria in excelsis*," to which either the choir responded with "*Et in terra pax hominibus*," or the congregation sang the hymn "Allein Gott in der Höh"— "To God alone on high." Then followed the blessing in Latin, "*Dominus vobiscum*," with the answer of the choir, "*Et cum spiritu tuo*." The Collect was now read, likewise in Latin—that is to say, sung to the proper tone; and after the chanting of the Epistle for the day from the reading-desk—in Advent and Lent the Litany was sung in such a way as that the congregation not only repeated the responses, but joined in the petitions—the Litany was chanted in St. Thomas' by four boys specially appointed and called "*Altarists*," the choir responding.[128] Then came a congregational hymn suited to the Gospel, while on other Sundays the congregational hymn followed immediately after the Epistle, and the Litany was omitted. The Gospel was then intoned from the desk, and after it the minister whose weekly turn it was intoned the "*Credo*" before the altar, and on the three last Sundays in Advent and in Lent, as also on the Festivals of the Apostles, the whole of the Nicene Creed was sung in Latin by the choir. On other Sundays the prelude to the principal piece of music came immediately after the minister's intoning, and then the music itself.[129] When it was ended the Creed was sung in

---

[128] This is shown by the regulations of the Leipzig Consistory of Jan. 31, 1810, which abrogates the former custom. In the Ephoral Archives at Leipzig.

[129] In Bach's list of the order we find under No. 8, after the words "the Gospel read," the following addition scratched through—" and *Credo* intoned." Hence Bach must, by mistake at first, have noted down the order of the service as it was told him for the first Sunday in Advent.

German by the whole congregation. (Wir glauben all' an einen Gott.) As has been said, on ordinary Sundays part-singing was performed in the two principal churches alternately. Thus when it was not performed the German Creed was sung immediately after the intoning of the *Credo* in Latin.[130] Then came the sermon on the Gospel, which was once more interrupted before the Gospel itself was read by the congregational hymn, " Herr Jesu Christ dich zu uns wend." The sermon lasted an hour, from eight till nine, since the whole service was throughout adapted to an exact division of time, and to keep it strictly to these limits a sand hour-glass was used.[131] After the sermon the general confession was recited with the usual church prayers, and after the customary ascription of praise from the pulpit, and the Lord's Prayer, the blessing of St. Paul, " The peace of God which passeth all understanding," &c., closed the service. While the preacher was descending from the pulpit a few verses were sung of some suitable hymn.

The Communion celebration formed the principal part of the service. It is not quite clear whether on ordinary Sundays it was preceded by the introductory words of Luther's paraphrase of the Lord's Prayer and the admonition which follows; it may however be considered probable from Duke Heinrich's decree. During the communion German hymns appropriate to the service were sung; but before these, even on ordinary Sundays and holidays, another harmonised composition was sung, as we learn from the list made by Bach and from a note written on the second portion of the *continuo* part for his Trinity Cantata " Höchst erwünschtes Freudenfest "—" Feast of joy so long desired." The end of the whole service— without a final collect as it would seem—consisted in the benediction " God be gracious and merciful unto us and give us His blessing." The length of the Morning service varied with

---

[130] It was thus the custom took its rise which Rochlitz speaks of (Für Freunde der Tonkunst, S. 4, 2nd ed., note to p. 278) as a hardly credible monstrosity: that the Latin *Credo* was sung in a " lively " manner just before the German creed.

[131] " 12. ggr. for repairing the sand-clock on the organ." Account of St. Thomas, 1739-1740, p. 62.

the number of communicants; often it was not over till eleven o'clock, thus lasting four hours. This of itself accounts for its beginning so early, and the consequence was that, in winter, the music before the sermon was for the most part performed by candlelight.[132]

Mid-day service, which began at about a quarter before noon, was very simple in character. It consisted of a sermon with two congregational hymns to precede and one to follow it. The choir were not employed in it.

At about a quarter past one Vespers began with a motett, followed by a congregational hymn. One of the deacons intoned a psalm from the desk, the Lord's Prayer, and a collect. Then came another hymn and the sermon, which treated of the epistle, or in Advent of the Catechism, and in Lent of the history of the Passion. After the sermon the *Magnificat* was sung in German to the four-part melody by Joh. Hermann Schein,[133] and, after a collect recited at the altar and the Blessing, the service concluded with the hymn "Nun danket alle Gott."

As regarded the Holy days, three great Festivals were specially celebrated, each being honoured with a particularly full performance of the service for three days running. Matins in St. Nicholas' remained unchanged, excepting that they began at five instead of at half-past. Morning service and Vespers were begun on each day by a hymn sung by the choir; these hymns were: at Christmas, *Puer natus in Bethlehem*; at Easter, "Heut triumphiret Gottes Sohn"; at Whitsuntide, *Spiritus sancti gratia*. After this, at Morning service, came the organ prelude and motett. The collect, which preceded the epistle, seems to have had reference to some introductory verse from the Bible, appropriate to the day.[134] After the hymn that followed the sermon the complete Latin pre-

---

[132] When part-music was performed the Cantor and the Conrector had to provide the lighting for the organ choir. The Cantor received for this purpose eleven thlrs. fifteen ggr. per annum out of the revenues of St. Thomas, and for St. Nicholas', where the choir was much smaller, seven thlrs. twenty-one ggr.

[133] See Vopelius, Neu Leipziger Gesangbuch, p. 440.

[134] This I infer from Vopelius, who has verses appropriate to every festival in his hymn book.

fation was sung as introductory to the Communion, and then the Sanctus was sung in parts; besides which during the Communion either a motett or a concerted piece was performed. The whole service closed with a festal hymn sung by the congregation after the Benediction.

At Vespers the collect before the sermon was omitted, and after the sermon the *Magnificat* was sung in Latin and in parts. Church cantatas were performed at Morning service and at Vespers in both churches on the first two days of each festival; at Vespers one took the place of the omitted offertory collect. Still only two cantatas were performed on each day, since the one which had been sung in the morning in one church served for the other in the afternoon, and *vice versa*. The church of which the Superintendent for the time being was the minister—in Bach's time therefore that of St. Nicholas—had the precedence—that is to say, the first and best choir sang there in the morning of the first Holy day under Bach's own direction, while the *principal music*, as it was called, was performed in the morning of the second day at St. Thomas'; on the third day music was given in only one of the churches.[135]

The festivals on which this double performance was given were Christmas, Easter, and Whitsuntide, also at the New Year, Epiphany, Ascension, and Trinity, and the Annunciation of the Blessed Virgin. The hymns specially appointed were, for the New Year and Epiphany, *Puer natus in Bethlehem;* for Ascension, "Heut triumphiret Gottes Sohn"; for Trinity, "*Spiritus sancti gratia*," the same as at Christmas, Easter, and Whitsuntide. At the New Year and Epiphany the mid-day sermon was preached in only one of the churches.

---

[135] See, "Texte zur Leipziger Kirchen-*Music*, auff die heiligen Oster-Feyertage, ingleichen auff Jubilate, Cantate, und das Fest der Himmelfarth Christi, *Anno* 1711. LEIPZIG, gedruckt bey Jmmanuel Tietzen," und "Texte zur Leipziger Kirchen-*Music*, Auf die Heiligen Weyhnachts-Feyertage, und den Sonntag darauf, 1720. Ingleichen auf das Fest Der Beschneidung Christi, den drauf folgenden Sonntag, Das Fest der Offenbahrung, und den Sonntag darauf, des 1721 sten Jahres. Leipzig, gedruckt bey Jmmanuel Tietzen." These are both to be found in the Library of the Historical Society (Verein für die Geschichte) at Leipzig.

Of the three festivals in honour of the Virgin, the Purification and the Annunciation had a specially ecclesiastical character; on the first, both morning and afternoon service were begun and ended alike with the hymn *Ex legis observantia*. The Festival of the Annunciation was held on March 25; but if this date fell on Maunday Thursday, Good Friday, or Easter Day, it was kept on Palm Sunday, and, in spite of its being in Lent, part-singing and the organ were allowed.

The Reformation Festival was kept only as a half-holiday and always on October 31; but if this fell on a Saturday or a Monday it was celebrated on the previous or the following Sunday.[136] The Morning service began with an organ piece and a motett; the Kyrie, which followed the Introit, was set in a concerted form. The Epistle—II. Thess. ii. 3-8—was intoned, and then for the Gospel, Rev. xiv. 6-8, was read. After the sermon the choir sang the *Te Deum* to the accompaniment of trumpets and drums, and, as a close to the service, the congregation sang "Nun danket alle Gott." The services for St. John's and St. Michael's days were essentially the same. All these festivals had this in common, that a festal hymn was sung in the middle of the sermon; apparently the collect before the Epistle was on all of these days founded on a suitable text, and that for the Reformation Festival seems to have been the Latin one *pro pace*, "for peace."[137]

The order of service for Passion week had this feature in common with the whole of Lent, that neither the organ nor concerted music were employed. This rule was not without exceptions, and it has already been said that it was broken through if the Annunciation fell on or was transferred to Palm Sunday. When Palm Sunday was celebrated as such the order of service was as follows: An organ

---

[136] Gerber, Historie der Kirchen-*CEREMONIEN* in Sachsen. Dresden und Leipzig, 1732, V. 4, p. 227.

[137] At least it was so in the year 1755, when, moreover, the order of the Reformation Day services underwent several changes. Compare also Schöberlein, Schatz des liturgischen Chor-und Gemeindegesangs, Part I., Göttingen, Vandenhoeck und Ruprecht, 1865, p. 487.

prelude led at once into the *Kyrie*. After the intonation of the *Gloria*, the hymn "Allein Gott in der Höh sei Ehr" was sung by the congregation. Instead of the Gospel, the Archdeacon chanted the history of the Passion, in German, according to St. Matthew, before the high altar, with the assistance of a choir of scholars. This was followed by the motett, *Ecce quomodo moritur justus*, by Gallus.[138] The celebration itself was not preceded by the whole of the Præfation, but as an introduction to the Lord's prayer, the so-called *praefatio orationis dominicae*[139] was read, as it seems, in Latin, and this was usual also on the preceding Sundays from the third Sunday in Lent onwards. Then a motett was again sung, and hymns on the Passion or the Communion, by the congregation. At the early service, which began at about half-past five,[140] it may be noticed that the choristers sang the hymn *Gloria, laus et honor tibi sit, Rex Christe*. Also that on Maunday Thursday the service began with an organ prelude. The passage from Phil. ii. 8, *Humiliavit semet ipsum*, was sung as an Introit after the Epistle, the hymn *Crux fidelis inter omnes*, and, during the Communion, the Motett, *Jesus Christus Dominus noster*. The same Introit was sung on Good Friday, but the organ was altogether silent. For the Epistle and Gospel, Psalm xxii. and Isaiah liii. were used interchangeably year and year about. Instead of the chanted passage from the Gospel, however, the Passion according to St. John was sung, as that according to St. Matthew had been on Palm Sunday. The accounts of the Passion in St. Mark and St. Luke were taken no cognisance of in the Liturgy. This part of the service was followed, not by the motett *Ecce quomodo*, but by the congregational hymn "O Traurigkeit! o Herzeleid!" The celebration was again

---

[138] Vopelius, p. 263 ff. Johann Adam Hiller, in the preface to his Vierstimmige lateinische und deutsche Chorgesänge—"Four-part choral hymns in Latin and German"—(Part I., Leipzig, 1791), says that this motett was always sung in churches on fast-days.

[139] On this see Schöberlein, loc. cit., p. 371 and 373.

[140] The authorities give the time as half-past six, but this must be a mistake, for the Morning service began at seven o'clock, and the early service must have been celebrated on Palm Sunday at the same time as on other Sundays.

preceded by the *Praefatio orationis dominicae*, but no motett was sung in the course of it, only hymns on the Communion and the Passion were sung by the congregation. Until the year 1721 no performances of the Passion were known in Leipzig excepting in the chorale form. The influence of operatic music had gradually become so strong that it had at last penetrated even into Passion week. At the date just mentioned, the modern " madrigal " or polyphonic style of Passion music, consisting wholly of part-writing, first found a place in the vesper service for Good Friday. Kuhnau, who so often bewailed the destructive influence of the opera on church music, was obliged to yield, nay, even to compose a work in this style himself. This still exists, though only in the sketch, and will be described more fully farther on. The new performances of Passion music took place in the Churches of St. Nicholas and St. Thomas in alternate years. The Thomaskirche was chosen first, perhaps because it had more suitable accommodation for musical performances on a large scale.[141] This arrangement remained in vogue for nearly half a century until the old performances in a chorale style were entirely abolished by an order from the Consistory on March 20, 1766. The deacons were for the most part unmusical, the performances sounded ill and had not tended to edification.

From 1766 onwards Passion music in the madrigal style was employed also in the Morning service, being one and the same in the two churches, but so arranged that in one year the performance took place in the Thomaskirche on Palm Sunday, and in the Nikolai-Kirche on Good Friday, while next year the order was reversed. In time these performances disappeared altogether from the service. The vesper service on Good Friday was performed in the following order: It began with a motett, after which the congregation sang the hymn " Da Jesus an dem Kreuze stund." Then came the Passion music. The sermon

---

[141] That this was the case is seen from Bach being obliged to arrange for the first performance of his Passion music in St. Nicholas'. Rathsarchiv "*Acta die Kirchen-Musik.*"

II.

which followed was always on the subject of Christ's entombment—that is to say, before 1721, up to which time the Good Friday vesper service was held only in the Thomaskirche. There is no doubt that this was the subject, as a rule, even after that date, although it would have involved an anomaly if the Passion music, like both of those by Bach, were in two parts. In such a case, the first part would have come before, and the second after the sermon, and so the continuity in the idea of the service would have been destroyed. After the sermon, the motett *Ecce quomodo* was again sung. The congregation once more sang the hymn "O Traurigkeit! o Herzeleid!" A collect and the blessing were chanted, and for the close the hymn "Nun danket alle Gott" was sung.[142]

Something still remains to be said with regard to the order of service in the University church on Sundays and Fastdays. Its course was considerably simpler. The Morning service began at nine o'clock with an organ prelude and the congregational hymn "Allein Gott in der Höh sei Ehr." Then followed another hymn proper to the day, then the creed, and after that the sermon, by one of the University professors of theology, upon the Gospel. After the sermon another hymn was sung, which concluded the music. In Kuhnau's time music was only performed on the high festivals and during mass. After Bach and Görner took part in the direction it became more frequent; but whether it took place every Sunday, or, if not, what other arrangement was made, we cannot say. After the music the service was brought simply to a conclusion by the blessing "Gott sei uns gnädig"—"God be merciful to us"—spoken by the preacher. In the afternoon a short vesper service was performed, lasting from a quarter-past three till four, and consisting only of a hymn, a sermon by one of the "candidates," and the closing hymn "Ach bleib bei uns Herr Jesu Christ." A vesper service for Good Friday, with organ and sacred music (probably a sort of Passion music), was begun in

---

[142] Documents with regard to the arrangement of public service for Palm Sunday and Good Friday, &c., 1766. Ephoralarchiv at Leipzig.

1728.[143] This improvement was gradually achieved by Görner. Of the number of week-day services we need only mention those held in the Thomaskirche. In the church of St. Nicholas the liturgy was in the charge of the choristers, and in the University church there were no week-day services. The days on which the choir was employed in St. Thomas' were Tuesday, Thursday, Friday, and Saturday. On Tuesday, at a quarter-past six, there was a sermon, preceded by a hymn. The choir sang several Latin psalms, the canticle of *Zacharias*, a German psalm, and at the end the *Benedicamus Domino*. On Thursday, at the same time, there was a sermon and the communion; the service was identical with that of Tuesday, except that the creed was sung before the sermon. On Friday, the usual day of penitence, there was a full service at half-past one, when the Litany was sung. On Saturday, at two o'clock, a confessional service was held especially for those who were to receive the Sacrament on the Sunday. The Thomasschule choir began with a vocal work (whether it were a motett, hymn, or psalm cannot be said); after a sermon on repentance, the *Magnificat*, a collect, and the blessing were chanted. While the private confessions were being heard, the chief practice of the music for Sunday took place in the organ loft. No composition by Bach can have been performed at these week-day services; at most there was only a possibility of it on Saturday. The mention of these services is, however, not unimportant, because they serve to complete the picture of the church life of which Bach's sacred music is an integral part. By this means alone, however, the picture cannot be presented in all its details. To do this it would be necessary to give an account of the services in the New Church and St. Peter's Church, with which Bach stood in some relation, if only a slight one, as well as those in St. John's and St. George's Churches.

---

[143] Antonius Weizius, Verbessertes Leipzig. Leipzig, 1728, p. 12. In the Paulinerkirche, "This year, 1728, the first Good Friday vesper sermon was delivered, on which occasion the organ was used to accompany German hymns, and other instruments were combined with it." These last words evidently indicate some sacred music of a concerted kind, since no hymn but the *Te Deum* was ever accompanied with instruments as well as the organ.

Their mere mention must here suffice; it will be readily imagined that there was an unusual variety of religious exercises at Leipzig. And this multiplicity of services had not been handed down from past times of different manners, so that they were kept up only out of respect; it was the last generation who, in 1699, had restored the service in the New Church, and in 1711 that in the Peterskirche, and who had changed the service in the University Church, which had hitherto been only occasionally held, into a regular service, to the delight of the whole town.[144] Thus it is beyond all doubt that at this time there was a very active devotional life, and that Bach was right in looking forward to a great effect in this place from any important sacred work.

The order of service, which was strictly prescribed in every detail, regulated even the share to be taken in it by the people. In the majority of cases not only were the places where the congregation were to join in prescribed, but the particular hymns to be used. In the ordinary Sunday services, as was said before, the hymns " Allein Gott in der Höh," " Wir glauben all," " Herr Jesu Christ dich zu uns wend," " Nun danket alle Gott," and " Ach bleib bei uns Herr Jesu Christ" had their own place assigned to them, while for the hymn before the Gospel, and that after the sermon, a choice, although always a limited one, was given. For festivals the specifications were more exact. Between the Epistle and the Gospel the following hymns were sung in the morning: at Christmas, " Gelobet seist du, Jesu Christ"; at Easter, " Christ lag in Todesbanden"; and at Whitsuntide, " Komm heiliger Geist, Herre Gott." After the exordium of the sermon: at Christmas, " Ein Kindelein so löbelich"; at Easter, " Christ ist erstanden"; and at Whitsuntide, " Nun bitten wir den heilgen Geist." For the evening service, after the anthem, the following were sung: for Christmas, "Vom Himmel hoch da komm ich her," " Vom Himmel kam der Engel Schaar," or Lobt Gott, ihr Christen allzugleich"; for Easter, " Christ

---

[144] Kuhnau begins one of his petitions to the Council with a statement that "the whole town had rejoiced at the new service being held regularly in the Church of St. Paul."

lag in Todesbanden"; for Whitsuntide, "Komm heiliger Geist, Herre Gott." For the New Year the Christmas hymns were repeated in the same places, and the New Year Hymns were sung also; the Christmas hymns were also used for the feasts of the Epiphany and the Purification of the Virgin, but besides them the hymn "Was fürcht'st du, Feind Herodes, sehr" was sung at the former, and " Mit Fried und Freud ich fahr dahin" at the latter, but only for the Evening service. For the Annunciation it was ordered that the hymn before the Gospel should be " Herr Christ, der einig Gott's Sohn," and that half-way through the sermon "Nun freut euch lieben Christen g'mein." The same hymn was sung on Ascension Day, before the Gospel. The one between the sections of the sermon being " Christ fuhr gen Himmel." On Trinity Sunday, before the Gospel, " Gott der Vater wohn uns bei" was sung, and on St. John's Day in the same place " Christ unser Herr zum Jordan kam." On Michaelmas Day, the hymn " Herr Gott dich loben alle wir" was indispensable. The festival of the Reformation was characterised by the hymn before the Gospel, " O Herre Gott dein göttlich Wort," and the sermon hymn "Erhalt uns Herr bei deinem Wort," besides the fact, which has been already mentioned, that the *Te Deum* was followed by the hymn "Nun danket alle Gott." For Palm Sunday, as ushering in the Passion Week, the hymn before the Gospel was " Aus tiefer Noth schrei ich zu dir," and for Maunday Thursday it was " Jesus Christus unser Heiland." On Good Friday, in the Morning service, the Passion music was preceded by the hymn " Da Jesus an dem Kreuze stund," and followed by " O Traurigkeit! o Herzeleid!" The hymn "O Lamm Gottes unschuldig" being sung between the sermon and the Communion. The hymns for the Evening service, which in the form used in Bach's time has a special interest for us, have already been described in connection with the general features of that service.[145]

---

[145] The following sources have been used, besides those already mentioned in the above description of the order of the Leipzig services: "Leipziger Kirchen-Staat, &c., Leipzig, 1710. Neo-Annalium Lipsiensium Continuatio II., &c., C. E. Sicul," 2nd edition, 1719, p. 565 ff. Bruno Brückner, Betrachtungen über die Agende der evangelisch-lutherischen Kirche. I. Leipzig, Edelmann. 1864. 4.

There was no general hymn-book in use in the Leipzig churches, nor was there any great necessity for one. The hymn-book of Vopelius was in use, but seems chiefly to have been used only for general purposes of reference.[146] Gesner wished every scholar to be in possession of a Dresden hymn-book and always to bring it into church; this was to be looked after by the Cantor and the Conrector, as the inspectors of their conduct in church.[147] This must have been the "Neuauffgelegte Dressdnische Gesang-Buch, Oder Gottgeheiligte Kirchen- und Hauss-Andachten"—"The newly-compiled Dresden Hymn-book, or Sacred Devotions for public and private use"; it was provided with tunes and appeared in a quarto form at Dresden and Leipzig in the year 1707.[148] In the Church of St. George's Orphanage, the hymn-book printed by Johann Montag in Halle was used.[149] A hymn book expressly for Leipzig use was published in octavo by C. G. Hofman in 1747; the arrangement was copied from that of Vopelius, but the tunes were omitted. This handy volume, however, appears not to have been much to the taste of the Leipzig church-goers, for, five year afterwards, Hofman had his hymn-book brought out in quarto. Bach himself made use of the rich collection of hymns made by Paul Wagner, and published after his death by Magister Johann Günther, Deacon of St. Nicholas, in the year 1697, at Leipzig, in eight octavo volumes under the title of "Andächtiger Seelen geistliches Brand- und Gantz-Opfer"—"The whole spiritual burnt-offering of devout souls."[150]

The custom, which was becoming more and more general, of accompanying the congregational singing throughout on

---

[146] A copy of this was procured in 1722 for the scholars in the Neue Kirche; vide the accounts of the Neue Kirche from Candlemas, 1722 to 1723, p. 34.

[147] See ante p. 232.

[148] The first edition appeared in 1694. In 1741, a Dresden hymn-book was published in octavo.

[149] Apparently "Glaübiger Christen Himmel-aufsteigende Hertzens- und Seelen-Music," a Halle hymn-book in octavo, of which the fifth edition appeared in 1710. Sicul, loc. cit., p. 572, says, that as all were accustomed to one hymn-book the practice arose of inscribing only the numbers of the hymns on black boards.

[150] See the inventory of Bach's effects given in Appendix B., XIV.

the organ, had not yet come into use at Leipzig. In times of national mourning, on days of penitence and fasting, the partial or entire cessation of the organ had the effect of intensifying the gloom of the services. Even on festal and ordinary Sundays the "sermon" hymn at least was always sung without accompaniment.[151] The variety thus produced gave greater richness and colour to the services. The same object is apparent in the combination of the organ and the choir singing. In the early service in the Nikolai-Kirche, the choristers sang the *Te Deum* in such a manner that they alternated with the organ at every verse. From this it may be gathered that, as a rule, the choral dialogue between the priest and the choir was without accompaniment. But on this point practical considerations had due weight and no universal rule prevailed in any of the churches. For instance, in the Litany the organ was constantly employed in the churches of Saxony in order to prevent the choir falling from the original pitch.[152] In part-music the *a cappella* style was falling more and more into disuse. In 1717 Mattheson said, with regard to his quarrel with Buttstedt: "Where are the vocalists who used to sing without instruments, even without a bass, whether of clavier or organ? (the trios which occur in the middle of the piece are differently constituted). Where are the singers, I ask, who can sing a whole aria without accompaniment and can keep in tune? My opponent probably preferred the ordinary style, for I know no example of such singers elsewhere."[153] And in agreement with this is what we learn from Bach's pupil Kirnberger, whose testimony as to the Leipzig order of services is of especial importance. "Performances of church-music, even when sung in four, eight, or more parts without instruments, were always accompanied

---

[151] For the feast of the Reformation in the years 1755 and 1757, this custom was broken through. Vide "*ACTA* die Feyer des Reformations-Festes betr. *Superintendur* Leipzig, 1755."

[152] Gerber, Historie der Kirchen-Ceremonien, p. 230.—Ephoralarchiv at Leipzig.—Scheibe (Critischer Musikus, p. 421) speaks of an arrangement whereby, in the responses, the priest sings alone and the choir with organ accompaniment.

[153] Mattheson, Das beschützte Orchestre. Hamburg, 1717, p. 83.

on the organ, which served to support and to keep up the pitch of the voices; or, at least, the manual was employed when a performance of music for the Passion or some other occasion was sung below in the church, for which double basses were used according to the number of the singers. Another arrangement was to accompany each voice part with trumpets and cornets,[154] but never without due reference to the employment of at least one organ manual."[155] The custom of combining instruments with the voices in the motett became usual at the beginning of the seventeenth century, when the pure form of the motett began to degenerate and to borrow from other forms.[156] Although in this way it approached very near to the usual form of concerted music, yet there was this essential difference, that the accompanying instruments were never allowed to play obbligato;[157] an exception was occasionally allowed in the case of the instrumental bass. So thoroughly did this kind of accompaniment become understood as essentially part of the idea of the motett, in the eighteenth century, that choral music with organ accompaniment used to be called simply "mere vocal music" (blosse Vocalmusik), or "*A cappella*-Musik."[158] Mattheson's statement probably goes too far, but is not very exact. The Thomasschule scholars, who never practised without the support of a string-bass, took with them, whenever they had to sing out of doors beyond their usual circuit of streets, a Regal belonging to the school.[159] That the

---

[154] "In former times the music was usually accompanied with cornets and trumpets, and this was particularly the case in motetts." Ruetz, Widerlegte Vorurtheile von der Beschaffenheit der heutigen Kirchenmusik. Lübeck, 1752, p. 27.

[155] Kirnberger, Grundsätze des Generalbasses. Berlin, 1781, p. 64.

[156] Compare Vol. I., p. 55.

[157] Scheibe, Critischer Musikus, p. 182.

[158] Ch. G. Thomas, of Leipzig, in the programme of a church concert arranged by him on the 19th of May, 1790, in Berlin, calls the fifth number "Den 149 Psalm, für zwey Chöre, blosse Vokalmusik," but adds afterwards that the organ was to accompany. Zelter calls the opening chorus of Bach's Cantata "Sehet welch eine Liebe" (B.-G., XVI., No. 64, P. No. 1652), "*a capella* gearbeitet*"*—written in the *a capella* style; see his catalogue of the Amalienbibliothek in the Joachimsthal Gymnasium in Berlin.

[159] Vide App. B. VII. B., under 1 and 2.

custom of accompanying the motetts with cornets and trombones, which was such a favourite one in the seventeenth century, had not yet been given up in Leipzig, is shown not only by different original compositions by Bach, but also by accompaniments to a Mass by Palestrina written with his own hand for cornets, trombones, and organ. An attentive consideration of the order of the services shows that it was the general rule here to accompany the motetts on the organ. On Good Friday morning, when the organ was entirely silent, the motett was also excluded, both before the sermon and in the Communion service; and in the evening when, on account of the Passion music the organ had to be used again, the motett was also there; and the same plan was followed on Palm Sunday and Maunday Thursday.[160] On ordinary Sundays the motett was preceded by a prelude on the organ, which would have been quite senseless if the motett itself were unaccompanied. On festivals the prelude came between the hymn and the motett, so that the former must have been without accompaniments and the latter with it. In all this the motive is clearly to employ the musical means which were usual in the service in alternation and contrast, and so to make even this a kind of work of art. It must be remembered, too, that it was usual to strengthen the voice parts with stringed instruments playing the same notes.[161]

The organs of the two principal churches, which, it is true, Bach in his capacity of Cantor was not required to play upon, answered only to the most limited expectations, for they were old and worn out. There were two in the Thomaskirche. The larger had been put there in the year 1525, having previously been in the Marienkirche of the Monks of St. Anthony at Eiche, not far from Leipzig. It was twice repaired in the seventeenth century, and in 1670 was also enlarged. In the year 1721, however, it was again renovated, as has been already said. The work, which consisted not only of a thorough improvement but also of the addition of

---

[160] Sicul, op. cit., p. 569, says, in general terms that when the organ was not played the motett was also omitted.
[161] Vide Vol. I., p. 61.

400 new pipes and the mixture stops,[162] was done under the direction of Johann Scheibe, the cleverest Leipzig organ builder of the time, who moreover was highly appreciated by no less strict a judge than Bach himself.[163] Although, during the time of Bach's holding office, the organ builders, David Apitzsch and Zacharias Hildebrand were still employed to keep the organ in repair, it was Scheibe who had charge of the principal improvements which were necessary during this period. The first of these was in 1730, and consisted in making the *Rückpositiv* separate from and independent of the *Hauptwerk* and providing it with a keyboard of its own.[164] The second was in the summer of 1747 when the organ had got so much out of order that it was hardly possible to use it at all. Bach and Görner together superintended the repairs, the cost of which was estimated at 200 thalers, and pronounced that Scheibe had done them all suitably and well.[165] The specification of the organ was as follows:—

Oberwerk.		Brustwerk.	
1. Principal	16 ft.	1. Grobgedackt	8 ft.
2. Principal	8 „	2. Principal	4 „
3. Quintatön (dble. diap.)	16 „	3. Nachthorn	4 „
4. Octave	4 „	4. Nasat	3 „
5. Quinte	3 „	5. Gemshorn	2 „
6. Superoctave	2 „	6. Cymbel	2 ranks
7. Spiel-Pfeife	8 „	7. Sesquialtera	
8. Sesquialtera	doubled	8. Regal	8 ft.
9. Mixture of 6, 8 to 10 ranks.		9. Geigenregal	4 „

Rückpositiv.

1. Principal	8 ft.	8. Mixtur	4 ranks
2. Quintatön	8 „	9. Sesquialtera[166]	
3. Leiblich Gedackt	8 „	10. Spitzflöte	4 ft.
4. Klein Gedackt	4 „	11. Schallflöte	1 „
5. Querflöte (Fto. traverso)	4 „	12. Krummhorn (Cremona)	16 „
6. Violine	2 „	13. Trompete	8 „
7. Rauschquinte	doubled		

[162] Accounts of the Thomaskirche from Candlemas 1721 to Candlemas 1722.

[163] A detailed account of various of Scheibe's ingenious inventions in the mechanism of organ building is to be found in the Leipzig "Neue Zeitung von gelehrten Sachen," XVIII., p. 833 f. Scheibe died September 3, 1748.

[164] Vide App. A., No. 15.

[165] Accounts of the Thomaskirche for 1747-1748, p. 52. According to this Bach and Görner were the makers of the contract, which still exists and is given in App. B., VIII. [166] On this stop see App. A., No. 15.

Pedals.
1. Sub-bass of metal - 16 ft.
2. Posaune - - - 16 ft.
3. Trompete - - - 8 „
4. Schalmei - - - 4 „
5. Cornet - - - - 3 „

} of metal pipes tinned over.

The organ stood close to the west wall of the church. In 1773 it was first moved into a better position and brought farther forward, by which process the Rückpositiv got shifted out of its proper place. The organ loft in which the choir stood had a different form in Bach's time to that which it has now. It was very much smaller, and besides that the seats for the Thomasschule scholars were in it. Hiller turned them out and put the trumpeter and the drummer in instead.[167] But even with this arrangement the space was not sufficient after a time, and so, in the year 1802, at the instance of the Cantor Müller the organ loft was altogether rebuilt, being heightened and provided with an ornamental railing to the balustrade. Another enlargement was made in the year 1823. The smaller of the two organs in the Thomaskirche was the older, having been built in 1489. When the larger organ was put into the church the smaller one was just beside it by the west wall. But it did not stay there long. In 1638 the gallery which still exists was built over the raised choir, and the small organ was placed in it, so that it stood opposite the large one. At Easter, 1639, after being repaired, it was played in this place for the first time, and it stood there until Bach's time.[168] In 1727 it was once more put in order by Zacharias Hildebrand;[169] but it was of very little use, and in 1740 Scheibe had to take it quite away. Such parts of it as were still available he used in building the organ in St. John's church, which he constructed in 1742-1744, to the entire satisfaction of the overseers Bach and Hildebrand.[170] This is the specification of the small organ:

---

[167] Manuscript note by the Rector Rost on p. 35 of the school regulations of 1723, in the copy kept in the Thomasschule library.
[168] Vogel, Leipziger Chronicke, loc. cit., and his *Annales*, p. 562.
[169] Accounts of the Thomaskirche, 1727-1728, p. 41.
[170] Accounts of the Johanniskirche from 1740-1744. Agricola, quoted in Adlung's Mus. Mechan., p. 251.

Oberwerk.

1. Principal	- - - - 8 ft.	5. Rauschquinte, 3 and 2 feet.[171]
2. Gedackt	- - - - 8 ,,	6. Mixtur, of 4, 5, 6, 8 and 10 ranks.
3. Quintatön	- - - - 8 ,,	
4. Octave	- - - - - 4 ,,	7. Cymbel 2 ranks.

Brustwerk.

1. Trichter-Regal 8 ft.
2. Sifflöte - - - 1 ,,
3. Spitzflöte - - 2 ,,

Rückpositiv.

1. Principal	- - - - 4 ft.	5. Octave - - - - - 2 ft.
2. Lieblich Gedackt	- 8 ,,	6. Sesquialtera - - doubled
3. Hohlflöte	- - - - 4 ,,	7. Dulciana - - - - 8 ft.
4. Nasat	- - - - - 3 ,,	8. Trompete - - - - 8 ,,

Pedal.

1. Sub-bass of wood 16 ft.
2. Fagott - - - 16 ,,
3. Trompete - - 8 ,,[172]

The organ was only used on the high festivals. It was not unusual, where there were two organs, to employ them for double chorus motetts or arias, in such a manner as that each choir had its own organ accompaniment, in which case it was necessary to separate the two choirs by a wider space than usual. In Wismar at the beginning of the 18th century, Keimann's hymn, set to a melody by Hammerschmidt, "Freuet euch, ihr Christen alle,"[173] was sung in this way at Christmas. The introductory Hallelujah was given out by the whole choir with accompaniment of cornets and trombones. Then the beginning of each verse was sung by a single voice supported by one organ, and answered by one of the full choirs accompanied on the other organ, to the words, "Freude, Freude über Freude," all joining

---

[171] *I.e.*, a Quinte of 3 feet and an Octave of 2 feet.

[172] When Scheibe broke up the organ several stops were taken away—namely, the Dulciana in the Rückpositiv organ and the Fagott and Sub-bass on the Pedal. Several ranks were also removed from the Mixture and Sesquialtera. The Lieblich Gedackt is now named "Grobgedackt," and the Trichter Regal "Ranquet." Rathsacten, book IX., A. 2, Vol. I., fol. 96.

[173] It is given by Winterfeld, in his Evang. Kirchenges. II., Musical examples, p. 102 ff.

together in the final Hallelujah.[174] The distance, by no means an inconsiderable one, between the two organs, in the Thomaskirche, rendered it indeed a matter of difficulty to keep the choirs exactly together. Everything, however, was done to overcome the difficulty, and if ever any confusion occurred it would be compensated for by the devotional effect which would be produced by the floods of sound streaming together from different parts of the church. For the celebration of the Reformation Festival in 1717, in the University Church, Kuhnau performed a festival work for three choirs, which were stationed in three different places in the church: one was put in the space in front of the newly built organ, and the other two in roomy pews by the side of the organ, and apparently behind the pulpit, where two loud sets of organ pipes were placed, and also instrumentalists.[175] In former times in the University church the very strange custom obtained of placing the singers at a great distance from the organ; that being behind the pulpit, the position of the choir was opposite, close to the church wall by the altar. Notwithstanding this, music had been performed successfully, although Kuhnau and Vetter rejoiced at the new arrangement, because it would be easier to avoid those differences between the choir and the organ which we have alluded to.[176]

---

[174] Ruetz, Widerlegte Vorurtheile, &c., p. 86 f.—Ruetz says that the full choir came in each time with the words "Freuet euch mit grossem Schalle." It is self-evident that this is a slip either of the pen or the memory, since the refrain does not begin with these words. Whether what he calls the single voice ought not to be really three, as is prescribed in the original, must be left undecided.

[175] Sicul. Die andere Beylage zu dem Leipziger Jahrbuche auf 1718, p. 73. In the year 1716 again Kuhnau had a Latin Ode set for three choirs performed in the same place; see Sicul, Beylage zu des Leipziger Jahrbuchs Dritten Probe, 1717, p. 11; compare p. 20.

[176] Archives of Leipzig University. Ch. G. Thomas, himself a Leipzig Musician, arranged, in 1790, a concert of compositions for three and four choirs in the garrison church at Berlin; the first choir was in the gallery opposite the organ, the second in the organ loft, the third on the right, and the fourth on the left, in the middle of the gallery. In his account he states positively that the music went well together in spite of the distance. (Sammelband der königl. Bibliothek zu Berlin, Abtheilung *Bibliotheca Dieziana*. Quarto 2900.)

The Nikolaikirche contained an organ dating from the year 1597-1598. The last repairs before Bach's time had been done in 1692. It then consisted of the following stops:—

Oberwerk.

1. Principal - - - 8 ft.
2. Sesquialtera - - 1⅕ „
3. Mixtur - - - - 3 ranks
4. Superoctave - - 2 ft.
5. Quinte - - - - 3 „
6. Octave - - - - 4 „
7. Gemshorn - - - 8 „
8. Grobgedackt - - 8 ft.
9. Quintatön - - - 16 „
10. Nasat - - - - 3 „
11. Waldflöte - - - 2 „
12. Fagott - - - - 16 „
13. Trompete - - - 8 „

Brustwerk.

1. Schalmei - - - 4 ft.
2. Principal - - - 4 „
3. Mixtur - - - - 3 ranks
4. Quinte - - - - 3 ft.
5. Octave - - - - 2 ft.
6. Sesquialtera - - 1⅕ „
7. Quintatön - - - 8 „

Rückpositiv.

1. Principal - - - 4 ft.
2. Gedackt - - - - 8 „
3. Viola da Gamba - 4 „
4. Gemshorn - - - 4 „
5. Quinte - - - - 3 „
6. Quintatön - - - 4 ft.
7. Octave - - - - 2 „
8. Sesquialtera - -. 1⅕ „
9. Mixtur - - - - 4 ranks
10. Bombart - - - 8 ft.

Pedal.

1. Cornet - - - - 2 ft.
2. Schalmei - - - 4 „
3. Trompete - - - 8 „
4. Octave - - - - 4 ft.
5. Gedackter Subbass 16 „
6. Posaune - - - - 16 „ [177]

In 1725 the organ was renewed by Scheibe. The improvements were very thorough and cost 600 thalers.[178] It is unfortunately impossible to know what alterations were made with respect to the disposition of the stops. In this state it remained till the year of Bach's death, when it was again repaired by Zacharias Hildebrand.

In both the Thomaskirche and the Nikolaikirche the organs were tuned to "chorus" pitch. This was then the usual pitch in Leipzig as well as in other places. The organ in the New Church was in the same pitch.[179]

---

[177] Vogel, Leipziger Chronicke, p. 97.
[178] Accounts of the Nikolaikirche from Candlemas 1724—1725, p. 49, and for 1725—1726, p. 53. The contract was concluded on Dec. 11, 1724, and the whole work finished on Dec. 22, 1725.
[179] See App. A., No. 16.

In contrast to these old organs, which were of only moderate capacity, and liable to get out of order frequently, there had been in the University church since Nov. 4, 1716, an organ which fulfilled the highest expectations, and which Bach must have chiefly employed when he played for his own pleasure or before other people. On this account it is of particular interest to become acquainted with its constitution.

Hauptwerk.

1. Great Principal (of pure tin) - - - 16 ft.
2. Great Quintatön - 16 „
3. Small Principal - 8 „
4. Schalmei - - - 8 „
5. *Flûte allemande* - - 8 „
6. Gemshorn - - - - 8 „
7. Octave - - - - 4 „
8. Quinte - - - - 3 ft.
9. Quint-Nasat - - 3 „
10. *Octavina* - - - 2 „
11. Waldflöte - - - 2 „
12. Great Mixtur - - 5 & 6 ranks
13. *Cornetti* - - „ 3 ranks
14. Zink (cornet) - - 2 „

Brustwerk.

1. Principal, of pure tin (in front) - - - 8 ft.
2. *Viola di Gamba naturelle*[180]
3. Grobgedackt with a wide mouthpiece - 8 ft.
4. Octave - - - - 4 „
5. Rohrflöte - - - 4 ft.
6. Octave - - - - 2 „
7. Nasat - - - - 3 „
8. *Sedecima* - - - 1 „
9. Schweizer Pfeife - 1 „
10. Largo - - - - [181]
11. Mixtur - - - - 3 ranks
12. Helle Cymbel - - 2 „

Unter-Clavier.

1. Lieblich Gedackt - 8 ft.
2. Quintatön - - - 8 „
3. *Flûte douce* - - - 4 „
4. *Quinta Decima* - - 4 „
5. *Decima nona* - - - 3 „
6. Hohlflöte - - - 2 „
7. Viola - - - - 2 ft.
8. *Vigesima nona* - - 1½ „
9. Weitpfeife - - - 1 „
10. Mixtur - - - - 3 ranks
11. Helle Cymbel - - 2 „
12. Sertin - - - - 8 ft.[182]

Pedal.

Six stops, which by a new and special invention were brought into connection with the great bellows of the manuals:—

---

[180] Scheibe distinguishes between the real and the so-called Viola di gamba; the former had a narrower mouthpiece. In 1722 he turned a so-called Viola di Gamba in the organ in the New Church into a real one, with great success.
[181] The number of feet is not given.
[182] What kind of stop this may have been, or whether it may not be a mistake or misprint for Serpent, I cannot say.

1. Great Principal of pure tin
   (in front) - - - - 16 ft.
2. Great Quintatön - - - 16 „
3. Octave - - - - - - 8 „
4. Octave - - 4 ft.
5. Quinte - - 3 „
6. Mixtur - - 5 & 6 ranks

These stops on the small Brust-Pedal bellows :—

7. Great (clear) Quintenbass
   (in front) - - - - - 6 ft.
8. Jubal - - - - - - - 8 „ [183]
9. Nachthorn- 4 ft.
10. Octave - - 2 „

And these on the great bellows on both sides :—

11. Great Principal of pure tin (in front) - - - 16 ft.
12. Sub-bass - - - - - - 16 „
13. Posaune - - - - - - 16 „
14. Trompete - 8 ft.
15. Hohlflöte - 1 „
16. Mixtur - 4 ranks

Extra Stops (Couplers, &c.).

Ventils
{
to the Hauptwerk
to the Brustwerk
to the Side Basses
to the Brust and Manual[184]
to the Stern
to the Hinterwerk
}

A Bell to call the blower.[185]

As has been before mentioned,[186] the honourable task of trying this organ after its completion, which must have been superintended by Vetter, was entrusted to Bach. Just at that time he had come from Weimar and had taken up his abode in Cöthen. He laid his opinion before the University, which we subjoin :—[187]

"Since at the desire of his excellency Herr Dr. Rechenberg, at present chief Rector of the honourable Academy at Leipzig, I was charged with the examination of the organ in the Pauliner Kirche, which has been partly renewed and partly repaired; I have fulfilled the task according to my power, have remarked any defects, and have prepared the following statement with regard to the whole work :—

---

[183] An octave stop. See Adlung, *Musica mechanica*, p. 107.

[184] *Sic.* The meaning must be that there was a Ventil between the Pedal and Manual of the Brustwerk.

[185] Sammlung einiger Nachrichten von berühmten Orgel Wercken in Teutschland mit vieler Mühe aufgesetzt von einem Liebhaber der Musik. Bresslau, 1757, 4, p. 54.

[186] See ante, p. 9.

[187] Archive of the Leipzig University "*ACTA*. Den Orgel- und andern Bau, ingl. Verschreibung der Capellen, Verlosung der Stühle und was dem mehr anhängig, in der Pauliner Kirche betr. *De aõ.* 1710. *Volum.* III." Repert. $\frac{\text{II.}}{\text{III.}}$ No. 5. Litt. B. Sect. II., Fol. 63—64.

1. Touching the whole structure, it cannot be disguised that it is in a very contracted space, so that it is a matter of difficulty to get at such parts of it as may at any time require to be repaired; this must be Herr Scheibe's excuse, because he was not the original maker of the organ, but found the case ready-made to his hand, and had to adapt himself to it as best he could; besides that the extra space which he required in order to make the structure more commodious was not granted to him.

2. The ordinary constituent parts of an organ, as the wind trunks, the bellows, the pipes, the sounding boards, and the other parts, have been repaired with great skill, and it need only be remarked that the wind must be caused to come more equally so that the unequal blows of the wind may be avoided; the sound boards ought to have been encased in frames to avoid all noises of wind in bad weather, but Herr Scheibe according to a method of his own, made them with panels, assuring us at the same time that the effect would be the same as that produced elsewhere by frames, and in consequence of this explanation it was let pass.

3. The parts included both in the description and in all the contracts are right both in quality and quantity, with the exception of two reed stops—namely, the Schallmey, 4 ft., and the Cornet, 2 ft.—which may have been omitted by an order from the honourable College, but in their stead a 2 ft. Octave has been introduced into the Brustwerk, and a 2 ft. Hohlflöte into the Hinterwerk.

4. The defects which still remain, such as inequality of intonation, must and can be done away with immediately by the organ builder; in particular, the lowest pipes in the Posaune and the bass Trompete should not speak so roughly and harshly, but should begin with and retain a pure and firm tone; besides this the other pipes which are unequal in tone, must be carefully corrected and equalised, which by means of frequent and thorough tuning of the whole instrument, and also in better weather than there has been of late, it will be quite easy to do.

5. The management of the organ ought indeed to be somewhat easier, and the keys ought not to have so great a fall, but this indeed cannot be otherwise, because of the excessive narrowness of the structure, so that it must perforce be left; yet notwithstanding it is still possible to play in such a manner that there need be no fear of coming to a sudden stop.

6. As the organ builder had to make a new wind trunk to the Brustwerk over and above what had been contracted for, and as the old wind trunk which was to have been used instead of a new one possessed in the first place a *Fundament Brete* (*Qy*. a fixed board supporting the wind trunk?) in itself incorrect and objectionable; and secondly, as the manual had the short compass peculiar to the old style so that there was no possibility of adding the keys which were requisite in order to bring the three manuals to an equality, had the old one been

employed a *deformité* would have been caused; it was, therefore, highly necessary to substitute a new one for it, so as to avoid the defects which were dreaded by the maker, and to preserve a satisfactory conformity. I consider, therefore, that the organ maker is entitled to the value of those parts which have been renewed over and above the terms of the contract, and that he ought to be indemnified.

Seeing that the organ builder has requested me to represent to the honourable College that, as certain parts were not allowed him, as, for instance, the ornamental woodwork, the gilding, and the other ornamentation which Herr Vetter had to superintend, these and whatever else may be necessary may be allowed for in payment, and that he be not held liable for them, since it is not the custom elsewhere to hold the builder liable for such things—and had it been the custom he would have made better terms—he begs humbly that he may be brought into no extra expenses on this account.

And finally it must not be left unmentioned (1) that the window behind the organ should be protected as far as to the top of the organ on the inside by a small wall, or a strong iron plate, so as to prevent any possible damage by weather; and (2) that it is customary, and in this case most necessary, for the organ builder to give a guarantee for one year at least to repair thoroughly any defects that may arise, which he moreover is perfectly willing to undertake to do, if his requests with regard to his expenses over and above the contract be granted speedily and completely.

This then is all that I have found necessary to remark upon in my examination of the organ, and recommending myself in all possible services to his most noble excellency, Herr D. Rechenberg, and to all the honourable College,

I remain,
Your most humble and devoted servant,
Joh: Seb: Bach.
Hochfürstlich Anhalt Cöthenische Capell Meister."[188]
Leipzig, Dec. 17, 1717.

When Scheibe undertook the work in 1710, Kuhnau and Vetter had no great opinion of his skill, but took note of him as an honest, cheap, and industrious workman; thus it was all the more praiseworthy that his completed work should come so well through a trial of so thorough and practical a kind. The organist who had the charge and care of the great instrument after Görner's departure was Johann Christoph Thiele, a man of whose artistic attainments nothing is known beyond this fact.

---

[188] Throughout in Bach's own hand. The address is wanting.

With the preludes and postludes which were then usual, the organ became an independent constituent part of the service. The technical expression for all such "voluntaries," as we should now call them, was "preludes," without regard to the place in the service when the organ was played by itself. From this it is evident that the postlude at the close of the service was not customary in all churches. Its use was not so much practical or liturgical as general and artistic, whereas the prelude was chiefly and expressly intended to prepare the congregation for the hymns which they were to sing, and especially in the case of a less known melody to make them acquainted with it. With the advancing development of the art of the organ, and with the growth of the chorale preludes into independent and organically shaped compositions, the custom of playing a concluding voluntary, in which the organist could exercise his talent at will in free fantasias or fugues, became more and more general. In none of our sources of information is anything said with regard to "playing-out" in the Leipzig churches. From this it does not follow that it was not the custom; indeed it may be inferred from the remarks of Johann Adolph Scheibe, that it was usual here.[189] With regard to the preludes, properly so-called, it was an universal rule to introduce the longest and most elaborate before the congregational hymn sung between the Epistle and the Gospel, and before the Communion hymns.[190] Here, again, the object was a practical one, for there was a greater freedom in the choice of these hymns, so that sometimes hymns with less known tunes were sung, whereas the other congregational hymns were always more or less the same. These preludes were, of course, formed upon the melody of the hymn that was to follow. The organ prelude which, as a rule, preceded

---

[189] He was born in Leipzig and worked for some time in his native town. Critischer Musikus, p. 428. On the other hand, Petri, in his "Anleitung zur Praktischen Musik," Leipzig, 1782, p. 297, says that the playing of a concluding voluntary only became "customary in several towns" at the end of the eighteenth century.

[190] Petri, op. cit., p. 299. Türk, von den wichtigsten Pflichten eines Organisten. Halle, 1787. P. 121 f.

any concerted church music, had a character of its own. Its practical purpose was to enable the instrumentalists to tune without disturbing the devotions of the congregation. For this reason the organist was not allowed to play in a strict style, but had to keep to a free fantasia style, and also to remain chiefly and for the longest time in those keys which corresponded to the tuning notes of the various instruments. When the instruments were tuned the conductor gave a sign for the organist to stop playing; in addition to this, his prelude had to prepare the way for the composition that was to follow, and also to have a kind of finish and roundness of its own. Thus the organist had no easy task if he was minded to make his prelude according to the rules. This, however, was by no means the case, owing to the roughness and carelessness of so many of the performances of church music, and on most occasions all that was heard was a confused and ugly medley of sound.[191]

During the music itself the organist had to play the *basso continuo* from a figured bass part, above which, in the case of recitatives, the vocal part was sometimes indicated.[192] The organ in this case took the part which in chamber music was taken by the harpsichord.[193] Since concerted music in churches had come into general use, this accompaniment became one of the regular duties of the organist, and that Bach viewed the matter in this light is clear from his report of the circumstance of his dispute with the University, in which he says that the organist had not only to accompany the music before and after the sermon, but also to play all the hymns until the very last. Occasionally, and in an

---

[191] Voigt, Gespräch von der Musik zwischen einem Organisten und Adjuvanten, p. 92 f. Adlung, p. 731. Orders from the Council for preventing the disorder and confusion of this prelude are given by Petri, p. 176 ff., and Türk, p. 136 ff.

[192] Voigt, loc. cit., p. 27, says, however, that no cantor would take the trouble to write out the voice part; but it occurs in Bach occasionally, for instance, in the cantata "Christus der ist mein Leben," B.-G., XXII., No. 95. See also App. A., No. 17.

[193] In Voigt, op. cit., p. 107, the Assistant says "I thought that it must be much harder to play in church than in a musical college, and that the mistakes would be much more noticed on the organ than on a harpsichord."

exceptional way, he may have handed over to some one else the task of accompanying from the figured bass, especially when Görner was officiating, since his discourses satisfied him but little. His manner of accompanying has been already spoken of at length.[194] It would be sufficient only to allude to it here, were it not that a document hitherto unknown has lately been discovered which places the circumstances in a still clearer light, and the importance of the subject justifies our returning to it. Heinrich Nikolaus Gerber must have learnt the art of playing from figured bass from Bach, and have practised it on the violin sonatas of Albinoni; subsequently he handed down to his son (the author of the Lexicon) the method of accompanying which he had learnt from studying these sonatas with Bach. The son informs us that an especial feature was that no one part was ever more prominent than the other, and that this accompaniment was of itself so beautiful, that no solo part could have added anything to the pleasure which it gave him. A specimen of an accompaniment of this kind by the elder Gerber still exists; it is throughout written by himself, and contains autograph corrections by Bach.[195] The corrections are comparatively few in number, so that the teacher was apparently satisfied with the work; still, having Bach's alterations, we are sure that this is such an accompaniment to a solo as he liked and approved.

Its state, however, proves that Gerber's work was unsatisfactory and forced, and that Bach's famous method was certainly not polyphonic. The term "polyphonic," it is

---

[194] See ante, p. 102.

[195] I obtained this in the Spring of 1876 from the musical collection of Musikdirector Rühl of Frankfurt a. M. who had died shortly before. Rühl had got the MS. from the bequest of Hofrath André, and he from that of the younger Gerber. A note by the younger Gerber on the MS. runs thus: "Made by Heinrich Nic. Gerber, and corrected by Sebastian Bach." The solo part and the figures of the bass are wanting, as is also all information as to the work of Albinoni from which the sonata is taken. It is, however, the sixth of Albinoni's *Trattenimenti Armonici per Camera divisi in dodici Sonate. Opera sesta*, and thus can be completely restored; it is given as Musical supplement I. to this work. A copy of the *Trattenimenti*, printed by Walsh in London, was bought some time ago at my instance for the Royal Library in Berlin, by Herr Dr. Espagne.

true, is capable of extension; but, when applied to Bach, a thematic and independent treatment of the parts is understood. Gerber's accompaniment, however, has no imitative use of motives, whether taken from the solo parts or freely invented. It is simple—a flowing movement in several parts, in which we never for a moment lose the feeling that the motive power is external to itself. It is only in so far as an unconstrained progression of harmonies which never encounters any obstacle must of itself have a certain melodic effect that we can speak of melody at all in this accompaniment, and this exactly suits the definition of a good accompaniment given by the younger Telemann: "a good flowing song (ein guter Gesang), *i.e.*, a well-proportioned and pleasing succession of sounds."[196] Four out of the small number of Bach's corrections owe their existence to the desire for such "pleasing sequence" (Nos. 1, 2, 9, and 10). Besides this we may observe that Bach even allows his pupil to pay no regard to the simple harmonies suggested by Albinoni, partly in order to make the harmonic motion more connected and flowing, and partly to make it more interesting. Accordingly we find a very independent treatment, only it has no melodic character, as we might suppose from the words of the younger Gerber; the limits which divide homophony and polyphony, or, which here is the same thing, the chief from the tributary subject, have been defined with the most accurate taste. Only a passing allusion can here be made to other instructive features in this accompaniment: to the constancy of the four-part writing, which corresponds exactly to the definition which Kirnberger gives of the characteristics of Bach's accompaniments; to the way in which in the second and third movements the themes which begin without accompaniment are strengthened by the other parts in unison—a method of treatment to be explained in the same way as the separate full chords in the three-part sonatas for violin and clavier.[197] Now that we are in a position to determine the exact mean-

---

[196] Georg Michael Telemann, Unterricht im Generalbass-Spielen. Hamburg, 1773, p. 17.
[197] See ante, p. 102.

ing of Gerber's words, we can judge of the precise import of Mizler's statement, that Bach used to accompany any given solo with figured bass in such a manner that it sounded like a concerto, and as if the melody which he made in the right hand had been written down before-hand. Mizler himself in another place gives a key to the right understanding of these words; but hitherto there has been no such certain example from which we might learn the practical application of it. In an account of Werckmeister's method of playing from figured bass, he deduces from it the general statement that, for a pleasing treatment of harmonies in accompanying, something more must be necessary than merely avoiding fifths and octaves. This "something more," says Mizler, is melody; and by melody he understands such a variety in the succession of notes that they could be sung with ease and would be pleasant to listen to. But since in the best melodies we notice the fewest leaps, it follows that the figured bass player must make no unusual leaps if he wishes to preserve intact the melodic element.[198] Apparently his description of Bach's method of accompanying tends the same way as what we already know from Heinrich Nickolaus Gerber's MS., namely, to a smooth combination of harmonies, the result of which is the production of a kind of melody in the upper part. The fact of Bach's scholar laying such stress upon this individuality is quite explained if we consider how irregular and tasteless the generality of accompaniments on a figured bass was at that time. Löhlein, a well-known teacher of music in the eighteenth century, tells us that soloists, even violoncellists, did not care to be accompanied on the clavier, but preferred a childish accompaniment on a viola, or even a violin; so that the bass part obtruded itself above the melody, and the effect was like a man standing on his head.[199] And even in the case of an accompaniment on the clavier or on the organ, many players

---

[198] Musikalische Bibliothek. Pt. II. Leipzig, 1737, p. 52.
[199] Georg Simon Löhlein, Clavierschule, &c., 4th edition. Leipzig, 1785, p. 114.

were quite content to take the chords just as they happened to suit the fingers, without respect to their position or connection.

Still, what has been said above must not lead us into supposing that Bach altogether forbade his pupil on the organ the use of all figured bass accompaniment that was adorned by imitations. Even if no great weight be laid on the fact that Kuhnau, Heinichen, Mattheson, Schröter and other authorities of that time valued an imitative treatment confined within proper limits, and considered it the highest art of the accompanist, there are still other evidences to prove that in special cases Bach sometimes used polyphony in accompanying, and even if it were not so, we have the right to conclude it from what we know of his musical nature in general. Gerber's work, then, confirms the difference, which we have before pointed out, that a distinction exists between what Bach allowed in this respect and what he required. With regard to the last, no doubt can any longer be entertained now that we possess Kirnberger's accompaniment to a trio by Bach, and that of Gerber to a solo sonata by Albinoni. It will not diminish the knowledge and power of the composition last mentioned if we allow that such an accompaniment could not have been put to an original composition by Bach, and that he would have required a very different sort of accompaniment for a work of his own. Every competent artist derives his standard from what he considers beautiful and fitting—namely, his own works. If the polyphonic enrichment of a solo composition by means of a figured bass accompaniment had seemed to Bach desirable on all occasions, he would have insisted on his pupils employing that method in this case also, where a work of an exclusively educational purpose was concerned. This is all the more certain, because the simple breadth of harmonic structure in Albinoni's sonata was particularly well adapted for the reception of elaborate detail, and, however much regard he may have had to the simple style of the whole work, some traces of polyphony must have been noticeable. Thus, when Bach's figured bass player went beyond what was required of him, he did it on his own responsibility. If some clever piece of ingenuity attempted here and there succeeded, he produced a feeling of pleasant surprise in the ear of the connoisseur, but if he failed he ruined the whole, and drew down upon him the wrath of the conductor. Löhlein says in one place, "The artistic or adorned style of accompaniment, that namely, in which the right hand plays some sort of melody with ornamented *agrémens* and imitations, is for those who have got beyond simple things; it demands great care and knowledge of composition. Herr Mattheson has given many examples of it in his Organistenprobe. Since, however, this kind of pretty decoration was found to spoil more than it improved, it has happily gone out of fashion."[200] This opinion meets the question very fairly; the adorned accompaniment was entirely a thing of musical fashion, as were also the *agrémens* and *manieren* of the clavier player, or the *fioriture* of the singer. It is the same view of the question that Adlung brings forward when, after speaking of the accents, mordents, trills, &c., which a player from a figured bass had at his command, he adds: "But the best *Manier* is melody."[201] Like all fashions, this also serves as one of the characteristics of the time.

Bach worked in a period of wide-spread and abundant musical creativeness. The fit presentment of a musical composition depends on the performance, which cannot possibly take place without the admixture of a certain subjective element, and this element generally took the form of arbitrary ornamentation of the written melodic phrases, even going so far as to distort their very shape by alterations. The composer submitted to this arrangement because the disfigurement his ideas might suffer was atoned for by the individual vitality with which a composition might be performed by a player capable of taking an independently creative share in the work. It is certain that Bach neither would nor could hold entirely aloof from this, an universal feature of the time in which he lived. But

---

[200] Löhlein, loc. cit. p. 76.
[201] Adlung, Anleitung zu der musikalischen Gelährtheit, p. 653.

in proportion to the individuality of an artist will be the care with which he will guard against the introduction of any foreign element into his creations. Of all the great musicians who appeared in this period, Bach is, without doubt, the most subjective, and stands the most apart from all others. He deserves the complaint made by Johann Adolph Scheibe, that "he indicated with actual notes all *manieren*, all the small ornaments, and everything that is understood by the word *method* in playing,"[202] so that he effectually closed the door upon any approach to individuality on the part of the performers. What Scheibe says of Bach's vocal and instrumental works holds good also of his figured bass parts, which he used, unless hindered by circumstances, to figure in the most elaborate way. His minute care extended even to the ritornels and short ritornel-like passages, in which, as a general rule, the accompanist was left to follow his own instinct as to what was most suitable. Thus, in the terzet in the cantata "Aus tiefer Noth schrei ich zu dir,"[203] he shows with the greatest exactness, by means of figuring, the counterpoint to be played in the right hand to the chief subject of the ritornel, which lies in the bass part. His pupil Agricola had a better appreciation of the case; he approved the exactitude which Scheibe blames, for he holds it to be no fault in a composer if, in order to obviate disfigurement, he expresses his ideas with the greatest possible clearness.[204] Bach's careful forethought is fully accounted for when we read the laments of earnest musicians over certain vain and trivial organists who, in contrast to those who never add more than is absolutely requisite, take every opportunity of "shaking out their sackfull of ornaments all at once, in fanciful tricks and runs, and, when the singer has to execute a passage, think it necessary to vie with the singer in ornamentation."[205] It is no less plain why the school of Bach always insists on using the polyphonic, or, indeed, any adorned style

---

[202] Scheibe, Critischer Musikus, p. 62.
[20] B.-G., VII., p. 296 ff. P. 1694.
[204] In the addenda to Tosi's Anleitung zur Singkunst, p. 74.
[205] Kuhnau, Der *Musicali*sche Qvack-Salber, 1700, p. 20 ff.

of accompaniment, as little as possible, and only as an exception,[206] as Philipp Emanuel does, often discarding it entirely. Kirnberger held that as the accompanist from a figured bass had only to add the harmonies, he ought to refrain from all ornaments which were not essential, and always aim at simplicity.[207] Johann Samuel Petri, a friend and pupil of Friedemann Bach, who made a special study of the art of accompanying under this master, forbids the organ-player to introduce shakes, or to play any melody with the right hand; he requires him to keep to the chords intended by the composer, and lays it down as a general principle that the organ accompaniment serves only to fill up the harmony and to strengthen the bass.[208] The same thing is said by Christian Carl Rolle, the intimate friend of Bach's sons and pupils.[209] But other musicians too, besides those of Bach's own circle, shared this opinion, and expressed it in the most decisive way. Johann Adolph Scheibe, with his well-known acuteness of æsthetic judgment, condemns the polyphonic accompaniment of a solo as contrary to good taste, and destructive to the composer's intentions.[210]

The question of the manner of accompanying also includes the way of employing the sound-material offered by the organ. In this respect there were firmly-fixed traditions. The bass was as a rule played by the left hand alone, while the right took the complementary harmonies. When an organ with several manuals was used, the left hand usually played on an independent manual with powerful stops. The pedal generally played the bass also, and then the left hand. could

---

[206] Versuch über die wahre Art das Clavier zu spielen. Pt. II., pp. 219 f and 241.

[207] In Sulzer's Allgemeine Theorie der schönen Künste, Second edition, Pt. I., p. 194. On Kirnberger's share in this work, see Gerber's Lexicon IV., co . 304 f.

[208] Anleitung zur praktischen Musik, p. 169 ff. Petri tells of his connection with Friedemann Bach in this very work, which is one of the best educational works on music of the eighteenth century, pp. 101, 269 (see also 268) and 285.

[209] Neue Wahrnehmungen zur Aufnahme und weitern Ausbreitung der Musik. Berlin, 1784, p. 49.

[210] Critischer Musikus, p. 416.

help to bring out the inner parts.[211] When the bass was strongly orchestrated it was considered sufficient to mark only the essential points in the harmonic progressions by short notes on the pedals.[212] In such cases, however, it was not unusual to leave the bass out altogether in order to avoid confusion, since it was generally taken by other instruments.[213] In the case of slightly instrumented works, some composers only used this accompaniment in the ritornels.[214] For accompanying arias and recitatives the eight-foot Gedackt was generally used alone, and from this circumstance got the name "Musikgedackt."[215] In the case of "recitativo secco," even when the bass notes were directed to be held, the chords on the organ, which used to be played *arpeggio* as on the harpsichord, were, as a rule, not held long, in order to give due prominence to the words sung.[216] Petri, however, holds that when there is a very soft stopped-flute, the chords may be held in the tenor register and that then each change of the harmony is to be indicated by a short pedal note. In order to give support to the singer, the organist was sometimes to hold out the bass note alone, and take off the rest of the chord quite short. Greater freedom was permitted in the case of accompanied recitative; he might either play both the chord and the bass together, or the latter alone, as it seemed good to him.[217]

---

[211] Adlung, p. 657. "At one time much attention was paid to divided playng, that is, when some of the inner parts were played by the left hand. It is also quite possible, if the notes can be given out on the pedals, for both hands to remain on one manual. But in the case of rapid basses, this arrangement fails, because such are best played on two manuals." Petri, p. 170.

[212] Rolle, p. 51. "The management of the pedals is attended with unusual difficulties, such as that of always knowing which are the right notes to play and which may be left out with impunity for the sake of rapidity." Schröter, Deutliche Anweisung zum General-Bass, Halberstadt, 1772, p. 188, § 348. Türk, p. 153. The same holds good, as a rule, for the double-bass also; see Quantz, Versuch, &c., p. 221, §. 7.

[213] Türk, p. 156.   [214] Petri, p. 170.

[215] Scheibe, Critischer Musikus, p. 415. Adlung, p. 386. Rolle, p. 50. Compare Gerber, Historie der Kirchen-Ceremonien, p. 280.

[216] (Voigt), Gespräch, &c., p. 29. Petri, p. 171, compare p. 311. Türk, p. 162 ff.

[217] Schröter, p. 186, § 344. Türk, p. 174.

It should be noticed that the staccato style of playing, now universally considered unsuited to the nature of the organ, was not considered so by the musicians of that time. The formation of fugue themes from reiterated notes, and the repetition of full chords served, in the opinion of organ masters of the Northern school, to produce a peculiarly charming effect—Christoph Gottlieb Schröter of Nordhausen, one of the most perfect organists of his time, always played staccato. By this method, indeed, he provoked the opposition of the scholars of Bach, who followed the example of their master in considering the *sostenuto* style as the finer,[218] and to their influence is to be ascribed the fact that the other style of playing gradually died out. But it was only for independent organ pieces that they insisted so definitely on the sostenuto style. For accompanying, the "lifted" style remained in use even within the circle of Bach's scholars. Kittel, who during a period of fifteen years' activity as a teacher spread Bach's systems among the organists of Thuringia, inculcated this method, and persons are still living who heard one of his best pupils, Michael Gotthardt Fischer of Erfurt, accompany the Church cantatas in this manner: he followed the harmonic course of the movement with short chords in the right hand while he played the bass *legato*, and with considerable power.[219] This agrees with Petri's direction, that the organist is to accompany in as short a style as possible and to withdraw the fingers from the keys directly after striking the chord.[220] From this it must on no account be concluded that Bach always accompanied in this way and in no other. As the style of his compositions was more *sostenuto* than that of Kittel's and Petri's time, he distinctly requires in the majority of cases a *legato* accompaniment, and what he directs to be accompanied in a "melodic" manner can generally be performed correctly in no other way. It must be remembered, however, that he was equally at home in

---

[218] Gerber, Lex. II., col. 455.
[219] It was heard by Herr Professor Edward Grell of Berlin, who was so kind as to narrate the circumstance to me by word of mouth.
[220] Petri, p. 170.

the other style, and employed it upon occasion; examples are offered in the A major aria in the cantata "Freue dich, erlöste Schaar," and the G major aria in the cantata "Am Abend aber desselbigen Sabbaths."[221] In general all customs of this kind depend more or less upon the circumstances of the time. The original manuscripts of Bach's Matthew Passion, and his cantatas "Was frag ich nach der Welt," "In allen meinen Thaten," shows that he also permitted the organist to employ short chords in *recitativo secco*. It appears, too, that in passages where the instrumentation is lighter, some of the double-basses must have been left out. In the bass aria of the fifth part of the Christmas Oratorio, one of the strengthening bass instruments is silent throughout.[222] In the B minor aria of the cantata "Wir danken dir Gott," the whole body of the basses scarcely take any part except in the ritornels.[223] We may notice something of the same kind in a passage from the first chorus of the cantata "Höchst erwünschtes Freudenfest." Here, in addition to the organ bass, string basses and bassoons are employed, and, according to the orchestral parts, have to play throughout without intermission. In the score, however, at bar 86 in the bass line, there is this direction: *Organo solo*, and again at bar 97, *Bassoni e Violoni*. The chorus is cast in the form of the French overture; this passage corresponds to the stereotyped trio passages which are of a softer character, and in the real overtures are given to the two oboes and bassoon, and which are contrasted with the pompous and complete effect of the full orchestra. In order to make the contrast really effective, Bach makes the bass of the orchestra cease, and the organ play alone. He must have deemed it sufficient to inform the instrumentalists of his intentions by word of mouth at the rehearsal. We know, moreover, that he considered the Gedackt as peculiarly adapted for purposes of accompanying, from what he himself says in the specification of the repairs for the Mühlhäusen organ.[224] We may

---

[221] B.-G., V.,¹ p. 352 ff.　B.-G., X., p. 72 ff.　P. Vol. 1017, 2144.
[222] B.-G., V.,² p. XVI.　　　　　[223] B.-G., V.,¹ p. 307 ff.　P. 1289.
[224] Vide Vol. I., p. 356.

therefore conclude with certainty that in similar passages he
frequently dispensed with a part of the instrumental basses,
and, especially in arias, employed the whole of them only for
the ritornels; that the *recitativo secco* was usually accompanied
in a short style by his direction, and that, as a rule, the
Gedackt was used for recitatives and arias. We cannot,
however, venture to deduce a rule which shall hold good
for all cases, but must rather conclude that Bach, disregarding the practice of others, kept himself perfectly
free in all matters of art; thus, in accordance with the
character of the piece he would alternate short chords with
sostenuto in recitatives, or the Gedackt with some other stop
of especial fitness for accompaniment, and in other ways
deviate from what was generally accepted, to the advantage
of the particular instance. Such deviations were the result
of his nature, the time, and the subject.

Schröter and Petri lay down the law that in accompanying
church music no use whatever must be made of reeds or
mixtures.[225] By this they only mean to lay stress upon
the fact that the organ ought never to drown the voices and
instruments. Besides this, the task of the organ was not
only to support and hold together the whole body of sound,
but also to give it unity of colour. In a certain sense it
occupied, with regard to the other instruments, a position
similar to that taken in the modern orchestra by the string
quartet. Just as the wind instruments group themselves
round this as a centre, so all the instruments grouped
themselves round the organ. The relations were different
however in this way, that the organ remained always in
the background, its effect being merely that of power, and
that on this background the other instruments were seen,
not so much as solo instruments, but rather as choric
groups. One of these groups was the quartet of strings,
another the oboes and bassoon, a third the cornet and
trombones, and a fourth the trumpets (or sometimes horns)

---

[225] Schröter p. 187 ff, where precise directions are given as to the management of the stops in the different parts of a cantata, and also as to the various
characteristics of such parts. Petri, p. 169.

and the drums. The flutes occupied a less independent place in Bach's orchestra, but in the seventeenth century they formed a group by themselves. Any individualisation of separate instruments such as is exhibited in the orchestra of Haydn was by this means excluded; the effects were produced rather by means of the juxtaposition and contrast of the great masses of sound, a method which perfectly corresponded to the character of the fundamental instrument, the organ. The sense of style in Bach's church music results partly from his having left these relations of the groups one to another, which had become fixed in the seventeenth century, unaltered both in outline and detail. In this, as in other respects, he had stronger sympathy with a bygone time than his contemporaries, who were more sensitively alive to the approaching development of concert music, and to whom, for that very reason, these traditional requirements were antipathetic; in their church cantatas we hardly ever entirely lose the feeling of a deep artistic anomaly. Besides this, to return to the comparison between the organ and the string quartet, an essential difference lies in the relation of the two bodies of sound to the voices. In a combination of voices with instruments, the natural condition is that the former rule and the latter serve; so that the former fix the character of the piece while the latter only give support and adornment. Now the vocal music of the sixteenth century had attained greatness, notwithstanding that each part was often sung by a single voice. These insignificant choruses had remained, with few exceptions, in universal use throughout the seventeenth century, and far on into the eighteenth, while on the other hand the treatment of the instruments continued steadily to increase in fulness and variety of colour; so that in Bach's time even what we should call an orchestra of weak strength outnumbered the singers by more than a third. In the Neue Kirche under Gerlach there were only four singers to ten instrumentalists.[226] Bach himself, in the memorial of August 23, 1730, fixed the number of singers at twelve

---

[226] Vide App. A., No. 14.

and that of the instrumentalists, besides the organist, at eighteen—in the ratio, therefore, of two to three, so the vocal parts certainly did not preponderate; thus the natural proportion was exactly reversed in consequence of an individual development. Handel and Bach, the two culminating centres of music at that time, sought, each in his own way, to rectify this state of things. The choir with which Handel performed his oratorios in England was indeed numerically smaller than the orchestra, but it consisted of singers of much greater technical ability than those of the German church choirs, and consequently the tone was much fuller; besides, Handel made a much more limited use of the organ. The characteristic feature of giving the vocal parts more importance than the instruments is very prominent with him, and pervades his music so strongly that, in the performances of his oratorios within a few years of his death, it was settled in England that the voices were to outnumber the orchestra. In Germany the change did not come so soon. In the festival performance of the Messiah, got up by Johann Adam Hiller in the Domkirche of Berlin, on May 19, 1786, the old proportions were adhered to; there were 118 vocalists, and 186 instrumentalists.[227]

This change, which was gradual in Germany, is to be ascribed to the influence of England. But it was only suited to the oratorio proper, not to German, or, which is the same thing here, to Bach's church music. In the case of most of Handel's oratorios, although the chorus is seldom or never to be regarded as representing persons in the drama, yet, for the proper understanding of the artistic idea in its entirety, the consciousness that it is constituted of human voices is of the greatest importance. In Bach the

[227] Hiller. Account of the performance of Handel's Messiah in the Domkirche in Berlin, on May 19, 1786, 4. The orchestra, strengthened by Hiller by the addition of flutes, oboes, bassoons, horns, and trombones, consisted of thirty-eight first, and thirty-nine second violins, eighteen violas, twenty-three violoncellos, fifteen double-basses, ten bassoons, twelve oboes, twelve flutes, eight horns, six trumpets, two trombones, drums, organ, and harpsichord. The choir, which comprised all the singers of the schools of Berlin and Potsdam, and all the opera singers, male and female, numbered thirty-seven sopranos, twenty-four altos, twenty-six tenors, and thirty-one basses.

use of the voice is of a much more abstract character; it is regarded rather as an instrument having the property of uttering words and sentences with and on the notes it gives forth. Handel's oratorio style tended towards laying a stronger and more decisive emphasis on the vocal factor, while Bach's chorus admits of strengthening additions only within narrowly-defined limits, and, from the first, never bore an indirect ratio to the instruments. For the practical side of German music, it has been a fatal error, although easily accounted for by historical fact, to reckon the oratorio as a branch of church music on the one hand, and on the other to regard church music from the point of view of the oratorio. This is one of the principal causes of the hybrid state of the German oratorio in the latter half of the eighteenth century; outward circumstances, it is true, contributed to this result, but so deeply imbued were the German composers with this amalgamation that, even after the practice of performing oratorios in the concert room had become usual, its influence long remained evident.

In Bach's church music the ruling or dominant factor is not the chorus or the voices—if there be any such factor, it can only be said to be the organ; or, to put it more decisively, the body of sound used in performing Bach's church music is regarded as a vast organ of which the stops are more refined and flexible and have the individuality of speech. Still, this organ is not to be conceived of as a dead mechanical instrument, but as the conveyer and the symbol of the devotional sentiment of the church, which is what it had indeed become in the course of the seventeenth century, and by the aid of Bach himself. While assigning it this place in his church music, he succeeded in effacing, so far as he was concerned, the disproportion which existed between vocal and instrumental music, and in combining them to form a third power higher than either; he could do it only in this manner in his position and sphere. Handel and Bach, the fundamental sources of whose genius were in part the same, had arrived at directly opposite results in this as in many other problems in art. This is obvious from a study of their works even without regard to comparison or analogy. It is, however, always interesting to have evidence that Bach was conscious of the individuality of his work. In the latter half of the century, as the influence of the Protestant church decreased, the spiritual meaning of Bach's church music became less understood. Kirnberger watched with anger the gradual and increasing disuse of the organ in church music, while a secular and theatrical style was demanded on all sides which lowered this whole branch of art. In his opposition to these tendencies he was joined by the school of Bach and many other musicians, who devoted themselves to the music of the better times that had gone by. Rolle, whom we have frequently mentioned, has formularised and handed down to posterity the verdict of these men. He says: "In theatrical performances, in serious operas, and particularly in operettas, and also in concert rooms where solo cantatas, great dramatic vocal pieces, and so forth are performed, we are accustomed to distinguish the voices in concerted pieces in the plainest manner possible, as they are not checked, obscured, and disturbed by any organ or other powerful accompaniment. We are misled by this into demanding the like delicacy of sensuous pleasure even in church music. Many practical musicians, however, judge quite differently. They say we must never mistake the right and true form of church music. We must treat that splendid instrument, the organ, rather as the ruling power than as passive or as a mere accompaniment, and this more especially in choruses, even though the ornamental details of both vocalists and instrumentalists may thereby be lost. We indeed desire good and beautiful melodies, which each separate part can and must have, but above all we require noble, complete, and splendid harmony."[228]

---

[228] Rolle, Neue Wahrnehmungen zur Aufnahm und weiteren Ausstreitung der Musik. Berlin, 1784. This book was severely criticised and soon forgotten. The style and arrangement are no doubt confused, but the work is notwithstanding full of practical observations and useful facts. The author, who was Cantor at the Jerusalem and New Church at Berlin, was a son of Christian F. Rolle (mentioned in Vol. I., p. 520), and familiar with Bach's school of music. The passage here quoted expressly refers to Bach's pupils, for the heading in

The vocal part of the church music was performed by boys and men. In Thuringia and other districts of central Germany the church choirs were strengthened by so-called "Adjuvanten," or assistants—*i.e.*, amateurs from the neighbourhood, who voluntarily took part in the performances. In Leipzig this custom seems not to have obtained to the same extent; we find it once mentioned that in Kuhnau's time an "advocate in law" had frequently accompanied the church music on the organ. The *Collegia Musica*, under the direction of Schott, Bach, and Görner, consisted almost exclusively of students, who certainly must have taken part in the church music. The solos for soprano and alto were given, as a rule, to the boys of the Thomasschule choir. In the case of pieces composed by Bach himself, their performance was no easy task, for in his arias, as is well known, great demands are generally made on flexibility of voice, and the art of taking breath; a boy's voice rarely lasts long enough for him to acquire a thorough technical education. His singers are said, indeed, often to have complained of the difficulty of this music.[229]

Still, it may be pointed out that a certain skill in technique was at that time more common than at present; it was in the air, so to speak, so that it would be more easily acquired. During all that period the Italian art of song was in full bloom and was known and admired throughout Germany. Little as the German school-choirs were capable of turning this art to account in its entirety, yet a certain superficial brilliancy found its way among them, and with some degree of success. To this, for example, is to be ascribed the study of the shake, which was enforced with great gravity and zeal in the school singing lessons. Wolfgang Caspar Printz, Cantor of Sorau, in his Gesangschule which appeared so early as 1678,[230] gives

---

the table of contents mentions Agricola, Graun, Hasse, Kirnberger, &c., and some of those who are related to the families of Bach and Rolle, as being "famous musicians, as distinguished in church music as in theatrical music."

[229] Forkel, p. 36.

[230] *Musica modulatoria vocalis*, oder manierliche und zierliche Sing-Kunst, 1678. He calls the shake "Tremolo," while he gives the name Trillo or Trilletto to the tremolo proper, which he also treats of (p. 57 f).

instructions as to the shake, and the same was done one hundred years later by Petri, who, like Printz, was a school Cantor,[231] and by Hiller, one of Bach's successors in the Thomasschule.[232] Both direct the study of the shake to be begun early and to be diligently practised every day. It is clear, from Bach's compositions, that he demanded and expected from his singers facility in executing shakes. The German style of vocalisation at this period was a mixture of roughness and over-refinement, which a great musician such as Bach could only make available for his ideal by merging it in the style of instrumental art, which then was at an incomparably higher grade of development. In these days even, a boy's voice seems to us to be utterly inadequate to the task of giving expression to the abundance of feeling contained in the arias of Bach; their depth and passion seem to demand before all else, and as an indispensable condition, a high degree of maturity of artistic feeling. Since it was impossible for Bach to reckon upon this condition being fulfilled,[233] the conclusion is unavoidable that it was not his intention to bring this feature of passionate depth prominently forward. Indeed, throughout his music the subjective emotions are rather suggested than fully developed; and this is the true explanation of the phenomenon that Bach's music has begun to be so deeply felt since Beethoven's time, for during this period men's feelings have been particularly open to such emotions. In Bach's own time an aria of his composition was, as it were, a lake frozen over; the boy's voice glided over the surface, careless as to the depths which lay below. Moreover, the suppression of all personal feeling was required by the very nature of church music; nor is this true only in the case of the soprano and alto voices, but for Bach's music as a whole; it is the deepest law of its individuality. Boys' voices were at least capable of

---

[231] Loc. cit., p. 203.
[232] Anweisung zum musikalische- richtigen Gesange, Leipzig, 1774. p. 38.
[233] "The more refined and expressive kind of singing is not to be expected of choir boys." Forkel in his admirable dissertation on Church music. (Allgemeine Geschichte der Musik, Vol. II., p. 37.)

fulfilling the requirements of this law. We cannot, however, venture to assert that the performance of the solos was as yet always assigned to boys alone, for the art of falsetto singing by men was still diligently cultivated. This art, the practice of which has now so completely disappeared[234] that even the rudiments of its *technique* seem to have become a secret, was quite an ordinary thing in Bach's time at Leipzig. In the musical societies, where cantatas were performed every year with the full number of parts, men alone were the performers; the names of the students to whom the four-part singing was generally entrusted, under Hoffmann's direction, have been given. And later, Gerlach had only four students at his disposal for the concerted music in the Neue Kirche; and the choristers of the Nikolai Kirche, when they had to sing in four-parts, must have been capable of doing it by themselves. By means of the falsetto a tenor voice was changed into a soprano and a bass into an alto. It is expressly stated that this style of vocalisation was employed not alone in choruses, but also with a particular effect in arias, and that a falsetto soprano could sing up to the astounding height of e''' and f'''.[235]

In speaking of customs in singing, the way of performing the recitative must not be forgotten. The singers of the present day are accustomed to deliver Bach's recitatives as they are written, and this they do with a view of giving them a solemn character, distinct from anything theatrical. It is a question, however, whether our present practice has not come to be directly opposite to that of the earlier time. The free alteration of separate notes and intervals in phrases of recitative was seldom if ever employed in theatrical recitative in Bach's time; it occurred more frequently in chamber music, but almost constantly in church recitative.[236] The reason was that the rule at that

---

[234] [In Germany, at least. Tr.]

[235] Kuhnau, Der *Musicali*sche Qvack-Salber, p. 336: "When he played the clavier, and let his alto falsetto be heard (his proper voice was a bass) in some favourite arias, the girl was quite captivated." Petri, p. 205 f, gives full directions as to the cultivation of the falsetto.

[236] Tosi-Agricola, p. 150 ff.

time universally followed was to treat church recitative in a melodious rather than in a declamatory manner,[287] whereas in opera it was to be exactly the reverse. These alterations, however, serve for the most part the purpose of increasing the melodious flow of the phrases. As to the cases in which they are to be introduced as a regular practice, we are given exact directions by Telemann and Agricola. Telemann, in the preface to a collection of his own cantatas which appeared in 1725, illustrates these uses by examples.[288] On the one hand, they refer to the downward skip of a fourth, especially common in final cadences. Phrases like these—

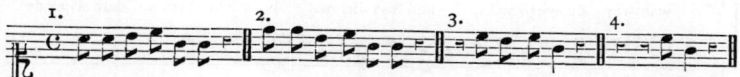

according to him, should always be sung thus:—

On the other hand, they treat of the employment of the so-called accent—*i.e., appoggiatura*, or prefatory note, consisting of the next note above or below the principal one. To make this clear, Telemann gives a recitative from one of the cantatas which occur in the work, both in the usual notation and according to the actual performance:—

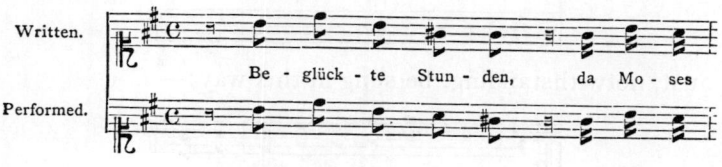

---

[237] Scheibe, Critischer Musikus, p. 163.

[238] Georg Philipp Telemann, " Harmonischer GOttes-Dienst, oder geistliche *CANTATEN* zum allgemeinen Gebrauche, &c." Service of Harmony, or sacred cantatas for general use, which are intended for the furtherance of devotion, as well private and in the house as public in the church, on the ordinary Epistles for Sundays and Holy Days throughout the year, &c. Folio. The preface is dated " Hamburg den 19. *Decembr*, 1725."

He remarks farther that it matters not though the accent should sometimes come in collision with the harmony, and that a phrase like this:—

must, notwithstanding, be sung in this way:—

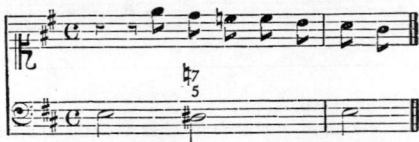

The list of "*Manieren*" which are illustrated in the longer examples, does not, as Telemann himself says, exhaust all those that are possible, but comprises only the most usual of their kind. Agricola gives a certain number of these as well, but adds several others, particularly in an ornamental style. In order to understand Bach's position with regard to the vocal treatment of recitative, it is necessary to consider two points with regard to it. First, that of all church composers he undoubtedly is the one who strove most earnestly after melodiousness in his recitative; and, on the other hand, that he would be most unwilling to give undue license to the arbitrary caprice of the performer. The first consideration must have led him to regard a free use of these ornaments as desirable, and the second to express them with the greatest possible precision by written notes. Viewed in this two-fold aspect, Bach's recitatives give the result which we should expect. The skip of the fourth in the cadence is always intended to be performed as Telemann directs, but it is always fully written out. Where it is found written in the ordinary manner, it must be sung in strict accordance with the notes. One of these cases—which are of the greatest rarity—is found in the second part of the Matthew Passion, where Jesus says: "Hereafter shall ye see the Son of Man sitting on the right hand of power, and coming in the clouds of Heaven."[239] In this case the harmony of the accompanying violins shows plainly that at the close of the phrase the notes to be sung must be *b* twice and not *é b*. If the interval is to be sung in a florid manner, Bach writes it down as such. This florid treatment was accomplished by filling up the interval with the notes that lay between, by which means *mordents*, shakes, and similar ornament might be introduced.[240] In the Christmas Oratorio, when the crafty Herod sends the message to the wise men from the east, commanding them to seek diligently for the young child, the phrase which Bach here gives at the close of his speech would under ordinary circumstances have been written thus:—

that I may come and wor-ship Him al - so.
dass ich auch kom - me und es an - be - te.

---

[239] B.-G., IV., p. 159, bar 1. Novello's edition, p. 102, lines 2 and 3.
[240] Agricola, loc. cit., p. 151 f.

In order to give expression to Herod's malignant scorn, Bach wished to signify that the skip of a fourth was to be slurred over; accordingly, he writes it completely out:—[241]

dass ich auch kom - me und es an - be - te.

The introduction of the accent or appoggiatura, which, as a rule, was only employed to precede an emphasised note, was possible in upward or downward direction. When it was used in a descending passage it was permitted in the interval of a third or a second. When only one emphasised note follows the skip of a third, as occurs at the end of the longer example of Telemann, Bach very frequently writes out the notes as they are to be sung when they deviate from the rule. It should be remarked, by the way, that he does not disdain the use of the accent after the skip of a third, even on a note which has no emphasis. Passages like the following:—

zu dem wer - de ich ein - ge - hen

or this—

so kann mich nichts von Je - su schei - den.

and many others exhibit this clearly. When two notes follow, the first of which is emphasised, as in bars 4—5 of Telemann's example, Bach not unfrequently writes in the notes to be sung; *e.g.*, in the cantata "Barmherziges Herze":—

ver - bess - re dei - ne Män - gel.

---

[241] B.-G., V.,² p. 236, bar 9 ff. Novello's edition. Another example occurs in the Cantata "Brich dem Hungrigen dein Brod," B.-G., VII., p. 335. P. 1295.
[242] B.-G., XVI., p. 15, bar 7 f ("Nun komm der Heiden Heiland"). P. No. 1668.
[243] B.-G., VII., p. 44, bars 3 and 4 ("Der Himmel lacht," &c.) P. 1695. A passage where this accent is wanting in the original score, but is written out in the autograph *part*, occurs in a duet B.-G., VII., p. 79, bar 3.
[244] Kirchengesänge von Johann Sebastian Bach. Berlin, Trautwein & Co., III., p. 19, bar 10.—See also B.-G., V.¹, p. 30, bar 1.—II., p. 27, bar 10.

When the voice descends only by a second, and one emphasised note follows, the accent may also be introduced; it occurs too in Bach several times written out, as in the cantata "Komm du süsse Todesstunde":—

zu mei - ner See - len Qual

If two or more notes follow (Telemann, bars 1—2), the first, supposing it to be emphasised, would be raised one note higher, exactly as Bach has written it in the "Himmelfahrt" Oratorio:—

der da hei - sset der Öl - berg

If, however, the second note has the emphasis it takes the accent, and thus gives rise to a melodic sequence, as in Telemann, bars 2—3 and 6; I have found no certain example to prove that Bach wrote this out in notes. With regard to its use in ascending passages, the accent when introduced *extempore* seems only to have been employed in the case of the interval of a second. A particularly expressive phrase resulted when two notes followed this interval, of which the first was emphasised (Telemann, 5—6 and 6—7); hence this kind of accent is found very frequently written out in Bach. See the passage from the Matthew Passion:—

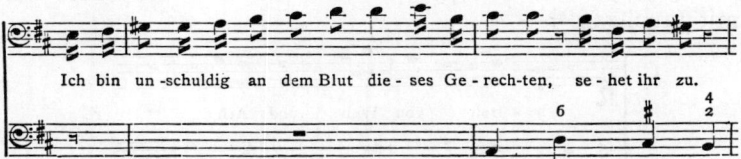

Ich bin un-schuldig an dem Blut die - ses Ge - rech-ten, se - het ihr zu.

I am innocent of the blood of this just person; see ye to it. The notes referred to have been altered in the English

---

[245] B.-G. II, p. 34, bars 13—14. P. 1279.

version to suit the words.[246]   Or at this from the Christmas Oratorio:—[247]

So, too, when only one emphasised note follows, as in the cantata "Was Gott thut, das ist wohlgethan":—[248]

When two notes followed, and the second was emphasised, the accent might be introduced as is done by Bach in one passage in the Matthew Passion:—

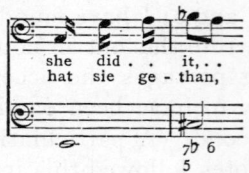

Finally, in the case of an ascending second followed by one emphasised note, the accent may be formed by taking the note below the emphasised one as the prefatory note. (Telemann, bars 3—4.)   Accordingly in the cantata "Gott fähret auf mit Jauchzen":—

---

[246] B.-G., IV., p. 200, bar 8.   Novello's edition, p. 131, bottom line.
[247] B.-G., V.,² p. 246, bars 12—13.   Novello's edition, p. 160.
[248] B.-G., XXII., p. 241, bars 7—8.   Compare bars 10 and 11.   P. 1669.

Taking these details into consideration, it seems safe to conclude that since Bach in general preferred to write out all the ornaments he wished used, and, as has been shown, wrote them out so very often in the recitatives, he must mean that they were not to be introduced in all the other places where they might perhaps be deemed suitable. I fear, however, that this summary method of proceeding would hardly fulfil his intention. Agricola, on whose works we have just been founding our opinion as to Bach's custom of writing out these embellishments, adds, after he has mentioned this as his custom, that we must at the same time be careful to distinguish between the places where supplementary notes are essential and those where they are only accidental and non-essential, and that a passage which is beautiful may have the possibility of becoming by many degrees more beautiful. When Bach wrote out an ornament in any passage, he regarded it as essential to that passage. We can see this from the fact that very often he supported the ornament by the harmony of the accompaniment, while the general custom was to leave the accents, which must always be dissonances, quite free. Proofs of this are afforded by the last two examples. It cannot be denied that the ornament thus ceases to be a mere ornament; many passages lose their recitative character, and the tendency towards the arioso, which is the general characteristic of Bach's recitatives, is made still more prominent. Still, he can hardly have abandoned the general custom so completely that he would not leave to a trustworthy singer the task of improving upon an already beautiful work, as Agricola would say, by using even the simplest and most ordinary means of adornment. Scheibe's statement that Bach wrote down in actual notes "all" the *Manieren* and "all" the little ornaments, gives us most valuable information as to one of Bach's fundamental rules; but it would scarcely be right to take it literally and apply it to each individual case. Besides, it must be remembered that in the case of a singer not pleasing Bach by his performance, he could always correct him by word of mouth, and we have already shown in one case quoted above how he availed himself of this resource in his relations with his musicians.[249]

To me it seems undeniable that very many passages in Bach's recitatives become more flexible, more expressive, and altogether correspond better to their inner meaning and nature by the use of the accent; so that it may be assumed that Bach himself conceived them as forming part of his idea. In such cases, indeed, the final decision must be left to taste. Before following taste, however, it must be considered in every case whether positive grounds cannot be found for singing the passage as it is written, and not unfrequently will such grounds appear. In the St. John Passion, when Peter denies Jesus, Bach sets the words the first time thus:—

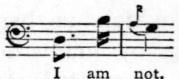

and the second time thus:—

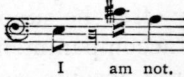

To introduce the accent in the second of these passages would be to destroy the composer's idea, since the growing excitement with which it ought to be sung is plain to every one.[250] Bach was very fond of this kind of psychological refinement; the example given above from the Matthew Passion, where, after an ascending second, the first and emphasised note of the two which follow is raised, corresponds to a passage that has gone before, in which the accent is also written out.[251] In both it is Pilate who speaks, the first time is a doubtful query, "Why, what evil hath He done?" And the second time distinctly deprecating the deed: "See ye to it." The identity of his feeling and opinion is expressed by the phrase being of similar construction in both cases. When the same cadence is repeated

---

[249] Compare also Rust's preface to B.G., XXII., p. xxi.

[250] B.-G., XII., p. 29, bars 14—15, and p. 33, bar 7. Novello's edition, p. 30, line 2 and 3; p. 36, line 3.

[251] B.-G., IV., p. 192, bar 13. Novello's edition, p. 125, bottom line.

twice with a short interval between, having an accent in one case and not in the other, we must assume that there is at least a reason for it. On this account it is right to sing the following passage from the Christmas Oratorio:—

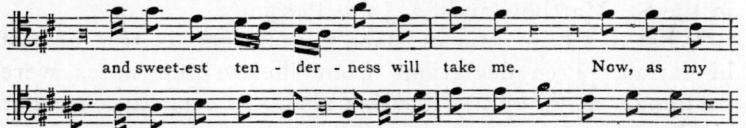

exactly as the notes stand, and without raising the last note but one.

All this, however, as has been said, is only true of accents, the simplest and commonest of the embellishments of recitative. The rarer and more elegant adornments Bach always—and this may be definitely affirmed—wrote out in notes; among these would be included the practice, usual in the case of several repetitions of the same note, of adorning one of the most emphatic of these with a mordent or something of the kind.[252] Passages in which Bach gave in to this custom in his own way, are for instance:—

Also all prolonged *fioriture*, like this:

---

[252] Agricola in Tosi, p. 155 f.
[253] Matthew Passion, B.-G., IV., p. 29, bars 9—10. Novello's edition, p. 23, line 3.
[254] "Cantate Ich hatte viel Bekümmerniss." B.-G., V.,[1] p. 30, bars 1 and 2, p. 31, bars 6—7. Novello's edition, "My Spirit was in heaviness," p. 28, lines 1 and 4.

and especially long florid passages [*Melismata*] at the close, which were very common in church-recitative.[255] These are all in the same category, as well as the two celebrated *Melismata* on the words "And he went out and wept bitterly," in the St. Matthew and St. John Passions.

What has been said of Bach's recitatives is true also of his arias. Free deviations from the written notes were customary at that time in arias also, and even to a greater extent in certain passages. They consisted partly of little embellishments of the melody by accents, mordents, and the like, partly of prolongation of the cadences, and partly of actual varying of the passages of the melody. This last method most commonly occurred in the third part of the aria. Its object was to prevent the hearer feeling wearied by a repetition note for note, or if the musician thoroughly felt his work, to heighten the effect of the first part and so to bring the sentiment of the piece to a higher pitch of passionate emotion. Of such variations there were again three kinds : to a passage of few notes more could be added, a passage of more notes could be simplified into fewer, or a certain number of notes could be exchanged for as many others.[256] In Bach, this last method could only be employed in the case of a true *da capo* aria, the first part of which comes to a full close in the principal key. This form is, however, by no means the rule in Bach's arias. He, who made it his first object to develop the materials at his command out of themselves, and to combine them one with another, could not long be contented with the cut and dried pattern of the *da capo* aria. Its form therefore appears in his case in manifold modifications, of which the most important consists in making the first part close in the

---

[255] Tosi-Agricola, p. 151.

[256] Agricola loc. cit. p. 235. In instrumental music the adagio movement especially used often to be played in this way. Examples are in Quantz, Versuch einer Anweisung die Flöte traversiere zu spielen. Tab. XVII.-XIX., and in Witting's edition of Corelli's Violin Sonatas. Wolfenbüttel, Holle. Two examples of varied arias, although of a somewhat later period, are given in Hiller's Anweisung zum musikalisch zierlichen Gesange. Leipzig, 1780. p. 135 ff.

dominant or some other nearly related key, so that the third part is not an exact repetition of the first but for its close contains another sequence of modulation; and besides this it very often happens that another aspect is given to the first part by new phrases and combinations. Thus, the tendency which led to the third part of the aria being altered and varied is by him intensified and endowed with meaning; not only is this part altered by means of outward adornment, but its inner nature is altogether changed. But in the case of a true *da capo*, we must remember Emanuel Bach's statement, that it was always understood that the accompaniment might be varied by simple alterations of the inner parts.[257] The polyphonic form of Bach's arias, the importance of each individual note of the melody, and the wealth of harmony allow, in most cases, of no alterations worthy the name. If a rudimentary knowledge of the rules of composition was always necessary for the proper execution of such variations, here it would surely be quite indispensable; but among the Thomasschule boys there can very seldom have been any who were capable of satisfying this demand. It is scarcely credible that Bach should have allowed his singers to do just what they liked with these profound and strictly-written compositions. With regard to the cadences the case may have been rather different. The adornment of these was confined, in the older periods of the art, to a shake on the second of the three notes which form the cadence proper. Subsequently a little ornament was introduced on the note before the shake, without however exceeding the proper length of the bar. It afterwards was carried farther, the last bar was sung more slowly, and at length began to be adorned with all manner of runs, skips, and other possible figures. This way of executing the cadence must have arisen between the years 1710 and 1716, and was in Bach's time in general use.[258] We can get a clear idea of it from a place where Bach, following his usual plan, has written it out; as, for example,

---

[257] Versuch über die wahre Art das Clavier zu spielen. Part II., p. 1.
[258] Agricola, loc. cit., p. 195 f.

II.

at the close of the second part of the first bass aria in the cantata "Freue dich, erlöste Schaar."[259] Where this has not been done Bach often gives the voice part such expressive passages at the close that there is not much room left for the exercise of the singer's fancy. That he permitted however a certain slackening of the time and a sparing introduction of ornaments appears very plainly from the numerous passages where all the accompanying instruments leave off before the final bars, with the exception of the figured bass, which goes on alone to the end. This cannot possibly have any other object than to give the voice opportunity and space for unrestricted and arbitrary movement. Finally, with regard to the small ornamentations in the course of the melody, the fundamental rule still holds good that Bach wrote them out whenever he considered them essential to the furtherance of the desired expression. In other cases he allowed his singers more or less liberty, since in the rehearsal he always had the power of directing them by word of mouth as he wished. That he relied upon this power, but also that in many places he did not consider it a matter of great importance whether an ornament was introduced or not, appears very clearly from a comparison of the instrumental ritornels with the voice part; frequently the one part exhibits ornaments which are wanting in the others, and this not only when one comes in after the other, but even when they are together. Whether any agreement can be restored between them, and if so, in what cases, can in our time be decided by taste alone.

These considerations extend also to the rendering of the choruses. It may appear strange that there can be any question of arbitrary ornamentation in them at least. But it is a fact that the so-called "Manieren" were introduced also into choral singing, and Petri even gives directions, as we learn, for extemporising inner parts in a four-part chorus."[260] This phenomenon is explained by the small constitution of the choirs and the slight difference that there was between solo and chorus singers. The singers

---

[259] B.-G., V.,¹ p. 347. P. 1017.   [260] Petri, loc. cit., p. 211 f.

were, indeed, divided into concertists and ripienists (*i.e.*, nearly the same as into soloists and chorus); there was no impassable distinction between them, however, but the former, besides taking the solos, sang the *tutti* movements as well, and thus constituted the proper nucleus of the chorus, the ripienists joining in to strengthen it. It is, nevertheless, a fact that the free introduction of " Manieren " often resulted in wild and inharmonious confusion, for which reason true musicians would have nothing to do with them. The only exception was when one part led off a theme adorned with " Manieren "; then the part that imitated it had to sing it in the same way, without the adornments being expressly specified; for the composer must have imagined his theme the second time to be the same as the first.[261] Whoever considers Bach's choruses knows that they are not wanting in all kinds of adornments, whether written out in full or only implied by the context. That these last had to be really and completely executed is, after all that has been said above, just as certain as that the singers were not permitted, as a rule, to introduce ornaments of their own invention.

The instruments added by Bach to complete and enrich the body of sound which before consisted simply of the organ and the voices, can only have been provided for in part from the town musicians. The rest must have been filled up by the scholars who could play. It was so both before and after his time, until Hiller succeeded in forming the entire orchestra from among the Thomasschule boys.[262] The limit thus set on the material, a limit regulated at that time solely by chance, must have proved a great hindrance to Bach in bringing out his works, had he treated the orchestra and chorus in the modern style. The older style of orchestral treatment had, it is true, defects of its own of a different kind. All wood wind instruments have a tendency to rise in pitch after playing some time. The strings can accommodate themselves to this, but not so the organ. Moreover, in the older style of treatment the wind

---

[261] Petri, p. 210.   [262] Gerber, N.L. II., col. 674.

was much more constantly kept in activity, and consequently the instruments got out of tune much sooner. Attempts were made in different ways to overcome this inconvenience. Johann Scheibe invented mechanism by which, by means of greater or less weight on the bellows, the pitch of the organ could be made higher or lower, and he attached it first to a small organ made in 1731, which had 12 manual stops, 2 manuals, and a 16-foot Fagott stop on the pedals.[263] It appears, however, that his invention did not meet with great success, since there was a simpler means by which the same object could be obtained; several of ach kind of wood wind instrument were kept in readiness, so that when one got out of tune by overblowing, another could be taken up.[264] The "chorus pitch," to which organs were generally tuned at that time, brought a difficulty with it, since most of the other instruments were tuned to the ordinary or so-called "high chamber pitch." This difficulty was usually overcome by transposing the organ part. If, however, it fell out that the only available wind instruments were those which were tuned to the "low chamber pitch," which was a semitone below the ordinary chamber pitch,[265] and that no transposed parts had been prepared for them, the strings were obliged to tune afresh. In Kuhnau's time the flutes and oboes which were used for the church music in Leipzig were of this pitch; and in Bach's cantata for Trinity Sunday, "Höchst erwünschtes Freudenfest," which dates from the beginning of his Leipzig period, they also occur. There can be no doubt that it was to prevent the continual retuning of the strings, which was detrimental alike to tone and purity of pitch, that Kuhnau often employed violins tuned to "chorus pitch" when no wood wind instruments were taking part. The trumpets stood as a rule in "chorus pitch," but were capable of being lowered to "chamber pitch" by an additional piece put on to the mouthpiece, so that the same

---

[263] Leipziger Neue Zeitung von gelehrten Sachen. XVIII. 833 f.
[264] Petri, loc. cit., p. 183.
[265] Adlung. Anl. zur Music Gel., p. 387.

METHODS OF CONDUCTING. 325

instrument could be used for D major and C major according to "chamber pitch."[266] And, lastly, the constant use of the whole group of trombones made it very often necessary to employ several players to relieve one another, because the bass trombone, in particular, when used, as was customary, to strengthen the bass part of the chorus, demanded an expenditure of physical strength for which a single player was incompetent. The treble trombone, when used in this way, was even more trying; this certainly was one reason for its being supplanted by the cornet, which is less fatiguing.[267]

The mode of conducting church music demands our special consideration, since in this respect, too, customs have much changed since the time of which we are speaking. They differed, moreover, among themselves, even in Bach's time. Johann Bähr, who was in his time Concertmeister at Weissenfels, says that one man conducts with the foot, another with the head, a third with the hand, some with both hands, some again take a roll of paper, and others a stick. Every ordinary director will know how to regulate his method according to place, time, and persons; whoever would give rules for general acceptation deserves to be laughed at. "Mind your own business, and let another man conduct as he likes, and do you conduct as you like; so there is no wrong done to any one."[268] All the styles mentioned by him have this in common with the modern practice that throughout the piece the time is visibly marked by a person who leads or conducts the rest. Pictorial representations, dating from the first decade of·the last century, which represent bands of musicians with conductors, make the matter quite clear. In a collection of engravings published before 1725, by Joh. Christoph Weigel, in Nuremberg, in which are depicted different kinds of musicians playing, there is a figure of a music-director who stands with a roll of music in each hand, directing a four-part

---

[266] See Appendix A., No. 17.
[267] Petri, loc. cit., p. 184.
[268] Joh. Bähr's Musikalische Discurse. Nürnberg, 1719, p. 171 ff.

motett, *Laudate Dominum*, from a score in front of him; beneath may be read these words:

> Ich bin, der dirigirt bey denen Music-Chören,
> Zwar still, was mich betrifft, doch mach ich alles laut,
> Erheb ich nur den Arm, so lässet sich bald hören,
> Was unsern Leib ergötzt, und auch die Seel erbaut.
> Mein Amt wird ewiglich, dort einsten auch, verbleiben,
> Wann Himmel, Erd und Meer in pures Nichts verstäuben.

> 'Tis I who lead and guide the tuneful choirs here;
> Silent myself, I cause the music I control.
> I do but raise my arm, and lo, at once ye hear
> Tones that enchant your sense and edify your soul.
> My sway survives the grave, and shall create delight
> When sky and earth and sea are sunk in endless night.

In other representations, the conductor, armed with his roll of music, stands near the organist and the bass player, at the organ, or separated from the organist and the trumpeters in the front of the organ loft, in the midst of the singers and fiddlers, who are grouped around him.[269] But the director was not always so "silent himself." Many cantors made use of a violin in conducting, so as to come to the rescue of the singers if necessary.[270] From the thirtieth year of the century onward the practice became different. It ceased to be the fashion for the conductor to stand and beat time all through the piece, and in time it became more and more usual to conduct from a harpsichord, that is, now to mark the time with the hand, and now to play the piece with the others, according as it was necessary, so that order was preserved, not merely by mute signs, but by audible musical influence. In very large performances alone, with a great number of executants, the older method remained in vogue, because it was indis-

---

[269] In the plates in "Der Durch das herrlich-angelegte Paradis-Gärtlein Erquickten Seele Geist-volle Jubel Freude bestehend In einem Kern auf allerley Anliegen und Zeiten angerichtete Lieder. Nürnberg, 1724." Also in "Altes und Neues aus dem Lieder-Schatze, Welcher von GOtt der einigen Evangelischen Kirchen reichlich geschencket," &c., published by *M.* Herrmann Joachim Hahn. Dresden, 1720, and also in Walther's Lexicon. Leipzig, 1732.

[270] (Voigt), Gespräch von der Musik, p. 36. Compare Petri, loc. cit., p. 172.

pensable. Besides this, the quiet unobtrusive style of conducting came to be a famous characteristic of musical performances in Germany, which compare very favourably in this respect with those in France, where they beat the time audibly with a large stick, and yet says Rousseau: "The opera in Paris is the only theatre in Europe where they beat the time without keeping it; in all other places they keep time without beating it." The practice of conducting from the harpsichord was widely imitated, because Hasse employed this method in Dresden with such happy effect that the opera performances there rose under his direction to a perfection very rarely surpassed. Rousseau has handed down to us a sketch of the disposition of the band under Hasse's direction.[271] From this we see what was indeed implied by the nature of the case, that the Capellmeister at his instrument did not undertake the task of playing the figured bass as well. For this a special *Clavecin d'accompagnement* was placed on the left side parallel with the front of the stage, while the Capellmeister and his clavier occupied the middle of the orchestra.

But the employment of the harpsichord as an instrument for direction had not been unknown even in earlier times. At Torgau a spinet (an instrument of the same genus as the harpsichord) was used in the Easter performances of 1660,[272] and in Leipzig there was in Kuhnau's time a harpsichord in the organ loft in both the Thomaskirche and Nikolaikirche, which he sometimes used; he preferred indeed to use the Italian lute, for he considered the penetrating tone of that instrument especially suitable for keeping the music together. When Bach entered on his duties he had the harpsichord in the Thomaskirche, which had become useless, set in order forthwith, and got the Council to expend the sum of six thalers a year upon keeping it regularly tuned, but it was out of use again in the year

---

[271] In his *Dictionnaire de Musique*, Planche G., Fig. I., reprinted by Fürstenau, Zur Geschichte der Musik und des Theaters am Hofe zu Dresden. II., p. 291.

[272] Taubert, Die Pflege der Musik in Torgau. Torgau, 1868, p. 18, note 3.

1733, till Easter, 1734, and also from Michaelmas, 1743.[273] In the Nikolaikirche, the organ loft of which was smaller, he used the harpsichord which was there, as it appears with a still longer interval during which it was not used; he first had it put in order for Good Friday, 1724, when he had a performance of the Passion Music in that church, and after that traces of its use begin to appear in the year 1732 and continue till 1750.[274] From the New Year, 1731, until the same date in 1733, Bach's son Philipp Emanuel, who entered at the university of Leipzig on Oct. 1, 1731, took charge of the tuning of the harpsichord in the Thomaskirche. With reference to this an opinion of his on the cembalo as an instrument for direction will be especially interesting. "The Clavier," he says,[275] "to which our practice entrusts the direction of the music, is of all instruments the best fitted to keep, not the basses alone, but also all the musicians, in the necessary equality of time; for even the best musician may find it difficult to preserve this equality, even though he may generally have his powers under control, or he may flag through fatigue. This being the case with one, the precaution is all the more necessary when many musicians are together, and the more so that by this means an excellent substitute is provided for the beating of the time, which is in our day only usual in the case of performances on a large scale. The notes of the clavier, which stands in the middle, surrounded by the musicians, are clearly heard by all. For I myself know that even performances on a large scale, where the performers are far apart, and in which many very moderate musicians take part voluntarily, can be kept in order simply by the tone of the harpsichord. If the first violinist stands, as he should, near the harpsichord, it is difficult for any confusion to ensue. In vocal arias, in which the measure is arbitrarily varied, or in which all the parts (of the accompaniment) sound together, and the voice part alone has long notes or triplets in which a very clear beat is

---

[273] Accounts of the Thomaskirche.
[274] Accounts of the Nikolaikirche.
[275] Versuch über die wahre Art das Clavier zu spielen. Part I., p. 5 f, note.

required, on account of the divided time, the singers' task is greatly facilitated by this method of conducting. The bass will find it easiest to keep up the equality of time when his part is least burthened with difficult embellished passages; and this often gives rise to the circumstance that a piece is begun with more vigour than it ends with. If, however, anybody begins to hurry or to drag the time, he can be corrected in the plainest possible way by means of the clavier; while the other instruments have enough to do with their own parts because of the number of passages and syncopations; and especially the parts which are in *Tempo rubato*, by this means get the necessary emphatic up-beat of the bar marked for them. Lastly, by this method—since the musicians are not hindered by the noise of the clavier from perceiving the slightest *nuances* of time — the pace can be slightly lessened, as is often necessary; and the musicians who stand behind or near the clavier have the beat of the bar given out in the most evident and consequently the most emphatic way before their eyes by both hands at once." Here we have a comprehensive statement of the advantages offered to a conductor by the harpsichord, from one well versed in the matter, and at the same time it is an open testimony that Bach availed himself of this method. For the words " our practice entrusts the direction to the clavier " can only have this meaning: that the person who directs the music, that is, the conductor, does so from the harpsichord and by its help. If we take the beautiful and lively description of Bach's conducting given by Gesner, who, by the way, represents him as sitting at the harpsichord, we shall get a clear and correct idea of him. Consistent with this picture is what Emanuel Bach and Agricola say when they are praising Sebastian Bach's facility in conducting: " In conducting he was very accurate, and in time, which he generally took at a very lively pace, he was always sure:"[276] for the use of the harpsichord did not by any means exclude an occasional beating of the time; the object of the instrument was only

---

[276] Mizler's Necrology, p. 171.

to keep the thing going, and quickly and imperceptibly to restore any defaulters to the right way.

Of course, for the most part the score was used in conducting, but sometimes, if this was for any reason impracticable, a simple conductor's part was used. The form of this part explains very clearly the task of the director. It has two staves, the lower for the bass and the upper for marking the points where the conductor's support was necessary. Generally the upper part alone is written, and in fugal movements the entrances of the parts are generally expressed by means of the clefs corresponding to the parts, generally in the bass stave, but sometimes in the upper one too. Besides this everything is given which is needful for information concerning the disposition of the work and the materials used in it; whether a particular movement is executed by an instrument or a singer, and by which, and whether for several voices or for all; whether they sing with or without instruments; when the ritornels come in, and so forth. Kuhnau, in leading the performance of his Passion according to St. Mark, used such a "*Directorium sive Quasi-Partitura*"; and such a director's part, which seems to have been used also for a figured bass part, to his cantata "Welt ade, ich bin dein müde," is still in existence.[277]

The method of using the clavier as an instrument for direction proved itself so good that it remained in vogue down to our century, for instance, in the performances of the Berlin Singakademie. It was also employed in purely instrumental works, and Haydn conducted his symphonies at Salomon's concerts in London from a harpsichord.[278] But at that time, besides the harpsichord director, there was a special conductor as time-beater. In the performance of Haydn's *Creation*, which took place at Vienna, in 1808, Kreuzer sat at the harpsichord, and Salieri conducted the whole.[279] In the year 1815, when Beethoven's *Christus* was

---

[277] In the Stadtbibliothek in Leipzig.
[278] Pohl, Mozart and Haydn in London. Second part, Vienna, Gerold. 1867, p. 119.
[279] Gerber, N.L. II., col. 557.

given at the same place, Wranitzky conducted, and Umlauf was at the clavier.[280] In the Berlin Singakademie, Zelter in his later years let one of his pupils, Rungenhagen or Grell, play the harpsichord, while he himself only beat time.[281] Here the clavier player had also to accompany the *recitativo secco* from the figured bass, which in Bach's performances was the duty of the organist. The harpsichord as an independent instrument crept gradually into church music, but yet the school of Bach used it only to strengthen the tone in such recitatives and arias as the composer had intended to be performed without organ accompaniment. Emanuel Bach at least recommends this practice; but Rolle, on the other hand, considers that such a strengthening of tone in the church is more or less an illusion, and that the constant new-quilling and tuning which were necessary in the winter months made the use of the harpsichord in the church difficult and expensive.[282] This, perhaps, was the reason why Bach gradually gave up its use in his later years. From 1730 onwards, he used instead the independent and practicable Rückpositiv organ in the Thomaskirche, and with greater convenience, as he could play the figured bass himself without being obliged to turn the organist out of his seat.[283] How he used to manage when the performers were so numerous as to require him to beat time throughout—whether in such cases he made some other persons, for example, his sons Friedemann or Emanuel, play the clavier, or whether he left it out and managed the whole thing by his beat alone, cannot be certainly known. If the first alternative seems to be probable, from internal con-

---

[280] Nottebohm, Beethoveniana, p. 37.
[281] From the verbal description given us by Herr Professor Grell.
[282] Rolle, loc. cit., p. 56. "The introduction of the harpsichord into church music has been strongly advised [alluding to the statement of Emanuel Bach given in Vol. I., p. 831]. Since, however, the sound rises in theatrical music, while in churches, on the contrary, the sound must come down from the organ loft, the introduction of the harpsichord would give no particularly emphatic support [E. Bach, loc. cit., had said that even in the loudest music, in the opera, and even in the open air, the harpsichord can be heard, if it be put on a raised place]."
[283] See Appendix A., No. 18.

siderations, the last is more likely when we remember that he could dispense with the clavier altogether for a whole year; thus he must, by his independent conducting, have developed great energy, certainty, and clearness, as indeed is expressly testified by Emanuel Bach and Agricola. What can be said as to the position of the performers has been for the most part given above in Emanuel Bach's words. We must, however, remark that the instrumental basses—or at least some of them, since, to ensure precision and time, a good many were employed—always used to be placed near the conductor, sometimes behind him as he sat at the harpsichord, so as to be able, in cases of necessity, to play with him from the score. Besides, the trumpets and drums were always put rather at a distance from the rest, so as not to drown the voices.[284] A favourite place for them was close to the organ, right and left of the organist.[285]

V.

KUHNAU.—THE CHURCH CANTATA.—TEXTS BY NEUMEISTER AND PICANDER.—COMPARISON OF THEIR MERITS.—BACH'S CHURCH CANTATAS.—THE "MAGNIFICAT."

KUHNAU, during the whole period of his being Cantor of the Thomasschule, waged an unequal, and, as far as he was concerned, an altogether unsuccessful warfare with the opera and all connected with it. In all his complaints one thought is uppermost: namely, that if he only had at his command a sufficient number of performers to take part in the church performances he would be able entirely to nullify

---

[284] Scheibe, Critischer Musikus, p. 713 ff.

[285] See the plate in the collection of songs by Hahn mentioned above, and also in Hiller's description of the performance of the *Messiah* in 1786, the appended ground-plan of the places of the performers. Compare the ground plan of the Dresden orchestra where two platforms were built for the trumpets and drums on the right and left sides; and lastly, Petri, p. 188. Hiller's ground plan is very interesting, both historically and musically, even though it is of little use for our purpose, since Bach did not dispose his great numbers of performers in the ordinary way, and in Handel's oratorios the harpsichord played quite a different part.

the pernicious influence of the theatre, and to promote the triumph of a more earnest musical feeling in Leipzig. He was mistaken: the fact that his influence on the public musical taste constantly diminished was only partly due to outward circumstances; the blame was equally due to his own character and that of his musical talents. In a time when the old and the new are striving together for the mastery, he alone is a successful leader of the public taste who is capable of understanding and recognising the rights of both. The power that was struggling for expression in the opera forms was unintelligible to Kuhnau and foreign to his whole nature. He had more than once attempted to enter the domain of dramatic music as a composer. In earlier life he himself translated a libretto on the story of Orpheus from the French and set it to music; it is not known with what degree of success.[286] But another opera of his, which must have been written very late in life, made a distinct *fiasco*.[287] It was his weakness not to perceive that the most versatile cannot do everything; otherwise he might have avoided giving such practical proof that his aversion to the opera was the result of his failure. Kuhnau was a master in the sphere of clavier music, and many considered him equally great in church music. There is no question that he showed abilities in this branch of art, which raised him above his contemporaries. He was better versed in the technicalities of vocal writing than most other German composers of the time. His five-part motett for Holy Thursday, *Tristis est anima mea usque ad mortem*,[288] may be reckoned among the more prominent works of the kind; if it is not of equal merit with the motetts of Joh. Christoph and Joh. Ludwig Bach

---

[286] "The foolish fellow now got upon the subject of the opera of 'Orpheus,' which I had formerly translated from the French into German poetry, and likewise composed the music." Kuhnau, Der *Musicali*sche Qvack-Salber, p. 456, compare p. 458 ff.

[287] Scheibe, Critischer Musikus, p. 879, note: "But notwithstanding his (Kuhnau's) great merit, we know well how badly he succeeded when he undertook to set an operetta to music, and put it on the stage."

[288] It exists in the separate parts in the library of the Leipzig Singakademie and is numbered 362.

even in technical qualities, it has a breadth of conception which betrays the study of the classical Italian models. A chamber cantata, *Spirate clemente, o zephyri amici*,[289] also shows that Kuhnau endeavoured to form himself on the style of the Italians. There exist seventeen church cantatas, written at different periods of his life.[290]

Scheibe, who considers Kuhnau, Keiser, Telemann, and Handel the greatest German composers of the century, says: Kuhnau "is now and then carried away by the flood of harmonic ideas; hence, he is often dull and devoid of the requisite poetic beauties and ornaments of expression, and consequently here and there he becomes too prosaic. That he was aware of this himself, however, and that sometimes he succeeded in writing deep and poetic music, is shown by his things for clavier, and by his last sacred works, especially his Passion oratorio, which he finished a few years before his death. In these works we see how clearly he understood the employment and laws of rhythm. We see too, that he was always careful to make his sacred works melodious and flowing, and in many cases really affecting, though he was not so happy in his dramatic work."[291] These words prove—what is plain from the works themselves—that by degrees Kuhnau tried even in his church music to make his style similar to that which prevailed in opera. The statement that he was "carried away by the flood of harmonic ideas" is not to be taken so much in the positive sense, *i.e.*, that Kuhnau immersed himself to too great an extent in polyphonic obscurity, as in the negative sense: namely, that he did not always give due consideration to the importance of melody and of variety of rhythm. In his earlier works, *e.g.*, a cantata "Christ lag in Todesbanden," dating from the seventeenth century, he keeps entirely to the style of the so-called "older" church

---

[289] In the Royal Library in Berlin.

[290] Ten are in the Royal Library in Berlin, and seven in the Town Library in Leipzig. Among the latter is a Christmas cantata "O heilge Zeit, wo Himmel, Erd und Luft," which however is certainly not by Kuhnau, but is the work of a younger master.

[291] Critischer Musikus, p. 764.

cantatas, of which illustrative examples by Buxtehude were considered in an early part of the present work. His style never radically altered from this, even in later life; though he adapted himself to the operatic style in many ways, he still composed to words by Neumeister, or in Neumeister's manner, so that a compromise was the result.[292] This is quite clearly seen in the recitatives. Bach's recitatives have always a strongly marked melodious character, but this style was invented by him, and founded on the dramatic recitative of his time. Kuhnau's recitative still retains the arioso form of the older church cantatas, varied, however, with the new-fashioned recitative phrases. Of the aria form in three sections he has left several excellent examples; one duet for alto and bass, in the cantata for Ascension, "Ihr Himmel jubilirt von oben," the polyphonic writing of which is very flowing and ingeniously developed, must be allowed to be a masterpiece. But Kuhnau felt himself much more at home in the old simple form of hymn, with its short and pleasing staves of melody and time-honoured ritornels.

Of the choruses, it can only be said that here and there they show attempts at a broader and more artistic development, but generally in a tentative manner; as a rule, they alternate between homophonic vocal movements and meaningless interludes, or between little solo portions and *tutti* movements. In the construction of the chorale, again, there is nothing more than an attempt at development. There is no free contrapuntal movement of the parts, and we have to content ourselves with a few meagre imitations, as, for instance, in the final chorus of the cantata "Christ lag in Todesbanden" or at the beginning of the cantata "Wie schön leucht't uns der Morgenstern." When the orchestra joins in with the chorale in figurations, it is very far from being employed in an independent way; in the Christmas cantata "Vom Himmel hoch" the chorus sings this chorale in the

---

[292] Neumeister's cantata "Uns ist ein Kind geboren" (see Vol. I., p. 486) was not performed in Leipzig till Christmas, 1720, but the words of church music even in the year 1711, are completely in Neumeister's style. For example Kuhnau's cantata "Und ob die Feinde Tag und Nacht" is written on a poem in this form.

simplest style of homophony, and the instruments support it, only the first and second violins having ascending and descending semiquaver passages quite in the style of the final chorus of Bach's early work, "Uns ist ein Kind geboren." The chorale "Gelobet seist du, Jesu Christ," in the cantata "Nicht nur allein am frühen Morgen," is treated in a similar way, but with the addition of a bass. Everything he writes is clever and in a flowing style, and hence the effect is pleasing; agreeable and even pathetic passages are frequently to be met with; but depth of feeling and grandeur of form are wholly wanting. Kuhnau must rank with the group of writers of the older church cantatas, because he had nothing in him to say which could not perfectly well be said in the forms of those cantatas; and this is equally true of the introductory instrumental symphonies, though his independent instrumental music was very admirable. His Passion according to St. Mark, composed for Holy Week of 1721, to which Scheibe gives especial praise, exists only as a sketch.[293] It can, however, be very plainly discerned, even here, that the prevailing characteristics of Kuhnau's church music have left their mark perhaps more strongly than ever, and that the composer endeavoured to assimilate the more emotional style of operatic music "by poetic beauties and ornaments of expression," *i.e.*, by the invention of such turns and phrases as had a certain innate dramatic value. His inmost nature, nevertheless, remained absolutely unchanged even in this. Kuhnau did not understand the world, nor did the world understand him; it was time that they should take leave of one another.

The position in which Bach found himself with regard to theatrical music was quite different. He had mastered the principles of all its forms, and had turned them to account for his own art. However unlike operatic music his compositions may seem, he was by no means devoid of an inner sympathy with that style, but rather held that it was

---

[293] "*Directorium sive Quasi-Partitura Passionis ex Evangelista Marco.*" A copy made by Burgmeister, in 1729, is in the Royal Library at Königsberg in Prussia, Department "Gotthold's Library."

a just demand of the time that respect should be had to it even in church music. The condition of music, as he clearly explained to the Council in his memorial on the improvement of church music, was quite different from what it had formerly been; the art had made considerable progress, taste had altered in a remarkable manner, the old-fashioned style of music in which Kuhnau was still writing had ceased to have any charm for the ears of his contemporaries. The interest which he took in the Dresden operatic performances, for instance, is well known, and a number of secular works in dramatic form by him still exist. So far as we know, however, he never wrote an actual opera. For although he may not perhaps have agreed with the opinion of Gottsched, who about this time gave out in Leipzig that the opera was the most preposterous absurdity that had ever been invented by the human mind,[294] it is yet conceivable that the glitter and glamour of this branch of art, which only serves for the entertainment of an hour, must have been antagonistic to his earnest, true, and deep artistic nature. Whenever he conceived the wish to go to Dresden he would say to his favourite son, "Friedemann, shall we go to Dresden again and hear their beautiful little songs?"[295] If the actual words of the expression have been handed down correctly, Bach's relation to the opera is characterised in them with striking brevity. Scheibe says, incidentally: "There are some great geniuses, who use the word 'song' (Lied) as a term of abuse; when they want to speak of a piece of music which is not sufficiently pompous and intricate for them, they call it a 'Lied.'"[296] Since a pompous and intricate style is just what Scheibe accuses Bach of,[297] it is more than probable that the words we have quoted were intended for Bach. The simple construction of the operatic forms seemed to him quite inadequate for the realisation of his art-ideal, and in this sense he may often have spoken disparagingly of them. But when employed by a Hasse and interpreted by a Faustina he could yet think them "beautiful,"

---

[294] Gottsched, Versuch einer Critischen Dichtkunst. Leipzig, 1730, p. 604.
[295] Forkel, p. 48.
[296] Critischer Musikus, p. 583.   [297] Idem. p. 62.

II.

and he knew just as well as his critic did that without operatic music he would not have been what he was.[298]

Bach's historical position in art can only be fully understood by regarding him not as opposed to this music, but as accepting all he could from it. The opera in Germany was incapable as yet of becoming a living musical drama, nor could such a change be effected at that time by the hand of a German alone. Under freer, broader conditions, such as were offered in England, it became the oratorio under Handel, while in Germany it was developed into Bach's church music. In England the result was attained by combination with the forms of Italian sacred music, and in Germany by the complete purification which it acquired by means of the national art of the organ. The history of the development of German opera can be clearly traced. After it had risen, by about 1700, to be a considerable musical power, it sank rapidly from its height, and in the next thirty years it had almost ceased to exist, until it was revivified in the latter half of the century by an impetus proceeding chiefly from France, and shown in the operettas of Hiller and Weisse. In Leipzig itself, it was all over with German opera by the year 1729. What came to replace it was Bach's music. This contained what Kuhnau had vainly striven after: namely, the spirit of the time, in so far as it could find fitting musical expression in operatic forms, and at the same time the genuine church style. Whether Bach did anything consciously and directly to extinguish the flickering flame of the opera at Leipzig is not known. But that his art tendencies speedily became predominant there cannot, under existing circumstances, be any doubt. It has been generally accepted and stated as a matter of fact, that Bach's church music was not understood, and soon forgotten, and that the mighty works of his creative genius failed to meet with due appreciation. I believe that too much stress is laid on certain expressions of disapproval, on some

---

[298] Idem., p. 591. Note.—"If we in Germany insist on bringing about the total banishment of operettas from the stage, we may be quite sure that we shall never again see a Hasse, a Graun, a Telemann, a Handel, or a Bach."

measures taken by the magistracy which are not properly understood, and on the partially insufficient means which Bach had at his command.[299] With regard to the last point, it may well be asked whether Handel was always so much better off for his oratorios, or Beethoven for his symphonies; and whether an eminent genius ought not to be capable of doing wonders with small means? The high respect in which Bach's name and music remained throughout the century at Leipzig, the extensive influence which he exercised upon the music of Northern and Central Germany, and the fact that many of his sacred vocal compositions found their way into Saxony and Thuringia, may serve to show that we are justified in concluding that his work in Leipzig made its mark.

There had never been any good writer of words for cantatas in Leipzig. In 1716 Gottfried Tilgner had collected five annual series of Neumeister's poems, by permission of the author, and published them under the title of "Five-fold church devotions."[300] These poems, which had hitherto been disseminated privately, were now brought within the reach of every one, and had such a sale that a new edition was demanded in the following year. Tilgner, a young literary man, lodged in the house of *Magister* Pezold, a colleague of Kuhnau, whom we have frequently mentioned.[301] Kuhnau had undoubtedly set many texts by Neumeister, and besides he was very capable of writing texts himself on Neumeister's pattern. Bach, however, had no such skill in verse making; therefore he was at once obliged to look about for a poet, and he chose Franck in preference to Neumeister.

But he had not to wait long before he found in Leipzig

---

[299] Bierey, the well-known Music Director at Breslau, made the acquaintance, as he told Julius Rietz, of an old church servant in Leipzig, who had been employed in Bach's time. He fully agreed with Bierey in his admiration of the master, as far as concerned his power as an organ or clavier player, and his opinion of the cantatas was thus expressed: "Ah! but you should have heard them!" It is not known, however, whether this opinion was shared by the general public.

[300] See Vol. I., p. 474.

[301] Sicul, Die andere Beilage zu dem Leipziger Jahr-Buche, for the year 1718, p. 187 ff. This gifted and industrious man was driven by melancholy and overwork to commit suicide in 1717.

itself an adequately skilled and always willing collaborator. Christian Friedrich Henrici, born at Stolpe in 1700, had studied at Wittenberg, and had lately settled in Leipzig, where he was living for the present in poor circumstances chiefly by writing "occasional" verses. In two of his poems he petitioned the King-Elector to grant him free board; and in 1727 dedicated to him, through Count Flemming, a cantata for his *fête* day on August 3, beginning " Ihr Häuser des Himmels, ihr scheinenden Lichter."[302] In the same year he obtained a situation in the Post-office, and in the Leipzig Directory of 1736 he figures as *Ober-Postcommissarius*.[303] In 1743 we again come across his name as a tax-gatherer and exciseman, and in this capacity he died in 1764. The higher officers of the churches and schools received a certain annual sum as compensation for the general tax on liquors; this was a special favour.[304] To this circumstance a small document in Bach's own writing owes its origin; it is a receipt given to the tax-gatherer at Easter, 1743, for a compensation for three casks of beer.[305] But the intercourse of the two men was not simply on matters of business; it had been for a long time of a friendly and artistic character. Among the sponsors to one of Bach's children, born in 1737, was the wife of Henrici, who had then been working for more than twelve

---

[302] State archives of Dresden.

[303] Das jetzt lebende und florirende Leipzig, 1736, p. 14.

[304] "Fruuntur nostri privilegio potus a collectis cerevisiae exemti." (They enjoy a privilege of having liquors exempted from taxation.) Kuhnau, Jura circa musicos ecclesiasticos. Leipzig, 1688. 4, Cap. VI., § 1.

[305] "Having duly received from Herr Christian Friedrich Henrici, appointed exciseman of the district, from *Quasimodogeniti* (*i.e.*, first Sunday after Easter), 1742, until same date 1743, by his Majesty the King of Poland and Serene Elector of Saxony, according the Electoral decree of November 9, 1646, for three casks, each at forty ggr. making altogether five thlrs.

Five Thalers

for the two taxes in money and kind, I hereby acknowledge the receipt of the same most thankfully. Season of *Quasimodogeniti*, 1743, at Leipzig.

"*Joh. Sebast: Bach.*"

And beneath are the autograph signatures of Deyling and a certain Johann Andreas Vater. The impression of Bach's seal represents a rose surmounted by a crown. This document, in October, 1870, formed part of a collection of autographs in the possession of the late General-consul Clauss of Leipzig.

years in collaboration with Bach. Henrici began his literary
career in 1722, as a satirist; in this respect we may call
him a disciple of Christian Günther, though, as he is not
to be compared to the last-named poet in talent, his mean
style and bad taste are all the more repulsive. He was
incapable alike of Günther's free and picturesque imagery,
and of his audacious licence. When his satirical poems
created ill feeling he was frightened, and declared that he
had only the best intentions in writing such productions,
but that the unfortunate results had spoilt the fun, and the
threats of the evil-disposed had deprived him of all his
pleasure in it. He now only wrote poems from time to
time to please his patrons and friends, but he does not
deny that he used "a sharpened pen." In the year 1724
he turned his hand to sacred poetry. He thought it
incumbent upon him to give a public explanation of this
sudden change. "He imagined that many people would
laugh to see him assuming a devotional attitude. He wished
to guard, however, against the imputation of having been
quite unmindful of heavenly things, and considered it only
right to offer to his Creator the fresh fruits of his youth,
and not the worn out remains of his old age. Let not
anyone blame him for making poetry his chief employment
and troubling himself little or nothing about other branches
of learning. If necessary he could produce credible testimony to his academic diligence. Besides, verse-making
was easy to him, and took him very little time. Why
should he not then employ this natural gift and turn it to
account for his living?" The work to which these and
other remarks served as preface was entitled: "Collection of
profitable thoughts for and upon the ordinary Sundays and
holidays"; he uses here the pseudonym of Picander, which
he adopts from this time. The work consists of meditations
in rhyme, mostly in Alexandrines, to which a set of verses
to the melody of some church hymn is usually appended.
They did not at first appear in a collected form; for a year
the poet was in the habit, on Saturdays and Sundays
after vespers, when other people were enjoying themselves
in unseemly dissipation, of putting into rhyme his thoughts

on the Gospel, and of bringing them out week by week on a half-folio sheet. To this custom he adhered; the first piece was written for the first Sunday in Advent, 1724, and the second and last for the same day in 1725.[806]

Considering that he had now made ample atonement for his sins, and that his reputation was firmly re-established, he once more threw himself into the arms of the secular muse. In 1726 there appeared, written by him, three German plays: "Der akademische Schlendrian," "Der Erz-Saüfer," and "Die Weiberprobe," "designed for amusement and instruction." They are low, repulsive farces,[307] but the tone which pervades them was his natural element, and he owns that it would be pleasanter and easier to him to sing four bridal songs than to grind out even one dirge. His chief object in writing these things was not attained, however: that of gaining enough to live on. He therefore brought out in the year 1727 a new collection, consisting of "Grave, gay, and satirical poems," and dedicated them "To fortune who will surely yet be kind to him, and grant him something more." Fortune was kind, and he got his situation in the Post-office. In 1728 there followed a collection of texts for cantatas in Neumeister's style.[308] This was the only collection of the kind which he brought out, and he subsequently incorporated it with his "Grave, gay, and satiric poems," to which four more parts were gradually added.[309] Henrici considered himself an original genius

---

[806] The title of the first part runs thus, differing slightly from that of the complete edition: "Sammlung | Erbaulicher Gedancken, | Bey und über die gewöhnlichen | Sonn- und Festtags- | Evangelien, | Mit | Poetischer Feder entworffen | Von | Picandern. | Leipzig, | Gedruckt bey Immanuel Tietzen." | 8. The Preface, which is wanting in the complete edition, is dated November 30, 1724. In the complete edition it is replaced by a short notice to the "kind reader" dated First Sunday in Advent, 1725; it is dedicated to Count Sporck. A copy is in the Royal Library at Berlin.

[307] On these and their connection with Weise's plays, compare Gervinus, Geschichte der deutschen Dichtung. Three Vols. (5th Ed.) p. 600 f.

[308] "Cantaten | Auf die Sonn- | und | Fest-Tage | durch | das gantze Jahr, | verfertiget | durch | Picandern. | Leipzig, 1728." The preface is dated June 24, 1728. A copy is in the Royal Public Library in Dresden, *Lit. Germ. rec. B.* 1126.

[309] The second in 1729, the third in 1732, the fourth in 1737, and the fifth in 1751. The series of cantatas is reprinted in the third part.

and a pioneer of public taste. He "foresaw," he says in the preface to the Sammlung erbaulicher Gedanken, "that imitators of his style would speedily arise; he wished them better success than he himself had met with. For his part, he would rather try an unbeaten track than follow in the footprints of another." And then he goes on alluding to the strong influence of English literature, which was then beginning to make itself felt: "Everybody at Leipzig wishes to be critics, patriots, and moralists, without having tested their powers, and it can only be regretted that the world-famous name of Leipzig should be used as a vain shelter for such unworthy productions. Nothing is more praiseworthy than the foresight with which the authorities have suppressed wretched trash of this kind, and these measures will go far to free Leipzig from the present polluted condition of its literature; Leipzig, so renowned even in foreign lands for good and refined taste. Well may such abortive productions hide their heads at last for very shame of their imperfection." He himself shines most in his satirical writings. They exhibit a certain keenness of perception, and a knowledge of human nature unusual in one so young, and they are very skilfully rhymed.[310] His giving up this line of work so soon is a sign that his satires were not so much the spontaneous production of his mind as imitations of others, suggested by outward circumstances. In the Epithalamia, the most numerous of his poems, a few pretty ideas are entirely overpowered by the dulness of the rest, while the plain-spoken improprieties are all the more odious from the weakness of the whole tone of the work. In spite of this, the fact remains that for a whole generation his poems enjoyed great popularity; they reached four editions before the year 1748, and no doubt they reflect with considerable truth the poetic taste of Leipzig at the time.

In his sacred poems, Picander shows even less original talent than in the satires and the secular occasional verses.

---

[310] The best specimens are to be found in the first edition of the first part of Ernst-, scherzhaffte und satyrische Gedichte, p. 477—566.

This branch of his art was utterly foreign to his nature, and probably he never would have attempted to write poems for cantatas had not Bach been in want of a versifier, and had it not been important to him, striving as he was for the mere necessaries of life, to be brought into connection with the celebrated composer. It is clearly perceptible too that Bach fashioned him for his own purpose. Many indications point to the fact of Bach having employed him in the beginning of 1724, or perhaps even for the Council-election of 1723. The first sacred poem by Picander to which he is known to have composed the music—the Michaelmas cantata "Es erhub sich ein Streit," was written in 1725. The "Sammlung erbaulicher Gedanken," begun in the previous year, afforded no opportunities whatever to the composer; but in this cantata Neumeister's form of poetry is used with success, at least in the recitatives, though the texts of the arias betray the hand of a beginner. As Picander had previously written words for several occasional cantatas, he cannot have found it hard to acquire the knack of getting the right form in the sacred cantatas. Many turns of expression show the influence of Neumeister, and especially in one particular, that he tries to give his diction an ecclesiastical tone by a free admixture of Scriptural expressions, and of allusions—often extremely far-fetched—to Biblical events. In the year 1725 he wrote for the first time a Passion poem, taking Brocke for his model. From this time the intercourse between Bach and himself became closer. Picander was himself not without musical talent and knowledge, and in this respect he had one advantage over Neumeister, to whom indeed he was not inferior in his easy use of language. In his secular poems the allusions to musical matters are of frequent occurrence, and they go into such detail that we may conclude that he not only took a lively interest in it, but practised it also himself.[311] In one place he even gives two very pretty dances as a musical

---

[311] Compare "Das Orgel-Werck der Liebe," written in 1723, in "Ernstscherzhaffte und satyrische Gedichte," 1727, p. 303 ff., and "Die Vortrefflichkeit der Music," written 1728, in the collected edition of 1748, Vol. II., p. 662. ff. and the same, p. 621 f.

appendix,[312] and from a poem of the year 1730 we learn that he was a member of a musical society—which must have been the one conducted by Bach.[313]

In the preface to the year-book of cantatas written in 1728-1729, Picander says: "To the glory of God, and actuated by the requests of many good friends, and by much devotion on my own part, I resolved to compose the present cantatas. I undertook the design the more readily, because I flatter myself that the lack of poetic charm may be compensated for by the loveliness of the music of our incomparable Kapellmeister Bach, and that these songs may be sung in the chief churches of our pious Leipzig." This cycle of poems was thus directly intended for Bach, and it seems to owe its origin to an unexpected wish expressed by him, since it does not correspond with the ecclesiastical year, but begins with St. John's Day, ending with the fourth Sunday after Trinity. For Good Friday, 1726, Picander wrote the text of the St. Matthew Passion, this time, however, not imitating Brocke's plan, but keeping the Bible words unchanged. Here again Bach's influence is easily perceptible, and it may be also traced in the circumstance that Picander seems to have borrowed some of his ideas, at least, from Franck. It is impossible to give any positive evidence of this, since Franck's skill in cantata writing was never anything but moderate. Still, a comparison of certain portions of the German text will seem sufficient proof to the reader who cares to search into the matter, and as Bach certainly loved Franck's work for the sake of his fervent and rapturous sentiment, it can only have been he who referred Picander to Franck's works.

After 1729 Picander published only a very few sacred poems, but it must not be supposed that he ceased writing them altogether. We may indeed feel certain of the contrary, for a lasting intercourse remained between him and Bach, and he was the only person in Leipzig who could undertake tasks of this kind with adequate skill. If we are not to

---

[312] Poems of 1727, p. 540.
[313] Collected edition, Vol. II., p. 819 ff.

regard Picander as the author of most of the other cantatas which Bach wrote in Leipzig, it is inexplicable that Bach should never have set to music a single one (so far, at least, as we know at present) out of the numerous collections of cantatas which appeared at that time, and which must have been accessible to him.[314] After what has been said, Picander's omission of these cantatas from his collected works is easily explained; he put no value on these manufactured compositions, which were put together hastily and to please his friend. Franck wrote out of a true poetic inspiration; Neumeister as an active theologian and preacher; while Henrici did not feel himself impelled to writing sacred poems by any genuine or hearty interest in such things. His impulse came solely from Bach, and this explains the pains he took in turning to account the productions of others and remodelling them for Bach's purposes; and of this procedure his treatment of Franck's hymns is not the only example. Bach also took an interest in the writing of Johann Jakob Rambach, several of whose devotional works he had in his library. He never used one of Rambach's cantata texts, however, although they are as good as anything of their kind. But he seems to have drawn Picander's notice to a pretty madrigal by Rambach ("Erwünschter Tag"):—

> O wished-for day,
> To be engraved on marble,
> Or metal that will ne'er decay—

for the same idea occurs in the text of one of Picander's Christmas cantatas ("Christen ätzet diesen Tag"):

> Christians, grave ye this great day
> In brass and stone that will not perish.

In a cantata by Rambach, for the Feast of the Purification, one of the arias begins thus ("Brechet, ihr verfallnen Augen"):—

> Rest, O eyes so dim and weary,
> Close in slumber's sweet repose.

---

[314] *E.g.*, the poem written by J. D. Schieferdecker and E. G. Brehme for music, in the castle churches at Weissenfels and Sangerhausen, in the years 1731-1735. Bach was Capellmeister at Weissenfels, and in that capacity had to perform certain duties at Court.

And in one of Bach's cantatas on the same feast there are these lines ("Schlummert ein, ihr matten Augen") :—
> Slumber now, O eyes so weary,
> Close in soft and sweet repose.[315]

Bach had composed music to an ode by Gottsched for the obsequies of Queen Christiana Eberhardine, and afterwards wished to put the music to some other purpose; Picander gave his assistance and wrote a new poem, so that the funeral ode became a Passion according to St. Mark.[316] He was just as willing and ready to put new words to compositions on his own texts, if the music could thereby be made available for other purposes.

It is worth while to draw attention to the difference which existed between the sister arts of music and poetry as regards their art-ideals. Music as applied to religion had attained a height which must be allowed to be unapproached before or since, in respect both to depth and richness of substance, and to variety and breadth of form. The art of sacred poetry, however, so far from rising to a similar level, had sunk, under the successors of Neumeister, to be a false and hollow mockery. It would not be too much to say that the influence of the cantata-poem upon the development of poetry at that time was really ruinous. For these collections of texts, although made for the special purpose of musical treatment, ere long asserted their claim to be regarded as independent creations. Originally printed separately, in order to enable the congregation to follow the words during the music, they soon came to be considered as devotional works on their own account, and were, in fact, sold as such in large numbers. Very many

---

[315] "*M* Joh. Jacob Rambachs, | *HALLENSIS*, | Geistliche | Poesien, | Davon | Der erste Theil | Zwey und siebenzig *CANTATEN* über | alle Sonn- und Fest-Tags-Evangelia; | Der andre Theil | Einige erbauliche Madrigale, | Sonnette und Geistliche | Lieder | in sich fasset." (Magister J. Jacob Rambach of Halle's Spiritual Poems, of which the first part consists of seventy-two cantatas on all the Gospels, for Sundays and Festivals; the other part contains several madrigals, sonnets, and sacred hymns.) Halle, 1720, pp. 218 and 257. B.-G., XX.,[1] p. 37.

[316] This matter has been cleared up in a conclusive manner by W. Rust, in the preface to B.-G., XX.,[2] p. VIII. f.

of these texts were never set to music; for Picander wrote cantata-poems even for the Sundays on which, as he knew very well, there was no music in church—viz., for the Sundays in Lent and the three last in Advent. Those who chiefly threw themselves into this branch of poetry were persons who either had no poetic faculty at all, or in whom whatever talent there might be inclined to another kind of work; to the last class belonged Picander, as well as another writer well known at that time, Daniel Stoppe, of Silesia. The gigantic advance made by Brocke in his "Irdisches Vergnügen in Gott;" and by Gellert in his odes and songs—not to mention Klopstock, who came somewhat later—can only be perfectly estimated by a comparison of their works with the great mass of the cantata-poems in vogue in Bach's time. Still, it must be admitted that the new spirit which appeared in the works of these men was unsuited and opposed to musical setting; their poems were to attain their end by their own poetical inspiration, independent of other aids. During Bach's period, artistic feeling and emotions in the domain of religion found vent almost exclusively in music, and from the moment when sacred poetry made itself felt as a prominent influence religious music began to decline. The overpowering predominance of the musical factor in this kind of work is very clearly seen in Bach's relations with his Leipzig poet. It might have been fatal to him, for it is not possible that church music can be genuine and good which utterly disregards the particular religious sentiment or emotion called up by the poetry; indeed, Keiser, Telemann, and Stölzel, although their gifts were by no means small, had succumbed to this very danger. Bach triumphed over it, because, however fully and comprehensively he represented all the musical inspirations of his time, he yet remained faithful to the foundation of all Protestant church music, namely, the chorale.

Bach composed in all five complete "year books" of church cantatas for all the Sundays and holy days.[317] Now, by

---

[317] Necrology, p. 168.

deducting the six Sundays in Lent and the last three in Advent, and adding in the three feasts of the Virgin, and also the festivals of the New Year, Epiphany, Ascension, St. John, St. Michael, and the Reformation festival, we get a total of fifty-nine cantatas to the ecclesiastical year at Leipzig. It follows, then, that Bach must have written on the whole 295 church cantatas. Of these at least twenty-nine belong to the period before his coming to Leipzig; then the maximum of those written at Leipzig must be put at 266. As Bach was twenty-seven years working at Leipzig, he would have written on an average ten cantatas in a year. Telemann, who was four years older than Bach, wrote nearly seven "year-books" of cantatas in 1718 alone;[818] and in 1722-23 Johann Friedrich Fasch wrote a double set of church compositions, for the morning and afternoon, and when Saints' days occurred, often four cantatas in one week.[819] This comparison of mere numbers is only given to confute the opinion of Bach having been a voluminous and rapid writer. The number of his works is indeed very large, but it is spread over a long life. The talents of Telemann, Fasch, and other of his contemporaries who were prolific in production, were of a shallower kind, and their work is therefore no proper standard of measurement for that of Bach. But even when compared to geniuses of nearer equality with himself, such as Handel and Mozart, Bach appears as a more deliberate worker, though at the same time a more clear and certain one. His scores do not give the impression that he made many preparatory sketches or experiments with the chief subjects, as Beethoven did, for example. They seem to have been written down when the work had been completed in its outline and general features in the composer's mind, but yet not so far that nothing could be added to it during the actual process of writing it down. The cases in which he discarded and altered the entire original design of any piece are comparatively rare,[820]

---

[818] Mattheson, Grosse General-Bass-Schule, p. 176.
[819] Gerber, N. L. II., col. 92.
[820] The first sketches of the cantatas "Die Himmel erzählen die Ehre Gottes" and "Man singet mit Freuden vom Sieg" met with this fate.

but alterations of detail are of frequent occurrence. When, after a lapse of some time, he took up one of his works, he never omitted to try it afresh, and, when occasion required, to improve it; compositions on which he laid great importance he used to rewrite again and again, refining and adorning them into perfection. This careful forethought is illustrated also by the fact that he very frequently took a share in the work of transcribing the parts. The Cantors always had among their scholars one or more musical copyists, and Bach did not omit to keep his pupils very hard at work at this employment on occasion; and even in the year 1778 Doles recalls to mind the quantity of notes that Bach made the Thomasschule boys transcribe.[321] His sons, too, had to give their assistance, but his best copyist was his wife; her handwriting, which helped in the production of the parts of all the first Leipzig cantatas, meets us even in the church compositions of Bach's later years, larger and stiffer, but as firm and accurate as ever. Where so many hands were helping, many musicians would have spared themselves the mechanical labour of writing out the parts, which must have taken much of his time from composition. Those 266 Leipzig cantatas are the work of the truest artistic industry devoting itself solely to the subject in hand; a mighty monument built up stone by stone. It has not remained until our day in an unimpaired condition, for all that are at present known of Bach's cantatas, even including the six cantatas of the Christmas Oratorio and the more important fragments, only reach the number of 210.

Bach's trial piece was performed on the Sunday next before Lent (February 7), 1723; it was the cantata "Jesus nahm zu sich die Zwölfe."[322] The cantata "Du wahrer Gott und David's Sohn"[323] seems to have been at first intended for this occasion; it is recognisable as having been written at this period, and while he was yet at Cöthen.

---

[321] In a written apology addressed to the Council on July 15th, 1778. See Council deeds " Die Schule zu St. Thomae betr. Fasc. II."

[322] B.-G., V.,¹ No. 22. P. No. 1290.

[323] B.-G., V.,¹ No. 23. P. No. 1551.

It consists of a duet for soprano and alto, a recitative for tenor, and two choruses, one free and one on a chorale; it shows the master at a height that none of his former works had approached. The text is not particularly adapted for musical setting—Scriptural words are wanting altogether, and the rhymed verses are in "madrigal" form, occasionally changing into Alexandrines. Here, more than in most cases, Bach had to give form to the whole, and it is quite peculiar. The Sunday (Estomihi, as it is called) is the one next before Lent, when the minds of the people are to be prepared for contemplating the sufferings of Christ. In the gospel for the day it is related how Jesus, with His disciples, is drawing nigh to Jerusalem, there to await His Passion. A blind man sits by the wayside, and appealing to Christ, as He is passing by, to have mercy upon him, receives his sight. These ideas form the germ of the religious feeling of the work: the fervent cry for help, and the presentiment of the tragedy that is approaching. The duet (in C minor *Adagio molto*) is accompanied by two oboes, which with the bass have an independent and more animated movement, surrounding the broad phrases of the melody and the sustained tones of woe in the vocal parts, which are in imitation. There is less emotion in the recitative, except at the end where it rises to passionate intensity. The prominent feeling of this piece is not however given out by the voice, but by the first violin and the oboe, which play the melody of the chorale "Christe du Lamm Gottes" very softly above the vocal recitative. This bold combination of chorale and recitative is a new effect, even in Bach, but it followed almost naturally from his melodic manner of treating recitative, and after he had ventured upon a recitative-duet.[824] Its effect is very powerful; the whole feeling of the Passion, so tragical and yet so full of comfort, comes over the hearer like a flood. The chorus (in E flat major) now rises in a perfect cycle, its chief section being repeated in a solid homophonic style, while the central portion consists of imitative two-part

---

[824] See Vol. I., p. 557.

interludes between the tenor and bass. Yet this chorus, with its mighty pathos, is not the climax of the emotional progress of the work. The expectation of Christ's passion and death is brought out as the strongest element; the E flat chorus is immediately followed by the chorale chorus in G minor on " Christe du Lamm Gottes." On account of the shortness of the melody, all three verses are gone through, which is rare with Bach. The first verse, sung more or less in homophony by the chorus, has an independent instrumental accompaniment of a mournful, wailing character; in the second verse the course of the lower parts is more animated, the upper part (with the melody) being imitated in canon both by the oboe and the first violin ; by the former, at the third note of the melody in the fourth below, and by the latter, at the sixth note in the third above; in the third verse, which closes in an Amen, the lower parts have a different, but quite as vigorous a counterpoint, while above the *Cantus firmus* the oboes give out a new melody of the highest intensity of expression, and sharply contrasted in its rhythm. The feeling rises throughout from sadness to intense grief, and at last is changed into a pious prayer, pleading for reconciliation with God. A new treatment of the chorale form is shown in the first verse, and one subsequently much used and largely developed by Bach ; here, however, it is only used in very modest proportions. In the second verse enormous skill is shown in polyphonic combination; it is evident that this form comes directly from organ-music, and Bach treated this very chorale, together with many others, in his " Orgelbüchlein."[325] The last verse, on the other hand, is in the manner of the older composers, and is only superior to their works in the constantly melodious character of its instrumentation. Moreover, it must not be overlooked that between the second and fourth numbers of the cantata deep poetic connection subsists which has been already noticed in different cantatas of the Weimar period;[326] the chorale melody, which in an earlier movement was given to the instruments alone,

---

[325] See Vol. I., p. 600.   [326] See Vol. I., p. 544.

is sung at the close, being used at the beginning to give a
tone to the emotional aspect of the work, and at the end to
bring out clearly the devotional feeling. In every respect
this cantata is a trial piece well worthy of Bach; compared
with the one which he actually employed for that purpose,
it is easy to see why he judged it necessary to keep this
cantata in reserve for a time. It was too grave, deep, and
elaborate. Bach knew the taste of the Leipzig public,
accustomed as they were to varied styles of operatic music
and to Kuhnau's soft and tender tunes.

In the cantata "Jesus nahm zu sich die Zwölfe," he
adapted himself more to this taste. The fugal chorus in the
first number shows a simplicity of counterpoint which is of
very ordinary occurrence in the works of Telemann, but
which surprises us in Bach. An agreeable tenor aria is
followed by an easily intelligible chorale number, in which
the chorus in four simple parts is accompanied by the uppermost instrumental parts in counterpoint of semiquavers,
which is continued in short interludes. This was a form
much in vogue at the time; Bach himself had formerly made
use of it,[327] and in this instance it is only enriched so far that
passages, mostly of an independent though not very characteristic order, are allotted to the second violin and the
viola. The other numbers are deeper, especially the opening
of the first, an unison chorus for tenor and bass; the most
important part of this is in the instrumental portion, which
is very delicately interwoven and is much less like an
accompaniment than an independent piece. From this
number alone we can see in truth that the composer understood his business. The work as a whole is well suited
to its purpose, but it can in no respect be considered of
equal merit with the first one, which seems to have been
first performed on the Sunday before Lent, 1724.[328]

We cannot now be sure what was the music for Whitsuntide with which Bach began his labours in the University
church.[329] The only cantata which is likely to have been

---

[327] See Vol. I., p. 491.   [328] See App. A., No. 19.
[329] See ante, p. 214.

used is "Erschallet ihr Lieder, erklinget ihr Saiten," but there are convincing reasons for placing the date of composition two or three years later. He began his church work as Cantor of the Thomaskirche on the first Sunday after Trinity. It has not, it is true, been expressly recorded that the first church music conducted by him and favourably received by the congregation was of his own composition; but this may be taken for granted according to the customs of the time.[330] There exist two cantatas for the first and second Sundays after Trinity, the last of which is dated by Bach himself 1723. The first is so exactly similar to it as regards the text and the musical form, both in outline and in detail, as well as in feeling, that no doubt can be entertained that they both belong to the same period, especially when we remember Bach's characteristic way, before alluded to, of producing several examples at once of any new form previously unused by him.[331] The chief novelty of form in this case is that we here meet, for the first time in Bach's work, with the church cantata in two divisions. We know already that on the high festivals part-music was sung before as well as after the sermon, and that it was allowed in this latter place even on ordinary Sundays and holy days. But to cast one single work in so large a mould, as that it could be divided into two independent parts, was at variance with previous customs, in Leipzig, at least. It seems to me that Bach, in this innovation, followed the method of the oratorio composers whose works, destined for the Catholic form of worship, were so arranged that the sermon came between the two parts.[332]

---

[330] *ACTA LIPSIENSIVM ACADEMICA.* 1723., p. 514.—"On the 30th of the said month (namely May), as on the first Sunday after Trinity, the new *Canto*r and *Collegii Musici Direct.*, Hr. Joh. Sebastian Bach, who came from the Ducal Court at Cothen, performed his first *Music* here with great applause." (See *Ante*, p. 215, Note 64.)—In Sicul, *ANNALIVM LIPSIENSIVM MAXIME ACADEMICORVM SECTIO* XX. Leipzig, 1726, p. 479, it is stated that this performance took place in the Nikolaikirche.

[331] See App. A., No. 20.

[332] The division which was usual, even in the 17th century, of oratorios into two parts, in contradistinction to the triple division of operas, is explained by this custom, which obtained, in Italy, at least, throughout the whole of the 18th century. See Burney, Present State of Music in France and Italy, p. 365.

The Gospels for the first two Sundays after Trinity are more than usually rich in deep and beautiful thoughts and in striking contrasts. The librettist unfortunately did not understand how to take advantage of this in the interests of music. In both he devotes his energies to didactic trivialities, which have but a loose connection with the Scriptural narratives, and might just as fitly have been written for many other Sundays. In the first cantata, "Die Elenden sollen essen, dass sie satt werden"[333]—"The poor shall eat and be satisfied"—(in E minor), reflections of a common-place kind are made, with reference to the story of the rich man and Lazarus, upon the worthless and transient character of earthly riches, upon Jesus as the essence of all good, upon a good conscience and contentment. Bach triumphs over this heterogeneous medley in four well contrasted arias, the first of which is ingeniously and fancifully built on the following motive :—

and in six recitatives. The chorale "Was Gott thut, das ist wohlgethan"—"That which God doth, is still well done"—plays an important part in the work. It appears first at the close of the first part, and in a new combination of Pachelbel's style, with that especially peculiar to Georg Böhm.[334] This movement is repeated again at the close of the second part, and a chorale fantasia[335] on the same melody serves as an introduction to the same part, not set for the organ, but for the string quartet with the trumpet—the quartet taking the independent and polyphonic accompaniment and the trumpet the chorale melody. The thread of the cantata, broken off by the sermon, could not be re-united in a more skilful or artistic manner. The introductory chorus, however, arouses the liveliest interest of all, speaking as it does in affecting tones of consolation in sorrow. The emotion of sorrow, as might have been

---

[333] B.-G., XVIII., No. 75. P. No. 1670.
[334] See Vol. I., p. 206.   [335] See Vol. I., p. 610.

expected in Bach, is more or less prominent both in the orchestra—which in the first movement of the chorus progresses independently in faltering pulsations—and in the agonised and sustained melodies in the voice parts. But a promise of joy gleams brightly through the sorrowful sounds (compare especially bar 53 ff.) and in the broad theme of the fugal movement which follows :—

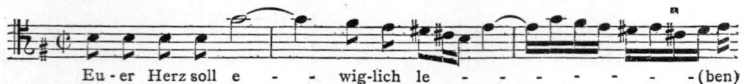
Eu - er Herz soll e - - wig-lich le - - - - - -(ben)

the sick and the weary soul inhales full draughts of the air which is to bring new life and health. But yet a certain veiled character predominates, and in the interrupted cadence (in bar 13 of the fugue) it has even a tone of piercing grief. From this point, however, all the forces of the music combine, and finally pass by a lovely modulation into D major; the major sixth which occurs in the soprano part, an interval strictly forbidden by the old rules for the formation of melodies, but used moderately often by Bach, has here a powerful effect of light and liberty.

Not only in the division into two parts and the insertion of an instrumental movement, but also in the number of its choruses and chorales, its recitatives and arias, and the order in which these forms are arranged, the cantata (in C major) for the second Sunday after Trinity, "Die Himmel erzählen die Ehre Gottes"[336]—"The heavens declare the glory of God"—agrees with the one just spoken of. Nay, the arias, whether intentionally or not, are in the same keys as those of the other work, their order alone being different. Here, again, the most important part of the whole is the first chorus, although scarcely any connection subsists between its words and the Gospel for the day. As before, the chorus consists of an opening section with free imitations and an independent instrumental accompaniment, agreeing with the other even in the time—and of a fugue; there, however, the entry of the full choir is preceded by a movement for soprano

---

[336] B.-G., XVIII., No. 76, · P. 1677.

LEIPZIG CANTATAS, 1723. 357

and alto, while here some of the basses begin alone.[337] The resemblance is seen even in similar phrases (compare bars 24 ff of the C major, with bars 26 ff of the E minor cantatas). The whole piece is very full and brilliant, and the fugue theme is strong and vigorous. An effect noticed in the cantata "Ich hatte viel Bekümmerniss"[338] is repeated here, and once again in a new work that we shall consider next in order to the present ("Ein ungefärbt Gemüthe"); the fugue is begun by solo voices, the tutti parts coming in gradually at the recurring entrances of the theme, giving the effect of a slow *crescendo* such as that produced on the organ by the gradual drawing out of more and more stops. This fugue also bears a resemblance to some of the earlier fugues, in that the trumpet is used as a fifth part in the working out. The chorale sung at the close of each part is set in a simple and familiar form. The instrumental movement at the beginning of the second part is a trio for oboe d'amore, viola da gamba, and double-bass, one of the forms of chamber music transferred to the church, of which we have already met with several examples in Bach. Bach, not long after, used this piece in one of his organ sonatas,[339] not, however, without considerable alterations. Many of these alterations are absolute improvements, in various ways, upon the older work; the calmer basses and the generally lower position of the upper part reveal its devotional purpose, for the second part of the cantata was performed during the Communion.[340]

---

[337] Bach had at first intended to begin the chorus immediately with a fugue with an independent accompaniment on the instruments. This was to have been the theme:—

Die Himmel er - zäh - - - - - - - len die Eh-re Got- tes.

[338] Compare vol. I., p. 534.

[339] B.-G., XV., p. 40 ff.

[340] Both cantatas were well known in an altered and abridged form, the first beginning with the first recitative, under the title "Was hilft des Purpurs Majestät," and the second, as "Gott segne noch die treue Schaar," beginning with the opening of the second part. The latter was also used as a Reformation cantata. See Breitkopf's catalogue. Leipzig, Michaelmas, 1761, p. 20.

It appears that Bach came again before the public with a new work fourteen days after this, on the fourth Sunday after Trinity (June 20).[341] The librettist of the two former cantatas could no longer satisfy him, so he turned back to Neumeister's poems, which he had tried and found successful. The text "Ein ungefärbt Gemüthe"—"A pure and guileless spirit"—taken from the fourth of the five year books of church cantatas, is elegant in versification but is anything but a model of poetic feeling. Bach viewed this didactic, prosaic dissertation in the light of his own musical imagination. To the two arias which the cantata contains he set fine and affecting music; the first of these (F major), with which the work opens, delights the hearer with its pleasant and contented expression. This may be called a trio for violin, viola, and bass, and the second a quartet for two *Oboi d'amore*, viola, and bass, so little prominence is given to the voice part. This method was here necessary from the inferior quality of the text; and from the relationship between the chief subject of the first aria and that of the last movement of the B minor sonata for violin and clavier,[342] it may be concluded that Bach was devoting himself now more especially to instrumental writing. The third number, a chorus on the text in Matt. vii., 12, "Alles nun, das ihr wollet"—"Therefore all things whatsoever ye would that men should do to you, do ye even so to them"—is very animated, energetic, and striking, especially in the double fugue, which forms its second part; on the other hand, the short phrases in which the chorus and the orchestra answer one another at the beginning, remind us of the style of the older church cantatas, while bars 11—18 have a resemblance to bars 11—25 of the second chorus in Bach's earlier cantata "Ich hatte viel Bekümmerniss." The closing chorale is simple in form, like that in the cantata "Jesus nahm zu sich die Zwölfe."[343]

For the seventh Sunday after Trinity (July 11) in this year,

---

[341] B.-G., Vol. I., No. 24. [342] B.-G., IX., p. 80 ff.
[343] See App. A., No. 21.

Bach chose a text from Franck's "Evangelische Sonn- und Festtags-Andachten," which had provided him with the groundwork of two compositions in Weimar.[344] Franck had intended it for the third Sunday in Advent, so that its adoption for another occasion involved the alteration of the first two arias. Recitatives, which do not usually occur in Franck's texts, were inserted, and in them we seem to discern the hand of the prolix compiler of the cantatas for the first and second Sundays after Trinity; their insertion renders the text sufficiently long to serve for a cantata in two parts. The first chorus "Aergre dich o Seele nicht" (G minor)—"Fret thyself no more, my soul"—is remarkably short, dispensing with any full working out, such as we are accustomed to in Bach. After a prelude the chorus starts with a phrase of three bars, which by means of striking suspensions is intended to produce the impression of vexation and anger. In Bach's vocal pieces it may often be noticed that on the one hand their character is derived not so much directly from the text as from the church teaching for the Sunday, or some other more general poetic idea, often merely from the necessity of musical contrast; while, on the other hand, certain special ideas provide the basis for characteristic musical subjects which direct and pervade the working out, giving outward colour rather than intrinsic purpose to the sentiment—an intimate union of the general and the particular. This is the case here; for as the idea of vexation is only viewed negatively, it could give no character to the whole, but it did give occasion for an opening of great musical interest. In contrast to this, there now comes in a declaimed fugato on the same words, full of Bach's emphatic intensity; the instruments are given an elaborate fugal movement on another theme in contrast to the air; the double-bass alone going on in its own course. After one working out, the close follows in D minor, and in a short homophonic phrase, the words which give the reason for those that have gone before are given out; then the whole process is repeated in C minor, leading back to the original

---

[344] See Vol. I., pp. 570 and 642.

key; six bars more by way of epilogue, and the piece is at an end. In spite of the unusual and concise form, the emotional picture is complete; we are sensible of the sure hand of a master not only in the elaborate detail, but in the whole structure. Altogether, the cantata, generally, stands above most of those that preceded it, and is worthy to rank with the cantata, "Du wahrer Gott." The recitatives are throughout very expressive, and at times deeply touching, especially the *arioso* endings; particularly that of the second recitative with the anticipations of the harmony in the bass, so ingeniously contrived to impart colour to the accompaniment. The three arias vie with one another in depth and intensity of expression, but the gem of the solo numbers is a duet (in C minor) for soprano and alto, "Lass Seele kein Leiden von Jesu dich scheiden"—"My soul, let no sorrow e'er part thee from Jesus." It is in the rhythm of a gigue, and has that melancholy grace which is peculiar to so many of Bach's pieces in dance-form. Lastly, the chorale with which the first part of the cantata closes is of great interest. Here, for the first time, we meet with the chorale fantasia since it was transferred to vocal music. In the cantata "Du wahrer Gott," Bach had indeed shown a predilection for this new production, but in the works which followed had gone back to the simpler forms. An instrumental piece, of a pious and innocent character, is played by the violins and flutes, with bass, answering each other; and the melody of "Es ist das Heil uns kommen her," steals in with the verse :—

> Ob sichs anliess, als wollt er nicht,
> Lass dich es nicht erschrecken.
>
> Although He seems to will it not,
> Yet let not this affright thee.

The *cantus firmus* is in the soprano, and the other voices have counterpoint, mostly imitation; the melody in diminution. In the score there is no chorale at the close of the whole work, but, by analogy with the cantatas for the first and second Sundays after Trinity, it may be assumed with certainty that the chorale of the first part was to be repeated.

Another of Franck's texts was set for the thirteenth Sunday after Trinity (Aug. 22). It is, however, not quite certain whether this work was written before the following year, when it was performed first on the 3rd of September. But I consider the first to be the more probable date.[345] For this work Bach returned to the "Evangelische Andachts-Opfer,"[346] the poetry of which was more suitable for music. The work is altogether without chorus, except at the very end, where a simple chorale is sung to the fifth verse of the hymn, "Herr Christ der einig Gott's Sohn"—"Lord Christ, the only Son of God." The Gospel narrates the parable of the Good Samaritan, and the whole musical work is full of Christian tenderness and compassion. In the first aria in G minor, "Ihr, die ihr euch von Christo nennet, wo bleibet die Barmherzigkeit"—"Ye upon whom Christ's name is called, where is your love and charity"—we seem to hear the Saviour Himself speaking, with divine tenderness and true human warmth of love and sympathy, yet not without a touch of sadness:—

Ihr, die ihr euch . . . von . . Chri - . sto nen - net,

In the second aria (in D minor), "Nur durch Lieb und durch Erbarmen werden wir Gott selber gleich"—"Only by our love and pity are we made like God Himself"—the Redeemer's teaching is applied by man to himself; and in the third it is delivered with joyous conviction as a duet, and in a strain that seems to have originated from the melody of the first movement:—

"Hands not grudging, ever ready." These two themes are worked out with a persistency unusual even in Bach, for the most part in canon on the unison or the octave, and in the duet also in contrary motion; they are heard unceasingly, now in the voices, now in the instruments, like those words of St John, "Little children, love one another."

---

[345] See App. A., No. 22.
[346] See Vol. I., p. 540.

The re-modelling of the beautiful Weimar cantata, "Wachet, betet,"[347] can with moderate certainty be assigned to this year. Since no church music was performed in Leipzig on the last three Sundays in Advent, Bach employed it for the twenty-sixth Sunday after Trinity, as it fits the Gospel for that day very well. The alteration consists essentially in this, that by the insertion of recitatives and the chorale "Freu dich sehr, O meine Seele"—"Now rejoice, O soul and Spirit"—the work is enlarged into a two-part cantata. To one recitative, one of the finest ever written by Bach, a chorale is added on the instruments, which raises the passionate personal emotion into the province of religious worship, while, on the other hand, it is enriched and individualised by it. Bach himself thought very highly of this cantata, and in later years performed it again and again.[348]

In addition to these eight cantatas there are yet two works, dating from the first year in Leipzig, which owe their existence to church ceremonies of an exceptional character. On the 24th of August in each year—St. Bartholomew's Day—the election of the new Town Council used to take place, and, on the Monday or Friday next following, a festival service was held before the members of the new Council took their seats. In the year 1723 this Monday fell on August 30,[349] and for this occasion Bach wrote the festival cantata "Preise, Jerusalem, den Herrn"[350]—"Now praise the Lord, Jerusalem"—a work equally remarkable for its vigorous and brilliant choruses and for its fervent and melodious solos. That the form of the work was intended for a festal occasion is shown very plainly by the character of the separate numbers. It opens with an overture in the French style, for the performance of which an orchestra

---

[347] See Vol. I., pp. 571 and 643.

[348] See Appendix A., No. 23.

[349] It was held on the Monday after St. Bartholomew's Day in the years 1724, 1725, 1726, 1727, 1731, and 1739; and on the Friday in the years 1728, 1729, "and 1730. See the deeds of the Council "Rathswahl betr. 1701, Vol. II.," and Nützliche Nachrichten von Denen Bemühungen derer Gelehrten und andern Begebenheiten in Leipzig. In the year 1739." P. 78.

[350] B. G., XXIV., No. 119. P. 1684.

consisting of four trumpets, drums, two flutes, three oboes, string quartet, and organ is requisite. The splendidly pompous *Grave* movement is played by the instruments, till at the *allegro* (12-8) the chorus enters with the words of Psalm cxlvii. (12—14), worked out not so much in fugal style, as with free imitation and episodical use of the chief subject which was first given out by the bass; this goes on until the recurrence of the *Grave* movement, played as before on the instruments alone, and filling the part of a postlude. It is not the first time that we meet with this bold transference of a thoroughly secular instrumental form into church music by Bach; for the same thing is done in the cantata "Nun komm der Heiden," written for Leipzig in 1714.[351] An essential difference is observable, however, in that the overture form in the earlier work is united with a chorale, and also that the voices take part in the *Grave* movement. The number of which we are speaking in this "Rathswahl" cantata inclines more strongly towards the secular side, because it comprises a chorus in free style, and more distinctly towards the instrumental side, in that the chorus does not take part in two of the chief sections into which the work is divided. This was allowable, since its purpose was not in the strictest sense devotional; besides, a fundamental religious sentiment, if nothing more, is preserved throughout the *allegro* by the polyphonic style displayed in it, which is all Bach's own. No more sacred character than this would befit the recitatives and arias which follow, of which the words set forth the happy circumstances of the town of Leipzig. But for his amazing gift of pure musical invention, Bach would have found it almost impossible to produce such a charming piece as the second aria " Die Obrigkeit ist Gottes Gabe, ja selber Gottes Ebenbild"— "Authority is God's ordaining, yea, and His image here on earth"—for he could hardly have got his inspiration from the words. But whenever it was possible to get any poetic impulse from the text, he availed himself of it, as is shown by the first aria "Wohl dir, der Volk der Linden, wohl dir,

---

[351] See Vol. I., p. 507.

der hast es gut "—" O dwellers by the lime trees, O blest, O favoured race." There are few pieces of so agreeable and sunny a character as this; the low oboes (*Oboi da caccia*) here employed by Bach lend it an idyllic character, yet a grave one as befitted the circumstances, so that it differs perceptibly from the aria of *Pales* in B flat from the cantata " Was mir behagt,"[352] which in other respects it closely resembles in feeling. This aria (in G major) is joined to the second (in G minor), which we have already mentioned, by an entirely separate movement; a piercing trumpet call introduces a bass recitative "So herrlich stehst du, liebe Stadt "—" So fair thou art, beloved town "—which is then accompanied by sustained harmonies on the two flutes and two English horns, in addition to the figured bass, the jubilant trumpet call recurring at the close; the strings are silent during this movement. After the G minor aria the whole body of instruments and voices reunite immediately in a splendid movement cast in the form of the *da capo* aria. The first section consists of a fugue, whose theme:—

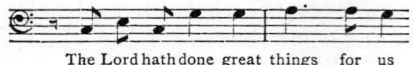

The Lord hath done great things for us

is apparently found from the first line of the chorale " Nun danket alle Gott." Such an independent use as this of parts of a chorale melody is extremely rare in Bach (in the motett " Nun danket alle Gott " he once did something similar to this); the chorale was to him consecrated, as it were, to the church, and where he introduces it he is wont to set it unaltered as the central point of his own composition. In departing from his usual method in this case, he doubtless felt impelled by the mingled sacred and secular character of the whole work. The second section of the cantata forms, in its homophony, a contrast to the first; a motive of vocal character given out by the trumpets in the instrumental ritornels:—

---

[352] See Vol. I., p. 566.

is worked out in this section with great skill. The strictly devotional feeling finds its first expression at the close of the cantata in a few lines from "Herr Gott dich loben wir."

During the years 1722-3, the Church of Störmthal near Leipzig had been restored, and at the same time a new organ had been erected, the building of which had been undertaken by Zacharias Hildebrand, a pupil of Gottfried Silbermann, for 400 thalers. A certain Kammerherr von Fullen, who was resident at Störmthal, had provided the requisite sum of money, and after the completion of the work requested Bach to try the instrument. On November 2, the Tuesday after the twenty-third Sunday after Trinity, a public service took place for the dedication of the organ, and Bach wrote a cantata for the occasion: "Höchst erwünschtes Freudenfest"—"Hail, thou longed-for feast of joy"—and conducted the performance himself. The organ, "certified" by him "as an excellent and durable instrument and highly commended," is in existence in all its essential parts to this day; it underwent thorough repair in 1840 at the hands of the organ builder, Kreuzbach, who at the same time expressed himself favourably with regard to the organ. It says a great deal for the respect in which Bach's person and name were and are still held in Saxony, that the tradition of his having been at Störmthal has remained there through more than a century and a half.[353] The cantata, which Bach must have taken his forces from Leipzig to perform, is written with especial care, owing, perhaps, to the high position of the personage who had given him the order.[354] It is, moreover, of intrinsic importance, and Bach, who always liked to turn the compositions written for special occasions to account for his regular duties, subsequently arranged it for Trinity Sunday, and often performed it in its altered shape on that day. We may learn much from comparing this work with the

---

[353] This was told me by Pastor Ficken, of Störmthal, to whose kindness I also owe the extract given above from the accounts of the church there.

[354] Both the autograph score and the original parts are in the Royal Library in Berlin. To the parts are appended the text printed on a folio sheet.

"Rathswahl" Cantata. Here, again, there is no strictly devotional purpose, and the fundamental forms are even more thoroughly secular, although purified and adapted to church use. The opening again consists of an overture in the French style, and the way in which the voices take part is similar, excepting that when the *grave* recurs at the close, they are introduced in the last bar, to heighten the whole effect. Besides this the *allegro* is here a true *fugato*, and the interrupting trio-passages are not omitted, so that the overture style is more strictly adhered to, even in detail. By the continual recurrence of the chief subject:—

the first aria has in some degree the character of a rondo, though the aria-form is retained. The second aria is entirely in the rhythm of a gavotte:—

The third keeps up the character of a gigue against the following theme used as a bass ritornel:—

Finally, the fourth has the movement of a minuet:—

Thus we have here the remarkable phenomenon of a cantata in the form of an orchestral suite, except that recitatives are introduced and that each of the two sections closes with a chorale. Bach probably intended in this way to suit the taste of the noble patron of the church at Störmthal, just as he had suited that of the Leipzig people with his trial cantata; for, under Augustus the Strong, French music was in great favour at the Court of Dresden. In so doing Bach did nothing whatever against his natural instinct, which was to weld together in his own style all the musical forms of the time. It was, therefore, in no way unsuitable when he subsequently performed this "organ dedication" cantata on Trinity Sunday. It is not undevotional, and yet it lacks that highest degree of sacredness which can only be given when the composer's imagination is set in motion by an event of universal importance to the Christian church.

Bach had entered on his post in the ferial portion of the ecclesiastical year. Remarkable as was the activity displayed by him as a church composer during this period, yet he had no opportunity of showing himself in his full greatness until the beginning of the ecclesiastical year 1723-1724. It is self-evident that he would celebrate the first occurrence of the festal days as far as possible with music of his own. On this assumption we can, with moderate certainty, assign most of the festal cantatas to this year. On his first Christmas Day in Leipzig, the chief music was his cantata "Christen ätzet diesen Tag in Metall und Marmorsteine"—"Christians, grave ye this great day In brass and stone that will not perish."[355] The particular emotion of the composition is revealed in a

---

[355] B.-G., XVI., No. 63.

duet occurring in it for alto and tenor. The words are as follows:—

> Ruft und fleht den Himmel an,
> Kommt ihr Christen, kommt zum Reihen,
> Ihr sollt euch an dem erfreuen,
> Was Gott hat anheut gethan.

> Come, ye Christians, praise and pray,
> Come with singing, come with dancing,
> With your joy and praise enhancing
> That which God hath done to-day.

The universal feeling of Christmas joy seems thus dramatised by the image of a festal procession mingling in sacred dances, and, as it were, obeying the commands of the Psalmist: "Praise Him in the cymbals and dances; praise Him upon the strings and pipe." The festal dance, represented by a graceful waving figure in the duet, comes out powerfully and splendidly in the ritornels of the first chorus. Without deviating from this character the chorus itself is formed from quite different melodies which are worked out very effectively in canon; it contains exhortations to Christmas rejoicing, and as each section is concluded the jubilant dance again breaks forth immediately. The last chorus, too, is influenced by the same idea, which, however, is soon changed; for the festal procession crowding together seems bowed in devout adoration before God; this is represented by a very characteristic double fugue full of fervent and intense expression. In the second part of the chorus another double fugue corresponds to this on the words "Lass es niemals nicht geschehn," &c.—" Let it never come to pass that Satan shall torment us"—which is coloured by the dramatic expression suggested by the word "torment." The other numbers of the cantata, two recitatives and a skilfully worked duet for soprano and bass, are of a more general religious character; their subjects, which are peculiar to themselves, being prolonged and developed in the oboe and instrumental bass parts. For the rest, the style of the work unmistakably approaches that of oratorio, and this it is which makes it especially remarkable among Bach's cantatas, although it has a very considerable amount of intrinsic musical value. It is noticeable that there is no chorale throughout.[356]

On high festivals, after the sermon, the *Sanctus* was sung in a florid style as an introduction to the Communion (ante, p. 270). There are a number of such Sanctuses composed by Bach, one of which I believe may be identified as that composed for Christmas Day, 1723. Like the cantata, it is in C major, and the instrumentation is almost the same; it is remarkable for its bright festal character, which culminates in the *Pleni sunt coeli et terra*.[357]

The performance of the cantata "Christen ätzet diesen Tag," with its attendant *Sanctus*, took place during the morning service, and was sung by the first choir in the Nikolaikirche. In the evening the cantata was repeated by the same choir in the Thomaskirche; and after the sermon the hymn of the Virgin was sung, set in its Latin form and in an elaborate style. For this purpose Bach wrote his great *Magnificat*, which, since it has become generally known, has rightly been reckoned one of the highest inspirations of his genius.[358] For the occasion of the festival of Christmas, for which it was intended, Bach expanded the Bible text by inserting four vocal numbers in suitable places, the words of which bore especial reference to the Leipzig form of service. These are (1) The opening verse of the hymn "Vom Himmel hoch"; (2) the verse " Freut euch und jubilirt, Zu Bethlehem gefunden wird Das herzeliebe Jesulein, Das soll euer Freud und Wonne sein"— " Rejoice with pious mind, To Bethlehem go now and find The fair and holy new-born Boy, Who is your comfort, peace, and joy "; (3) the *Gloria in excelsis Deo*; (4) the lines *Virga Jesse floruit, Emanuel noster apparuit, Induit carnem hominis, Fit puer delectabilis. Alleluja*. Although these words are some German, some Latin, and have no outward connection of form, yet Kuhnau had built upon them a

---

[356] See App. A., No. 24.
[357] B.-G., XI.,[1] p. 69 ff. See App. A., No. 19.
[358] B.-G., XI.,[1] p. 3 ff, is the form in which it was afterwards cast by Bach. It was published in its original form by N. Simrock, at Bonn, in the year 1811. P. 40.

Christmas cantata, retaining the exact order given above.[859] They are very suitable in feeling, for they give a kind of dramatic character to the events of the Christmas night. Several points of detail go to prove that Bach took the words direct. from Kuhnau's cantata. One of these is in the form of the angel's song *Gloria in excelsis*. The Greek words καὶ ἐπὶ γῆς εἰρήνη, ἐν ἀνθρώποις εὐδοκία are wrongly rendered in the Vulgate by the words *et in terra pax hominibus bonæ voluntatis*. The meaning of this will be either "Peace on earth to men of good-will," or "Peace on earth to the men with whom God is well pleased," while the true meaning is (as in our English version) "Peace on earth, and good will towards men," as Luther rightly translated it (Friede auf Erden und den Menschen ein Wohlgefallen). The Vulgate version, however, is nearly always retained, even in Protestant churches, when the words are set to music. The cantors, who for the most part built their texts on the model of the Romish composers, were severely blamed by the theologians for thus introducing errors or stamping them with their approval, whether from ignorance or negligence; but it was justly answered that from the musician's standpoint the Vulgate version had a more melodious rhythm.[860] Kuhnau, however, had amended the reading of the Vulgate in the cantata referred to, replacing it by the words *bonæ voluntes* in accordance with the Greek; and Bach, who in all his other *Glorias* retained the words *bonæ voluntatis*, here follows the correct reading. So far, however, as we may gather from the musical phrasing, either the sense was not quite clear to him or he felt himself raised above and beyond it by the greatness of his musical ideas; for he phrases the words thus: *et in terra pax hominibus, bona voluntas*, approaching near to the original sense, but not quite reaching it; so that we may perceive that the Vulgate version was running in his head, and that he set the texts given to him as a whole, without paying any especial heed to the theological

---

[859] The cantata exists in the separate parts in the Town Library at Leipzig.

[860] S. Rango, Von der *Musica*, alten und neuen Liedern, &c. Greiffswald, 1694, p. 22 ff.

battles which raged round him. A second piece of evidence is found in the lines *Virga Jesse*, &c. This is a fragment of a longer Christmas hymn, given by Vopelius in its entirety.[361] It runs thus:—

Virga Jesse floruit,	The stem of Jesse hath flourished,
Emanuel noster apparuit,	Our Emanuel hath appeared,
Induit carnem hominis,	And hath put on human flesh,
Fit puer delectabilis.	And become a lovely child.
Domum pudici pectoris	The home of the Virgin chaste
Ingreditur Salvator et	Receives the Saviour and
Autor humani generis.	Creator of the human race.
Ubi natus est Rex gloriæ?	Where is the King of glory born?
Pastores, dicite!	Ye shepherds, say?
In Bethlehem Juda.	In Bethlehem Juda.
Sause,[362] liebes Kindelein,	Slumber, little baby dear,
Eya, Eya,	Eya, Eya,
Zu Bethlehem Juda.	In Bethlehem Juda.
Virga Jesse floruit,	The stem of Jesse hath flourished,
Emanuel noster apparuit,	Our Emanuel hath appeared,
Induit carnem hominis,	And hath put on human flesh,
Fit puer delectabilis,	And become a lovely child,
Alleluja!	Alleluja!

Kuhnau only used the last four lines, and it would have been more than strange if Bach had hit upon the same fragment quite independently. But the order in which Bach inserted the four passages into the *Magnificat* is precisely that adopted by Kuhnau. This is not the last time that we shall see Bach treading in Kuhnau's foot-prints, as in this instance. Just as he cherished and revered the traditions of his family, and was always ready to learn whatever he could from the works of older and contemporary masters, so now, when a great master had preceded him, it was foreign to his nature to study to appear as an innovator, or, trusting in his own power, to leave the work of his predecessor disregarded. It is true that his inmost nature was different from Kuhnau's, and his adherence to his method of working could only be superficial; it is, however, none the less

---

[361] Neu Leipziger Gesangbuch, 1682, p. 77 ff.
[362] Meaning to "sing in sleep," or "sing one's self asleep," still remaining in ordinary use in the low-German "susen."

important for the understanding of Bach's character. In this instance Kuhnau's Christmas music was connected with certain Leipzig church customs which must have been interesting to Bach, for their own sake. The Latin-German Christmas hymn above quoted is, in part at least, a lullaby sung at the cradle of Christ, and, since it was included by the Leipzig Cantor, Vopelius, in his Leipzig hymn book, it must have been in ordinary use there. It was an old Christian custom to place a manger in the church, and to perform the events of Christmas night as a drama or mystery. Boys represented the angels and proclaimed the birth of the Saviour, and then priests entered as the shepherds and drew near to the manger; others asked what they had seen there (*Pastores, dicite*); they gave answer and sung a lullaby at the manger. Mary and Joseph were also represented: Mary asks Joseph to help her to rock the Child; he declares himself ready, and the shepherds sing a song.[363] This custom of "Kindleinwiegen," as it was called, traces of which have remained into the present century, was in full force in the beginning of the previous century in Leipzig. It was one of the customs which the Council wished to abolish in the year 1702.[364] That their proposal met with but little favour has already been told. This particular custom of "Kindleinwiegen" actually survived to Bach's day, for we find a *lullaby* in his Christmas Oratorio. As it is spoken of in connection with *Laudes*, it can only have taken place at the close of the service in the Nikolaikirche. The cradle hymn which used to be used on this occasion is the old and favourite one, "Joseph, lieber Joseph mein, hilf mir wiegen mein Kindelein."[365] It is evident that the hymn *Virga Jesse floruit* must have been intended for the same purpose. That it was actually so

---

[363] Compare Weinhold, Weihnacht-Spiele und Lieder. Graz., 1870, p. 47 ff.

[364] "That sundry fanciful Latin *Responsoria, Antiphonæ*, Psalms, *hymni* and *Collects*, as also the generally so-called *Laudes* at Christmas time with the 'Joseph lieber Joseph mein,' and 'Kindlein wiegen,' be henceforth discontinued in public service." See ante, p. 264.

[365] It is reprinted in Schöberlein, Schatz des liturgischen Chor-und Gemeindegesanges. Part II., p. 164 ff.

used in the Leipzig Christmas service is not, indeed,
expressly stated; but we may assume it with certainty. So
much at any rate is clear; the words put together by
Kuhnau for a Christmas cantata and adopted by Bach reflect
the simple old custom, still popular at that time in Leipzig,
of representing dramatically in church the angelic message
and the adoration of the shepherds; but in a more ideal way,
being, as it were, its poetic and musical counterpart.
Viewed in this light, Bach's *Magnificat* gains a special
meaning, which lends a greater depth to the feeling of this
mighty composition. If we compare the four pieces he
inserted—which form a contrast to it both by their partly
German text and by their substance—with the materials
and forms used in the principal body of the work itself, we
shall perceive a striking difference. With the exception of
the *Gloria*, which to a certain degree requires the assistance
of the instruments, all these pieces are accompanied only
by a figured bass.[366] It had long been a favourite custom
in the Lutheran church to sing the music on Christmas
night with antiphonal changes, whether between the choir
and the people or between a large and a small choir, these
being usually placed at a distance from each other.[367] This
custom, originally followed with only short sections of hymns,
gradually grew as time went on, until longer pieces were
performed in this way. In Leipzig itself performances with
responsive choirs were of no unusual occurrence (comp. ante,
p. 284). Now the Thomaskirche, where the *Magnificat* was
first performed, contained a smaller organ built above the
"high choir," so that it was opposite to the great organ.
We know that this was only used on high festivals. It
evidently then was used for no other purpose than for these
alternating antiphonal performances of vocal music. From
this I gather that Bach, when he brought his *Magnificat* to
a hearing, made these four Christmas hymns sound down
from the smaller organ loft, the contracted dimensions of

---

[366] In "Vom Himmel hoch" the bass is not once figured, showing that the organ took no share in the accompaniment.
[367] Examples are given in Schöberlein, loc. cit., p. 52 f. and 96 f.

which admitted only a small band of singers and few instrumentalists. This is the easiest way of explaining the arrangement of the score and his subsequent treatment of these numbers. They certainly seem to have been composed at the same time as the *Magnificat,* and intended to form one with it, but Bach did not write them in the score in the places they were intended to occupy. In the autograph score they run consecutively, from the twelfth page, along the bottom stave, below the *Magnificat,* and are marked with references to the points at which they are to be inserted. If they were to be sung from the smaller organ-loft they could only be performed in the Thomaskirche; in the Nikolaikirche, where the great musical performance took place in the afternoon of the second day of Christmas, they had to be omitted altogether, since there was no similar situation in the church from which to sing them. And further: since an elaborate *Magnificat* was sung both at Easter and Whitsuntide, and Bach intended his work to be available for these feasts also (besides this, so far as we know, he only wrote a small *Magnificat* for soprano solo, which is not known to exist), and since these interlude numbers were only suitable for Christmas time, it is conceivable that in a later recension of the work he should leave them out altogether.[368]

The *Magnificat* is written for five-part chorus with accompaniment of organ, strings, two oboes, three trumpets, and drums, to which two flutes are added in the later recension. The opening chorus is set only to the words *Magnificat anima mea Dominum.* Its external form is that of the Italian aria, but is only thoroughly intelligible when reference is made to the concerto form. The instrumental introduction has less the character of a ritornel than that of a concerto tutti.

---

[368] The autograph scores of the work, both in its first conception and in its altered form, are in the Royal Library at Berlin. In the latter the work is transposed from E flat to D, besides which it is enriched in respect of orchestration; the part-writing is elaborated here and there, and in several places it is made more complete in other respects, or else altered. This alteration took place about 1730. On the lost smaller *Magnificat,* see Rust in the Preface to B.-G., XI.,[1] p. xviii.—And Appendix A., No. 24.

We have already seen how Bach developed and re-modelled the concerto form,[369] and how it was one of his chief objects of study in Cöthen, and from this circumstance it is very natural that having just come from thence he should use the form which had become so familiar to him there. There are not two contrasting subjects, but the whole material is exhibited in the tutti, the jubilant character of which is indicated by this motive :—[370]

But in the treatment of the chorus with regard to the orchestra, the concerto idea is shown in the plainest way possible. The first entry of the chorus follows this like that of a solo, accompanied only by a figured bass, and its melodic form accords with the opening of the tutti just as is the case in the real concerto. Two bars later the instrumental tutti comes in anew, but breaks off after two bars to allow the choir to be heard alone in another subject taken likewise from the tutti.[371] It then continues its course until the sixteenth bar of its united progress with the chorus, but so that the latter is the subordinate element as regards musical importance. It follows the course of the instrumental parts, now in unison and now in octaves ; not indeed renouncing its independence altogether, but freeing itself and then becoming re-united to the instrumental parts in passages of the most wonderful lightness ; but the chorus is here by no means a musical factor of equal importance with the orchestra, as it is, for example, in the opening chorus of the cantata "Ärgre dich o Seele nicht"; it only gives definiteness to the poetic emotion—though it indeed does this in the most decisive way possible. In the second section the conditions are changed. Here the chorus predominates even in respect of the music, the subject

---

[369] See ante, p. 125.
[370] I quote in the key of the later recension.
[371] Compare with this the first movement of the violin concerto in E, B.-G., XXI.,¹ p. 21 ff.

of its second entry being diligently worked out, while the orchestra comes in again in concerto style in little snatches taken from the great tutti. The third section is like the first, excepting that the working out leads from the subdominant into the original key; the last fifteen bars of the instrumental tutti form a postlude.

To appreciate Bach's power of building new forms, we must compare with this the chorus in the cantata "Wer sich selbst erhöhet" in which concerto-like elements are worked in, as they are here.[872] The aria for second soprano which now follows—*Et exultavit spiritus meus in Deo salutari meo*—is in the same key—a rare thing with Bach. It carries out the feeling expressed in the opening, but transfers it from the region of general joy and exultation to that of the quieter and more childlike joy of Christmas. We know the way in which this feeling is expressed by Bach too well not to recognise it here.[873] In order to leave no doubt of the composer's intention, this aria is immediately followed by the chorale "Vom Himmel hoch da komm ich her," the first of the inserted numbers. The *Cantus firmus* lies in the soprano; it is in four parts, in Pachelbel's form, and treated with evident love and delight; the counterpoint is throughout formed from the lines of the melody in diminution, and treated imitatively with great art, except that in a few cases the harmonies are somewhat daring. This number is also in the principal key, forming one group with the two that preceded it. We must bear in mind, too, that the chorale "Von Himmel hoch" was one of the hymns usually sung before the sermon, on Christmas Eve, by the congregation, if we are to appreciate the effect it must have produced when it came in, sounding down from the top of the church in the midst of the Latin songs, which were inaccessible to the direct sympathies of the congregation.

Now another style begins. An aria for the first soprano

---

[872] See ante, p. 13.
[873] See Vol. I., pp. 510 and 561, and in the Christmas oratorio (B.-G., V.²) especially the chorale, p. 37 ff. In the Simrock edition of the score of the *Magnificat*, the last four bars of the opening ritornel are omitted, it would seem by oversight.

gives out the words *Quia respexit humilitatem ancillae suae. Ecce enim ex hoc beatam me dicent (omnes generationes)*. The fact that the church, from the earliest times, included the Song of the Virgin in the liturgy, divested it of its personal character, and Bach, as his work shows, viewed it in its broadest signification. On the other hand, he was tempted by the operatic forms of church music to give it individuality. Whenever this occurs—and we shall often have occasion to notice it—the treatment is never dramatic throughout, but is confined to a single number of the work. This subjective tendency was one of his chief motives in musical composition. Bach always pierced deep into all the Biblical and ecclesiastical relations of his texts, in order to gain suggestions for new forms in art; but it was his sense of musical connection or contrast which ultimately ruled his choice. In Bach's compositions, even though we may not perceive the inmost motives of each separate part, we always receive the impression of a musical organism rounded, complete, and intelligible in itself; interpolations, which are otherwise than obvious and easily understood, are never so important as to disturb the harmony of the whole, like an unsolved enigma. We saw in Bach's solos that an outward smoothness of form, moulded by a master's hand, conceals a passionate and ever-varying emotion; and here again, in the midst of a work made up of many parts, and whose aim appears to be simple and clear, we discover a store of varied powers; it is only by the discovery of these powers that we can thoroughly appreciate Bach's spirit, with its own individual impulse and activity. The feeling of the first aria was innocent Christmas rejoicing; in the second the composer is inspired by the idea of the Mother of God. Scarcely ever has the idea of virgin purity, simplicity, and humble happiness found more perfect expression than in this German picture of the Madonna, translated, as it were, into musical language. Somewhat allied to this in fundamental feeling is the aria in B minor from the cantata "Alles was von Gott geboren,"[874] but

---

[874] See Vol. I., p. 563.

the direct reference to a particular individual and the effect of the *oboe d'amore* in the accompaniment give this aria of the *Magnificat* a greater intensity of feeling and a quite different character. The words *omnes generationes* are not sung by the soprano, but are taken up by the whole choir; thus the dramatic fiction, which gave the aria its peculiar character, is here again discarded. The choral movement composed on these two words alone is not less deeply felt, though it is conceived in another way, as representing the entrance of an innumerable company of people moved by one and the same idea. It is very characteristic of Bach that this idea is not cast in the form of a hymn of praise, as the preceding words might have suggested. This perhaps might have been Handel's method; but Bach, with his less subjective nature, represents only the idea of a great and universal movement. By the theme not entering alone, but being always surrounded by three moving parts, we are flung, as it were, into the midst of the throng. The working out is not fugal: at first the different parts seize upon the chief subject as it comes within their range; they overtop one another in gradually ascending entries; and at last they build themselves up in canon on a dominant pedal. A gifted editor of the *Magnificat* has expressed his opinion that Bach meant here to represent the all-conquering power of Christianity driving the nations against one another in deadly battle.[375] I cannot go so far as this, chiefly because a movement with this character does not seem to me to suit the object of the whole work, which is to represent the joy of the Christmas festival. And I have found that the character of this chorus in performance, although grave and mighty in its rush and flow, is yet not properly speaking wild or vehement. Bach, following his musical nature, gives to movements of this kind a more excited character than others—as, for instance, Handel—would have done. Certain alterations made in this particular movement, in the later recension, seem to me to be especially

---

[375] Robert Franz in his thoughtful little pamphlet: Mittheilungen über Johann Sebastian Bach's "Magnificat." Halle, Karmrodt. 1863.

instructive. In bars 6 and following from the end, the bass was originally this :—

on the pause the second soprano remained on d″, the instruments keeping silence until the entry of the last bar but one. And in bar 3 from the beginning the expression was harsher, this being the original passage in the bass part :—

We cannot suppose that Bach meant to express any different feeling by the two versions. Nor can technical reasons account for the alterations. He must rather have found that the expression had, here and there, exceeded its limits and required moderating.[376]

The bass aria, constructed on a *basso quasi ostinato*, that reminds us of Böhm—*Quia fecit mihi magna qui potens est et sanctum nomen ejus:* "He that is mighty hath magnified me and holy is His name"—breathes a fresh joyfulness into the feeling of the music, which in the last movement had become graver, and leads up to the Angels' song "Freut euch und jubilirt"—"Rejoice and be ye glad"—which ends a second group. This hymn, for only two soprani, alto, and tenor, is full of imagery; it seems to

---

[376] I take this opportunity of drawing attention to a clerical error in this chorus, on Bach's own part in the writing out of the later recension, which error has been transferred to the edition of the Bach Gesellschaft. In bar 6, the second half of the alto part runs as follows:—

But as is shown both by the accompanying first flute and by the older score in the alto part, it should be:—

float in a realm of light above the dimmer earth; a fifth part is given to the *continuo*. Bach may perhaps have had in his mind the old custom when boys, dressed as angels, sang Christmas hymns in the festival service. This beautiful piece bears a remarkable resemblance to Kuhnau's work; in Bach the theme of the first section is as follows:—

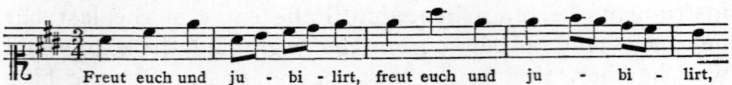

in Kuhnau:—

The intermediate subject in both is full of deep feeling— Kuhnau indicates it as *affettuoso*—and the finale in both is a regular fugato, though Bach has carried it out the more thoroughly. The mercifulness of God is praised in a very melodious duet for the alto and tenor in E minor, almost homophonous throughout, with an accompaniment of violins *con sordini*, and flutes; the words *timentibus eum*—"On them that fear Him"—offer an opportunity for closing with an elaborate and picturesque treatment of a very interesting character. As a contrast to this, the chorus *Fecit potentiam* represents the power of the Almighty bringing the pride of man to nothingness; and then, for the third time, the treatment leads up to a hymn sung by celestial beings, in the Latin *Gloria in excelsis Deo*. The chorus *Fecit potentiam*—"The Lord hath shewed strength"—displays in its principal theme a sort of sweeping and irresistible energy, with a peculiar crushing force in the accompanying chords and the rhythm of the instrumental bass. There is a poetical detail worthy of note in the way in which the instruments, divided into two portions, imitate the contrapuntal melody of the upper part:—

at the third note in contrary motion, as if to represent

the idea that there is no escape from the hand of the Lord. Among other striking details, the attention is especially rivetted by the closing *Adagio*, where the idea of arrogance is expressed with forcible verisimilitude by the pompous, widely spread notes of the dispersed chord. Here again we have an instance where the separate words of the text are taken advantage of to serve as the motive for a peculiarly effective musical close. In after years Bach composed a *Magnificat* in German[377] (in part paraphrased) for the Festival of the Visitation, in which the same portion of the text gave him the opportunity for an equally picturesque setting. It is in recitative and contains a very elaborate Melisma, which was a favourite way of concluding recitatives.[378]

A fourth group is composed of two arias and the verse "*Virga Jesse*" set as a duet. The first aria has a very vigorous, nay resolute, character; it has undergone many alterations in the process of re-arrangement, resting partly on practical grounds only, and by no means invariably striking as improvements. In the second aria, which was conceived in a softer vein of feeling, flutes were introduced by Bach, even in its first state. An expression of longing is stamped very beautifully on the principal idea by a striking turn of the rhythm:—

E - su - ri - en - tes

The abrupt close—an illustration of the words *dimisit inanes*—"sent empty away"—is an innovation in the second treatment; originally the flutes were carried on to the last note. The final bars of the duet are lost; the voices rise and fall

---

[377] B.-G. I., No. 10.

[378] Franz suggests (op. cit. p. 19) that Bach had by mistake attributed the words of the Vulgate *mente cordis sui* to God, for it would have been a great fault in taste to emphasise the idea of arrogance by the display of the sublimest means of expression at his command. Such a mistake is, however, hardly conceivable in a man so versed in his Bible as Bach, and even in classical Latin *sui* would be right. To me the feeling of the extended chord is quite clear; the impulse which led Bach up to this Adagio close was not derived from the individual passage, which only serves to colour its expression, but from the fundamental idea of the whole.

above a slow and lulling *continuo* which, as in the aria *Quia fecit*, is *quasi ostinato*.

Up to this point a simple Christmas joy has been the ruling idea through all the various movements, but when the last Christmas hymn, properly speaking, has died away a new element of feeling comes into play. At Vespers a sermon was to be preached on the Epistle, which enlarges on the work of Redemption as being the final aim of Christ's incarnation. The only words which distinctly refer to this in the *Magnificat* are in the verse *Suscepit Israel puerum suum recordatus misericordiae suae*. Their suggestiveness did not escape Bach's notice. He takes them for a chorale arrangement, in which two soprani and an alto voice have the counterpoint, while a trumpet (or in the second arrangement two oboes) softly plays the old church melody to the *Magnificat*. Much has been said as to the idea conveyed by this particular form.[379] Bach employs it when he desires to give utterance to a mystical and abstract emotion. In his German *Magnificat* he has treated the same passage in the same way, with evident reference to this previous composition. But the effect produced is not precisely similar, because the chorale has already been introduced and re-appears again at the close. In the Latin *Magnificat* the chorale does not occur anywhere but in this place; the unexpected melody, therefore, sounds doubly suggestive, pathetic, and melancholy; though the whole number, from its high position, is singularly translucent and visionary. It steals into the midst of the happy triumph which has hitherto possessed our feelings, as a shadow glides across a sunlit meadow, but the original sentiment is immediately restored by a powerful five-part fugue without concertante instruments. The *Magnificat* is closed, according to the custom of the church, with the usual Doxology, *Gloria Patri et Filio*, &c. Bach has set the threefold *Gloria* to grand rolling passages in triplets, towering up in repeated imitations, and sinking to rest in broad harmonies—a composition of exceptional grandeur

---

[379] See Vol. I., pp. 459 and 544.

and power. In *Sicut erat* he returns to the opening chorus, reconstructing its principal motives into a gorgeous and glowing whole.

The part played by the *Magnificat* in the evening service determined the form of this composition. A cantata was performed before the sermon, and, if the service was not to be made too long for custom and convenience, Bach could not allow himself to expatiate too largely on his musical ideas; all the more so because he proposed to extend the text by introducing four Christmas hymns. The *Magnificat* is consequently emphatically distinct from the rest of Bach's grand church compositions by the compactness and concentrated power of the separate numbers—particularly of the choruses—by the lavish use of the means at command, and by its vividly emotional and yet not too agitating variety. It stands at the entrance of a new path and a fresh period of his productivity, at once full of significance in itself and of promise for the future development of the perennial genius which could always re-create itself from its own elements.

If the cantata for the first day of Christmas had seemed to resuscitate that side of Christianity which rejoices with mirth and dancing before the altars of the healing Godhead, and if in the *Magnificat* the popular feeling of Christmas festivity had found its expression against a background, as it were, of simple festal usages, in the cantata " Dazu ist erschienen," which was probably composed for the second day of Christmas, 1723, we see Christ represented as the radiant hero sent into the world to conquer the powers of darkness.[380] "That was the true light which lighteth every man that cometh into the world. . . . And the Word was made flesh and dwelt among us, and we behold His glory, the glory as of the only begotten Son of the Father, full of grace and truth." The cantata is made to illustrate these fundamental ideas of the Gospel, and some of these words serve as the text of the first recitative; the grand chorus which follows is on the words of I. John iii., 8:

---

[380] B.-G., VII., No. 40.

"The Son of God was manifested that He might destroy the works of the devil." Bach has concentrated the feeling of the work in this one chorus, more than he has done in the case of any of his other cantatas, for chorales are used much more freely than usual, no less than three, each in four parts, being introduced in it; but, with one exception, they are new melodies not in general congregational use.[381] For novelty, boldness, and breadth of structure this is far superior to all the choruses of the *Magnificat* or of the first Christmas cantata. The instrumental introduction is not a proper ritornel, but has something of a concerto character, like the first chorus in the *Magnificat*, in which Bach seems to have tried his wings before a wider flight. Then the chorus works out the motive which the instruments have given in a compressed form. At a first glance it appears as though the instrumental prelude were wrought out of independent materials, but on closer examination we may convince ourselves that the subject of the chorus is in fact contained in it, as a kernel in its shell. The first phrase of the chorus is almost exclusively homophonous in character, then it occurs alternately with the orchestra in an interchange of the very shortest phrases; but on the words "destroy the works of the devil," uttered in defiant declamation, the answering bodies unite and combine in unison—which in Bach is a very rarely used device—radiating from thence like a sheaf of rays from the focus of a lens, chasing the shades of night and pouring a flood of brightness on all around. From this latter portion a double fugue is worked out, as the first portion of the chorus is from the Prelude. The first theme is constructed by augmentation, while the second preserves the original time. It sets out with wonderful boldness on the unprepared

---

[381] Erk has proved that the melody "Schwing dich auf zu deinem Gott"—"Lift thy spirit up to God"—was not, as Winterfeld says, composed by Bach; see his admirable essay on Bach's Chorales, Part I., p. 121, under No. 114. The final hymn, the fourth verse of "Freuet euch ihr Christen alle," by Keimann and Hammerschmidt, had become popular less as a congregational hymn than as a sacred chorus tune. The first chorale is the third verse of the sixteenth century Christmas hymn "Wir Christenleut."

seventh, and, as if this were not enough, the first horn follows it in sixths, and a wild triumphant battle turmoil is worked out. Then, exactly as in the *Magnificat*, it passes from the subdominant back to the first part. When it is said that this chorus gives us the true heart and root of the whole composition, this is so far true that its musical elements continue to leaven the cantata throughout. After a haughty and scornful air for the bass " Höllische Schlange wird dir nicht bange "—" Serpents of hell will not affright thee "—we have a recitative with an accompaniment, of which the rhythm is derived from some of the instrumental figures in the chorus, and the motive of the last aria is also traceable to certain elements of that grand composition.[382]

For the third day of Christmas we have another cantata, which was probably composed in this year, or, at any rate, during the earlier part of Bach's residence in Leipzig, and performed there and then.[383] This again offers a strong contrast to the Christmas music previously described. It has little to do with the incidents of the festival itself; the previous compositions had sufficed in this particular. It alludes in an encouraging manner to the love of God, who is fain to be the Father of His human creatures, and the imperishable grace which Jesus wrought for them by His incarnation. This gives rise to a tone of grave collectedness and devoted faith which prevades the whole work. As in the cantata "Dazu ist erschienen," a more extensive use of the simple church hymn is made here than in Bach's other works, and this, in itself, points to its having been composed at the same time and to words by the same author. Three different chorales are introduced in the course of the work : the first, which is the last verse of "Gelobet seist du Jesu Christ," immediately follows the introductory chorus, a fine and solemn fugue on the words " Sehet, welch eine Liebe hat uns der Vater erzeiget, dass wir seine Kinder heissen "—" Behold what manner of love

---

[382] See Appendix A., No. 25.
[383] B.-G., XVI., No. 64.

the Father hath bestowed upon us, that we should be called the sons of God"—in which the voices are reinforced by the strings and the trumpets with the cornet. The effect of this chorale depended in great measure on the circumstance that it was sung by the congregation before the intoning of the Gospel, and consequently the sentiment of the Gospel was brought vividly before them in a glorified foreshadowing. The second chorale consists of the first verse of the hymn "Was frag ich nach der Welt"—"What care I for the world"; it follows closely on an alto-recitative, in which rapid rising and falling passages in the bass figure forth the transitoriness of all earthly things that pass away "like smoke." It was in the same spirit as that in which Bach had on a former occasion given a contrapuntal setting to the organ chorale "Ach wie flüchtig, ach wie nichtig," [884] and subsequently composed the introductory chorus to a cantata on this same chorale,[885] that he now worked out the soprano aria which here follows. The last chorale, which forms the last number, is the fifth verse of "Jesu meine Freude," a favourite chorale with the master.[886]

As in this year Christmas Day fell on Saturday, there was only one Sunday after Christmas; the next music had to be composed for the New Year's festival of 1724. Among Bach's New Year's cantatas there is but one of which we can assert with any certainty that it must have been written between 1724 and 1727, this is "Singet dem Herrn ein neues Lied"—"Sing to the Lord a new song."[887] It may, therefore, be mentioned in this place. The first chorus—in D major 3-4 time—is founded on words taken from Psalms cxlix. and cl. At first the voices are used in a rather homophonous and massive way, but on the words "Alles was Odem hat lobe den Herrn"—"All that have voice and breath, praise ye the Lord"—a fugue begins. In two places the grand composition is interrupted by all the voices de-

---

[884] See Vol. I., p. 602.
[885] B.-G., V.,¹ No. 26.
[886] See App. A., No. 26.
[887] A fragment of the original score and some of the original parts exist in the Royal Library at Berlin.

claiming in mighty unison the first two lines of the chorale "Herr Gott, dich loben wir"—"Lord God, we sing Thy praise." In the second number these lines are used for a four-part chorus interwoven with recitatives. The third movement is an alto aria, of a gay and almost dance-like character, and compact in form (A major, 3-4); it is accompanied only by a quartet of strings. A bass recitative then leads into a duet of a deeply emotional character (D major, 6-8) for the tenor and bass, with violin *concertante*, "Jesus soll mir alles sein"; and after another recitative for the tenor the piece concludes with a second verse of the New Year's hymn "Jesu nun sei gepreiset." This work is in no respect inferior in importance to the Christmas compositions just discussed, and Bach himself thought it worthy to serve, after some revision, for the first day of the Jubilee in memory of the Augsburg confession, on June 25, 1730. Picander undertook the necessary alterations in the text. It is not improbable, indeed, that he had also written the text for the New Year's cantata, for it is included in his works;[388] indeed, the last recitative of the older form of it follows out much the same train of thought as we find at the end of a piece in Picander's "Erbaulichen Gedanken,"[389] written for the New Year of 1725. This last recitative also forcibly reminds us of the first recitative in the *Rathswahl* cantata "Preise Jerusalem den Herrn," and it would seem from this that the same poet wrote this also.[390]

Bach came forward with another grand new composition for the Feast of Epiphany, on January 6. It refers to the Epistle for the day, and not to the interesting Gospel narrative, and so is better adapted for Vespers than for the first service. The eye of the Prophet sees the crowd of nations over whom the light of the new doctrines dawns and spreads, the thousands across the seas that are converted to Christ, and the might of the heathen that are gathered in to Him; unfortunately the poet has not sufficiently

---

[388] Vol. I., p. 207, of the collected edition of his poems.
[389] Leipzig, 1725, p. 78.
[390] See p. 173 of this Vol., and Appendix A., No. 27.

concentrated these grandiose images, and has availed himself of the greater portion of the space afforded to him for a moralising homily, to the great detriment of the work as a whole, though it was perhaps suggested by the doctrinal purpose of the Epistle in the service. From the very first recitative this cantata "Sie werden aus Saba alle kommen"—"They shall all come from Sheba"[391]—has a grave and deliberate character, but no remarkable originality; and even the closing chorale does not lead us up to a coherent and definite festal feeling, but carries on the general sentiment which is suggested by the last aria. It is difficult to understand why, in this place, Bach did not introduce some change. It would almost seem as though he on his part had wished to insist on the feeling of the Epistle in the cantata, though it had to be performed at morning service. But the beginning—a chorus on the last verse of the Epistle with a chorale immediately following—is of lofty and peculiar beauty. Dense masses seem to come crowding on to do homage to the Saviour, "Bringing gold and frankincense, and to declare the praise of God." First on the key note and then on the dominant, with close imitations in canon, the pilgrims seem to come pouring in, almost treading on each other's heels; a few small gaps occur in the tumult of the fugue which follows, and then in the last bars they sing, as with one voice, the Glory of the Lord. A solemn and mystical brilliancy is given to the whole picture by the use of the horns, flutes, and *Oboe da caccia*. The introduction, immediately after, of the short chorale "Die Könige aus Saba kamen dar"—"Kings from afar have come to Thee"—sounds strange. Though in poetic purpose it follows what has gone before as the fulfilment of a prophecy, in musical effect the small and simple form is swamped by the broader and richer one. Here, again, it is the connection it bears to the service which explains and justifies it. The hymn appointed for use at the Epiphany was *Puer natus in Bethlehem*, and this chorale is the fourth verse of that hymn (*Reges de Saba veniunt*). The hymn

---

[391] B.-G., XVI., No. 65. P. 1280.

was sung at the beginning of the service by a chorus *a cappella*, and the return to it here has a symbolical significance which would suffice to counterbalance the great chorus; but it is indeed more than justifiable, it is indispensable in this place to give the work a thoroughly sacred and festal character. The recitatives and aria bear, it is true, the stamp of church use, but they are but slightly connected with this particular festival. The first chorus gives a wonderfully artistic form to the leading idea of the Epiphany; but the mode in which an incident has been musically embodied in it approaches very remarkably the oratorio style, of which the distinguishing mark is the way in which it directly depicts the emotion an event gives rise to, without the modifying stamp of church use. This chorale is treated on a method which, lying between the two, defines them both; it strictly confines the human emotion within the limits of church use, while it concentrates the devotional feeling on the festival of the day. Its oratorio style gives this first chorus an affinity with the Christmas cantata "Christen ätzet diesen Tag"—"Christians, mark this happy day." They have other resemblances in detail. When the first chorus in the Christmas cantata runs thus:—

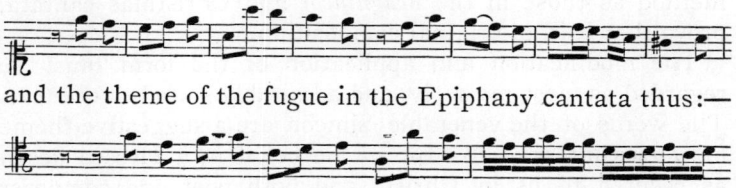

and the theme of the fugue in the Epiphany cantata thus:—

the same principal idea can be traced in both, though dressed in a different garb; indeed, here the same idea is used again, treated in canon (from bar 27).[392]

The next festival of the church year was the Purification of the Virgin. This was held on February 2, which, in 1724, fell upon the Tuesday after the fourth Sunday after

---

[392] See Appendix A., No. 28.

the Epiphany. The music which Bach seems to have composed for this day, "Erfreute Zeit im neuen Bunde,"[898] is an off-shoot from the first chorus of the *Magnificat* and the cantata for the second day of Christmas, and a pendant to the cantata for dedicating the organ, of November 2 of the previous year. There we had a cantata in the form of an orchestral suite; in the first chorus of the *Magnificat* and in the Christmas cantata we had an imitation of the first movement of a concerto. The music for the festival of the Purification has assumed the form of a complete Italian concerto; and, as if Bach was resolved to force this on the hearer's consciousness, he adds a violin part *concertante* to both the first and third movements, one being an aria for the alto and the other for the tenor. The resemblance to an instrumental concerto is indeed so conspicuous that we might almost be tempted to think it had been founded on one, and remodelled from its first form for church purposes. This, however, is not the case. The use of the aria form, it is true, would not suffice to disprove it, since it occurs in real concertos by Bach, as, for instance, the violin concerto in E major. But the regular and normal ritornels, though worked out on the same method as those in the *Magnificat* and Christmas cantata, prove that it is an original composition.

The modification and application of the form must be regarded as most masterly, and admirably suited to the text. The words of the venerable Simeon are a suggestive theme for verses intended to express the belief in a blessed death as secured to us by Christ; and both poet and composer have treated the motive grandly and broadly. It is conspicuous in the text, but is not the predominant idea; at least, as much importance is given to the feeling that faith is effectual in bestowing strength and happiness in this life; but Bach has worked exclusively on the former ground, and to embody it musically the powerful and, at the same time, plastic style of an opening concerto movement offered a form as admirable as it was novel. This was not quite the case

---

[898] B.-G., XX.,[1] No. 83.

with the third or *giga* movement, here represented by the tenor aria " Eile Herz voll Freudigkeit "—" Haste my heart with eager joy"—for in the first place the common factor in the conception of the text and of the music is, in this case, the rather superficial one representing the idea of "haste"; however, even in this, an irrepressible and elastic nature asserts its power. The piece which divides the two arias, in the same way takes the place of the concerto *adagio;* more, however, by the contrast it offers to its surroundings than by its inherent nature. The song of Simeon was sung in the Leipzig liturgy as preliminary to the Collect.[394] In the second section of the cantata Bach takes up the first three verses of it, gives them to the bass, and has them accompanied by an independent instrumental subject in two parts, worked out now in strict canon and now in free imitation. This is interrupted by recitatives. The severely sacred character of the elaborate counterpoint to the old melody puts the listener into the fitting vein of feeling for the first and third movements. A fine effect is produced by the melody of the three verses being repeated, not in the same position, but each time in a lower one, each time more calmly and dimly, as it were, like the dying thoughts of one who "departs in peace."

Another short recitative follows the tenor aria, and then comes the fourth verse, in four parts, of the chorale " Mit Fried und Freud ich fahr dahin," which was the appointed congregational hymn for this festival, for Vespers only. It is not without interest to note the arrangement of the keys in the different movements of this cantata. The first movement, the alto aria, is in F major; the second, the chorale arrangement, in B flat major; the third, the tenor aria, is in F major again, and the final four-part chorale is in D in the Doric mode. When a cantata included only two arias, and the second was not the final movement of the whole composition, Bach usually avoided composing them both in the same key, and its occurrence here is farther proof that the type of the Italian concerto was floating in his

---

[394] According to Vopelius, p. 112.

mind. The first three numbers form of themselves a musical entity, and what follows is an external addition.[395]

It is not known what Bach composed this year for the festival of the Annunciation; for the present we will pass over the Passion music which he had performed on Good Friday and go on to the first day of Easter. Here a truly grandiose work meets our view, the cantata "Christ lag in Todesbanden"—[396] "Christ lay in bonds of darkness." It is certain that Bach composed this in the early years of his life in Leipzig, and we discern, moreover, that his ideas reverted to Kuhnau in this work in the same way as in the *Magnificat* at Christmas. It is, therefore, highly probable that it was composed for April 9, 1724.[397] A MS. of the year 1693 has been preserved containing a sacred composition by Bach's predecessor, in which the same chorale constitutes the central idea.[398] The first and last stanzas of the hymn form the beginning and the close, and between them there are independent lines in verse-form, themselves a paraphrase of the ideas of the older hymn. It begins with an instrumental sonata in the old form; the first verse is sung by the soprano alone with an accompaniment of two cornets and continuo in this rhythm—

and this is immediately followed by a *vivace* Allelujah in four parts. In comparing this first section of Bach's cantata—the short introductory symphony, the composition of the second verse, and the Allelujah at the close of the first— we at once detect that he has allowed himself to be inspired by Kuhnau's work, which he must have found among the music in the St. Thomas' Library. However, this special musical impulse has not carried him beyond the first movement, though the whole cantata betrays a constant endeavour to recover the earlier forms of expression which

---

[395] See App. A., No. 29.
[396] B.-G., I., No. 4. P. 1196.
[397] See Appendix A., No. 30.
[398] In the Royal Library at Berlin.

he had in fact long since left behind him; and in this respect it evades comparison with any other of his works. We may be very sure, from the profound nature of the artist, that in doing this, what he aimed at was something more than a mere rivalry with his esteemed predecessor. The melody of the chorale is one of the most ancient in existence; it is easy to recognise it as a modification of a hymn already well known in the twelfth century, "Christ ist erstanden"—"Christ is risen." If the high antiquity of this tune was known to the composer—as is certainly very probable—he would no doubt feel the fitness of stamping on the whole composition he developed from it a correspondingly antique character, and this he thought could best be done by the adaptation and revival of forms which were not yet wholly cast off by the modern time, but which yet had some flavour of antiquity. Since, too, in the morning service, both these old hymns were sung, and at Vespers "Christ lag in Todesbanden" was again used by the congregation in the churches both of St. Nicholas and St. Thomas, the melody gave utterance to the festal feeling of this special day above all other festivals, and guided the emotional side of the whole service into the right path.

An antique character is impressed on it merely by the constitution of the orchestra. It is well known that in the seventeenth century harmony in five parts was almost invariably preferred to four, and for this reason two violas were frequently added to the two violins. Bach himself had followed this custom in some of his earlier cantatas, as in the Advent music "Nun Komm der Heiden Heiland"—"Come, O Saviour of the nations"—written in 1714, and the Easter cantata "Der Himmel lacht"—"The Heavens laugh"—in 1715. The cantata for Sexagesima, which was written still earlier, "Gleichwie der Regen und Schnee vom Himmel fällt"—"Like as the rain watereth"—has four violas, the violins being altogether absent. In the Leipzig cantatas it is an exception when the two violas are employed, and this is one of the exceptions. None but stringed instruments are introduced; the trumpets and cornet belonging to them are only used in a few passages to support the voices. The composer has carefully avoided all the "madrigal" types of music, as likewise the arioso and all solo singing strictly speaking. The seven verses of Luther's hymn serve exclusively for the text, and he works out the melody in seven numbers, each different from the other, so that this is the only work by Bach which is literally and thoroughly a church cantata in the sense in which Buxtehude, Pachelbel, and Kuhnau used the word.[899]

The introductory *Sinfonia* is quite in the style of Buxtehude's sacred music, and it must remain doubtful whether Bach purposely returned to the forms of expression of an earlier period, or used a work of his youth as the foundation of it. This melody, played by the first violin:—

—the feeling of the first two bars; the repetition of the same phrase; the interrupted progressions; the episodical dismemberment of the first line of the chorale, which is, as it were, only caught in passing; and, finally, the brevity of the piece, which altogether contains but fourteen bars—all this is so foreign to Bach's later style of writing, that the second hypothesis seems the more probable of the two.

Each of the seven verses undergoes a special treatment. The first and fourth are in the style of Pachelbel; the full choir is employed in them, and in both without any independent instrumental accompaniment. In the first the *cantus firmus* is given to the soprano; in the fourth to the alto. In the second verse, for the soprano, alto, and *continuo*, the lines are dissected and worked out in Böhm's manner. The third is constructed on the principles of an organ trio, and the only voice employed is a tenor, which has the *cantus firmus*; the fifth verse, on the contrary, is sung by the bass alone, the melody lying in the first violin of the accompanying strings; but the lines of the air do not follow each other

---

[899] See Vol. I., p. 305.

immediately, but are separated by interludes, in which the first violin has an independent part. These interludes, as well as the opportunity afforded by the prelude, are taken advantage of by the bass voice, which sings each line in anticipation, whereas in the passages where the melody is given to the instruments, a counterpoint is allotted to it; thus each line is repeated twice, and both times on notes of the same value, and the fact that the instruments are the true exponents of the chorale, and not the voice, is only recognisable from the place in the scale in which it appears on the instruments. In this number also we find in certain pregnant passages an extension of the melody in Böhm's manner. The sixth strophe is given to the soprano and tenor; the delivery of the chorale is distributed between them, the tenor taking the first two lines and the soprano the third and fourth; the fifth again is for the soprano, the sixth for the tenor, and they both sing the last two. This alternation, however, only applies to the lines of the chorale itself; the two voices are for the most part employed together throughout; the voice which is not singing the melody sings in counterpoint, and also leads the two first couplets with the melody sung in the manner of a prelude, but on the fourth above or the fifth below the chief part. In the seventh verse, the whole chorus sing the simple finale.

These treatments of the chorale bear abundant traces of the earlier style. They lie partly in the numerous imitations of Böhm's effects and partly in certain combinations of the instruments in the first chorus. While the second violin and the viola for the most part support the voices, the first violin goes on its own way high above the general body of sound; at the same time it drags the second violin into its own rhythm, and so develops a movement such as we have often met with in Bach's earliest cantatas.[400] The entrance of the strings, too, in the second bar, reminds us of Buxtehude's tendencies towards mere fulness of tone, irrespective of the thematic value of melodic phrases. On the other hand, again, the cantata displays a wealth of chorale forms which the old

---

[400] See Vol. I., pp. 444 and 453. B.-G., XXIII., p. 169.

masters were far from having at their command; nor had
they any intuition of the dramatically sacred sentiment
which we here meet with at every line. The type of the
first chorus is, it is true, that of Pachelbel; still, this is not
perceptible in the first two lines, since the *cantus firmus*
starts on the very first note, and scarcely any dependence on
the theme is perceptible in the parts which have the
counterpoint. But when, as an introduction to the following
lines ("der ist wieder erstanden und hat uns bracht das
Leben"—"Who hath risen again and brought us life"—) they
strike in with a broad fugal subject which is at last crowned
by the soprano with the expected melody—when all the
parts begin to extend, and spread, and overflow with
independent vitality—then we discover what a deep poetical
intelligence has here pervaded and animated the whole. The
extension of the lines in the second stanza is, in the first
instance, a recurrence to the standard of the old type of
chorale, but it is also subservient to the poetical idea. On
examination of the separate phrases, it is easy to perceive
that they consist for the most part of five bars or of five
half-bars. The text speaks of the impotency of men against
spiritual death, which has overcome them and holds them
captive; hence this broken and abrupt rhythm, which
seems to hold the music spellbound. In the sixth verse we
find the same artifice, but with what a different aim!

> So feiern wir das hohe Fest Come let us keep the holy feast
>   Mit Herzensfreud und Wonne,   With joy and exultation,
> Das uns der Herre scheinen lässt, Our Sun is risen in the east,
>   Er ist selber die Sonne,   He is our soul's salvation,

says the poet, and after each section of the melody a long
train of light seems to fall across the path. I have spoken
of the treatment of the fifth strophe as the expression of
a mystical emotion. It is so here; a mysterious parallel
is drawn between the Paschal Lamb of the Passion and
its saving power, and the sacrificial death of Christ. The
instruments, like an invisible choir, glorify the mystery
which is proclaimed by the bass; but he does not speak
of it as a Catholic priest would, but with a personal
and Protestant participation in it; this arises from the

fervency with which Bach throws himself at once into the purport of the text. The image of the Cross is vividly given by a dislocated *melisma*, full of anguish ; that of Death by a leap of a diminished twelfth down to the darkest depth ; that of Death overcome, by a d¹, held through several bars, and almost boastful in its effect. In bars 43 and 52, the voice imitates, as it were, the movement of the mystic type.

Indeed, the cantata is full of picturesque details throughout. In the third verse, the " Form of Death," which alone remains when all living powers have been overcome by death, is represented by a peculiar counterpoint, which seems to shrink away humbled and confounded. " Die Schrift hat verkündigt das Wie ein Tod den andern frass, Ein Spott aus dem Tod ist worden "—"The Scripture hath declared to us that as One Death hath swallowed all, death is now mocked at." Thus begins the fourth verse; the parts enter in counterpoint on the first line, impressively and powerfully, like the shout of a herald ; on the second they entwine in a close maze in canon, in which the parts seem to swallow each other in turn ; in the third they dance gaily and victoriously, disdaining the *cantus firmus*; in the fourth stanza, the *cantus firmus* derives a peculiar effect from the circumstance that it is carried on, not in the fundamental key, but in that of the fifth above. This bold and striking combination obviously serves a poetic purpose, for " it was a great and fearful fight that death and life were waging." If we listen to the cantata all through, as a whole, the effect is at first somewhat monotonous, in consequence of the persistency of the chorale melody and of the key of E minor, and from the uniformly low and gloomy pitch of feeling throughout. A dim and mournful light, as of the regions of the north, seems to shine upon it ; it is gnarled and yet majestic, like the primeval oak of the forest. From the total absence of all Italian forms, it bears a German and exclusively national stamp. Such a product of art could never have matured under a southern sun—a work in which the Spring festival of the church, the joyful and hopeful Easter-tide, is celebrated in tones at once so grandiose and so gloomy.

The cantatas written by Bach for the second and third

days of Easter, 1724, are lost, as well as that composed for Ascension Day; those for the first and third days of Whitsuntide we have, on the contrary, and also one for Trinity Sunday, all of which seem to have been written in this year, or, at any rate, early during his residence in Leipzig.

The composition for "Erschallet ihr Lieder, erklinget ihr Saiten"—"Sing out, all ye minstrels, your lutes now be sounding"—belongs to the first day of Whitsuntide.[401] The verses are probably by Franck; they are not, to be sure, included in any of the printed collections of his poems, still, we are not forced to conclude that Franck published all his cantata texts. A glance at the "Geist- und weltlichen Poesien" shows us that he was fond of setting Bible verses, not merely at the beginning and end of a text, but of introducing them here and there, and enlarging upon them in words of his own. An uniform musical composition was hardly to be expected from this process, but Franck never very well understood how to work hand in hand with the musician. In the cantata, "Erschallet ihr Lieder," the Bible text, "He that loveth Me, keepeth My word," which ought properly to have crowned the whole, takes quite a subordinate place. As is very frequent with Franck, the words for the arias are not in the *da capo* form, but like the verses of a hymn; there are no recitative passages at all. Franck's peculiar manner is most conspicuous in the duet, and if we compare the dialogue it contains with some of the verses of a Whitsuntide cantata in his "Geist- und weltlichen Poesien" we cannot but recognise the same hand in both. As we shall presently see, again, this is not the only unpublished text by Franck which Bach made use of in Leipzig.

The duet, which reminds us of the music of the chorus, "Himmelskönig," is the most important movement of the cantata. The soprano and alto sing together above a *basso quasi ostinato*, while an independent instrumental part works out the Whitsuntide chorale "Komm heiliger Geist, Herre Gott"; in a subsequent rearrangement, he gave the bass

---

[401] The original parts are in the Royal Library at Berlin.

and chorale to an *obbligato* organ. The artistic elaboration of this complicated movement is enhanced by Bach having introduced a variety of highly-coloured detail into the chorale in Buxtehude's manner; at the same time, he has not used the whole long tune, but only the first three lines and the last two, treating the repeated "Hallelujah" at the end as a single line. It is worthy of remark that he used the same abridged form in the original arrangement of his chorale fantasia for the organ on the same melody, but there the "Hallelujah" also is omitted.[402] The whole movement is instinct with a fervency and ecstasy which astonish us even in Bach. Buxtehude's mode of treatment was peculiarly suited to express such emotions; the liberty it gave the imagination facilitated the carrying out of such intricate combinations, by allowing small deviations from the strict order of the time, as in the "Hallelujah," or even a break in the melody, as in bar 9. The solo songs are equally full of characteristic beauty, as the tenor aria, where the passages for the united violins float by like airs of spring, and the magnificent bass aria, accompanied only by trumpets, drums, and bassoons. The chorus, which is repeated at the close, recalls that in the Christmas cantata "Christians, mark this happy day"; it is, however, less important in character. It must be observed, as reminding us of the Easter-day music—"Christ lay in bonds of darkness "—that we here again have two viola parts.[403]

The music for the second day of Whitsuntide, 1724, again is wanting, while that for the third may have been preserved in the cantata "Erwünschtes Freudenlicht,"[404] which at any rate was written at about this period. At the same time there can be no doubt of the fact that it is a remodelling of a secular cantata. For what occasion the original was composed is not known; not a trace of it survives. But

---

[402] The omission of the "hallelujah" justifies us in assuming that Bach had arranged the first four lines, and wished to give the last of them a closing tune which should lead back into the original key, for the fourth line is almost exactly the same as the one before it.

[403] See App. A., No. 31.

[404] The original score and parts are in the Royal Library at Berlin.

that it was actually founded on a secular cantata is evident from the popular dance-like character of the duet, and still more from the gavotte measure of the final chorus. Another particularly strong evidence is that it recurs with a different text as the closing chorus of a secular composition written in 1733—that is to say, the first twenty-four bars of it do.[405]

This proves that it was originally written for a secular purpose. Bach himself must have had a particular liking for the subject, for often as it occurs with him to transform a secular into a sacred piece, it is equally rarely that he transfers a number from one secular cantata to another. The chorale which precedes the gavotte was not inserted till later, when the work was adapted to church purposes, but it was insufficient to alter essentially the tone of perfectly worldly cheerfulness of the whole piece.

To the same period, and probably the same year, belongs the Trinity cantata " O heilges Geist- und Wasserbad "— " O fountain of the Spirit's grace "—of which the text is again taken from Franck's "Evangelische Andachts-Opfer"[406]—a flat and empty poem to which Bach has written a pleasing and graceful though not very important composition. The opening subject, an aria for the soprano, is a remarkable piece, an extremely artistically wrought fugue with strettos, inversions, and countersubjects. The two other airs proceed more simply, but exhibit throughout the same finish of detail. The chorus only comes in in the simple final chorale, and the feeling of the whole is mild and temperate.

This closes the series of the festival cantatas which can be assigned with more or less certainty to the first complete ecclesiastical year in Leipzig. We may now consider the works composed for ordinary Sunday use, belonging—or appearing to belong—to this year.

---

[405] Lasst uns sorgen, lasst uns wachen," written for the birthday of the Elector of Saxony, September 5, 1733. The text is in Picander, Part 4, 1737; the autograph score and parts are in the Royal Library at Berlin. The circumstance that the beginning of the final chorus is a clean copy, proves that it was not first written for this cantata.

[406] See App. A., No. 32.

I.—The Sunday after New Year's Day, January 2, 1724, I should place here the cantata "Schau lieber Gott, wie meine Feind"; a grand and boldly planned tenor aria, "Stürmt nur, stürmt ihr Trübsals-Wetter," and an alto air of wonderful melodious charm, are associated with three simply set chorales, one of which is placed at the beginning, the second after a bass arioso included in a recitative, and the third at the end. There is no chorus on the chorale nor on any independent subject, and the arrangement of the movement is different from what is usual with Bach.[407]

II.—First Sunday after Epiphany, January 9, 1724, "Mein liebster Jesus ist verloren." As in the former cantata, the chorus has only two simple chorales, one of which forms the close and the other is the third number. The arias are of great beauty; the tenor begins in a mournful and genuinely Bach-like longing strain:—

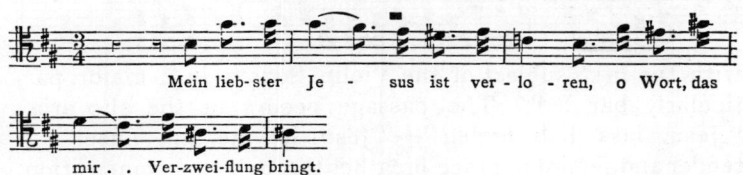

Mein lieb-ster Je - sus ist ver-lo - ren, o Wort, das mir . . . Ver-zwei-flung bringt.

The following lines: "O Schwert, das durch die Seele dringt"—"O sword, that piercest through the soul," &c.— are evidently suggested by Joh. Rist's hymn, "O Ewigkeit, du Donnerwort, O Schwert, das durch die Seele bohrt." Bach composed two cantatas beginning with this chorale. In one of these, that for the twenty-fourth Sunday after Trinity,[408] the anxious expectancy with which man watches for the coming of the Judge is expressed by a light quivering motion of semiquavers on the strings. Bach here employs precisely the same image to the analogous words, and in composing the cantata for Trinity Sunday, 1732, this earlier work may have recurred to his mind. Another

---

[407] The original parts are in the Royal Library at Berlin. See App. A., No. 33.
[408] B.-G., XII.,² p. 171.

striking resemblance occurs in the tenor aria which precedes the last chorale with this passage :—

In a duet in the cantata "Liebster Jesu, mein Verlangen," which also belongs to the first Sunday after Epiphany,[409] the same passage occurs elaborately worked out, and we cannot doubt that the repetition was intentional. It may also probably be recognised in the fugue in F sharp major of the second part of the Wohltemperirte Clavier, where the polyphonic progression is interrupted twice by a long homophonic section worked out almost without any thematic connection with the subject quoted above; and again this cantata reminds us of a third composition: compare the following passage :—

with the first subject of the Violin Sonata in A major, particularly bar 8.[410] The passage occurs in the alto aria: "Jesu, lass dich finden"—"Jesu, let me find Thee." A tender and feminine grace breathes through this composition. It is felt from the first in the principal subject :—

and the piece acquires throughout a soft ethereal brilliancy from the accompaniment, consisting only of violins and violas with two *oboi d'amore*, and the rocking movement of the principal parts :—

which is carried on by the violas, gives it a peculiar character.

The parents of Christ, so runs the Gospel narrative, went up with Him to the Passover at Jerusalem. There

---

[409] B.-G., VII., No. 32, p. 55, P. 1663.
[410] B.-G., IX., p. 84. See ante, p. 113.

they lost their Son, and after long seeking "found Him at last in the temple sitting in the midst of the doctors both hearing them and asking them questions." With gentle reproach, Mary says to Him "Son, why hast Thou dealt thus with us? Behold Thy father and I have sought Thee sorrowing." And the Child justifies Himself in words almost of reproof. "And His mother," we are told, "pondered these words in her heart." The writer of the text has treated the incident symbolically as representing the longing of the soul for Christ, and Bach has adopted this as the fundamental feeling of his composition. Indeed, we know how ready he always was to find some special purport in the Bible text itself, or in the ecclesiastical significance of the Sunday, and here, when we see moreover that the above quoted aria is immediately followed by the words from the Bible in which Jesus reproaches His mother, "Wist ye not that I must be about My Father's business," we cannot for a moment doubt that it was the image of the blessed Mother—so vividly set before us in the Gospel narrative—which floated before the fancy of the deep-souled composer. It is an instance similar to that of the B minor aria in the *Magnificat* (see ante, p. 377).[411]

III.—Fourth Sunday after Epiphany, January 30, 1724, "Jesus schläft, was soll ich hoffen"—"Jesus sleeps, what can I hope for?"—Jesus is sailing across the sea with His disciples—a storm rises, but He is asleep. They wake Him in their terror, He reproves them for their little faith, speaks to the sea and it is still. The hearer, whose mind is full of this picture, when he listens to this cantata will be startled by the first aria, for it is quite beside the situation as thus depicted. It is a dusky night piece, the Saviour sleeps and some weird apparition wrings cries of terror from the lonely watchers. It is not till the second aria that the musical work corresponds somewhat more closely to the Gospel narrative, and here all the means at command, even the voice part, are engaged in representing the sea surging in a storm. In the following movement the connection

---

[411] See App. A., No. 34.

is even closer; we hear Christ speak, "O ye of little faith wherefore are ye so fearful?" and in the magnificent E minor aria we see Him rise majestically to rule the winds and waves with words of might. It is vain to seek here any thread of dramatic purpose which might give unity to the whole. It is indeed the very privilege of the composer that he may treat the Gospel narrative itself, in its simple or its symbolical meaning, as the amalgamating factor; he may view his subject from different sides, and need regard none but musical requirements in arraying and ordering the pictures he sets before the hearer. In this cantata Bach has shown how with the smallest means he could produce the grandest results. It is beyond question one of the most stupendous productions, not only of his art but of German music at any time. In every bar it may be said that his genius reveals its full power. No one can listen without deep emotion to the chorale "Jesu meine Freude" which comes in with consolatory effect.

IV.—Quinquagesima Sunday, February 20, 1724, "Du wahrer Gott und Davids Sohn." This cantata, composed a year before, has already been discussed (see p. 350).

V.—Jubilate Sunday (third after Easter), April 30, 1724, "Weinen, Klagen, Sorgen, Zagen"—"Sorrow, weeping, anguish, terror."[412] There are clear proofs indicating that this cantata was composed at the same time as that for Whitsunday, "Erschallet ihr Lieder"—"Sing out, all ye minstrels"—though at the same time we are tempted to trace in it, as in "Christ lag in Todesbanden," a remodelling of an earlier work. The autograph score, which is preserved, is a beautifully executed, fair copy. If it is founded on an older composition, this must have been written during the Weimar period, perhaps about 1714; the spirit and style of the words betray the hand of Salomo Franck. The symphony is one of those broad and richly harmonised *adagios* in which Bach rose superior to Gabrieli's church

---

[412] B.-G., II., No. 12. The statement in the preface that in the autograph score only the symphony has a figured bass is not wholly correct. The figuring, though it is imperfect, extends throughout the first recitative and first aria. As to the chronology, see App. A., No. 31.

sonatas in one movement, not without assimilating in some degree the purport and feeling of the introductory *adagio* of the Italian chamber sonatas. Now this form occurs most frequently in his earlier sacred works.[413] The first chorus is in three divisions in aria-form, and the first and third sections are a *passacaglio* adapted to the chorus and orchestra. A parallel to this also exists among Bach's earlier works: the *chaconne* at the end of the cantata "Nach dir Herr verlanget mich."[414] In both, difficulties of form are got over in a masterly manner; still this *passacaglio* is the more interesting from a musical point of view, and more thoughtfully harmonised. Its pathetic and tearful feeling, revelling in melancholy, is also characteristic of that period of Bach's life when he was still engaged on church music of the older type, and was developing his own line of feeling in the forms it offered. I have spoken in another place of the internal connection of this *passacaglio* with a chorus in the Mühlhausen Raths-wechsel cantata, and with an air by Erlebach (Vol. I, p. 351). Its bass part is one of Bach's favourite motives; it occurs also in the cantata "Nach der Herr verlanget mich"; there, however, it is the theme of a fugue for the first chorus; again we find it in the cantata "Jesu, der du meine Seele," where other details also remind us of the chorus "Weinen, Klagen."[415] The arias follow each other, as was Franck's wont, without any recitatives inserted between. The alto aria is remarkable because the instrumental ritornel has a different idea in it to the voice part, and the same is the case in the Weimar Easter cantata "Der Himmel lacht, die Erde jubiliret." Finally, this work resembles the cantata "Nun komm der Heiden Heiland" in the circumstance that it closes in a different key from that it begins in, and falls naturally into two divisions of the first four and the last

---

[413] For instance, in "Ich hatte viel Bekümmerniss" (see Vol. I., p. 531), "Himmelskönig sei willkommen" (Vol. I., p. 539), and in "Der Herr denket an uns" (Vol. I., p. 371). Compare also Vol. I., p. 124.

[414] See Vol. I., p. 443.

[415] Compare bar 83 of "Weinen, Klagen" with bar 25, &c., and bar 57, &c., of "Jesu, der du meine Seele."

three movements (see Vol. I., p. 507). Thus we find throughout traces which connect it with an earlier period of Bach's writing. It is further noteworthy that in bar 17 of the alto aria an imitation in canon of the voice part, beginning on the fourth quaver, implies the use of an organ accompaniment; this proceeding frequently occurs, not only in Bach, but in other music of the time, in closing cadences and particularly in recitatives. In evidence we may quote the end of the recitative in the cantata "Gott der Hoffnung erfülle euch":—[416]

VI.—Second Sunday after Trinity, June 18, 1724, "Siehe zu, dass deine Gottesfurcht nicht Heuchelei sei"—"Take thou heed, that thy fear of God be not hypocrisy." This cantata is evidently a companion piece to that for Trinity Sunday, "O heilges Geist- und Wasserbad," but it is worked out in grander forms, and is more important and full of meaning. Still, an extraordinary resemblance is perceptible in the structure of the first movements, which are quite exceptional and highly complicated. What in the Festival cantata is a soprano aria, in this is a four-part chorus, with an independent bass in some passages. The theme of the fugue is worked out in both pieces *in motu recto et contrario*, and in each a second theme is introduced, which afterwards combines with the first to form a double fugue. A certain relationship is also traceable in the themes,

---

[416] See App. A., No, 21.

at any rate, in the principal themes. In the Festival cantata the theme is as follows :—

in the other :—

But the progression of the parts is in the former more intricate and close, almost overwrought indeed; the fugal aria might almost be called a study for the choral fugue which is so broad, free, and imposing. But it is not in this only, but in the two arias and the closing chorale that Bach has fully displayed his unsurpassed powers of combination; particularly in the second aria, in which the soprano and bass have a real quartet with two *oboi da caccia*, which is quite admirable in its lavish use of harmonies and modulations. The cantata resembles the former one in ending in a different key from that in which it begins.[417]

VII.—Twelfth Sunday after Trinity, August 27, 1724, "Lobe den Herrn, meine Seele"—"Praise thou the Lord, O my soul." This cantata owes its sumptuous and festal character less to its appropriateness to this particular Sunday than to the circumstance that it is evidently intended also for a Rathswahl cantata. The service usual on this occasion took place, in 1724, on the 28th of August, and the cantata was then performed a second time. A third performance took place about 1735; it no doubt did not serve for Sunday use, but only for the Rathswahl service. For in the re-arrangement which it underwent for this occasion the reference to the city government was brought out more distinctly in the text, so that its application to the twelfth Sunday after Trinity is altogether lost. The solo movements of this cantata are of no conspicuous importance. But in the first chorus Bach has put forth all his powers: it is a double fugue worked out

---

[417] The autograph score and some of the original parts are in the Royal Library at Berlin.

on broad proportions with an aria-like opening, and it is one of the most brilliant and powerful pieces of the kind that remain of his writing.[418]

Having followed Bach through the first complete church-year of his residence in Leipzig, we may henceforth study all that he wrote in the way of cantatas in larger groups, according to periods. Such a group may be very properly limited by the date of the death of the Queen Christiana Eberhardine, September 7, 1727, when a public mourning of four months began, during which all church and organ music was silenced.[419] Of four of the cantatas composed until this time, the year and day of their performance can be exactly determined, and it can be approximately ascertained as regards several. As has been said, Picander published for the years 1724—1725 a "collection of edifying thoughts," which appeared in weekly parts. If we may hazard a guess that he had written for Bach so early as for the Rathswahl of 1723, and New Year's day, 1724, we can point out in this collection the first texts which Picander can be proved to have written for Bach at all. The "Erbaulichen Gedanken" had no pretensions to be fitted for music, excepting in so far as they included hymns in stanzas, and the introduction of a hymn in verses into the musical setting of a sacred "Madrigal" involved many anomalies. For this reason Picander found himself obliged to give a "Madrigal" form to the poems in stanzas contained in the "edifying thoughts" for St. Michael's day, by shortening, compressing, and transposing them with certain additions to some of the lines.

On comparing, as I have been able to do, the text of the cantata as it stands with that in Picander's collection, we find that the first two verses have been re-written to suit the ideas suggested by the Epistle appointed for the Festival. But the words of the soprano aria correspond exactly with the third verse in the "Erbaulichen Gedanken," while the tenor recitative is the first verse transformed by a facile

---

[418] B.-G., XVI., No. 69, P. 1667. See Appendix A., No. 37.
[419] See Appendix A., No. 33.

hand into the "Madrigal" form; this is the case, too, with the last two strophes, which have been adapted to the soprano recitative, while the tenor air is a new and independent verse. The closing chorale consists of the eighth verse of the hymn "Freu dich sehr, O meine Seele." There is not the smallest doubt in my mind that the cantata text and the verses in the collection were alike written for the Feast of St. Michael of 1725, and that while the former was intended only for Bach to compose to, the latter was intended for publication as a separate work. Picander wrote too quickly and readily in the Madrigal form for it to be possible that he should in later years have worked again on a poem which, even in its original form, was, on the whole, but meagrely adapted to the requirements of a cantata.

I have already taken occasion to point out that Bach derived the impetus to this composition from a work by his uncle, Joh. Christoph Bach.[420] We know that he had that composition performed in Leipzig, and that it produced a great effect. Its influence, in his mind, betrayed itself in the first place in the construction of the text, in which Picander certainly followed the prompting of Bach; and which, contrary to custom, refers more to the epistle than to the Gospel for the day, so far as it was possible to reconcile this with the use of the verses he had at hand. Besides this, the effort to produce dramatic and oratorio-like tone-pictures is conspicuous throughout the composition. Even in the first chorus, where the text offers an idea which suggests a figurative movement in the music, it is eagerly seized upon. And yet the result is not properly an oratorio chorus. It is not the objective—or, may we say, epic—concentration on the matter in hand which is so effective, but a torrent of feeling, roused by some stupendous event, which roars and rushes by, reflecting the quavering picture in vague and broken outlines. On comparison with Joh. Christoph's composition the chorus plainly reveals the characteristics and limitations of Sebastian's

---

[420] See Vol. I., p. 51.

genius. A composer of oratorio, like Handel, would have made nothing of it, even if the external conditions had offered themselves; but Bach's style was precisely what was required in church music. In the soprano air, to the quaint words:—

Gott schickt uns Mahanaim zu,	God shields us with the Mahanaim,
Wir stehen oder gehen	Whether we stay or go
So können wir in sichrer Ruh	We walk in safety all the same,
Für unsern Feinden stehen.—	Nor fear our ghostly foe.—
	(Gen. xxxii., 2.)

the image of "Mahanaim"—the angelic hosts which guard humanity at every step—is beautifully set before us by a close tissue of music woven out of the principal melody. The tenor air is even more profoundly significant in its combinations. The orchestra accompanies the voice with a *Siciliano*, which is almost independent of it, while the trumpet rings out the melody of the chorale "Herzlich lieb hab ich dich o Herr"—"Thee, Lord, I love with all my heart." This calls our thoughts, not to the first verse, but to the last:—

Ach Herr, lass dein lieb Engelein	Thy ministering angels send
Am letzten End die Seele mein	O Lord! my parting soul to tend;
In Abrahams Schooss tragen,	To Abraham's bosom take me.
Den Leib in sein'm Schlafkäm-merlein	My body in the grave shall spend
Gar sanft ohn einge Qual und Pein	The days in quiet, till the end
Ruhn bis am jüngsten Tage!	When the last trump shall wake me.
Alsdann vom Tod erwecke mich,	And grant me then with gladden'd eyes
Dass meine Augen sehen dich	To see Thy glory in the skies
In aller Freud, o Gottessohn,	O Lord! in perfect joy and peace;
Mein Heiland und Genaden — Thron!	My Saviour, fount, and throne of grace!
Herr Jesu Christ!	Lord Jesu Christ!
Erhöre mich, erhöre mich!	O hear Thou me! O hear Thou me!
Ich will dich preisen ewiglich!	Thee will I praise eternally.

The length of the stanza resulted in an unusual expansion of the aria. It is requisite to conceive of the chorale melody as forming the nucleus of it in order not to feel the movement altogether too long. However, we can perceive how this combination serves as a preparation for the closing chorale, which leads our thoughts away from the warlike images of the commencement to the peace of the blessed dead.[421]

The second cantata text which Picander arranged in part from the collection of his poems belongs to the seventeenth Sunday after Trinity, 1725, and begins with the Bible text "Bringet her dem Herrn Ehre seines Namens" (Psalm xxix., 2). In 1725 the seventeenth Sunday after Trinity fell on September 23, so this cantata must have been performed six days before that for Michaelmas. Thus, if we had closely followed the chronological order we ought to have studied it first. Since, however, the text is a less evident instance of the method on which Picander worked, and the composition generally would give rise to few observations of any interest, it is more fitly placed after the St. Michael's cantata. The subject of it is the feeling of rejoicing in the Lord in His Sanctuary, and though it is far from being so grand a work as the St. Michael's cantata, it is nevertheless of such high merit as to hold its place well by the side of it. The first chorus has a particular soaring swing, with a stamp of vigorous nationality; it is an effective union of homophonic sections and fugal subjects on very productive themes. The alto aria, accompanied by three oboes and figured bass, breathes of the solemn and joyful Sabbath feeling of the worshipping Christian.[422]

The third of the four cantatas written between 1724 and 1728, of which we can exactly fix the date, belongs to the beginning of February, 1727, and the text is by Picander. This, like the cantata "Lobe den Herrn, meine Seele," of 1724, has a two-fold purpose; it was intended both for a church cantata and for occasional music. In its first application it belonged to the feast of the Purification, February 2; and in the second it was adapted to a mourning ceremonial held only four days later. Johann Christoph von Ponickau, the elder, Lord of Pomssen, Naunhoff, Grosszschocher and Winddorff, Chamberlain, *Hof- und Appellationsrath*, had

---

[421] B.-G., II., No. 19. The autograph score and original parts, which are in the Royal Library at Berlin, offer no special evidence as to the date of the work.

[422] See Appendix A., No. 38.

died in October, 1726, in the 75th year of his age, and was buried October 31, in the family tomb at Pomssen. He had acquired many honours in Saxony, and had become a highly respected and important personage. Picander himself had good reason to be grateful to him, and gave expression to this feeling in his mourning ode. On February 6, 1727, a solemn mourning service was performed in memory of the deceased, in the church at Pomssen; Picander wrote for the occasion the text beginning "Ich lasse dich nicht"—"I will not leave Thee except Thou bless me"—and Bach composed the music.[423] It would seem as though he had done so less from an impulse of his own than to oblige the poet who was his friend, and for his part was more engaged in considering how the music could be made at the same time to serve the purposes of the church service. The text avoids all personal allusions—probably by Bach's desire—and without the alteration of a single word could be used for the Purification a few days previously. The composition throughout has no specially solemn character; it is a grave and meditative composition in the strain of feeling of the words of the aged Simeon—"Lord, now lettest Thou Thy servant depart in peace." There is no chorus whatever but the last chorale.

Fourthly, we must mention the cantata "Herz und Mund und That und Leben," which was probably written for the Fourth Sunday in Advent, 1716, in Weimar; but it had undergone a very comprehensive revision in Leipzig, and, as music was not used in the Advent season, it was now adapted to the Visitation of the Virgin Mary. Though it is not absolutely certain, it is extremely probable that this was not done before 1727.[424]

There now follows a series of church cantatas, of which all that can be said with any certainty is that they were written between 1723 and 1727.

---

[423] S. Schwartz, Historische Nachlese Zu denen Geschichten der Stadt Leipzig; Leipzig, 1744, 4, p. 33. The autograph of Bach's cantata is not known to exist. I only have seen it in a copy preserved in the Royal Library at Berlin.

[424] See Vol. I., pp. 574 and 643.—Appendix A., No. 31.

I. For the New Year, "Herr Gott dich loben wir." In speaking of the cantata "Jesus nahm zu sich die Zwölfe"—"Jesus called to Him the twelve"—it has already been pointed out that Bach by no means disdained to accommodate himself on occasion to the taste of the Leipzig public, and of all his church compositions that are known to us, this, for the New Year, is the one in which an intentional return to Telemann's mode of writing is most manifest. Not, to be sure, in the first movement, which is a splendid chorale for the chorus on the first four lines of the *Te Deum*; in this form Bach could borrow nothing from Telemann, nor, indeed, could Telemann have followed in his steps, even at a remote distance. But a quite different spirit seems to speak to us in the second chorus, "Lasst uns jauchzen, lasst uns freuen"—"Come rejoicing, come with gladness." The alternation of the bass and the full choir, the pleasing style of the melody, the incisive style of expression, the handling of the chorus—all this is deceptively like Telemann's choral subjects,[425] though on a closer inspection we at once find traces of Bach's more powerful mind. Bach's connection with Telemann did not rest merely on personal friendship; he by no means undervalued him as a composer, and transcribed with his own hand one of Telemann's cantatas for use in his Leipzig performances.[426] Again, in the very tuneful and fervent tenor air "Geliebter Jesu, du allein," we cannot fail to detect a reflection from the solo airs by Keiser and Telemann. Comparing this with the tenor aria of the cantata discussed, No. IV. (pp. 404 and 350), we at once perceive a certain affinity in their vein of feeling.[427]

II. Third Sunday after Epiphany, "Herr wie du willt, so schicks mit mir"—"Lord, as Thou wilt so let it be."[428]

---

[425] Only to give one example, I would refer the reader to Telemann's Whitsuntide cantata "Ich bin der erste und der letzte"—"I am the first and the last"—and particularly the chorus "Auf, lasst uns jauchzen."

[426] "Machet die Thore weit"—"Open wide the gates"—which exists in Bach's autograph in the Royal Library at Berlin.

[427] B.-G., II., No. 16. P. 1286. See Appendix A., No. 40.

[428] B.-G., XVIII., No. 73. P. 1676. See Appendix A., No. 19.

The chorale chorus at the beginning, which refers, though only incidentally, to a passage in the Gospel, displays a form which in many respects is new. It is intersected throughout by recitative portions which carry out the ideas suggested by Melissander's hymn. We often meet with such a scheme in Picander's sacred texts; the last aria of "Ich lasse dich nicht" is on a similar plan, and so is the beginning of a text written for the third Sunday after Epiphany, 1729; we may, therefore, conclude that the words of this cantata are by him. The chorale is in four parts, and so far homophonic as that the voices are never interfered with by imitative elaborations. The instruments are perfectly independent. Before and between the lines a ritornel comes again and again, always the same, but differing in key after each line, and it does duty as an accompaniment to the recitatives. A singular use is also made of one of the subjects of the chorale. A horn, which supports the *Cantus firmus*,[429] and at first plays a little prelude, asserts itself now and then as a subsidiary supported by the strings, giving out the first lines of the melody in a dismembered form and in diminution. This is particularly the case with the notes that fall on the words "Herr wie du willt":—

which are brought in in diminution over and over again.

At last, long after the chorus has done its part, it seems suddenly to understand the idea that the composer has been incessantly suggesting to it by the instruments; three times again, at long intervals, we hear it briefly ejaculating "Lord, as Thou wilt," breaking off at last on the dominant seventh, when the instruments come in with a rapid closing cadence. But the whole purport of this singular tone-picture is not disclosed till we come to the bass aria, which precedes the final chorale. The text consists of three verses of four lines each, all beginning

---

[429] Bach afterwards allotted the horn part to the Rückpositiv of the organ of St. Thomas'.

with the words "Lord, as Thou wilt." Bach designates this movement as an *Aria*, though it is in fact a form of his own invention. He works out in it the idea of the first chorus in such a way as plainly shows that this hymn was in his mind when he wrote the chorus. The theme is given out by the voice without any preliminary notes in the instruments:—

Herr, so du willt,

they immediately take it up and repeat it in diminution, but the rhythm only:—

and they work it out with pertinacity. The accompaniment in semiquavers is also borrowed from the first chorus. The feeling of a man who bows in humble submission before the incomprehensible counsels of the Almighty is here expressed with the deepest fervour. From the passage where the strings *pizzicato* imitate the tolling of a knell, a sort of vision of peace seems to be revealed through the dismal gate of death.[430]

III. Third Sunday after Epiphany, "Alles nur nach Gottes Willen"[431]—"Lord, Thy will alone obeying." This text is from Franck's "Evangelisches Andachtsopfer," and certainly one of the most suggestive. It follows out the same line of thought as the former cantata, and it would almost seem as though one had influenced the other; at the same time, the feeling in the former is emphasised rather in the sense of a pious resignation to the sufferings of life, while in this work that blissful contentment is praised which has its root in the assurance that the hand of a loving father is to be traced in everything. The imaginative features and gloomy colouring of the former cantata do not exist here, but a trustful and childlike devotion of most touching

---

[430] Compare Vol. I., p. 553.
[431] B.-G., XVIII., No. 72, P. 1299. See App. A., No. 41.

power. This feeling finds its strongest expression in the soprano aria "Mein Jesus will es thun"—"This will my Saviour do, my sorrows He will sweeten"—one of the most lovely vocal pieces Bach ever wrote; but the remainder is no less delightful in its way: the less tranquil alto aria "Mit Allem was ich hab und bin"—"With all I have, and as I am"—the arioso which precedes it, in which rhythms of two and three time respectively are intermingled with such wonderful effect, and the preliminary chorus which marches on in such magnificent breadth, and overflows with fervent feeling.

IV. Septuagesima Sunday, "Nimm was dein ist und gehe hin"—"Take that thine is, and go thy way."[432] The cantata begins with a fugue, in which the resolute, nay stern, repudiation of the fancied claims of the workmen for fairer payment is treated with almost dramatic force. It is a pity that the writer of the text should have so inadequately grasped the deeper meaning of the Gospel parable of the householder who hires labourers for his vineyard, that he could find nothing better to say about it than the praises of contentment; this must necessarily impair the interest of the music. What Bach did with the trivial couplets is of course full of ingenuity and purpose, but it does not stir us deeply; still, this cantata seems to have become very popular, and the text was used by several composers.[433]

V. Sexagesima Sunday. "Leichtgesinnte Flattergeister"—"Empty thoughts of worldly folly." The bass aria at the beginning, figuring the empty and foolish thoughts that scatter the blessings of the Divine word to the winds, is a characteristic composition, full of individuality. The finale is an independent chorus in the Italian aria form; it is a

---

[432] The autograph score in the Royal Library at Berlin. Published in "Kirchengesänge für Solo- und Chor- Stimmen mit Instrumentalbegleitung von Joh. Seb. Bach." Berlin, Trautwein, No. 1. See App. A., No. 19.

[433] The first chorus of this work of Bach is quoted as a model by Marburg, for its admirable declamation, "Kritische Tonkunst," Vol. I., p. 381. He speaks of a public performance of it. The beginning of the alto aria "Murre nicht, lieber Christ," is quoted by Sulzer, "Allgemeine Theorie der schönen Künste," Part IV., p. 267, Leipzig, 1779, but to another setting. I have not been able to make any farther inquiries in this direction.

fugue, but simple and popular in subject, and must originally have belonged to a secular work.

VI. First Sunday after Easter, "Halt im Gedächtniss Jesum Christ"—"Keep in remembrance Jesus Christ." Here again we have a work which must satisfy every requirement, even as regards the text. The beautiful Gospel narrative of how, after His resurrection, Jesus appeared among His disciples, bidding them "Peace be with you," and strengthening their faith, is clearly reflected in the text, which is made up of well-chosen passages from the Bible, suitable chorales, and melodious verse. It reminds us of Franck's manner, and if Picander wrote it he surpassed himself. There is but one solo in this cantata—irrespective of a few short recitatives—a beautiful tenor aria. In the first chorus—a splendid fugal movement freely worked out on two principal themes—Scheibe would have found a perfect example of that "poetic embellishment and graceful expression" which he demanded as the conditions of adequate and expressive church music. About half-way through the work, the Easter chorale "Erschienen ist der herrlich Tag"—"The glorious day has now arrived"—is brought in with great effect. From the words of the recitative which leads up to it we see that the same chorale must have been previously sung by the congregation. This was not prescribed by rule, so Bach must have expressly determined that this hymn should be used on this occasion.[434] The next chorus is extremely peculiar; it is designated as an *Aria*, from the verse form of the text. The bass sings the words of Christ, "Peace be on you all," to a tender and deeply felt melody below a soft floating accompaniment on the flute and oboes. Against this, the three upper voice-parts express by word and note their faith and confidence in Christ's protection and aid; the whole dies away in the lingering benediction of the bass, and then the chorale "Du Friedefürst Herr Jesu Christ"—"Thou Prince of Peace, Lord Jesu Christ"—once more briefly concentrates the leading idea of the cantata.[435]

---

[434] See ante, p. 231.
[435] B.-G., XVI., No. 67, P. 1293. See App. A., No. 19.

VII. Second Sunday after Easter, "Du Hirte Israel"—"Shepherd of Israel." A sacred pastoral, which exhibits a beautiful combination of tenderness and gravity, grace and depth. The rhythm of the first chorus is, properly speaking, not to be regarded as in 3-4 time, but as 9-8, since the beat throughout is chiefly in triplets, and the rhythmical figure ♩♪ according to the custom of that time should be read ♩, ♪ (see Vol. I., p. 563). By this means, and by the heavy droning bass like a bagpipe, the stamp of pastoral music is delicately impressed on the composition. Besides its wonderful melodic charm, this chorus is at the same time a masterpiece of artistic structure. In the vocal parts, homophonic sections alternate with fugal workings out; three *schalmei* (oboes) are added to the three upper parts to give them support and colour. The stringed instruments meanwhile involve them in a glittering network of rising and falling figures, always however perfectly independent. Bach had never before composed such a work; it is a fresh evidence of his inexhaustible imagination. We may compare the pastoral symphony of the Christmas oratorio as the only worthy pendant to this movement in feeling and in the delightful magic of its harmonious development, as well as in the combination of different qualities of tone. The bass air has in many respects a similar character, while, on the other hand, the tenor aria which comes between it and the chorus expresses that dejected sentiment which is naturally aroused by such passages in the Psalms as "Though I walk through the valley of the shadow of death, I fear no ill," and "As the hart panteth for the waterbrooks, so longeth my soul for Thee, O God"; and it is certain that Bach had these passages in his thoughts, since the text is to much the same effect, and a stanza of Psalm xxiii. versified forms the close of the cantata.[436]

VIII. Second Sunday after Easter, "Wo gehst du hin?"—"Where goest thou?" The text, like those of many of Bach's cantatas, reveals a lamentable incapacity on the part of the writer for grasping the idea of the Gospel and giving it a

---

[436] B.-G., XXIII., No. 104, P. 1680. See App. A., No. 19.

poetic form. After a slight reference to it we find ourselves once more in the beaten track of exhortation to think of heaven, and reflections on the transitoriness of all earthly things. It is a source of constant astonishment how Bach was always equal to the occasion, and could always produce new and still new forms and styles to give life to this poetical monotony. The work is a solo cantata, for, excepting the closing chorale, no chorus—in many parts, at any rate—is employed. The first and third sections claim our particular attention; in the former Bach has found it possible to compose an aria on a text of four words, which, moreover, merely put a question, "Wo gehst du hin?" It has a singular effect from the phrases, in three bars each, of which it is constructed. In the third number the soprano sings the third verse of Ringwald's hymn "Herr Jesu Christ ich weiss gar wohl," while the instruments carry on a two-part counterpoint. In this piece we meet for the first time with a complete and deliberate transfer of the organ trio to vocal music; in the cantata "Erfreute Zeit im neuen Bunde" we found scarcely more than an attempt at it. It was during the early period of his residence in Leipzig that he composed his six great organ sonata trios; hence we here see the outcome of his labours in this direction.[437]

IX. Rogation Sunday, "Wahrlich, wahrlich, ich sage euch, so ihr den Vater bitten werdet"—"Verily, verily, I say unto you, that ye shall pray to the Father." This is a work which bears the most obvious and close relationship with the next preceding one, both in purport and date. The arrangement of the text is precisely the same; first a text taken from the Bible, then an aria, chorale, recitative, aria, and final chorale. The means of which Bach has availed himself are also identical, even the arias are for the same voices, but in reversed order; each, however, begins with a solo in the bass and ends with a simple four-part chorale. The chorale subject in the middle, on the last verse of

---

[437] The original parts of the Cantata are in the Royal Library at Berlin. See App. A., No. 19.

"Kommt her zu mir, spricht Gottes Sohn," reveals in both cases a form borrowed from the organ; in this cantata three parts are employed in the counterpart, but the melodic character of the parts is nearly related in the two. In the cantata "Wo gehst du hin?" it begins thus:—

In "Wahrlich, wahrlich," it is thus:—

Bach was wont to give Bible words in the arioso form to one single voice, because any adaptation of the opera forms might so easily seem profane; and when, for once, he deviated from this rule in the cantata "Wo gehst du hin?" it certainly arose from the peculiarities of the text. It is true that the movement which stands at the beginning of "Wahrlich, wahrlich," is not an arioso; but the dignified attitude which befits the composition of a Bible text is attained in another way—a way, it is true, in which none but such a master as Bach could venture to walk; he has entangled the song in a regular four-part instrumental fugue, and so worked it out that for the most part it constitutes an independent fifth part; it is only now and then that it flows in unison with the instrumental bass, and thus distinctly recalls the usual type of the old-fashioned sacred *arioso*. A companion piece to this original composition is not wanting; we find it in the first movement of the cantata "O heilges Geist- und Wasserbad"; it therefore seems probable that the two works were written within a short time of each other.[438]

X. Sixth Sunday after Easter, "Sie werden euch in den Bann Thun"—"They shall cast out your name as evil" (G minor). This text again is precisely similar in its construction to those of the last two cantatas, and the composition also displays a general resemblance to them,

---

[438] B.-G., XX.,[1] No. 86. See App. A., No. 19.

particularly in the two arias. The opening piece is a duet for the tenor and bass, afterwards taken up by the chorus. The greatest polyphonic skill prevails throughout the duet, which may be compared with that in the duet in the cantata "Du wahrer Gott und Davidssohn" (see p. 350). The chorus, on the other hand, with its popular style of arrangement and easily understood polyphony, reminds us of the choruses in "Jesus nahm zu sich die Zwölfe" (see p. 353) and "Ein ungefärbt Gemüthe" (see p. 358). Something perfectly new is revealed to us again in the middle chorale movement "Ach Gott, wie manches Herzelied." It is a free imitation, not of any organ trio or quatuor but of Böhm's type of chorale treatment.[439] The simple chorale itself is sung by the tenor, accompanied only by a figured bass, and that in phrases developed by diminution from the first line of the chorale, which by their chromatic dislocations are intended to convey the idea of "heart-sickness." As the verse consists of only four lines the subject is soon over, almost too soon for its whole significance to be grasped and understood. In a cantata written much later, Bach worked out this melody in the same way,[440] only, instead of one voice, he employs a four-part chorus. But in the later composition the effect produced is greater, because recitatives are inserted between the lines of the chorale, and the hearer consequently has time afforded him to take in the original structure of the form. The closing chorale of the present cantata is the last verse of Flemming's hymn, "In allen meinen Thaten."[441]

XI. First Sunday after Trinity, "O Ewigkeit, du Donnerwort"—"Eternity! that awful word." Rist's hymn is the basis for the text, the first, second, and sixteenth verses being transcribed bodily; the remainder, excepting verses seven and eight which are omitted, is cast in the madrigal form. We must recognise Picander's hand in this work, for it is worked out in precisely the same

---

[439] Compare Vol. I., p. 206.
[440] "Ach Gott wie manches Herzeleid," A major; B.-G., I., p. 84.
[441] B.-G., X., No. 44, P. 1659. See App. A., No. 19.

way as the Michaelmas cantata "·Es erhub sich ein Streit." The work is in two portions, each concluding with a verse of a chorale set to the same harmonies; thus quite in the manner frequently found in Bach's earlier Leipzig cantatas. It is evident that in this one he worked *con amore;* in it a passionate agitation is combined with a mighty and imposing solemnity. Through four arias and a duet the image of the awfulness of the Divine Judge and of eternal torment is brought home to the personality of the hearer, and displayed with all the dramatic vividness the limitations of church music admit. The separate numbers are in strong contrast to each other, so that although it is the same idea which is varied in them, our keenest interest is kept up to the end; and it adds to the vigour of the expression that, as a whole, it is brief and abrupt. Even in the opening chorus we are already made to feel that personal and passionate conception of the subject which, to a certain extent, we always meet with in Bach, but which is the especial stamp of this work. We need only study the phrases given in the counterpoint on the first and third lines of the chorale to the three deepest parts; the tremor which takes possession of the instruments in bars 13, 17, 23, and 27, and the terrified flight, as it were, of the combined rhythm in bar 90. Besides, this chorus affords the third instance in the course of Bach's early years at Leipzig of the adaptation of the French *ouverture* to church music. In the cantatas " Preise Jerusalem den Herrn " and " Höchst erwünschtes Freudenfest " it is not, however, founded on a chorale; here, as a chorale is amalgamated with the overture form, the cantata affords a striking pendant to the Advent cantata " Nun komm der Heiden Heiland," written in 1714.[442]

XII. Feast of St. John the Baptist, " Ihr Menschen rühmet Gottes Liebe "—" O men, declare God's loving-kindness "—a cantata of less importance, and which offers hardly any opportunity for special observation. Its character is cheerful and pleasing, the forms simple and easy to under-

---

[442] See Vol. I., p. 507.

stand. It is only in the middle duet with *oboe da caccia* that the master has displayed his higher art; and this even is marred by a certain dryness, which we cannot wonder at when we consider the vapid emptiness of the words. A graceful feature may be mentioned in the way in which the bass recitative which precedes the final chorale appears as a sort of prelude to the first line. The form of the chorale itself is that already known from the cantata "Jesus nahm zu sich die Zwölfe."[443]

XIII. Eighth Sunday after Trinity, "Erforsche mich Gott und erfahre mein Herz"—"Search me, O God, and know my heart." It would be possible to select from among Bach's cantatas a group which might be designated "orthodox compositions," and this one would be included in it. It is full of a stern zeal verging on severity, which is peculiar to Bach among the sacred composers of the time, and which, in this cantata, is most conspicuous in the important opening chorus. The conception of the Bible words, Psalm cxxxix., 23, to which it is composed, was suggested by the Gospel for the day, which is directed against false prophets—one of the most fertile themes for an orthodox preacher.[444]

XIV. Ninth Sunday after Trinity, "Thue Rechnung! Donnerwort"—"Day of reckoning! awful word." The text is from Franck's "Evangelischem Andachtsopfer," hence, we must suppose that the work is an early Leipzig composition. Apparently Bach wrote it in the same year as the cantata "Ihr, die ihr euch von Christo nennet" (see p. 361).[445] It has no chorus but the final chorale. Of the solo pieces, the tenor aria is unsatisfactory by reason of its truly amazing text, "Capital und Interessen"—"Capital and interest of my sins, both great and small; I must soon account for all!" The bass aria, on the contrary, and the duet for soprano and alto are splendid examples of force and of characteristic treatment.

---

[443] The original parts are in the Royal Library at Berlin. See Appendix A., No. 19.
[444] The original parts in the Royal Library at Berlin. See Appendix A., No. 19.
[445] See Vol. I., Appendix A., No. 26, p. 640.

XV. Ninth Sunday after Trinity, "Herr, gehe nicht ins Gericht"—"Lord, enter not into judgment." This has the character of a fervent and supplicatory penitential prayer, and the orchestra begins in G minor with the two upper parts worked out in canon; in the closing cadenza the four-part chorus comes in with "Enter not into judgment with Thy servant, O Lord," Psalm cxliii., 2. It derives nothing from the instrumental subject but the motive of the canon treatment, the other parts are constructed out of freshly introduced ideas over a ground bass. At the end of six bars the voices are silent again, while the instruments repeat their inarticulate penitential hymn on the fifth above. It then is worked out in double counterpoint, and the same is done in the chorus which comes in again eight bars later, the soprano answering the tenor, the alto the soprano, and the tenor the alto. The instruments then borrow a rhythmical motive from the chorus, and work it out into an independent picture which is welded with the chorus into a masterly whole. Once more there is a brief pause in the chorus, while the instruments carry on their motive, at the same time referring distinctly to a certain passage of the opening subject—bar 5. Now, for the third time, the chorus comes in, emphasising and freely working out the principal subject; at last it gives out its own penitential cry, going through it completely in the middle range of compass, as if it flowed straight from the hearts of the singing host. The chorus ended, the feeling is allowed to die softly away on a long organ point on the dominant. An admirably constructed *adagio* subject is immediately followed by an animated fugue "For in Thy sight shall no man living be justified." We might place at the head of this movement, as an appropriate motto, the words "I, the Lord thy God, am a jealous God"; the treatment of the resolute theme leads in many places to passages which rage and roll like angry billows. Once we suddenly come to a long *piano* passage—an extremely rare device with Bach—and this presently sinks even to *pianissimo*, as though man were cowering to hide from the dreadful eye of God. The remainder of the cantata is in no way inferior to the impressive effect of this opening chorus.

Wie zittern und wanken,	The sinful must languish,
Der Sünder Gedanken,	In torment and anguish;
Indem sie sich unter einander verklagen,	They turn on each other with impotent railing,
Und wiederum sich zu entschuldigen wagen.	Or plead their temptations with bitter bewailing.
So wird ein geängstigt Gewissen	Thus racked by its self-accusation
Durch eigene Folter zerrissen.	Their guilt works its own condemnation.

So runs the text of the first aria, to which Bach has set a composition of the greatest originality. A tremulous semiquaver figure on the violins goes on throughout, while the soprano, with an oboe concertante, sings a boldly constructed and impressive melody. No figured bass is added, the lowest part is given to the viola in a steady slow *tremolo* of quavers. A secret terror, and at the same time a profound grief pervades the whole air. A change comes over it with the accompanied recitative for the bass which follows: "In Jesus ist Trost, er öffnet uns einst die ewigen Hütten." None but Bach could have found tones of such deep pathos to express these words; they introduce an artistically constructed aria for the tenor, overflowing with the sentiment of restored calm. Its rhythm is particularly noteworthy. The words it begins with: "Kann ich nur Jesum mir zum Freunde machen"—"If only Jesus be my Friend and Saviour"—Bach has adapted to a phrase of which the periods fall into a half and a whole bar in common time:—

and the three notes of the first half-bar are subsequently used very ingeniously for episodic phrases. The closing chorale combines the two main ideas of the cantata—dread and reassurance are both expressed in it. The vocal part speaks of reassurance, and the violins which re-echo the soprano air in trembling semiquavers keep up the feeling of fear. But by degrees the beating heart becomes calmer and more peaceful; the semiquavers sink to triplets of quavers, then simple quavers, then to tied ($\sharp\,^3\,\flat$) triplets, and finally in the last bar to crotchets. It is clear that Bach intended to keep the memory of the soprano air alive to the very last, because at the end the chorale is played on the instruments without any figured bass, but with the viola for the deepest part.

It is impossible sufficiently to admire the successful combination of antagonistic feeling in this composition; it opens before us a realm of music utterly unknown to any of Bach's forerunners and contemporaries. The song of Orestes in "Iphigenia in Tauris," "Le calme rentre dans mon cœur," is justly praised as Gluck's *chef d'œuvre* in dramatic music, but he was not the first to disclose under a musical aspect the inmost depths of the inarticulate complication of human feeling; half a century before him Bach had solved the problem with no less mastery.

I must not here omit to notice a singular reminiscence of Handel's Passion music to Brocke's text, part of which Bach had copied with his own hand, probably in the later Cöthen period. The progress of the melody of the soprano air is free and animated, but one passage startles us by its studied and almost forced style, which does not duly harmonise with the rest. It is in bar 28, &c., which is as follows:—

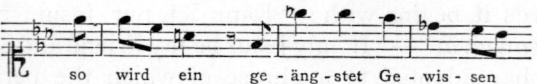

In Handel's Passion this passage occurs:—[446]

I feel convinced that Bach in this place was carried away by his remembrance of this very beautiful and impressive passage; that Handel's work was exerting an influence on him throughout this composition it is fair to conclude from the use of the oboe in imitation, and the feeling all through is very similar. This bears important testimony to the

---

[446] German Händel Society's edition, part XV., p. 80, bar 3.

interest which Bach took in the works of his great contemporary, and the reader will remember that it is not the only instance.[447]

XVI. Tenth Sunday after Trinity, "Schauet doch und sehet, ob irgend ein Schmerz sei, wie mein Schmerz"—"Behold and see, if there be any sorrow like unto my sorrow." This cantata is a companion piece to the previous one, and they were undoubtedly composed at the same time; they are alike in structure and in feeling too, so far as the different characteristics of the subjects admit. In the Gospel for the day, Jesus prophesies with weeping the coming destruction of Jerusalem; the text, as adapted to this conception, is very unskilfully managed. It alludes first to the former destruction by Nebuchadnezzar, and then makes an awkward transition to the coming fall under Titus, drawing from it an application to the judgment impending over all mankind, of whom Jesus will nevertheless tenderly shield the pious. As the musical treatment gives the greatest weight to the beginning, the work as a whole lacks directness and clearness of dramatic purpose. This is much to be regretted, for in the conception and working-out of individual parts the cantata is one of the most striking and thrilling works that Bach ever created. The whole essence of the lamentation of Jeremiah is compressed into the first chorus (in D minor) with quite incomparable force; every note is tearful, and every interval a sigh. As in the cantata "Herr gehe nicht ins Gericht," the chorus consists of a slow subject developed in canonic imitations by the voice parts, and of a lively fugue. The first has sixty-six bars, the second seventy-six, in 3-4 time—thus it is of some considerable length; and yet it is not more than enough to convey the sentiment which it has to express. In certain carefully chosen passages the chorus is supported by a trumpet and two *oboi da caccia*, while the stringed quartet and two flutes play round and about the *adagio* subject in graceful and appropriate arabesque. Certain sobbing passages on the flute remind us distinctly of the chorale

---

[447] See ante, p. 175. B.-G., XXIII., No. 105. See App. A., No. 19.

for chorus in the Passion according to St. Matthew, "O
Mensch, bewein dein Sünde gross";[448] but in the Passion the
sentiment is qualified by thankfulness for the Redeemer's
death, while in the cantata there is no comfort for the
burning anguish that torments the soul; these two wonderful
compositions stand in contrast like the Old and New Covenant.
Next to the chorus our attention is most rivetted by the grand
bass aria " Dein Wetter zog sich auf von weitem." The
main idea, rising in thirds from the B of the instrumental
bass :—

has in it something mysteriously terrible which scarcely any
other composer could have found means of expressing; this
feeling is enhanced by the long-drawn f" high above all in
the trumpets,[449] like a shaft of light piercing the dark storm-
clouds and giving, as has been aptly observed, " a red hue
as of blood."[450] The chromatic rise and fall in the middle,
bar for bar (45 to 54 and 67 to 76), of the instrumental bass
is highly effective. The alto air, G minor, which is accompanied only by the flutes and oboes without any figured bass,
paraphrases the words of Christ, " How often would I have
gathered thy children," &c., Matt. xxiii., 37, and has a
soothing though solemn character, as befits the context. In
the closing chorale—the ninth verse of Meyfart's hymn,
" O grosser Gott von Macht "—we again hear the flute
passages of the opening chorus in brief interludes ; this is a
reference to the beginning, similar to that in the final chorale
of " Herr, gehe nicht ins Gericht."

---

[448] This chorus was originally written for the Passion according to St John,
and was already in existence when Bach composed the cantata in question.
More will be said on the subject when we deal with the Passion music.

[449] Bach here indicates, as in the opening chorus and final chorale, *Tromba
o Corno da tirarsi*. The *Corno da tirarsi*, which is frequently put in by him—
for instance, again, in the cantata " Halt im Gedächtniss Jesum Christ "—was
the same instrument, or a similar one, as the *Tromba da tirarsi*, in which a
combination was attempted of the trumpet with the trombone. Kuhnau speaks
of it in the " Musikalischer Quacksalber," p. 82.

[450] Lindner, *Zur Tonkunst*, p. 124. Berlin, Guttentag, 1864.

It has already been explained (ante p. 2) that even in a composition for choral singing, all use of ornament need not be excluded; two passages of the first chorus here are examples of this. In bar 37 in the tenor, and bar 51 in the alto, the falling intervals of thirds are filled up by the addition of "accents"; in the voice parts these are not indicated, though they are in the accompanying oboes.[451] Of course they must be used in every part, or no intelligible harmony could result, though it is true they thus cease to be embellishment and become part of the tune; but the limit line was often overlooked by Bach (see ante p. 316).[452]

XVII. Thirteenth Sunday after Trinity, "Du sollst Gott deinen Herren lieben"—"Thou shalt love the Lord thy God." The style of this cantata is conspicuously different from all we have hitherto discussed. The arias are considerably simpler than we are accustomed to find them in Bach. In one of them two instruments are introduced *concertante* (probably two oboes) with the soprano: this they carry on almost uniformly in parallel thirds or sixths, and we nowhere find that they work out any polyphony worth mentioning. The sentiment verges on that quiet ecstacy which is peculiar to Bach's earliest church compositions, when he was still lingering on the borders of the old cantata. The soprano aria even reminds us very plainly of "Jesu dir sei Dank gesungen," from the cantata "Uns ist ein Kind geboren."[453] The condition of the autograph offers no evidence in support of the idea that Bach has here remodelled an older composition, unless we detect it in the haste with which it has been written, and which seems to indicate lack of leisure. As regards the first chorus, however, the difference of style consists in its presenting itself in a perfectly new form, ingeniously

---

[451] In the second passage the oboe part is as follows—

which seems to imply that the shake should begin on the passing note above.
[452] B.-G., X., No. 46, P. 1660, App. A., No. 19.
[453] See Vol. I., p. 491.

conceived, and worked out in a masterly manner. The text is taken from the Gospel, " Du sollst Gott deinen Herrn lieben "—" Thou shalt love the Lord thy God with all thy heart, and all thy soul, and with all thy strength, and with all thy mind, and thy neighbour as thyself" (Luke x., 27). Now it was not unknown to the composer, who was well versed in his Bible, that the incident which called forth this injunction is reported in a more extended form by the Evangelists Matthew (xxii., 35—40) and Mark (xii., 28—34); and the precept which follows, "On these two commandments hang all the laws and the prophets," was full of significance to him. He brought in the melody of Luther's hymn, " Dies sind die heilgen zehn Gebot," in the bass in minims, as a *Cantus firmus*, working out the chorus in quavers from the first line of the chorale, and finally gave out the chorale in crotchets on the *Tromba da tirarsi*. Thus its very essence pervades every portion of the composition, and closes it in on every side; and the thought that all God's laws are embodied in these two precepts acquires the most figurative musical presentment which is in any way possible.[454] It is clear that the form, regarded from the purely musical stand-point, is that of the organ chorale; we have, indeed, two real organ chorales by Bach on this same melody. One is in the third part of the " Clavierübung,"[455] and belongs consequently to the latest period of Bach's work. The other is in the " Orgelbüchlein,"[456] and so must have been written at Weimar. The later one treats the melody in strict canon on the octave in the inner parts. In the earlier organ chorale the melody lies in the upper part, and the counterpart is worked out from the first line. This chorus, therefore, as regards its musical treatment, holds a middle place between the two organ chorales. A working out in strict canon form between the instrumental bass and trumpet was inadmissible, since, in the first place, neither the value of the notes nor the

---

[454] W. Rust has given a sympathetic interpretation of the deep meaning of this chorus. B.-G., XVIII., p. xv.
[455] B.-G., III., p. 206.
[456] P. S. V., Vol. V. (No. 244). See Vol. I., p. 600.

intervals are the same; and, in the second place, the trumpet repeats the first line after each of the others in order to emphasise very expressively the words "These ten are God's most holy laws"; finally, the whole melody is repeated once more straight through, above an organ point on G. This playing with fragments of the melody, so to speak, rather points to the influence of the Northern school. So, indeed, does another circumstance. Bach divides the fourth line—

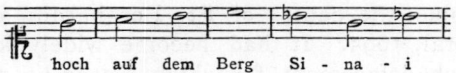
hoch auf dem Berg Si - na - i

in a singular manner into two sections, treating the first four notes separately, and connecting the last three with the *Kyrie*. No reason for this, either poetical or musical, is discernible; it is simply a whim of that capricious art which the Northern composers were so ready to yield to; and, in fact, something similar occurs in a chorale arrangement by Buxtehude.[457] Bach has followed the same course again in the organ chorale in the "Clavierübung," whence we may conclude that he had the cantata chorus in his mind when he wrote it.[458]

XVIII. Sixteenth Sunday after Trinity, "Liebster Gott, wann werd ich sterben?"—"Ah, Lord God, when shall I see Thee?" This cantata seems to have been composed very nearly at the same time as that for the Thirteenth Sunday after Trinity, "Ihr, die ihr euch von Christo nennet."[459] Its subject consists in meditations on Death, only very remotely suggested by the narrative of the widow's son, of Nain. A verse of Neumann's hymn "Liebster Gott, wann werd ich sterben," is used both at the beginning and end in the original form; verses two, three, and four, on the contrary, are so paraphrased on the "madrigal form" that verse two serves as the text for a tenor air, verse three

---

[457] *Te Deum laudamus.* See Spitta's edition of Buxtehude's Organ Compositions. Vol. II., p. 53, bar 5.
[458] B.-G., XVIII., No. 77, P. 1675. See App. A., No. 19.
[459] See ante, p. 361.

for an alto recitative, the first half of verse four for a bass air, and the second half for a soprano recitative.

The aria-like form of the hymn was due to Daniel Vetter, who has been frequently mentioned in this work (and who died in 1721) as organist to the church of St. Nicholas, in Leipzig. Vetter had been a pupil of Werner Fabricius, and at his death, January 9, 1679, he had succeeded him as organist (on August 11 of the same year).[460] He was a native of Breslau and composed this hymn at the request of his friend Wilisius, the cantor of St. Bernhardin at Breslau, for his funeral, 1695. It had become widely known and suffered much defacement, for which reason he republished it in 1713, in the second part of his Musicalischen Kirch- und Haus-Ergötzlichkeit[461] set for four parts. Bach must have known this four-part aria, for it is the same which appears at the end of his cantata, in a somewhat altered form, but easily recognisable. Here again we perceive that Bach held his Leipzig predecessors in due honour. In the first chorus the melody is treated in the form of a chorale fantasia (compare p..361). This is a very remarkable composition—the sound of tolling bells, the fragrance of blossoms pervade it—the sentiment of a churchyard in spring time. The character of the piece may no doubt have been largely determined by the fact that it is not strictly a chorale but a sacred *aria* which is under treatment, but this does not sufficiently account for it. Rather might we suppose that the tender encouraging tone of the Gospel story had suggested the feeling, particularly when we regard the whole cantata; for its gentle grace, not unfrequently passing into a blissful childlike playfulness, contrasts strangely enough with the stern gravity of Bach's other funeral cantatas. A knell is imitated in exactly the same way as in the Weimar cantata "Komm du süsse Todesstunde" by *pizzicatos* on the strings, and rapidly reiterated high notes on the flute. Two *oboi d'amore* float along above the strings, now crossing each other in flowing melody,

---

[460] Archives of the University and of the Town Council of Leipzig.
[461] Winterfeld, Ev. Kir. III., p. 487, and Musical Supplement, p. 140.

and now united in soft passages of thirds and sixths; the piece of music thus evolved almost suffices of itself to fill our souls with peacefulness. Indeed, the musical impression of the whole rests upon it; it consists of sixty-eight bars, while the homophonic chorus, which comes in interruptedly and gives us the original melody with no embellishments but a few delicate *Melismata*, includes altogether no more than twenty bars. Nevertheless its words of death attune our feelings to that peculiar vein of melancholy which we experience beside the bier of a child or a youth. The tolling of the bell goes on in the basses all through the highly strung aria given to the tenor, and sometimes even appears in the voice part, bars 29-31. The melodious and elaborate bass air and the two recitatives fully correspond in beauty to the other pieces.[462]

XIX.—Sunday after Christmas, "Gottlob, nun geht das Jahr zu Ende"—"All praise to God, the year has gone." This is the last we have to mention of Bach's cantatas that are composed to texts by Neumeister,[463] and as regards the use of the chorus it is the finest. The principal chorus is the second number, but such is its weight, that the finished beauty of the preceding soprano air hardly asserts itself, and all that comes after sinks into nothingness. Bach had taken the composition of the chorus in hand earlier than the rest of the work, and had sketched it first separately, for in the complete score it shows hardly any corrections and has all the appearance of a fair copy. At the conclusion of this gigantic work the master himself looked back on it with proud satisfaction—he has done what he scarcely ever did—counted up its 174 bars, and noted them at the end. It is a chorale for chorus on "Nun lob mein Seel den Herren"—"My soul now praise the Lord"—and resembles a motett in so far as that the instruments—strings, three oboes, cornet, and three trombones—work with the voices, and it is only the figured bass which is here and there allowed a way of its own. The type is that of the

---

[462] B.-G., I., No. 8. P. 1199. See App. A., No. 43.
[463] See Vol. I., p. 487.

Pachelbel organ chorale, elaborated to the highest degree of which it was capable within the limits of the motett form. Particularly we may note, as belonging to this form, the picturesque musical rendering of the separate lines of the verses by the use of contrapuntal parts, which interpret the forgiveness of "us miserable sinners" by acute chromatic passages, or pour out the consolations of God as it were in a stream over wretched humanity, and then soar up "like to the eagle." Bach subsequently wrote several pieces of this kind,[464] and they are worthy of the first-born, but not one surpasses it.[465]

On September 7, 1727, a general mourning of four months began for the Queen Christiana Eberhardine. The interruption this occasioned of course made a break in the long series of Bach's Leipzig compositions; this is, therefore, a suitable place to pause for a retrospect. Our final judgment as to the Weimar cantatas was much to the effect that in them the ideal of church music in Bach's hands had already been found, excepting in the one particular of the treatment of the chorus.[466] In spite of the occurrence in them of many important choral numbers, these are on the whole outweighed by the solo pieces, in which the form gives us an impression of perfect maturity; and the student who has thoroughly examined them will find very little that is new as regards form in Bach's later compositions for solo voices. Then it became evident from the Cöthen cantata, "Wer sich selbst erhöhet,"[467] that Bach, from his having long occupied himself in composing for the organ and other instruments, had acquired a complete mastery of the art of working out independent choral compositions in the most grand and elaborate forms. In the cantatas written during the first four years of his life at Leipzig, we again find that unlimited wealth of invention which the artist derived from his power of applying the

---

[464] "Ach Gott vom Himmel sieh darein," B.-G., I., No. 2. P. 1194, "Aus tiefer Noth schrei ich zu dir," B.-G., VII., No. 38. P. 1694.
[465] B.-G., V.,¹ No. 28. See App. A., No. 41.
[466] Vol. I., p. 565.
[467] See ante, p. 12.

forms of instrumental music to his sacred compositions in a way previously undreamt of; we find him unhesitatingly adapting parts of the chamber sonata, and utilising it as an instrumental opening to the second part of the cantata, "Die Himmel erzählen die Ehre Gottes." He blends the elements of the first movement of the Italian concerto with true choral forms, as in the Magnificat and the Christmas piece, "Dazu ist erschienen der Sohn Gottes"; he casts whole cantatas in the concerto form, as "Erfreute Zeit im neuen Bunde," or in that of the orchestral suite, as "Höchsterwünschtes Freudenfest." He combines the French *ouverture* with an independently conceived chorus or even with a chorale, as "Preise Jerusalem, den Herrn," "O Ewigkeit, du Donnerwort"; he makes the *giga* serve its turn as a sacred duet, "Aergre dich, o Seele, nicht," and the *passecaille* as the basis of a chorus of lamentation, "Weinen, Klagen."

In the cantata "Die Elenden sollen essen" he uses the instruments of secular music for a chorale *fantasia*, and in the Michaelmas music, "Es erhub sich ein Streit," he takes a *Siciliano* for counterpoint to a chorale melody; he avails himself of everything that he or his predecessors had ever invented in the whole realm of the organ chorale for his sacred vocal music. We meet once more with the type created by Pachelbel and with those of Buxtehude and Böhm in new and figurative modifications, sometimes pure, as in "Erschallet ihr Lieder"; sometimes mixed, as in "Die Elenden sollen essen," "Christ lag in Todesbanden," "Du sollst Gott deinen Herrn lieben." The chorale, trio, and quartet which Bach constructed in so masterly a way for the organ we find again in the cantatas, "Wo gehst du hin?" and "Wahrlich, wahrlich, ich sage euch," but now in a vocal form. He welds the orchestra and chorus together with a mighty hand to unite in the chorale fantasia; he calls upon the instrumental chorale to accompany the irregular figures of the recitative ("Du wahrer Gott und Davidssohn"); he inserts the appealing phrase of the recitative, which has a personality of its own, between the sections of the chorale chorus ("Herr, wie du willst, so schicks mit mir"), and he impresses on the solo voice the polyphonic form of the instrumental fugue ("O heilges Geist- und Wasserbad," "Wahrlich, wahrlich, ich sage euch"); and among them all we find the old well-known forms of the aria, the arioso, the recitative, and the simple chorale, but always filled with new meaning from a perennial fount of inexhaustible inventiveness, made deeper, broader, and grander, and either linked together by a deep and inherent poetical purpose, or connected with one of his newly invented forms. All this may be detected on a narrower and less ambitious scale in his Weimar cantatas, but what distinguishes the Leipzig compositions from these in a very conspicuous manner is the lavish introduction of powerfully and boldly outlined choruses. Only a small proportion of the cantatas hitherto discussed are devoid of such numbers. It need hardly be said—for it is evident from the descriptions given above—that a variety worthy of Bach is to be found in them; at the same time— and this is characteristic of this group of cantatas—the freely invented choruses are decidedly the more numerous class. The chorales for chorus which occur in the cantatas, "Du wahrer Gott und Davidssohn," "Christ lag in Todesbanden," "O Ewigkeit, du Donnerwort," "Du sollst Gott deinen Herrn lieben," "Liebster Gott, wann werd ich sterben," "Gottlob, nun geht das Jahr zu ende," and in a few other places, are beyond a doubt thoroughly thought-out subjects, some of them very grand, and each and all such as Bach alone could compose; nor must it be forgotten that the final chorus of the first portion of the Passion according to St. Matthew is also to be attributed to this period. But when we set them all in the scale against the mass of independent choruses written at the same time, we see at once that Bach's inclinations tended towards the latter. Their form varies, but on the whole the fugue is evidently preferred to any other, and is often prefaced by an *adagio*. It is highly significant as indicating Bach's attitude of mind towards the chorale that there are among these cantatas, some—and these by no means unimportant ones—in which a chorale is altogether wanting ("Christen ätzen diesen Tag"), and not a few in which it plays quite a secondary part. Bach found in Leipzig a public which next to Kuhnau, preferred Telemann's music above all other. Telemann's strength lay in a certain style of brilliant chorus, superficially graphic and highly effective to the general public from its obvious and picturesque imagery. Now, though Bach may never have thought of taking him for a model in this, still, the tendency of popular taste, which he had already taken into account in his examination cantata, may have been an incentive to him to occupy himself chiefly with the composition of independent choruses, while he did not disdain to copy with his own hand a piece written for Advent by Telemann. There are indeed features in his choruses and solo pieces which have a certain air of Telemann about them; this is most conspicuous in the cantata, "Herr Gott dich loben wir." But we have also seen that he derived something from Kuhnau, and once made use of a composition by Vetter that had become popular. This open mind as regarded the works of his contemporaries and his anxiety to learn from them as much as possible, or at least to show his respect for them, and through them for the public, is a trait in his character which has not till now met with due recognition, though it is as characteristic of his art as of his nature.

VI.

BACH'S CANTATAS—(CONTINUED).

WE are now entering on the year 1728, and approaching the period when the work was written which to all appearance Bach himself valued most highly among his sacred compositions. We must however defer (as he did) giving our attention to the Passion according to St. Matthew, and must first occupy ourselves with the church cantatas composed between this and 1734. It may here be pointed out, however, that this great work must have prevented the composition of any other church music, at any rate after the last months of 1728; all the more so since for the new year of 1729 he had to compose the music for the great mourning ceremonial at the obsequies of Prince Leopold von Anhalt-Cöthen, and to conduct it at Cöthen in person. It is therefore no wonder that we can only indicate one single cantata which may be attributed with tolerable certainty to 1728: "Wer nur den lieben Gott lässt walten," for the fifth Sunday after Trinity. In this again we find clear tokens that Picander must have written the text, though it is true that it is not to be found in the cycle of cantata texts which Picander began on the Feast of St. John the Baptist, which fell immediately before the fifth Sunday after Trinity in 1728. Still, such poems were not invariably written solely with a view to composition, and still less with the idea that they would one and all be set to music and performed within that same year.[468] Besides this, Bach exercised considerable influence over the poet who for a long time lived in his immediate neighbourhood;[469] it never occurred to him to set everything Picander put into rhyme as soon as it was written, and he expected something more than the details to be adapted to his wishes; no doubt he generally sketched the foundation lines of the purport and feeling of the whole. The cantata "Wer nur den lieben Gott" affords an instance in support of this. It is not wholly devoid of reference to the Gospel for the day, but its general tendency leads us tolerably far from it. Bach, in the first place, desired to make Neumark's consolatory hymn the central point of a composition, and Picander has used all the seven verses for the text; the first, fourth, and last, in their original form. He also preserved the words of the fifth, and almost all those of the second, only he has woven in with them recitatives in madrigal form, and he has dealt freely with the meaning of the third and sixth, though he has preserved some of the original phrases. The attitude taken up by Bach with regard to the separate verses of the melody is precisely analogous. In the sixth, a few fragments of the tune are incidentally introduced into a soprano air (bars 23-25, 28-30, 35-37, and again 31-32 in diminution); the tenor aria, for which the third verse is used, reminds us generally of the chorale by the retention of the verse form, and besides this, at the beginning of each section of the

---

[468] See on this point Appendix A., No. 45.
[469] In the Burgstrasse; see Das jetzt lebende und florirende Leipzig, 1736, p. 14; 1746-47, p. 11.

verse, reminiscences of the corresponding parts of the melody are brought in transposed, the first time into the major key. Verses three and six are given to a solo voice which alternates between the long-drawn phrases of the chorale tunes and the more animated lines of the recitative. The rest of the stanzas give us the chorale form complete; the last verse being a simple four-part subject; the fourth is on the model of the Pachelbel organ chorale—the soprano and alto singing counterpoint motives to the lines of the melody, which is at the same time played by all the instruments, and the first is set to a modified form of chorale fantasia.

Such an undeviating reference to the same chorale melody throughout a whole work has only once before come under our notice, in "Christ lag in Todesbanden." But the difference is evident at a glance. We there meet only with regular church chorale forms; however great their freedom and variety of treatment, a strict *Cantus firmus* is present throughout. Here, on the contrary, the chorale appears as the general starting point of personal devotion. This is not the case however in all the numbers, for the fourth and seventh verses are within the strict limits of congregational feeling; but the rest express a frame of mind which strives to give to religious consolation a form that may answer to subjective needs. In each the chorale serves only as the nucleus—the motive and incentive to the aria, but in such a way as that this motive is not concealed, but must be felt and understood by the hearer, or the piece will fail of its due effect. In the recitatives to each line of the melody appropriate reflections are added, by which means the chorale as a whole is dissevered and lost; for in the bass recitative all its parts are not even brought in, and in the tenor recitative each line appears in a different key. Even in the opening chorus this character is plainly discernible. If the form of the chorale fantasia was to be successfully transferred to the chorus and orchestra, their relations had in the nature of things to be so adjusted that the delivery of the independent tone picture which expounds the fundamental feeling of the chorale should be given to the instruments, while the chorus filled the rôle of *Cantus firmus*. This could be effected in various ways, for instance, by a simple four-part treatment, but contrived in such a manner as that some of the parts give out the calm flow of the melody, while others surround it with more rapid figures, in the course of which they may, of course, sometimes approach or coincide with the parts given to the instruments; it is only necessary to maintain the general principle of contrast. In the present instance we see that, irrespective of the independent instrumental accompaniment, the voices prelude each line with an introductory subject in which they have a fugal arrangement of the line that follows. The impression produced is that the subjective sentiment, after having first dwelt upon the meaning of each separate line, rises to the level of the lofty general feeling of a congregation of worshippers. I confess frankly that I find it difficult to comprehend and enter into this chorale chorus, as a whole, but the composer's purpose does not seem to me to admit of a doubt. It need scarcely be pointed out that the musical character of the cantata is thus throughout contemplative. The fervency which pervades each separate portion of it acquires from this a peculiar colouring which is most easily and plainly discernible in the beautiful and touching aria in E flat major. In Bach's time musicians had already begun to write compositions in the grand style for household worship.[470] Although the cantata "Wer nur den lieben Gott lässt walten" was used as church music, in feeling it borders on the domain of private devotion. As his letter to Erdmann tells us, and as is proved by the large number of various kinds of instruments which he possessed, Bach had musical performances in his own house. It is very possible that he conceived and composed this cantata more with a reference to this than for its church purposes.

So far as we can judge from our present knowledge of Bach's church music, he composed music to nine of the cantata texts by Picander which first appeared in 1728-29. Four of these he probably wrote in 1731, the other five I assign to 1729 and 1730, of which years no cantatas can

---

[470] For instance, Telemann in his Harmonisches Gottes-Dienst, Hamburg, 1725.

be proved to exist excepting a few pieces for festivals, while a considerable number remain of the following years. A Christmas Cantata, "Ehre sei Gott in der Höhe," survives only in a fragment,[471] but the chief part of it was transferred to a later piece composed for a wedding ceremonial.[472] The alto air "O du angenehmer Schatz" is one of those lovely cradle songs, one of which we have already met with in the cantata "Tritt auf die Glaubensbahn."[473]

We have a cantata for the New Year which is remarkable for a dignified and powerful fugue at the beginning with this theme:—

Gott, wie dein Name, so ist auch dein Ruhm bis an der Welt En - de.

However, Bach seems to have set aside this cantata in an unfinished state, and not to have worked it up till a later period when he added a soprano air from the cantata "Der zufriedengestellte Aeolus" and the final chorale from the New Year's piece "Jesu nun sei gepreiset."

A cantata for the third Sunday after Ephipany, "Ich steh mit einem Fuss im Grabe"—"With one foot in the grave I stand"—is full of the solemnity of death, and at the same time of believing expectation. It opens with a symphony which proceeds in the style of the first adagio of a chamber sonata. This is immediately followed by an aria in the form of a chorale quartet, in which the soprano delivers the chorale "Machs mit mir Gott nach deiner Güt"—"Do with me, Lord, just as Thou wilt"—other words in "madrigal" form being added by the tenor—a highly poetic composition from the sinking motion and halting rhythm of the contrapuntal parts.[474]

---

[471] In the possession of Herr Professor Epstein. See App. A., 46.
[472] B.-G., XIII.,[1] No. 3.
[473] See Vol. I., p. 560.
[474] I only know this cantata through a score by Franz Hauser, who found the parts in the Thomasschule in 1833; they are no longer there. It was probably composed in 1730, and, if so, was first performed on January 22, 1730. In 1729 Bach would have been too busy with the mourning music; and in 1731 there was no third Sunday after Epiphany.

The cantata for Quinquagesima Sunday, "Sehet, wir gehen hinauf nach Jerusalem"—"Behold, we go up to Jerusalem"—is no less meritorious. In every portion of it the affinity to the Passion music (St. Matthew) is conspicuous, it is full of the same sentiment that pervades that work. The cantata begins with the same words as that for Quinquagesima of 1723: "Jesus nahm zu sich die Zwölfe"—"Jesus called unto Him the twelve"—and the words in which Christ declared to His disciples His approaching passion. Here too they are set to an expressive arioso which derives its peculiar character from a wandering motive in the bass; recitatives for the alto carry on the feelings called up by the Bible words. A very beautiful and softly flowing chorale trio on the sixth verse of Gerhardt's Passion hymn "O Haupt voll Blut und Wunden"—"O Thou whose head was wounded"—an indescribably pious and deeply felt Bass aria "Es ist vollbracht das Leid ist alle"—"It is finished, Thy pain is over"—and the simply set chorale "Jesu deine Passion"—"Jesus, Lord, Thy Passion"—with a short recitative for tenor, constitute the remainder of this cantata, which though of no great extent is very beautiful, and exhibits to the full the genius of Bach.[475]

The music for the third day of Easter, "Ich lebe, mein Herze, zu deinem Ergötzen"—"I live, O my heart, for thy joy and thy gladness"—is one of the freshest and brightest of the master's works; the bass aria "Merke, mein Herze, beständig nur dies" is full of swing, and even has something dance-like about it; we almost fancy we can see sturdy and happy figures dancing in the spring. Like the two preceding cantatas, this one also is written for solo voices only, not counting the usual final closing chorale in four parts.[476]

---

[475] The autograph is wanting. A MS. copy by Christian Friedrich Penzel, Cantor of Merseburg, who was foundation boy at St. Thomas' from 1751 to 1756, is in the possession of Herr Joseph Hauser, of Carlsruhe.

[476] This work is known to me only from a recent MS. in Zelter's collection. In this the chorale "Auf, mein Herz, des Herren Tag" and a chorus "So du mit deinem Munde bekennest Jesum" are introduced before the opening duet.

We now may go on to the cantatas of the years 1731 to 1734, and will first consider those compositions which again are based on Picander's series of texts. The cantata for Septuagesima "Ich bin vergnügt"—"I am content"—(for Jan. 21, 1731, or Feb. 10, 1732), occurs in Bach's composition in an altered form, which Picander himself arranged however. In this delicately treated composition we find, what rarely happens with Bach, that it is set throughout exclusively for a soprano voice. We shall learn when speaking of another cantata "Ich habe genug"—"I have enough"— which belongs to this period, that this was done with special reference to Anna Magdalena Bach, and that the piece was actually written for her; for it was at this time that Bach's household band and singers, of which his wife and children were the mainstay, were in the prime of their powers. Anna Magdalena had kept up her practice as singer in the Church services, and she had never appeared in public in any other way since, from being a court singer at Cöthen,[477] she had become Sebastian's wife.[478] Still, she exercised her talents in private music, and her husband took care in his compositions to give her the opportunity. The cantata "Ich bin vergnügt" has more distinctly the stamp of music for domestic performance than the former one. It is only by supposing that Bach had this purpose immediately in view that we can explain the modification made in the text, in which the original arrangement is exactly preserved, and almost the same ideas are worked out; and it is of the same length. But, excepting the final chorale—which is the last verse of the hymn "Wer weiss, wie nahe mir mein Ende"—all the phrases in which God is directly addressed are altered or omitted. The whole, as

---

There can be no doubt as to the genuineness of the chorale, but I have serious hesitation in accepting the chorus: the mode of working out the melody and the fugue is not Bach's, but more like Telemann. From these two pieces at the beginning, the cantata is sometimes designated by the first lines of them; see Mosewius, J. S. Bach in seinen Kirchen-Cantaten und Choralgesängen, p. 21, where they are assigned to the first Easter day, or to Easter generally.

[477] That this had been the case is proved by a notice found in the Baptismal register of the Cathedral Church of Cöthen, of Sept. 25, 1721.

[478] See Gerber, Lex., I., cap. 76.

it stands, is a devout meditation culminating in the chorale as a prayer.[479]

The cantata for the second day of Whitsuntide, "Ich liebe den Höchsten von ganzem Gemüthe"—"I love Thee, my Saviour, with all my affection"—(May 14, 1731, or June 2, 1732), begins with a symphony borrowed from the first subject of the third Brandenburg concerto,[480] which Bach has enriched with much art by the addition of two horns and three oboes. Such an application of secular music to church purposes was not new to him; we have already seen an attempt made in this direction in the Weimar cantata "Der Himmel lacht" and "Gleichwie der Regen."[481] At this time, when he was acting as director of the Telemann Musical Union, it must have seemed to him a very obvious course, and we shall soon meet with several instances of it. In this cantata we have no complete chorus, but in the aria for bass "Greifet zu! Fasst das Heil, ihr Glaubenshände"—"Grasp and hold. Hold it fast, it is salvation"—we have a composition of the highest class. The accompaniment is given to violins and violas in unison with a figured bass. The ritornelle is in two sections: a broad melody of four bars—

and a more animated and vigorously marked subject of eight bars; from these materials—figuring on the one hand the blessings of divine grace, and on the other the eager reception of it through faith—the whole air is developed. The four-bar melody is taken up by the voice part, but occasionally the instruments are employed, and a splendid effect is produced where here and there they come in unexpectedly with the air in a rich body of sound; or again, as in bar 94, follow the voice in canon. In the recitative which precedes it we observe that Bach has left three lines without music; the first of the omitted lines is exactly like

---

[479] B.-G., XX.,[1] No. 84. App. A., No 46.
[480] See ante, p. 133.
[481] See Vol. I., pp. 541 and 492.

the first of those that follow, so that here we probably have merely an oversight due to haste. Bach often went hurriedly to work on recitatives and arioso passages; of this we find an interesting example in the arioso of the cantata "Gottlob, nun geht das Jahr zu Ende."[482] He here wrote the words under an empty stave and without any repetitions; when he did this he was not yet quite clear as to the music he should set it to, for when he set to work to compose it he scratched through, shifted the words, and added repeat-marks until the text fitted the music.[483] In the Michaelmas cantata, "Man singet mit Freuden vom Sieg," Sept. 29, 1731, the Saturday before the Nineteenth Sunday after Trinity, we have again a complete chorus; this is not a new composition but taken from the Wiemar cantata "Was mir behagt ist nur die muntre Jagd," the secular cantata which he had composed for such a very different purpose, and which he had also utilised for the church cantata "Also hat Gott die Welt geliebt."[484] Here it forms the finale, set in F major, and it must be admitted that the music is admirably suited to the fifteenth and sixteenth verses of Psalm cxviii.

A comparison of the revised form with the original is highly instructive and interesting, as it always is in such cases with Bach, although we cannot say that the remodelling has been very thorough. It is transposed into D major, the horns are exchanged for trumpets, a third trumpet and a drum are added, some sections are enlarged, the parts are worked out rather more briskly, and certain alterations have resulted from the new text and the new key; finally, towards the close we have a very effective *unisono* in the chorus, and this is about all. The stroke of genius lies in the keen perception which discerned in the old piece its fitness for the new purpose. Bach, however, had at first intended to compose an altogether new chorus to the verses of the Psalm. The beginning of this exists on a sheet of paper which he afterwards made use of for a secular cantata. This cantata, called "Der

---

[482] B.-G.,[1] p. 266.
[483] The original score and parts of the cantata "Ich liebe den Höchsten" are in the Royal Library at Berlin. See App. A, No. 46.
[484] Vol. I., p. 568.

Streit zwischen (the contest between) Phöbus und Pan," was composed in 1731, and it is from it that we know the date when this Michaelmas cantata was written. It is clear that other occupations caused Bach to leave this sketch for the chorus incomplete, and that then, when pressed by time, he fell back on an earlier work.[485] Among the solo subjects the soprano aria in A major, "Gottes Engel weichen nie," is remarkable for its sweetly melodious and gently floating character. The last number but one, too, a duet between alto and tenor with bassoon obbligato, is as artistic as it is expressive, with its penetrating but simple treatment of the melody; it is in G major, and the final chorale is in C major. As this cantata, like the former Michaelmas cantata "Es erhub sich ein Streit," turns entirely on the desire that the angels may bear the souls of the departed to the abode of the blessed, the persistent descent towards the subdominant has undoubtedly been meant by Bach to have a mystical and poetical significance, and it does in fact produce that effect. The last chorale consists of the third verse of Schalling's hymn "Herzlich lieb ich, o Herr" ("Ach Herr, lass dein' lieb' Engelein")[486]

The fourth cantata on the remaining text by Picander: "Ich habe meine Zuversicht"—"I have a perfect confidence"—is for the twenty-first Sunday after Trinity, Oct. 14, 1731, or perhaps Oct. 29, 1730. The composition is interesting because it is the first among those composed in Leipzig in which an organ obbligato is introduced; this, indeed, could not be done before 1730, since it was not till that year that the *Rückpositiv* of the organ of St. Thomas' was fitted with a manual of its own, and so could be used independently of the great organ. The cantata "Ich habe

---

[485] This sketch is also in D major; the trumpet begins:—

Of the voice parts only the first note of the bars is written with the word "Man" under it. The score of "Phöbus und Pan" exists in the Royal Library at Berlin.

[486] The autograph of this cantata is not known to exist. A copy by Penzel is in the possession of Herr Joseph Hauser, of Carlsruhe.

meine *Zuversicht*" was intended to be introduced by the clavier (or violin) concerto in D minor,[487] which underwent a special re-arrangement for this purpose. A similar use made of chamber music has already been mentioned in "Ich liebe den Höchsten." The fact that the whole concerto is placed at the beginning, and not merely a movement from it, reveals an intention of letting the congregation hear to full advantage the improvement in the organ; and we may infer from this that the cantata must have been performed in 1730, immediately after the alterations in the *Rückpositiv*, but we do not know exactly when they were finished. The obbligato organ is silent during the first air, but in the second it combines with the alto voice in a trio of conspicuous beauty.[488]

As soon as Bach found himself enabled to introduce the organ obbligato in church music we find him using it tolerably often. He arranged older cantatas for it, as "Erschallet ihr Lieder"; but he also composed several new ones with an organ part *concertante*. In a few of these—as in the one just discussed—he adapted chamber compositions for instrumental symphonies. This is evidently the case in the cantata for the twelfth Sunday after Trinity, "Geist und Seele wird verwirret"—probably August 12, 1731.[489] Though the instrumental symphonies of the first and second sections are neither of them extant in their original form, we can detect that they have been transferred to the cantata from the fact that here they are written in fair copy.[490] In form they constitute the first and third movements of a concerto, and the middle movement Adagio may have been the A minor aria, a Siciliano such as Bach has used for the middle of a concerto in other cases besides this. A clavier concerto in E major has, between two allegro subjects, a Siciliano in C sharp minor.[491]

---

[487] B.-G., XVII., No. 1.
[488] See App. A., No. 47.
[489] B.-G., VII., No. 35. See App. A., No. 46.
[490] Only a fragment of the first movement is preserved in the original form. It is B.-G., XVII., p. xx.
[491] B.-G., XVII., p. 45.

This concerto too has been utilised for church music, and in all its movements. The two first are included in the cantata for the eighteenth Sunday after Trinity, "Gott soll allein mein Herze haben," Sept. 23, 1731, or Oct. 12, 1732; the first, transposed into D major, and enriched by the addition of three oboes, being the introductory symphony. The vocal piece that follows is Arioso in 3-8 time, interspersed with recitative in common time. In the following aria, D major common time, the obbligato organ comes in again; this, however, is a perfectly new composition. On the other hand, in the second aria, in B minor, 12-8, the Siciliano of the concerto undergoes a remodelling and extension (from thirty-seven to forty-six bars) which is stamped with genius. The voice part, which begins as follows:—

Stirb .. in mir, .. stirb .. in mir, Welt . .

is newly inserted.[492] A recitative and the chorale "Du süsse Lieb schenk uns deine Gunst" (the third verse of Luther's hymn, "Nun bitten wir den heiligen Geist")

---

[492] The editor of the concerto in the B.-G. edition has failed to detect the relationship between this air and the Siciliano of the concerto. After pointing it out it is hardly possible to regard his assertion as accurate, that the cantata is the earlier and the concerto the later work, B.-G., XVII., p. xv. So much as this is, at any rate, self-evident, that the E major concerto could not well have preceded the cantata as it now stands, even allowing for such simplifications as the different nature of the organ would demand. The melody of the B minor air, in spite of all the mastery displayed in it, gives us too closely the impression of a supplementary composition for us to think it possible that the whole number, as we find it in the cantata, is in its original form, and that it was subsequently simplified for the concerto. A middle course, which relieves us from this dilemma, is afforded by the supposition that the clavier concerto in E major now exists complete only in a later re-arrangement, as can be proved to have been the case with the D minor concerto, and that, in adapting it for the cantata, Bach followed the original casting of it. Such an earlier form of the Siciliano exists, and is to be found B.-G., XVII., p. 314. There can be no reason for doubting that the first six bars of the clavier part of the older form were filled up by pauses, since here the clavier has only the accompaniment in figures, which would sound badly on the organ. In later years Bach collected his clavier concertos in a volume, and may probably have remodelled this one at that time. See App. A., No. 48.

conclude the work. It may be assumed that in the A minor air of the cantata "Geist und Seele" we have a similar case of adaptation, with all the more reason because the resemblance in structure of the two cantatas is obvious at a glance; it extends even to the vocal melodies made use of, since both are for an alto voice, only the final chorale is wanting. We meet with the last subject of the E major concerto in the cantata for the twentieth Sunday after Trinity, "Ich geh und suche mit Verlangen,"[493] which must, therefore, have been written in the same year as the former one, and have been performed for the first time only a fortnight later. It here forms the introductory symphony, and is filled out by the addition of an *oboe d'amore;* but with respect to the solo part it is treated more simply, which must be partly ascribed to the different conditions of the organ. In the course of the work the obbligato organ plays its part in newly composed and engrafted pieces, some of these being solos for the soprano or bass voices, some dialogues for the two, as representing the Soul and Jesus. The cantata includes no chorus. However, in the last piece for the soprano we hear the final verse of the chorale "Wie schön leucht't uns der Morgenstern," while the bass sings, in a vein which may almost be called fervid, this paraphrase of Bible words:—

Dich hab ich je und je geliebet,	From all eternity beloved
Und darum zieh ich dich zu mir.	I, loving thee, have longed for thee.
Ich komme bald	Now, lo! I come
Ich stehe vor der Thür,	And say "prepare for me,
Mach auf, mach auf, mein Aufenthalt!	Open thy heart to be my home."

A stringed quartet with *oboe d'amore* and organ concertante complete this highly significant picture.

I have already mentioned, in another place, that Bach, who when he conceived his three violin sonatas was quite as much possessed by the feeling of the clavier or the organ as by that of the violin, subsequently re-arranged them in fact as clavier and organ pieces.[494] He adopted

---

[493] B.-G., X., No. 49.
[494] See ante, p. 80.

the same course with the prelude to the suite for violin alone, in E major; he has transposed it for the organ to D major, and furnished it with an orchestral accompaniment of strings, two oboes, three trumpets, and drums; and in this state it forms the instrumental symphony to the Rathswahl cantata performed on Monday, Aug. 27, 1731, "Wir danken dir Gott, wir danken dir"—"We thank Thee, Lord God, we thank Thee, Lord."[495] Though this is a less remarkable work than the transformation of a clavier solo into a clavier concerto for the clavier, violin, and flute, with a stringed orchestra (see Vol. I., 420), this symphony nevertheless sets Bach's powers of combination in a very clear light. As in a clavier concerto with orchestra there must be a special instrument for the figured bass to be played upon, so, usually, when Bach uses an organ obbligato in church music, the great organ is also introduced to support and connect the whole. In the cantata "Wir danken dir Gott" this certainly cannot invariably have been the case, for it was performed twice again (Aug. 31, 1739, and in 1749) in the church of St. Nicholas; during the performance of Divine service on the occasion of a change of council, and the organ there had no independent *Rück-positiv*.[496] The symphony has a festal and lively character, and leads up very fitly to the cantata, which gives expression to gratitude towards God in gushing songs of triumph, and solemn, majestic choruses, emphasizing this feeling rather than the occasional purpose of the composition. The principal chorus is on the words from Psalm lxxv., and we must take this verse—"But I will declare for ever; I will sing praises to the God of Jacob"—as the starting point of the sentiment of the whole great work. Bach

---

[495] B.-G., V.,[1] No. 29. W. Rust, in B.-G., V.,[1] p. xxxii., and B.-G., VII., p. xxvii., expresses an opinion that the violin prelude was arranged from the symphony. But this is contradicted plainly by the circumstance that the violin suites were completed long before 1731 (See App. A., No. 4). The date of the performance is given from the documents of the Leipzig Rath "concerning the Rathswahl (election of councillors) 1701."

[496] A. Dörffel has noted the first repetition, Musikalisches Wochenblatt. Leipzig, 1870, p. 559. The second we know of from a text-book of 1749 accompanying the autograph score.

works out two themes in fugue, one after the other, without their ever coalescing to form a double fugue. On the contrary, the response always follows in artistic stretto movement, which beyond a doubt—as in the cantata "Sie werden aus Saba alle kommen" (see ante p. 217)—was founded on a dramatic feeling.[497] The first of the two themes is formed on an ancient type of the old church chorale, of which Handel also made extensive use.[498]

Hardly two weeks later, September 9, 1731, Bach must have conducted a cantata for the sixteenth Sunday after Trinity, in which again an organ obbligato is introduced: "Wer weiss, wie nahe mir mein Ende"—"Who knows how near my end is?"[499] In this we find no revised chamber composition; the alto aria accompanied by organ obbligato and *oboe da caccia* is a perfectly fresh invention.[500] If we are not altogether deceived, Bach's hand is also to be traced in the text, which is recognisable as a modified version of a poem by Neumeister. The words set by Bach are as follows:—

Willkommen! will ich sagen,	When Death shall come to call me
Wenn der Tod ans Bette tritt.	I will hail him as a friend,
Fröhlich { folg ich / will ich folgen } in die Gruft,	When he leads me to the tomb
Wenn er ruft.	I will come.
Alle meine Plagen	Still shall pain and grieving
Nehm ich mit.	Go with me;
Willkommen! will ich sagen,	When Death shall come to call me
Wenn der Tod ans Bette tritt.	I will hail him as a friend.

---

[497] The indication of the time suggests a somewhat quick *tempo*. Bach says in his Generalbasslehre ch. iv., that the way of indicating a (rapid) tempo in common time by a 2, was used by the French in pieces which are to go quickly and gaily, and the Teutons have imitated it from the French. So that he himself felt that he had but imitated the French in this matter. See App. B., XII., Cap. 4.

[498] See Chrysander, Handel I., p. 393.

[499] B.-G., V.,[1] No. 27. See App. A., No. 48.

[500] In the B.-G. edition it has escaped notice that in the score the aria has the title *Aria à Hautb. da Caccia e Cembalo obligato*. On the other hand, on the autograph organ part we find *Organo obligato*, and on the cover for the parts, likewise in Bach's hand, *Organo oblig*. It is certainly surprising that the organ part is not transposed, but in E flat major. This was probably done to please any player who might perform the concertante part on the Cembalo, while Bach himself played at the performance in the church and transposed it at the time.

The metrical construction of the text is unskilful, and the expressions are awkward, since, if the tomb is to be considered as the longed-for goal of rest, the pains of the deceased cannot follow him thither; it is clear that the poet, in altering the verses, has only written what occurred to him as fitting the rhyme to "tritt"; Picander with his knowledge of form would scarcely have done this. But we find in the composition that expression, peculiar to Bach, of a fervent longing for death, reaching a pitch which, so far as my knowledge goes, is not to be discerned in any other work by Bach, whether earlier or later. The Gospel for the day refers to the youth of Nain, and it may be remembered that we spoke before of another cantata on the subject, for the same Sunday, "Liebster Gott, wann werd ich sterben." A more striking contrast than that between the two is scarcely conceivable. The older cantata is pervaded by a youthful sentiment, painful though sweet, while in the later composition we have the feeling of one who, having bid the "false world farewell," longs to depart. There can be no doubt that Bach made this strong contrast intentionally, for the finales of the two cantatas plainly show that when he wrote the later one he had the older one in his mind. The close of the first consisted of an air by Daniel Vetter; here we have a five-part composition by Johann Rosenmüller "Welt ade! ich bin dein müde." They were Leipzig musicians to whom Bach in this way paid a kind of homage.[501] The whole work is composed with the same devotion as the aria. In this frame of mind Bach was more at home than in any other. If we try to picture to ourselves what the most concentrated form might be of the feeling which lies at the base of the two Passion-musics, and of the funeral ode, we shall find that it is incorporated in this cantata. A token that Bach actually *lived in* the sentiment he embodied in this composition is to be found in certain points of resemblance to those greater works. Only one or

---

[501] Rosenmüller's composition is to be found in Vopelius, p. 947. Bach has reproduced it unaltered; a small difference in the sixth bar arose beyond a doubt from a slip of the pen. See Rust, B.-G., V.,[1] p. xxvii.

two need be here pointed out. 1st. The beginning of the final chorus of the Passion according to St. Matthew on the one hand, and on the other the beginning of the first movement of the cantata, and again bar 66 of the bass aria. 2nd. The motive for the two flutes in the first chorus of the mourning ode 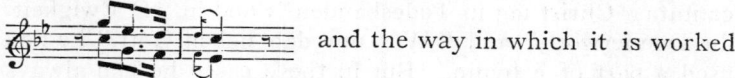 and that for the oboes in the first chorus of the cantata (from bar 13),

and the way in which it is worked out.[502] 3rd. The first bar of the aria " Es ist vollbracht "—" It is finished "—in the Passion according to St. John, and bars 35 and 78 of the bass aria in the cantata. Still, the whole speaks even more clearly than the details.

Bach must have been so greatly pleased with the effects of the organ concertante that his inventive genius found more and more ways of turning it to account. For the sixth Sunday after Trinity, probably of 1732 (July 20), he wrote a solo cantata for an alto voice, " Vergnügte Ruh, beliebte Seelenlust "—" Contented rest, with sweet and heart-felt joy "—in which he introduced an obbligato organ subject for two manuals, as an accompaniment to the second aria. This however has to be performed on the great organ: all is subsidiary to the figured bass, and violins and violas in unison have the lowest part. This most original combination led to the production of a composition which is not only remarkably artistic, but also deeply emotional; and it stands among worthy surroundings, for the whole cantata is one of the most beautiful of its kind.[503]

---

[502] The mourning ode was worked up again for the Passion according to St. Mark, and was then first performed in 1731.

[503] See App. A., No. 48.—The autograph score is in the Royal Library at Berlin. There also is a later recension of the work (from Fischhoff's bequest) in which the cantata is transposed from D major to C major. Besides this the first movement alone remained intact: it is followed by a new recitative, and then by a great final chorus, which is none other than the opening chorus of " Herz und Mund und That und Leben " in ¾ time, and with slightly altered text. I know not for what occasion this recension was made, nor, consequently,

It happened at this time that Bach had become sickened with the trivial rhyming of the "madrigal" cantatas, on which for nearly ten years he had been almost incessantly engaged, and begun to yearn for stronger poetic diet. The old Protestant church hymns supplied him with what he required. Consequently, he occupied himself in the musical treatment of a number of the best sacred verses of the sixteenth and seventeenth centuries. He had already made use of the whole of the words of the hymn in the Easter cantata "Christ lag in Todesbanden"; and in "O Ewigkeit, du Donnerwort," and "Wer nur den lieben Gott," he had used a part of a hymn. But in these cases he had always worked up the melody side by side with the words, by which those works were rendered quite distinct from the group of cantatas now to be characterised. In these Bach adopts the text of the hymn rather as a mere poem on a sacred subject, to serve him as the germ from which he may develop an independent composition. He does not wholly ignore the melody proper to the hymn; on the contrary, these cantatas always include at least one movement in which it appears in its fullest form, but they also contain others in which it is almost or altogether set aside. Johann Adam Hiller, one of Bach's successors in office in Leipzig, says: "Old and new spiritual songs have also been treated as cantatas by certain composers; I confess that this kind of composition seems to me one of the fittest for church use, only the composer should refrain entirely from the use of recitative. Recitative has none of the characteristics that are proper to this kind of song, no symmetrical periods, no lines of perfectly equal length, no jingle of rhymes, and it is therefore extremely displeasing when we hear four-lined verses, for instance, which rhyme line for line, delivered in recitative."[504] Since the works of the great cantor were at

---

whether Bach was the originator of this transformation. It is quite possible that he was, for, since the cantata "Herz und Mund" was used in Leipzig for the feast of the Visitation, the amalgamated work may have been performed in 1742, when the Visitation (July 2) followed immediately upon the 6th Sunday after Trinity (July 1).

[504] Hiller, Beyträge zu wahrer Kirchenmusik. Zweyte vermehrte Auflage. Leipzig, 1791, p. 7.

hand, we cannot doubt that these words were aimed at him. But what he blames—namely, the treatment in recitative of verses of hymns—Bach has certainly not avoided, any more than he thought proper to adhere to the rules laid down in 1754 by the Musical Society of Leipzig as to the construction of cantata texts, although he was a member of the society. Hiller's reflections were justified, however; the madrigal form in sacred texts had been retained principally for the sake of the recitative. It is clear that Bach's chief object was to have texts of fuller meaning to work upon, but it never occurred to him to give up any part of the musical form. As his style of recitative differed essentially from that in common use, he had no need to consider that the poetical and musical forms were in perceptible contradiction to each other.

He composed two hymns of the sixteenth century in the cantata form of his time. The paraphrase of Psalm xxiii. written by Wolfgang Musculus, "Der Herr ist mein getreuer Hirt"—"The Lord my faithful shepherd is"—was adopted for the second Sunday after Easter, and was first performed on April 8, 1731, or April 27, 1732.[505] The hymn "Ich ruf zu dir, Herr Jesu Christ"—"On Thee I call, Lord Jesu Christ"—was arranged for the fourth Sunday after Trinity, July 6, 1732.[506] The setting of the fine hymn by Johann Olearius, "Gelobet sei der Herr, mein Gott, mein Licht, mein Leben"—"All praise to Thee, O Lord, my God, my life, my glory"[507]—for Trinity Sunday, 1732. Joachim Neander's hymn, written in 1679, "Lobe den Herren, den mächtigen König der Ehren"—"Praise ye the Lord, the Almighty King of glory"—was composed for the twelfth Sunday after Trinity; it may have been performed for the first time in 1732, on August 31.[508] Paul Flemming's hymn, on the

---

[505] B.-G., XXIV., No. 112. P. 1682. See App. A., No. 46.

[506] The autograph score in the Royal Library at Berlin has these words written at the end, "*Il fine SDGl. ao* 1732." The original parts are in the Library of the Thomasschule at Leipzig. See App. A., No. 44.

[507] The original parts are in the Library of the Thomasschule, and a opy by Penzel, dated 1757, is in the Royal Library at Berlin. See App. B., No. 48.

[508] The original parts are in the Library of the Thomasschule. See App. A., No. 46. This cantata, and "Lobe den Herren meine Seele" (see ante, p. 407),

occasion of starting on a journey, "In allen meinen Thaten," written in 1633, was set by Bach in 1734, but for what particular event is not known.[509] Martin Rinckart's hymn "Nun danket alle Gott" (1644)—"Now thank we all our God"—and one by Jacob Schützen, "Sei Lob und Ehr dem höchsten Gut" (1673)—"All praise and glory be to Thee"—have also no trace of any special occasion or purpose. It can only be said, as regards the date of their composition, that this must have taken place at about the time we are now considering, while Johann Heermann's "Was willst du dich betrüben" (about 1630)—"Why art thou so dejected?"—and Rodigast's "Was Gott thut, das ist wohlgethan" (1675)—"That which God doth is still well done"—appear to be of later date.[510] It is common to all these cantatas to begin a grand chorale in chorus treated in the form of a chorale fantasia. In "Nun danket alle Gott" only, this form has been so far altered that freely invented subjects for the chorus introduce and interrupt the treatment of the chorale and come in again as a finale. In the cantata "In allen meinen Thaten" the chorale fantasia assumes the aspect of a French *ouverture,* and the chorus is built into its fugal allegro with admirable art. This piece reminds us perceptibly of the opening chorus of "O Ewigkeit, du Donnerwort," which, however, we concluded we must assign to an earlier period.

The finale chorales display a greater variety—they sometimes appear as simple four-part compositions; in "Lobe den Herren," and in "In allen meinen Thaten" they are expanded and made splendid by the addition of three independent instrumental upper parts; again, we have them

---

were evidently connected with the change of Council. The twelfth Sunday after Trinity in 1732 was the first after St. Bartholomew's day, when the change by rotation commonly took place, so the service for the occasion was performed on the 25th.

[509] B.-G., XXII., No. 97. P. 1674. See App. A., No. 49.

[510] "Sei Lob und Ehr," B.-G. XXIV., No. 117. P. 1690. "Was willst du dich betrüben," B.-G., XXIII., No. 107. P. 1685. "Was Gott thut, das ist wohlgethan," B.-G., XXII., No. 100. P. 1669. "Nun danket alle Gott," is in the original parts, in the Royal Library at Berlin, unfortunately not complete. See App. A., No. 44.

TREATMENT OF THE CHORALE. 457

in the chorale fantasia form; or, as in "Gelobet sei der
Herr" and "Was Gott thut," they are treated in a manner
which reminds us of Böhm, a certain phrase being given
to the orchestra and repeated after each line, besides being
played with the lines when possible. But, excepting in the
opening and final movements, the melody of the hymn recurs
only twice in its complete form—namely, in the second and
fourth verses of "Lobe den Herren"; in the latter instance
it is given to a trumpet concertante. In most other cases no
heed is given to it at all; this is the case throughout the
cantatas "Der Herr ist mein getreuer Hirt," "Ich ruf zu
dir," "In allen meinen Thaten," the rest of the verses
serving simply as a text for independently invented arias,
duets, and recitatives; often, too, the composer has indulged
in a whimsical sporting with the melody, allowing it to
come out now more strongly and then more softly. In this
way Bach developed a new phase of church music. Up to
this time we have not met with anything like it, excepting
a slight example in the cantata "Wer nur den lieben Gott,"
and I there pointed out its æsthetic importance.[511] From
the musical point of view, the Suites offer us something
analogous, where the beginning of each dance of the series
has a certain connection with that of the Allemande. The
florid opening phrases commonly have reference to the first
line or the first two lines of the chorale. For example:—

Chorale Melody.

Ge - lo - bet sei der Herr, mein Gott, mein Licht, mein Le - ben.

Third aria of the Cantata.

Ge - lo - bet sei der Herr, mein Gott, der e - wig le - bet.

---

[511] The resemblance to the chorale "Nun danket alle Gott" in the Rathswahl cantata "Preise Jerusalem" (see ante, p. 364), need not be dwelt upon, as this chorale occurs nowhere else in the cantata. Its only meaning is a symbolic or poetical one.

Chorale Melody.

Fourth verse of the Cantata.

In the cantata "Lobe den Herren" the melody—

is once transposed into the minor, and worked out as a duet, thus:—

In the fifth verse of "Sei Lob und Ehr," which begins with a recitative, we come suddenly on a phrase which is directly connected with the first four notes of the melody:—

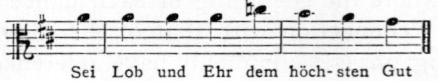

the bass imitates it, and an arrangement grows out of it which reminds us vividly of the chorale treatment of Bach's early time.[512] But other lines besides the first are sometimes touched on lightly, and, as it were, in passing by; in this respect the fourth verse of "Was Gott thut" is worthy of attention. In verse five of "Was willst du dich betrüben," which begins with a very ingenious fantasia on the first lines, and then develops it freely as an aria, it occurs that at the close the last line of the chorale comes in quite unadorned and simple, as though the fancy of the composer had sunk back to its source. Again, the second verse of "Nun danket alle Gott" is formed on a free

---

[512] See Vol. I., p. 214.

utilisation of the chorale melody in such a way as that it is heard throughout this grand and wonderfully brilliant work, sometimes full and clear, and sometimes veiled in sound.

Meanwhile the master must have been fully aware that this method of dealing with the hymn and chorale had its objectionable side. Though he continued during several years to return to it from time to time, still a glance at his later works as a writer of cantatas reveals that we must regard those that have now been described as the creations of a period of transition or digression, outside and beyond which lay a still more perfect type. Even during the first Leipzig period we come upon a few works in which the form is already plainly discernible, which became to Bach in his latter years the very ideal and type of the church cantata. In these, too, a hymn with its appropriate tune forms the nucleus, but the hymn text is not made use of for airs or recitatives, nor, on the other hand, is the hymn tune sacrificed to fanciful embellishments. On the contrary, words and compositions which, though independent, are developed out of the church hymn, are used to serve the more personal emotion which is aroused by the congregational feeling; the chorale preserves its unapproachable and unalterable nature, though it still pervades the whole as a unifying power, even where neither the original words nor the original music are to be heard.

Very closely approaching to this ideal form is the magnificent composition which Bach prepared for the twenty-seventh Sunday after Trinity of 1731, November 25. This Sunday, as is well known, but rarely occurs in the ecclesiastical year; and for this reason, and because of its poetically and mysteriously solemn Gospel, Bach felt himself prompted to compose for it a creation of the very highest order. Nicolai's three-verse hymn "Wachet auf, ruft uns die Stimme"—"Wake, arise, a voice is calling"—has, with just feeling, been selected as the basis of the work; this has an obvious connection with the Gospel story of the ten virgins (Matthew xxv., 1—13), and it leads on and up to the beatific contemplation of the Song of

Solomon and of the Revelation of St. John, chap. xxi.[513] Between the stanzas are inserted recitatives and dialogues between Christ and the Bride, duets of the highest art, which breathe of chaste fervency without ever trenching on the domain of personal passion. The three verses of the chorale are precisely at the beginning, middle, and end, and figure the mystical tone that pervades the whole work, and which is required by the ideas of the solemn silence of the night when the Heavenly Bridegroom is looked for, and the unspeakable joys of the glory of the New Jerusalem. The first verse is a chorale fantasia: this motive—

which comes in on the fifth bar, infuses a feeling of mysterious bliss into the majestic rhythm of the orchestra, and this feeling overflows again and again in happy and expressive passages. The soprano has the melody, while its dramatic purport is indicated by the other voices in figures of wonderful significance. In the second verse, which is a trio for tenor voice, violins, and bass, the mystical sentiment is most fully brought out. It is like the dance of souls in bliss, swaying to and fro with a strange and mysterious expression on the low notes of all the violins—all Zion and the faithful have passed with Christ into the joys of the heavenly banquet. The last verse, in which "Gloria, mit Menschen und Engel-Zungen"—"Glory, with tongues of Men and Angels"—is sung, appears in unadorned simplicity. The splendid melody has here once more an opportunity of producing its effect by its own beauty.

The cantata "Was Gott thut, das ist wohlgethan," of which I spoke above, is, so far as its opening and final subjects are concerned, only an extended and embellished remodelling of earlier pieces. The closing movement occurs in the cantata "Die Elenden sollen essen"—"The poor

---

[513] The cantata, the original parts of which are preserved in the Thomasschule Library, is as yet published only in Winterfeld, Evang. Kirchenges., III, Appendix, p. 172 ff, and in P. 1691. On the date of composition, see Appendix A., No. 44.

shall eat and be satisfied"[514](1723). The first movement—but in a simpler form—also introduces a work which may have been written two or three years earlier, about 1733. (?) It repeats the type of the cantata "Wachet auf" in a somewhat simpler development, making use of only two verses of the chorale. We find them at the beginning and at the end in the same form as in that cantata, and between them come a number of vocal movements in "madrigal" form.[515] A third composition beginning with the same chorale must have served for the twenty-first Sunday after Trinity, and have been performed for the first time October 21, 1731, or November 2, 1732.[516] It contains only one verse of the chorale, which is at the beginning and set in the form of a chorale fantasia; still the instrumental tone-picture is not in the first instance strictly homogeneous, and during the first section of the tune it has more of the character of a ritornel. In the remainder of the work the chorale melody is nowhere used again, not even at the close, for the cantata ends in a bass aria. Hence it is imperfect as to form, a mere sketch which has never received the final touches; even the separate movements are inferior to the other two cantatas in musical value. A cantata for the sixth Sunday after Trinity, on the other hand, "Es ist das Heil uns kommen her"—"Now is salvation come to us"—gives us perfect satisfaction by its masterly completeness and fulness of form. If similarity in the aspect of details is any evidence of a similar period of composition—and with Bach this is certainly the case—this cantata must have been written in the same year as "Wachet auf"—viz., 1731. Only the first and twelfth verses of Paul Speratus' hymn are used and placed at the beginning and end of the work, while between them, among other pieces in madrigal form, an admirable duet in canon finds a place; but the treatment of the first verse resembles that of the first verse of "Wachet auf" in a surprising degree; particularly in the two-part imitations,

---

[514] Composed in 1723. See ante, p. 355.
[515] B.-G., XXII., No. 99. P. 1670.
[516] B.-G., XXII., No. 98. See Appendix A., No. 48.

and partly in the accompanying rhythm of the instrumental subject.[517]

A new and deeply thoughtful composition for the sixteenth Sunday after Trinity (probably September 28, 1732) "Christus der ist mein Leben"—"Christ, who is my life"—must be mentioned in this place, although as to form it hardly belongs here. It begins with a chorale chorus on that hymn, but besides this three chorales are used in it, all among the most beautiful and best known funeral hymns of the Protestant church. By this means the sentiment of the cantata, poetical and musical alike, is enhanced to powerful intensity; still, we cannot fail to observe an absence of unity. The importance of the chorale in Bach's cantatas is different and greater than what could arise merely from the combination of a well thought-out poem with a fine melody; it has to serve as the common centre for the more subjective arias and recitative. It is even justifiable to introduce a different chorale at the end from that with which the cantata opens; in such cases the feeling is developed from the starting point of the strictly church sentiment in such a way as that it returns again to that narrow and confined province of religious art. It can only prove a disturbing element when, instead of this one starting point, two or three are taken. We know how Bach could revel in images of mortality and death; he has here again given the reins to this sombre mood, and has gone so far in the first chorale chorus as entirely to destroy the proportions of the four lines of the chorale by a long extension for the sake of the word "Sterben"—"Dying." The way in which this chorus passes into the solo without any musical break is very delicate and imaginative; first into an arioso, then a recitative, which is immediately followed by a fresh chorus in the simple style of Pachelbel, on the words "Mit Fried und Freud ich fahr dahin"— "With joy and peace I pass away." The third chorale "Valet will ich dir geben"—"Farewell to thee addressing" —is treated as a trio, and it is impossible to overlook its

---

[517] B. G., I., No. 9. P. 1281. See Appendix A., No. 44.

affinity with the chorale trio in "Wachet auf." At the close comes the fourth verse of the hymn " Wenn mein Stündlein vorhanden ist "—" When my last hour is close at hand."[518]

Though from this time forth it became more and more firmly established that the first movement should assume the form of the chorale fantasia, still the craving for variety in Bach was sometimes so strong that he assigned the first place to other forms; this is the case with a piece written for the first Sunday in the year, January 4, 1733. In this the vocal portion associated with the instrumental work is in two parts only, the soprano singing the first verse of the chorale, "Ach Gott, wie manches Herzeleid"—"Ah! Lord, how many a pang of heart"—while the bass sings against it in a fashion of its own, admonishing us to patience and endurance. The same kind of contrast is once more brought out musically in the final movement, which again is not a chorale fantasia, but a bass air, and the chorale is worked into it by the soprano voice. The text of this chorale consists of the second verse of Martin Böhm's hymn, "O Jesu Christ, meins Lebens Licht"—"O light of life, my Saviour dear"—while the melody is the same as at the beginning. From the character of alternate complaint and consolation which are worked into this cantata—though only in the principal movements—this work has been styled the *Dialogus*.[519] It did not remain alone of its kind; Bach wrote another *dialogus* for the twenty-fourth Sunday after Trinity (probably November 23, 1732), in which the characters are "Hope" and "Fear." The bass alone appears as opposed to them, and we may suppose it to represent that " Voice from Heaven" which, in Rev. xiv., 13, utters the words here given to the bass to sing. The first movement is constructed in a way exactly analogous to that of the former cantata. The alto sings the melody " O Ewigkeit, du Donnerwort"; this, however, is not repeated at the close, where we have Joh. Rudolph Ahle's expressive aria "Es ist genug" in a four-part setting.[520]

---

[518] B.-G., XXII., No. 95. P. 1673. See Appendix A., No. 50.
[519] B.-G., XII.,² No. 58. P. 1195. See App. A., No. 48.
[520] B.-G., XII.,² No. 60. P. 1285. See App. A., No. 51.

We have now done with the chorale cantatas, and, before concluding, must glance briefly at a group of church pieces which are based entirely, or chiefly, on free invention. In most of these a text from the Bible has supplied the poetical motive, and in accordance with the custom which was general, and which Bach himself by his practice acknowledged as well-founded, the idea is usually embodied in the chorus form. Among these, two works stand pre-eminent. First, a cantata for the tenth Sunday after Trinity (probably July 29, 1731), "Herr deine Augen sehen nach"—"O Lord, are not Thine eyes upon the truth?" (Jer. v., 3)—belongs to that class of works which I have ventured to designate as orthodox compositions (see ante, p. 423).[521] If Picander compiled the text he must have surpassed himself; in the recitative, it is true, the commonplace method prevails, but the airs have a strict metrical structure and vigorous phraseology. Besides the text for the principal chorus, another passage is quoted from Rom. ii., 4-5, while Picander usually contented himself with one Bible text. Since Bach was planning an exceptional work of art he may, perhaps, for once have stirred up his colleague to unwonted vigour, while in the cantata "Schauet doch und sehet"—"Behold, now, and see"—he revels in insatiable lamentation as if he desired to do full justice to the words of the prophet, "Mine eye runneth down with rivers of water for the destruction of the daughter of my people" (Lam. iii., 48-49). He here appears as the ardent, almost fanatical, preacher of repentance. As usual, the first chorus gives the key to the prevailing sentiment; it is arranged in a masterly and quite new way. A prelude expresses in a connected whole the principal ideas which afterwards and separately form the basis of the chorus, transposed in various ways. We often meet with this method of procedure in Bach, but it is astonishing to find two elaborate fugue subjects interwoven in the working out.

---

[521] It is one of the many services rendered by Rust to have restored this cantata to its original form; previously it was only known to the general public through the edition of A. B. Marx, Kirchenmusik von J. S. Bach, No. 2, Bonn, Simrock. It is now published B.-G., XXIII., No. 102, P. 1679. See App. A., No. 48.

Their themes—

are of extraordinary boldness and overwhelming energy, and it is impossible to overlook the oratorio-like stamp of this stupendous chorus. It is conspicuous in the aria and in the bass arioso, often rising almost to dramatic energy. The tenor aria is characterised by certain peculiarities of form. The text is as follows:—

Du allzu sichre Seele	Thou all too boastful spirit
Erschrecke doch!	Be dumb with fears,
Denk, was dich würdig zähle	Remember the due merit
Der Sünden Joch!	Of sinful years!
Die Gottes-Langmuth geht auf einem Fuss von Blei,	Though God's long-suffering may linger with foot like lead,
Damit ihr Zorn hernach dir desto schwerer sei.	'Tis only that His wrath may crush thy guilty head.

The first of these lines was not graphic enough for the vein of agitation which Bach desired should predominate throughout this cantata; he therefore began on the second, and with a musical phrase which paints terror with emphatic significance. The aria offers us one of those cases—rare in Bach—where the principal musical thought, which is introduced as a prelude in the ritornelle, remains throughout conspicuous in the instrumental part, and hardly appears at all in the voice part. The chief melodic phrase

---

522 The objections raised by Hauptmann to the passage in bars 37—44 have been removed by Rust, who refers it to the general plan of the whole; still, it cannot be denied that there is something strange in the anticipation of the words " Du schlägest sie," &c.—" Thou hast stricken them but they felt it not; Thou hast plagued them but they amended them not "—which, besides, are not once treated separately with the expression which they suggest.

here seems to have been devised to give us the words in their right order—

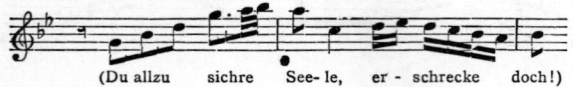

(Du allzu sichre See-le, er-schrecke doch!)

and if this suggestion is the right one, this aria is an instructive example of the way in which, to Bach's mind, the vocal and instrumental aspects of a work presented themselves as an inseparable unity.

The other cantata, which I rank as equal to this one, is based on the words of Psalm xxxviii., 3. Its purport is much the same as that of the former cantata, and we now must imagine that the repentance, preached with such fervency in that, has penetrated and filled the soul of the sinner. "Es ist nichts gesundes an meinem Leibe," &c.— "There is no soundness in my flesh because of Thine anger; neither is there any rest in my bones because of my sin"— is his cry.[523] The chorus is a double fugue, complete in itself, and full of contrite expression. From bar 15 onwards, the four-part chorale "Ach Herr mich armen Sünder"[524] is given out at regular intervals by the flutes, cornet, and three trombones, and is heard through the fugue, which in parts is accompanied by particular phrases on the strings. The four-part subject is equally complete in itself, and, like the fugue, might be performed independently with a satisfactory effect; nevertheless, the two bodies of sound amalgamate as if they had grown from one and the same root. The depth of the effect produced when the sacred penitential hymn comes in, sung, as it were, by invisible voices above and beyond the abased multitude entreating from the dust, is indescribable and unfathomable. It appears in augmentation in the instrumental bass before each couplet of the first section of the tune.

---

[523] B.-G., V.,[1] No. 25. P. 1650. See App. A., No. 44. The water-mark is distinguishable only on the wrapper of the original parts.

[524] No other hymn can possibly be intended. The melody also belongs, as is well known, to the words "O Haupt voll Blut und Wunden"—"O Thou whose head was wounded."

Even now all is not told. The two themes of the fugue are derived from two lines of the chorale—the former from the second and the latter from the first. The way in which this is done serves to remind us of the construction of the melody of the cantata, which is founded on the entire text of a church hymn; for this reason it appears to me probable that the cantata "Es ist nichts gesundes" belongs to the same period—and that a limited one—as this. The composition evolved out of all these musical and poetical motives is extremely remarkable and unique in its way. The movement "Es ist der alte Bund," out of the cantata "Gottes Zeit" (see Vol. I., p. 457), cannot be compared with it, because in that the chorus subjects appear rather as mere interludes in the chorale; but here we see the forms borrowed from the chorale fantasia and transferred to the chorus and orchestra in an inverted position: the chorus fulfils the duty of the instruments, and a body of instruments takes that of the chorus. It is neither a freely worked-out chorus on a Bible text, nor is it a chorale chorus; it is something compounded of the two and superior to each, which we vainly strive to comprehend. A beautiful bass aria with an instrumental bass independently worked out, full of character, brings the overwhelming conception of the first movement home to the personality of the hearer; this air still lingers in the same domain of feeling, but it passes gradually into a consolatory vein, in which the work is brought to an end.

A brighter picture is put before us in the cantata for Ascension Day, "Wer da glaubet und getauft wird"[525]— "He that believeth and is baptised." In the principal chorus two melodies are worked out in contrast—one broad and calm, and the other animated and eager; they are treated throughout under the same conditions as are displayed in the beautiful bass air of the Whitsuntide cantata "Ich liebe den Höchsten" (see ante, p. 444). Even the poetic sentiment is the same, and the two cantatas must belong to about the same period. The particularly *singable* chorus

---

[525] B.-G., VII., No. 37. P. 1693.

is associated with a rich six-part accompaniment, and the whole body of sound has a very splendid and ample effect, while the words of the Gospel, " Go forth into all lands, and preach the Gospel to every creature," are delivered in an animated rhythm that flows fully and freely onwards. In the middle we have a two-part chorale—the fifth verse of " Wie schön leuchtet der Morgenstern"; the lines extended by *melismata* are given out now by the soprano and now by the alto, with imitations in the other parts, while in the bass a characteristic episode is wrought out of the first line of the chorale.

A very remarkable chorus opens a cantata for the twenty-first Sunday after Trinity, " Ich glaube, lieber Herr "—" Lord, I believe; help Thou mine unbelief" (Mark ix., 24).[526] It expresses the sentiment of doubt and wavering in a way which is as unmistakable as it is masterly, for the parts wander about separately and, as it were, aimlessly, and only combine into compact figures now and then, and for a short while. The same idea is given by other means in the tenor air, of which the text serves to give us the key to the meaning of the chorus. Here we have a chorale fantasia at the close, on the seventh verse of " Durch Adams Fall ist ganz verderbt."

The cantata on the words from Psalm xcvii., 11-12, " Dem Gerechten muss das Licht "—" Light is sown for the righteous, and gladness for the upright in heart "—was written neither for a Sunday nor a festival, but for a betrothal. It has a superlatively festal and brilliant stamp, and reminds us of the style of the cantata " Lobe den Herrn meine Seele," of the year 1724. It opens with a couple of splendid fugues (in common time and 6-8), in which, as in some few cantatas of the early Leipzig period, the fugal treatment is begun in a small chorus and is gradually transferred to the great chorus.[527] The closing chorus, which is homophonic, is broad and powerful, and between them

---

[526] B.-G., XXIII., No. 109. P. 1686. See App. A., No. 44. The water-marks do not occur in the autograph score, but are very plainly in the original parts.
[527] See ante, p. 187.

stands a bass aria in the Lombardic style,[528] so-called, combining a flow of melody which reminds us of Italian grace with a conspicuously festive character. A wealth of warmth hangs over the whole work, which, however, in its present form can hardly be regarded as in its original state, since it must have resulted from a revision of a composition which belongs indeed to the very earliest Leipzig period.[529] The history of its origin seems to have been the same as that of the funeral cantata "Herr Gott, Beherrscher aller Dinge," which, in part at least, is founded on an older Rathswahl cantata, "Gott, man lobet dich in der Stille," and, in part, on an adagio from the violin sonata in G major.[530] The alliances which have arisen between different works of Bach, through remodelling and transferring, are often extremely intricate. The Rathswahl cantata just mentioned must, to all appearance, have also served as the second Jubilee cantata for the centenary anniversary of the Augsburg Confession, June 26, 1730, and may finally have been once more worked up into the form in which we now have it for its original purpose. The Bible text, Psalm lxv., 1, is not here set for a chorus—this seems to have been prohibited by the purport of the words—but as an alto solo, of which the florid character most expressively conveys the feeling of festivity. The principal chorus follows, on words in madrigal form.

It may here be observed that the triad of Jubilee cantatas of which Bach conducted the performance on three successive days in 1730[531] are to be traced to earlier works from which they have been remodelled. The first, "Singet dem Herrn ein neues Lied"—"Sing to the Lord a new song"— is the New Year's cantata for 1724,[532] and the third "Wünschet Jerusalem Glück"—"Wish thou joy to Jerusalem"— is a Rathswahl cantata for August 25, 1727, which was

---

[528] See Vol. I., p. 419.
[529] B.-G., XIII.,[1] p. 3. P. 1662.
[530] B.-G., XXIV., No. 120. Compare B.-G., IX., p. 252; also ante, p. 117; and see App. A., No. 52.
[531] See ante, p. 243.
[532] See ante, p. 386.

repeated on a similar occasion, August 18, 1741;[533] this, and the piece in the original state, have both been lost. There is yet a fourth re-arrangement which we may assign to 1730. The assertion is no doubt well founded that in this year the celebration of the Reformation festival was considered of special importance and kept accordingly; and it is evident that the cantata "Ein feste Burg ist unser Gott"—"A stronghold sure is our God"—must have been intended for some such extraordinary solemnity.[534] Bach took the music he had composed at Weimar to "Alles was von Gott geboren"—which, as it was intended for the third Sunday in Lent, he had not yet been able to use in Leipzig—and he added new movements to the first and fifth numbers. These are chorale choruses on the first and third verses of Luther's hymn. The bold spirit of native vigour which called the German Reformation into being, and which still stirred and moved in Bach's art, has never found any artistic expression which could even remotely compare with this stupendous creation. The first number, including 228 bars, is in the Pachelbel form, excepting that the *Cantus firmus* is carried on in canon by the trumpet and the instrumental basses; it stands up like some impregnable giant fortress. The second chorus, No. 5, is a chorale fantasia with episodical treatment of the first line of the tune; the whole chorus sings the *Cantus firmus* in unison, while the orchestra plays a whirl of grotesque and wildly leaping figures, through which the chorus makes its way undistracted and never misled—an illustration of the third verse (" Und wenn die

---

[533] With a few omissions and additions, see Nützliche Nachrichten von Denen Bemühungen derer Gelehrten, &c., p. 82, Leipzig, 1741.

[534] B.-G., XVIII., No. 80. P. 1012, and in English translation published by Novello. See Vol. I., p. 641. It is certainly possible that this cantata may have been written for the year 1739, with reference to the Jubilee of the two-hundredth anniversary of the adoption of the Evangelical doctrines in Saxony. The service took place in the principal churches on Whitsunday, May 17, but by express order without any special ceremonial. The University commemorated this event on August 25; Görner had composed a Latin ode for the purpose; see Gretschel, Kirchliche Zustände Leipzigs vor und während der Reformation im Jahre, 1539, p. 292. Leipzig, 1839.—It will be presently shown that the Reformation cantata "Gott der Herr ist Sonn und Schild" was composed neither in 1730 nor 1739.

Welt "—" If all the world with friends were filled ")—as grandiose and characteristic as it is possible to conceive.[535]

Before quitting the subject, there are two other revised arrangements of chorales, of which we can only say that they seem to have been brought out about this time—that is to say, about 1730. Bach made use of his beautiful serenata written at Cöthen, "Durchläuchtger Leopold," for a cantata for the second day of Whitsuntide ("Erhöhtes Fleisch und Blut"), and for the same purpose turned the final duet into a chorus.[536] The birthday cantata for the Princess of Anhalt Cöthen he adapted for the first Sunday in Advent, enriching it with two admirable arrangements of the chorale "Nun komm der Heiden Heiland" and two simpler chorales. One of these arrangements is a chorale fantasia with a *Cantus firmus* given to the tenor. In the other, on the contrary—a rare occurrence with Bach—the chorale tune is treated in the manner of a motett, line for line, for a soprano and an alto. We are here reminded of the beautiful two-part chorale subject in the cantata " Wer da glaubet und getauft wird "—" He that believeth and is baptised "—but the development is more completely worked out in the Advent cantata.[537]

When words from the Bible were used for solo songs, the established form was the bass arioso, and, in general, Bach did not deviate from this practice. Still, his observance of it was often merely superficial—that is to say, he avoided the designation *aria*, but actually wrote pieces of music which must be regarded as independent arias. This is the case with the solo of the cantata " Herr, deine Augen sehen nach dem Glauben," which bears the title *Arioso;* with the solo in the Rathswahl cantata "Gott man lobet dich in der Stille," which has no designation at all, and which is exceptionally given to an alto voice, and again with the song which opens a cantata for the twenty-second Sunday after

---

[535] The cantata was published about 1822 by Breitkopf and Härtel, of Leipzig, and was the first engraved and published after Bach's death. Rochlitz gives a section to the subject (Für Freunde der Tonkunst, Vol. III., p. 229).

[536] See ante, p. 7, and App. A., No. 44.

[537] B.-G., VII., No. 36. P. 1292. See ante, p. 158.—See App. A., No. 44.

Trinity "Was soll ich aus dir machen, Ephraim"—" How shall I give thee up, Ephraim?" (Hosea xi., 8). On the other hand, the introductory solo of a cantata for the fifth Sunday after Trinity, "Siehe, ich will viel Fischer aussenden"— "Behold, I will send forth many fishers" (Jerem. xvi., 16)— is simply entitled an aria. Both these works belong to this period, still the exact date of composition can only be given with any approach to certainty in the case of the second (July 13, 1732).[538] From the fact that the solo arias are among the most beautiful that Bach ever wrote, we can see that he must have worked at them with the devotion due to the sacred text on which they are founded. They have certain peculiarities of form which are not always accounted for by the structure of the text. The cantata "Siehe, ich will viel Fischer aussenden," which is altogether a very remarkable work, contains two such solo pieces on Bible words. The second begins with two bars, common time, in G major, as introductory, while the remainder is in D major. The first consists of two grand pieces in D major and G major, but the whole cantata is very remarkable as regards the arrangement of the keys and the distribution of the orchestral accompaniments. But in this instance there is no covert meaning to be found in the singularity. The peculiar feeling which finds expression in the first movement will have due justice done to it, in connection with a similar phase of sentiment, when we have occasion to discuss a portion of the Passion according to St. Matthew.

These two last cantatas are, with the exception of the closing chorales, mere solo cantatas. We still have to add to them a few of the same kind, which are however composed to madrigal poems: "Ich armer Mensch, ich Sündenknecht"—" I, wretched man, the child of sin!"—for the twenty-second Sunday after Trinity (October 21, 1731, or November 9, 1732), "Ich will den Kreuzstab gerne tragen"— "How gladly will I bear the cross!"—for the nineteenth

---

[538] B.-G., XX.,[1] Nos. 88 and 89. As to the former, see App. A., No. 48; to the latter, No. 44. The watermark is only distinguishable in the horn part. This cantata cannot have been composed in 1733, because in that year the Reformation festival fell on the twenty-second Sunday after Trinity.

Sunday after Trinity (October 7, 1731, or October 26, 1732),[539] "Jauchzet Gott in allen Landen"—"Sing to the Lord in every land"—for the fifteenth Sunday after Trinity,[540] and, finally, "Ich habe genug"—"I now have enough"—for the Purification (1731 or 1732).[541] The cantata "Jauchzet Gott," a fiery solo cantata for a soprano voice, has no connection with the Epistle or Gospel for the day, it passes into a chorale fantasia on "Nun lob, mein Seel, den Herren," and closes it with a Hallelujah subject fugally treated. It must have been intended properly for some other occasion which we can no longer even guess.[542] A somewhat sombre feeling is common to the other three cantatas, and all three bear the stamp of having been composed by Bach at the happiest period of his mature powers and fullest contentment. If one of them is inferior to the others it is "Ich armer Mensch," in which the succeeding movements are not quite worthy of the first aria with its fervent and contrite emotion. In the other two the glowing and thoroughly dignified character of the texts particularly deserves to be pointed out. The expressive passage, at the close of the first aria in "Ich will den Kreuzstab," stands out in beauty, both of rhythm and melody, like a sigh of deep happiness after final relief, and it returns with wonderful effect at end of the last recitative. This evidently was never intended by the author: it is Bach who here has once more outdone the poet. We also trace an unmistakable poetical purpose in the scheme by which the work is made to die away with the sixth verse of the chorale "Du O schönes Weltgebäude" on the subdominant of the principal key.

The music for the Purification is based on the prevailing sentiment of the hymn of Siméon. It is interesting to

---

[539] B.-G., XII.,[2] Nos. 55 and 56. The second is P. 1664, see App. A., No. 44.
[540] B.-G., XII.,[2] No. 51. P. 1655. See App. A., No. 44.
[541] B.-G., XX.,[1] No. 82. P. 2149. See App. A., No. 46.
[542] The indication *et in ogni tempo* shows this. The text was altered at a later date, see B.-G., XII.,[2] p. 9. From the alterations in the first aria we might conclude it was intended for the festival of Michaelmas, according to this it may have been repeated in 1737, when St. Michael's Day and the fifteenth Sunday after Trinity fell on the same day.

compare it with the earlier cantata "Erfreute Zeit im neuen Bunde" (see ante, p. 390), as showing us how Bach could conceive of the same sacred occasion from two such different points of view. In the former the flow of feeling is cheerful and hopeful of life; in this it is weary, and happy only in the contemplation of approaching death. The aria "Schlummert ein, ihr matten Augen"—"Close in rest, ye weary eyelids"—expresses this sentiment with indescribable beauty. This cantata was originally composed for Anna Magdalena Bach; then the master arranged it for a mezzo-soprano or alto voice, and finally for a bass. With reference to its use in church, the last must be considered the most proper, since the text is a paraphrase of Simeon's words, and the connection with the Gospel for the day is thus made most obvious from the musical point of view;[543] but it was no doubt really intended for sacred chamber or domestic music. It is very significant that Bach has expressly designated this a *cantata*, which he never did with his church music, properly so-called.

This brings me back to a circumstance already touched upon. A vein of feeling which points rather to family than to congregational worship is conspicuous in certain works of the middle Leipzig period. It is evident in the cantata "Wer nur den lieben Gott lässt walten," and it could only be this which prompted the master—quite contrary to the practice of both his earlier and his later years—to introduce into some of his cantatas the whole of certain church hymns, but with the different portions of the chorale tune modified with the utmost ingenuity, while giving them an entirely new musical connection. Formerly, he had dealt with the chorale as with a dogma from whose impeccable unity it was permissible to borrow portions—that was strictly orthodox; now he merged it in subjective emotion—this was proper to the spirit of private worship. Naturally, this would find expression principally in the form of solo song. Bach had already written many solo cantatas at an earlier time; but, with the single exception of the Weimar cantata, "Ich weiss, dass mein Erlöser lebt," in not one of them

---

[543] See App. A., No. 53.

could it be pointed out that one and the same voice was to perform the whole piece. This constitutes an essential difference, for by the mere fact of several voices taking part in the same subject—even though it be only one after the other—the subjective feeling is abstracted from the expression of the sentiment. In the cantatas of the middle period it was to a certain extent different.

The first which can be called a solo cantata in the strictest sense of the word, since only one voice is employed in it, was the Septuagesima cantata "Ich bin vergnügt" —"I am content"; and that this was in the first instance intended for private use seems very clear from the alterations made in the text. By degrees, seven others were added to it, among which the composition "Ich will den Kreuzstab" was actually marked by Bach himself as *Cantata à Voce sola è stromenti*; it is striking, too, that a considerable number of them ("Geist und Seele," "Gott soll allein mein Herze haben," and "Vergnügte Ruh") are written for an alto or mezzo-soprano voice. We also find three for a soprano ("Ich bin vergnügt," "Jauchzet Gott," and "Ich habe genug"), the last being also arranged for alto and for bass. If we may suppose that the soprano cantatas were composed in the first instance for Bach's wife, the alto cantatas were perhaps intended for his daughter Katharina, of whose proficiency as a singer her father himself bears favourable testimony. Besides the arrangement of "Ich habe genug," we meet with only one cantata for the bass and one for the tenor—"Ich will den Kreuzstab" and "Ich armer Mensch." But I do not believe that these can have been all the compositions written at this period. Three pieces certainly come into the same category: one a composition for a soprano voice for the twenty-third Sunday after Trinity, " Falsche Welt, dir trau ich nicht"— "World so false, I trust thee not"—which has, as an introduction, the first movement of the first Brandenburg concerto, just as the cantata "Ich liebe den Höchsten" has the first movement of the third Brandenburg concerto;[544] an

---

[544] B.-G., XII.,² No. 52.

alto cantata for no special occasion, "Widerstehe doch der Sünde"—"Stand and fight against temptation"[545]—as well as the well-known beautiful aria for an alto voice, "Schlage doch, gewünschte Stunde"—"Haste to strike, oh, longed-for hour"—in which I believe the style of Franck is to be detected in the text.[546] It is self-evident that this aria cannot have been intended for church use, for there is no part of the service where it could have been introduced; it is too short for the regular church music, which had to last from twenty-five to thirty minutes, and the text is not suited for any extraordinary occasion of mourning. It may be regarded as certain that Bach, though as much inclined as ever to introduce a musical imitation of the sound of bells, would never have brought a real bell into the church to produce the effect, while in the family circle no one would have objected.[547] All the other cantatas for solo voices may have been used in divine service; this would no more have been a breach of good taste than the adaptation of secular cantatas to church purposes. It has already been said that Bach's style was penetrated throughout by a spirit of sacred feeling and that it remained sacred, in a certain sense, even when he meant to be secular; but within these broad limits we can detect many subtle differences by which Bach certainly distinguishes his secular music from what he wrote originally for church use, and such distinctions also exist between that and those compositions which give rise to the comparisons here drawn.

---

[545] B.-G., XII.,[2] No. 54. This work is inserted in Breitkopf's list for Michaelmas, 1761, not under the heading of Church music, but at p. 10, under the heading of "Smaller sacred cantatas and arias."

[546] B.-G., XII.,[2] No. 53. Also in Peter's "Alt-Album."

[547] In Breitkopf's list previously quoted, we find on p. 23, "Trauer-Arie: Schlage doch gewünschte Stunde, à Campanella, 2 *Violini, Viola, Alto solo, Basso*," and not "*Organo*." It is singular that this composition, which is so undoubtedly Bach's, has no original warranty for his name; in Breitkopf's list even there is no author's name. Forkel's opinion that the mention of the *Campanella* of itself proves it to belong to a period when Bach's taste was still imperfect is thus justified. Forkel, Uber Joh. S. Bach's Leben und Kunstwerke, p. 61. Still, it is very certain that in its full and mature state it is not a youthful work.

Moreover, the period, which in point of time fell about half-way in Bach's labours as a writer of cantatas, is also a middle stage in the character of his work; since here the different tendencies meet which, before and afterwards, we see the master following with remarkable distinctness and singleness of aim. Still, all such limitations have, of course, only a relative value, and require that we should allow them a certain elasticity and mutability. Even in this middle period we find cantatas with freely invented choruses on both Bible and verse text in considerable numbers, and in examples which for grandeur and profundity are inferior to no earlier works of the same character. But, by the side of these, the chorale cantata, properly so-called, is already conspicuous, and it was to this that in his later years Bach gave more and more preponderance. The years between 1727 and 1734 are the richest and most fruitful of Bach's life; this is amply proved by a consideration of the church cantatas only, and it will be fully confirmed as we study other domains, in which his inexhaustible genius laboured during the same period.

VII.

PASSION MUSIC BEFORE BACH.—THE ST. JOHN PASSION—THE ST. MATTHEW PASSION.

It has been customary, in investigating the history of Bach's Passion Music, to rest satisfied with referring its origin to a dramatic interpretation of the Gospels with a side glance at the oratorio and at the opera. But the matter is not so simple; many and diverse elements must have co-operated for the evolution of such creations as these, which we must admire as the culminating efforts of Protestant Church music.

The custom of singing the history of the Passion according to each of the four Evangelists on the four days of Holy week, seems to have been established in the church tolerably early in the middle ages. The object was to set the story as clearly as possible before the intelligence of the people, since the Latin words were understood by very few. One priest sang the narrative portion, a second the words of Christ, a third those of other individuals; while the utterances of the populace—the crowd or *turba*—were repeated by the choir. The Protestant church kept up this peculiar form of Passion service. Luther, it is true, considered it unnecessary that all four Gospel narratives should be sung, and attributed no particular importance to the performance.[548] Nevertheless Johann Walther, as early as in 1530, had composed the Passion according to St. Matthew and that according to St. John, with a German text for church use; the first being intended for Palm Sunday and the second for Good Friday.[549] The same composer, in 1552, arranged a Passion music in four parts to a German text compiled from the four Gospels.[550] In the course of the sixteenth century the German text of the Passion came into universal use by the Protestants. So early as 1559 we again meet with a St. Matthew Passion at Meissen;[551] in the year 1570 we find the first printed edition so far as is known;[552] others followed in 1573 and 1587.[553] From this the Passion performances must have continued in vogue in the Protestant churches throughout the seventeenth century. In 1613 Melchior Vulpius brought out a St. Matthew Passion; two Passions by Thomas Mancinus— St. Matthew and St. John—were published in 1620; in 1653 Christoph Schultz, cantor at Delitzsch, had a St. Luke Passion printed at Leipzig; and in Vopelius, Neuem Leipziger Gesangbuch, of 1682, we find Passions after St. Matthew

---

[548] Deutsche Messe und Ordnung des Gottesdienstes 1526. Vol. XXII., of Luther's collected works, p. 243, Erlangen Edition, by Heyder, 1833.

[549] O. Kade, Der neuaufgefundene Luther-Codex, p. 126. Dresden 1871.

[550] S. R. Eitner in den Monatsheften für Musikgeschichte, 1872. P. 59 of the Supplement.

[551] In a MS. in the Court Library at Vienna and which came thither from Meissen. Ambros, Geschichte der Musik, Vol. III., p. 416.

[552] Also according to St. Matthew, composed by Clemens Stephani, printed at Nüremburg. Chrysander, Händel, Vol. I., p. 427.

[553] In the hymn books of Keuchenthal und Selneccer: Winterfeld, Ev. Kir., Vol. I., p. 311.

and St. John.[554] And no less a musician than Heinrich Schütz treated the history of the Redeemer's sufferings in four compositions, after the four Evangelists, which still exist, though it is true only in MS.[555]

This custom continued in practice until late in the eighteenth century. In Leipzig, as has been already said (p. 273), the singing of the Passion music was first discontinued in 1766; but at Dosdorf, near Arnstadt, the cantor, Kramer, composed a work of this kind so late as 1735, and this exactly corresponds with the Passion text printed in the Arnstadt Hymn-book of 1745, above which it is stated that "in most of the districts and villages belonging here (*i.e.*, to Arnstadt) every year on Good Friday, it is customarily chanted *in stylo recitativo.*"[556]

As to musical form these Passions are so absolutely stereotyped that it would be difficult to imagine why they were printed so often and so often newly composed—if we may so call it—if it were not that the custom had struck such deep root into the life of the church. Almost all of them are in the transposed Ionic mode—F major. The separate voices recite the plain-song, and the phrases, which have very little melodic movement, resemble each other with scarcely any variation. The narrator—the Evangelist—has the tenor part, Christ sings in the bass; the remaining personages are represented by an alto voice, even Pilate's wife and the two maid servants; although as a rule, in the sixteenth and seventeenth centuries, the alto parts were sung by men. As an exception,

---

[554] I myself possess original impressions of the Passions by Vulpius and Schultz. Those of Mancinus appeared in the *Musica Divina* (Wolffenbüttel, 1620) and are reprinted in Schöberlein's Schatz des liturgischen Chor-und Gemeindegesangs. Pt. II., p. 362.

[555] They are to be found in a MS. copy in the Library at Leipzig, written probably about the end of the seventeenth century by Zacharias Grundig, the date 1666 on the title page of the St. Matthew Passion can at any rate refer only to this; for the St. John Passion, which is still extant in autograph, is dated April 10, 1665. See Chrysander, Jahrbücher für musikalische Wissenschaft, Vol. I., p. 172.

[556] Herr Stade, town cantor of Arnstadt, possesses the autograph of Kramer's Passion. The words of the Passion according to St. Matthew, here spoken of, are in the Arnstadt Gesang-Buch of 1745, p. 679.

we occasionally find these persons represented by a treble solo, for instance, in the Passions by Walther, 1552, Schultz, and Kramer.[557] In the part writing rather more development and variety are to be found. In some, even the *turbae* are treated in such a simple style of pure recitation that they could hardly claim to be called part music if some more definite melodic phrase or characteristic series of harmonies were not here and there observable. Melchior Vulpius makes the two false witnesses—whose words were usually sung by the chorus in four parts—sing together in two part imitation. In very critical situations he repeats the words several times. When the people call for Barabbas, the chorus repeats his name six times in a passionate syncopated rhythm, and when they cry for the last time "Crucify him," the choir divides into two parts, the lower voices calling after the higher ones; then they re-unite and, whereas throughout the rest of the work a four part treatment prevails, we here have a setting in six parts. The same occurs in the corresponding passage in Schultz. The close commonly consisted of a short act of thanksgiving, called *Gratiarum actio*; and throughout these Passion compositions the Latin names of the persons were preserved —a token of their early church origin—*ancilla, servus,* Pilati uxor, *latro, centurio* or *miles.*

The words of the thanksgiving were "Thanks be to our Lord Jesus Christ, who hath redeemed us from the torments of Hell." Even when the music was of corresponding brevity this offered an opportunity for expressing a purely lyric sentiment, which gave rise to the development of more lavish musical resources. In the seventeenth century, indeed, these meagre words often proved inadequate for the utterance of fervid and eager emotion, and we find in their place verses from hymns with a kind of motett setting, as in Schütz; or even an original verse in hymn form, as in a very remarkable manner in Schultz. The opening consisted, in

---

[557] I have not myself seen Walther's Passion; in that the treble is said also to sing the High Priest's part (see Eitner, Op. cit. p, 61), but I think there must be some error here.

the same way, of a devotional chorus to correspond to the finale. For this the text would consist merely of the announcement "The sufferings of our Lord Jesus Christ, as they are declared to us by the holy Evangelist," or "The sufferings—sometimes the bitter sufferings—and death of our Lord Jesus Christ, according to the holy Evangelist," or to that effect. The importance of the announcement was intended to be emphasised by the choral singing. However, the introduction of the choir was not immutable either at the beginning or the end; it even occurred—as in the Leipzig St. Matthew Passion by Vopelius—that the beginning and end were in unison plain-song.

Music being generally introduced to grace the services of Passion week and to represent those events of the life of Christ from which they took their rise, this form could no longer be considered adequate so soon as music had reached a higher development. Consequently, at the earliest period of composition for several voices, side by side with the old plain-song Passion, there arose settings of the Latin texts which were treated throughout in the motett style, and in part writing. This style was also adopted by the Protestant composers with the words of the German Bible. The oldest German Passion music of this kind known to me is by Johann Machold, published at Erfurt in 1593.[558] In his preface the composer refers to a Passion by Joachim von Burck "which had been written a few years previously," and had served him as a model; that, therefore, must also have been in the motett style.[559] Machold composed the Passion according to St. Matthew, which Burck had not done, and lived in the hope that it might sometimes be performed alternately with Burck's, so as "not always to fiddle on one string." From this it may be inferred that

---

[558] In the Royal University Library at Königsberg. It is in five parts; the alto part is wanting to this copy.

[559] Walther, Lexicon, p. 119, mentions a German Passion by Burck, printed at Erfurt in 1550. As Burck was born in 1540 or 1541 this cannot be accurate. If for 1550 we read 1590 this would agree with the music alluded to by Machold. A Passion by Burck is said to exist in the Rathsbibliothek at Löbau.

such settings of the Passion had an established place in the service as well as the plain-song Passion. The beginning consists, as in these, of the chorus announcing the subject, and at the end there is a hymn-subject on the words—

> O Jesus Christ, God's only Son,
> We humbly sue before Thy throne,
> That through Thy bitter cross and pain,
> Our souls may life and comfort gain.

No very high degree of artistic merit can be attributed to Machold's unpretending work; but more can be said for a St. John Passion in the same style, written in six parts by Christoph Demantius, and brought out at Freiberg in Saxony in 1631.[560] This also begins with the announcement "Hear the sufferings of our Lord Jesus Christ from the Gospel of John the Evangelist," which plainly indicates that Demantius contemplated not merely a setting of the Gospel narrative, but actually a musical remodelling of the old chanted Passion; the words "newly composed for six voices" inform us of this. The detailed character of the text naturally precluded the broad motett treatment; the narrative portions are rapidly got over, the speeches of the persons stand out in dramatic and animated phrases, in which we find repetitions both of the text and of musical subjects; still, solo singing is wholly excluded, and the introduction of persons is indicated by smaller groups of voices, the body of the chorus being divided. The end, here again, is a general reflection referring to the words in John xix., 35, "We believe, Lord; do Thou give increase of faith. Amen."

A Latin St. John Passion exists by Gallus, of 1587;[561] it is composed in the same manner, and the consideration of the works by Machold and Demantius naturally leads us back to it. The subject matter is divided into three sections, and the case is the same in the Passion by Demantius,

---

[560] A copy in the Church Library at Pirna; the bass part is unfortunately wanting.

[561] *Secundus tomus Musici operis Authore Jacobo Händl. Pragae, Anno* MDLXXXVII. There is a copy in the University Library at Königsberg. Compare Winterfeld, Joh. Gabrieli, Part II., p. 204.

indeed, the divisions occur at nearly the same passages; the first ends with the words of Christ "Why smitest Thou me?"; the second begins with the leading of Christ before Caiaphas and ends with the high priests' words "We have no king but Cæsar"; the third includes the crucifixion and death of Christ. It is to be observed that Machold's Passion is also in three sections, and, so far as the difference in the text allows, is divided at about the same places, so the assumption seems well founded that this was a form that had become established by custom; and though the number of German Passions in the motett form at present known to us is but small, we may, nevertheless, suppose that at one time they were in very general use, and the Passion by Demantius shows that this was not alone in the sixteenth century. It was the first method adopted for dealing musically with the history of the Passion, with all the means available at the period. As such it has an important historical interest, both as regards later church music and the oratorio.

As a middle stage between the chanted and motett forms of Passion music, we may mention those compositions in which the Evangelist's narrative and the speeches of Christ were alike recited in plain song, while all the rest were in parts. This form too—which was employed, among other Catholic composers, by Orlando Lasso and Jakob Reiner— found imitations among the Protestant musicians. To this class belong the well-known Passion by Bartholomäus Gese, written in the year 1588.[562]

Thus, we find three forms existing at the time when concerted church music found its way into Germany from Italy, and their elements are easily recognisable in the works of Heinrich Schütz, the greatest German Protestant composer of the seventeenth century. His "Seven words of Christ on the Cross" treats, it is true, only a section of the history of the Passion, still the work must be considered as belonging to that category.[563] Of the plain-chant Passion,

---

[562] Published by F. Commer. *Musica sacra*, Berlin. Vol. VI., p. 88.
[563] In parts in the Library at Cassel.

it retains the one-part recitation for the Evangelist and for the speeches of the persons; while of the motett type it preserves the use of the repetition of the four-part setting in the narrative portion; and like the Passions of the older type, has the devotional choruses at the beginning and end. A verse of the Protestant hymn "Da Jesus an dem Kreuze stund"—"When Jesus stood before the cross"—is used for the opening chorus, and another verse for the close, but only as having appropriate words, and without any reference to the tune. But what is new, and a result indeed of the concerted style, is the instrumental accompaniment, and the symphonies so full of meaning which occur after the first and before the last chorus; new too, and an outcome of the dramatic style, is the abandonment of the plain-song, which has given way to the recitative invented in about 1600.

But this work, which, as we see, combined such very different elements in a new form, had already had a predecessor, resembling it in artistic treatment though not in its materials. In the "Historie der fröhlichen und siegreichen Auferstehung unsers einigen Erlösers und Seligmachers Jesu Christi" (1623)—"History of the joyful and triumphant resurrection of our only Saviour and Benefactor, Jesus Christ"—the plain-song is still preserved, although it shows a marked disposition to pass over into the mode of the newly devised monologue. The first and last chorus are in the old manner, as are also a dramatic chorus of youths; the speeches of persons are some of them in several parts without any dramatic emotion, as in the Passions in motett form; while the figured bass accompaniment has sprung from the soil of the new style of music.

It is only at the first glance that the four Passions by Schütz seem to keep wholly within the limits of the old liturgical form. The Evangelist recites, while in the speeches of individuals or groups of persons single singers come in; there is the devotional chorus at the beginning and end, and no instrumental accompaniment. But on closer examination we perceive that, at any rate in the St. Matthew Passion, the separate singers no longer use the old plain-song, but actually sing the more modern recitative, and that the old style of writing is only apparently adhered to. A very limited amount of melodic movement is proper to the plain-song; it is only at the beginning and end of a phrase that melodic intervals are introduced, and these consist of a few frequently-recurring formulas, while the end of the phrase falls almost without exception on the key-note or the fifth. In those portions of Schütz's St. Matthew Passion which are intended to be recited, we detect an unexpected and varied melodic movement, which almost seems to demand a figured bass accompaniment, and in which, indeed, only a few turns remain to remind us of the plain-chant. It is no longer an equable sacred calm, but an eager personal emotion that pervades it. This is perceptible in the repetition of words and phrases, which is wholly incompatible with chanting. For instance, Judas sings—

In the St. Luke and St. John Passions we likewise find the modern recitative concealed under the mask of the plain-chant; still, it is more old-fashioned in character, and in the St. Mark Passion the old style of chanting is preserved in all its simplicity. The St. Matthew and St. John Passions again are not in the usual mode; the first is in a transposed Doric, the latter in the Phrygian. If we turn from the solo songs to the dramatic choruses, here again a vivacity and keenness of expression are to be remarked such as could only have become possible under the development of concerted music. In the St. Mark Passion, which is richest in passionate choruses, these are singularly contrasted with the unisonous monotone of the solo portions; while in the other works—and most completely in the St. Matthew Passion—the contrast is softened down by the old form of plain-chant being enlivened to the more modern monologue; even the choruses at the beginning display that free and bolder style of part progression which was introduced by the concerted style. At the opening of all four of these Passions the announcement is set to music, but the customary thanksgiving occurs only at the close of the St. Mark Passion, which thus adheres more closely to the older type; for the others Schütz has chosen verses of church hymns: for the St. Matthew Passion the last verse of "Ach wir armen Sünder"; for that according to St. Luke the ninth verse of "Da Jesus an dem Kreuze stund," with a slight variation in the text; and for that according to St. John the last verse of "Christus der uns Selig macht." But his aim in so doing was not—or was in a very small degree—to effect a closer alliance with congregational singing; it was in the last of these three instances only that he made use of the chorale melody, working it out as a motett. He set the other verses to original music, and we find the same in the Seven Words and in his other works. Church hymns at that time always afforded, next to Bible words, the most available text for German church music; but Schütz's efforts were already tending towards the introduction of the Italian madrigal style into Germany, and he hailed with delight Caspar Ziegler's first attempts of the kind.[564] In these Passions there is a peculiar mixture of old and new, but the new predominates, and this, as Schütz understood it, was not church music nor the church sentiment; it was partly sacred, but partly secular—the Oratorio.

In the farther course of the seventeenth century, the musical transformation of the German Passion music, and the gradual advance through its various phases of Protestant church music in general, continued to progress side by side at an equal pace. The vocal recitative preserved in all essentials the *arioso* character it had acquired in Schütz's hands; even in church cantatas down to 1700, recitative, properly speaking, does not occur. The instrumental accompaniment became universal; some modest attempts at

---

[564] See Vol. I., p. 469.

making it more independent are perceptible, and in suitable places short symphonies appear. In the choruses, which were by preference in five parts, the same feeble and etiolated character is perceptible, as we have seen in the cantatas—homophony preponderates; here and there we have short imitations, insignificant sub-divisions, and frequent ritornelles.

The characteristic form of this period is the sacred aria; it is now introduced into the Passion; I first find it used under its own name in a St. Luke Passion of 1683 by Funcke, Cantor of Lüneburg.[565] A Passion setting from Rudolstadt, of 1688, is already abundantly supplied with arias in several verses.[566] The simultaneous appearance of the same form of development at two places, tolerably far apart, allows us to conclude that the custom of inserting sacred songs into the Passion music had already been for some time widely spread. Among the arias we may include the "little hymn of thanksgiving for the bitter sufferings of Jesus Christ," which forms the close of the St. Matthew Passion, by Giovanni Sebastiani, written in 1672.[567] For although the *gratiarum actio* in a song form had long since ceased to be new, up to this time it had lacked the instrumental accompaniment peculiar to the aria. Besides this, in Sebastiani's work, a number of chorales are inserted "for the arousing of greater devoutness"; these, like the thanksgiving—all but its last verse—were to be performed by the treble only, with an accompaniment of four "deep viols" and figured bass; thus they were to be sung as arias. This is a new phase in the development of the German Passion music. Writers have hitherto been content to establish the fact without explaining it, for which reason we must dwell on it for a moment. The mode in which Sebastiani adopted the chorale into his Passion cannot of course have been original; it presupposes a stage of development in which the

---

[565] A copy of the text is in the Library of the Johanneum (school) at Lüneburg.

[566] A copy of the text is in the Royal Library at Sondershausen. On the title page is added "as in use for musical performance from day to day in Holy week," thus indicating a custom already some time established.

[567] A copy of this work, which, since Winterfeld's investigations, has been frequently mentioned, exists in the Royal Library at Königsberg.

chorales, here treated as arias, were sung by the whole congregation, as their very nature required that they should be. And the congregation did in fact participate in this way in the old chanted Passion. The regular order of the divine service provided, indeed, that the congregation should sing, both before and after the Passion, a hymn having reference to it; for the chanting of the Passion in Holy week took the place of the lessons from the Gospels read on Sundays. But this was not thought enough. Since the chanting was a long affair, in order to keep the interest of the congregation alive and to enhance the edifying effect, pauses were made at certain places where the assembled Christians struck in with a suitable hymn. Of this fact we have evidence, of which the value is rather increased than diminished by its dating from a period later than Sebastiani's Passion. For all that could avail to uphold true church music under the revolutionary movement which, in the beginning of the eighteenth century, threatened to annihilate it, undoubtedly rested on old and tried foundations. A little volume of Passions brought out in 1709, at Merseburg,[568] contains narratives of the Passion abridged from the four Gospels as they were at that time in use at Merseburg. It is immediately clear that these were sung in the old plain-song form, I might say in the oldest; for the thanksgiving is lacking from the end, and the introit, which is introduced into the St. Matthew Passion only, "Höret das Leiden"—"Hear ye the sufferings"—is not sung by the choir, but intoned by the Evangelist *choraliter*.[569] Arias are wholly excluded, but not chorales; these, however, are not printed in their place, but the first lines of the verses are given in brackets and usually there is an indication in the margin with the words "Here shall be sung verse * * of hymn * *." They mostly consist of one or more verses of Stockmann's hymn

---

[568] Auserlesene | Passions- | Gesänge, | wie auch | Die Historie vom blu- | tigen Leiden und Sterben | unsers Heylandes Christi | JEsu, | und wie solche von dem Chor, | nach dem *MATTHÆO, MARCO, LUCA* und *JOHANNE*, | abgesungen wird, Merseburg, 1709. In the Count's Library at Wernigerode.

[569] As it is in the St. Matthew Passion by Vopelius.

"Jesu Leiden, Pein und Tod"—"Jesu! Suffering, pain and death"—in which the whole history of the Passion is told in verse; in this way the congregation is enabled to follow the course of the events. But other hymns of five, six, seven, and even ten verses are indicated, which were to be sung at full length, and towards the end of the St. John Passion a hymn containing twenty-one verses "Nun giebt mein Jesus gute Nacht"—"And now the Saviour bids good night"—is ordered to be sung. Besides these chorales we also find in certain places one or two verses of hymns printed complete; possibly the chorus alone were to sing these for the sake of variety.

The same kind of interest was sometimes shown by the congregation in other ways. At the first performance of a Passion music, to words in madrigal form, in a town of Saxony, part of the people present sang quite calmly and devoutly in the first chorale, but presently expressed their displeased surprise at the rest being so different from what they were accustomed to.[570] Sebastiani himself published a small book in 1686, "Kurze Nachricht, wie die Passion . . . in einer recitirenden Harmonie abgehandelt und nebst den darin befindlichen Liedern gesungen wird" (A short account of how the Passion is to be treated in harmonised recitative and sung, together with the hymns to be found herein), from which we may conclude that the congregation were at liberty to join in the hymns.[571] And even as the Passion acquired an increasingly rich musical treatment the custom was here and there adhered to, and the congregation sang in the chorales.[572] But the more the sacred aria gained ground the less this practice could be kept up; Sebastiani's music is quite on the border line. His chorales no doubt are all of early origin, but the mode of their performance

---

[570] Gerber, Historie du Kirchen-Ceremonien in Sachsen, p. 284.
[571] To be seen in the Royal Library at Königsberg.
[572] Gerber, op. cit., p. 283. "As at this time folks have begun to perform even the Passion story—which was formerly so fine *de simplici et plano* and chanted simply and devoutly—with all sorts of instruments in the artistic manner, and sometimes to mix in a Passion hymn, so that the whole congregation sings with the instruments."

is no longer that of the ancient chorale. The whole character of chorale singing had assumed the aria type; all the new tunes, which were composed at this time in considerable numbers and of great beauty, were, in fact, arias, and no objection was made even to calling them so; as, for instance, Paul Gerhardt's chorale "Ein Lämmlein geht und trägt die Schuld"—"A white Lamb comes to bear our guilt"—to the old melody of "An Wasserflüssen Babylon."[573] Chorales for a soprano solo with an instrumental accompaniment occur not only in Sebastiani's Passion, but in the church cantatas of the period, as in those of Buxtehude; but here they appear in such a connection as entirely to exclude the idea of any co-operation on the part of the congregation. That this should be the case was but natural, since the aria is the expression of individual feeling, and thus, at last, in Seebach's Passion (to be fully described presently) even the separate *dramatis personæ*, "Jesus, John, the Virgin Mary, and the Bride of Christ," sing now in recitative, now arias, and now chorales.

It is undeniable that, in the last quarter of the seventeenth century, chorales play a considerable part in the Passion music. This, however, may not prove that the congregation had any more intimate share in it, nor that it had become more closely amalgamated with the service. On the contrary, like the introduction of the aria, it seems to prove a gradual evanescence of the animating spirit of church community, and it was the aria which helped the chorale to assert and hold its place in church music. Certain forms of music, in which formerly the congregation had taken enthusiastic part, they now were satisfied to hear sung, and could listen to them with calm sympathy. The Protestant chorale began at the same time to find in the organ the importance it was losing in vocal music.

Wherever the chorale is set in several parts in these Passion works, it is always in the simplest possible form. Though Eccard and Hassler set chorales in *contrapunctum*

---

[573] In the Rudolstädt Passion of 1688, at the beginning of *Actus* III., the indication stands "to begin with Paul Gerhard's *Aria*."

simplicem and gave the melody to the upper part, still the others had always some intelligent and tuneful progression, and there was an abundance of strong and original harmony. Now the most meagre progression was deemed sufficient, and the most commonplace harmonies. In the first instance this may have arisen from a desire to make it convenient to the congregation to join in. But that this was very soon lost sight of is proved by the parts as well as the printed texts; for the chorales—like the portions ascribed to the Jewish populace, the youths, and the high priest—have the simple indication *Chorus;* and soon the case was the same with the Passion music as with the church cantatas; for no composer ever took the co-operation of the congregation into consideration in writing a chorale, but, on the contrary, conceived of it in a spirit absolutely opposed to congregational singing.[574] The chorale now appeared in the place of the old announcement and thanksgiving, or if these were piously retained, there was at least one chorale into the bargain. As a thanksgiving "Nun wir danken dir von Herzen"—"From our hearts we thank the Saviour"—(the last verse from the hymn "Jesu meines Lebens Leben") was in almost universal use; as an introduction any appropriate hymn might be selected. The voices were supported by the instruments, which at most had some brief and insignificant solo interludes. Rarely we find a paraphrase of a hymn with an original setting, as at the close of a St. Matthew Passion, by J. C. Rothe, of 1697.[575]

The most perfect Passion music that remains to us in the style of the older church cantatas is no doubt that by Kuhnau, according to St. Mark. It was written for Good

---

[574] In the preface to a series of cantata texts which were printed in 1696 for the Hofcapelle at Gotha under the title: "Erbauliche Uebereinstimmung der Sonn- und Fest-Tags-Evangelien," and compiled by Witt (in the Library at Wernigerode), which consist chiefly of chorales and Bible texts, it is written "it is necessary too that those who sing should themselves devoutly ponder on what they sing, and that the hearers should not give heed to the sweet music only, but still more to the noble matter of the song."

[575] The autograph is in the Library of the Castle Capelle at Sondershausen. This paraphrase is founded on the usual verse "Nun ich danke dir."

Friday, 1721, and so about twenty years too late, but nevertheless it won the approbation of connoisseurs (see p. 336). The eighteen arias it includes are for one, two, three, and four voices. They all have the character of hymns in verse and, with one exception, have ritornels. This single exception is the four-part aria "O theures Blut, du dienst zum Leben"—"O precious Blood, of life the fountain"—in which the close affinity of the sacred aria to the true chorale, as it was at that time sung, is very conspicuous; there is nothing in it to make it less proper for congregational use than, for example, the hymn "O Traurigkeit, o Herzeleid" —"O bitter ruth and grief of heart." Besides these eighteen arias we find twenty chorales, the last of which is that just named, "O Traurigkeit"; however, it does not strictly speaking belong to the work, which ends with "Nun ich danke dir von Herzen." Kuhnau, who opened the performances of Passion music in part writing in Leipzig with his St. Mark Passions, evidently intended to bring them into closer connection with the customary services there, and this hymn had, as has been said (p. 274), its fixed place in the Good Friday service. For the same reason he inserted, after the alto aria which follows the words "And Jesus cried out and gave up the Ghost," the hymn "*Ecce quomodo*" of Gallus—a highly suggestive idea which no one seems to have had before him, at least I have not met with this motett in any other Passion music.

With the year 1700 began a period during which the theatrical music of the Italians became infused into church music in an ever spreading flow. Under the name "theatrical music" I include the Italian oratorio, which, however, was widely distinct from the opera, not only by the subject matter and a few broadly developed choruses, but by its presentment on a stage never having become an established custom. It was of course self-evident that the narrative of the Passion offered admirable material for a German oratorio on the Italian pattern, and Ch. F. Hunold at once set to work and wrote the words of the "Bleeding and dying Jesus" ("blutigen und sterbenden Jesus"), in which not the recital of the Evangelist alone, but the Bible words and even the chorale were dispensed with, and the whole material cast in one mould like an Italian oratorio. Keiser set this poem to music, and it was performed at Hamburg in Holy week, 1704. But it is an injustice to regard Hunold as the pioneer in this new line of work; he simply transferred to the Passion a process which had previously been tried with success in the church cantata. The gifted originator, therefore, of this new form was not Hunold but Neumeister, who had published his first cycle of "madrigal" cantata texts in 1700. As Hunold was his great admirer, and in 1707 published, without his leave, Neumeister's work "Die allerneueste Art zur reinen und galanten Poesie zu gelangen," there can be no doubt that he simply imitated him. Neumeister's first annual series—and indeed his second also, published in 1708—are quite independent alike of Bible texts and chorales, and are wholly original verse. We have seen what angry reprobation this practice incurred from various sides as a profanation of the church; and the same vehement antagonism arose against the new form of Passion writing, but without any result worth mentioning. It is indeed very true that, besides those in the Italian oratorio form, many Passion texts continued to be compiled, in which the recital of the Evangelist and the Bible words were retained as well as the chorales. But this could hardly be regarded as a general concession to the opponents of the new form. Still, among the poets of the time there were certainly not too many who could go even half way towards the satisfactory accomplishment of such a task. The impulse towards new texts of compositions of the narrative of the Passion proved a strong one, and it was partly for convenience sake only that music was set to the older form. It certainly was not respect for the words of the Bible, as something reverend and sacred, which prevented its utter deposition, or else the traditional arioso in solo singing might have been retained for Bible words, as was done in the church cantatas. No one will assert that the perfectly secular—not to say flippant—way in which most of the composers after this date set the deeply pathetic history of the Passion to a sort of recitative betrays any feeling for the sublimity of the language of the Bible.

On the other hand, both the Bible words and the chorale offered undoubted advantages to the musician; a regard for these induced Neumeister to make use of them again in the third and fourth series of his cantatas. Still, this combination of Bible words, chorales, and original verse could not henceforth have become the normal and immutable form if musicians had not recognised in it the ideal form of that period. For the time, however, the Italian oratorio remained the fundamental prototype of all who wrote and arranged Passion-texts, and who had any pretensions to be considered poets. Postel's Passion according to St. John, which Handel composed, still stood on the debatable ground, though it was written certainly not later than Hunold's work, and probably one or two years earlier; and a few years later Benjamin Neukirch and Johann Ulrich König still sailed in the wake of the Italians. Neukirch in his "Weinende Petrus"—"Weeping Peter"—treated, not the Passion properly speaking, but, as it were, a reflection of it. Peter and Judas Iscariot have both sinned against Christ: they are tortured by conscience, and Judas is driven by it to kill himself. Belial and spirits of hell would fain ensnare Peter too into despair, but the comforting words of John and of Mary Magdalene, as they remind him of the infinite love of God, give him fresh courage to suffer, to endure, and to conquer. This work is in fact nothing else than a sacred opera; it even falls into three acts, or incidents, like an opera; while the oratorio usually has but two sections, and each act has a different scene. Scenic effects are throughout taken into account; thus at the beginning we are told: " Peter goes, in melancholy thought, to a desert place, and presently begins—" and at the end of the first act, " Peter goes off sadly on one side, and Judas, full of despair, on the other." Besides these two, the following persons appear: Philip the younger, Zion, Belial, and the allegorical figures of Despair and Faith. The first act ends with a chorus of youths, the second with a chorus of the demons of hell, and the third with a chorus of angels and the righteous; the work consists of recitatives, arias, and one duet. There are no chorales; all the verse is original. When and by whom the text was composed I know not; on the title page it is said to be "for devotion on the Passion."[576] König called his poem honestly an oratorio, and farther bestowed on it the designation of "Tears under the cross of Jesus." We are told that these words were set by Keiser and performed in 1711, "on Monday, Tuesday, and Wednesday in Holy week, at vespers." The contents actually consist of the history of the Passion ; chorales are introduced, and in part the words of the Bible narrative are retained, though certainly in a very singular manner, for they are used as stage directions between the songs. For instance, after an aria sung by Mary the mother of Cleophas, it is added in brackets, "They that passed by blasphemed Him," whereupon, Mary goes on in recitative till soon after the words given to her are again interrupted in a similar manner :—

> Great Heaven! and can this really be,
> Can Christ while hanging on the tree
> Be mocked by all the scum of earth?
> (And one of the malefactors blasphemed him.)
> The malefactors now begin
> To blaspheme Him who knew no sin.
> (Then answered him the other.) &c.

These notes were intended for those who were following the performance in the book of words, supplying the place of dramatic action, and König considered this subordinate service quite good enough for the Bible words.[577]

---

[576] The "Weinende Petrus" must have been written in 1711 or 1712 at latest, for the third act alludes to the Emperor Joseph I. in such a way as to show that he must have lately died; Joseph I. died in the spring of 1711. The text was first printed as an appendix to the "Adachts Ubung | Zur Kirchen Music. | In Cantaten, Oden und | Arien. Franckfurt und Leipzig, 1721." It is to be found in the Count's Library at Wernigerode.

[577] "Theatralische, geistliche, vermischte und galante Gedichte von König." Hamburg and Leipzig, 1713, p. 307. "Thränen Unter dem Creutze JESU, In einem *ORATORIO* Montags, Dienstags und Mittwochs zur Vesper-Zeit In der stillen Woche Musicalisch aufgeführt. M.DCC.XI." In the Town Library of Leipzig. Winterfeld mentions a "verurtheilten und gekreuzigtens Jesu "— "Judgment and crucifixion of Jesus "—by Johann Ulrich König, as performed in Hamburg with Keiser's music in 1714. I have not seen this poem.

He found imitators; in 1719 Joachim Beccau brought out a poem of the Passion after the four Evangelists, in which all the events are inserted in this way; for instance:—

Jesus. Whom seek ye?
The Host. Him that is called Jesus.
Jesus. I am He. (And they fell to the ground.)

In Beccau, indeed, as may be seen by this quotation, the whole is dramatised. Peter, Jesus, Mary, and Judas sing arias; there is an animated dialogue between them, with choruses of various kinds. Besides this, there is a lyrical devotional song for Sulamith, who often takes an eager part in the proceedings and speaks between the speeches of the others. The piece begins with a *canzonetta* of the disciples at the last supper, and closes with a chorale of believers, "O hilf Christe, Gottes Sohn"—"Help, O Christ! thou Son of God."[578]

The Passion oratorio of Johann Georg Seebach, Erlebach's son-in-law, is on the whole constructed on the same lines; it was brought out in 1714.[579] The speeches of the actors are a paraphrase of the Gospel text; the narrative is partly given, as in König and Beccau, as a kind of stage direction, but in part is supposed to be known to the hearer. Besides this, the personages sing independent arias, as well as chorales—as if it mattered not which—and there are several scattered throughout the work. The action begins with the institution of the Lord's Supper, and the oratorio opens with a hymn of praise by the disciples, in the aria form, as in Beccau.

It is impossible not to see that Beccau and Seebach wrote under the influence of an older work, of which I have intentionally reserved all mention till this place. In 1712, Barthold Heinrich Brockes, a member of the Town Council

---

[578] *Beccau*, Zulässige Verkürtzung müssiger Stunden | Hamburg, 1719, p. 83. "Heilige Fastenlust | oder: das Leyden und Sterben unsers Herrn JEsu Christi | nach der Historie der Vier Evangelisten." Royal Public Library at Dresden. (Sacred Lenten Diversion on the sufferings and death of our Lord Jesus Christ.)

[579] Johann Georg Seebach, Der leidende und sterbende JESVS. . . . In einem *ORATORIO* und in geistlichen Liedern zur Erweckung heiliger Andacht ans Licht gestellet. Gotha, 1714. In the Library at Wernigerode.

of Hamburg, compiled what was, according to the views of
that time, a model text: "Jesus tortured and dying for the
sins of the world."[580] Keiser was the first to set it to music,
and it was performed in Holy week both in 1712 and 1713.
He was followed by Telemann and Handel in 1716, and
Mattheson in 1718; a remarkable setting of this much
admired work also exists by Stölzel,[581] and finally, as we
shall presently see, Sebastian Bach was not unacquainted
with it. What Brockes' capabilities were as a poet,
he subsequently showed in his "Irdisches Vergnügen
in Gott" ("Earthly joys in God"). The Passion text, it
cannot be denied, is skilfully arranged and well fitted for
music; but the turgid language, overloaded with glaring
imagery, by which the sublimity of the subject is sensibly
depreciated, makes the poem distasteful to us, though it was
precisely this which delighted his contemporaries. It has
long since been estimated at its true value,[582] and I may, therefore, restrict my observations to these few words. Brockes
adopted the independent oratorio form which had now for
some years been popular, but, intimidated perhaps by his
opponents, he did not want to abandon entirely the older
customs. He, therefore, retained the recital of the Evangelist,
but gave him a paraphrase of the Bible words—a kind of compromise which could not satisfy either party, and in which,
so far as I know, he found but one single imitator. The
admiration excited by this poem was founded on other
characteristics. It frequently occurred that a selection of
aria texts from this work was introduced into Biblical
Passions; a St. Mark Passion by Telemann, composed
before 1729, contains—besides the Gospel words, a number
of chorales, and a few arias by an unknown author—seven
passages derived from Brockes.[583]

Thus the German Passion had developed through the

---

[580] "Den für die Sünden der Welt gemarterten und sterbenden Jesus, aus
den vier Evangelisten in gebundener Rede vorgestellt."

[581] The autograph is in the Royal Library at Königsberg. Stölzel divided
his work into four parts, and the second and fourth are unfortunately wanting.

[582] By Winterfeld, Ev. Kir. III., p. 128; and Chrysander Händel, I., p. 429.

[583] This Passion is in my possession in a copy made in 1729 by J. P. Hasse.

influence of the Italian oratorio into a very singular compound. The most ancient and modern forms stood side by side; the simple and the ornate, the sacred and the secular were worked out in juxtaposition. It had become possible by these enhanced means of art to appeal to the emotions from the most opposite sides. But it had not yet been given to any man to co-ordinate and amalgamate this mass of elements from a high stand-point, not even to Handel; and by subsequently making use of all that was best in his Passion music for other works he showed that he was conscious of this. The chorale here plays the most perplexing part; its uses from the musical and poetical side alike were not to be undervalued. Seebach says very astutely in the preface to his Passion oratorio "a well-known and well-designed poem has in truth no small effect; it has a very lively one, and quickens the hidden heart of man like a refreshing balm. It is as though our faith took breath once more, our love grows fervent and our hope blossoms out in sacred admiration of God's glory, grace, mercy, long-suffering, and kindness. Yea, in so far as the poet is able to connect the chorale with his own thoughts, a well-composed oratorio or cantata will often wring tears and sighs from the heart of a child of God, particularly when the music is of equal weight with the words of the poetry." But it was just this—the musical treatment—which the composers could not succeed in. They had borrowed the chorale from the Passion form which had existed before 1700; and the simple guise in which they had found it there was suitable to the music among which it stood; but it no longer matched with the varied and passionate character of the new type of Passion. However, as the composers could make nothing else out of it they had to leave it as they found it. To stand still is to go back, and we are positively startled at the more and more wretched garb in which the noble figure of the chorale was condemned to appear in their works. Thus the original church element, which had existed in the modern Passion form, dwindled away; and the elements of oratorio it contained did not fare much better. When

MISUSE OF THE WORD *ORATORIO*.

Scheibe gives the name of *Oratorio* to Kuhnau's St. Mark Passion, which had its root in an older and very different period, this shows that he had hardly a shadow of an idea of what is indispensable to the oratorio, and yet he was one of the most keen and intelligent men of his time. The conception of the oratorio was so completely lost to the multitude that most people ceased even to understand the meaning of the name. Count Heinrich XII. of Reuss established at Schleitz a scheme of church music by which, on Sundays, first a chorale was sung, then the Gospel was read (or more probably intoned), then an aria was performed, and the whole concluded with a chorale; and the text compiled and printed for this purpose he called an *oratorio*.[584] Gottfried Behrndt published in 1731 a collection of sacred poems, some of which had, however, been written much earlier, and these were called on the title page "*Oratorien* so-called." These "so-called oratorios" are very simple church cantatas in Neumeister's style. But it is to be noted that Behrndt does not use the Latin word *Oratorium* or the Italian *Oratorio*; he connects its etymology with the Orator—*i.e.*, the speaker or preacher.[585] Accordingly, the Passion necessarily degenerated into a religious cantata, although it long retained, superficially, much in common with the older and purer form. "The death of Jesus" ("Der Tod Jesu") by Ramler and Graun is such a cantata; this was written in 1756, and must be regarded as affording

---

[584] *ORATORIUM* | Welches, | nach Anleitung | derer Sonn- und Fest- | tägigen | Evangelien, | zu Erweckung | einer Christlichen Andacht, | aufgeführt wird | In der Schloss-Capelle | zu Schleitz. | Schleitz, *s. a.* Auf der gräflichen Bibliothek zu Wernigerode.

[585] Gottfried Behrndt, *Zitt. Lusat.*, Poetische Sonn- und Fest-Tags- | Betrachtungen | über die verordneten Evangelien | durch das gantze Jahr, | In so genandten Oratorien | bestehend. Magdeburg, 1731. In the Library at Wernigerode. In explanation of his misuse of the term, one more circumstance may be mentioned, namely, that works of the oratorio class were at that time not unfrequently performed on the regular school speech-days by the choirs of pupils. This was the case, for instance, with a poem and music by Constantin Bellermann, "Die himmlischen Heerschaaren," performed in 1726 at the Gymnasium at Göttingen (see Marpurg, Kritische Briefe, Vol. III., p. 13). Heiland speaks of similar performances in the programme of the Wiemar college, 1858. P. 16.

a fair standard of the Holy week music of the second half of the eighteenth century. The chorales it contains, the misapplied survivals of an earlier period, might still have some edifying effect, but their importance as church music is wholly lost. In 1759 Doles conducted a performance of an original Passion on Ramler's text in St. Thomas' church at Leipzig; I even fancy that a few portions of Graun's music were transcribed bodily into it.[586] In 1766 the primitive chorale Passions were abolished once for all, and with reason. When Graun's style of art began to be pre-eminent all true intelligence must have vanished to the last trace.

From the very first the old church form of the Passion had contained a dramatic germ capable of development. In the middle ages the Passion miracle play had grown out of it, without its ceasing to exist in its simple identity. The reading of the Gospel with distributed parts—and not of the history of the Passion only—survived the ruin which overtook the so-called "mysteries" of the sixteenth century, and had preserved enough of their characteristics to bring them to a second bloom, so far as was possible under totally altered circumstances. It is evident, from the Latin text of Carissimi's oratorios, and still more from the *Historicus* (the Historian or Evangelist) who appears in them, that the semi-dramatic musical style of the Italians in the seventeenth century did not remain outside all connection with it. But the attempts of a similar character made in Germany in the seventeenth century must not be regarded in this connection as an imitation of the Italians. It is quite certain that the Italian oratorio was not unfrequently performed with scenery and action; but long before any Italian oratorio existed at all, performances of the Passion with action had been customary in the churches of Germany. I do not of course refer to the medieval Passion plays—though they too originally had a liturgical purpose, and were performed in the churches until they freed themselves from ecclesiastical control, and then rapidly became secularised and degenerate—but to the simple Gospel

---

[586] The text of this performance is preserved in the Library at Wernigerode.

readings with distributed parts. It is a fact that has only recently become known, that dramatic representations of the Passion took place in Protestant neighbourhoods after the Reformation. Such a performance took place on the fifth Sunday in Lent, 1571, at Zittau. The composition was by a certain Paurbach, and had long been in existence. Ten years previously, Christoph Bornmann, tavern-keeper of the "Freiberg cellar" at Dresden, had made it a present to the Church of St. John at Zittau. The stage was erected near the altar, and the actors were the three lowest masters of the school and two treble choristers.[587] It is not stated whether the performance took place before or after divine service; but this is of no importance. The real interest of the matter lies in this, that evidently the feeling still survived of the connection between the old dramatised Gospel readings and the Passion play, for it is only under this supposition that such an attempt is explicable.

Because the sacred plays of the people during the sixteenth and seventeenth centuries were of but small importance in the development of dramatic art in Germany, but little attention has been paid to them, with the single exception of the Oberammergau play. And yet a due regard to them is indispensable to a complete comprehension of the German Passion music and its allied forms; and even the first beginnings of the German opera must undoubtedly be referred to it.[588] In Thuringia, Upper Saxony, and Silesia the popular sacred drama throughout the seventeenth century was universally delighted in, even when its progress was checked by the scholastic drama. It was not merely venerated as a piece of respectable antiquity, and allowed as such to subsist, but it grew and lent itself to the needs of the time. A Christmas drama from Arnstadt in Thuringia, of which I possess a copy, made in 1700, reproduces the general outlines common to most of the popular Christmas

---

[587] Reinhold Zöllner, Das deutsche Kirchenlied in der Oberlausitz, Dresden, 1871, p. 31.

[588] See Chrysander, Händel, I., p. 78, and Weinhold, Weihnacht-Spiele und Lieder, p. 290.

dramas, but betrays itself in language, versification, and music as a new arrangement of well-known motives which cannot have originated long before 1700; and yet its essentially popular character is plainly to be seen from the fact that two peasants and two shepherds who figure in it speak the Thuringian dialect.[589] In the same way, in the Zuckmantler Passion play, we have one of the re-arrangements of an older work which were made in the seventeenth century; this is even more clear from the music than from the language. There are seven arias altogether bearing the stamp and feeling common in the middle of that century.[590] No doubt, however, it was not so much the specially dramatic element, as a general popular tone and treatment, that the Passion music and its allied forms at their most flourishing period had borrowed from the contemporary sacred drama. A time when the creative instinct, which flooded and obliterated every other artistic tendency, was so purely musical was especially unfit for the development of a new type of dramatic poetry; such germs of it as could assert their existence could, for a time, only expand in Handel's oratorios and Bach's church music—to forms indeed of transcendent beauty, but to the total exclusion of scenic representation and with a general tendency to resolve everything into lyric forms.

In short, the position held by the medieval "Mystery" in the life of the people was not very different from that of the Passion music of the seventeenth and eighteenth centuries—making due allowance for the progress and changes that had meanwhile taken place in art. In both we see an obvious dependence on the Church; each had for its principal subject the Life of Christ, and was performed at fixed church festivals. In both, the liturgical and sternly ecclesiastical element acquired by degrees a greater admixture of the secular. Just as in the "Mysteries" we trace, first the explanation of the Latin Bible and sacred

---

[589] In a volume of collected pieces in the Ministerial Library at Sondershausen.

[590] Zuckmantler Passionsspiel, edited and elucidated by Anton Peter. Troppau, 1868-9.

words by a context in German, then the interpolation of independent verses, and, finally, an entire suppression of the church element and complete secularisation; so in the Passion music we first find the interpreting congregational hymn side by side with the Gospel narrative, then the introduction of the independent sacred aria, and by degrees of the various richer forms of modern music; till at last the church element is felt as a disturbing one, and is as far as possible set aside. But the latter process of development is distinct from the former; for a man had now appeared on the scene, capable at the right moment of giving worthy utterance at once to the sacred and the secular feeling, and of restoring the church element to all its dignity, without in any way detracting from the fulness, splendour, and richness of the secular adjuncts.

It has been said above that from the seventeenth century onwards, the development of the German Passion performances ran parallel with that of Protestant church music, and this is equally true as applied to Bach: for the musical style of his Passion music is identical with that of his church cantatas. But this was new altogether, evolved from organ music and the chorale, and hence truly church-like; while in it, at the same time, all the musical forms of the period were purified and re-invigorated. By transferring them to the Passion, he achieved the amalgamation and natural coalescence of all the heterogeneous elements which in the lapse of time had become agglomerated under that name. At the same time, the style was eminently German; for if ever anything was a national product, the organ music of the seventeenth and eighteenth centuries and the popular hymns of the reformed faith were essentially German. Thus, when we find that the influence excited by the German sacred plays on the Passion performances and other allied forms was, properly speaking, only fully revealed in Bach, the internal process and connection are very intelligible. The more his contemporaries ran after the Italian opera and Italian oratorio, the farther they found themselves from the national spirit; Bach would not yield to the momentum given by foreign art, though he did not ignore it; but the point from which he started and the work he produced were invariably conceived from the essentially German views of art. The ineradicable hold these had on his mind is the evident result of his inheritance from a long race of artists, and here again we see how important to a thorough comprehension of Bach's individuality is a just estimate of his ancestry. None but a man who could take his stand on a tradition founded in an unbroken intimacy with the life of the people, and in whom their feelings, deeds, and thoughts had become an innate and inseparable part of his being, could, at such a period of general ferment as the beginning of the eighteenth century, have restored to a form of art so vacuous as the Passion had become all the dignity and mastery of the German genius. Bach's Passions and kindred works are indeed a revival of the medieval sacred drama of the best period, but on an immeasurably higher level—nay, it may be said they are the very culmination and crown of their kind. Bach was careful not to attribute the title of oratorio, at any rate, to his Passion music. His compositions for Christmas, Easter, and Ascension-tides have, it is true, kept that name—in fact, at that time there was no other or more suitable term in use; I shall venture in the following pages to revert to the more comprehensive designation of "mysteries," alike for the Passions and for these three oratorios.

According to the Necrology in Mizler (p. 168), Bach left five compositions of the Passion, and we cannot doubt the statement of this, our best authority, because the number corresponds with that of the annual series of cantatas (see ante p. 348). After his death, his sons, Friedemann and Emanuel, divided these cantatas between them, and the Passions were no doubt included. Emanuel had the original scores of the St. John and the St. Matthew Passions. He treasured them faithfully and they still exist. The original manuscript of the other three fell into the hands of the dissipated Friedemann, who now grew wilder than ever; they were sold for a trifle, and two have entirely disappeared. A St. Luke Passion in Bach's handwriting may be the third;

but whether this is a genuine work by him is a question still unsolved. I hope to bring it a little nearer to a solution, but first I have a few words to offer as to the lost settings of the Passion.

On Good Friday, 1731, a Passion music was performed in St. Thomas' Church on the text of the Gospel of St. Mark; Picander, who in 1729 had compiled the St. Matthew Passion text for Bach, had also written this one.[591] It is in two parts, the first being sung before the sermon and the second after. Besides the narrative from chapters xiv. and xv. of the Gospel, which was sung in recitative and which includes twelve dramatic choruses, this text contains a lyric chorus at the beginning and end, six arias, and sixteen chorales. Now it is almost inevitable that, if Picander wrote this text for St. Thomas', and if it were really printed and set to music, Bach must have written the music; this is nearly certain when we consider Bach's official position and his relations with Picander. And the hypothesis is raised to a certainty by the identity we can trace between the text of the opening and closing choruses, and of the second, third, and fourth arias, with the corresponding choruses, and the three arias in the funeral ode for Queen Christiana Eberhardine (1727). Although the feeling and purport of those is different, there is a similarity in the metrical structure which leaves no room for doubt that Picander adapted his text to the music already composed by Bach; and this is particularly conspicuous in the last line of the final chorus.[592] Picander has here exercised his skill, which was often turned to similar account, in the most commendable way. Hence we may regard the musical portion of the St. Mark Passion as not wholly lost, since five lyrical pieces of it are preserved in the funeral ode,

---

[591] It is to be found in the third part of his poems, p. 49, with the title: "*TEXTE* Zur Passions-*Music* nach dem Evangelisten Marco am Char-Freytage, 1731."

[592] Rust deserves the credit of first detecting this resemblance (see B.-G., XX.,² p. 8). In Breitkopft's list for the New Year, 1764, p. 18, we find "*Anonymo*, Passions-Cantate, *secundum Marcum*. Geh. Jesu, geh zu deiner Pein." The beginning of the text is the same as in the mourning ode, and so is the music down to the employment of two viol-di-gamba and a lute.

though this is indeed but a meagre substitute for a complete work rich in recitatives, dramatic choruses, and chorales.

Besides this there exists a third Passion text by Picander, hitherto unknown: the first in point of order, since it was written for Good Friday, 1725.[593] A remark made some pages back applies to this work; it is constructed entirely on Brockes' model, and, so far as I know, Picander in this stands alone. The Bible narrative is brought into the madrigal form, and is to be recited by the Evangelist. The *dramatis personæ* are John, Peter, Jesus, and Mary. Still, Peter has rather the more important part; that of Jesus is, by comparison, meagrely worked out, and of all the important and pathetic situations in which the Gospel story represents Him, hardly one is done justice to—indeed, the rhymed narrative of the Evangelist touches on the most important events with a superficiality that is hardly credible. For instance, all the incidents between the seizure of Peter on the Mount of Olives and his denial of Christ are summed up in the feeble line, "And Jesus went out calmly, and came to the high priest, Caiaphas." There are no dramatic choruses whatever, while, on the other hand, the lyrical reflections given to the allegorical figures of Zion and the Soul occupy a wide space. There are only two chorales introduced, fewer therefore than in Brockes' text; the second is Picander's original composition, and vapid enough. Neither of them are skilfully brought in; he calls the whole thing an *oratorium*. A comparison with the prototype is very unfavourable to Picander. In that of Brockes we cannot but recognise that all the musical factors are skilfully brought out, and the separate incidents depicted with vivid, though too often painfully glaring, colours; the flat lyrical monotony of Picander's poem closely borders on the religious Passion cantata of a later date. It is noteworthy, though from a quite different point of view, how many resemblances it bears to the text of the St. Matthew Passion; the words of the closing chorus are almost identical in both.

---

[593] Sammlung Erbaulicher Gedancken über und auf die gewöhnlichen Sonn- und Fest-Tage. Leipzig, 1725. For the complete text in German the reader is referred to the original German of the present work, Vol. II., App. B., x. 1.

Now, did Bach compose the music for this text? It is not possible to give any certain answer, but it is not improbable, if we suppose that it was intended to be set to music. The composers in Leipzig at the time who might have done it were Görner, Schott, and Bach. Görner, at any rate, is out of the question, since he did not institute Good Friday performances till 1728, and as regards Schott, so far as is known, Picander had never any connection with him; but it can be proved that he had written other texts for Bach in the same year as this very Passion. We must also pay due regard to the following remarkable circumstance: Bach's St. John Passion was probably projected during the last months of his residence at Cöthen, and was certainly first performed on Good Friday, 1724, as will presently be amply proved. It began originally with that grandiose chorus "O Mensch, bewein dein Sünde gross"—"O man, thy heavy sin lament"—which Bach subsequently placed at the end of the first part of the St. Matthew Passion when he passed it under revision in 1740. It has hitherto been supposed by those who have paid any attention to the manifold uses made of that chorus that it was transferred direct from the St. John to the St. Matthew Passion, but a closer examination of the MS. materials show that this is not the case; on the contrary, Bach had already had the St. John Passion performed in 1727, with the introductory chorus as it now exists, and the chorus "O man, thy heavy sin," had therefore been removed at a time when the St. Matthew Passion (written in 1729) had not yet been thought of. It is certain that when he changed the chorus, which remains one of the grandest of his compositions, it was not because it appeared to him unsuitable, but because it asserted its claim to some other use. The chorale "O man, thy heavy sin," is a Passion hymn, and can have found its use in the church only in the Holy week. Now the only concerted church music performed at that season was the Passion, at vespers on Good Friday; hence this chorale chorus must necessarily have found its place here, and consequently Bach must have written another Passion between the years 1724 and 1727, in which it could be introduced. This hypothesis coincides very well with the date of Picander's first poem, written in 1725. It is quite conceivable that Bach, dissatisfied with the meagre amount of sacred purpose in the poem, wished to give it a more characteristic style by opening it with this masterpiece of chorale writing. In this case he may simply have set aside Picander's text for the chorus, "Sammelt euch, getreue Seelen"—"Gather together, true believers"—still, it was not unusual in the Passions to allow a composition on a madrigal text to follow immediately after the chorale at the beginning, and in the same way a funeral hymn, set as a choral aria, not unfrequently preceded the closing chorale at the end. The fact that the poetical form of this Passion is unlike that of Bach's other Passions need not rouse our suspicions. Certainly, Brockes' type of Passion, with its paraphrases and modernising of the Bible words, was not Bach's ideal; but that he could nevertheless condescend to it, is proved by his Christmas oratorio. If we remember, moreover, that it was precisely in his first year at Leipzig that, to please his public, he now and then lent himself to the Hamburg vein of writing, many things combine to render it probable that Bach supplied the music to Picander's text of 1725, and that we may consider this to have been one of the lost settings of the Passion.

If I here venture to interrupt the course of my narrative with some critical observations, I only do so on the assumption that the question as to the number of Bach's settings of the Passion, their date and fate, is one of very general interest. Four can be proved with more or less certainty, and the fifth may be the St. Luke Passion above mentioned, of which, however, the genuineness has hitherto been doubted. I shall here take occasion to discuss this more fully. I have already said that a MS. of this music exists in Bach's writing.[594] It bears no express notice that it is his own composition, but on the other hand we find on the title the letters J. J. (*Jesu Juva*), which Bach was accustomed to add only to his own works, and not to copies from those of other composers. Besides

---

[594] In the possession of Herr Joseph Hauser, private singer to the Grand Duke at Carlsruhe. See App. A., No. 54.

this, in a list of MS. musical works which was printed by Immanuel Breitkopf at Michaelmas, 1761, on page 25 we find "Bach, J. S. Capellmeisters und Musicdirectors in Leipzig, Passion unsers Herrn Jesu Christi, nach dem Evangelisten Lucas, *à 2 Traversi, 2 Oboi, Taille, Bassono, 2 Violini, Viola, 5 Voci ed Organo.*" The list of instruments corresponds exactly with those in the Passion under discussion, but not the vocal parts, as there are nowhere more than four; but any one who had occasion to note the numerous errors in Breitkopf's list[595] will find no difficulty in thinking that 5 may be a misprint for 4. A doubt as to the authenticity of the work can hardly be founded on these details; but the music itself is strange and puzzling: its very simple forms reveal a tender and soft expressiveness, but it is far away from the power, fervency, and solemn grandeur of the St. John and St. Matthew Passions. Judging from this alone we should be disinclined to regard the St. Luke Passion as genuine. But we have become familiar with a mass of works of Bach's youth, and if we compare the St. Luke Passion with them it appears in quite another light. Although the score which exists was undoubtedly written at Leipzig, nothing compels us to assume that it was composed there. In Weimar, where he displayed no small industry as a composer of cantatas, he also occupied himself greatly with the whole class of Passion Music. He wrote out at Weimar, with his own hand, the parts of a St. Mark Passion by Keiser "Jesus Christus ist um unsrer Missethat willen verwundet"—"He was wounded for our transgressions"—and must therefore have had it performed there.[596] We must unquestionably assign the St. Luke Passion to the first half of the Weimar period. Though in many respects it is evidently inferior to the cantatas "Nach dir, Herr," "Aus der Tiefe rufe ich," and "Gottes Zeit," we must consider that in this, his earliest work of the kind, Bach would naturally cling closely to the type of Passion

---

[595] Thus, for instance, on p. 20 in the cantata "O Wunderkraft der Liebe" ("Herr Christ der einge Gottessohn," B.-G., XXII., No. 96, P. 2142), we find "*à* 3 *Voci*" instead of "*à* 4 *Voci*."

[596] See App. A., No. 55.

music prevalent at the time, while in the province of the cantata he could already move with freedom. We cannot justly estimate the St. Luke Passion unless we assign it a place about half-way between these and his earlier Weimar cantatas. Then its weak points will seem explicable, and our comprehension will be opened to its more important features.

As regards the text, the early date of its composition is stamped on the face of it. The independent lyric pieces in it are but eight; two arias for the soprano, one for the alto, three for the tenor, an important opening chorus, and a chorus for the women who followed Christ to His Crucifixion. The Gospel narrative, Luke xxii. and xxiii. to v. 53, is all brought in, and no less than thirty-one chorales are introduced. Such a lavish use of chorales almost always occurs in the older Passions of central Germany which had not yet come within the influence of the Italian oratorio, and it was subsequently adhered to in many places where the new style found no favour. In the Rudolstadt Passion of 1729 there are twenty-eight chorales; in the Gera setting twenty-five; in the Gotha Passion of 1707 nineteen; in the Schleiz Passion of 1729 twenty-seven; in the Weissenfels Passion of 1733 thirty; Seebach, in 1714 —who was strongly influenced by Brockes—inserted no less than forty-nine in his setting. Original verses stand in inverse proportion: sometimes they are altogether wanting, but as time went on they increased in numbers, and finally supplanted the chorale altogether. Bach's St. Luke Passion, which is but slenderly provided with lyric verse, still shows in places clear traces of Italian influence; as when, after the words "And there followed Him a great company of people and of women," a chorus of soprani and alti sing :—

Weh und Schmerz in dem Gebären   Our pangs and woe in travail
Ist nichts gegen deine Noth.   Are as nought beside Thy pain.

Here the women who are brought in as the speakers have original words to sing, while all the rest of the dramatic music is set to Bible words alone. From these tokens we may assign the text to about the year 1710. Certain important Passion hymns recur naturally with more or less regularity in every setting; and besides this, in certain neighbourhoods some chorales seem to have been particular favourites, with passages from the Litany and the *Te Deum*. In the Rudolstadt Passion, after the narrative of Judas hanging himself, come the lines from the Litany "From the crafts and assaults of the devil, from murder and sudden death, and from everlasting damnation, good Lord deliver us." Farther on, after the passage telling how John took home with him the Mother of Jesus, come the lines "To defend and provide for the fatherless children and widows, We beseech Thee to hear us, good Lord! To have mercy upon all men, We beseech Thee to hear us, good Lord!" Again, after Pilate's question "What is truth?" the words "Thou art the King of glory, O Christ!" and two other suitable verses are quoted from the *Te Deum*. In the Gotha Passion according to St. Matthew, of 1707, after the history of the crucifixion, ch. xxvii., v. 38, the chorus comes in with "By Thy cross and passion, in our utmost need, help us, O Lord God!" In the same way in this St. Luke Passion, in two places we find lines from the Litany, and three times words from the *Te Deum*—the last, it may be observed in passing, exactly to the same melodic setting as exists in the organ arrangement of that hymn by Bach, which is still extant.

I have not succeeded in discovering who can have been the writer of the lyrical portions of this work; it was not Franck, in spite of some resemblance to the poems in his sacred and secular verse. Texts of such uniform metre and shallow purpose were not in his line. Still, there is a recognisable resemblance between these and the aria texts of the cantata "Nach dir, Herr, verlanget mich," which also belongs to Bach's early Weimar period.

Any one who is familiar with the glowing style of melody which characterises Bach's recitatives will at once recognise it in those of the St. Luke Passion. It cannot be denied that the harmonic sequences, above which the recitative is carried on, are sometimes rather loose and halting, and instead of the usual final cadence proper to the recitative, an arioso close occurs more often than usual, without the melody having assumed the arioso form early enough to confirm it. We get an impression that it is the work of a composer who has had but little practice in writing recitatives; and this agrees with the facts of the case, for in the earlier church cantatas to which Bach had hitherto devoted himself there is, as we know, no recitative. The style of the Biblical dramatic choruses is equally unskilful. In a few we still recognise that more general church-feeling which prevails in the Passions of the seventeenth century; others—indeed the greater number—are marked by a dramatic vividness which, it is true, does not rise to the mark of the choruses in the St. John and St. Matthew Passions, but still is conspicuously higher than that of contemporary composers. The chorus of lamenting women set to the poem above quoted is not very important, but it is well thought out, and its peculiar instrumentation—two flutes, violins, violas, and no bass—gives promise of later masterpieces, such as the soprano arias in the Ascension Oratorio, and the cantata "Herr, gehe nicht ins Gericht"—"Enter not into judgment, O Lord." The opening chorus, too, reminds us of this cantata; of course no one will expect to find in it anything like the stupendous chorale figures of the later Passions, but those who find much to admire in the four-part aria of the cantata "Denn du wirst meine Seele"—"Thou shalt not leave my soul in hell"—and the C major chorus of "Uns ist ein Kind geboren"—"Unto us a child is born"—cannot hesitate to accept this also.

The *Da capo* form predominates both in the two choruses and in the arias; the length of the pieces is small, but not smaller than in the cantatas "Nach dir, Herr, verlanget mich," or "Uns ist ein Kind geboren." The first two, it is true, show small trace of Bach's essential characteristics, they rather remind us of Handel's earliest work; and the soprano aria "Selbst der Bau der Welt erschüttert"—"Nay, the earth itself is quaking"—must be called meagre and insignificant. The rest of the arias, on the other hand, are so full of power and individuality that no one but Bach can be named who could have written them. Particularly noteworthy is the use here made of the bassoon. In his Weimar period Bach showed a special predilection for this instrument; it is possible that he had at the time a remarkably good player at his command. In the cantata "Nach dir, Herr," it is introduced obbligato, and again later, more artistically, in " Mein Gott wie lang, ach lange." The chorales throughout are more simply harmonised than we are accustomed to find them in Bach. But, do we not find the utmost harmonic simplicity in the closing chorale of " Denn du wirst meine Seele"? and it cannot be denied that the treatment is careful. Moreover, Bach's deep sense of fitness is so unmistakably revealed in the way in which they are introduced that in view of this alone all remaining doubts as to the authenticity of the work must surely vanish.

In the St. Luke Passion Johann Flittner's hymn " Jesu meines Herzens Freud "—" Jesus, Thou my heart's delight " —constitutes, as it were, the focus of the church sentiment, recurring in the most important situations in fresh developments, as the melody "O Haupt voll Blut"—"O Thou whose head was wounded"—does in the St. Matthew Passion; and Stockmann's hymn "Jesu Leiden, Pein und Tod"—"Jesus, suffering pain and death"—in the St. John Passion. It recurs no less than four times—first, after Christ's words, "Where is the guest-chamber where I may eat the Passover with my disciples?" xxii., 11 (with the third verse " Weide mir "—" Pasture me "); then after the words " With desire I have desired to eat this Passover with you before I suffer" (with the fourth verse "Nichts ist lieblicher"—" Nothing fairer is than thine"); again, after the scornful address of the soldier, "If Thou be the King of the Jews, save Thyself!" (with the fifth verse "Ich bin krank"— "I am sick, O heal thou me"); and lastly, after the prayer of the repentant thief, "Lord, remember me when Thou comest into Thy kingdom" (with the second verse "Tausendmal gedenk ich dein"—"Often have I cared for thee"). The tender grace of this hymn, which can hardly be termed a chorale, is characteristic of the spirit of the Passion music of the time,[597] and particularly of this St. Luke Passion. It was known and loved in many places, but it can never have been a church hymn, properly speaking, and in general use, for the reason that hardly any tune underwent so much variation as did this one. It was originally written in the minor, but was soon transposed into the major key. Each of these forms, which held their place side by side in many parts of Germany, shows many variations in the details both of rhythm and melody.[598] In the St. Luke Passion Bach used only the major tune, and, as we might expect, has given it a different form each of the four times he has introduced it. But, whereas in his later works he was wont to alter the harmony only, here the changes are chiefly in the melody and rhythm. This is simply explained by the early date of the work; when he composed the St. Luke Passion, Bach was not yet wholly free from the influence of those organ composers—principally the North Germans—who, with their taste for arbitrary ornamentations, obeyed their own immediate musical promptings rather than the permanent needs of the congregation.[599] It is not merely in the tune of this hymn of Flittner that melodic and rhythmic alterations of the chorale are to be found. The Passion hymn "Herzliebster Jesu"—"Jesus, beloved"—occurs in the second part of the St. Luke Passion in common time and its usual melodic form, while, in the first part, it is in 3-4 time and has two alterations in the tune. Arbitrary deviations from the normal tunes occur in other chorales in the same work. The deeper Bach penetrated into the church feeling and

---

[597] Two verses of it are also used in the Rudolstädt Passion of 1688.

[598] Dretzel, Des Evangelischen Zions Musicalische Harmonie, p. 316; Nürnberg, 1731; gives three forms of the major and two of the minor. In Mühlhausen the minor form was in use in Joh. Rudolf Ahle's time, Winterfeld, Ev. Kir. II., p. 467, and at Gotha, as in Witt's Cantional, 1715, p. 203, at Sondershausen, as we see from Gerber's Chorale book, which was never printed, but exists in my possession in a MS. of 1745. Freylinghausen's hymn book, on the other hand, in the Halle edition of 1741 has the major tune; but it is not to be found in the earlier edition of 1710. Bach himself has the minor melody under No. 696 of the Naumburg-Zeitzer hymn book (known as Schemel's) of 1736, and in his Choralgesängen, Vol. III., 264, of 1786, the major tune. In no two of these editions do the tunes exactly agree.

[599] See Vol. I., pp. 312 and 592.

value of the chorale, the more strictly he adhered to that form of the melody which he had once adopted; and then it was only in exceptional cases that he allowed himself to alter a single note of the *Cantus firmus*. Such an exception occurs in the seventh line of the seventh stanza of his chorale cantata "Christ lag in Todesbanden."[600]

In reviewing the instances in which Bach has introduced chorales into the St. Luke Passion, we are first struck by the close of the first part: Peter has denied his Lord, he goes out and weeps bitterly; an important tenor aria prolongs the emotion, and then, as a closing chorale, comes the sixth verse of Schwämmlein's hymn "Aus der Tiefe rufe ich"—"From the depths I cry, O Lord!" But it is not in four parts; it is for one voice only and is sung by Peter. This is evidently a deeply considered feature and full of sentiment. But here again we recognise the connection between Bach's work and the Passions composed on the pattern of the early church cantatas; it reminds us of a similar passage in the Passion composed by Sebastiani of Weimar. The subjective character is rendered prominent by the free transformation of the two last lines of the text. Here we see the hand of the poet who compiled the text for Bach, and to him we must also refer a few other chorale texts which belong to no well-known hymns. Bach had a farther opportunity of revealing his deep religious feeling in a still more striking manner, in the course of the story of Peter's denial—in the passage where we are told how that Peter, who had previously avowed himself ready to follow his Lord to prison and to death, followed Jesus into the High Priest's palace; here we are to discover whether he is strong enough to keep his pledge. At this point the chorus sings in unison, and in the "Collect" tone, the words of the Lord's prayer "And lead us not into temptatoin, but deliver us from evil." The solemn and impressive effect of this short movement is intensified when we turn to the place in the Gospel of St. Mark xiv., 38, which Bach must

---

[600] B.-G., I., p. 124., P. 1196. The expressive Melismata with which Bach loved to ornament his chorales of course do not here come into the question, since they neither change nor conceal the fundamental form of the melody.

certainly have had in his mind: Christ, praying in the garden of Gethsemane, finds His disciples asleep; He turns first to Peter—"Watch and pray," He says, "lest ye enter into temptation. The spirit is willing, but the flesh is weak." But the most deeply felt adaptation of the chorale Bach has reserved till near the end of the work. The Evangelist sings "And having said thus, He gave up the ghost." Here we have, in four parts, the chorale " Ich hab mein Sach Gott heimgestellt "—" All care for me I leave to God "—performed only by wind instruments, oboes and bassoons. After this symphony is ended the chorus sings, still in four parts, the twelfth verse of the same chorale :—

> Derselbe mein Herr Jesu Christ
> Für all mein Sünd gestorben ist,
> Und auferstanden mir zu gut,
> Der Höllen Gluth
> Gelöscht mit seinem theuren Blut.

> My Saviour, Christ the crucified,
> To purge my sin has bled and died;
> And He is risen for my good.
> Hell's fiery flood
> Is slaked by His most precious blood.

It is this which first suggests the real meaning of the instrumental subject which is repeated after the singing. Then the narrative goes on as far as the descent from the cross, interrupted once more by a chorale. A tenor aria, full of the stricken fervent feeling which is so essentially Bach's, goes on :—

> Lasst mich ihn nur noch einmal küssen,
> Und legt dann meine Freud ins Grab.

> Ah, grant me but once more to kiss Him,
> And then the grave must hold my Joy.

Nine bars after, the oboes and bassoons begin once more the same chorale, and carry it on with a few breaks till the end. This combination of independent verse sung by voices, and a chorale for instruments only—the chorale having already acquired a peculiar dramatic chiaroscuro from its connection with what has gone before—could have been conceived and worked out by no composer of the eighteenth century excepting Bach himself. A farther and strong evidence of his authorship is to be found in the fact that a very similar combination occurs in the cantata "Gottes Zeit"—"God's time is the best"—and that in both cases it is the same chorale that is so treated, and the reader will recollect that this cantata was also written in the early Weimar period.[601] As it is altogether more mature in style we may conclude that the St. Luke Passion was written earliest. It cannot fail to strike any one who compares them that the form of the chorale is different in the two. In both it varies from the original melody; but both the forms were in use together with the original tune in that neighbourhood. The melody in the cantata at first is identical with that of "Warum betrübst du dich, mein Herz," and I presume that Bach selected it on that account.[602] This is not the only indication of an internal connection between the St. Luke Passion and the *Actus tragicus*. The celebrated treatment of the closing words "Ja komm, Herr Jesu"—"Yea come, Lord Jesus"—has its counterpart at the close of the chorale "Stille, stille! ist die Losung"—"Silence, silence! is our watchword." Here all the voices are silent excepting the soprano, which has the words "Silence! Silence!" alone, above the bass, having previously begun the chorale in the same manner and with the same words.

If we acknowledge the St. Luke Passion as a genuine composition by Bach, the number attributed to him by the Necrology is complete. For, though in a list of the compositions by Seb. Bach, published in 1790 by his son, Philipp Emanuel, we read, on p. 81, "A Passion according to St. Matthew, *incomplete*," we may consider this to mean merely the first sketch of the great St. Matthew Passion to

---

[601] See Vol. I., p. 456.

[602] The use of this tune in Thuringia is not only established by Witt's Cantional, but also by Michael Bach's motett "Unser Leben ist ein Schatten." Rust very justly regards Bach's use of it as a fresh proof that the cantata "Gottes Zeit" was written in Weimar, B.-G., XXIII., p. xl. The form adopted in the St. Luke Passion is to be found in Freylinghausen Gesangbuch, 1741; but in the tenor aria Bach deviates from this in one passage.

Picander's words. Otherwise Emanuel Bach must have possessed three Passions and Friedemann only two, which is contradicted by the perfectly trustworthy account of the way in which the property left by their father was divided. At the same time it must remain uncertain whether the MS. spoken of was actually incomplete, or whether, as the grand closing chorus of the first part was not yet inserted, it may not have misled the writer of the catalogue into thinking so. We must pay due honour to the St. Luke Passion as the first attempt in this direction of a great genius. Bach himself thought it worthy of revision even after he had written his greatest work of the kind. In the course of that revision, which was probably carried out during the years 1732-1734, he may have improved it in some details, but the whole aspect of the work tells us that it can have undergone no radical change; it is true, however, that it would hardly lead us to anticipate the height to which Bach was able to rise as a composer of the Passion subsequently, at Leipzig. This is revealed in the St. John and St. Matthew Passions, to which we will now turn our attention. It is by a singular and happy chance that these three Passions have been preserved to us, for the process of Bach's development is broadly stamped upon them. As has been already said, the narratives of St. Mark and St. Luke had no place in the liturgy used at Leipzig. Bach, who loved to establish the most intimate alliance between his own music and the services of the church, of course did not overlook this circumstance. It is therefore highly probable that he himself attributed the highest importance to the two later works, and we may indulge in the belief that in them is preserved to us the very best that he thought himself capable of in this branch of his art. The other two Passions seem not long to have survived their author, even in the scene of his labours. In the year 1780, Doles, the Cantor, had only three performed; the St. Matthew, St. John, and possibly the St. Luke Passions.[603]

---

[603] Rochlitz, Für Freunde der Tonkunst, Vol. IV., p. 282. "As a boy, under Doles, I only made acquaintance with three Passions." This he says when

If the hypothesis is a correct one that Bach did really compose the music to Picander's Passion text of 1725, the Passion music which he conducted on Good Friday, April 7, 1724, can only have been that according to St. John; and this assumption is so far confirmed by the state of the MS. that it acquires the certainty of an established fact. This was the fourth year in which concerted Passion music had been performed in one or another of the Leipzig churches. When Kuhnau first introduced it, in St. Thomas', in 1721, the Council agreed that these Passion performances should take place every year in the two chief churches alternately, so in 1724 it must have come to the turn of St. Nicholas'. Since, however, the organ there was of very limited compass, Bach preferred to remain at St. Thomas', and had made all his arrangements, and had sent out notices by issuing the printed books of words. The superintendent of St. Nicholas, however, would not forego the honour; he laid his protest before the Council, and within four days of the performance Bach had to submit to remove to the church of St. Nicholas, to carry out as quickly as possible the necessary preparations, and to print new programmes.

The form in which the St. John Passion was performed for the first time was not that in which we now know it. Before it finally crystallised it had undergone various vicissitudes. We have every reason to suppose that Bach had already written it in Cöthen, in fact at the time when he decided on applying for the post of cantor to St. Thomas' and expected to receive the appointment. His application was made at the end of the year 1722, and he no doubt reckoned that he should be in office in Leipzig by Good Friday, 1723, and must be prepared for that event. Hence, the work would have been composed in the early months of 1723. As we know, however, Bach's call was delayed till May, so he could not make use of his work till Good Friday, 1724. This history of its origin accounts for a

---

speaking of the St. John Passion, which must therefore have been one of the three, and of course the St. Matthew, his greatest and most famous work, was another. The St. Luke Passion was still procurable in the Leipzig music shops, at any rate, as late as 1760, as is shown by Breitkopf's list mentioned above.

number of peculiarities in the work. In the first instance, the composer had to hurry it on if he wished to have it ready in time, and, since there was no poet equal to the task in Cöthen, he had to do the best he could himself with the text. The Bible narrative was before him, and he was more competent than any one else could be to select the appropriate chorale verses; the original verses presented the only difficulty. To find these he naturally turned to the favourite Passion poem by Brockes. This supplied him with the texts for the arias "From the bondage of transgression," "Haste, ye deeply wounded spirits," "Beloved Saviour," "My heart, behold how all the world," *Arioso*, and the next song "Dissolve, O my heart," with the words of the last chorus "Rest here in peace."[604] Bach did not borrow any of these verbatim from Brockes; the last three are very independently dealt with, and the first two altered in details. The changes do not always seem to be for the better; though in some instances he has mitigated a few harsh expressions in the original. Still, they are not on the whole unworthy of the praises lavished on them at a later period by Birnbaum;[605] it is true they do not satisfy our highest requirements as to the co-ordination of idea and suggestiveness of expression, but they show a delicate taste for musical verse. A few comparisons of the two works will show this more plainly. After the death of Christ is accomplished, in Brockes' text the believer is made to ask in an indirect way whether the work of redemption is indeed finished. "Thus," as he puts it, "asks the daughter of Zion, and the Saviour bows His head in sign of assent." As Bach has it, the question is directly addressed to the Redeemer: "Beloved Saviour, wilt Thou answer, as Thou hast now the cross endured . . . Shall all the world redemption see? Thou canst for anguish now say nothing, yet dost Thou bow Thy head and say in silence, Yea."

---

[604] "Von den Stricken meiner Sünden," "Eilt, ihr angefochtnen Seelen," "Mein theurer Heiland," "Mein Herz, indem die ganze Welt," "Zerfliesse, mein Herz," "Ruht wohl, ihr heiligen Gebeine." The English text is quoted throughout from Novello's edition.

[605] See ante, p. 238.

The final chorus is founded on Brockes' model, only so far as the second half is concerned; the first portion is on the usual pattern of those closing verses which Bach no doubt had in his memory; and the five remaining aria texts cannot be referred with certainty to any earlier poems, though the aria "Zerschmettert mich" may have been suggested by a verse of Franck. Bach may have relied for the most part on his own inventive powers. This, however, cannot be actually proved, nor is it indeed certain that it was he who wrote the text modified from Brockes. But we here meet with the same want of skill in the use of words as has been already alluded to, and the construction is that of a man whose head was well furnished with Bible texts and hymns, but who was not apt at combining them into a whole. But a man who was only slightly experienced in putting verses together could not have compiled this text; and as I shall presently bring forward evidence based on a MS. by Bach himself, that on another occasion he certainly made an attempt at poetry,[606] of quite the same stamp as that which we are now discussing, we have every reason to suspect this to be his handiwork.

In the setting of the Bible words, and even of the chorales, Bach subsequently found but little to alter essentially. But this was not the case with the pieces in madrigal form, in which the words of the text, hastily thrown together, had as it were run away with him. Three of these he dispensed with entirely, and not, certainly, because they were of inferior musical merit; their character must have seemed to him unsuited to their place. In lieu of two of them he inserted new solos of a different stamp, one of which—"Consider, O my soul," with the aria that follows it—is, again, a paraphrase from Brockes. This he seems certainly to have written himself, and it cannot be said that the attempt is any more successful than the former ones. Brockes' words are in the very worst taste, but Bach's verge on utter nonsense.[607] Another portion

---

[606] *A propos* of the cantata "Vergnügte Pleissenstadt."

[607] The text of the arioso, for bass, is moreover wrongly given in the edition of the B.-G., XII.,[1] p. 55. The original parts and score both give the words

of the madrigal verses were subsequently cut out by Bach
and replaced by an instrumental symphony ("Mein Herz
indem"—"My heart, behold"—with the preceding recitative and the soprano air following it), but he afterwards
restored them on the original plan. He must also certainly
have laid a revising hand on the opening and closing
choruses. Both these were grand chorale choruses. The
last "Christe, du Lamm Gottes," now forms the last
number of the cantata for the Sunday before Lent "Du
wahrer Gott" (see ante p. 351); indeed, it must surely have
belonged there from the first and have been transferred
from thence to the Passion, merely because the composer
laid that cantata aside for the time as beside his immediate
purpose; but after failing to bring out the Passion in 1723
the chorus was restored to its original position, and probably
was never once performed as a portion of the Passion.
The opening chorus, "O Mensch, bewein dein Sünde gross,"
was, we may imagine, made use of by Bach for his third
Passion, written in 1725, and finally incorporated in the
revised form of the St. Matthew Passion, as the closing
chorus of the first section, transposed from E flat major to
E major. Its place in the St. John Passion was taken
by a new chorus, "Herr unser Herrscher"—"Lord, our
Redeemer"—with a simple chorale to close the whole work,
"Ach Herr, lass dein lieb Engelein"—"Lord Jesus, Thy
dear Angel send." Most of the alterations here enumerated
were made for a second performance of this Passion, which
seems to have taken place on Good Friday, 1727. Bach
afterwards conducted it at least twice again, and each time
made some few farther alterations; the dates of these
performances cannot be exactly ascertained, at least one,
however, must have taken place in 1730.[608]

The old custom of giving the narrative of the Evangelist
and the speeches of individuals in recitative, while those
of a number of persons were sung by a choir in parts, put

"mit bittrer Lust und halb beklemmtem Herzen" and afterwards "wie dir aus
Dornen, so ihn stechen, die Himmelsschlüsselblumen blühn." It is equally
incorrect in Winterfeld, Ev. Kir. III., p. 368.

[608] See App. A., No. 56.

many and great difficulties in the way of any attempt to
create a well proportioned and homogeneous work of art.
The composer was absolutely dependent on a text which
was not written for musical treatment, and in which he
could not venture to alter any essential features. Many
portions of it compelled him to employ all the wealth
of musical means, while in others, of which the purport
was no less weighty, he had to confine himself to simple
recitative. Short, and even subsidiary passages, were dis-
proportionately prominent from having to be treated in
chorus, while some deep and vital utterance ran a risk of
passing disregarded in the modest delivery of a solo voice.
Thus it became a leading function in the inserted chorales
and madrigal pieces to restore the balance; they had to
rivet the attention of the hearer to the important points,
to define and concentrate the separate incidents of the
drama, and to distribute light and shade alike on the whole
and on the details. It cannot be concealed that in this
respect the St. John Passion leaves much to be desired.
In more than one passage the images mingle in confusion—
as Christ's examination before Annas, with Peter's denial;
Christ's appearance at the Judgment seat, with His crucifixion; and, again, the episode where Christ from the cross
commends His mother to the care of John, does not stand
out distinctly.[609] We cannot always consider the places
where the madrigal verses are brought in as well chosen.
The recitative "Simon Peter also followed Jesus" introduces
the description of Peter's denial, and when we find inserted
here a song of hopeful faith :—

> I follow Thee also, my Saviour, with gladness,
> And will not forsake Thee my Life and my Light—

while the incident it leads to is one of contemptible weakness and cowardly retractation, it needs no very subtle sense
of fitness to perceive that a transient feeling is made
prominent at the cost of the whole effect, and all the more
so since two arias are thus made to follow each other almost

---

[609] B.-G., XII.,[1] pp. 29, 31, 83, 102, and 104. Novello's edition, 29-32, 85, 98, 111.

without interruption. Even where the place seems appropriate for the insertion of a lyric piece the verses themselves are not satisfactory. After Jesus has been smitten by one of the High Priest's servants, the feeling of the situation is very finely concentrated in the chorale "O Lord, who dares to smite Thee." But this was immediately followed by a second portion of a chorale, on the seventeenth verse of "Jesu Leiden, Pein und Tod," in which the bass sang below the *Cantus firmus* of the soprano an independent text, which, though it referred to the Saviour scourged and crowned with thorns, was by no means suited to this place. The words "went out and wept bitterly" were originally followed by a passionate aria "Zerschmettert mich"—"Fall on me, O ye hills"—which gave the sentiment a turn quite different from that which was led up to by the setting of the foregoing words from the Bible. All these things point to the perplexity in which Bach found himself as to the necessary poetical additions, when he undertook the composition of the St. John Passion. He had actually no text for the more important places, and for the others he had none that were appropriate. The numerous alterations which the work constantly underwent prove that it did not satisfy him. Much he certainly improved in time: the first aria here named he eliminated, in spite of its great musical merit, and the second he replaced, with much taste, by a song of lamentation. Still many imperfections remain.

It is very evident that it was this difficulty as to the lyrical antithesis which led Bach to a quite peculiar treatment of the dramatic choruses. Of these, the St. John Passion includes a considerable number, highly artistic, vigorous, and characteristic compositions. The breadth and solidity with which most of them are worked out are very striking—almost as if they were oratorio choruses; and it need hardly be said that the words themselves, so far as any connection with the narrative is concerned, do not involve any such treatment, since they suggest movement rather than passive strength; indeed, a glance at the St. Matthew Passion shows that Bach was quite clear on the subject. But the need for some broad lyric forms was imperative and had to be satisfied, as well as might be, with the means at hand. From this point of view it becomes intelligible how such comparatively tame and unemotional words as " Let us not divide it, but cast lots for it who shall have it," or " It is not lawful for us to put a man to death," can have been set to choruses so full of musical purpose. There is still another proof that Bach, when he wrote the choruses of Jews, ranked as first in importance the musical form he required to give, rather than any striking dramatic rendering of a mass of people. It is that he has reset almost every one of these compositions once or oftener to other words, with small alterations and such amplification or abridgement as the text may have required; thus, the chorus " If this man were not an evil doer " to the words " It is not lawful"; the chorus " We salute Thee, King of the Jews " to " Write not the King of the Jews "; the chorus " We have a law" to the words " If thou let this man go." Besides this, Bach has constructed a short subject of four bars which he introduces in no less than four choral passages, which are sung respectively by the soldiers who take Jesus prisoner, by the populace outside the palace of Pilate, and finally by the high priests, under the most diverse conditions; and the accompaniment given to these in semiquavers on the upper instruments is also used with episodical extensions for the grander chorus " It is not lawful for us to put any man to death." It cannot be asserted that the music is equally well suited to all the different texts. The repetition of the phrase to which the reiterated cry of "Crucify Him" is set is founded on internal fitness, and it is not positively repugnant to the poetic sense, even in other instances; but we feel it quite unsuitable when we find the music to the words " We salute Thee, King of the Jews," afterwards adapted to the text " Write not," &c. For in this instance the fundamental feeling is entirely different; in the first instance it is malicious scorn— which, indeed, the music admirably depicts so far as its means allow; in the second it is remonstrance or a secret anxiety; Bach has here sacrificed suitability of character to musical solidity and unity.

Of the four Gospel narratives of the sufferings and death of Christ, that of St. John is the least detailed and animated. A mass of significant circumstances, well suited for musical treatment, is passed over in silence; as, for instance, the institution of the Lord's supper, the agony in the garden of Gethsemane, and the convulsions of nature at the moment of Christ's death. Bach felt this as a serious deficiency, and to compensate for it in some degree he transferred, at the right place, the description of the earthquake from St. Matthew, and set the words which close the episode of Peter's denial with true musical feeling: "Then Peter thought upon the word of Jesus, and he went forth and wept bitterly."[610] On the whole, he could not of course remedy the meagreness of the narrative; but this circumstance must be taken into consideration to explain the fact that the St. John Passion is far inferior to the St. Matthew, or even to the St. Luke. Its highest permanent value does not lie in the general construction; as a whole, it displays a certain murky monotony and vague mistiness. This must be confessed even though we are fully conscious that such a work must be viewed from quite a different standpoint to an oratorio or a musical drama. It is true in considering a Passion setting we must never forget that it is church music, which does not admit of any radical differences in the treatment of the epic, lyric, and dramatic elements; but, on the contrary, must express everything, even the most opposite ideas, within the spell-bound circle of impersonal generalities. Hence it ought not to be regarded as want of style when the recitative of the speaking persons is not sharply distinguished from that given to the Evangelist by its greater animation, if the Evangelist himself seems impressed by them; the words of the Bible have the same importance, no matter who it is that speaks them. The composer must remain at full liberty to render the expressions of the soldiers or high priests by a full chorus as well as

---

[610] The words of Jesus were "In this night before the cock crow thou shalt deny me thrice." They are wanting in St. John's Gospel, and as Bach has omitted to borrow them from St. Matthew the passage he has inserted has no connection or sense.

those of the populace; it is sufficient that the contrast between individuals and numbers should be generally indicated. Nor can any objection be made where a grand chorus is constructed on a short sentence, such as "Crucify Him," since, even from the church point of view, it can be full of high symbolical significance. All the dramatic elements which underlie the text of the Passion music can only approximate a true dramatic utterance, but cannot assume its most vivid and natural expression. Meanwhile, though granting all this, we find, remaining within the limits thus imposed, ample opportunity for subtle distinctions and gradations; this has been incontrovertibly proved by Bach himself in the St. Matthew Passion; but the St. John Passion cannot be admitted to be of the highest perfection in this respect.

The adaptation of Brockes' poetry in the St. John Passion has given it a certain resemblance to the Italian oratorio, inasmuch as the aria "Haste, ye deeply wounded spirits," represents a dialogue between the Daughter of Sion and the souls of the believers. The division into two portions which, in the same way, we find in the St. Luke Passion, must also be referred to the Italian oratorio, in that the first portion was intended to be performed before the sermon and the second after it. The custom of the Protestant church did not correspond to this; hence, when the narrative of the Passion was divided, it was into six sections, one for each day of Passion week; or sometimes even more, some being performed on the previous Sundays in Lent.[611] Of course, when we consider the actual process by which the text of the St. John Passion was evolved, any considerable influence from the Italian oratorio is out of the question; it is far more conspicuous in the St. Matthew Passion. I mention this merely as characterising the difference between the earlier and the later works, and not as a merit or a demerit

---

[611] The Rudolstadt Passion of 1688 is divided into six parts (Actus). Count Heinrich XII. introduced at Schleiz a Passion in twelve sections, the first to be sung on the first Sunday in Lent and the last on Good Friday. The text exists in the Count's Library at Wernigerode. Both these are composed to "harmonies of the Gospel"—*i.e.*, texts compiled from all the four Gospels.

in either. But as a result the St. John Passion has a less modern air according to the standard of the time.

This becomes more than negatively evident in the chorales which are used to introduce and close each part, and which (excepting the first, "O Mensch, bewein dein Sünde," which was afterwards cut out) are set simply in four parts. If Bach had been minded to follow out the old custom in every particular, he would, at the end of the whole work, have introduced either the usual thanksgiving, or with a nearer allusion to the burial of Christ, Rist's chorale "O Traurigkeit, o Herzeleid!"—"O anguish sore, O grief of heart." The choral aria over the sepulchre, which precedes the final chorale—and which occurs very similar in form in many of the Passion settings of Central Germany at the end of the seventeenth and beginning of the eighteenth centuries—owes its existence to an ancient ceremonial—*i.e.*, the performance of the interment in front of the altar (auf dem Chor), at which stage of the proceedings a motett was sung on the chorale "O Traurigkeit." I have not discovered that this was still customary at Leipzig in Bach's time, but that the ceremony had left its record in the close which continued in use in the Passion music is very certain. The remembrance of it must, therefore, have survived in Leipzig, since it was an established thing that after the Passion music was ended the congregation sang the hymn "O Traurigkeit," and this was obviously the reason why Bach would not introduce it again at the end of the St. John Passion. Much later, at a time when Passion settings were already performed in concert halls, it became common to designate the finale by the name "By the sepulchre."[612] Nowadays, if this dirge is not omitted and, at the same time, the closing chorale is retained, two choral pieces must be sung in succession, as Bach has planned in the St. John Passion; and the same thing is observable in many other Passions of the time of the old church cantatas. The close relationship which the St. John Passion bears

---

[612] Israel, Frankfurter Concert-Chronik of 1713 to 1780. Frankfurt a. M 1876, p. 33 (a concert programme of March 26, 1743).

to these is also discernible in the recitation of the Bible text, which renounces all attempt to give prominence to the speeches of individuals—Christ, for instance—by the use of additional effects, and is throughout content with a simple fundamental bass.

But it is only by this reticence that Bach adheres to the early practice, and not in the form of the recitatives themselves. After pointing out those peculiarities of the St. John Passion which cannot entirely satisfy our highest demands, we must all the more emphatically insist that in everything which relates to musical style, in invention, and in the elaboration of the separate compositions, Bach proves himself to have reached the heights of ripe and perfect mastery. The treatment of the recitatives, as in the cantatas, is of Bach's best period; but any one who should look for a difference in manner corresponding to that between a contemplative passage, a narrative, or a dramatic speech will be disappointed. The composer has not neglected such opportunities as the Gospel text affords for more impressive phrases or more incisive accentuation; and, besides frequently giving strong relief to important passages by special melodic or harmonic combinations, he often depicts the idea of movement—such as going backward, falling to the earth, the drawing or sheathing of swords, interment, smiting, scourging, or fighting—by graphic musical phrases, usually in the voice, but occasionally in the accompanying bass. Emotions also, which are only indirectly connected with an image or a statement, are expressed as opportunity offers; thus, the word "dying" is illustrated by a lingering, longing *melisma*, "Crucifixion" and "Golgotha" by a painful and dislocated turn. When the servants warm themselves by the fire Bach sets the words to a figure which is obviously intended to convey the sentiment of comfort, though we might not perhaps suspect the intention if the same phrase did not recur when Peter warms himself (pp. 30 and 34 of Novello's English edition). Much more, resembling this, is left to be divined. When on the words "Every one that is of the truth heareth my voice," spoken by Jesus, a wailing *melisma* is brought out,[613] there is certainly no direct instigation in the text itself. Possibly the composer had in his mind the passage in the Gospel of St. Luke, where Christ weeps over Jerusalem saying, "If thou hadst known the things that belong unto thy peace," xix., 42. Now and then Bach even diverges into the theatrically dramatic, and introduces declamatory figures which seem to require the addition of gesture for their full effect; thus, in Peter's reiterated cry "I am not"; the phrase "Now Barabbas was a robber," and the *a-parte* passages which are carefully indicated by the directions *piano* and *forte*.[614] Still, from all these examples it is not possible to deduct any fixed principle of treatment, for Bach quite as often leaves words and sentences, which would well bear some characteristic accent or melisma, to be chanted in commonplace recitative. The supreme and sole principle of form that governs throughout is the result of his own innate tendency towards vigorous melodic movement. Everything else is merely a means to this end; if it were not so it would be inconceivable why Bach should sometimes introduce picturesque details which have no dramatic musical purpose, and which, in themselves, are mere sports of the composer—*e.g.*, when the word "the Pavement" (in German the word "Hochpflaster") is declaimed in such a way as that the first syllable is sung on G sharp, reached by a leap of a sixth, and the other two on C sharp.[615] A stronger development of the melodic element suited the style of the church recitative, and it was a favourite method to let the closing fall die away into an *arioso* passage, in which it is impossible not to recognise an affinity to the character of the cadence in the Gregorian church mode. It was from such cadences that the elaborate —nay, too elaborate—*melismata* had been derived, and Bach employed them in setting the phrases "and wept bitterly" (p. 37), and "Pilate took Jesus and scourged Him" (p. 57). But even a reference to the style of the church recitative

---

[613] B.-G., XII.,¹ p. 53, bar 9. Nov. ed., p. 56, bar 3.
[614] B.-G., XII.,¹ p. 33, 54, 119, 122. Nov. ed., pp. 36, 57, 121, 123, &c.
[615] See B.-G., XII.,¹ p. 78, bars 4-5. The word "Hoch" is thus strongly emphasised in the German.

is insufficient to explain the matter as it lies before us. Bach's recitative—I can only repeat my former expressions,[616] for I find none more apt—has in it something of the character of a prelude or an independent fantasia. The composer wanders through the realm of musical imagery and gives himself up to realising it, now in one way and now in another—prompted to do so now by some important factor, and now by something wholly unimportant, and without ever renouncing his right to absolute free-will. We can find no reason in the nature of things why in one place he should devote all the means at his command to an exhaustive illustration of some emphatic word, while in another he passes it over with complete indifference: it was his good pleasure, so he did it.

I have already alluded to the oratorio character of the dramatic choruses in the St. John Passion; still there is something in them which does not bear the true stamp of the oratorio. They are all characterised by a polyphony of exceptional complexity and a certain compactness of structure, so far as is compatible with broad working out. It may be admitted that by this means the fanatical Jews, in their craving for a victim, their wild accusations and threats, are graphically depicted. But this is not the real reason which made Bach write in this particular manner, for the chorus of the soldiers who cast lots for Christ's coat is constructed in the same way. Though Bach had been forced, from the lack of lyric text, to find a substitute for it in the deeper musical treatment of the dramatic chorus, he must, on the other hand, have seen very clearly that he could not deviate very widely from the narrow dimensions of the forms properly associated with the poetic essence of these choruses without detriment to the whole. It is the mixture of breadth and concentration, arising from a compromise between the oratorio and the dramatic styles, which gives the Biblical choruses of the St. John Passion so peculiar an aspect. These grandiose forms, which are full to overflowing of musical sense and

---

[616] Compare Vol. I., p. 497.

significance, testify to a stupendous creative power, but at the same time have something oppressive and sultry about them. The large space they fill in the whole work stamps the greater part of it with their peculiar character.

The chorales are almost all set simply in four parts, and are such as Bach could write at the height of his powers. By a marvellous pliancy in the treatment of the parts, and an inexhaustible wealth of harmonic resource, he was able to distribute over the whole a fresh and varied vitality with a subtle and significant illustration of details; and at the same time to work out the chorales, each as a whole, in most effective contrast to the rest. The best example of this is offered by the chorale "Ach grosser König"—"Thy bonds, O Son of God"—which is permeated by a vein of ecstatic love; and that wonderful hymn, touching in its extreme simplicity, "In meines Herzens Grunde"—"Within our inmost being." Indeed, the choice of the chorale is worthy of the great master, both as regards the text and the melody. Stockmann's hymn on the Passion "Jesu Leiden, Pein und Tod" forms, as it were, the pivot of the church feeling of the work; originally it recurred four times with four different verses. One verse Bach subsequently took out, as well as the aria "Himmel reisse"—"Rend, ye heavens." It is now twice repeated in a simple form, and finally, after the words "And He bowed His head and departed," as a chorale fantasia; for the bass sings an aria to the organ, while the soft and meditative chorale in four parts is woven in with it.

As regards the solo songs, they probably all (with the exception of the aria "Ach windet euch nicht so" and that which subsequently took its place "Consider, O my soul") are among the best that Bach ever wrote. How they can ever have been supposed to betray the manner of an earlier period[617] it is hard to discover, for they almost all depart, more or less, by their grand, free, and novel form, from the traditional type of aria. The very first "From the bondage of transgression" is interesting as to its form, and

---

[617] Winterfeld, Ev. Kir., III., p. 364.

important from the episodical transition from the second to the third part. The aria "Zerschmettert mich" derives, from the frequent changes of time and the bold close of the voice part on the dominant, a character of personal passion devoid of all conventionality; in novelty and captivating ingenuity it is superior to that which subsequently took its place "Ah, my soul!" though this too, its other merits apart, is distinguished by its ingenious construction. We are insensibly led on from the second part into the third, which here consists only of the opening ritornel, while the song continues its course independently. The construction of the aria "It is finished" is again quite peculiar, with its *adagio* subject, accompanied only by the organ and viola di gamba, contrasted with an *allegro* to the whole orchestra of strings, returning at the close with startling effect to the *adagio*. The bass aria "Haste ye deeply wounded spirits," with its stressful rhythm and the impressive questioning of the chorus, is so full of dramatic force that scarcely any other solo piece out of Bach's Mysteries will bear comparison with it. The two *arioso* numbers also: "Consider, O my soul" and "My heart behold," are compositions of the most striking character and deep musical feeling.[618] The highly artistic, and, indeed, over-refined, setting which characterises most of these pieces no doubt prevents their having that simple charm and popular effect which is the specialty of almost all the solos in the St. Matthew Passion. The impression they leave is profound and grave, and their prevalent feeling is gloomy; they reveal a near affinity to the duet in the cantata "Du wahrer Gott und Davidssohn," which was, indeed, composed at the same period. There is hardly any interest to be found in comparing Bach's settings of Brockes' text with those of other composers. Not only do they surpass them immeasurably in richness and

---

[618] The former is to be accompanied by two *Viole d'amore*, lute, and bass. In the absence of the lute Bach has given the part to a cembalo, and an autograph part for that instrument is in existence. Later, after 1730, he transferred it to an organ obbligato; the autograph part exists—it is in D flat major and has the indication in Bach's handwriting "To be played on the organ with the eight and four foot Gedackt." The harpsichord part is in E flat major.

depth, but, in consequence of the profound church feeling that prevades his style, they stand forth as something totally distinct when compared with the operatic religionism of the other German masters. The solitary grandeur in which Bach dwells as a composer of church music is only rendered more clear by a comparison with similar works by Keiser, Telemann, Mattheson, Stölzel, or even by Handel.

The St. John Passion includes only two madrigal choruses. The dirge, "Rest here in peace," which immediately precedes the finale chorale, is an aria for chorus which, however, is not in the stanza form, but in that of the Italian *da capo* air. A very simple form of structure was here an accepted law of style; the upper part has the melody accompanied by the lower voices in free and graceful movement; the ritornel comes in at the usual places. The length of this number is all the more surprising. It is in five instead of three sections, for the second is repeated as a fourth in a different key; the whole thus attains the enormous length of 172 bars. It is an inexhaustible lament and leavetaking over the grave; flowing passages in quavers on the strings sink softly down to the lowest depths, mingling with the tearful tenderness of the vocal parts like the dull, slow fall of clods on the coffin. Bach evidently did not trouble himself with the fact that the burial of Christ was effected in a manner different from ours, for there can be no doubt that he intended to represent a descent into the grave. A very remarkable composition lies before us in the opening chorus, which Bach wrote for the second performance of the St. John Passion. The text is as follows:—

> Lord our Redeemer, Thou whose name
> In all the world is glorious,
> Show us in this Thy Passion
> That Thou, the true and only Son.
> For evermore.
> E'en from humiliation sore,
> Dost rise victorious.

The first lines resemble the eighth Psalm, and were, no doubt, borrowed from it; the key is G minor. The voices roll onward above long-held pedal points—now in impressive outcry and massive homophonic subjects, piled up in figures of semiquavers; now in proudly-mounting themes or broadly flowing passages, worked out in canon, in fugue, or in free imitation: a mighty picture of divine power and glory. The orchestra is quite differently treated; a vague and incessant rushing movement is heard throughout the work, generally in three parts on the violins and violas, and always in the medium or lower notes, with brief excursions into the basses and rapid passages in the lower flutes and oboes, while above this the wind instruments hold long-drawn notes of lamentation almost uninterruptedly. What Bach's general intention was is obvious at first sight; he proposed to depict at once the· majesty and might of the Son of God, and His utter humiliation under the bitterest sufferings of which humanity is capable.

Nay, we can hardly err even in our interpretation of the semiquaver movement in the instruments. It was a favourite trope with the sacred poets of the time to figure the miseries of human life as the waves of the sea which threaten to surge up and overwhelm man; the narrative of Christ crossing the sea was what proximately suggested this. A great part of the cantata previously discussed, "Jesu schläft" —"Jesus sleeps"—is founded on this conception, and in the cantata, "My spirit was in heaviness,"[619] we find an aria with the same purport. It was, of course, an easy and pleasing task to represent the motion of the waves by musical means; in the last-named aria this is done by a figure on the violins, which bears considerable resemblance to that which predominates in the first chorus of the St. John Passion. We therefore can have no doubt as to the meaning which was conveyed in this heavily-rolling instrumental figure. Bach has given the antagonistic images and sentiments which he desired to render to antithetical bodies of sound; the element of divine supremacy was to be expressed by the voices; the instruments were to lend the expression of human suffering. Nor is it contradictory to this interpretation when we find that in some passages the

---

[619] See Vol. I., p. 532. English edition by Novello.

voices unite with the stringed instruments; for, when sung, the semiquaver figures acquire a different character, and all the more so when, as here, they occur only in a transient form and in alternation with a rich variety of other figures. This is not, indeed, the first time that we have met with such a combination in Bach; in the final chorale of the cantata "Enter not into judgment," the same idea prevails. Of course, we have not here two different conceptions merely superficially connected; both have arisen from one common idea, and are only contrasted for the sake of greater effect. The power of imagination which could enable Bach to amalgamate two such opposite images into one is truly marvellous; but there is yet another thing which excites our astonishment in no less degree. I have already pointed out again and again how Bach loved, in his cantatas, to epitomise and concentrate the whole domain of feeling in the opening chorus, and so to define the limits to which the development of the work in hand might extend. The introductory chorus of the St. John Passion serves the same end; it is, as it were, the prologue to the drama. Beyond a doubt, it was only after deep consideration that Bach decided on selecting this text, which bears neither on the sentiment of lamentation, on a feeling for the sufferings of Christ, nor on our blessedness as won by this sacrifice of Himself, but merely on the contrast between the eternal omnipotence of the Son of God and His temporary humiliation. In point of fact, even in the setting of this chorus we nowhere find any expression of warm sympathetic emotion; it is full of a dark unapproachable grandeur, and remains, in that respect, unique among Bach's works.[620] But it rehearses with impressive distinctness the sentiment of the whole work—a work which is not always synthetically constructed nor elaborated with all the charm of variety, but which stands out in a grandiose outline, though in a dim and lurid chiaroscuro which contrasts strangely enough with the

---

[620] Even Winterfeld (Ev. Kir., III., p. 366) has pointed out this feature in the introductory chorus to the St. John Passion, and has subjected it throughout to a very sympathetic analysis. His objections to the views of Rochlitz seem to me ill-founded; however, their differences might well be reconciled.

ideas of tenderness and love that we are accustomed to associate with the character of the writer of this Gospel.

At the time when Bach decided on composing a Passion according to St. Matthew, he had at his disposal what he had conspicuously lacked in the case of the St. John Passion; the assistance, namely, of a skilled and sympathetic poet, with whom he had already co-operated for some years. Picander's talent was less than mediocre, but Bach asked for nothing more than facility of outward poetic form, and this he could command. The main outlines of the scheme of the Passion were long since established by tradition, and, in spite of the tendencies of his time, Bach was fully convinced that the Biblical core of the work could not be tampered with if the church feeling of the significance of the sufferings and death of Christ was to be fully brought out in the highest and most comprehensive form. He himself possessed, in the highest degree, that knowledge and sentiment which were needed to give the utmost intensity to the fundamental Protestant tone of the work by the interpolation of appropriate chorales at suitable places. Besides this, he was sufficiently imbued with the spirit of the popular sacred drama to require of his poet all possible regard for it. How much of the construction of the text is due to Bach and how much to Picander is hardly capable of demonstration in detail, but we may regard it as certain that Bach had a very considerable share in it. The mere fact that Picander's text was printed again and again, not only without the Bible text, but without the chorales, proves that he felt no particular interest in the form it had assumed by the addition of the chorales.[621] But Bach in some degree influenced the independent poetry; it must have been he who insisted on the adaptation of Franck's hymn "Auf Christi Begräbniss gegen Abend"—"Towards evening at

---

[621] He had the St. Mark Passion printed with the Bible words and chorales, although it is to be presumed that Bach himself also selected these. But in this Picander's poetical contribution was so small that it would not have been worth printing by itself. The St. Mark Passion is not included in the collected edition of Picander's poems.

Christ's burial"—and it seems as though he had not been content to make use of Franck's text only once in the Passion; for the recitative "Thou blessed Saviour, Thou," agrees in feeling, and even in certain turns of thought, with one of Franck's madrigals, in the same way as the recitative "It was in the cool of eventide" corresponds to the above-named dirge. It is incontestable that their co-operation resulted in a text which satisfied Bach's requirements in every particular, and which we must recognise as being in every way suited to its purpose, whatever our opinion may be of Picander's trivial rhymes.

The story of the Passion according to St. Matthew was sung in Leipzig every year on Palm Sunday, treated chorally. The close connection with divine service which it thus acquired must have been a special incitement to Bach to bestow on it a consistent and thorough artistic treatment; though, indeed, a sufficient inducement lay, no doubt, in the subject itself, for the narrative of St. Matthew far exceeds those of St. Mark or St. Luke in fulness and vividness. Bach divided the story into two parts, but not according to the break in the chapters xxvi. and xxvii.; he ends the first section with the taking of Jesus and the flight of the disciples (xxvi., 56). The hearing before Caiaphas, Peter's denial, the judgment of Pontius Pilate, with the episode of the death of Judas, the progress to Golgotha, Crucifixion, Death, and Burial of Christ, are all included in the second part, while the first comprises only the conspiracy of the High Priests and Scribes, the anointing of Christ, the institution of the Lord's Supper, the prayer on the Mount of Olives, and the betrayal by Judas. This division alone proves the judgment exercised in the treatment. In a work planned to represent the most stupendous events, and engaged throughout with none but the saddest emotions, every possible contrast had to be made the utmost use of. The first section is contrasted with the second as a prologue with the crisis; in one a solemn stillness reigns, in the other a passionate stir; in the former the lyric, in the latter the dramatic element. The number of independent poems is considerable, no less than twenty-eight if we consider the recitatives in verse as separate pieces. Besides this we have fifteen chorales. With so abundant a supply of lyric compositions, Bach found himself in a position to treat all the separate events in a quite different way from that he had used in the St. John Passion, defining them clearly and dividing them with a satisfactory close. Only in two places do we find an incident without a concluding lyric. The narrative of the death of Judas is followed by the aria "Give, O give me back my Lord," in connection with the restoration of the "price of blood," since the Bible words which follow, no doubt, seemed less proper as an introduction to a suitable meditation. Bach follows up the description of Christ's death very appropriately with the confession of the heathen set to guard Him: "Truly this was the Son of God," and he connects the proceedings of the women who had been present at the Crucifixion with the scene of the Burial, which is quite fitting, because the two Marys have a further part to play in the narrative of the Resurrection.

The grandeur and breadth of the poetical matter are adequately met by the musical means made use of.[622] Bach has arranged it for double chorus, and given to each chorus its own orchestra and its own organ accompaniment. He has made a truly astounding use of these two main masses of sound, both to emphasise all that has poetic value, alike from the lyric and dramatic point of view, and to express in music the many elements which compose the mighty picture. In the unanimous utterances meant to characterise the vehemence of the fanatical persecutors of Christ, the choruses commonly sing together in polyphony, only coalescing in a compact four-part structure at the culminating points of their passion. In less emotional portions, Bach was satisfied to employ only one choir, as when the servants of Caiaphas address Peter "Surely thou also art one of them, for thy speech bewrayeth thee." The scornful words at the foot of the Cross: "He calleth for

---

[622] The St. Matthew Passion is contained in Vol. IV. of the B.-G. edition; and it has been published with English words by Messrs. Novello & Co. This edition is quoted throughout.

Elias," are given to the first choir, and those which immediately follow: "Wait, let us see if Elias will come to save Him," are sung by the second. The disciples are represented only by the first choir; but in all the chorales, except where supplementary lyrics are interwoven, and except in the deeply significant dramatic chorus "Truly this was the Son of God," the choirs combine in one mass of sound. In the grand "madrigal" tone-pictures at the beginning and end, and towards the close of the first part, they work together as a double chorus with a grandiose progression of parts, and in the opening number a third chorus of one part only—*soprano ripieno*[623] —is associated with them. Turning to the solo songs we find all the Biblical personages, with the exception of the false witnesses, supported by the first choir. The "madrigal" recitatives and arias are pretty equally divided between the two choirs, with a trifling preponderance in favour of the first. This detail, which is commonly neglected in performances of the work at the present time, is, nevertheless, not unimportant, since the choirs, of course divided into two bodies, were placed on the right and left sides of the organ choir.

Bach did not write for solo singers, who should stand out in contrast to the chorus. His *concertists* sang with the chorus, and only came forward for the moment when it was their turn to sing; it must, therefore, have had a characteristic effect when the solo voices rang out in the spacious church from one side and then from the other, and this would be all the more striking when a single voice was heard above its own proper accompaniment from one side, while on the other the whole chorus joined in with its full orchestra. For instance, when in the scene on the Mount of Olives one voice in the first choir sings "O grief! Now pants His agonising heart," and from the other side the chorale rises up, like the penitential prayer of a kneeling congregation, "O Saviour, why must all this ill befall Thee?" the effect must have bordered very closely on the dramatic.

---

[623] P. 4 of Novello's edition.

Bach has succeeded in subtly characterising the various poetical elements which, in the course of time, had become infused into the form of the Passion music, without injuring the fundamental unity of the style. He did not hesitate to introduce a certain number of madrigal recitatives side by side with those from the Bible. He has even added to their colour by the addition of instrumental effects, and yet left it possible that each class shall be immediately recognisable. The Evangelist and the other persons who are introduced as speakers sing in *recitativo secco;* when Jesus speaks, to distinguish Him from the others, a stringed quartet is brought in, but the accompaniment is mostly restricted to held chords, though in certain passages it adds a graphic illustration of the words, and in the institution of the Lord's supper it works round the voice part, which gradually develops into a long arioso in a highly artistic four-part subject. The accompaniments to the words of Christ are chiefly used to give colour; the "madrigal" recitatives, on the other hand, have an obbligato accompaniment in which a motive is usually worked out, which bears a figurative reference to some important image in the words. These recitatives are thus a degree nearer to an organic composition and constitute a very natural transition to the form, complete in itself, of the arias which they lead up to. The accompaniments are frequently given to wind instruments, so as to produce a distinct contrast to the words of Christ; a sufficient variety is thus produced, and a musical antithesis corresponding to the different importance of the recitatives. And this applies to the chorale portions as well as to the solo parts. Bach, who had at his command an incomparable wealth of forms of chorale treatment, has, nevertheless, in the St. Matthew Passion refrained from using any other than a simple choral setting. By far the greater number of chorales are set in a severe style, and accentuate the congregational feeling in all its modesty and force. Twice, however, in the chorales at the opening and close of the first part, he has extended this simple form to a chorale fantasia, thus giving free expression to personal sentiment. Even these two numbers again are quite dissimilar; since in the final chorus of the first part the subjective element finds its chief expression in the instrumental picture, while in the introductory number it lies with the two principal masses of the chorus, each having its own text. While Bach thus bridged over the gap between the strictly sacred and the independent chorus, he succeeded, by combining the lyric solo and the simple chorale for chorus ("O grief now pants"), in effecting a union between these two antagonistic elements; and in the opening chorus of part II. ("Alas! now is my Saviour gone!") between the aria and the chorale treatment of Bible words in general, without effacing the peculiar essence of each. The choruses in madrigal form roll by in the broadest waves of sound, but in the simplest possible treatment; in the dramatic chorus he indulges in the contrapuntal profundity which beseems the Bible words, and yet he resolutely confines himself to the concisest expression possible. Apart from a few certainly quite new and peculiar developments the musical forms in the St. Matthew Passion differ in no respect from those in the cantatas. But what is amazing is the evident economy of artistic means in their application. A distinct organisation prevails throughout, and at the same time a delicate treatment of the outlines and a tender toning down of contrast which are far above anything that can be mentioned in the St. John Passion. This circumstance is in the highest degree characteristic of the whole impression produced.

There is scarcely anything left to be added with regard to the Bible recitatives in the St. Matthew Passion to what has been said in discussing the St. John Passion. But as one critic has tried to detect in them a clear and definite characterisation of the various Biblical personages,[624] it may here be expressly insisted that nothing of the kind was done by Bach, or even intended by him. The Passion has, it is true, borrowed certain features from the drama, but it is not really a drama for all that. All that he proposed to do by way of distinguishing the actors is effected by the

---

[624] Mosewius, Johann Sebastian Bachs Matthäus-Passion, musikalisch-ästhetisch dargestellt. Berlin, 1852, p. 70.

distribution of the speeches to different voices, and by the addition, when Christ speaks, of a more highly coloured and sometimes even quite independent accompaniment; but from this to dramatic individualisation is a long step. This can only consist in the development of a special basis of feeling for each of the different *dramatis personæ*, on which the mode of expression for each shall be modelled in every particular. If this cannot be proved to exist, all the supposed characterisation dwindles down to a suitable accentuation of the words and phrases of the actors. And this, even, is by no means so constant as to justify us in regarding it as a ruling principle, always before his mind. The essential character of his recitative can only be perfectly understood by considering it as a musical improvisation under dramatic conditions, and, even then, as an improvisation within the pale of strict church forms and style; for it is only by considering its church use that we can account for the fact that the narrative of the Evangelist is given with the same fervent utterance as the speeches of individuals, a fact which is sufficient by itself to contravene all notions of dramatic treatment. Throughout the Evangelist's narrative we may note an emotional unction which often is nothing more than this, but often, on the contrary, is concentrated in a special sentiment. The agony and terrors of Christ in Gethsemane, Peter's bitter tears of repentance, and Christ's Crucifixion are not so much related by the Evangelist as experienced by him, with all the devout fervour of the sympathising Christians. He has not translated the cry " Eli, Eli, Lama sabachthani," as Handel has done in setting the Passion[625] compiled by Brockes; but he lets the feeling of the appeal find an echo, as it were, in his own breast ; and we see too that the melody was invented for the German rather than the Hebrew words, for it does not perfectly agree with these in accent. When Peter denies the Lord for the third time, the Evangelist reiterates the phrase of recitative a fifth higher, on the words "And

---

[625] Handel-Gesellschaft, XV., p. 134.

immediately the cock crew," thus reminding him of Christ's prediction by a mocking echo of his own pitiable weakness. Now such a treatment was only possible to Bach because he regarded the words of the Evangelist and those of the actors from the point of view of the Protestant believer, and not from that of the dramatist.

Even the stringed quartet accompaniment which, as a critic has elegantly said, floats round the utterances of Christ like a glory,[626] does not proceed from a dramatising tendency. The notion of *characterising* the omnipotent God by mere human means would certainly have seemed to Bach a blasphemy in itself; besides, the style of music given to Christ is precisely the same as that of the other persons. But, just as in earlier times Christ's words had been delivered in several parts with the idea that the use of this fuller effect gave them, musically speaking, a higher value; reverence for the person of the Redeemer has here prompted the composer to attune the minds of the hearers to special devotion when He speaks. This, however, was not in itself a novel mode of proceeding. It is not necessary to go back to the "Seven Words" by Schütz for an example, and it is hardly likely that Bach can have known the work; if Bach had a model in his mind he is more likely to have found it in Telemann's St. Mark Passion (in B flat major). Still, Telemann frequently set the words of Jesus to an arioso, a means which Bach has only used to give greater prominence to the words spoken at the institution of the Lord's supper. Both composers have dispensed with the string quartet in one place, different, however, in each; Telemann at the brief reply made by the suffering Saviour to Pilate: "Thou hast said." In this place Bach, it is true, ceases the continuous accompaniment, but throws in short chords before and after it; on the other hand, Bach stops the accompaniment on words uttered in the depth of crushed humiliation, "Eli, Eli, Lama sabachthani." Both are highly graphic, but Bach's feeling, which makes him

---

[626] Winterfeld, Ev. Kir., III., p. 372.

extinguish, as it were, the glory round the Redeemer's head, at this instant is by far the most figurative.[627] As far as regards the general management of the accompaniment, Telemann, of course, cannot measure his strength against Bach by a long way; with him it is no more than a superficial means of emphasis, while in Bach it really rouses a feeling of higher devotion;[628] and if Bach was not altogether original in his idea, he certainly was in his method of dealing with it.

Though the treatment of the Biblical recitative in the two Passions according to St. John and St. Matthew is essentially the same, in the choral numbers on Bible texts we see a conspicuous dissimilarity. No form of composition used in them shows more convincingly that Bach's Mysteries form a class of art apart from the dramatic choruses—so called for brevity—in the St. Matthew Passion. Consider the passage where the Jewish people, prompted by the high priests and elders, demand the release of Barabbas. The Evangelist makes them reply to Pilate's question with the single word "Barabbas"; the situation is no doubt full of emotion, and an oratorio writer might have been prompted to let the electric tension of the moment discharge itself in a chorus. But it must necessarily be embodied in a form in which the chorus could have its full value as a musical factor, in a broadly worked out composition and on a text of somewhat greater extent. The dramatic, or operatic, composer would have given it the utmost brevity, since it stands midway in the critical development of an event; he would have to con-

---

[627] Bach has, at this passage, given held chords to the organ which has a particularly solemn effect, as it elsewhere has only short chords. That the unsupported recitative of the St. Matthew Passion was intended to be accompanied on the cembalo and not on the organ, is an unfounded hypothesis of Julius Rietz, B.-G., IV., p. xxii; for the organ part which still exists contains the recitative accompaniments complete. I need only refer the reader to p. 299 of this volume to prove that the idea rests on an altogether false assumption.

[628] The principles which guided Telemann are rendered more obvious by his St. Mark Passion of 1759 (in G major), in which he gives a stringed quartet accompaniment not merely to all the words of Christ, but to other important and significant passages.

sider the progress of the action as well as the expression of feeling: an excited populace thronging wildly and tumultuously round the governor. A sudden roar and brief turmoil of voices would be the kind of movement best suited to his purpose. Bach, composing a devotional Passion, makes the whole chorus groan out the name of Barabbas once only, on the chord of the minor seventh, led up to by a false close. This of course is not oratorio style, but even in a dramatic work such brevity would be inadmissible at such a crisis. Bach commonly pays no regard to the scenic situation, a freedom of which, in this case, he availed himself to concentrate the utterance even beyond the limits of dramatic usage. He depicts in the strongest manner the savage feeling of the populace by giving them a dramatic identity, and at the same time suggests the sudden horror which seizes the believing Christians at their answer. It is a master-stroke, equally admirable for the decisive rightness of the feeling for form which it reveals and the overwhelming force of its utterance. Although it is directly suggested by the text itself, it never occurred to any composer of eminence before Bach; all have composed a longer choral movement by repeating the word "Barabbas" several times.

The chorus "Let Him be crucified" which, after a few bars of recitative, follows this soul-shaking cry, and its repetition later on, is a fresh example of the style peculiar to the Passion music. It is a fugue subject of eight bars, in which the parts, from the bass upwards, come in with strict regularity on the closing note of the subject in the previous part, thus giving us the impression of a coherent musical whole artistically worked out.[629] The shortness of the movement is sufficient to prevent our feelings lapsing into calm, and this is enhanced by the fact that the response to the theme is not constructed by strict rule, for the subject starts in A minor and we are ultimately flung aground, as it were, on the dominant of E minor. This of itself

---

[629] Marx (Kompositionslehre II., p. 276) finds expressed in this the solemnity of popular justice. Solemnity indeed there is, but it is that of Protestant church music.

accentuates a dramatic element, which is brought out in full force when the chorus is repeated; the words of the Evangelist, "But they cried out all the more and said," require an increased emphasis. In the oratorio style, where musical principles are paramount, this would have had to be brought out by a more complicated and intensified effect in the music itself. Bach simply repeats the chorus, but a tone higher. He actually depicts the populace in its natural excitation. The sense of their utterance remains the same—they only shout in shriller pitch, starting on B, with which he ended the first time; he now closes the chorus on C sharp (in the bass), the dominant of F sharp minor. In this passage a comparison with the St. John Passion is particularly instructive. In that also the chorus "Crucify Him" occurs twice. The composition is very similar on both occasions, but the first time it is in G minor and the second, half a tone lower, in F sharp minor. Here no enhanced effect has been attempted. The secret of this strange proceeding lies in the order of the modulations of the whole of the second part of St. John. For the sake of the general principle of musical construction Bach has not only sacrificed all dramatic emphasis, but even counteracted it, in the same way as a short time previously he has repeated the chorus "We have a law" to other words, but a semitone lower.

The influence of the dramatic aspect is shown in a peculiar way of closing the choruses in other passages of the St. Matthew Passion, besides those at the Crucifixion. Not unfrequently they end on the dominant of the key, even when the text does not contain a question, and thus produce an impression of something unfinished and leading on to a sequel. Bach has also used this effect almost throughout the St. John Passion.

But besides this, again, in the St. Matthew Passion, choruses occur where the progression of the modulation does not return on itself, but leads to some new end—where, consequently, a psychological progression is depicted and not a single feeling. The chorus "His blood be upon us and upon our children" carries us on, from the gloomy surging of hatred and blind impulse to murder, to an insolent defiance of the law of divine retribution. In the choral number "Sir, we remember," the picture of the chief priests and Pharisees talking themselves into vehement zeal has led Bach to modulate at last quite out of the key of E flat major, which is undoubtedly the original. The mixture of breadth and conciseness, which is characteristic of the choruses of the St. John Passion, we must regard as a not perfectly successful amalgamation of the oratorio and dramatic styles. In the St. Matthew, the problem is throughout triumphantly solved; an elaborate musical scheme for a long chorus is nowhere to be traced in the dramatic portions, and where the choruses are of any length it is the result of the length of the words to be sung. They are marked throughout by the utmost concentration, but combined with a severity and artistic treatment of the musical texture which always bears a direct proportion to the importance of the poetic sense to be interpreted, while both are in keeping with the purest church feeling.

As regards the different expressions of emotion which depend on individuals or situations, to do them full justice we must study them through the double medium of church feeling on one hand, and a conception midway between the drama and the oratorio on the other. Even then they exhibit a very sufficient differentiation. The choruses of disciples are often distinguished by a feeling of humble devotedness which, in the passage where they desire to prepare the Paschal Lamb for their Lord, has a vein of solemnity, and which, when they hear that there is a traitor among them, is tinged with anxious sadness. In the choruses of the persecutors, fanatical hatred is revealed in all its varying shades; it rises to terrific rage when they demand that Christ shall be crucified—none but a transcendent genius could come so near to naturalistic expression without overstepping the limits imposed by the style of the whole work.[630] As soon as the Jews are certain of their victim

---

[630] It has long been supposed that the dislocated form of the theme and the crossing of the parts in their progression might be intended to have a graphic effect. The similarity of the theme with that of the Crucifixion chorus in the

their fury turns to fiendish triumph; and a solemn conviction is stamped on the short chorus of the heathen: "Truly this was the Son of God."

As in the Passions according to St. Luke and St. John, so in that according to St. Matthew, Bach has distinguished one of the chorales introduced from the rest by frequent repetition, thus making it the centre of the church sentiment of the whole work. Among the fourteen simply set chorales included in the work in its original form, the melody "O Thou whose Head was wounded" occurs five times; it was a favourite melody with Bach, and there is no other that, throughout his long life, he used so frequently or more thoroughly exhausted as to its harmonic possibilities for every variety of purpose. It comes in three times in the second part: first when Jesus silently bows to His fate at Pilate's decision. Here it is used with the words of the first verse of Gerhardt's hymn "Commit thy ways to Jesus," one of the few passages in this incomparable work to which perhaps exception might be taken. It was a beautiful idea to associate the pious submissiveness of Jesus with a congregational meditation on it; but the soft and heartfelt tone of Gerhardt's hymn, with its admonitions to patience under human affliction, is not commensurate nor appropriate to the solemn pathos of the situation and the awfulness of the sufferings of the Son of God. Apparently Bach felt chiefly the need for bringing in the melody; in the hymn "O Haupt voll Blut und Wunden" there was no suitable verse, and he seized, not very happily, on another well-known hymn by the same writer. The second time the

---

St. John Passion is striking. The disposition to a picturesque treatment at that time sometimes found expression in a kind of writing for the eye; thus Mattheson once figured the "rainbow," marked by the scourging on the Redeemer's back, by a series of notes, which, as written, showed the form of an arch (see Winterfeld, Ev. Kir. III., p. 179). The leading idea of Bach's Crucifixion theme—

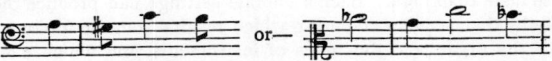

make the sign of the cross when the first and last notes are joined by a line and the two middle ones by another; nor is there anything displeasing in the conceit, since the musical idea is a sound one.

chorale is sung is in the second section, immediately before the progress to the Cross, when the soldiers have crowned the Saviour with thorns and mocked Him and smitten Him; and we here have the first two verses of the hymn addressed to the Head of Christ.[631]

Nothing more suitable could be found for this place, and the effect is consequently deeply touching. The third time it is the last chorale in the work, and it comes in after the words "But Jesus cried with a loud voice and departed," with the ninth verse of the hymn "If I should e'er forsake Thee, Forsake me not, O Lord." This climax has always been justly regarded as one of the most thrilling of the whole work. The infinite significance of the sacrifice could not possibly be more simply, comprehensively, and convincingly expressed than in this marvellous prayer. Bach has chosen for it a particularly low pitch, and while he has in other places treated the mode as Ionic, as it was originally, he has here, certainly after full consideration, worked out the solemn and twilight effects of the Phrygian.[632] The three moments marked by the entrance and recurrence of this melody, are the three decisive crises in the treatment of the second part.

---

[631] The English version slightly alters the sense. The following is rather more literal:—

"O wounded Head, and bleeding,
Weighed down with pain and scorn,
Thou'rt crowned by cruel mockers
With piercing boughs of thorn!
O Head! thy radiant glory,
With honour crowned of yore,
Is now sweat-stained and gory—
I hail Thee and adore!"—*Translators.*

[632] It is an error which has grown to an almost universal custom to have this chorale movement sung *a cappella*. Setting aside that it is unfaithful to the original, as it certainly is, it decks out the passage by the use of an effect quite unlike Bach, and gives it a touch of sentimentality which could nowhere be in worse taste than here. Bach's chorale settings can produce their special effects only by that peculiar colouring which results from the mixture of human voices with the organ and the tones of instruments, and which cannot find a substitute in anything else. The instruments have besides, in Bach's hands, so much to say of their own individuality that they constitute an intelligible symbolism when they come in unanimously with the four voice parts.

The passages in the first part where it is introduced are in the scene on the Mount of Olives. Jesus says "Ye shall all be offended because of Me this night: for it is written, I will smite the shepherd, and the sheep of the flock shall be scattered abroad." The figure of the shepherd leads to the introduction of the fifth verse :—

> O Lord, Thy love's unbounded,
> So full, so sweet, so free.[633]

Then follows the prediction of Peter's denial, and Peter and the other disciples pledge themselves to be faithful; this is followed by the sixth verse—

> Though all men should forsake Thee,
> Yet will not I, O Lord.

In the St. John Passion—after the narrative of Peter following Jesus when He is led captive to the palace of the High Priest, where he subsequently denies Him, we have the aria "I follow Thee also, my Saviour, with gladness," and it has been already pointed out that this sequel can hardly be thought satisfactory. On a comparison with the corresponding passage in the St. Matthew Passion, it seems quite clear that Bach could not carry out the idea he had in his mind for lack of adequate verse. It was not the reflections arising out of this particular incident, but those suggested by the course of events generally that he desired to express. This is made quite clear by the original closing chorale of the first part, which, after the statement "Then all the disciples forsook Him and fled," begins "Jesum lass ich nicht von mir":—

> Never will I quit Thee, Lord,
> Walking ever at Thy side.[634]

The Christian congregation stands in conscious contrast to the disciples: they fled while it remains unalienated from Jesus. Thus, through the whole of the events on the Mount of Olives, the atoning sentiment prevails that the

---

[633] "Erkenne mich, mein Hüter, Mein Hirte nimm mich an"—"Know me for Thine, O Saviour, My Shepherd keep me safe."

[634] This composition, never hitherto published, is given as a Musical Supplement, No. 3, to this work.

very sufferings endured by Jesus at the hands of those who had once loved and honoured Him are what bind Him the more closely to His people. Now we begin to understand what is signified when the chorale is introduced for the second time with precisely the same setting as at first, and only a semitone lower (E flat major instead of E major). The contemplation of human weakness revealed in the disciples fills the people with humility; but even in humiliation they remain firm.

Next to this chorale, the one which fills the most important part is Johann Heermann's hymn, "Herzliebster Jesu, was hast du verbrochen?"—"O blessed Jesus, what is Thy transgression?"—with the first, third, and fourth verses,[635] and we have also Gerhardt's hymn, "O Welt, sieh hier dein Leben," twice, with the fifth and third verses. The other three chorales, like the original closing chorale of the first part, are not strictly Passion hymns, but of more general import, and all chosen with a fine sense of fitness.[636] At the end of the first part Bach subsequently put what had been originally the opening chorus of the St. John Passion. The intention of the two verses of the chorale "O Haupt," which are introduced into the first part, is certainly thus rendered rather less evident. However, the grand proportions of the whole work acquire a more impressive culmination, and Bach must certainly have desired not to omit from a work of so exhaustive a character the best known and most venerable of the hymns on the Passion; such a hymn is Sebaldus Heyden's "O Mensch bewein dein Sünde gross," which was frequently sung even in the old chanted Passions, though in Bach's time it was no longer used on Good Friday in the chief churches of Leipzig. The composition which Bach has set to the first verse of this hymn is not a simple four-part chorale, but a chorale fantasia on the grandest

---

[635] "Mysterious act of God's Almighty mercy," and "O Saviour, why must all this ill befall Thee?"

[636] "O Father, let Thy will be done," verse 1 ("Was mein Gott will, das g'scheh allzeit"); "The world accuses falsely, Lord," verse 5 of the hymn "In dich hab ich gehoffet, Herr"; and "If I should e'er forsake Thee," verse 6 of "Werde munter, mein Gemüthe."

scale and of the richest style. The tone-picture given to
the instruments is woven out of a motive which seems to
owe its origin to an illustration of "weeping"; and while this
represents the general feeling, the alto, tenor, and bass voices
of the combined chorus are busied with working out the
individual shades of sentiment, the *Cantus firmus* being given
to the soprano. This mighty movement, saturated with the
most intensified feeling of the divine Passion, has so long been
universally regarded as an integral portion of the St. Matthew
Passion, and it so admirably fits its place, musically speaking—
particularly as it is a splendid finish to the first part—that
we are scarcely conscious of the fact that, in one respect, it
cannot conceal its original purpose; the poetical purport of
the words refers almost exclusively to things which are
antecedent to the Passion properly speaking. The verses
ought to form an introduction to a rhymed narrative of it,
and are therefore out of place here, where the history has
already begun; and the contrast must have been still more
perceptible at a time when the purport and feeling of
Heyden's hymn were more familiar to every one than
they are now. Bach did not transfer the chorus directly
from the St. John Passion; he seems to have used
it first for a fifth Passion setting, now lost. It was not
till near the end of his career that he added another
embellishment to his greatest Protestant church work, by
adding this; thus framing the first part between two chorale
choruses of the very grandest type; for the opening chorus
also is nothing more, as to musical structure, than a treatment of the chorale "O Lamm Gottes unschuldig," though
it is, to be sure, a treatment in which the counterpoint is
carried on by two choirs with their own texts, and two
orchestras, while a third choir in a single part sings the
Cantus firmus. When we compare it with the chorale in the
Little Organ Book, constructed on the same melody (twenty-
seven bars long), no words are too strong for admiration of
the colossal power which was equal to such an elaboration
of form, and which, within the development of one man's
mind, was able to produce what in other cases required
the evolution of generations—for instance, consider the

history of the symphony. We shall return to a closer analysis of this chorus; but, for the present, it may lead us from the study of the chorale numbers to that of the numbers in madrigal form, since it combines both of these in itself.

The madrigal words have been given by Picander to the Daughter of Sion and to the believers; in this we may perhaps trace his dependence on Brockes. At the same time the idea is not systematically carried out as it is in Brockes and, indeed, in Picander's own Passion text of 1725. It remains conspicuous only in six numbers—important ones it is true — and it would here have resulted in a distinctly oratorio-like effect if Bach had consistently throughout given the words of the Daughter of Sion—a strictly allegorical personage—to a solo voice, as Handel did in his setting of Brockes' Passion. But he has not composed them all as songs for a single voice, but also in the choral and duet forms; and in the solo songs he varies the voice, thus generalising the personal feeling. From a few trifling changes made also in the text it is made additionally clear that he was throughout inclined to express the feelings of the congregation.[637]

From a poetic point of view no separation can subsist between the individual believers and the body of the Christian church,[638] for the believers have church hymns to sing as well as independent verses ("Lamb of God" and "Mysterious act"); it is only in a musical sense that the distinction is marked. A general observation has already been made as to the style of the madrigal choruses; while conciseness and an intricate treatment are combined in the dramatic choruses, in these breadth and simplicity have joined hands; they are intended for sacred arias and are so entitled in the text. It was, of course, impossible to give

---

[637] At the beginning of the second recitative "Although mine eyes with tears o'erflow" Bach changed "mine" into "ours." In Novello's edition the singular is restored in the English. In the second recitative of Part II. (p. 98) Bach substituted the words "to show *us*" for Picander's text which simply said "to show" ("Um uns damit zu zeigen," for "Um damit anzuzeigen").

[638] Winterfeld, Ev. Kir., III., p. 372, assumes this to be the case.

them the verse form in such a work, and the poet has therefore disregarded this consideration: they are in the Italian aria form. Bach has, however, succeeded in keeping up a connection with the older form by adhering to the homophonic method of setting and the simplest possible phrasing. In this way a new style has been originated in this department, bearing fresh witness to Bach's inexhaustible genius. In the whole realm of church music there is nothing that can be compared with these choruses; they have the broad dimensions of Bach's cantata choruses, and at the same time are as simple and as intelligible as a ballad.

The first time such a chorus comes in is when Jesus is praying in the garden of Gethsemane. It is introduced by a recitative for a tenor of the first choir, full of sympathy for the Agony of Jesus, and which is interrupted by chorale lines sung by the second choir, "O Saviour! why must all this ill befall Thee?" (p. 43) and then the anguish is merged in the resolution to cling to Jesus alone and find rest in Him from the torments of the consciousness of sin. Here again a tenor comes in with the soothing, lulling passages, from the second chorus, which represent the "falling asleep of sin." We listen with rapture to the flowing *sostenuto* melody of the upper parts, and scarcely perceive that the same melodic charm is equally present in all the others. They might equally well fill the place of upper parts and, in fact, become so by various changes of position in the course of the double counterpoint.

When Jesus has been taken another grand madrigal choral piece comes in. Two voices in the first choir begin their lament "My Saviour Jesus now is taken." It is supported by a lower part on the violins and violas which certify their connection with the notion of "being led away captive" by syncopations, and then by slowly progressing quaver passages—one of the many examples of Bach's manner of extracting a musical idea from the external factor in an event contemplated from within, and of then freeing himself from its influence and working out only its lyrical features. The deep melancholy of the two solo voices is broken by short and vehement cries from the second choir, "Leave Him!

leave Him, bind Him not." Then both choirs burst out in a *vivace*, and with righteous indignation call upon Heaven to hurl down thunder and lightning on the traitor and his accomplices. The music wars and raves like the wind and storm; nevertheless the form is very simple: an aria of which the second and third divisions are compressed into one—the choral treatment is perfectly homophonic apart from the fugal beginning; the two choirs sing in parts, each by itself, in a compact body, and only unite at the close. Any one knowing Bach only from his cantatas would hardly trust in his capacity for producing from such simple combinations an effect which is one of the most thrilling in the whole realm of music. And how miserably do the undignified execrations, which in other Passion settings have to be sung by a solo voice in this place, compare with this picture, traced as it were in flames of fire![639]

The third madrigal-choral number of the St. Matthew Passion is the final song by the tomb of Christ. This also has an introduction, a brief recitative for different voices in turn, with short choral subjects full of yearning, and of that deep feeling which checks all but subdued utterance as we stand by the grave. The intimate connection of the final chorus with that of the St. John Passion is self-evident. The character has, however, become far calmer and simpler, the contrast in style with the dramatic choruses more decided, the sense of woe more pious, and more confident in the work of redemption wrought by Christ—a combination of beatitude and sorrow which none but Bach could ever achieve and which is most conspicuous in the words " Hope returned, Here the weary close their eyes." And although we gain from this chorus an impression of breadth which is precisely suited to the rest of the work, it is considerably shorter than the corresponding one in the St. John Passion. The inexhaustible flow of lamentation there leaves a remnant of unsatisfied feeling which is only relieved by the chorale which succeeds it. In the St. Matthew Passion this is not

---

[639] Picander evidently took as his model the text of Peter's aria in the Passion by Brockes.

needed; the sentiment soars slowly upwards, and is amply worked out within the established limits of the aria form. To close a Passion-setting with an aria instead of a chorale was quite opposed to all tradition, and we must not underestimate this fact. It contributes in an important degree to give to the St. Matthew Passion that tenderly human stamp which distinguishes it from the St. John Passion. It seems to me that the simplicity of form of all these madrigal choruses does not arise merely from a general necessity for musical contrast—I see in it also a sign that Bach desired to have them recognised as an extended form of sacred aria; and there is fresh and striking evidence in favour of this in the introduction to Part II. of the St. Matthew Passion. The alto solo of the first choir sings with the whole of the second choir, but the solo sings a lyric text and the choir sings Bible words.[640] Another style of choral writing is here at once perceptible; subjects of a motett character, beautifully complete, come in with a consolatory effect, answering the Daughter of Sion, while her strains drop one by one, like hot tears, with an expression quite different from that deep and tearless grief with which, in the first part, she gazed after Jesus as they bound Him and led Him away.

One more of the solo arias included in the St. Matthew Passion claims the co-operation of the chorus. In the figure of Jesus, hanging with out-stretched arms upon the Cross, the Daughter of Sion beholds the image of Love ready to embrace in His mercy those who crave redemption. She calls on the "forsaken broods"[641] to shelter themselves in Him, and the believers interrupt her with ejaculated questions. The musical idea is of the same character as "Haste, ye deeply wounded spirits," in the St. John Passion. It was Brockes who introduced this style, and he found many followers, among them Rambach, and Picander also, in many parts of the St. Matthew text. However, in the opening chorus, and again after Christ is

---

[640] Song of Solomon, VI., 1.
[641] Compare Matthew xxiii., 37, "As a hen gathereth her chickens."

apprehended, the short choral ejaculations participate in a connected manner with the regular chorus, and have a definite place in the development and completing of the composition. The impression given when the number is considered solely as a musical work—which is an indispensable condition with a form of art which is to produce its effect apart from action—is one of urgent longing, exactly like that of the bass aria in the St. John Passion.

The grades of feeling traversed by Bach in the solo songs of the St. Matthew Passion are all the more impressive, because every sentiment of joy in its various shades is wholly excluded; they are all based on the emotions of sorrow. The most fervent sympathy with the sufferings of the Son of Man, rising to the utmost anguish, childlike trustfulness, manly earnestness, and tenderly longing devotion to the Redeemer: repentance for the personal sins that His sufferings must atone for, and passionate entreaties for mercy; an absorbed contemplation of the example offered by the sufferings of Jesus, and solemn vows pronounced over His dead Body never to forsake or forget Him—these are the themes Bach had to treat. And he has solved the difficult problem as if it were child's play, with that inexhaustible wealth of resource which was most at his command precisely when he had to depict the sadder emotions. In no other of his works (unless it be in the Christmas oratorio) do we find such a store of lovely and various solo airs, nor did Bach even ever write melodies more expressive and persuasive than those of the arias in the St. Matthew Passion. If here and there the stream of tunefulness flows somewhat less richly, as in the tenor aria "Rejoice, rejoice," it is because the contemplative character of the text has led to it. A critical note is, however, demanded by the bass aria "Give, O give me back my Lord." It comes in after Judas has restored the "price of blood" to the Chief Priests and testified that Jesus was innocent, and the words express the desire that Jesus should be set free, since even a Judas must acknowledge His innocence. Now, all the other madrigal texts express sentiments and reflections which, it is true, are connected with certain crises of the history of the Passion, but still contain permanent truths of Christianity. This one alone is not the utterance of a member of the Christian communion—who has not lost Christ through His captivity and death but who, on the contrary, has not till now truly found Him—it is that of a person who has stood by during the events recorded; a disciple perhaps or some other follower of Jesus. The influence of the dramatised oratorio is here perceptible, and not quite satisfactory. The hearer is obliged to change his point of view; in the other arias it is based on the general facts and bearing of the work of redemption; here, on the contrary, it is limited to the feeling aroused by the immediate incident.

All the madrigal recitatives have obbligato accompaniments in which, as has been said, wind instruments are employed in preference to the strings, in order to distinguish them as vividly as possible from the Biblical recitatives of Christ. Space will not admit of an analysis here of all the picturesque features and ingenious phrases in which they are so rich; where he thought the work would gain by it Bach did not shrink from the boldest strokes, such as the startling enharmonic modulation at the end of the recitative "All gracious God!"

The recitative "'Twas in the cool of Eventide" (p. 167) requires, however, some more detailed explanation; the text is on the model of Franck's. It is a well-known trait in the Teutonic mind that it loves to regard nature, external to man, as sympathising in the joys and sufferings of humanity, and to think of its phenomena as animated by a spirit akin to the human soul. The nations of antiquity and the Romans, who in their metaphysical conceptions had borrowed from them largely, were strangers to this characteristic; their attitude towards nature was determined solely by the pleasures she offered, or the burdens she imposed. This vein in German poetry is prominent or subordinate in proportion as the poem moves chiefly on national or on foreign ground. While the medieval poems, particularly those of the Minnesänger, are full of this sentiment of natural symbolism, and while in popular songs it was still preserved, in the artificial poetry of the seventeenth century it disappeared more and more, not to revive till the second half of the eighteenth century, and then to shoot forth once more in full vigour from the same soil—the national ballad. A parallel process is observable in music. So long as Italian ideas were paramount in the music of Europe we find hardly a single attempt at expressing the romantic moods and aspects of nature. Suggestions of this kind occur in Gluck—in the "Orfeo" and in "Armida"—and again in Cherubini, who had certain German proclivities —in "Elisa"—but it is not till we come to the accompanied vocal music of Weber and Schubert that this feature is fully revealed, while in Haydn's Oratorios the antique view of Nature is still sensibly felt. It is, therefore, especially noteworthy that here and there in Bach's music the romantic feeling for nature is already unmistakably perceptible; nay, this may even be said of some of his ancestors.[642] An instance of such imagery—revealing the core of Bach's thoroughly German spirit—is offered in this very recitative. The softened violins float and linger like the tender mists of twilight; there are no figured bass chords to overweigh the dreamy web of sound, only the long drawn tones of the organ and stringed bass lend it support. The feeling that is breathed out from this composition is not, in the first instance, the religious sentiment of peace and redemption spoken of in the text. It is the sense of evening; the deeper sentiment is only brought out musically through this medium, and the emotions that stir the heart of the Christian are infused into the picture of nature. An instance, in many respects similar, is offered by the opening chorus of the cantata "Bleib bei uns"—"Bide with us"[643]—in which the sombre tones of the united violins and violas alternating with three oboes, which at first move very softly through several bars and then are long held, give us a feeling as of falling shadows. But here it is less the peaceful sense of evening calm than the unearthly tremor which accompanies the gradual fading of daylight.

---

[642] See Vol. I., p. 71.
[643] B.-G., I., No. 6. Also as "Bide with us," with English words by Rev. John Troutbeck, D.D. (Novello & Co.)

I have already taken occasion to point out another instance of the same kind; a cantata for the fifth Sunday after Trinity begins with the verse from the Bible "Behold I will send for many fishers, saith the Lord," &c. (Jer. xvi., 16). Bach has not used these words, according to his usual custom, for a simple dignified arioso; he has worked them up into a grand tone-picture in two divisions, in which we are first led to see the tossing surface of the sea, and then to listen to the horns echoing through the wood. These nature-pictures do not merely supply him with a definite musical motive; they constitute the romantic aspect of the feeling he desires to depict.[644] In the tenor aria of the Whitsuntide cantata "Erschallet ihr Lieder,"[645] and the soprano aria of another Whitsuntide cantata "Also hat Gott die Welt geliebt" (transferred from the secular cantata "Was mir behagt"[646]) we feel, as it were, the breath of May. And the Easter cantata "Der Himmel lacht" contains passages in which we almost seem to feel the airs of Spring.[647] A threatening and stormy mood breathes through the cantata "Schauet doch und sehet," and in the bass aria "Dein Wetter zog sich auf von weitem."[648] Even in several instrumental pieces this romantic vein is very perceptible; conspicuously, for instance, in the Pastoral Symphony of the Christmas Oratorio. As this spirit has something pantheistic in it, it is quite intelligible why in such works as these—which stand on the foundation of the old Lutheran doctrine—a very small space can be allotted to it. Still, the occasions are numerous when it is impossible to ignore its presence. They suffice to show us the contrast between Bach and Handel as to their views of art, both sacred and secular. A poem like "Allegro e Pensieroso," which in its most beautiful passages takes for its subject the harmonious sympathy of nature with humanity, could not have failed to call forth a romantic response in the region of music, if any such conception had ever existed in the sphere

---

[644] See ante, p. 472.   [645] See ante, p. 398.   [646] See Vol. I., p. 568.
[647] See Vol. I., p. 541.   [648] See ante, p. 428.

of Handel's ideas; but Handel stood too firmly on an Italian basis for this to be possible. He has derived a multitude of musical motives from the realm of nature, but he is far from knowing how to vivify them with any mysterious reflex action of the human soul. His music is worked out in clear and classical lines; the composer stands face to face with nature, gazing at creation with a steady eye, as its master and not as an integral portion of it; the mystic absorption into the great universal mind which is essential to the romance of nature has left not a trace. To interpret its moods in musical forms demands more complicated lines than Handel loved to work in, and above all a greater wealth of colour. Everywhere in Handel we find a blooming and abundant sweetness of melody, but no very great variety of handling; Bach is the greater colourist, and the effect of his works results in a high degree from the richer effects he has lavished on them.

These reflections lead us back to that great tone-picture in madrigal form in Part I., which follows closely on the capture of Jesus:—

> My Saviour Jesus now is taken.
> Moon and stars
> Now refuse to yield their brightness.

Since the events narrated occurred on the day of the Passover, the full moon must have shone down on Christ when he prayed by night on the Mount of Olives. We must suppose Him to have been taken towards the end of the night, since the moon had already set. By taking this into consideration Picander has introduced a popular trait into his poem; the Teutonic mind would recognise a deep symbolism in the coincidence of a natural phenomenon with the transcendently important occurrences in Gethsemane—nature mourning over the sufferings of Jesus.[649]

---

[649] This or similar ideas were frequently worked out by the German poets. The pretty popular song is well-known:—

> " Our Saviour Christ went forth to pray,
> And soon began His agony,
> The greenwood mourned and grassy sward,
> That Judas could betray his Lord."

That the desolate feeling that pervades Bach's composition was determined by this idea in the poem, I cannot venture to assert. We may rather suppose that it was Bach's influence which led to this application of the popular *motif*. It is impossible not to see that the opening chorus also reflects some popular usage. The text, which treats of the progress to the Crucifixion, where Jesus bears His own Cross, amid the lamentations and tears of the Daughters of Jerusalem, Luke xxiii., 27, does not seem particularly suitable as an introduction to the whole history of the Passion. It gives prominence to a single incident, and that not the most important, in a drama full of movement; and we are led on through Part I. and half of Part II. before we arrive at the progress to the Cross. Usually the Passions were introduced by a chorale, or else with an exhortation in verse to contemplate the sufferings of Christ; and a text like the present must have seemed altogether strange if it had not borne a reference to the ancient custom of the Good Friday processions. The Passion plays were so conducted, in many parts of Germany, that only a preliminary portion was performed in the church, while the principal action was played in a procession, arranged to go to a raised spot outside the church, called the Calvary or Hill of the Cross. This procession was planned on the Biblical narrative of the progress to the Cross; the different personages, distinguished by their clothes or by emblems—among them a representative of Christ with the Cross—marched in tra-

---

"Christus, der Herr im Garten ging." Des Knaben Wunderhorn. Part I., Ed. II., Heidelberg, 1819.

Friedrich Spee (1592—1635) in a "Lamentation for Christ's agony in the garden," makes the Saviour Himself say:—

> The lovely moon will soon be gone,
>   And rise no more for sorrow;
> The stars this night have hid their light,
>   In grief for Me to-morrow.
>
> No song is heard, no piping bird,
>   The woods in silence languish;
> The wild beasts grieve in hole and cave,
>   And share in all My anguish.

"Der schöne Mon, wil vndergohn," in Trutz Nachtigal. Cöllen, 1649, p. 227.

ditional order, chanting hymns of lamentation. At certain spots the procession halted and performed the more dramatic scenes. These, however, always included those incidents which had occurred before the progress to the Cross, so that the whole thing might suffice as a concentrated representation of the history of the Passion, ending with the Crucifixion on Calvary and the Burial. Remains of these processions continued to be kept up in Schleswig as late as the beginning of this century, and in the lower Rhine provinces till the end of the seventeenth.[650] We have no present means of knowing how far any traces of the Good Friday processions had survived in Thuringia and Saxony. But it is quite clear that the text of the introductory chorus of the St. Matthew Passion could only have been based on a view which attributed to the procession of the Cross, and its sequel, everything that was essential in the history of the Passion. What Bach constructed musically is a stupendous picture of a surging crowd moving onwards with hymns of lamentation. But, at the same time, he has clung to the custom of opening the Passion with a chorale. A third chorus in unison joins in with the intricate maze of the other two choirs and the double orchestra, with the chorale "O Thou, begotten Son of God." As it was a favourite method in Leipzig to place separate choirs in different parts of the church to sing together, it is extremely probable that the chorale was sung by the small choir, which was placed opposite the grand organ. It must also be said, to Picander's credit, that he has shown great skill in adapting the different portions of the madrigal text to the lines of the chorale. The chorale is made to stand out both outwardly and logically as the power which governs the whole, and in this way this grand composition was most intelligibly brought within the limits of Protestant church music.

It is not improbable that Picander, who in other portions of the poem took Franck and Brockes for his models, here

---

[650] Peter, Zuckmantler Passionsspiel. Troppau 1868, p. 10. Also Rein, Vier geistliche Spiele des 17, Jahrhunderts. Crefeld, 1853, p. 9.

actually imitated certain popular hymns. Setting aside the
form of question and answer, the text of the opening chorus
has quite the stamp of the hymns sung in the Good Friday
processions.[651] It also contains one expression which clearly
betrays its primitive and popular origin. According to old
German custom, it was the part of the relatives of a deceased
person to "help" in the wailing over the corpse; and the
expression is frequently used in the composition of Passion
texts, particularly in the Marien Klage, "Mary s Lament."[652]
Picander makes the Daughter of Sion sing "Come, ye
daughters, help me to lament." That he had in his mind
some poem of the kind I have no doubt, and all the less
because a similar poem certainly—though probably unconsciously—influenced his mind when writing the words
for the aria "Blute nur, du liebes Herz"—"Break and
die, thou dearest heart." Picander is always clear and
lucid in his own verse, but where he imitates or paraphrases
he is apt to lose himself. In the words:—

> Break and die, thou dearest heart!
> Ah! a child which thou hast trained,
> Which upon thy breast remained,
> Now a serpent has become,
> Murder is the parent's doom—

the wildest confusion certainly prevails. We do not immediately see whether it is the heart of the believer, or of
the Daughter of Sion, or of Jesus Himself that is spoken
of; judging from the last two lines, Picander can only have
meant the heart of Jesus, though the expressions in the
second and third do not correspond to this idea. The
phraseology is in fact in great measure that of a "Mary's
Lament," and made use of in this very sense by Picander
himself in the soliloquy he puts into the mouth of the

---

[651] Compare in Rein, the hymn on p. 20. "Schaue Sion deinen König."
[652] "*Maria cantat:* Johann's beloved cousin mine, Help me bewail my grief
and thine," from a MS. of St. Gall of the fifteenth century printed in Mone,
Schauspiele des Mittelalters, Vol. I., p. 200. "Weep, ye faithful sisters, and
help me, a poor troubled soul; help me to bewail my grief, my loss which
is grown great, and my heart's pain." In a Low German Marienklage in a
MS. from Wolfenbüttel published by Schönemann, Hanover, 1855.

Virgin Mary, in the Passion of 1725. It is difficult to answer the question as to what Bach with his keen intelligence can have understood by these meaningless words, for which he, nevertheless, evolved one of his finest arias.

There is still another passage full of naïve simplicity in the St. Matthew Passion, which, however, is attributable solely to the composer. After Peter has denied the Lord, the Evangelist tells us "And Peter remembered the words of Jesus, which said unto him 'Before the cock crow thou shalt deny me thrice,'" and he has set the words to a musical figure which mimics the crowing of a cock. In the scene on the Mount of Olives, too, to which these words refer, there is the same graphic figure, and in the corresponding passage in the St. John Passion there is a bass figure in semiquavers which is inexplicable but on the supposition that it represents the crowing of a cock; the intention is, therefore, unmistakable. Those who regard such an interpretation, as beneath the dignity of Bach's genius,[653] may be asked to remember that Schütz, in his three Passions, and Sebastiani also have made use of a similar though less assertive kind of imagery. We may also ask whether it is less extravagant when Bach represents the idea of a "high" pavement, and even of a "high" priest by the use of high notes, and finally point to the character of Bach's recitative in general, such as I have attempted to show it. In spite of all this, however, we must admit that there is something rather startling in the mimicry of an animal's cry, and that the contrast it offers to the solemn gravity of the whole work threatens to fall into absurdity. But the direct connection with popular traditions of art gives the proceeding a deeper meaning, and even a certain justification, if we rightly comprehend the principles on which Bach bases his work. The crowing of the cock in the sacred plays formed a favourite crisis, with the populace at any rate, by reason of its simple naturalism,[654] and it is no wonder that they objected to its

---

[653] As Mosewius, Johann Sebastian Bach's Matthäus Passion, p. 6.
[654] Mone, Schauspiele des Mittelalters. Vol. I., p. 108.

omission even in the Passion music. Scheibe, to give an instance of certain ancient Franconian modes of expression, gives us the narrative of a musician from Schleswig: "He once conducted the performance of a Passion music, and in order to render the crowing of the cock very plainly and naturally he hid one of his musicians behind the organ, who at the proper time imitated on nothing but the pipe of the hautboy the crowing of the cock with such naturalness that all the hearers were put into the utmost astonishment, and gave due praise to his happy idea."[655] Certain performances of the sufferings of our Lord existed in Saxony down to the twentieth year of the present century, the remains of the Passion plays. They were given by perambulating choirs of singers, and in these the three crowings of the cock, mimicked on a pipe by one of the singers, was always looked for with excitement, and received with great delight and applause.[656]

Bach's St. Matthew Passion as a whole is, therefore, in a remarkable degree a popular work; and this character does not rest solely on its connection with certain national aspects of thought, or in the faithful preservation of church traditions that had grown dear, but on the whole character of the music. With all its profundity, breadth, and wealth, and in spite of all the art lavished upon it, it never belies the lucidity and simplicity which are its mainstay, and at the same time seizes and grasps with amazing certainty that leading sentiment which pervades the whole history of Christ's sufferings and death—namely, atoning love. Though violent and thrilling emotions are not absent, they only serve to make the tender fundamental feeling stand out as all the more perfect and impressive. The contrast it thus presents to the gloomy St. John Passion is a marked one. It is needless to repeat once more how it surpasses that in many other respects, and how in it all the demands of the most complicated form of art that is conceivable are amply satisfied, and in the most masterly

---

[655] Scheibe, Critischer Musikus, p. 58.
[656] See a paper in the Musikalisches Wochenblatt, 1870, p. 337, by Kriebitzsch, who remembered having witnessed it himself.

manner. Favoured by a happy concurrence of circumstances, Bach has created, in the St. Matthew Passion, a masterpiece such as it is granted to the human race to have bestowed on them but rarely as the centuries grow and wane; and a monument at the same time of the German national character which will perish only with the spirit that gave it life.

The first performance of the St. Matthew Passion took place on Good Friday, April 15, 1729, at afternoon service. It would be rash to suppose that the hearers could at once appreciate it in all its significance. At the same time its reception seems to have been even less favourable than Bach was justified in expecting, since the Town Council did not feel moved to grant certain modest requests of the composer with regard to matters musical (see p. 241). Time alone brought due recognition of the St. Matthew Passion, and though it may have never become very widely known—the difficulty of performing it, and the insufficient means of most of the church choirs at the time, must have hindered that—it took a firm root in the soil of Leipzig and was performed by the Cantors of St Thomas' till the end of the seventeenth century.[657] Vesper service began at a quarter-past one, and under ordinary circumstances lasted till about three; since at a quarter-past three the University Church service began. But with the St. Matthew Passion the Good Friday vespers must have lasted more than four hours, since the music alone demands about two hours and a half for its performance. Its abnormal length would therefore reverse the usual and due proportions of the Passion music to the service; while the original idea—founded, indeed, in the nature of things—pointed to the embellishment of the service by the introduction of music, here we have a sacred concert to be performed with the addition

---

[657] See note, p. 518. In Mizler, Mus. Bib., Vol. IV., p. 109, an incomparable Passion music is mentioned, which, by reason of the vehemence of the emotions it expressed, had a good effect as chamber music, but a bad effect in a church. It is possible that this may have been the St. Matthew Passion.

THE ST. MATTHEW PASSION. 569

of the church service. Actually, indeed, the St. Matthew Passion trenches closely on the limit line where church music ceases and concert music begins; nor is this by reason of its musical style, which is still throughout church-like, but from the self-sufficing sense of art, for art's sake, which is revealed in it. As far as possible it exhausts and completes its subject by its own methods, and, in attaining the effect it aims at, it hardly retains the church element, excepting as a background determining the feeling. It hovers so near the borderland of secular music that the attempt must certainly have been made occasionally to produce it in the hall or room used by the Musical Union, or even in Bach's spacious dwelling. This characteristic affords fresh support for the opinion, derived from experience, that a work of art, in which the ideal type of its special form is brought to absolute perfection, bears in it the germ, which, in growing, entails its destruction.

The reader may here remember that, irrespective of the Passion performances which had long existed in the New Church since 1728, Görner had instituted Good Friday music at Vespers in the University Church. It follows that the latter service would have clashed with the performance of the St. Matthew Passion on account of the great length of Bach's work, so that after 1730, when Görner became organist at St. Thomas', he could not be available for the organ accompaniment of all the great Passion performances. But Bach conducted the performances of his own great works in St Thomas', preferring it for the sake of the greater space. Thus in 1731 he produced the St. Mark Passion; then, probably in 1734, the revised St. Mark Passion; and, possibly in 1736, a third performance of the St. John.[658] The St. Matthew Passion, in the altered and extended form in which we now possess it, could only have been brought out at the earliest in 1740.[659]

---

[658] See Appendix A., No. 54 and No. 56, towards the end.
[659] See Appendix A., No. 57.

VIII.

BACH'S COMPOSITIONS FOR CHRISTMAS, EASTER, AND ASCENSION.

THE deep congregational feeling which lies at the root of Bach's Passions underlies also his grand compositions for Christmas, Easter, and Ascension; at the same time the conditions of its manifestation are somewhat dissimilar. There is certain evidence that, in the medieval church, Gospel recitations were performed, with a distribution of parts, at Christmas, Easter, and Ascension, as well as during Passion week; but the Protestant church did not avail herself of them in the same way. At least, I have found no instance of their use at Christmas or Ascension during the first century and a half after the introduction of the reformed faith. It is not till the end of the seventeenth century that they occur in Thuringia,[660] and we can hardly suppose them to have had then any recognised connection with the customs of the older church; they must rather be regarded as merely suggested by the ancient and favourite Passion form. This explains why Bach should have given the works in question the title, then so foreign, of Oratorio. There was no designation at his service derived from church tradition and custom, so he borrowed the name of the form which seemed to have a comparatively close affinity to that which he had evolved. In this one respect the connection of the oratorios with the liturgy for the three festivals is far less deeply rooted than is that of the Passion music with its corresponding liturgy; on the other hand, the affinity to certain other national

---

[660] Rudolphstädtischer | Christ Abend, &c. (the Rudolphstadt Christmas, that is to say, the joyful narrative of the incarnation and birth of our Lord and Saviour Jesus Christ, brought together out of the Evangelists Matthew and Luke, and mingled with suitable hymns. As it is performed [*music*irt] at the meeting of the preparation or eve of the Christmas festival in the Hof Kappelle of the Counts of the Schwartzburg, at Rudolphstadt; printed at Rudolphstadt) Heinrich Urban. Im Jahr, 1698. In the Ministerial Library at Sondershäusen. From this title we gather that it was not a solitary, but a customary performance.

THE ORATORIOS. 571

church usages is all the more conspicuous in parts. The sacred dramas, in so far as they comprehended the whole work of Redemption, went back beyond the Incarnation, to the Old Testament types and preparatory events; but as they did not extend beyond the Ascension of Christ, the occurrences of Pentecost were not included. Bach, again, has left no Whitsuntide "oratorio," and from that we might infer that he had followed the popular tradition; but we must guard against too confidently assuming this to have been his intention, since at the end of the seventeenth century even Whitsuntide music on the pattern of the Passion was not unknown.[661]

I have already observed (p. 499) that the part-reading of the Passion gradually developed into the Passion plays of which the church was originally the theatre. The Christmas, Easter, and Ascension dramas had the same origin; but as the requirements were far simpler for the ceremonies and processions needed in their dramatic presentment, it is easy to understand that they could be included, entire, within the limits of the liturgy; while the Passion liturgy, to counteract the degeneracy of the drama as performed outside the church, was restricted again to the merest indications of dramatic purpose. The Christmas story offers only four incidents: the Birth of Christ; the announcement of the angels to the shepherds; the visit of the shepherds to Bethlehem; and the adoration of the three wise Kings. The symbolical dramatisation of these events had for a long time wholly or partly fallen into disuse in the Protestant service. What survived at Leipzig in Bach's day I have already spoken of when discussing the *Magnificat* (p. 372). Besides the custom of lulling the holy child, a symbolic ceremony representing the angels' message must also have been known there. It consisted in placing boys, dressed as angels, and divided into four choirs, in four parts of the church—for instance, in front of the altar, in the

---

[661] In the Ministerial Library at Sondershäusen a printed text exists (Rudolphstadt, 1690, printed by Johann Rudolph Löwe) of the same character as the Christmas music mentioned in the preceding note.

pulpit, in the officials' seats, and in the organ-loft—when they sang the Christmas hymn, *Quem pastores laudavere*, line for line alternately.[662] That this was done in Leipzig we know from Vopelius,[663] but this does not prove that it was customary in Bach's time. But, besides these liturgical ceremonies, there were the Christmas plays outside the church, which also represented these four principal events in a popular style, and with the utmost simplicity and freedom. Since, too, the festival of Christmas had absorbed into itself many primeval German heathen customs and ideas, it was rooted more deeply and firmly in the national life than the Passion plays; it consequently flourished more vigorously, and had been familiar to the children of the Thuringian race from the earliest times. Circumstances were, therefore, particularly favourable there for the institution of a Christmas mystery as part of the church service, which should prove an exhaustive means of expressing the ideas and sentiments of the people with the highest artistic perfection.

The Christmas Oratorio was written in 1734.[664] The Biblical text is from Luke ii., 1 and 3—21; Matthew ii., 1—12. This text is not divided into two sections on the plan of the Italian Oratorio as the Passions are, but falls into six sections for the three days of Christmas, New Year's day, New Year's Sunday, and the festival of the Epiphany; each division thus constitutes a complete composition for one of six days, and this is how it was usually performed. It has indeed been asserted that the Christmas Oratorio is merely a series of superficially connected, but really independent, cantatas. In our day, when church usages are neglected or forgotten, such an opinion cannot seem surprising, and the title of "Oratorium," though given by Bach himself, is misleading. But the church regarded the whole period till

---

[662] 1589, in Brandenburg, as we learn from Schöberlein, II., p. 52, corresponding to the German version of the hymn *Nunc angelorum gloria*, in the Arnstadt hymn-book of 1745, p. 22 f.

[663] Neu Leipziger Gesang-Buch. 1681. p. 44.

[664] B.-G., V.² Novello's octavo edition is here quoted for the English text.

Twelfth night—from Christmas Day, that is, till the Epiphany —which had been held as a feast even in heathen Germany —as one festival season of which the Birth of Christ was the central idea. Though the Catholic church had given certain Saints' days a place within these twelve, the Protestants endeavoured to remove them so as to devote them solely to the person of Jesus. Thus it was now in Leipzig. On the three days of Christmas the Birth of Christ was illustrated: on New Year's day, the Circumcision; on the first Sunday in the year and the feast of the Epiphany, the persecution by Herod and the adoration of the Kings; and on all these days the Christmas hymns were sung.[665] Thus, irrespective of the fact that the six portions of the Christmas Oratorio deal with a progressive series of events, they must be held, according to church views, to constitute a whole, as being intended for the six consecutive days. This church idea is indeed indispensable to the proper comprehension of the work; we must not approach it, any more than the Passions, with the expectation of finding a concert-room oratorio. What Bach did here for Christmas he had already done for the usages of Holy week, as I have already observed (p. 527).[666]

The madrigal pieces are for the most part transferred from secular occasional music. The opening chorus of the third part and two arias are derived from a *Drama per Musica* written by Bach, and performed by the Musical Union for the Queen's birthday, on December 8, 1733; four arias, a duet and a chorus are transferred from a work of a similar character composed for the birthday of the heir-apparent, September 5, of the same year, and one aria from a complimentary cantata to King Friederich August III., on his visit to Leipzig, October 5, 1734. Of the remaining six pieces it is probable that four were transferred from other

---

[665] See ante, p. 277. The fact that the Christmas hymns were again sung on the feast of the Purification (Feb. 2) must be regarded as a last echo of Christmas.

[666] As the Christmas Oratorio was composed for a year when there was no Sunday after Christmas till after New Year's day, in after times so long as Bach lived, it could only be completely given in years when this occurred again—three times, namely, 1739—40, 1744—45, 1745—46.

compositions which no longer exist in their original form.[667] The text for the cantata on the Queen's birthday was evidently written by Bach himself,[668] while Picander wrote the text of that for the Electoral Prince, and probably that of the complimentary ode.[669] Hence it follows, almost of course, that Picander and Bach must each be responsible for the alterations in his own portion of the text of the Christmas Oratorio; this view is confirmed by the text of the first chorus which is partly founded on Ps. c., 2, and partly bears an evident resemblance to two arias by Georg Ahle, which can hardly have been known to Picander, though they would be to Bach, Ahle's immediate successor at Mühlhausen.[670] Whether the secular compositions which have found their way into the Christmas Oratorio may not already have been altered from still earlier works, at any rate in part, is a question that cannot be decidedly answered in the negative. At any rate the final chorus of the *Drama* for the young prince—which to be sure is not adopted into the Christmas Oratorio—had already been used in the Cantata. "Erwünschtes Freudenlicht," of course with other words, and even this was not its original place and use.[671]

It might seem as though, under these circumstances, the Christmas Oratorio must fail in unity of treatment; and it must be granted that many details in the madrigal portions betray the fact of their having been originally set to other texts. The words of the first chorus were at first "Tönet ihr Pauken, erschallet Trompeten!"—"Sound all ye drums,

---

[667] No such transfer is probable in the case of the alto aria of the third part, "Keep, O my spirit." In the autograph it is conspicuous for the multitude of corrections of the original conception; and that it was Bach's intention to introduce something entirely new in this place is evident from a previous sketch of another aria in B minor, 3-8, which, however, remains a fragment and was subsequently struck through by Bach. The opening chorus of the fifth part also, "Glory be to God Almighty," has, in the autograph, no appearance of being taken from a former work. Compare Rust's Preface to B.-G., V.[2]

[668] As will be shown presently when the occasional cantatas come under discussion.

[669] Although this text is not included with the other in Picander's poems, the style prohibits our attributing it to Bach.

[670] See Winterfeld, Ev. Kir., III., p. 344.

[671] See ante, p. 399.

and shout out ye trumpets"—and the music suits these words, beginning with kettledrums, immediately followed by the fanfare of trumpets — while it scarcely suits those now substituted "Jauchzet, frohlocket, auf!"—"Christians, be joyful." The opening chorus of Part IV. had for its text "Lasst uns sorgen, lasst uns wachen"—"Let us be wary, let us be watchful"—and the idea of watchfulness is figured by two staccato quavers, but the word "loben"—"praise Him"—in the altered text, does not seem to require this mode of utterance. In the cantata for the Crown Prince we find the words "Auf meinen Flügeln sollst du schweben"— "Thou on my pinions shalt be carried"—and the music which represents a floating and soaring motion with graphic dignity is set in Part IV. of the Christmas Oratorio to the text "Ich will nur dir zu Ehren leben," &c.—"'Tis Thee I would be praising ever, My Saviour give me power and skill, and all my heart with ardour fill." In the same cantata Hercules sings "Ich will dich nicht hören, ich will dich nicht wissen":—

> Mine ears may not hear ye,
> My heart shall not know ye,
> Prohibited joys ye shall never be mine;
> Wreathing serpents, long since crushed and sore defeated,
> Seek not my spirit to lull and entwine.

The accompaniment (p. 103) at suitable places depicts the twining of the snakes, and to the altered words, "Strengthen me, that Thy mercy worthily to praise I may endeavour," the significance of the phrases is lost.

Still, the mere fact that Bach himself never hesitated before such trifling incongruities is extremely instructive for the true comprehension of his music. Ready as he was to sprinkle his works with picturesque figures, he did not do so as a result of fundamental principles based on a sense of the graphic power of music. Those figures are transient flashes, and their presence or absence cannot alter the value or intelligibility of the composition in its integrity. In studying Bach, when we meet with some conspicuously melodious line or some strikingly harmonious tune, that happens to coincide with an emphatic or emotional word, we are too ready to attribute to them a much closer and deeper connection than can ever have dwelt in the purpose of the composer; if in Bach's recitatives we often find that the relation of the music to the words is quite secondary—almost accidental—how much more is this the case in the consecutive pieces! Any one who has listened to the first alto aria of the Christmas Oratorio must have been delighted with the tender expression given to the isolated cry, "The fairest, the purest." (p. 15, l. 4.) The music and text, as they stand, seem inseparable; and yet in the original form these notes belong to the words "I will not, I may not," and suit them, too, extremely well. Often enough the rendering of these subtle associations is left to the interpretation of the singer—Bach's music neither forbidding nor requiring it. Nay, the dramatic feeling suggested by the text is allowed to influence the construction of the musical composition as a whole less by Bach, on the average, than by other masters of his own or of a later time. The aria here quoted seems in the secular cantata to express, in the words of Hercules, his contempt for the delights of pleasure, and in the Christmas cantata the desire of the Daughter of Sion to receive her longed-for Bridegroom with due honour; and yet all will agree in feeling that there can hardly be a more characteristic piece of music than this, only the character lies in the music itself.

Now, when Bach transferred a piece complete and bodily from one work to the other, the question, with him, was chiefly whether it fitted the musical context and—since its original purpose was secular—whether its style was not antagonistic to a sacred work. As to this I must refer the reader to what has already been said in Vol. I. (p. 569). Bach's whole mode of expression was built on true church feeling; whether he wrote sacred or secular music, whether he composed organ fugues or chamber sonatas, the fundamental church sentiment developed directly from the nature of the organ pervades all his works, and he consequently could write nothing that jarred with it. On the other hand, his secular occasional pieces were not genuinely secular; as such they scarcely fulfilled their aim, and the composer only restored them to their native home when he applied them to church uses. In the face of the inexhaustible inventive wealth, and the profound sense of artistic responsibility, testified in Bach's works, no one would dare to assert that such transfers were made for mere convenience sake, or for lack of time. They were made with a perfect feeling that the compositions in question would not be seen in their right place till they were set for church use. At the same time we cannot deny to Bach a certain distinction between his sacred and his secular styles. Irrespective of the general vein of feeling which would govern him in the act of composition, Bach must naturally have felt moved in a very different way when he was writing for the delectation of a musical union, or for the edification of a band of Christian worshippers, when singing in honour of the Electress of Saxony, or celebrating the glory of the Infant Christ. The character of the secular cantata is lighter and more sportive; but even within the cycle of church life there were suitable opportunities for striking this key of feeling, and Christmas afforded one above all others. The festal cheerfulness and innocent sweetness which are the principal features of the Christmas oratorio, and exhaustively express the very essence of the Christmas festival, are most happily attained by this very transfer of secular compositions. It is true that pieces by Bach are not wanting in which the style seems more decisively determined by the words; still, in changing the use of these due caution has been exercised, and the altered words have been skilfully adapted to characteristic pieces. The cradle song of Part II. was indeed, as originally composed for the birthday cantata, a lullaby to the Crown Prince.

No other work by Bach contains a richer collection of charming and easily comprehended melodies than the Christmas oratorio. Still, it is not alone in its musical aspect that its popular character lies; wherever it was possible, reference has been had in the words as well as in the music to the traditional ceremonies connected with the Christmas plays and hymns. The custom of cradling the Child was reflected in the cradle song of Part II., "Slumber, beloved," a composition of enchanting grace and the sweetest melody.[672] It does not certainly fill its proper place, which is in Part III.; but musical considerations must have prevented Bach's inserting it there. In Part II. its introduction is led up to by a bass voice, which delivers the call to the shepherds, and charges them when they are come to Bethlehem to sing "in sweet harmonious tone, and all with one accord;"[673] the following song, however, is certainly not appropriate in any way for choral singing. The only thing here needed was to supply a reason for its introduction; Bach has by preference given the song itself to the Virgin Mary, since it is written for an alto voice, while in the Birthday Cantata it was written for a soprano, a minor third higher. In the Christmas dramas and pastorals it was a stereotyped detail that after the appearance of the angels the shepherds should be encouraged to go to Bethlehem. Thus, in a play used in Schleswig we find:—

> Laufet ihr Hirten, lauft alle zugleich,
> Nehmet Schalmeien und Pfeifen mit euch, &c.

> Hasten, ye shepherds, rise, hasten away,
> Take with you reed-pipe and tabor to play,
> With joyful accord
> At Bethlehèm.
> And there, in the stable, worship the Lord.

And in a pastoral:—

> Up, O shepherds, rise and hasten,
> Jesus newly born to greet;
> He in yonder stall is lying,
> Fall in worship at His feet. [674]

If we compare these with the text of the tenor aria in Part II. of the Christmas oratorio we cannot doubt their association:—

> Haste, ye shepherds, haste to meet Him,
> Why should you delay to greet Him?
> Haste the gracious Child to see.

---

[672] The precursor of this lovely composition is to be found in the Weimar Cantata, "Tritt auf die Glaubensbahn." See Vol. I., p. 560.

[673] Reminding us of Luther's "Righteous Susanna," who sang "her heart's delight in sweetest tones."

[674] Weinhold, Weihnacht-Spiele und Lieder, pp. 119 and 434.

The matter is of some importance, because Bach himself wrote the words for the aria taken from the cantata for the Queen's birthday, and consequently it is probable that he undertook the other paraphrases; we must therefore ascribe the introduction of this suggestive reminiscence to him.

The bass aria of Part I. belonged to the same cantata; in this Jesus is praised as "Mighty Lord and King, all glorious": here, too, the words must be the composer's. The lowliness in which Jesus came into the world must always have offered a significant and obvious contrast to the idea of His divine omnipotence; it seems, however, as if Bach had had an ulterior view. The three wise kings—whose coming had been dramatically treated in the primeval church liturgy, and who had subsequently become very popular personages even in the processions outside the church walls —bring, as we know, three gifts to the Infant Christ: gold, frankincense, and myrrh; gold symbolising His sovereignty, incense His divinity, and myrrh His redeeming sufferings and death on the Cross.[675] In Part I. of the Christmas oratorio we hear nothing, it is true, of the three kings; but we have already seen that Bach seems to have cared more for giving general expression to certain popular and church aspects of the subject than for assigning to them any particular position. The premonition of Christ's death, immediately after His birth, has assumed a very striking form in Part I. by the adaptation of the melody of "O Haupt voll Blut und Wunden,"[676] to the words of greeting, "How shall I fitly meet thee?" which falls across the bright festal tone like a dim shadow. It must have seemed unnecessary to bring out

---

[675] Thus rendered from the Latin Christmas hymn in an old German version, reprinted (from the Andernacher Gesangbuch of 1608) by F. Böhme Altdeutsches Liederbuch, Leipzig, Breitkopft and Härtel, 1877, p. 640.

" Als ein könig brachten Gold,
weihrauch dass er opfern solt
myrrhen dass er sterben wolt."

" Gold they brought for Royalty,
Incense for Divinity,
Myrrh to show that He should die."

[676] That of the chorale, "O Lord! Thy love unbounded," in the St. Matthew Passion (p. 18 of Novello's Edition).

the idea of Christ's divinity by any specially suggestive subject, but it was all the more obvious to emphasise His royalty, and there was surely room enough for this in the whole work. And yet the feeling of Part I. may not unfitly be expressed in the saying of the Apostle, " He made Himself of no reputation and took upon Him the form of a servant." Framed, as it were, in festal joy, it is a picture of humility and self-abasement that meets our gaze; but the objection taken to the bass aria as not corresponding to this is surely much diminished when we point out the affinity which subsists between it and the first chorale, on the ground of popular feeling.[677]

In Part II., when the angels have proclaimed the birth of Christ, the aria just mentioned, "Haste, ye shepherds, haste to meet Him," is introduced by the following bass recitative :—

> What God to Abraham revealed,
> He to the shepherds doth accord
> To see fulfilled.
> To shepherds, lo! our gracious Lord
> His purposes unfoldeth;
> That blessing which, in days of old,
> He to a shepherd first foretold,
> A shepherd first beholdeth.

This text also seems to contain an obscure reference. The insipid antithesis of Abraham as a shepherd, and the shepherds of Bethlehem, can hardly have been the whole motive of the poem; the author must rather have had in his mind an idea of praising the shepherd's calling generally. In the Christmas plays it was the custom for the shepherds watching by night to sing a *Cantilena de laude pastorum*, to while away the time.[678] Thus, to be perfectly

---

[677] Winterfield, Ev. Kir., Vol. III., p. 347, raises this objection, and thinks the fact of the aria being taken from another work lies at the root of its inappropriateness; but his hypothesis as to how Bach could have allowed himself such an error of taste is answered if what is said above be correct.

[678] Such a piece exists in a Bavarian Christmas drama, beginning " Lasst uns singen von den Hirten ":—
Weinhold, op. cit., p. 176.
> " Let us sing the shepherds' glory,
> Who have been renowned in story."

in sympathy with the instrumental symphony which opens the second part, we shall do well to imbue our minds with the sentiment on which the scene of the shepherds by night was based in the Christmas plays. A combination of opposite factors—which presented no difficulties to the naïve minds of the people—of the grace of the Eastern idyl with the severity of the starlit boreal winter's night, gave the fundamental feeling of this symphony. This wonderful composition, woven as it were of silver rays, and enchanting us by harmony of hues, is full of calm rejoicing, and yet unutterably solemn, child-like, and overflowing with yearning. The romantic feeling for nature, which so unmistakably breathes from it, also prevades the magnificent chorus of angels, " Glory to God in the highest ! " where the sparkling accompaniment makes us feel as if we are gazing into the vault of stars.

The ground covered by the Gospel narrative of Christmas events is much narrower, and the incidents themselves are less tragical and vivid than those of the Passion ; hence the lyric element becomes more important, and, as a whole, it decidedly tends to the style of the church cantata. Bach was well aware of this ; indeed, he has brought this element into greater prominence than the circumstances required. Even connected portions of the Gospel, treated as such by Bach himself, are not unfrequently interrupted by reflections in verse, which never occur in the Passions. Bach has not by any means invariably availed himself of the opportunities offered for varying the sung recital by the introduction of a chorus. In Part II. he even lets the Angel's speech, begun in the soprano, be continued by the Evangelist, the tenor ; certainly this takes place after the announcement made by

---

And then it enumerates all the shepherds named in the Old Testament. And that this was a popular theme in more recent times is proved by the well-known shepherds' dance by Johannes Falk :—

> What can be of fairer fame,
> What can give a nobler name
> Than to come of shepherd lineage ?

the Angel has been interrupted by a recitative and an aria.[679] After this Bach may have thought it unsuitable to bring in the Angel again without the intervention of the Evangelist. But if it had been incumbent on him to carry out a dramatic fiction, he would undoubtedly have arranged the separate numbers differently throughout. In Part V. Herod calls together the chief priests and scribes, and enquires of them as to the birthplace of the Messiah. They answer, "In Bethlehem of Judæa: for thus it is written by the prophet, 'And thou Bethlehem, in the land of Juda, art not the least.'" St. Matthew ii., 6. According to rule these words ought to have been sung by a chorus, but they are given to the Evangelist. The reason is not, as it seems to me, that Bach undervalued the dramatic element. In the Christmas story the most popular personages, next to Mary, Joseph, and the Child, were the herald angels, the heavenly host praising God, the shepherds, the three kings, and Herod; and, so far as the Gospel text allowed, Bach has introduced them all. But he must have seen reason to fear that a chorus of chief priests and scribes would produce a puzzling and bewildering effect, since these persons were of no importance in the incidents directly concerning the birth of Christ, and his object was to remain strictly within the cycle of popular feeling. In this way, no doubt, he left a powerful vein of dramatic animation untouched; but it is the predominant lyrical character of the work which enables us to comprehend the easy freedom with which Bach inserts pieces, such as the cradle song, or the praises of Christ as King, in places where, in the due course of events, they do not belong. He took his stand on the ground of the church cantata, in which all the incidents of the Gospel are accepted as known, and the emotions raised by them are made use of from a purely musical point of view.

The Christmas Oratorio falls naturally into three divisions of the subject; the narrative of the birth of Christ and its announcement, which compose the story of Christmas Day strictly speaking, are treated in the first three parts. The fourth

---

[679] B.-G., V.,² p. 66. Novello's Edition, p. 34.

relates the naming of Jesus, and the fifth and sixth the visit of the three kings. Bach has rounded off the whole work into unity in a very unmistakable manner, by making the first chorale of Part I. recur at the close of Part VI. in the form of a brilliant and festal chorale fantasia. Otherwise each section has its own particular character. In the first three the Christmas feeling prevails most vividly; this is effected in great measure by the chorales, which are interspersed in far greater numbers than in the last three, and which are almost all familiar Christmas hymns.[680] Most of them are simply set in four parts with highly ingenious applications of the church modes. The melody of " O Haupt voll Blut und Wunden" is harmonised by Bach in the Phrygian mode, thus excluding all doubt as to his purpose in introducing it. Where the same tune occurs in Part II., with the words of a verse from Gerhardt's hymn, " Schaut, schaut, was ist für Wunder das?" the harmonies have Mixolydian chords, and the same mode is strongly stamped on the last chorale but one of Part III. (p. 78), where the last verse of the hymn " Fröhlich soll mein Herze springen" is sung to the melody—somewhat modified by Bach—of " Warum sollt ich mich denn grämen"; and in both these cases we feel that the character of fervent humility, which the chorales derive from this mode of treatment, is rendered especially beautiful and impressive from the places they fill.[681] Unlike the St. Matthew Passion the Christmas Oratorio contains one single-part chorale. When the Evangelist has recited " And she brought forth her first-born Son . . . and laid Him in the

---

[680] Luther's melody of " Gelobet seist du " comes in twice (Nos. 7 and 28), "Vom Himmel hoch " three times (Nos. 9, 17, and 23), " Fröhlich soll mein Herze springen" (No. 33, with another tune) and " Wir Christenleut" (No. 59) each once. Besides these the words of the Advent hymn are very poetically set to the tune of " O Haupt voll Blut und Wunden "; and the tune of Rist's hymn, " Ermuntre dich, mein schwacher Geist," has been adapted to " Break forth, O beauteous morning light " (No. 12), as has been observed by L. Erk. Bach's mehrstimmige Choralgesänge. Part I., p. 115, No. 27.

[681] See Winterfeld's well-considered comparison, Ev. Kir., V. III., p. 348, 350, and 302. It makes no difference that he has given a wrong text for the first of these two hymns.

manger because there was no room for them in the inn," two oboes with the basses and organs come in with an exquisite composition of the most tender and modest fervency, to which the soprano calmly sings the sixth verse of "Gelobet seist du"—"For us, to earth He cometh poor" (p. 19)—and between the lines the bass sings a devotional recitative. Chorale subjects, such as we usually find in Bach, close Parts I. and II. Here, again, the chorale appears in simple four-part treatment, but with interludes and a coda on instruments founded on motives or echoes of the first number of the respective sections. In the first instance we have verse xiii. of the hymn "Vom Himmel hoch"; in the second, verse ii. of Gerhardt's hymn, "Wir singen dir, Immanuel," but the tune is again that of "Vom Himmel hoch," and not that originally written to it;[682] to make it fit the text Bach has omitted the final "Hallelujah."

In Part III. we again detect the purpose of connecting the musical feeling of the end with the beginning; for it does not conclude with a chorale; it is the opening chorus that is repeated—a rare thing with Bach. Parts V. and VI., devoted to the history of the three kings, are in no respect inferior to the first three. The lyrical choruses are full of artistic beauty and swing. Among the solo pieces a terzet in dramatic style (p. 130); "Ah, when shall we see salvation," and a four-part fugal recitative are of conspicuous importance. On the other hand, Bach has neglected to avail himself of the events on which the work is based as completely as in his Passion music. The cantata character is more conspicuous here than in the first three sections, and the specially Christmas feeling resides more in the general tone of the music than in the chorales. The two Christmas hymns principally used in the Leipzig churches, "Vom Himmel hoch," and "Gelobet seist du," do not occur at all. Of the four that are introduced, indeed, only one is really a Christmas hymn (Gerhardt's "Ich steh an deiner Krippen hier"—"Beside Thy cradle"), one is an Epiphany hymn, the other two have no distinctively festal

---

[682] "Erschienen ist der herrlich Tag."

text; and even the tunes ("Gott des Himmels" and "O Haupt voll Blut") are not Christmas tunes. In the Epiphany hymn Bach has chosen to alter the words of the chorale (p. 122, "All darkness flies"), and in "This proud heart" he has had to repeat the last note of each of the last two lines in order to make the words and the tune fit each other.

Part IV. has least of the character of church festival music. The Biblical matter consists of a single verse from the Gospel of St. Luke, ii., 21, which relates the circumcision and naming of Jesus. Not much material could be worked out of this, and Bach has almost entirely set aside all adjuncts from the liturgy. No Christmas hymn, indeed no true chorale, is introduced in it. We find, it is true, two verses of hymns by Rist; the fifteenth verse of "Hilf, Herr Jesu," is used for the finale (p. 105) in a similar form to those of Parts I. and II., and the first verse of "Jesu du mein liebstes Leben" is divided into independent subjects for a soprano arioso, with a bass recitative in counterpoint, "Immanuel, beloved name" (p. 91). But in both these cases Bach has not used the regular church tune, but others of an aria stamp, which, as they are to be found nowhere else, have long been regarded with justice as of Bach's own invention.[688] This section, therefore, bears more strongly the stamp merely of a religious composition; it is full of grace and sweetness, and can only have derived its full significance for congregational use from its position in context with the rest of the work.

I have already taken occasion, more than once, to point out that towards the middle of his Leipzig period Bach showed a disposition to divert church music into the channel of sacred domestic music, or—as we may say on the other hand—to raise the home music he performed and loved to the dignity of the church style. Part IV. of the Christmas oratorio is fresh evidence of this tendency. One feature by which it is manifested in the details deserves notice, in spite of its inconspicuous nature. In the verse of

---

[688] Since Winterfeld; see Ev. Kir., Vol. III., p. 354.

the final chorale Rist begins every line with the name in whose honour the day is held sacred, "Jesu, guide my every action; Jesu, still abide with me," &c. Bach alters "Jesu" (the vocative) to "Jesus" (the nominative). Thus Rist addresses a supplication to the Redeemer, Bach merely utters a pious wish, and it is not till the last line that he uses the form of prayer. This, as we have seen, is the same frame of mind as that which guided the paraphrase in the cantata "Ich bin vergnügt."[684] Traces also occur elsewhere in the Christmas oratorio of a free mode of dealing with sacred property, which plainly betrays Bach's tendency towards a sentiment of family edification; it is for this reason that I have intentionally noticed the liberties, greater or smaller, which he has taken with the chorales: "Thee with tender care," p. 78; "With all Thy Hosts," p. 55; "All darkness flies," p. 122; and "This proud heart," p. 139. Whether Bach could have introduced such a composition as the soprano aria "Ah, my Saviour," with a double echo of a second soprano voice and an oboe, into any work other than this very peculiar Part IV. may well be doubted. Of course, childlike naïveté is in its place in the Christmas festival if anywhere, and it is touching to note how the grave and thoughtful master has allowed himself to yield so completely to this festal feeling as to admit such sportive movements into the work. Indeed, he has not simply copied the echo in its natural form, but has conceived of it in an independent manner as a subject motive. In the first instance the aria had its place in the occasional cantata for the birth of the Crown Prince. This *Drama per musica* deals with the narrative of Hercules choosing between Vice and Virtue. After Pleasure and Virtue have exercised their persuasive powers, Hercules, who has not yet decided what he shall do, says:—

> Faithful Echo of this glade,
> If I should be soon betrayed
> An I were content to go
> Where this sweet voice bids, say No.
> *Echo*—No.

---

[684] See ante, p. 443.

> But if then the warning voice,
> On through toil and weariness,
> Guides me to a better choice,
> Then, I bid thee, answer Yes.
> *Echo*—Yes.

Here the situation fairly allows the presence of the Echo; if Bach had transferred the aria into the Christmas oratorio merely for the sake of its pleasing music, and so disjoined it from the scenery that befits it, it would have been an obvious breach of good taste. But he did not do this. The person who sings the hymn tune in the oratorio is the Bride, who goes forth to meet her Beloved. To verify the correctness of this interpretation we must go back to the origin of such Echo songs. It is to be found in the Trutz Nachtigall of Friedrich Spee. In one of the sweetest poems contained in this collection, " the spouse of Jesus talks in the forest with an echo or reverberation."[685]

Der schöne Frühling schon begunnt,	The bonny spring had well begun,
Es war im halben Märzen,	And March was half way thorough,
Da seufzet ich von Seelengrund,	And from my deepest soul I longed
Der Brand mir schlug vom Herzen.	To cast away my sorrow.
Ich Jesum rief	Jesus! I cried,
Aus Herzen tief,	And deeply sighed,
Ach Jesu! thät ich klagen:	"*Ah, Jesu,*" humbly praying;
Da hört ich bald	And lo! I heard
Auch aus dem Wald	An answering word,
Ach Jesu! deutlich sagen.	"*Ah, Jesu!*" plainly saying.

This play of words is carried on in a very pleasing way through a multitude of appeals and answers till the last verse:—

> 'Tis well, sweet echo, to my call
> That answerest from thy bower,
> With thee I oft will play at ball,
> For many a happy hour.

---

[685] Trutz Nachtigall, 1649, p. 10. This is the second verse; there are twenty in all.

> And this the ball
> That ne'er shall fall
>   Between us, Jesus' name
> From me to thee,
> From thee to me,
>   'Tis ever Jesus' name.[686]

Echo songs soon became a favourite conceit with poets; Franck has several times availed himself of this *motif;* in a "Sacred echo to the cooing turtle dove"[687] the plagiarism from Spee is self-evident.

What has been said above as to the elements of Bach's secular cantatas, namely, that by transferring them to sacred ones he actually restored them to their natural sphere, is equally true of the echo song, but in a quite different manner: the source of these poems lay in the domain of sacred poetry. That Bach, and the poet who turned the words of the aria into their sacred form, must both have known Spee's poem I consider as probable, if only because it involves the glorification of the name of Jesus, and so gives the leading idea of Part IV. of the oratorio. Certainly Bach's composition is its worthy counterpart.

Dramatised Gospel recitations had from the earliest times held an important place in the Easter solemnities of the Protestant Church, though they were less general than in those of Holy week. This is to be accounted for by the fact that the Easter Gospels are the direct sequel of the narratives of the Passion. We have Easter compositions in the style of the older Passions by Scandelli and by Schütz, both Capellmeisters at Dresden; and in Vopelius' Leipziger Gesangbuch, of 1681, there is one which must still have been sung at Leipzig at that time, but which must have been written considerably earlier. The text is harmonious and the musical treatment remarkable, because the Evangelist is represented by a baritone and not by a tenor, while the

---

[686] The English reader will be reminded of George Herbert's *Heaven*, the last poem but one in "The Temple."

[687] Geist-und weltliche Poesien. Part I., p. 80. Compare the Song of Solomon II., 14.

speeches of the individuals are in several parts. Christ's words are given in four parts and the rest in two.[688] But these Easter readings had already fallen into disuse in Kuhnau's time, and it is impossible to say at what stage of divine service they may have been introduced. On the other hand, in the so-called sacred concerto of the seventeenth century a form was developed which, by the omission of all purely narrative portions, approached very nearly to the dramatic musical *scena*. Schütz and Hammerschmidt worked in this form with great success.

We have, by Hammerschmidt, a small Easter drama called a *Dialogus*, which is put together from the words of the Evangelists, and deals with the great event of the Resurrection.[689] Mary Magdalene, Mary the mother of James, and Salome (Mark xvi., 1) have come to the sepulchre to anoint the body of Christ. After an introductory symphony begins a three-part song, "Who shall roll us away the stone from the door of the sepulchre?" To which two men in shining garments answer, "Why seek ye the living among the dead? He is risen, He is not here." The women lament, "They have taken away the Lord and we know not where they have laid Him." The answer to this is a chorus on the Easter hymn, "*Surrexit Christus hodie Humano pro solamine. Alleluja.*" This constitutes, as it were, the first scene; the next is between Mary Magdalene and Christ (John xx., 13 and 15—17). She now laments and entreats alone, "They have taken away my Lord and I know not where they have laid Him. Sir, if thou hast borne Him hence, tell me where thou hast laid Him, and I will take Him away." The risen Saviour replies to her with the questions, "Woman, why weepest thou? Whom seekest thou?" And then sings slowly and significantly "Maria." She recognises Him and cries out "Rabboni!" several times repeated, while He charges her to announce His resurrection and approaching ascension to the disciples. The little work ends with

---

[688] Vopelius, pp. 311—365. As to the probable date of this "Auferstehung"—Resurrection—Winterfeld gives some information in Ev. Kir., Vol. II., p. 556.
[689] Hammerschmidt, Musikalische Andachten. Freiberg, 1646. No. 7.

a repetition of the chorus "*Surrexit Christus.*" It is one of those which led and prepared the way for the oratorio in Germany. Although it was undoubtedly written for church use it has nothing that bears any special indication of it; the "*Surrexit*" is not introduced in the usual manner, but in a newly devised way, and rather as being merely a suitable text, as is frequently done by Hammerschmidt and Schütz; it nowhere exhibits any striking polyphonic treatment, nor any well-composed melody, but a good deal of clear dramatic emphasis. This work is in many respects closely allied to Bach's Easter oratorio. In this, also, the narrative portions are wanting, only the two Marys, Peter, and John, appear in person; there are no chorales, and the portions given to the chorus are, when measured by Bach's standard, of remarkable simplicity. The text consists, not of Biblical quotations, but solely of poems in madrigal form, and the resemblance of the whole work to the Italian model is immediately perceptible. Of all Bach's compositions this has the fairest right to the name of "Oratorio," though not, to be sure, in Handel's sense of the word.

The text, of which the author is unknown, is meagre enough. All that is most beautiful and significant in the history of the Resurrection, and that has been given above in the outline of the *Dialogus,* has not been made any use of. It begins with a duet between John and Peter, who are informed of Christ's resurrection by the women, and who run joyfully to the sepulchre to convince themselves (John xx., 3 and 4). There Mary the mother of James, and Salome, reproach them with not having also purposed to anoint the body of the Lord and thus testifying their love for Him. The men excuse themselves, saying that their anointing has been "with briny tears, and deep despair and longing." Then the women explain that these, happily, are no longer needed, since the Lord is risen. They gaze into the empty tomb; John asks where the Saviour can be, to which Mary Magdalene replies—what the men have long known:—

"He now has risen from the dead.
To us an angel did appear
Who told us, lo! He is not here."

Peter directs his attention to the "linen cloth," and this leads him to recall the tears he had shed over his denial of Jesus—a very tasteless episode. The women next express their longing to see Jesus once more; John rejoices that the Lord lives again, and the end is a chorus:—

> Thanks and praise
> Be to Thee for ever, Lord!
> Satan's legions now are bound,
> His dominion now hath ceased,
> Let the highest heaven resound
> With your songs, ye souls released.
> Fly open, ye gates! Open radiant and glorious!
> The Lion of Judah comes riding victorious.

It cannot but surprise us to find that Bach could have been satisfied with such a text. He has embodied the history of the Passion in a stupendous work, and he knew that the Resurrection had been sung at an earlier period, for he knew and made use of Vopelius' hymn book. It might be supposed that this would have been reason enough for his treating the history of the Resurrection in a worthier and more dignified way. Nor is this a work of his youth; the forms show the handling of a mature master, and from the MS. we may see that the work must have been written about 1736.[690] I can only find an explanation in the regulations for divine service at Leipzig; there was, in fact, no opening for a comprehensive work in the style of the St. John or the St. Matthew Passions. The *Magnificat* was performed at Vespers, and in the morning there was only time for a piece of about the length of a cantata, which could not even be in two sections, since after the sermon the *Sanctus* had to be sung. It is clear that Bach, having written "Mysteries" for Christmas, Holy week, and Whitsuntide, simply wished not to omit Easter; and as he could not deal with the Gospel narrative in so extensive a form as he thought desirable—and as he found adopted by Vopelius—he preferred giving the form of an Italian oratorio, of which less was expected and demanded, to selecting a portion of it.

In the Easter plays the race between Peter and John to the sepulchre was a favourite event for representation; on this

---

[690] See App. A., No. 58.

occasion Peter appears as the weaker and less important personage. I do not think it merely accidental that Bach's work should begin with a lengthy duet between the disciples as they run to the sepulchre ; nor, again, that Peter's part is given to the tenor and John's to the bass, while in the St. Matthew Passion the reverse plan is adopted. Bach subsequently obscured the popular sentiment which lay at the root of this, the principal event treated in the work ; for he re-arranged the duet as a four-part chorus, though the circumstances, of course, allow us to suppose that, besides the two disciples and the two Marys, other of Christ's followers would have hastened to the sepulchre.[691] As regards the church feeling in a text which avoids both Biblical words and chorales, it certainly can only arise from the circumstance that it treats of an event of supreme importance in the church. Bach's music had to do its utmost and best to support it. Of course we cannot expect grandeur and depth, as in the cantata "Christ lag in Todesbanden," or even the triumphant spring-like joy of the Weimar cantata "Der Himmel lacht," since the words are absolutely devoid of any incitement to either. Bach has given to the whole a fresh and innocent character, suggested, perhaps, by the words :—

Lachen und Scherzen	Laughter and gladness
Begleitet die Herzen	Now drive away sadness
Denn unser Heil ist auferweckt.	For lo ! the Lord hath waked from sleep.

A symphony in the two movements, together with the first vocal number, constitute a complete instrumental concerto. Among the arias, that given to Peter is distinguished by being a soft lulling cradle song, such as Bach was fond of writing. Singularly enough the principal motive is the same as that at the beginning of the Coffee Cantata which Bach composed about 1732 ; this of itself is definitive as to the

---

[691] In the parts, as last written out for this alteration, Bach did not note the names of the four *dramatis personæ*, nor in the score, but only in the earlier parts ; but we need not conclude from this that he had altogether given up the dramatic scheme of the work. This would render the first and second recitation perfectly unintelligible.

cheerfulness of the feelings with which he composed the Easter Oratorio. The final chorus, freely worked out in the form of the French *ouverture*, is attractive from the breadth and splendour of the first subject; compared with this, the Fugato is surprisingly brief, and produces no profound effect.[692]

In the Ascension Oratorio Bach, on the contrary, has adhered to the old liturgical forms. The historical matter was, in this case, of such limited extent that it could easily be worked out within the narrow bounds allowed by the regular service for Ascension Day. The Biblical narrative is compiled and harmonised from St. Luke xxiv., 50—52; Acts i., 9—12, and Mark xvi., 19. Bach has made no use of the customary hymns "Nun freut euch" and "Christ fuhr gen Himmel"; after Christ's ascension has been related we have the fourth verse of Rist's hymn, "Du Lebensfürst, Herr Jesu Christ!" and at the end the last verse of Sacer's "Gott fahret auf gen Himmel." It is not known who wrote the texts in madrigal form, consisting of one chorus, two recitatives and two arias, nor can anything exact be stated with regard to the date of the composition. The style displays the fullest maturity in the master, hence we may assume that the Ascension Oratorio, like those for Christmas and Easter, was written in 1729.[693] Like the Christmas Oratorio, it has at the beginning a madrigal chorus in the Italian aria-form, and at the end a choral-fantasia. The opening chorus, which may have been transferred to this place from some occasional music, is a masterpiece in its way, a combination of tuneful simplicity with polyphonic elaboration; in the upper parts the jubilant melodies flow on without interruption with the most artless phrasing, and yet the lower parts evolve a rich and independent vitality.

It is remarkable that the solo pieces are not devotional, but dramatic in conception. In the alto aria, and the bass

---

[692] B.-G., XXI.³

[693] The autograph has the characteristics of Bach's later writing, but beyond this gives no data as to its chronology. The work is B.-G., II., No. 11. I may here observe that there is in the Royal Library, at Berlin, a figured organ part in autograph, which has not been made use of in editing the work.

recitative which precedes it, Jesus is implored not yet to depart from among those who believe in Him; in the soprano air consolation is derived from the reflection that Jesus, though He has ascended into heaven, is still present in spirit to the faithful. In the St. Matthew Passion, in one passage only ("Give, O give me back my Lord"), the same view is taken, and there seems hardly justified. The arias in the Christmas Oratorio, "Haste, ye shepherds" and "Slumber beloved," had their origin in popular tradition, and in the solos of the Ascension Oratorio we must recognise a dramatic expansion of the simple narrative in the same spirit of the sacred plays. In fact, we actually meet with such an expansion in a medieval Ascension-play, where first Peter (compare the bass recitative in Bach) and then the Virgin Mary (compare the alto aria) bewail the departure of the Redeemer.[694] Both the arias are full of splendid harmonies; the first breathes an overpowering fervency, the second a magical radiance of glory, and the contrast between the two shows Bach's power and poetical feeling for colour. The chorale, "Nun lieget alles unter dir"— "Now art Thou Sovereign over all"—is set in as low a range as possible; the aria, which is accompanied only by the flutes, violins and violas—dispensing with the basses—soars up to the realms of light. In one we have the image of the disciples left below, in the other that of the transfigured Saviour floating upwards.

IX.

BACH'S MOTETTS.

IN order to fill up the whole cycle of all Bach's labours as a church composer during the years 1723—1734, it is necessary, after having considered his "mysteries," to pass in review his motetts. A critical examination of these is rendered difficult by many external circumstances. Some of the motetts have been lost or have at any rate disappeared;

---

[694] Mone, Schauspiele des Mittelalters. Vol. I., p. 261.

many pass for Bach's which would seem to be spurious; of those which are undoubtedly genuine only the smallest portion are yet existing in Bach's writing, and the greater part of them have been handed down in a very incomplete and unsatisfactory condition; while of one alone do we know the date of composition.[695] This one belongs however to the section of his life between the years 1723—1734, and since most of the others bear the stamp of his full maturity of style they cannot be placed far from this one in regard to date.

At all events Bach did not begin to turn his attention to the composition of motetts for the first time in Leipzig. It was impossible that one who employed his energies in vocal church music so early in life, and so thoroughly as he had done both at Mühlhausen and at Weimar, could leave unnoticed the Motett—a form of composition which, in spite of transformation and disfigurement, remained so full of vitality, and which was an indispensable part of the church services at that time. It is also not improbable that an early composition of this kind by Bach is contained in a motett, " Unser Wandel ist im Himmel "—" Our conversation is in heaven." The text is taken from the Epistle for the 23rd Sunday after Trinity (Philippians iii., 20 and 21); the work is divided naturally into two fugues, the first of which is preceded by a short homophonic movement, while between the two is inserted the second verse of the chorale " Herr Gott, nun schleuss den Himmel auf," simply set in four parts. In the " Orgelbüchlein " Bach treated the same melody, though in a somewhat altered form; this is no sort of evidence against its Weimar origin, for Bach, even in his Leipzig time, employed varying forms of one and the same melody (e.g., " Warum sollt ich mich denn grämen," " Helft mir Gotts Güte preisen "). The four-part writing of the chorale somewhat resembles that of the Passion according to St. Luke. In the fugues the parts are now and then treated with too little care, and their working out is not always quite successful, but on the whole they are

---

[695] See Appendix A., No. 59.

flowing and animated musical pieces, not unworthy of Bach. This is especially true of the second fugue; it shows, too, an intimate connection with the fugal movement "Herr, höre meine Stimme," from the Weimar cantata "Aus der Tiefe rufe ich Herr zu dir."[696] Not only are the key and the indication of time identical, but the character of the theme and the movement which prevails throughout present a striking similarity, and in the motett the idea of agitation is represented in a realistic manner, like that of Telemann, as that of entreaty is in the cantata. Again, the close of the preceding number in the cantata agrees to some extent with the end of the first fugue in the motett. Several passages, too, where the vocal bass part lies above the tenor, and is yet to be considered as the root of the harmony, make it probable that it was intended to be accompanied by a 16-foot *Basso continuo*.[697]

In the services of the churches of St. Thomas and St. Nicholas the motett had its regularly appointed place at the beginning of the early service and of vespers, after the organ prelude. Besides this a motett was occasionally sung on the high festivals during the communion; this was always the case on Palm Sunday and Holy Thursday. The motett was only omitted altogether when the organ was not played. There were also occasions for the composition of motetts outside the circle of church ordinances, and especially in the case of funeral ceremonies; and one of Bach's motetts really owes its existence to the burial of the rector, Johann Heinrich Ernesti (d. Oct. 16, 1729). Bach however turned it to an ecclesiastical purpose, as was his wont with other "occasional" compositions. The position in the service occupied by the motett, limited it as regards length; nor could the motett be of great musical importance, since it

---

[696] See Vol. I., p. 450.

[697] In parts, which may have been written about 1800, in the Library of the Singakademie in Leipzig. Inscription: "*Motetto di Bach.*" On the same sheet are also written, "*Ecce quomodo moritur*" by Gallus, and "*Tristis est anima mea*," by Kuhnau. Although no Christian name is added, it cannot be doubted that the copyist of the parts, the exemplar of which must have been preserved among the archives of the Thomasschule, referred it to Sebastian Bach.

only served the purpose of an introduction. Many of Bach's motetts are, however, of such grand proportions and are worked out with such an exhaustive treatment of the special feeling suggested by the Church ordinances that they cannot possibly have served merely as an introduction to the service. They must be regarded rather as pieces to be performed before the sermon in place of the cantata, and we know from Bach's own words that he occasionally substituted the one for the other.[698]

The motett of Bach takes its rise from his cantatas, and, like them, primarily from his organ music. This explains the relation it bears to the motett form as that is generally understood. It is only indirectly connected with the motett of the seventeenth century. That was influenced by the concerted vocal music of the time, and reflects its half-developed forms with moderate completeness both of outline and detail.[699] In so far indeed as Bach's cantatas owe their origin to these, they have something in common; but Bach's motetts, like his cantatas, are absolutely free from the dramatic elements which appear in Schütz's and Hammerschmidt's sacred concertos and madrigals, and which also found their way into the motetts of the time. Whereas in the cantatas the forms of organ music have scarcely more than a subsidiary influence, they here make themselves felt in their full power. The organ-style governs the whole; it determines the characteristic formation of the melodies and the polyphony, always founded upon the laws of harmonic progression; and, above all, it has re-established the chorale in its full significance. Bach's cantatas were a central form of art, which included and made use of every musical element of a living and progressive character which existed at the time. Even the motett had been absorbed into the Bach cantata, and it was afterwards not so much born fresh from it as set free once more from its trammels. It re-appeared, not as an independent form of art, but as an offshoot of the Bach cantata.

---

[698] See ante, p. 247.
[699] See Vol. I., p. 55.

So far as it is possible to venture on an assertion, owing to the incompleteness of the materials, the history of the motett bears out this view. The name motett first occurs in Bach as applied to a piece of really concerted church music, namely, to the Rathswechsel cantata of 1708; the title is also borne, in the autograph by the cantata "Aus der Tiefe rufe ich Herr zu dir," the very work of which we are so strongly reminded in the motett "Unser Wandel ist im Himmel." The chief reason of this is to be found in the text of this work, but yet the changed application of the title would be inexplicable had not Bach been actuated by the idea of uniting the form of the motett with that of concerted music.[700] We first meet with a choral movement in genuine motett style in the Weimar cantatas "Ich hatte viel Bekümmerniss" and "Himmelskönig, sei willkommen,"[701] then in the first inserted piece in the great *Magnificat* "Vom Himmel hoch," then in the treatment of the fourth verse of the Easter cantata "Christ lag in Todesbanden," and again in the second movement of the cantata "Gottlob nun geht das Jahr zu Ende."[702] This last number was afterwards turned into an independent work, after having been remodelled in such a manner that the *Basso continuo* could be entirely left out.[703] Bach subsequently prefixed to the cantatas "Aus tiefer Noth schrei ich zu dir" and "Ach Gott vom Himmel sieh darein"[704] chorale choruses in motett style, which afterwards obtained celebrity as inde-

---

[700] See Vol. I., p. 345 and 449. The title "motett" for the Cantata "Aus der Tiefe" is given in Aloys Fuchs' catalogue of autographs, which is found in MS. in the Stadtbibliothek in Leipzig.

[701] See Vol. I., p. 531 f. and 539.

[702] See ante, pp. 376, 394, and 433.

[703] In MS. in the Amalienbibliothek in the Joachimsthal Gymnasium in Berlin, Nos. 24 and 31. Title: "*Choral.* | Sey Lob und Preiss mit Ehren. | vom Herrn | Bach." For the first verse of "Nun lob, mein Seel, den Herren," the fifth verse is substituted. Important changes have been made in several places; in particular, the closing symphony is entirely different. For this reason we might suspect Bach to be the remodeller, if some other features of the work did not make it improbable.

[704] B.-G., VII., No. 38, and I., No. 2. P. 1694 and 1194.

pendent works.[705] But texts of Scripture set in motett form occur also in the cantatas. The Christmas cantata "Sehet, welch eine Liebe,"[706] opens with a number of this kind, and another number in motett form, "Wenn aber jener der Geist der Wahrheit kommen wird," stands in the middle of a work for the fourth Sunday after Easter ("Es ist euch gut, dass ich hingehe"). All that Bach has left us in the motett form is throughout written in the same style. The texts consist either of Bible words only or of Bible words combined with a chorale, or else of a sacred aria-text joined to these two, or lastly of the aria-text alone. All the texts are in German. As the motetts sung at the beginning of divine service, even in Bach's time in Leipzig, were for the most part in Latin, he may very likely have been prompted to write some also in that language. In the year 1767 a certain one heard, at the early Christmas service, a Latin motett for two choruses by Bach, and "was thrilled to his inmost soul," and considered that nothing could compare with the lofty devotion and the beauty which pervaded it.[707] They are, however, all lost.

Of four-part motetts only one remains: the 117th Psalm, "Lobet den Herrn, alle Heiden"[708]—"Praise ye the Lord, all ye heathen"—a grand work, which flows onward uninterruptedly in one movement in the old style, except that the Hallelujah movement at the close is disjoined from the rest.

There are two motetts in five parts; both have a chorale for their groundwork, but are in all other respects very dissimilar. In one the words chosen are taken from Ecclesiasticus l., 24—26, "Now therefore bless ye the God of all, which only doeth wondrous things everywhere,

---

[705] In a MS. of Agricola's in the Amalienbibliothek, Nos. 37 and 38. The chorale movement "Aus tiefer Noth" is still followed by the terzet, which occurs in the same cantata.

[706] See ante, p. 385.

[707] Gerber, N. L. I., col. 222 f.

[708] Published in 1821, in score and parts, by Breitkopf and Härtel, in Leipzig. It must have been done from Bach's autograph, which however has not hitherto come to light again.

which exalteth our days from the womb, and dealeth with us according to His mercy. He grant us joyfulness of heart, and that peace may be in our days in Israel for ever: That He would confirm His mercy with us, and deliver us at His time!" Martin Rinckart, as is well known, paraphrased this passage in the two first verses of his hymn "Nun danket alle Gott"—"Now thank we all our God"—and this hymn forms the groundwork of the motett. It not only concludes with the third verse of the hymn set in simple harmony, but its first section is pervaded with reminiscences of the first line of the chorale melody. This mode of treatment may justify the supposition that the motett dates from about the year 1730, since in the cantatas also Bach liked to employ fragments of chorale tunes as free motives.[709] To what extent his thoughts, in the composition of the motett, ran on the words of the hymn is shown by the fact that in one passage, obviously unintentionally, he substitutes the words of the hymn for those of the Bible; he wrote, namely, thus: "Who wondrous things hath done, wherein His world rejoices," which occur indeed in the hymn, but not in Ecclesiasticus.[710]

In the other five-part motett Johannes Franck's devotional hymn "Jesu meine Freude"—"Jesus, my joy"—is set in all six verses. Its musical treatment displays of course great variety. The first and last verses are set in four-part harmony, "*simplici stilo*," as Bach himself used to designate pieces of this kind,[711] the harmony being the same for both verses, although the lower parts display rich animation and transcendental fervour. The same may be said of the second verse in five parts, and the fourth in four, although the lower parts, especially in the fourth verse, exhibit a still greater individuality and many graphic and characteristic touches. In the fifth verse the principal key of E minor changes to that of A minor, for the purpose of giving the *Cantus firmus* to the alto part; the two soprano parts and the tenor have counterpoint consisting for the most

---

[709] See ante, p. 457 f.
[710] The motett has hitherto remained unpublished. See Appendix A., No. 60.
[711] See Vol. I., p. 557, Note 271.

part of freely invented phrases, which however vary with
each line of the tune, so that the whole form is different
from that which we have termed the chorale fantasia, and
which we found employed so frequently in the cantatas.
The third verse is treated quite freely, taking the lines of the
chorale in only the most general way as its motives, without
strict regard to the exact intervals or length of these lines.
The imitative and homophonic styles are used alternately,
and unison even is employed to complete a picture which
for power, variety, and individuality has not its match among
Bach's motetts. In this work that resemblance which we
noticed before to Buxtehude's chorale cantata "Jesu meine
Freude" is more prominent than ever,[712] but the soft and
tame emotion of the older master does not show to advantage
beside this warlike strength and eager desire of conflict.
Between the six verses of the chorale Bach introduces
freely invented passages in five and three parts, on words
from Rom. viii., 1, 2, 9, 10, and 11, the first of which
agrees as regards musical material with the last.[713] In these
he discourses with the fervency of faith on the importance
of Christ's atoning work. The congregational feeling infused
into these subjects, as being appropriate to their general
dogmatic purport, is pointedly applied to the practical
Christian life by the intervening verse; and thus the germ
of Protestant Christianity is embodied in this great work.
Bach uses all the power of his inmost convictions to give
expression to the teaching of Luther in its utmost rigour
and purity. But with this keen dogmatic certainty he
combines the deepest personal devotion to Christ. In no
other of his works do we so plainly see how completely
the two parties into which the church of his time was
divided—namely, orthodoxy and pietism—had ceased to exist
for him. Even if we knew nothing more with regard to the
position taken by Bach in church disputes, an attentive

---

[712] See Vol. I., p. 307 ff.

[713] It has been already mentioned (Vol. I., p. 67) that Michael Bach also
used the chorale "Jesu meine Freude" as the nucleus of a motett. It is a
simple work for double chorus in one movement, which Sebastian may indeed
have had in his head when composing this work in this too the key is E minor.

consideration of this motett would suffice to guide us to the right opinion.[714] It is indeed a work "for all time," allotted to no particular day in the ecclesiastical year, though occasion for it may have been given by the eighth Sunday after Trinity, the Epistle for that day being taken from the eighth chapter of the Epistle to the Romans. Of course it was not intended as an introduction to the service, but as a substitute for the concerted music between the reading of the Gospel and the sermon.[715]

The four remaining motetts are set for double chorus. Among them is found the one mentioned above, which Bach wrote for the funeral of the Rector Ernesti. It consists of two sections, the words being taken from Rom. viii., 26 and 27; the first only is set for double chorus, the second being a four-part fugue. Bach afterwards added the third verse of the chorale "Komm heiliger Geist, Herre Gott"; judging from the words of the latter the motett was probably used for Whitsuntide, but it may also have served for the fourth Sunday after Trinity, the text being taken from the Epistle for that day. Another of these motetts begins with a number on Ps. cxlix., 1—3 ("Sing to the Lord a new song"), consisting of no less than one hundred and fifty-one bars. This is followed by the third verse of the chorale "Nun lob nein Seel," interrupted by fragments of the first chorus introduced as interludes, yet held together by a connected text, which is written according to the metric scheme of Rist's hymn "O Ewigkeit, du Donnerwort," although the melody which belongs to these words is not employed in the music. Next comes another double

---

[714] See Vol. I., p. 365, ff.

[715] This Motett is notified as Sebastian Bach's work in Breitkopf's Catalogue for the New Year 1764, p. 5. The MS. is in the Amalienbibliothek, Nos. 10, 12, and 30. There are three copies, which have, notwithstanding their many clerical errors, a great importance as regards authenticity. Published in score by Breitkopf and Härtel ("Motetten von Johann Sebastian Bach," No. 5). The text has here undergone much modernising; Bach set it correctly in its original form. Also the titles of the separate numbers, and the indications of time, with the exception of the word "chorale" over verses 1, 2, 4, and 6, and an "Andante" over the three-part number, "So aber Christus in euch ist," are modern additions.

chorus on Ps. cl., 2, and finally a closing fugue in four parts on v. 6 of the same Psalm. The work is apparently a composition for the New Year.[716] A third motett " Fürchte dich nicht, ich bin bei dir "—" Fear not, for I am with thee "—is set to Is. xli., 10, and xliii., 1 ; in the last verse the treatment is again in four parts, but in such a manner that the three lower parts work out the subject as a *Fugato*, while the soprano sings the last two verses of Gerhardt's hymn " Warum sollt ich mich denn grämen." The chief subject of the *Fugato*, on the words " For I have redeemed thee," referring to the crucifixion of Christ, is free and chromatic. From the third verse before the end onwards, Bach makes an arbitrary alteration in the expressive melody of the chorale ; its key now appears as E major, while it begins in A major. By this means it gains the mixolydian character much more distinctly than in the Christmas Oratorio.[717] Grief for Christ's sorrow and lowly devotion to His person are thus united into one emotional picture, which gives a deep and genuinely Protestant foundation to the trust in God which was before expressed. As regards the form, we again have a chorale fantasia in the style developed by Bach in his organ music. At the same time it must be understood that the soprano and the three other parts are not contrasted together as two dramatic factors, but that their poetic and musical import is merged in a more universal religious feeling. That Sebastian Bach, by this method of viewing and treating subjects of this kind, drew a sharp line of demarcation between himself and his

---

[716] The autographs of both motetts are in the Royal Library at Berlin. They are published by Breitkopf and Härtel as Nos. 1 and 6 of the collection before mentioned, but with spurious time-indications and many alterations of the text, which are probably due to J. G. Schicht, Cantor of the Thomasschule, 1810—1823. Below the second section of the last motett, in the autograph score, stand the words " The second verse is like the first, except that the choirs change about, the first choir singing the Chorale, and the second the *Aria*." The second verse must probably have been the fourth verse of the Chorale, but Bach must have found that in this way the work would be too long, for in the original parts the interchange between the choirs is not found. The connection of the Psalm-verse which follows is, in fact, not very obvious.

[717] Compare ante, p. 583.

uncle, Johann Christoph, who set part of the same text, and also interwove a chorale melody to form a motett, has been explained before.[718] For the last of the motetts for double chorus Bach used an aria text, and in this case dispensed with a chorale altogether. The text consists of two stanzas beginning "Komm Jesu, komm, mein Leib ist müde"—"Come Jesu, come, for I am weary." The poet, who is unknown, may have written these words with a special view to their being set to music. They cannot have been intended for congregational use, for their metre suits none of the Protestant chorale tunes that existed in 1750. The music to the second stanza in aria form is evidently Bach's original composition. The first stanza is fully developed for double chorus, and the picture which it presents of fervent longing for death is as majestic as it is deeply moving.[719]

In the treatment of an ordinary motett one fundamental principle held good: the separate sections or lines of the text were to be worked out fugally, not, however, to the exclusion of shorter homophonous passages. In motetts for double chorus, on the other hand, the working out was done by means of alternating the two bodies of the chorus, which were opposed to one another, as self-contained and complete entities, the two being usually only united at the chief cadences. By this means there was very little scope for thematic development; there exist, indeed, excellent motetts of the seventeenth century which exhibit nothing of that kind at all. Sebastian Bach followed out this principle. With that keen penetration which went to the very heart

---

[718] See Vol. I., p. 93. This motett is notified as Seb. Bach's work in Breitkopf's Catalogue of 1764, p. 5, and was published in Breitkopf and Härtel's collection as No. 2. The MS. is in the Amalienbibliothek, No. 15—17, to which is added a note in Kirnberger's writing with regard to the origin of the chorale employed.

[719] Designated as Seb. Bach's work in Breitkopf's Catalogue of 1764, p. 5, and published by Breitkopf and Härtel as No. 4, but with altered words. The MS. score in the Amalienbibliothek, No. 18—21, must have been put together from the separate parts, for above the beginning stand the words *Soprano Chorimi*, which the careless copyist apparently read for *Soprano Chori I mi*, and transferred to the score.

of all new forms and followed them out to their ultimate
consequences, he never once divided his body of voices into
a higher and a lower chorus, as was frequently done by the
older masters; and it was but seldom, and in a way of his
own, that he introduced a fugue in more than four parts.
In this method there was, indeed, a danger of the choruses
becoming united into one mass; but a fugue in which all
the parts took an equally active and independent share
would be a conscious transgression of the ruling principle
of the form, being not so much a union of all the parts to
one whole, as the resolution of two factors into eight.
Wherever Bach employs a fugal movement in which all
eight parts have a share, as for instance in "Komm Jesu,
komm," bars 44—57, and in "Der Geist hilft," bars 76—84
and 124 ff., it always happens that the separate parts of the
choruses are treated antiphonally. The individuality of
Bach's style is most prominent in the motetts for double
chorus, because here he was impelled most strongly towards
the use of homophony. I use this word only to denote the
absence of imitative writing. It is not intended to convey
the impression that there is any lack of melodic movement in the separate parts. It is just this movement
which is no less conspicuous in the homophonous than in
the fugal portions, and which testifies with the greatest
certainty to the origin of Bach's motett style. These
passages did not originate in the nature of the human voice,
which excels most in the simplest movements, and particularly in gradations of strength reached by imperceptible
degrees in colouring and *nuance*—they took their rise in
the conception of a musical instrument which has no means
of embodying the whole force of emotion but by varying
degrees of mere external movement within the limits
imposed by unalterable strength of tone. The church organ
is the parent of these passages that flow up and down and
across each other; and its influence may be traced even in
the minutest details of their form. In the organ the lack
of power to give proper expression to melody and rhythm
leads naturally to a greater prominence of purely harmonic
effects. Though Johann Christoph Bach, and others of his

time, strove to gain their effects by homophonous movements, a clearly recognisable melody, of however simple a form, is always heard persistently in the upper part. In Sebastian Bach there occur passages—as at the beginning of "Singet dem Herrn"—which can only be regarded as waves of harmony twisted into melody. The astounding boldness of the part-writing finds in this view its explanation and only justification. The grand harmonic portions, developed with the surest and truest instinct, afford firm points of support, or rather of suspension, at the beginning and the end, between which the separate parts can disport themselves at will. Contact and collisions between the parts, nay, even transgressions of the elementary rules of part-writing—notice the octaves between the extreme parts in bars 26—27 of the motett "Singet dem Herrn"—are not avoided, provided only that the harmonic progression is clearly intelligible as a whole. We cannot but notice that the whole is full of life and movement, but in many passages the effect of Bach's motetts in no way depends upon the exact perception of how that movement is obtained. Combinations of parts such as these :—

are not written by Bach in other places. In the same way he gets his effects of increased power in a manner similar to that employed on the organ. The four-part fugues, chorales,

---

[720] Motett " Singet dem Herrn," bar 72.
[721] Motett " Komm Jesu, komm," bar 150 ff.

or arias, at the close of the motetts for double chorus, represent, as it were, the full organ. When in the elaborate fugue "Die Kinder Zion sein fröhlich," from the first chorus of "Singet dem Herrn," more and more parts from the second choir are brought in to strengthen the entrances of the theme, the method may be compared to that of increasing the force in the course of an organ piece by drawing out one powerful stop after another.

It has been stated in another place[722] that at that time in Leipzig, as everywhere else, it was the custom to accompany the motett on the organ or some other supporting instrument. Bach's position relatively to this custom has been a matter of doubt. When he set free the motett from the church cantata, forming it, as is most evident, upon the style of the organ, never scrupling to let the bass part rise above the tenor, though it is always to be regarded as the continuous root of the harmony; when choral numbers, from cantatas like "Aus tiefer Noth" and "Ach Gott vom Himmel," could become widely known by the name "motett," even with a figured bass, which was partially, at least, independent of the vocal parts; and when his pupil Kirnberger expressly testifies to the co-operation of the organ, the doubt seems superfluous. Indeed, an organ accompaniment exists for the Psalm "Lobet den Herrn alle Heiden," and there is a figured organ part written by Bach himself for the motett "Der Geist hilft unsrer Schwachheit auf." Of course this can only have been intended for use in church. The performance at funeral ceremonies took place at the house of mourning; for this the instrumental parts which also remain to us—two violins, viola, violoncello, for the first choir, and two oboes, tenor, and bassoon—may have been employed, in which case it remains uncertain whether or no they, as well as the organ, took part in the church performances. The organ part is very interesting. On the one hand it offers an example rarely given by the composer of how agreeably a figured bass may be added to his motetts for double chorus without being independent in any way,

---

[722] See ante, p. 279.

since the organ part contains only one semiquaver which is not also in the vocal bass. On the other hand, the figuring, which in the passages where it is used follows the fundamental harmony, regardless of some rather serious false relations, shows clearly by what considerations Bach was led in writing his parts. If we consider the relation between the two vocal bass parts, it appears very similar to that which exists in Bach's cantata choruses between the vocal and instrumental bass parts. To write the two bass parts so that they frequently came together either in unison or in octaves was indeed usual before his time. But the way in which Bach often makes one of the bass parts separate from the other for a short time, in order to execute some rather important phrase or other, and then flow on again, united to the other (see "Singet dem Herrn," bar 122 ff.); the way in which one joins in with the other on the figured notes, thus simplifying the phrase (see the first of the examples given above, and also "Fürchte dich nicht," bar 34 and "Komm, Jesu, komm," bar 60); the way the bass of one of the choirs is brought in alone to complete the harmony of the other (see "Fürchte dich nicht," bar 57)—all this can be called neither unity nor duality of parts; it is a free method of treating the parts under his control according to the necessity of the moment, such as we also notice in the treatment of the figured bass, with regard to the vocal bass in the cantata choruses. Here, indeed, the figured bass is the dominant part, whereas in the motetts for double chorus this part is taken now by one of the basses, now by the other, sometimes even by both together. But Bach could only fall into this style of writing while the idea of his accompanied cantata choruses was in his head.

If we consider the fugal movements that occur in the motetts, not a few of them are so arranged that the one part which has the theme does not lead off quite alone, but supporting harmonies are supplied by the other parts. Thus it is in the second section of the motett "Jesu meine Freude" ("die nicht nach dem Fleische wandeln"), vocal harmonies surround the opening of the fugue from which real parts are gradually evolved. In the fugue "sondern der

Geist," &c., from the Funeral Motett, the theme given out by the soprano of the first choir is accompanied by all the other parts of the same choir. In the fugue " die Kinder Zion sein fröhlich " (in " Singet dem Herrn"), the whole second choir at first only accompanies the first in harmonic masses, until the separate parts are gradually drawn in by the irresistible flood of sound, and made to take part in the fugal development. But Bach usually constructs the fugal movements of his cantatas above or among the accompanying harmonies of the figured bass or of the instruments. If we compare the opening chorus of the cantata " Wer sich selbst erhöhet "[723] we shall have an almost exact counterpart to the motett fugue last alluded to. We see in this how the individual characteristics of Bach's concerted music are repeated again, even in detail, in his motetts.

In spite of this, the question as to the accompaniment of Bach's motetts is not yet dismissed. Though in Breitkopf's printed catalogue of 1764 some of these works are entered under the name " Motetts without instruments," this proves very little. For, as may be seen on the next page of the catalogue, compositions with organ accompaniment are also comprised under this name, so that the plural "Instruments" is to be taken in its strictest meaning. Nor is the fact of the organ not being expressly mentioned in Bach's pieces any reason for denying that it was used, for, as has been shown, it was very often taken for granted. The fact of its having to provide the necessary completion of the harmony has nothing to do with the point. The question is simply whether the quality of the vocal writing is not such as to be only fully explained and justified by the addition of the organ effect. A strong piece of evidence on the other side is, however, that among the original parts of the motett, "Singet dem Herrn," which are preserved in their entirety, no organ part is found. And more important still is the testimony of Ernst Ludwig Gerber, who in 1767 heard a motett for double chorus in Leipzig, and remarked that the Thomasschule boys " were wont to sing " these compositions by Bach

---

[723] B.-G., X., No. 47. P. 1698.

without any accompaniment. This expression cannot be misunderstood, and at the same time it points to a custom which had prevailed for a long time. Thus there exist contradictory testimonies with regard to this question.

The whole of Bach's creative work is pervaded by a tendency to cast off all that is unnecessary and redundant attaching to the forms, to limit the means of expression to what was indispensably necessary, and in the highest possible degree to spiritualise what was material. It was this tendency which prompted him to cast off the figured bass harmonies with the clavier obbligato in the sonatas for violin and viol da gamba, and to write sonatas and suites for violin and violoncello without any accompaniment whatever. His predecessors preferred to write motetts with rather than without an independent figured bass part. Only a single work of his is known in which the figured bass is necessary to its completeness, at any rate in a portion of it—this is the Psalm " Lobet den Herrn alle Heiden"; the motett movements, which were originally parts of cantatas, are not referred to here. For the rest, all that is requisite to the tone-picture was to be done by the voices alone. It is perfectly conceivable that Bach should have allowed himself to be led by that spiritualising tendency so far as to eliminate from his motetts even the admixture of that quality of tone, and of that colouring, which first gave them their individual style. This would have been no more than what he had previously done in the case of his solo sonatas for violin and violoncello, the style of which in the same way was derived less from the nature of those instruments than from that of the clavier and organ. We must constantly bear in mind that Bach's fancy lived and had its being in the tone of the organ, and that as he listened to the unaccompanied motetts—or to the sonatas—his mental ear would vividly supply, in addition to their audible effect, that quality which had originally given them life and which, though separated from them, yet hovered around them like a delicate perfume. It is certain that the motetts could only produce their full effect when accompanied on the organ. Without this their right appreciation would depend chiefly upon the ease with which their technical difficulties could be overcome, and upon the extent to which the voices could assimilate themselves in quality to the restful flow of the organ; also, upon the degree in which the hearer can succeed in saturating his imagination with the feeling of organ music, and so supplying what is unattainable by the instruments used.

For the rest Bach's motetts are the only compositions of his that have at no time vanished completely from the world of music. The cantors of St. Thomas' after him always held them in honour, and had them sung, though certainly not always *con amore*. Rochlitz, calling to mind his practice in the Thomasschule, remembered particularly the motetts "Singet dem Herrn," "Der Geist hilft unsrer Schwachheit auf," and "Sei Lob und Preis mit Ehren," with a certain mixture of feeling, however, for the difficulty of these works caused the boys much trouble.[724] When Mozart went to Leipzig in 1789, the Cantor Doles had the motett "Singet dem Herrn" sung for him, and he was so charmed with it that he studied all the motetts by Bach that the Thomasschule possessed. Then came a time when it seemed as though Bach would again meet with more general appreciation. This was the period when a transient admiration for Bach was created by Forkel's work; it passed, however, very quickly. But that the motetts were not forgotten, even later than this, is shown by their first appearance in print, in an edition prepared by Schicht in 1802 and 1803. Outside Leipzig, too, they seem to have been moderately well known; in Saxony, at least, as was natural from the fact that a great number of Saxon cantorates were held by men who had formerly been scholars in the Thomasschule. Marschner, when a boy at Zittau, sang Bach's motetts, under the direction of Friedrich Schneider, the choir prefect; and in a MS. choral-book preserved at Nieder-Wiesa, in Saxony, dating from the end of the last

---

[724] Rochlitz, Für Freunde der Tonkunst, II. (Edit. 3.), p. 134 f.—The Motett "Wie sich ein Vater erbarmet," mentioned by him, in addition to the above, is only the second part of "Singet dem Herrn." It seems that this comprehensive work was at that time only performed piecemeal.

century, is found the text of the motett "Komm Jesu, komm," adapted to a newly invented melody, and turned into a chorale.[725]

X.

"OCCASIONAL" COMPOSITIONS.

DURING the years 1723—34 an abundant wealth of compositions in the domain of church music was brought to perfection by Bach's genius. These works bear most certain and eloquent witness to the fact that Bach was here in his element. His obtaining the post of Cantor at St. Thomas's just in the ripest years of manhood must be regarded as no less happy a contingency than his having been summoned to Weimar and Cöthen, which served the purpose of giving to his nature its first free development and its contemplative depth. As compared with the result of this eleven years' work in Leipzig, the disappointments and unpleasantnesses which the office necessarily involved are seen to be of very small actual importance. Bach really valued his post and considered it a favourable one, and, as we shall soon see, felt himself bound to sing the praises of Leipzig. The plunge into the unknown which he had ventured upon "in the name of the Most High"[726] was really a fortunate one, although at times it seemed to him to be otherwise. In order to get a clear idea of his life and work, it is necessary that his productiveness in the way of church music should be dwelt on at full length, and shown to be the power which drove all other kinds of work into the background; by this means alone can light and shade be rightly thrown upon the path of his development. For this reason detailed reference will not be made here to the numerous instrumental compositions which date from about this time. It will be more suitable to leave the consideration of these to the end, and the more so because Bach never, even at the end of his life, ceased to

---

[725] See Jakob and Richter, Reformatorisches Choralbuch. Berlin, Stubenrauch. Part II., No. 918.
[726] See ante, p. 254.

produce works of high importance in this branch of art. The case is different with regard to the "occasional" compositions for voices and instruments. The greater number of these date also from the Leipzig period; besides which, they are in part, at least, so closely connected with the church compositions, that even on this account they would claim examination in this place.

The "occasional" compositions, which were intended to celebrate some specially important event in the life of an individual, or of some public institution, are both sacred and secular. To the first class accordingly belong the funeral compositions: "Dem Gerechten muss das Licht" and "Herr Gott, Beherrscher alle Dinge"; the funeral cantata for Herr von Ponickau, "Ich lasse dich nicht" (1727); the burial motett for the Rector Ernesti, "Der Geist hilft unsrer Schwachheit auf" (1729); and the cantata for the dedication of the organ at Störmthal (1723): while the compositions for the annually recurring election of the Councillors (Rathswahl-cantaten), and those written for special ecclesiastical festivals, cannot be included in this category, although in a certain sense they too may be called "occasional" cantatas. Mention has before been made of the works just alluded to, for some of them were employed afterwards as "Rathswahl" cantatas, while some were used in their entirety for church purposes.[727]

We know that a portion of the Passion according to St. Mark is nothing more than an adaptation of an "occasional" composition.[728] This Passion music no longer exists, but the composition which was thus adapted is completely preserved as the music for the death of Queen Christiana Eberhardine. It would be valuable from this cause alone; but it deserves recognition also from the fact that it has an ecclesiastical and political *raison d'être*. Queen Christiana Eberhardine, of the family of the Margraves of Brandenburg-Bayreuth, had been married to Frederick Augustus since 1693. When her husband had, in 1697,

---

[727] See ante, pp. 468—9, 411—12, 602, 365—67.
[728] See ante, p. 505 f.

ascended the throne of Poland and embraced the Catholic religion she had remained true to the Evangelical church. She had no desire for honours that were to be purchased at the price of her religion; and, although she could not persist in her refusal to bear the title of Queen, she never set foot in the kingdom of Poland. Even before her husband had cut himself off from her by his connection with the Countess Königsmark she had completely separated herself from him, and lived quietly at Pretzsch near Wittenberg. It was not long before her son, the successor to the throne, whose earliest education she had undertaken, abjured the religion of his father, which he did in 1717. A living religion of the heart, and firmly rooted convictions, were her Protestant doctrine, as she showed on the occasion referred to, in a letter in which, being well versed in Scripture, she sought to point out the errors of the Catholic Church, and adjures her son, "for his own poor soul's sake, and for his poor mother's sake, whom else he would bring down with sorrow to the grave, to return again to Evangelical truth."[729] By their perversion to Catholicism the Electors of Saxony lost the leading position which they had held in Protestant Germany since the Reformation, and which now passed into the hands of Prussia. Though the prince gave many pacifying assurances that his change of religion was a purely personal circumstance, and that everything would remain as it had been of old in the country, the people of Saxony were so zealously Protestant that they could not but regard with suspicion a movement which seemed to favour or at least to allow the introduction of Catholicism. At the Saxon diet of the year 1718, very violent accusations and petitions were brought on this question.[730] Protestant preachers, zealous for the faith, did their best to stir up excitement and hatred. The doubts raised by the Leipzig Council in 1702 with regard to the propriety of the services there being conducted in Latin, which made them resemble

---

[729] The letter is quoted entire by Förster, Frederick August II., Potsdam, Reigel, 1839, p. 245—249.
[730] Gretschel, Geschichte Sachsens, II., p. 589, ff.

the Catholic services too closely[731] appear in a new light under these conditions. A year before the queen's death religious fanacticism had even led to murder and insurrection. The archdeacon of the Kreuzkirche in Dresden, Magister Joachim Hahn, was stabbed on May 21, 1726, by a Catholic whom he had converted to Lutheranism, and who had afterwards been in doubt as to his spiritual condition. This deed created a fearful tumult, which had to be quelled by military power.[732] In these circumstances it was natural that the Saxon populace should regard the steadfast queen with special respect and love, and should be deeply moved by her sudden decease. The general mourning which was commanded lasted from September 7 to January 6.[733] On October 17 the town of Leipzig showed its respect for the departed princess by a grand public funeral ceremony; compared with this the indifference shown on the occasion of the king's death, six years later, is sufficiently significant. In the year 1697 when, after the election of the King of Poland, a *Te Deum* was appointed to be sung in the Saxon churches, the congregations, with unequivocal demonstrations, sang Lutheran and other Protestant hymns after it.[734] But the funeral ceremonies for the queen, who had remained true to her faith, were performed in a devotional Protestant spirit. The poem set by Bach avoids, and that designedly, all that might give offence to the king, but does not omit to celebrate the queen, " the pattern of a great woman," as the " defender of the faith."

The ceremony did not bear a strictly devotional stamp; it was arranged in the same way as the academic "speeches," and took place in the University church, the University having taken the principal part in its arrangement.[735] On

---

[731] See ante, p. 264.

[732] Gretschel, loc. cit., p. 592 f.—Picander celebrated the event in a dull and bombastic poem (Part I., p. 212—231).

[733] See Appendix A., No. 33.

[734] Gretschel, loc. cit., p. 475.

[735] The printed matter relating to this ceremony—viz., (1) the Latin introductory speech by the Rector of the University; (2) the text of Gottsched's Ode, as it was distributed among those present in the church; (3) the eulogy

occasions such as this it was usual to set to music, not cantata texts in the modern madrigal form, but Latin odes. Bach had previously set a Latin ode for the birthday festivities of Duke Friedrich cf Saxe Gotha.[736] The general interest taken in the funeral ceremony must have been the reason why the German language was considered more suitable on this occasion. Gottsched, who had been for several years lecturer in the University, and was Senior of the Deutsche Gesellschaft, wrote the ode; it cannot be praised for any great flow or wealth of ideas, but its general character is worthy of its subject, and its diction is correct.[737] The composition is divided into two parts, of which the first was sung before and the second after the funeral oration. This was delivered by Hans Carl von Kirchbach, "Assessor of the Royal and Electoral Upper Court of Mines (Oberberggerichts) at Freyberg."

Bach finished the composition on the 15th of October, only two days before the funeral. As the parts had then to be written out we may see in this case with how few preparations this kind of musical performance was usually given.[738] He gave the work the cantata form by ingeniously dividing the strophes of the ode into choruses, recitatives, and arias. This seems to have been an innovation as applied to poems of this kind, for a reporter mentions expressly that the ode was composed "in the Italian style."

---

and lament delivered by Hanns Carl von Kirchbach; (4) a funeral ode by *M. Samuel Seidel*—is all found together in one volume in the Royal Public Library in Dresden (*Hist. Saxon.* c. 232). The order of the ceremony is described by Sicul, Das thränende Leipzig, 1727.

[736] See ante, p. 215.

[737] It was reprinted in an elaborated form in "Oden der Deutschen Gesellschaft in Leipzig." Leipzig, 1728, p. 79 ff.

[738] At the close of the autograph score stand the words, "*Fine SDG. aõ 1727. d. Oct.* 15. *J S Bach*." On the title-page Bach gave the date of performance as October 18. That he was mistaken in the date is indubitable since the recovery of the original printed form of the ode, which gives the date as October 17 (see above, note 735). If the ceremony had been obliged for any reason to be put off for a day, Sicul would surely have mentioned it in his elaborate description. This rectifies and completes the statement made in the Preface to B.-G., XIII., 3.

The Italian style is also prominent in the use of the clavicembalo, which Bach himself played, and on which he must have accompanied the recitatives and arias, as was done in the Italian opera : the organ, probably, was only introduced in the choruses. The place which the work holds, half-way between secular and sacred music, by comprising continual references to God, and by the employment of chorales, while it is only a mortal person who is celebrated, but who is brought, as it were, into the region of the church, is explained by the circumstance that[739] the deceased queen had been very fond of music. Bümler, the Capellmeister at Anspach, was in her service from 1723—25, and in the summer she used frequently to invite members of the Dresden orchestra for her own pleasure.[740] This may have given a special incitement to the composer. The music to the Trauerode (funeral ode) is one of Bach's finest works. In the broad phrases of the first chorus, to which a full and florid colouring is given by the employment of gambas and lutes, a feeling is breathed forth which is closely related to that of the final chorus in the Matthew Passion. But in the chorus in the Trauerode, even though it is an introductory movement, there is a more vehement and passionate sorrow. This also comes out in the second recitative, which the instruments accompany with passages resembling the sound of bells ; the high flute begins, and is followed by a gradually increasing number of instruments going from the highest to the lowest, in figures of different kinds; last of all comes in the bass in long pulsations, and this sea of sounds flows on further in peculiarly sorrowful modulations, until it gradually ceases. The undeveloped sketch for this is found in the Weimar Cantata " Komm, du süsse Todesstunde," in describing which we spoke of the æsthetic justification of this graphic realism.[741] As the funeral ode goes on the feeling becomes calmer, particularly in the choruses, which are now smoother and more quiet,

---

[739] Sicul, loc. cit., p. 22 f. See Appendix A., No. 6.
[740] Hiller, Lebensbeschreibungen. Leipzig, 1784, p. 56 and 197.
[741] See Vol. I., p. 549 ff.

the last being quite in the simplest style. The prophecy here uttered by the poet :—

> Doch Königin du stirbest nicht,
> Man weiss, was man an dir besessen,
> Die Nachwelt wird dich nicht vergessen.

> But, noble Queen, thou diest not;
> We know what we possessed in thee,
> Posterity shall not forget thee.

has been fulfilled, although in a manner somewhat different from that which he intended. It is the art of Bach which has given immortality to the figure of the good queen, and I believe that the privilege of being so gloriously remembered will not be lost to her by the fact that the Trauerode, since it has become known again, is sung to a modern text of quite general religious import. No æsthetic reason, it is true, can be brought against this practice, and Bach himself employed the work for church purposes set to another text. The bitter reproaches which have been justly poured on a later and less intelligent period for a similar procedure with Handel's funeral hymn for Queen Caroline are not applicable here.[742] Handel's music can only be properly appreciated in its full depth and beauty by him who remains conscious throughout that it was dedicated to the pious memory of a noble personage. Instead of this free and broad human feeling, in Bach we find a strictly devotional sentiment; his music always speaks the same language. But the Protestant queen who remained true to her faith, and inspired Bach to such a masterpiece, might surely appear to subsequent generations worthy of an hour's remembrance.

A second funeral composition of yet greater length was written by Bach a year later, in honour of Prince Leopold, Anhalt-Cöthen. A lasting intercourse had been kept up between Bach and his cultured patron, even after he had

---

[742] Chrysander, Handel II., p. 445.—Among the German capellmeisters and cantors of that time it was the general custom to employ funeral compositions written for distinguished personages, with altered texts, for church cantatas. Johann Ernst Bach's funeral music on the death of Duke Ernst August Constantin, of Weimar (1758), also met with this fate.

left his court. The title of "Royal Capellmeister of Cöthen," which he continued to bear, bound him to perform certain honorary duties. Thus he composed a cantata for November 30, 1726, the birthday of the princess; and in the cradle of her first-born, the hereditary Prince Emanuel Ludwig (b. September 12, 1726), he laid an autograph copy of the first Partita in the "Clavierübung," which had just then appeared in print, together with a dedicatory poem written by himself, from which his relations with the princely house appear to have been of a most friendly kind. In May, 1727, at the time of the "Jubilee" fair, Prince Leopold was in Leipzig, where he heard the festival music which Bach had performed on May 12, in honour of the king, who was also present.[743] This seems to have been their last meeting. The death of the Prince took place unexpectedly soon after this, on November, 19, 1728. The funeral ceremonies at Cöthen, at which Bach conducted his compositions himself, did not take place till the following year; the exact date is not known. It has of late been made extremely probable that this work, which has for some time been lost, was for the most part made up of portions of the then newly composed St. Matthew Passion. Here, accordingly, the same relations subsist between the two works as between the Trauerode of 1727 and the St. Mark Passion; only in this case the church composition was certainly the older of the two.[744]

Turning now to the considerably larger number of secular "occasional" cantatas, I must first mention, by way of supplement to a former chapter, a work which belongs

---

[743] Sicul, *ANNALIVM LIPSIENSIVM SECTIO* XXIX., Leipzig, 1728.

[744] For the discovery of this adaptation we are indebted to W. Rust; see B.-G. XX.,² p. X. f. The only circumstance which can be adduced against the result of Rust's research is that Forkel, who possessed the funeral music in autograph, and also knew the St. Matthew Passion, did not remark on their identity. Certainly Forkel's knowledge of the St. Matthew Passion may have been only very superficial, see his work on Bach, p. 62. Rust mentions (loc. cit. p. XIV.), among the lost "occasional" compositions of Bach, a third funeral cantata ("Mein Gott, nimm die gerechte Seele") adducing Breitkopf's Catalogue of Michaelmas, 1761. The cantata is indeed mentioned there (p. 23), but without the name of the composer.

indeed to the Cöthen period, but which has only lately come to light.[745] Since both the title and the opening are lost, the occasion and purpose of its composition can only be guessed. Thus much, however, is clear, that it tends to the glorification of the whole princely house of Anhalt-Cöthen. It is possible that Bach dedicated the cantata, a work of considerable extent, which, to all appearance, was written before the end of the year 1721, to the honoured family at the close of the year, and that he himself wrote the words, which leave much to be desired in both substance and form. The words of the last recitative ("Ja sei durch mich dem theursten Leopold") show in the clearest way the purpose of the cantata. This delicate composition, which is full of lovely ideas, and in which a duet between alto and tenor (E flat major, common time) is remarkable for its especial beauty, is set throughout for solo singers, even the four-part number, at the close, being intended for single voices. In this the cantata agrees with the birthday cantata "Durchlauchtger Leopold,"[746] and shows, moreover, that Bach when in Cöthen had no available chorus at his service. As that had been used for a Whitsuntide Cantata, so this was afterwards turned into a work for the third day of Easter ("Ein Herz, das seinen Jesum lebend weiss") in Leipzig, without all its numbers being used however.[747] Here I must once more point out that the cantata mentioned earlier ("Steigt freudig in die Luft"), which Bach wrote for the first birthday which the second wife of Prince Leopold celebrated in Cöthen, was afterwards re-arranged several times.[748] The first of these re-arrangements was made for the birthday of a master,

---

[745] I discovered it in 1876 in the collection of autographs belonging to Herr W. Kraukling, of Dresden, who was kind enough to entrust it to me for a considerable time. See App. A., 61.

[746] See ante, p. 6.

[747] The autograph score of this Easter Cantata is in the Royal Library at Berlin.

[748] In Picander I., p. 14, it is expressly stated: "For the first birthday festival of the Serene Princess, at Anhalt-Cöthen, 1726." Now since her birthday was on Nov. 30, and her marriage took place on June 2, 1725, it follows that on Nov. 30, 1725, the Princess cannot have been at Cöthen. Compare ante, p. 158.

perhaps of Gesner;[749] the second transformed the work into a church cantata for the first Sunday in Advent.[750] There lived in Leipzig a lawyer, Johann Florens Rivinus (b. July 28, 1681), who was appointed *Professor ordinarius* to the University on June 9, 1723. He must at that time have been exceedingly popular, for on the evening of that day the students performed a serenade in his honour.[751] At least ten years after this the students again waited upon him with music on his birthday; for this the Cöthen Cantata underwent its third re-arrangement, and hereby regained its original secular character.[752] Bach seems to have had a special liking for this work, which, although pleasing, is not of great importance, possibly because such agreeable remembrances were connected with it.

Bach's secular "occasional" cantatas belong almost exclusively to the class of dramatic chamber-music. They are founded either upon some action, represented by means of a number of personages who come on and speak in character, or else upon a situation which is explained by the speeches of various persons. While, in general, vocal chamber-music stood half-way between sacred and dramatic music, this kind of composition leans more strongly towards the side of the opera. Of course such pieces were of much shorter extent, and rather resembled the last act of an opera. The plot and the characters were by preference taken from ancient mythology, although they were treated for the most part allegorically. Dramatic situations and style were demanded in as great a measure as for a work for the

---

[749] See ante, p. 262. Noticed in Breitkopf's Catalogue of Michaelmas, 1761, p. 33, among "Cantatas for promotions and days of honour" ("Promotions-und Ehrentags-Cantaten"). [750] See ante, p. 471.

[751] Vogel, *Continuation* Derer Leipzigischen Jahrbücher von *Anno* 1714 bis 1728. MS. in the Leipzig Town Library. Fol. 32b. The text of the cantata written on this occasion is there given, without mention of the composer. Winterfeld, Ev. Kirchenger III., 262, suspects Bach to be the composer, but hitherto the suspicion has received no kind of confirmation.

[752] The words begin "Die Freude reget sich, erhebt die muntern Töne." The original parts are in the Royal Library at Berlin. The parts are no longer complete, but, under existing circumstances, the loss is not very great. B.-G., XII.,[2] p. V., Note, is to be corrected accordingly. Bach was also a personal friend of Rivinus, and in 1735 asked him to be godfather to his son, Johann Christian.

stage; only costume, action, and scenery were wanting; and the last not entirely. It does not follow, from these compositions being reckoned as chamber-music, that they were always performed in the music-room. There is one essential difference between the musical performance of our time and of the past, in that a much greater freedom and variety prevailed then with regard to the place of the performance. At that time the ties connecting music with social life were much more numerous than they are now. Whereas now almost all the music which is performed outside the domestic circle—except it be either ecclesiastical or theatrical—is confined to the concert-room, where the public assembles for the special purpose of hearing it; at that time people chose the street, the garden, the wood, or even the lake or river, according to whether the music to be performed referred to a public ceremony, a wedding, a birthday, a hunting expedition, or any other festivity. If the work were of a dramatic kind the place of the performance was regarded as the scene, and the incidents of the piece had to be such as were suitable, though it was only in exceptional cases that such performances took place in costume. For a just appreciation of the character and effect of Bach's dramatic cantatas it is necessary to keep this mode of performing such works constantly in mind. It involved the procedure—so strange to us—of bringing in, not only the full chorus, but also solos in recitative and aria style, accompanied with instruments and the chorus in the open air. The usual weak and often simple treatment of the voices and instruments could not, of course, produce that far-reaching fulness of tone which is needed in an unlimited space under the open sky. Yet the skilled composer had to take care to suit his effects to the place of performance, and certain strange details in the Bach cantatas with which we are now concerned become intelligible in connection with this necessity.[753]

---

[753] Scheibe (Critischer Musikus, p. 450 ff.) goes so far as to give exact rules as to the different treatment of works of this kind, according to whether they were to be performed on the water or on land, in a room or in the wood, in a garden or in an open space surrounded with trees.

Apart from these peculiarities, the employment of certain rules of style underlies the dramatic cantatas, since they belong to the class of chamber-music. They had to unite the solidity of the sacred style with the ease and elegance of that of the theatre. In them we see the musical root and foundation from and on which the Handelian oratorio arose; its connection with the dramatic chamber cantatas is revealed even in external style by the circumstance that Handel had his "Acis and Galatea" performed in a suitably decorated scene, although in this case it was an artificial one.[754] But Handel, it must be owned, embodied the full import of the ancient myths in works which were intended for the whole cultured world, while in Bach an allegory supplies the dramatic motive, and the purpose of the work is to do homage to some particular person and to delight a small and select circle. While Handel is in his element here, Bach can hardly appear in any other capacity than that of a musician whose services are given to order. He would not have been the grand musician that he was if he could have contented himself with the composition of dramatic cantatas. For the public concert, which in England enabled Handel to elevate these and similar kinds of musical work to a branch of art of the first rank, was wholly wanting in Germany. The only place from which a man with Bach's comprehensive musical gifts could exercise an influence on the German people was the church choir. The influence of this, his own special province, pervaded all he undertook, consciously or unconsciously to himself. Accordingly he made use of the greater part of the secular cantatas for church music, without radical alteration, and they are seen to be in their right place. It follows from this that in their original form they cannot have been very suitable to the purpose for which they were written. Indeed, Bach's chamber cantatas, so far from keeping their proper central position, lean strongly towards the church style. The characteristic treatment of the persons and situations which is required in the more dramatic of them shows indeed in its great sharpness and contrast that the artist

---

[754] Chrysander, Handel, II., p. 266.

has worked with mature deliberation; still it exhibits nothing more than the versatility of his inventive faculty in his own province. It fulfils but one requirement of the dramatic style; it is very much too heavy for the circumstances; the light play of the emotions is turned into serious pathos, and the comic into the grotesque. Although many of the figures appear full of character when compared with one another, yet they in no way show that Bach's talent would have been suitable to opera; on the contrary, it cannot be doubted that he was wanting in the class of sentiment which is an indispensable condition of that sphere of art. It is consequently impossible to hold up Bach's chamber cantatas as models of their kind. They are works full of value and charm, but only for him who brings himself into sympathy with Bach's individuality. This by way of introduction to the examination which we are about to make of the separate works.

We have seen already how Bach always took trouble to assure to himself the goodwill of the student-world of Leipzig.[755] It evidently has to do with this when we find him, even in the first years of his residence there, before he had undertaken the conducting of Telemann's Musical Society, occupied frequently in helping the students' enterprises by means of his art. Among the popular teachers at the University was the Doctor of Philosophy, August Friedrich Müller, whose nameday fell on the 3rd of August.[756]

In the year 1725 his pupils wished to honour him with a musical work on this occasion, for which the ever-ready Picander made the text, which was set by Bach.[757] It was a dramatic cantata, evidently intended to be performed in the open air, and this is confirmed by the strong treatment of the accompaniment. The characters in the piece are

---

[755] See ante, p. 214.
[756] Müller was born in 1684 and died in 1761. On October 19, 1731, he became Professor extraordinary. His scientific works are noticed in the "Leipziger Neue Zeitung von gelehrten Sachen," III. 224, 519. VI. 654, 672. XVII. 760. XX. 103. In Ch. E. Hoffmann's "Geographischer Schau-Platz Aller vier Theile der Welt," Andere Theil., an engraving of him is given.
[757] Picander's Gedichte, I., p. 146 ff. B. G., XI.,[2] 139 ff.

DRAMATIC CANTATAS. 625

drawn from ancient mythology. Pallas wishes to celebrate a festival on Helicon with the Muses in honour of the learned man. It is feared that it will be interrupted by a violent autumn tempest (a fear, by the way, that appears somewhat premature at the beginning of August); the fierce winds are heard murmuring in their prison, and Æolus, their ruler, promises that they shall soon be set free; however, after Zephyrus, the god of the warm summer breezes, and Pomona, the protectress of the fruits of the earth, have both entreated in vain, he consents, on the representations of Pallas, to leave the pleasant repose of the season for a time undisturbed. The chorus of winds and the recitative of Æolus, with which the cantata opens, are taken by Picander from a well-known image in Virgil (Æn. I., 56 ff.). The winds are in a mountain dungeon, where they wrestle with one another, and vent their untamable fury upon the locks and bolts of their prison.[758] Æolus is rejoicing in the expectation that they soon will be set free:—

> When through the world they rush and roar
> The rocks and hills are firm no more,
> But all things fly before them.

By these figures and ideas Bach was moved to create tone-pictures of strange grandeur. In what follows, too, Picander prepared the composer's way very efficiently to some beautiful musical contrasts, even introducing some striking features into the more detailed character-drawing. The aria of Zephyrus, with its softly breathing figures and tender, charming colouring, is one of Bach's loveliest pictures of nature; and Æolus becomes in his hands a wild unmannerly churl. His demeanour, it is true, is more suitable for a tragedy than for a cheerful garden festival, and when he is induced, by the single name "Müller," to

---

[758] *Hic vasto rex Æolus antro*
Luctantes ventos, tempestatesque senoras
Imperio premit, ac vinclis et carcere frenat.
Illi indignantes magno cum murmure montis
Circum claustra fremunt.

hush the uproarious winds in their cavern,[759] the effect becomes irresistibly comic.[760]

Nine years later Bach brought his work again to a hearing on a much more important occasion. On January 17, 1734, Friedrich August II. was crowned King of Poland, as August III., in Cracow. For the festivities of the Leipzig University, which began on February 19, Görner had to provide a Latin ode.[761] Already in January, when the approaching coronation became known, Bach had prepared the Musical Society for a festival performance which was to take place so soon as news should arrive that the coronation was over. The necessary alterations in the text were undertaken by himself. Æolus was transformed into Valour, who, in the first recitative, instead of the raging of the winds, described the deeds done by her in the war with Stanislaus, the opposing king, and his party. Zephyrus had to take the *rôle* of Justice, and Pomona that of Mercy, Pallas alone remaining in person to beg the king's protection on behalf of the Muses. The name "August" now referred no longer to Professor Müller, but to the king of Poland. By this means the music, of course, lost all its dramatic character, but it is also evident that Bach did not consider the chief importance of the work to lie in this quality; if he had, the alteration of the text would have been a barbarism which could the less be justified because he was not obliged by any kind of command to celebrate this festal occasion.[762]

---

[759] *Celsa sedet Æolus arce,*
Sceptra tenens: mollitque animos, et temperat iras.

[760] E. O. Lindner (Zur Tonkunst, p. 129), gives it as his opinion that the aria, "Wie will ich lustig lachen," has a wild humour, compared with which Handel's lauded Polyphemus sinks almost to the level of the ordinary Italian *buffo* air. But I do not understand how the whole can so be lost sight of compared with the details. The greatness, never to be surpassed, of Handel's Pastoral arises from the perfect harmony existing between the subject and its musical treatment. Bach used stronger materials in his treatment, but it is not usual to adorn a summer house with church steeples.

[761] Mittag, Leben und Thaten Friedrich Augusti III. Leipzig, 1737, p. 333 f. Note.

[762] The original printed form of the text is in the Royal Public Library at Dresden, *Hist. Polon.*, 672, 17. That Bach is the poet is clear from the form

On another occasion, in the year 1726, Bach made himself serviceable to the students. Gottlieb Kortte, a scholar, born in the year 1698, at Beeskow, in Lusatia, who had studied first theology and philology in Leipzig, and afterwards jurisprudence at Frankfort on the Oder, was promoted on December 11, 1726, to be professor extraordinary of law at Leipzig. From the comprehensive knowledge he had acquired, and his personal merit, he was already considered an ornament to the University. During his few remaining years this opinion was confirmed, and his early death, which took place on April 7, 1731, "was greatly bewailed by the young students, with whom he was very popular, and by all who knew the thoroughness of his erudition."[763] The performance in honour of the ceremony of his promotion took place, of course, in the Great Hall of the University, and, as was suitable to the scene, the personages of the dramatic cantata performed on the occasion are only allegorical: Diligence, Honour, Happiness, and Gratitude. Though Bach gave a pompous character to this work he did not write it all afresh, but employed[764] the third movement of the first Brandenburg Concerto, which he adapted to the opening chorus with great skill, although with a certain license of genius. The second trio of the dance-numbers appended to the Concerto was introduced as the ritornel of the duet. The alto aria, which is independent, so far as can now be determined, of any other work, has the rhythm of a minuet, and, with its softly sounding phrases in the form of fanfares, reminds us of the united violins and violas in the second gavotte in the

---

of the title. The poet's name was never omitted on occasions of this kind, and it was much more likely for the composer not to be mentioned. From the fact that a blank space is left for the date, we see that the text was printed before the day of the coronation was known.

[763] Leipziger Neue Zeitung von gelehrten Sachen XVII., p. 264. The November number of the *Acta Eruditorum* of 1731 contains, in an *Elogium G. Cortii*, his biography, which has served as the source of later notices. See Sicul, Leipziger Jahr-Geschichte, 1720, pp. 92, 127, 212. Kortte's writings were mostly of a philological kind (Ausgaben von Ciceros Episteln, Sallust, Lucan, &c.).

[764] See ante, p. 130.

orchestral suite in C major.[765] The whole work opens with a march such as usually accompanied festival entrances and exits, that is to say, rather a fanfare prolonged into a march-form of two sections than a composition of purposeful and compactly articulated melody: the march in Erlebach's great Homage-music (Huldigungsmusik) of 1705 is cast in a similar form.[766] The cantata was afterwards performed again under Bach's direction, with an altered text, for the name-day festival of King August III. (August 3).[767]

The Cantata "Siehe der Hüter Israel" seems to have been written for a similar occasion as the music for Kortte. It was for four voices, string orchestra, and harpsichord, but has been lost.[768]

Five months after Kortte's promotion there took place a new student festivity with music by Bach. On May 3, 1727, King August II. arrived in Leipzig, meaning to stay a few days for the Jubilee fair. On May 12 was his birthday. As he had had a severe illness in the beginning of the year at Bialystok, in Poland, the patriotic Leipzigers purposed to make unusual efforts to celebrate this birthday with due solemnity. In the morning there was a festal ceremony in the University church, with a Latin ode of Görner's composition, and a Te Deum with firing of cannon and pealing of bells. In the evening, at eight o'clock, the students belonging to the academical common table performed a dramatic cantata outside the king's lodgings, which were then, as usual, in Apel's house in the Market (No. 17); for this Bach composed the music and conducted it in person. The music is lost: the poem was written by a

---

[765] See ante, p. 141.

[766] See Vol. I., p. 351.

[767] That it was really August III. and not August II. is shown by the tenor recitative, "Ihr Fröhlichen, herbei." The military events here alluded to only fit August III., and are evidently the tumults which prevailed during the first year of his reign. Besides, in the year 1733, the king's birthday was celebrated by a performance of Picander's Cantata "Frohes Volk, vergnügte Sachsen" (Picanders Gedichte, IV., p. 14 ff.), so that Bach's altered composition was not apparently used till later.—B.-G., XX.,² p. 73 ff.

[768] In Breitkopf's Catalogue of Michaelmas, 1761, p. 33, it is entered under " Promotions- und Ehrentags-Cantaten."

certain Christian Friedrich Haupt. Picander's indefatigable pen was not idle; he prepared a festival poem in Alexandrines. The town was illuminated. It has been already mentioned that Prince Leopold was present in Leipzig on the day.[769]

With a single exception, no "occasional" compositions written by Bach for the students especially can be pointed out during the following years. From the year 1729 he had, in the old Musical Society of Telemann, the solid assistance which he needed for his work, and at the same time suitable material for cases when he considered it advisable to come to the front with musical "attendances" (*i.e.*, occasional performances). In the first and second years of August III.'s reign this happened at least five times with reference to the royal family. It is evident that this assiduity was connected with Bach's personal position. On July 27, 1733, he had presented the first two movements of the B minor mass to the king in Dresden, with the petition that he would grant him a Court-title, which might ensure him against being further troubled by the Leipzig Council; and he no doubt thought to add to the urgency of his request by the festival performance got up by him in Leipzig. We have already spoken of the Coronation Cantata performed in January, 1734. But before this, on September 5, 1733, he had brought to a hearing a dramatic cantata "Hercules in indecision," set to words by Picander, for the Electoral Prince's birthday (b. September 5, 1722).[770] The convention of the *Collegium musicum* took place in the course of the summer out of doors, in Zimmermann's garden, and the Cantata, also, is evidently intended for an open-air performance. The characters in the drama—Hercules, Pleasure, and Virtue —are excellently drawn, and the contrasts are more vivid than in Handel's "Choice of Hercules";[771] the whole work glows with freshness and wealth of idea. But, as a whole, it again betrays a transcendental feeling which suits neither

---

[769] Sicul (Das frohlockende Leipzig, 1728) describes the festivities.

[770] Picanders Gedichte, IV., p. 22 ff. The autograph score of the music is in the Royal Library at Berlin.

[771] Händel Gesellschaft, XVIII.

the subject nor the purpose of the work. We may, therefore, congratulate ourselves that, with the exception of the final chorus and the recitative, it was all transferred into the Christmas Oratorio composed a year later.[772]

Three months later, December 8, 1733, the *Collegium musicum* again kept the birthday of the queen with a dramatic cantata, "Tönet ihr Pauken, erschallet Trompeten." On this occasion, too, Bach was the librettist, and he only finished the work the day before the performance.[773] Irene, Bellona, Pallas, and Fama are introduced. The lovely music, with the exception of Bellona's aria "Blast die wohlgegriffnen Flöten," and the recitatives, was made use of for the Christmas Oratorio.[774]

During the visit of the royal pair to Leipzig, which lasted from the 17th to the 26th of May, the musicians of that place seem to have been as inactive as they were on the occasion of the allegiance ceremony, held in Leipzig on April 21, 1733. But when the king and queen again visited the town, on October 2, 1734, the queen wishing to celebrate the anniversary of August III's election to the Polish throne, on the 5th of October, Bach had an opportunity of supplying the students with a serenade, to be performed to their Majesties. It was got up in haste. The unknown manufacturer of the text wrote it quite at random, his only object being to get in the necessary amount of words; in one place, in the last recitative, he wrote such nonsense that the composer was obliged to make some alterations in them. The production was a cantata, half lyrical, half dramatic; what personage may be intended at the beginning of the last number but one the poet has not revealed. It was impossible even for Bach to write entirely

---

[772] Comp. ante, p. 573 and 586.

[773] The autograph score, in the Royal Library at Berlin, bears at its close the note "*Fine DSGl.* 1733 d. 7. *Dec.*" The authorship is betrayed, as in the case of the altered Æolus-Cantata, even in the title of the original printed copy, which is in the Royal Public Library at Dresden, as well as in the awkward putting together of the text, and more especially in some Thuringian Provincialisms, such as "zum Axen" for "zu den Axen," "zum Sternen" for "zu den Sternen," "seyn" for "sind."

[774] Comp. ante, p. 575

new music in the given space of time—three days. He, therefore, made use in part, at least, of older works, of whose original purpose we cannot however speak with certainty. The original aria "Durch die von Eifer entflammten Waffen" was soon to find a place in the Christmas Oratorio. The opening chorus "Preise dein Glücke, gesegnetes Sachsen" was subsequently, and, as it seems, in its original shorter form, introduced into the B minor mass.[775]

Only two days after this, on October 7, was the birthday of the king, who was still staying in Leipzig. For this occasion Bach had, apparently some time before, prepared a dramatic cantata, and performed it with his Society on that day. In beauty and importance the music may rank with the "Hercules in indecision," and the cantata for the birthday of the queen. The rivers Vistula, Elbe, and Danube sing the praises of Augustus in turn, in whom they profess to have each an equal share; the Vistula and the Elbe, as being the chief rivers of Poland and Saxony, and the Danube, because the queen, Maria Josepha, was an Austrian princess. Their pretensions are very properly rejected by the Pleisse, the river of Leipzig, and the work, having opened with a four-part song, calling upon their fountains and streams to join in the king's praise, closes with their good wishes for the king's happiness, also sung in four parts; so they are permitted to return to their banks and cliffs, although it beseems them to flow with a gentle murmur "for wonder and fear." This first piece is a charming and romantic picture of nature; now it moves with a gentle swing, and now with rushing violence, with the loveliest play of colour. Bach repeated the work for the name-day of the king (August 3) in the year 1736 or 1737; as far as we know he never made it serve for church music.[776]

---

[775] The autograph score, with the title "*Drama per Muscia overo Cantata gratulatoria*," is in the Royal Library at Berlin; a copy of the text, written by Bach himself, is bound up with it. As to the ceremony itself, see Mittag, op. cit., p. 485 f., a passage which Bitter (II., 40 f.) was the first to point out. Compare also ante, p. 573.

[776] B.-G., XX.,² p. 3 ff. See App. A., No. 62.

When the king and queen came to Leipzig, in 1738, for the Easter fair, Bach waited upon them with a Serenade, which even gave rise to a printed criticism. A year afterwards Magister Birnbaum, in a work devoted to Bach's art, expressed his opinion of this composition in these terms: " That the Court Composer (der Herr. Hofcompositeur) composes in a moving, expressive, natural, and orderly style, and not in debased, but in the best taste, is shown particularly and undeniably in the Serenade publicly performed by him at the Easter fair, before our most Serene and Royal Rulers, on the occasion of their gracious visit to Leipzig, which composition met with thorough success."[777] The music has, however, been lost.

Lastly, Bach turned his attention to the old Weimar cantata " Was mir behagt ist nur die muntre Jagd," with the view of performing it with the Musical Society for the king's nameday. Whether the king was August II. or III. we cannot, however, be sure, but probably the latter, because Bach, as we have noticed, took all possible trouble to please this monarch.[778]

In the way of occasional cantatas written between the years 1723 and 1734, which became publicly known, as having been suggested by public personages or events, the dedicatory music for the rebuilt Thomasschule (July 5, 1732) only remains to be mentioned.[779] The private occasional compositions of Bach consist of some secular wedding cantatas; according to old custom they were intended to be sung during the wedding feast. They comprise only

---

[777] Scheibe, Critischer Musikus, p. 997.

[778] See Vol. I., p. 567. There were other festival performances of music after this: on April 30, 1741, when the students performed a serenade for the Electoral Prince and Prince Xaverius, on the occasion of their being in Leipzig for the first time (the text is in the Royal Public Library in Dresden), and in 1747, when, on the first coming to Leipzig of the Electoral Prince and his wife, an academic ceremony was held in the Paulinerkirche (the text, as before, in Dresden). But whether Bach composed the music for these cannot even be surmised. Since the publication of Vol. I. of this translation, the cantata "Was mir behagt" has been published in B.-G., XXIX. p. 1.

[779] See ante, p. 262.

WEDDING CANTATAS.

solos. The oldest of them is certainly the cantata "Weichet nur betrübte Schatten,"[780] belonging possibly to the Cöthen period. Johannes Ringk has preserved it to posterity in a MS. dating from the year 1730. Ringk, born at Frankenhayn, in Thuringia, was formerly a pupil of Peter Kellner, of Gräfenrode, a gifted admirer of Sebastian Bach, who himself wrote out many of the master's works. It was apparently through Kellner that Ringk became acquainted with the cantata. If Kellner was cantor at Gräfenrode when he first knew the cantata, it must have been between 1727 and 1730, for he entered on his post, at the earliest, in 1727.[781] He had formerly, however, worked at Frankenhayn, Ringk's home, where in 1726 he wrote out Bach's sonatas and suites for violin solo; and there exists a copy of one of Bach's organ-fugues made by Kellner in the year 1725.[782] And as he had in 1719 been a pupil of the organist, Schmidt of Zella, in the Thuringerwald, who seems to have been for a long time intimate with Bach,[783] the materials of the evidence lead us rather far back.[784] The text is a pleasing poem about spring, which subject naturally leads to the spring of love in two hearts; it is also remarkable for a refinement of expression not very frequent in poems of the kind. The work, set only for a solo soprano voice, is pervaded even in the first number by a gentle breath of romance. Evidence of its early origin is found not only in the concise forms of the arias, but also in another circumstance. The sixth of the violin sonatas with clavier obbligato[785] is known to have been twice remodelled by Bach, and at last furnished with an Allegro, the first subject of which is taken from the C major aria of this cantata. I have previously mentioned the bridal feeling which pervades this sonata,

---

[780] B.-G., XI.,² p. 75 ff.
[781] Gerber, N.L., col. 715.
[782] See Appendix A., No. 4.
[783] Probably the same Schmidt who, on November 9, 1713, wrote out a Præludium of Bach's. See Vol. I., p. 434; note 135.
[784] In the collection of Grasnick (now in the Royal Library at Berlin) there is a clavier fugue by Bach in B flat major, in Ringk's writing, which is evidently a work of his earlier period.
[85] See note, p. 116.

especially in the middle movement, *Cantabile, ma un poco Adagio*. It would seem that Bach, in remodelling the sonata for the last time, experienced this feeling, and was thereby prompted to use for the last Allegro a subject taken from an actual wedding composition which was written about the same time. A second agreement is also remarkable; the chief subject of the D major aria, given in the prelude to the oboe, is found again as the chief subject of the bass aria in the cantata "Liebster Gott, wann werd ich sterben."[786]

Although the musical portion of a secular[787] wedding cantata for February 5, 1728, has disappeared, an interesting circumstance attaches to Picander's poetry for it. The newly-wedded pair referred to were the Herr Johann Heinrich Wolff, of Leipzig, and the daughter of Hempel, the Royal and Electoral Commissioner of Excise, at Zittau.[788] The marriage took place at noon, being performed by Archdeacon Carpzov in Schellhafer's house, "by royal command."[789] This was situated in the Klostergasse, and was a favourite place for convivial gatherings; Görner used to hold the meetings of his *Collegium musicum* there, and the wedding banquet, at which Bach's cantata was performed, must have been given at the same place. The personages introduced in the cantata are the rivers Pleisse and Neisse, according to which it would appear that the bride did not live in Zittau before her marriage. What is of greater importance, however, is that Bach himself afterwards remodelled the text to do honour to the Town Council of Leipzig. The Pleisse became Apollo (tenor) and the Neisse Mercury (alto), representing respectively Learning and Commerce, the two pillars of the town's fame. The text, preserved in Bach's own writing, gives an

---

[786] See ante, p. 431 ff.

[787] It was formerly in the possession of Aloys Fuchs, in Vienna.

[788] Picander II., p. 379 ff. The printed text contains only the initials of the names; in order to complete the verses and the rhyme it is necessary to insert the words "Hempelin" and "Wolff." This is followed by a second poem on the same occasion: "Der Liebes-Congress zwischen dem Cupido, Wolff und Hampelmann."

[789] Marriage Register, St. Thomas' Church, 1728, p. 172.

interesting opportunity of seeing the musician engaged in the troublesome work of hammering out verses and rhymes. It is sufficiently remarkable that he could permit himself to sing the praises of Leipzig, and that of the Council which had caused him so much annoyance in his own person, with such hyperbole of expression as to write these words as the text of an aria :—

> Mit Lachen und Scherzen
> Mit freudigem Herzen
> Verleib ich mein Leipzig der Ewigkeit ein.
> Ich habe hier meine Behausung erkoren,
> Und selber den Göttern geschworen,
> Hier gerne zu sein.
>
> With glad jubilation,
> And true gratulation,
> I wish thee, dear Leipzig, perpetual youth.
> Here have I been content to fix my household fane,
> And sworn to the gods in all truth,
> Content to remain.

Even when he put the words into the mouth of Mercury, no one would ever conceive the idea of their referring to a place which he himself found unbearable. This parody is a complete and reliable testimony to the fact that Bach was very well pleased with Leipzig at this time.[790]

I would draw attention in this place, to a third secular wedding cantata, "O holder Tag, erwünschte Zeit," for soprano, although it was written much later, probably not before 1749. The young bridegroom to whom reference is made must have been a patron of music, for these words occur in the work :—

> Man, loved and honoured, prosper in thy way,
> Ever remaining true to noble harmony, &c.

Express allusion is made to the despisers and disparagers of music, which seems very appropriate to the particular year 1749, since, in the spring of this year, Biedermann the Rector, at Freiberg, had heaped gross insults upon musicians in his pamphlet, "*De vita musica.*" If we may venture to

---

[790] The autograph, a folio sheet written on three sides, was in the possession of Herr Grasnick, of Berlin. It is now in the Royal Library there.

draw a conclusion from the careful and even elegant style in which the manuscript was prepared by Bach, he seems to have attached a special value to this composition, which in truth is of great beauty.[791] With another text treating only of the praise of music (beginning "O angenehme Melodei"—"O melody, delightful power"), Bach brought this cantata to a hearing on three different occasions. Once it was done for a scion of the family of the Counts Flemming, in which music was loved and cultivated.[792] One aria is borrowed, and that is from a composition for the ceremony of allegiance of the year 1737. A protegé of Count Brühl, Johann Christian Hennicke, who had gradually risen from the position of a lackey to that of a Count of the Empire, received homage on September 28 of this year, on the occasion of his coming into possession of Wiederau, which had been granted to him as an hereditary feoff. For this ceremony Picander and Bach made a cantata, in which were introduced Destiny, Happiness, Time, and the river Elster; and in which the composer was extraordinarily successful, especially in the choruses at the beginning and the end, which are full of swing and freshness, and in the alto aria, which is remarkable for originality, elegance, and beauty of tone. We shall come across it again in the guise of a church cantata.[793] There is another

---

[791] The autograph is in the Royal Library at Berlin; the paper on which it is written is of the same kind as that of a harpsichord part to Bach's "Musikalisches Opfer," which is in the same place. That work was composed in 1747.

[792] For example, by General Field-Marshal Jakob Heinrich von Flemming (died 1728), and by the Governor of the town of Leipzig, Joachim Freidrich von Flemming (died 1740). Only the soprano part (some of which is in this autograph) of the altered form is preserved in the Royal Library, in Berlin. The writing and the paper alike are those of Bach's latest period, the cover alone, with the watermark, M A, being a remnant of earlier days. Since the music remains the same under all the alterations, with the exception of the last recitative, no loss has taken place, as is erroneously asserted in B.-G. XX.,[2] p. XIV. Since the publication of the original of this work, the Wedding Cantata has been published in B.-G., XXIX., p. 69. See also App. II. to the same volume.

[793] B.-G., V., No. 30 and Preface.—On Hennicke see Gretschel, III., p. 17. There is another poem on him in Picander, V., p. 350 f.

composition for a ceremony of allegiance, the so-called Peasant's Cantata (Bauerncantate) of the year 1742, which we cannot even touch upon in this place, for its style is so widely different from the other occasional cantatas as to demand an extended description in another place.

Besides these secular occasional compositions, Bach wrote some chamber cantatas, both in German and in Italian, of which, as they were not suggested by outward circumstances, the chief intention and purpose is to be found in the works themselves. Their number is not large. Impelled by other ideals, Bach could but rarely feel prompted to write independent music of this kind. When he did so, it is not surprising, considering his keen interest in the productions of other persons and nations, and the zeal with which he had formerly studied the instrumental chamber-music of the Italians, that he should also turn his attention to Italian chamber-music of a vocal kind. This tendency shows itself in the fact of his setting music to Italian texts. That Italian poetry, as such, was a thing he did not care for is proved by the fact that at least one of the texts is apparently from the pen of a German, who was by no means a perfect master of Italian. One of these cantatas, "*Andro dall colle al prato*," set for soprano with accompaniment of two flutes, string quartet, and bass, is lost.[794] In another, "*Amore traditore*," a bass voice is accompanied by the harpsichord, which is treated in parts as an obbligato instrument. This is not, we believe, an innovation of Bach's, but it is found moderately often, both in the Italian composers of that time and in the Germans who formed themselves on the Italian style; thus it occurs in Porpora, Conti, Heinichen, and others.[795] The fact is rather that even

---

[794] It was mentioned in Breitkopf's Catalogue for Michaelmas, 1770, p. 17.

[795] A cantata, "*La dove in grembo*," for a solo voice with a very brilliant harpsichord accompaniment, by Heinichen, is preserved in the king of Saxony's collection of music at Dresden. The first of the twelve cantatas by Porpora, which appeared in London in 1735, contains recitatives with an equally rich and elaborate harpsichord accompaniment.

Bach appears to form himself on the pattern of the Italians in the harpsichord acccompaniment which he appends to the second aria in his cantata. In other circumstances it was not his manner to write an obbligato part chiefly in broken harmonies, nor was he especially fond of using the obbligato treatment, except now and then. The breadth of form exhibited in the work points to the time of his fullest maturity; he first came to a thorough knowledge of Italian vocal music through the intercourse which was kept up between Dresden and Leipzig.[796] Traces of this familiarity with Italian forms, and especially resemblances to Lotti, are found here and there, even in his church cantatas. An aria for Marziano from Lotti's *Alessandro Severo* :—

agrees, as far as this opening is concerned, in all essential points with the opening of the bass aria in Bach's church cantata "Liebster Gott, wann werd ich sterben," and with the D major aria in the chamber cantata "Weichet nur, betrübte Schatten." In the opera of *L'Ascanio*, composed by Lotti, for Dresden, in 1718, the first aria begins thus :—

thus resembling the opening of the D minor aria in the fourth part of Bach's Christmas Oratorio. In the *Sicut erat in principio*, in the *Magnificat*, Bach, in repeating the subject of the first movement in a compressed form, was imitating

---

[796] "*Amore traditore*" is published in B.-G., XI.,² p. 93 ff. The autograph does not exist. It is noted as Bach's in Breitkopf's Catalogue for the New Year, 1764, p. 32.

the Italian church composers. The same thing is done in Leonardo Leo's *Dixit Dominus* for double choir and instruments (in C major), and in a five-part *Dixit*, by Lotti (in A major).[797] The latter work exercised an influence on Bach in other respects when writing his *Magnificat*. The grand close of a chorus, on the words *Dispersit superbos mente cordis sui*, has its model in Lotti's five-part chorus *Conquassabit in terra capita multorum*; the aria *Quia fecit mihi magna* is built on a bass theme which is similar to an alto aria in Lotti's work:—

[798]

It may here be noticed that a mass in G minor, by Lotti, was written out by Bach's own hand, in the middle of his Leipzig period, a circumstance which affords tangible evidence to confirm the fact of his having been much occupied with the music of the Italian master.[799]

A third Italian chamber cantata by Bach treats of an actual occurrence. This can, however, only be made out dimly from the text, which is evidently put together by a German, and consists of awkward and sometimes incorrect and meaningless Italian, with the admixture of scraps extracted from original Italian poems. A friend wishes to return to his native country, that is, from Germany into Italy. He is supposed to have been resident for some time in Anspach, and to congratulate himself on being once more able to be of service to his country, the more so that his work in foreign lands has not met with due recognition and support. But the poet tells us that the favourable opinion of some illustrious personage which he has gained while in Anspach will assist him in achieving great things in his

---

[797] Both works, in old handwriting, are in the king of Saxony's collection at Dresden. The autograph score of Leo's *Dixit* is in the Fitzwilliam Museum, at Cambridge, and the work, edited by Mr. C. V. Stanford, has been published by Messrs. Novello.

[798] See ante, pp. 379—81. Handel also seems to have known this *Dixit*. The theme of the "Amen" fugue in "The Messiah" is nearly related to the closing fugue in Lotti's work.

[799] The transcript is in the Royal Library in Berlin.

own country. Personal circumstances in connection with Bach seem to play some part in this. The Italian taste was in the ascendant in the band of the Margrave of Anspach, under Pistocchi and Bümler (1696—1745), and Torelli worked there also at the beginning of the century; but the words "*Tuo saver al tempo e l'età contrasta*" forbid us to apply the words to any Italian musician whom Bach may have befriended, if, that is to say, we may take the words of this bungling poet literally. The composition, set for soprano solo, flute and quartet of strings, betrays a thorough study of Italian chamber music. The fusion of Bach's original style with the Italian in this work makes it incomparably more interesting than the Cantata "*Amore traditore.*" The melodies have the breath of Italy about them; this is especially felt in the second part of the first aria and throughout the second, while the introductory symphony (in B minor), a piece in the style of the first movement of a concerto, and bearing a remarkable resemblance to the first movement of the violin concerto in D minor,[800] is in Bach's own style.[801]

Bach was probably incited to the composition of German chamber cantatas by the musical portion of his family circle. The very elaborate cantata for soprano about contentment ("Ich bin in mir vergnügt, ein andrer mache Grillen"— "Let others have their whims, I still will be contented"), which may have been written for Anna Magdalena Bach, claims attention less on account of any particular musical charm than from the fact of its being still in existence.[802] The music is pleasing and suitable to the words, and that is all. But the fact that he felt moved to set this text with its commonplace garrulity is characteristic of Bach, whose domestic nature knew how to prize most highly the tranquil repose of family life, notwithstanding the place in art which he had achieved, and all the honours showered upon him by princes and great people.

---

[800] See ante, p. 128.

[801] The work is in a MS. in Forkel's collection, in the Royal Library at Berlin.

[802] B.-G., XI.,² p. 105 ff.

Besides these pleasures, all the luxuries produced in various parts of the world occasionally introduced temptation into the contented family circle. The small differences which might arise between a father and children afforded Bach the material for a comic cantata. In the seventeenth century a new luxury was introduced into European society in the shape of coffee. As wine and tobacco[803] had already been extolled in song, the musician conceived the idea of doing the same with coffee. It would seem that the French had preceded him in this respect. In a collection of *Cantates françoises* (*Troisième livre*, No. 4) which appeared in Paris about 1703, the praises of coffee were sung in a very elegant style. The Germans were not long in following the example; in 1716 Johann Gottfried Krause wrote a text for a Coffee cantata.[804] But among all the German towns Leipzig was most remarkable for a particularly strong fancy for foreign productions. Although the luxury of coffee was limited, until the seven years' war, to the well-to-do classes, yet as early as 1697 the Leipzig Council sought to derive profit by taxing the "undue number of coffee houses"; and in 1725 Leipzig contained no fewer than eight licensed coffee houses.[805] In this Picander found material for satire. He published in the first volume of his poems (1727), under the title of "Von allerhand *Nouvellen*," a kind of journal in villanous rhyme, in which, under the form of correspondence from people of all countries, he satirised the existing state of things in this respect. Thus the news came from Paris: "A few days ago a royal mandate came to the parliament, which ran thus—'We have long noticed, and with sorrow, that many an one has been ruined simply by coffee. In order to stop this evil in time, we command that no one shall dare to drink coffee, the king and his court alone excepted. But permission may be obtained,' &c., &c. Then there arose a

---

[803] On Bach's Tobacco song; see ante, p. 151. A cantata by Stölzel, for a bass voice beginning "Toback du edle Panacee" is in the Royal Court Church Library at Sondershäusen.

[804] "Poetische Blumen von Joh. Gottfried Krausen. Erstes *Bouquet*. Langen-Saltza 1716," p. 129.

[805] Gretschel, loc. cit. II., p. 524.

long wail; 'Ah! cried the womenfolk, take our bread from us, for without coffee our life is dead.' . . But all this broke not the king's determination, but the people died like flies. As in time of pestilence they were thrown in heaps into the grave, and the womenfolk took on dreadfully until the mandate was torn down and destroyed, when the deaths ceased in France." Several years later he turned it into the subject of a comic cantata, and Bach set it to music in 1732. There was nothing new in treating circumstances and events of low life in this kind of comic vein. The "Jenaische Wein- und Bierrufer" by Nikolaus Bach was of the same class.[806] Other cantatas treated of the tooth-drawer, of the watchman in love, of the "female Magister," even of the "woman who makes cakes to cure the worms in Leipzig"; and the fun was not always of the most refined description.[807] In Picander's Coffee Cantata, one Father Schlendrian wants to cure his daughter Lieschen of the passion for coffee, which she shares with all the ladies of Leipzig. All his threats are in vain except the last, that she shall never have a husband; this seems at last to have an effect upon her; but she has laid a trap for her father, and while he goes to look after an eligible son-in-law she puts it about that "no lover need come to the house unless he will promise to me in person, and insert it in the marriage settlements, that I am to be allowed to make coffee as I like it." The last phrase is not by Picander, who concludes his poem with Lieschen's promise to give up coffee-drinking for a husband. Bach, probably, had something to do with the end, preventing the joke from becoming vulgar by the addition of the waggish ending; at all events it does credit to his taste that he set this, rather than the original form of the poem, to music. The two characters, besides which there is another, a narrating tenor, are kept clear and distinct, and drawn with great power. The old fellow grumbles and blusters about, Lieschen indulges in anticipations of enjoyment; Schlendrian takes council with himself, and seems of great importance in his own eyes, while the daughter rejoices over

---

[806] See Vol. I., p. 135.
[807] Breitkopf's Catalogue for Michaelmas, 1761, p. 34 f.

the bridegroom she is to expect; she has the liveliness and innocence of youth, he the heaviness and acerbity of age. This original couple seem to have delighted the world. In the *Frankfort News* of 1739, may be read the notice: "On Tuesday, April 7, a foreign musician will give a concert in the Kauffhauss under the N. Krämen, at which, among other things, will be performed a drama, Schlendrian and his daughter Lissgen; tickets 30 Kreuzer, the words 12."[808] It is not, indeed, expressly stated that the work to be performed by the "foreign musician" was Bach's composition; but who else would ever have put music to a poem treating of the state of things in Leipzig, and written by a Leipzig poet especially for Bach?[809]

A dramatic chamber cantata of larger calibre entitled "The strife between Phœbus and Pan"—("Der Streit zwischen Phöbus und Pan")—has a cheerful and partly satirical character. Picander wrote the words in 1731, and it was first performed at the summer meeting of the Musical Society in that year.[810] The old Greek myth, of course, is that Phœbus Apollo, as the god of the lyre, and Marsyas, the inventor of flute-playing, entered into competition; Apollo having gained the victory, by the terms of which he could do what he liked with his conquered rival, flayed Marsyas alive. The later myth changes Marsyas for the rural god Pan, who is subdued in the same way by Apollo; but the penalty is not inflicted on him, but on the Phrygian king Midas, on whom asses' ears are bestowed, because he thought Pan's playing the more beautiful. In this form Picander treated the myth, using as his classical authority Ovid's Metamorphoses, XI., 146—179. Besides the characters already mentioned, he

---

[808] Israël, Frankfurter Concert-Chronik von 1713—1780. Frankfurt am Main, 1876, p. 28.

[809] The Coffee cantata appeared, edited by S. W. Dehn, published by Gustav Crantz in Berlin, and in a second edition, thoroughly revised and corrected, published by C. A. Klemm, in Leipzig. See App. A., No. 63. It has been published in B.-G., XXIX., p. 141; and in an English form, edited by Mr. Samuel Reay, Mus. Bac., and published, together with the "Bauern Cantate," by Weekes & Co.

[810] See Appendix A., No. 46. Published B.-G., XI.,^{2} p 3. ff.

introduces the Lydian mountain-god Tmolus, as the arbitrator of the dispute, and—for which no authority is given by Ovid—Momus, the god of mirth, and Mercury to preside over the contest. The cantata opens with a fresh and graphic chorus of winds, which are very soon driven back into their confinement, in order to leave the music of the rivals free to echo and re-echo in nature's perfect stillness. Then the two competitors enter. Pan boasts of the mighty power and effect of his pipe, and charms Momus by singing a merry song. Mercury commands the contest, which is, however, carried out not with instrumental but with vocal music. Phœbus begins by singing an aria to his beloved and beautiful Hyacinthus; Pan thereupon sings a merry dance aria, in the middle of which he derides his opponent's grave and solemn tunes. This is followed by the verdict. Tmolus, in a congratulatory aria, declares Phœbus the conqueror, Midas, on the contrary, votes for Pan, on which he receives his punishment. Mercury and Momus point the moral, and a finale on Apollo's art closes the work, which is a masterpiece of characteristic variety, only granting that Bach's style is suitable for such things. As far as its poetic character goes, we might include it in the same category with Handel's "Acis" and "The Choice of Hercules," were it not for the fact of its being an allegory.

This time the purpose of the work is not to sing the praises of an exalted personage, but to do honour to a much maligned branch of musical art. Pan's flute delights the forest and the nymphs; he represents music of an agreeable kind, and easy to be understood by everybody; his light and untutored song pleased Midas so much that "he perceived it at once." The art of Phœbus unites beauty of melody with noble grandeur and depth; it is fit "to satisfy the gods." Both kinds of art are justified, but must always be opposed to one another; still, where the art is duly cherished, the last will be recognised as more important than the first. This would be the one pervading idea of the drama were it not that an intentional purpose is visible of mocking at Pan's music, and representing it as worthless, while extolling that of

Apollo in the hearing of his ignorant censurer. The introduction of Mercury also probably contains some reference to particular circumstances. The myth says nothing of his having taken any part in the transaction. He is indeed the father of Pan, but he could not, as such, take the opposite side, as he exclusively does. But as the god of commerce he personifies the Leipzig world of business, as Apollo does that of learning; in his remodelled version of the cantata "Vergnügte Pleissenstadt," Bach introduces them both in this signification. If the only object of the work was to exhibit Phœbus and Pan as contrasting elements in music, this might have been done in strict accordance with the myth, by performances on the lyre (or lute) and on the flute; or, if the parts were to be sung, it might have been expected that the instruments of the rivals would have played an important part in their songs. The fact that Bach did not do this, although he was generally so easily induced to take his musical motives from external circumstances, shows that his intention was not so much to depict the character of Apollo and Pan as to enforce the contrast between the grave, serious, and elaborate style on the one hand, and the light and merely pleasing on the other. He himself represented the former, as he could not be unaware, while the latter was represented by the opera composers and nearly all the rest of the musical world of the period. Pan is the patron of this class, while Apollo is Bach, who portrays himself in the beautiful aria in B minor which is written with evident purpose, and in the same way he laughs at himself with quaint irony in the middle section of Pan's song ("When the notes too doleful sound, And their mouths seem tightly bound, Strains of joy do not resound").

Who is Midas? Of course it must be some Leipziger, for Midas's aversion can only refer to Bach's vocal music, and this was not yet known outside Leipzig, although his instrumental music had already met with universal admiration. We only know one inhabitant of Leipzig who lifted up a voice of censure against Bach's vocal compositions, and that is Johann Adolph Scheibe. He was the son of

Johann Scheibe, the organ builder, whom we have so often mentioned, and was born in 1708; since the autumn of 1725 he had studied in the University, and at the same studied with a view to becoming a musician. On the death of Christian Gräbner, the organist of the Thomaskirche, in 1729, Scheibe, among others, applied for the situation. Bach was among the judges. Scheibe's trial playing seems not to have made a favourable impression; at all events, the post was given, not to him, but to Görner.[811] He remained, however, in Leipzig until 1735, giving lessons on the clavier, and having compositions of his own performed.[812] In 1737 he began to publish the "Critischer Musikus" in Hamburg, and in the sixth part of it he attacked both Bach and Görner, accusing the latter of being a conceited ignoramus,[813] and the former for his intricate and pompous style of writing, which was, according to him, both vain and tedious, as it was against reason. This opinion, which made a great commotion in certain circles, and caused a paper war, to which we must return later on, must be considered in connection with Scheibe's personal experience in Leipzig, although it is not unintelligible from a musical point of view alone. We must weigh it after knowing what he himself says, with commendable candour, of his earlier opinions. "Several years ago," he writes on July 28, 1739, "there lived in a certain famous town a certain man, whom I can the better describe from having been associated with him from youth, and have known him even as myself. I will call him Alfonso for the nonce. From certain circumstances he was obliged to devote himself to music . . . . . When he himself began to remark his own daily improvement, a secret jealousy of the excellencies of others appeared at the same time in him . . . When he heard praise bestowed on the merits of skilful men, he immediately became envious, for the simple reason that he did not possess such capabilities." Further on: "At last the jealousy that held him captive was changed into a

---

[811] See Critischer Musikus, p. 410.
[812] Gerber, Lex. II., col. 413.
[813] See ante, p. 211.

passionate desire of emulation, which impelled him to imitate the excellencies of great men. And gradually he so overcame his envy that he was able to hear the praise of able men without becoming red in the face, and at last he could even praise them himself with an honest heart, and give them the admiration they deserved." Scheibe expressly admits that this feeling had prompted him to action: "Perhaps my readers recognise me in Alfonso. And very likely my actions at that time bore a great likeness to those of Alfonso."[814] According to this, he must at that time have done all he could against Bach, and been sorry for it later; and those views of art which he still defended, even at Hamburg, when he had become more enlightened, he may have upheld and asserted with increased vehemence, especially as he may have been irritated at coming off so badly in the trial of organists. It is quite conceivable that Bach and his Musical Society were annoyed by these views, although they were only enunciated by a youth of three-and-twenty. Scheibe was an able fellow, and as such had influence with the students, and might have got up a party against Bach.[815] Now without the students Bach's *Collegium musicum* could not exist. Indeed, these motions of opposition made themselves felt in wide circles. Ludwig Friedrich Hudemann, Doctor of Laws in Hamburg, a capable musical amateur, and an old friend of Bach, as is shown by a canon dedicated to him in the year 1727, published "Proben einiger Gedichte," in 1732, at Hamburg, in which are included lines "To Herr Capellmeister J. S. Bach." From references in this poem it is probable that he knew all about Picander's poem on the goat-footed Pan, and Midas with the asses' ears.

The allegorical and polemic point spoils the harmonious impression of the cantata, in spite of its freshness, wealth of beauty, and drastic comedy (the pointed asses

---

[814] Critischer Musikus, p. 445 and 446.

[815] I notice that even George Friedrich Einicke, for instance, who did not go to the Leipzig University until 1732, went both to Bach and Scheibe in order to complete his musical education. See Marpurg, Kritische Briefe über die Tonkunst, II., p. 461.

ears on p. 52, bars 19 ff. and afterwards, are exceedingly funny), and causes it to rank on this account below Handel's work of similar kind. But on this very account it is of peculiar biographical interest. The differences which Bach had in 1714 with the church Elders at Halle and in 1725 with the University of Leipzig, and the lengthy dispute with the Rector Ernesti, soon to be related, betoken a certain combative nature, as well as a touch of the pettifogging orthodoxy of his time. We have here a case in which he believed himself capable of defending his artistic procedures and methods against his antagonist. He did not fly to his pen, like Mattheson; he was too genuine an artist for that. But he did not, like Handel, leave his compositions to speak for him entirely alone. He carried on his defence by means of a work of art, with a special tendency.[816] To this work threads attach themselves which connect it with a later time, and we shall see the feeling it betrays appearing again and again. The next period, the last of his life, has, however, an essentially different aspect from this, the richest and happiest of his creative life, which we conclude with the characteristic work " Phöbus und Pan."

---

[816] S. W. Dehn pointed out this explanation of the cantata, " Phöbus und Pan," in the October number of Westermann's Magazine for 1856. He went on partially wrong premises, and for this he incurred E. O. Lindner's censure (Zur Tonkunst. Berlin. Guttentag. 1864. p. 87 ff.). But yet in the main Dehn took the right view. Dr. E. Baumgart attempted to put him on the right track again, in a delightful little treatise, which also contains an elaborate and delicate musical analysis (" Uber den Streit zwischen Phöbus und Pan," Verhandlungen der schlesische Gesellschaft für vaterländische Cultur. Philosophisch-historische Abtheilung. Breslau, 1873).

APPENDIX (A, TO VOL. II.)

1 (p. 13). "**Wer sich selbst erhöhet.**" The preface to Helbig's series is dated March 22, 1720. Thus the cantatas must have been intended for the church year 1719 to 1720; they were at first printed separately, and then collected into a little volume with a preface, as was then usual. Bach, therefore, cannot have composed this cantata before 1720, and hardly later than 1722, since he was already living in Leipzig by the seventeenth Sunday after Trinity of 1723, and there had other texts at his command; besides, the character of the autograph is opposed to this assumption. There are no circumstances, it is true, to prohibit the idea of its having been written during the two intermediate years, still we do not know of his having undertaken any journey which might have prompted him to compose it. Moreover, it is to be observed that such series of cantata-texts, which at that time sprang up like mushrooms, were rarely remembered beyond the year for which they were written, unless they were the work of some distinguished poet, more particularly when, as in this case, they were not used in the place where they were written. The autograph, which is in the Royal Library at Berlin, is of such a character that we almost involuntarily associate it with the Carlsbad journey. Besides the difference in the writing and whole appearance from those of the Weimar and Leipzig periods, we are struck by the peculiarity of the paper, of which the last sheet and a half are quite unlike the others as to appearance and watermark, while the writing is the same; so it is evident the composer's paper came to an end while he was writing it, and he was forced to get more of a different kind. The watermarks of the first portion do not ascribe it undoubtedly to Cöthen (a kind of paper which he certainly used there is recognisable by the "Harzmann," as it is called—a figure of a savage clothed in a skin, and holding a fir tree in his hand; Bach used this for the birthday serenata for Prince Leopold). They are, on one sheet the eagle, and on the other all sorts of outlines, among which a D is plainly traceable; quite unusual marks. The last sheet and a half bear a shield with the crossed swords on one half of the field, thus betraying its manufacture in Saxony. Whatever view we take of these details, it is clear that the autograph was written under exceptional conditions.

I must add a further observation on the Bach-Gesellschaft edition of this cantata. The first aria is, in the score, accompanied only by an organ obbligato; but the part belonging to it is not for the organ but quite certainly set for the violin; indeed, the word "*organo*" has been written over it by a more recent hand. It would be inexplicable why the bass should be wanting, as it is, and why the right should play alone; why, again, the many ties and staccato marks, which have no sense as applied to the organ, should be marked in the part, though wanting in the score, why in the second portion of the aria so many double notes should have been altered, if it were not to render it feasible for the violin; for instance, in bars 128-129 instead of—

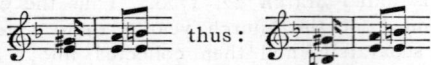

and, finally, why the parts should be written in D minor, and the part written out for the organ bass in C minor. If any performance ever took place with an obbligato organ accompaniment, Bach himself must have played it and transposed the part at sight. I, however, think it was a mistake to deviate from the score in favour of the part copy. The considerations adduced in the preface to the cantata (p. XXIV.) show it to have been quite unnecessary.

2 (p. 32). **The F minor toccata.** A MS. authority is Andreas Bach's book. An autograph which has now disappeared (see Griepenkerl's Preface) contains, besides the toccata in F sharp minor, another organ chorale on "Valet will ich dir geben" in B flat major, which certainly was written in Weimar; and Bach would naturally have written out together two pieces composed at the same time. It is against all probability and analogy that two toccatas constructed alike should not have originated at the same period; and, if for this reason alone, the assumption is obvious that it should have been at Weimar, even though the direct evidence of Andreas Bach's book is lacking. There is indeed a third toccata in F minor, and of similar structure so far as that it begins with an ornate subject, which however soon gives way to severer imitative passages, followed by a slow movement in 3-2 time with full broad chords and throughout sustained in character, while the finale is an animated fugue. This toccata is preserved in two old MSS., neither of which, however, names Bach as the author. One, in the possession of Dr. Rust, he inherited from his grandfather, F. W. Rust, erewhile Kapellmeister at Dessau; it is inscribed "*Toccata con Fuga in F moll*"; nothing more. The other is in the Royal Library of the Institute for Church Music at Berlin; it is only called a "toccata," and bears on it the name of "Dobenecker." But there was a third MS. in the possession also of F. W. Rust, of which now only a copy of the fugue remains; this Herr Schubring of Dessau had from him in his life-time, and he expressly said that the composer was Sebastian Bach. The question now is whether the internal evidence of the work itself is such that it can be unhesitatingly attributed to Bach, in spite of various doubts. I consider it to be so; and I see no particu-

lar difficulty in the name "Dobenecker" in the Berlin MS.[1] It may very well be the name of its owner at one time or another, who may have been all the more tempted to name himself on the title-page, because—as we see from Rust's MS.—the toccata was anonymous. Christian Friedrich Dobenecker may have been this owner, but nothing is known of any composer of that name. If the work is by Bach we must refer it back to the first Arnstadt period, and any resemblance of form we may find in it to the toccatas in F sharp minor and C minor can only be accidental and superficial. The polyphony of the first movement is, to be sure, skilfully managed, though it is not free from awkward and halting passages, easily detected by an ear familiar with Bach's masterpieces; it has besides a considerable resemblance in texture to that C minor prelude (P. S. V., C. 4, 243, No. 5) which, with its fugue, appears to me to belong to the earlier Arnstadt time. The middle movement, which, with its obscure imitations and heavy harmonic masses, reminds us in parts of certain toccatas by the southern masters—for instance, Georg Muffat—has various harsh phrases, and the fugue betrays so clearly, to any one who is accustomed to Bach's various manners, the traces of his beginner's hand that it is unnecessary to go into details. Still, it is Bach's genius that shines through it all; and we can but suppose that Kittel was mistaken in his statement as to the D minor toccata, which is a far riper work. As the matter at present stands, I could not venture to give the F minor toccata a place among Bach's authentic works; we must wait, and some new material may perhaps come to light to support the view. The MSS. themselves do not agree, and indicate certain changes by the composer. Rust's copy is the original form; in the Berlin copy the first bar of the fugue theme (of three bars long) is repeated, and a better phrase is thus constructed, not to speak of other small improvements; and the missing MS. agreed with this. In the Berlin copy it is obviously wrong that the movement in 3-2 should be placed at the end, and there are many other errors. It contains besides a fugue in G minor which must also be attributed to Bach, and it has in fact a conspicuous resemblance to the closing fugue of the E minor toccata, excepting that it is stiffer and less mature. Both the toccata and fugue were edited from the Berlin MS. by Fr. Commer (*Musica Sacra*, I., No. 9), and Dobenecker named as the composer; but they are not correct.

3 (p. 36). **Bach's fingering and execution.** What Quantz says in the passage alluded to is as follows:—" In performing running

---

[1] A merchant named Dobenecker lived in Leipzig in Bach's time. A son of his, named Christian Friedrich, went to the University there in 1728, and in 1735 was still living in the place. Compare also what Huttman says in the Musicalischer Quacksalber, p. 163: "He did just as much in the matter as certain village schoolmasters who sign their names to all their copies of music simply because they have written them out."

notes it is not right to lift the finger off at once; rather should the player place the tips of them on the farthest part of the keys, and withdraw them towards him till they glide from the keys. I refer this method to the example of one of the greatest of clavier-players, who used and taught it." That this meant Bach is proved by the index to the book, where, under Bach (Johann Sebastian), this passage is referred to. The description by Forkel of the action of his fingers corresponds with this, but Quantz restricts it to the execution of runs, while Forkel very rightly regards it as a general characteristic of Bach's *technique*. It is only in this sense that we can understand the application to the clavichord of this particular drawing inwards of the finger—it would be useless on the quilled harpsichord or organ, were it not that perfect equality of touch could only be acquired by this means. The real value of the method lay in its wide applicability. Its merit is most conspicuous in passage-work on the clavichord, and the most obvious to a superficial judge, because in the unavoidable jangling made by the tangents in striking the strings, the longest possible pressure of the keys is absolutely indispensable for the distinctness of the sequence of notes. It is therefore intelligible why Quantz, a flute-player, should speak of it in that sense only. But when Ph. Em. Bach confines this withdrawing action of the fingers to certain occasions—he calls it "the rapid touch, by means of which the fingers glide off the keys as quickly as possible, and which always must be used with a certain degree of strength" (II. 1, § 36)—when he directs it to be employed only in rapid changes of the fingers on one key (I. 1, § 90), at the last time of striking the upper note of a shake (II. 3, § 8), in the ornament of the "Schneller," or trill (II. 8, § 1), and in giving out quick subjects (III., § 1), not regarding it as a general rule of playing, we see that his style of fingering must have differed from his father's. This, already explained, will shortly be made clearer from other considerations. Forkel's surprise is therefore ungrounded. Much that Philipp Emanuel wrote has been taken for Sebastian's teaching, and thereby erroneous opinions have become widely spread.

4 (pp. 69, 72, and 81). **The dates of the sonatas for strings.** After what has been said about the thoroughness of Bach's nature, which never allowed him to relinquish a form of art when once he had taken it up till he had worked it out in every direction, it cannot be doubted that two works so similar in character as the six violin soli, and six for violoncello, must have been written in succession. But the second series was not, in fact, written for the violoncello, but for the *viola pompposa*, invented by Bach himself. What has been said above concerning this is taken from Gerber. He had heard from his father, who was Bach's pupil at Leipzig from 1724 to 1727, and who was still living, that his master had at times given the place of the violoncello to the *viola pomposa*, to facilitate the clearer per-

formance of the difficult and rapid bass figures, more particularly those that lie high, in his church compositions (see Gerber, Lexicon I., cols. 491 and 90). If we suppose the *viola pomposa* to have been invented in about 1724, as the younger Gerber does (Lexicon II., App., p. 85), the composition of these two works must have fallen within a period of Bach's life when his whole mind was taken up with the transition to a new and busy sphere of life, and with the effort to accommodate himself to it. That this is improbable is evident at a glance. Besides this, we may certainly assume that the first five violoncello soli, which Bach himself could not play, were written with reference to some skilled master of that instrument—such as Abel, of the Cöthen band—since many other of his most important pieces of chamber music can be proved to have been composed in Cöthen; and, on the whole, nothing can be more natural than that the composition of a musical specialty should have taken place in a town where the motive for writing them was stronger than in any other. The contrary and very remote possibility that the two works should have been composed at the later Leipzig period is disproved by the fact that the violin soli occur, copied out, in the often-mentioned collection made by Joh. Peter Kellner, now in the possession of Herr Roitzsch, with this note, " Franckenhayn den 3 Juli, 1736 " (with the exception of the B minor suite, and in this order: G minor, A minor, C major, E major, D minor); nay, that even so early as 1725 this same Kellner had copied for himself an organ fugue arranged from the G minor sonata (see B.-G., XV., p. 25). Finally, further proof is found in the state of an autograph of the violin soli. Any one who has studied the master's autographs can scarcely be mistaken in his handwriting, as it is known to us from the vast mass of his compositions at Leipzig; and the small variety in the sorts of paper he used, with their different watermarks, are another and tolerably certain standard. The Royal Library at Berlin possesses an autograph of the violin pieces in which the writing is altogether different from that of the Leipzig period, while it displays a decided affinity in its sharp and pointed character to the second copy of the "Inventionen und Sinfonien," mentioned in note 102 to p. 60. The watermark too, a double eagle, is different in Leipzig autographs: I have only met with the single eagle, and that but rarely. This may be the first fair copy which Bach made. Georg Pölchau, the professor of music at Hamburg, acquired this manuscript in 1814, from among the papers left by Palschau, the clavier-player of Petersburg, which were being disposed of to the butterman. It consists of twenty-three leaves, not all written over, for regard is always had to the necessity for turning over; hence it happens, too, that each piece, when the end of a page is reached without its being finished, is continued on another page in such a way that by placing them side by side it can be read straight on without turning over. The title is wanting in the superscription of each piece; Bach only names, with much precision, the pieces in

G minor, A minor, and C major, as *sonatas;* for in each of these two slow movements alternate with a quick one; while the others, which are made up of dances, he calls *partien.* The autograph of the last of these, in E major, is wanting; and the last twelve bars of the D minor chaconne are written in an unpractised, childish hand, perhaps Friedemann Bach's. A second and complete original MS., likewise in the Royal Library at Berlin, comprises a title-page and twenty-two other leaves, and the writing, watermarks, and title-page all refer it to Leipzig. The title, written by another hand, runs thus: "*Pars* 1. | *Violino Solo| Senza Basso | composée | par | Sr. Jean Seb: Bach.* | *Pars* 2. | *Violoncello Solo.* | *Senza Basso.* | *composée* | *par* | *Sr. J. S. Bach.*|*Maitre de la Chapelle| et* | *Directeur de la Musique* | *a* | *Leipsic.*"|; and below, to the right hand: "*écrite par Madame* | *Bachen. Son Epouse.*" The MS. is in the hand of Anna Magdalena Bach. Here, again, there are three sonatas and three partien. The title indicates that the violin and violoncello soli are united in one work in two parts; but the latter do, in fact, constitute a separate part, written out likewise by Anna Magdalena, with the title: "6 | *Suites a* | *Violoncello Solo* | *senza* | *Basso* | *composées* | *par* | *Sr. J. S. Bach.* | *Maitre de Chapelle.*" | There are nineteen leaves not written all over; the word "Suite" is added by Bach himself to each separate piece. The second "*autograph*" of the solo violin, sonatas and partien is therefore not so at all; it was written by Bach's wife, all but the title and a few directions, which have been added in a strange hand. So far as I can see, Sebastian has only added a few details to the sonata in C major.

5 (p. 103). **Bach's treatment of figured bass.** There are still extant the complete autograph scores of three chamber-trios with clavier obbligato. A sonata in G minor for viol-da-gamba and clavier has a figured bass (B.-G., IX., p. 203) where one of the above-described passages is to be found, but there alone. A sonata in G major, for the same instruments, shows no trace of figuring (B.-G., IX., p. 175), because no such passages occur in it. The six sonatas for violin and clavier exist in a MS. of which the last movement of the last sonata only was written by Bach, though he revised the whole; the figured bass is only indicated here and there, and he may have thought it unnecessary to complete it; the little there is, however, suffices to show that the same principle prevails. There is yet another very valuable ancient MS., the bass of which is more completely figured, and in it, too, the figures are only added to the bass at the first entrance of the theme in each part (*vide* B.-G.). It is certainly only an accident that the figuring is lacking in the E major sonata; and here the chords must be filled up in bars 1—4 and 35—49 (with the exception of the single bar in which the right hand is alone) and 120—123, of course in as decisive and simple a way as possible. But this cannot be required in places where the accompaniment makes a complete harmony of itself, like the beginning of the third movement of the A major sonata,

the beginning of the second section of the last movement in the C minor sonata (at least not for the first two bars), at the beginning of the first movement in the viol-da-gamba sonata in D major (B.-G., IX., p. 189); or where both parts are separated from one another by about an octave, and that their relative and interdependence must be plainly brought out, as at the beginning of the finale of the G major violin sonata, or the corresponding place in the B minor flute sonata (B.-G., IX., p. 15); which last, since it exists in autograph, confirms in the strongest possible way the justice of our remark, by the omission of figuring in the bass part at this point. A nearly complete autograph also exists of a trio for flute and clavier in A major (B.-G., IX., pp. 32 ff. and 245 ff.), the first movement of which has a figured bass in some passages. These passages are when the flute enters with one of the two contrasting themes (for the movement is written in concerto form) namely, at the beginning and for the introduction of the middle section; and here moreover the character of the form gives an accompaniment by itself. On this model the first movement of the flute sonata in E flat major (B.-G., IX., p. 22 ff.) should be played, where all passages are to be accompanied, even when the clavier part ceases in the right hand. Separate cases in which a single full chord should be struck find confirmation in the *largo* in four parts (and in that movement alone) of the violin sonata in F minor, by the direction at bar 8 " *accomp.*," and the natural indicated in bar 58, from which it is clear that in this and all ambiguous passages the violin is to be united with the bass by a few simple harmonies. The viol-da-gamba sonata in D major, which exists in a carefully written MS. of the Thomasschule pupil, Christian Friedrich Penzel, dating from the year 1753, is figured in bar 22 of its third movement, which may be a genuine indication, although the two parts harmonising is perfectly sufficient. Lastly, in the third movement of the G major violin sonata (according to the first arrangement in B.-G., IX., 252 ff.) we find, and in very unusual places, a figured bass, which is justified here by the unique plan of the piece. And this is the whole list of passages of this kind; its scantiness is clear evidence of how diligently and constantly Bach strove after pure and uninterrupted three-part writing. I repeat once more that all the instances consist only of passages where the obbligato part for the right hand is silent. To add parts and fill up the harmony in the case of these characteristic passages is quite inadmissible, in my opinion, and contrary to the intentions of Bach in outline and in detail alike. I cannot agree with the opinions put forth by Herr W. Rust on this subject (B.-G., IX., pp. 16 and 17).

6 (p. 110). **Bach's use of the organ and harpsichord.** Within the last few years an opinion has prevailed that in his compositions Bach used both the harpsichord and the organ; the harpsichord always for arias and recitatives; and the most wonderfully fanciful notions on the subject were given to the world with more assurance than truth by a writer in the "Allgemeine Musikalische Zeitung," Nos. 31

and 33, for 1872. It would seem that an essay by Franz Chrysander gave rise to them, in which he sets forth the manner in which Handel desired to have the organ used in "Saul" (Jahrbücher für Musikalische Wissenschaft, I., p. 408). The total outcome of these very interesting communications, based as they were on Handel's own private copy, is just what has already been said in the text of this work—that Handel did not regard the organ as a central feature of his orchestra in oratorio, but merely used it as he might any other instrument, where he thought it would produce its due effect—hence only to any great extent in the choruses and instrumental movements; and in these cases very frequently it is only for strengthening the bass. This is the sole object with which he introduces it in the few solo pieces in which it is brought in; while the proper instrument for such accompaniments is the harpsichord. The variety of ways in which Handel sometimes uses the organ, sometimes discards it—sometimes in one part only, sometimes in full chords—shows the sovereign master who knows the exact place for everything. But he had derived the principle of his treatment from the Italians, who transferred both their chamber music and their theatrical traditions to their church music. In Hamburg, the most important home of the opera in Germany, all that Handel justly claimed for his new ideal form of art had been simply copied from the Italians in the strictest limits of church music; and the harpsichord had soon so completely displaced the organ that in 1739 Mattheson thought he ought to put in a good word for it, and in the Vollkommene Capellmeister (p. 484, § 29) wrote that: "for various reasons it would not be bad if in churches small and neat organs, answering readily to the touch, and without harsh stops, could be combined with the clavicembalo." Bach, on the other hand, always adhered to the pure German principle, and would have nothing to say to a permanent harpsichord for church music, any more than he would have anything to do with theatrical music. I do not know that Chrysander has anywhere drawn any inference as to Bach's method from Handel. That any one else should have done so is the more incomprehensible because more than a hundred cantatas by Bach are now published in the Bach-Gesellschaft edition, and afford an ample mass of evidence from which any one can derive information on the subject. Since, happily, a great number of cantatas exist in autograph, or in copies revised by the composer, the organ parts—transposed a tone higher and figured from the first bar to the last—are beyond dispute so far as Bach is concerned. For why the busy composer should have given himself the trouble of transposing and figuring-in pieces which were not suited at all to playing on the organ it would indeed be difficult to say. When we find both a figured organ-part and another figured part in "chamber" pitch, this of course does not mean that both were played when the piece was performed, but that the harpsichord part was used for the rehearsals which did not take place in the church. And this regular co-operation of the organ was insisted on by Bach throughout his life, not merely at Leipzig, whence most of the examples are derived, but at Weimar also. The Advent cantata for 1714, in which a list of all the instruments to be employed is to be found on the outside page, specifies the organ and no other instrument; and the shifts that Bach was put to in writing out his scores, in consequence of the "cornet" pitch of the castle organ, have been described in Vol. I., App. A., No. 17. Now, in the cantatas which belong to the earlier phase of his work, there ought never to be any question of the harpsichord, even if the autograph score and printed parts of the Mühlhausen Rathswechsel cantata did not afford the fullest evidence as to the unrelaxed use of the organ. Nothing could have given rise to the introduction of the harpsichord into the church but the adoption of the Italian aria and recitative—forms that had grown up in opera music and were wanting to the older church cantata. Nor can even the semblance of a precedent be discovered for performing the cantata— for instance, "Gottes Zeit ist die allerbeste Zeit"—without an organ. If farther evidence were needed it might be derived from Bach's own words in the Mühlhausen specification (see Vol. I., p. 355 f.); he says of the 8-feet "Stillgedackt," introduced into the new Brustpositiv, that it "sounds well in combination"—meaning, with the church music as performed by singers and instrumentalists. The character of the stops introduced proves that solo movements were intended. It has been shown, in the proper place, that in Mühlhausen Bach started on the path from which all his life he never deviated. Very rarely does it occur that the figured organ-bass is absent from any single number in the cantata; among the published cantatas one such instance occurs (B.-G., V., 1., p. 200). But neither in the score nor in the parts do we find a trace of the introduction here of a harpsichord; even if this had been the case, a few unimportant exceptions could not avail against the rule; apparently, however, Bach himself undertook to accompany on the *positiv* organ, which was generally the task of the organist. Only in one single instance do we know of the introduction of the harpsichord, and this only serves to prove that Bach did not usually employ it. It is in the mourning ode on Queen Christiana Eberhardine, of which it is said in Sicul's Thränende Leipzig (1727, p. 22): "Soon after was performed the mourning music, which this time was composed by the Herr Capellmeister, Johann Sebastian Bach, after the Italian manner, with *clave di cembalo*, which Herr Bach himself played, organ, *violes-di-gamba*, lutes, violins, *fleutes douces*, and *fleutes traverses*." He had set the ode "after the Italian manner"—as is particularly stated on account of the exceptional circumstance—because, without having in itself any sacred element, it was to be performed in a church, and for these hybrid circumstances the hybrid style of the Italians seemed the fittest.

How far the unsuitable use of clavier music in churches had extended in Germany I am unable to ascertain with accuracy. As Bach was the only composer who wrote in the true church style, it ought not to surprise us if he alone had given the organ its due and proper share in it. But it must be said, to the honour of his Thuringian countrymen, that they at any rate have always duly valued the only really church instrument as such. Perhaps the harpsichord may never have found its way into their churches; at any rate, I have before me cantatas of the years 1768 and 1769, which have no accompaniment whatever but the organ! It may at any rate be considered certain for the first half of the last century; for the author of the "Gespräch von der Musik zwischen einem Organisten und Adjuvanten" (Erfurt, 1742), who was perfectly familiar with Thuringian life, takes an organ accompaniment for granted—as at p. 29, where he warns the player not to hold the chords in accompanying recitative, so that the hearers may understand the words. Even G. H. Stölzel, who had travelled through Italy, does not seem to have been unfaithful to the traditions of his native land; the word *cembalo* was, however, very loosely used. Thus Altnikol, in his copy of his father-in-law's cantata, "Ein feste Burg ist unser Gott" (B.-G., XVIII., No. 80), could write for the upper of the two bass parts in the first chorus, *Violoncello e cembalo;* for the lower, *Violone ed organo*. In the rehearsals the cembalo must have set off at once, strengthening the violoncello bass, and not have waited until the entry of the *canto fermo* in a lower register; in the church, the organ would fill the place of the harpsichord. On the other hand, we can now understand what Kittel means when, in the passage quoted in the text out of the "Angehende Praktische Organist," he says that a pupil always had to accompany on the harpsichord when Bach was conducting a cantata. It means of course, though inaccurately put, nothing more than a rehearsal; Bach's interruption as there described, and the feeling of the pupil, could only refer to a rehearsal.

But it would be quite unwarrantable to call in the evidence of Philipp Emanuel Bach in matters concerning his father. In studying the methods of fingering of the two men we saw clearly that their views were not identical. Nor could they be, for the son diverged into new paths different from his father's. He gave himself little more trouble with the organ; his whole endeavour was directed to the clavier, now becoming more and more independent; he had no more intimate connection with church music than his father's and his own contemporaries had; besides all this, it is a significant fact that he lived more than twenty years in Hamburg. He nevertheless made the most various demands on the harpsichord as an accompanying instrument. He says: "The organ, the harpsichord, the fortepiano, and the clavichord are the instruments most in use for accompaniments. The organ is indispensable in church use for fugues, strong choruses, and particularly in combinations; it gives grandeur and maintains steadiness. But whenever recitatives and arias are used in churches, and particularly when the middle parts form a simple accompaniment, and give the voice part perfect liberty for variations, there must also be a harpsichord. We hear only too often how bad the effect is in such a case without the harpsichord accompaniment" (Versuch über die wahre Art, &c., II., p. 1). The last sentence puts his meaning beyond a doubt. The harpsichord was not to be used instead of the organ in arias and recitatives to accompany from a figured bass; it was only to come in where an organ accompaniment was, in the nature of things, out of the question, and it was given to other instruments—no doubt the stringed quartet. Here, particularly in very simple accompaniments, a reinforcement by the harpsichord was needed, or the effect of sound was too meagre. It is quite in accordance with this—indeed, only intelligible on these grounds—that, at p. 259, he should give rules for accompanying a recitative on the organ in connection with other *tenuto* instruments. For the usual style of the church composers of the eighteenth century—if indeed we may speak of their style—the use of the organ in solo movements was highly inconvenient; it kept all expression and execution within the strict bounds of sacred feeling and of those forms and limitations which are the very essence of church music; all frivolity and sentimentality must vanish before its grave and dignified tones. But these writers had no suspicion even of this high ideal significance of the organ in the construction of a sacred cantata; and Kirnberger—who for two years had, as Bach's pupil, assisted in his church services every Sunday and had derived all his musical views from him—exclaims indignantly: " Hitherto always has church music been accompanied by the organ as its foundation and support. In these days of modern enlightenment, when a piece of sacred music must be just like a comic opera, the organ is considered quite unfit for the accompaniment, whereby a band is degraded from its dignity and the musical abortions go at once to the beerhouses."

7 (pp. 57 and 125). **A three-part sonata for violins and bass, in A minor,** exists in the Royal Library at Berlin, in Seb. Bach's own MS. It is written on the same particularly stout yellow paper as the autograph also preserved there of the six-part *ricercar* from the " Musikalisches Opfer " (press-mark, P. 226), consequently it dates from the last years of Bach's life, about 1747. That the piece was composed by him is certified by the title-page, written in another hand. But this certainly is an error; not a breath of Sebastian's genius animates any portion of it. Probably one of his sons may have been the author—it would not be the only instance of the father having copied one of their works in affectionate sympathy. The themes are given in A. Dörffel's thematic catalogue (App. I., p. 3, No. 5). The same library possesses a small brochure in oblong quarto, with " Inventionen " for violin and figured bass, without naming the composer. This too is written out by Bach himself. The character and shape are the same as in the second autograph of the " Clavier Inventionen und Sinfonien " (see note 102 to p. 60). It is, to say the least, very doubtful whether they can be

his original compositions, from the manner in which the pieces are named. The volume begins with "*Inventio seconda*"; it is in B minor, and consists of—*Largo, Balletto; Allegro, Scherzo; Andante, Capriccio; Allegro.* Then follow two blank pages, but for the most part ruled; and immediately after "*Inventio quinta,*" in B flat major: the first movement has no designation; *Aria, Giga; Presto, Fantasia; Amabile.* Then comes "*Inventio sexta,*" in C minor: a first movement; then *Lamentevole, Balletto; Allegro, Aria; Comodo assai, Fantasia.* This invention is on four leaves, the last page however is blank. Finally "*Inventio settima,*" in D major: the first movement has no designation; *Presto; Bifaria,*[2] *Largo; Andamento, Presto;* on four leaves, the last only partly written on. This ends the volume; the beginnings of the themes are given in the thematic catalogue (App. I., p. 3, Nos. 8 to 11). It is easy to see that here we have only a copy which was not even taken from an original collection; it is impossible, to be sure, to assert that the compositions absolutely cannot be Bach's; he may possibly have proposed to make a selection from an older MS. for some particular purpose, and the musical style is at any rate of his time. But it is highly improbable, if only from the elaborate designation of the movements, which was quite foreign to Bach's habit, and still more from the insignificance and meagreness of the pieces as to motive and form, and the character of the phrasing, which is quite unlike Bach's scheme of expression. The general form is that of the violin sonata in restricted proportions. The most interesting detail for us is the name "*Invention.*" It proves to a certainty so much at any rate as this, that Förkel's definition of the terms is radically inaccurate when he says (p. 54): " The name *invention* was given to a musical subject which was so constructed that out of it, by imitations and transposition of the parts, the course of a whole piece might be developed. The remainder was all working-out, and needed no new inventiveness when the composer knew the proper method of development." This definition is evidently founded on Bach's clavier *inventions*.

8 (p. 125). **Authorship of the C major Trio.** I can here supplement the MS. material on which the Bach-Gesellschaft have based their edition of the C major trio. In the Gotthold Library at Königsberg, in Prussia, there is a collection of preambles or preludes to chorales written out by Gotthold himself—No. 498 in J. Müller's catalogue. Among these, on folio 11, we find the C major trio, entitled " Trio von Gollberg "; the name being evidently a slip of the pen for Goldberg. He was the well-known pupil of Sebastian, for whom were written the thirty variations of the fourth part of the Clavierübung; he was a native of Königsberg. Since the authorship of this trio by Bach has long since been fully established, it is evident

---

[2] *Sic* (meant probably for Bizzarria, whimsicality). The term was sometimes applied to pieces of music, as we learn from Walther in his Musiklehre, 1708.

that we have here a fresh instance of a confusion which we so often meet with in studying MSS.; Gotthold must have copied from a MS. by Goldberg, and in the course of time the name of the copyist has slipped into the place of the composer. Goldberg, as a pupil of Bach's, would carefully have made his transcript from a very good copy, and therefore it is not without its value in the reproduction of the text, the autograph being lost. It is not figured. A few of the variants from the established text have some internal justification; but none of them bear on any essential point. As the MS. is accessible for comparison, I need say no more about it here.

9 (p. 143). **Writing and watermark of the " Partien."** The unpublished orchestral Partie, in D major, I only know from the copy which came into the Berlin Library from the papers left by Fischhoff. Its genuineness is, however, amply proved, irrespective of the character of the subjects, by the circumstance that Bach has worked up the overture in his most admirable manner in the cantata " Unser Mund sei voll Lachens." This fine work must, therefore, be promoted from its place among the doubtful works in the appendix to the thematic catalogue (Ser. VI., No. 3). As regards the time when the Partien were written, the autograph parts of the other D major Partie point, by their watermark, M A, to Leipzig. In those of the B minor Partie, on the other hand, the writing displays the Cöthen style of handwriting; it is preserved with the former in the Royal Library at Berlin; the character is sharp and pointed, and stiffer than that of the later Leipzig period. The naturals, which Bach subsequently wrote by the method of first making the connected strokes ⌐ and adding the other angle thus ⌐, often occur here in their old form, thus ♮. The later and peculiar form of sharp, which results from the vertical strokes not being drawn long enough, and so often failing to touch the lower horizontal strokes, does not yet prevail. Still, both these later methods of writing occur here and there among the other, so that a transition stage is plainly revealed. The watermark, too, is quite peculiar, and I have found no other like it in the Leipzig autographs. If we now duly consider the probability that Bach, as chief of the Prince's orchestra in Cöthen, may have written such a work, it seems very likely that the B minor partie should have been transcribed there. The autograph of the C major partie is wanting, but a certain simplicity of treatment would seem to indicate an earlier, rather than a later, origin. The two in D major, which are characterised by a richer instrumentation, may both have been composed in Leipzig.

10 (p. 152). **Authorship of " Willst du dein Herz," Giovannini.** In the larger of Anna Magdalena's music-books, on the inner pages of the two leaves which now immediately follow p. 111, we find the well-known song, " Willst du dein Herz mir schenken." On

the middle of the outer page of the first leaf are the words: "*Aria di G(i)ovannini*." The leaves are now loose, but they have belonged to the book, for on the last page we find begun for the second time the aria "Schlummert ein, ihr matten Augen," which is continued on the following page. A few pages must originally have been left blank on which the song was afterwards written, and these with a few other leaves torn or cut out. The music and words are neither in Bach's writing nor his wife's, the words being in Roman hand; and the lines for the music are more closely ruled than those commonly used. The book passed from the Bach archives of Ph. Em. Bach into the hands of KarlFriedrich Zelter, the director of the Berlin Academy of Singing; these leaves were then already loose, as he himself states. The idea that the music and poetry both were Bach's also is Zelter's, who, in a note which is still with the copy, hazards the following conjecture: "Giovannini may have been Joh. S. Bach's pet name turned into Italian; and the poem as well as the composition may have been written by him at the time when he was betrothed to Anna Magdalena, who is said to have sung very well. The copy, which is in a girlish hand, may be by the hand of his beloved. If this hypothesis is well-founded, such a memorial of the happiest period of the great man's life is not to be rejected, though Dr. Forkel will have it that Seb. Bach never wrote such a song." What Zelter only put forward as a possibility was considered proved when A. E. Brachvogel had adopted it as the theme of a romance in his novel, " Friedemann Bach"; and Ernst Leistner made this song—as a composition and poem by Joh. Seb. Bach—the turning point of a play in two parts (Leipzig, O. Leiner, 1870). It is also sold in the house at Eisenach, supposed to have belonged to the Bachs, as a "Memorial of Johann Sebastian Bach's birthplace"; it is sung in private circles and even in concert rooms and applauded for its touching antiquity; but, in spite of all this, unprejudiced hearers have always shaken their heads, for this music cannot be by Bach. That it is not his, any one may see who will study the MS. with an unprejudiced eye, and besides, the name of the composer is clear and plain on the title-page. Giovannini was an Italian gentleman of the middle of the eighteenth century, who resided for a long time in Germany, and who is mentioned with respect as a violinist and composer (Gerber, L. I., col. 510; N. L., II., col. 332). He was a master of the German language, and made many experiments in composing songs. In the third and fourth parts of the Odensammlung, edited by Johann Friedrich Graefe, 1741 and 1743, we find seven odes with music by him, and in the preface to the fourth part the editor gives us some personal details concerning him. Ernst Otto Lindner has included two of these compositions in his Geschichte des deutschen Liedes im XVIII. Jahrh (Leipzig, 1871—Musical Supp., p. 103, compare also p. 31 and 33 of the text); and we at once recognise in them the style of the composer of "Willst du dein Herz." How and when the song found its way into Anna Magdalena's book cannot, of course, be determined; probably not until after her death, when the book had passed into other hands, for the rest of the contents are of an earlier date, as has been said. From the circumstance that the four verses of the song are written in a Roman hand it may perhaps be inferred that the copyist, who certainly was not very expert, transcribed it from Giovannini's original, since he, no doubt, would have used this character for German as well as for his mother tongue. The poem, of which the playful grace ranks it above the music, I consider to be a translation from the Italian. The character of it, in my opinion, is not reconcilable with the state of German literature between 1750 and 1780. And it is quite incomprehensible to me how any one can seriously imagine that Bach could have written such a song,—Bach, whose poetical tastes had been formed on hymns and cantata-texts, by Neumeister, Franck, and Picander, and whose mode of expression, not to speak of the matter expressed, is known to us from his letters and official documents. I much regret that Dr. W. Rust (in the preface to the B.-G., Vol. XX., 1, p. xv) should have declared in favour of the genuineness of this song; all the more so because I had previously indicated to him Giovannini as its real composer, and because there was no particular reason for mentioning it there at all. He opines that in certain places the notes reveal Seb. Bach's handwriting; I, as I have said, am of the opposite opinion. I flatter myself I am sure of general agreement when I say that I have here settled once for all the question as to the writer of "Willst du dein Herz." In November, 1873, the Deutsche Zeitung of February 12, 1873, was sent me from Vienna, in which Herr Franz Gehring had already pointed out the similarity of style in "Willst du dein Herz," and in the songs by Giovannini, published by Ernst Otto Lindner.

11 (p. 159). **Autograph of the French Suites.** This is the most important autograph copy now extant of the whole work. A MS. containing only four of these suites is now in the possession of Professor Wagener, of Marburg. Whether this is written throughout by Bach himself, as used to be supposed, I must frankly doubt. It certainly only seems to me that it is older than the copy in Anna Magdalena's Clavierbüchlein. From this in the course of time a few sheets, being loose, have been lost; hence the copy is imperfect. The sixth suite, in E major, is altogether wanting, and several others are incomplete. But the usual order of the six suites is here already retained. The D minor suite has evidently been at the beginning, but the *allemande* is wanting and the first portion of *courante*, and from the second part of it we also miss 1 and 2-3 of a bar, the remainder is paged by Bach 4, 5, 6, 7, 8, 9. Since, however, the same suite recurs in Anna Magdalena's larger book, the gap can be filled up. The C minor suite must have come next, and here the *allemande* and the first part of the *courante* are again wanting, and only twelve bars and the preliminary

portion of a bar of the *gigue* are to be found; the remainder must have been on a lost leaf. Before the gigue Bach has noted: "N.B. Hierher gehöret die fast zu ende stehende *Men. ex. c. b*" ("N.B. Here belongs the minuet in C minor which is almost at the end"). The minuet is in fact in quite a different place in the book; from this notice I infer that the rest of the suite had to come after this beginning. Here again the larger book is of use to supplement the smaller one; only partially, it is true, since it contains only so far as the sarabande of the C minor suite, pp. 96 to 100. The B minor suite must have followed with this superscription :—

"*Suite pour le Clavessin par J. S. Bach.*"

The *Courante, Sarabande,* and *Anglaise* are wanting, excepting the last twenty-two bars of the second part of the Anglaise; the two minuets, on the other hand, are in the book, but farther on among other pieces, and were probably composed later. The Gigue, again, is perfect, after bars 10 and 28 of the first part of it, and bars 12 and 28 of the second, the next two bars are in each instance inserted in German "Tabulatur" on the upper or lower margin. Then follows "Suite ex Dis (D minor) pour le Clavessin," and then "Suite pour le Clavessin ex G♮." Both these are perfect.

12 (p. 166). "**The Wohltemperirte Clavier and Friedemann's Büchlein.**" In the edition of the Wohltemperirte Clavier, published by Hoffmeister and Kühnel, of Leipzig, 1801, Forkel, who edited it, has given the preludes in C major, C minor, C sharp major, C sharp minor, D major, D minor, E flat minor, E minor, F minor, and G major in a shorter form. He supposed them to be the final form in which they were projected by the composer (see his work on J. S. Bach, p. 63). But this is in direct contradiction not only to musical feeling, but to all the autographs of the Well-tempered Clavier. Nor has he based his views on a due study of the documents. A MS. which passed from his hands into the Royal Library at Berlin is, in the first place, somewhat incorrect, and, in the second place, contains only the Preludes in C sharp major and E flat minor of all those in question. At the same time it is difficult ever to imagine that such serious alterations, and in so many pieces, could have been made without any reference whatever to the composer. The Little Clavier Book of Friedemann Bach, which has not hitherto been utilised, gives us the wished-for solution. It shows that, in fact, a shorter form of several of the preludes was left by Bach. The question now arises whether this shorter form represents the original sketch or merely a re-arrangement intended not to overtask the powers of the younger pupils. But here again the little book helps us out to comparative, if not absolute, certainty. For, on the one hand, all the preludes which exist in an abridged form are not to be found in Friedemann Bach's book—that in G major is altogether wanting—and, on the other hand, several are here in the more extended form, thus C sharp major is

complete in one hundred and four bars (the beginning is thus in the right hand: [musical notation]) and C sharp minor in thirty-nine bars. E flat minor breaks off at the thirty-fifth bar, with the passage in semiquavers, on the chord of the diminished seventh; enough, however, is given to show that it is not in the form given by Forkel, and the same is the case with the F minor Prelude, which breaks off on the pedal C in bar 18. The C major Prelude also shows a considerable deviation from Forkel's version, with additions which are at once seen to be improvements; for instance —not to speak of minor changes—after each of the bars 4, 6, and 8, a bar is inserted by which the effect of the veiled melody is essentially enhanced. We thus see that in modifying the preludes none but purely musical motives have been acted on; and, as it is proved that in a few of the preludes the Forkel form, as we may call it, proceeded directly from the composer, from the character of the alterations it is not too bold to infer that this is the case in all. The Prelude in D major, which occurs in both the longer and shorter form, is, unfortunately, so fragmentary in Friedemann's book—ending at bar 19—that it is impossible to determine which form was intended. In the C major prelude, again, we have three forms to discriminate, that of Forkel, that of Friedemann Bach, and that finally arranged for the Well-tempered Clavier. Thus the case is precisely the reverse of what Forkel supposed; the shorter preludes were the earlier, and the longer were the later. Some MSS. of the preludes, in their first state, must have come into his hands, perhaps directly from Friedemann Bach, and he took them for later and revised arrangements, misled by having seen similar instances in other of Bach's works.

13 (p. 166). **The hitherto unknown autograph of the Well-tempered Clavier**, of which I am here enabled to give information, was formerly in the possession of Hans Georg Nägeli, of Zürich. So far as I can gain any information as to his acquisition of the Bach autograph, he seems to have procured it in 1802, through a friend, Professor J. K. Horner, of Hamburg, from the only daughter of Ph. Em. Bach, who was then still living there. This lady, Anna Karoline Philippine Bach, after her father's death, in concert with her mother, carried on a sale of the musical papers of Ph. Em. and of Sebastian Bach, and continued to do so alone after her mother's death in 1795. This I gather from a notice in No. 122 of the *Hamburger Correspondent* for 1795 (given by Bitter in Emanuel und Friedemann Bach, II., p. 127); and it is very probable that the two Eisenach autographs of Joh. Christoph Bach now in my possession, were first derived from this source (see Vol. I., p. 130, note 159).

This autograph of the Wohltemperirte Clavier was purchased of Nägeli's son in 1854, by Herr Ott-Usteri, of Zürich, and, thanks to the kind intervention of Herr Hofrath Sauppe, of Göttingen, he was persuaded to entrust it to me for a short time in the autumn of 1869. Since then Herr Ott-Usteri has died, in the summer of 1872, and, as I learn, bequeathed all his collection of autographs to the Town Library of Zürich.

This MS. has a wrapper, cover, and title-page also in autograph, which must originally have served for the two portions of the well-tempered clavier, for it runs as follows: "Zweymal XXIV | *Praeludia* 1ʳ Theil 24 2ʳ Theil 24 (then, in another hand, below the word *Praeludia*) und *Fugen* | aus | allen 12. *Dur* und *moll* Tönen. | vors *Clavier* | von | *Joh. Seb. Bach* | *Dir. Mus.* in Leipzig | ." The addition of " und *Fugen* " is in another hand, while in the work itself the fugues are written together with the preludes in such a way that at the end of the prelude "*Fuga seq.*" is frequently written. Both prelude and fugue are often written on the same sheet, so it is clear that this cover did not originally belong to this MS. but to another which contained the preludes only. This proves that Bach did not regard the preludes as inseparable from the fugues, but even at some period had collected them into an independent work. The same hand that added these words has written on the inside of the cover a paged table of contents, thus:—

"*Praelud.* 1. 2 Seiten *Fuga* 1. — 2 Seiten
2. 2 . . 2. — 2 . "

and so forth; and stated the total of the pages in sheets. It has also added, over the D minor Fugue, which is written by Bach, the words "bleibt weg." All the contents of the wrapper are not, however, in autograph; the first six preludes and fugues are written in another and much younger hand, probably that of a copyist; the writing is very round, the paper crisper, the staves ruled to a different scale and far more carefully. Bach's writing begins with the D minor Fugue, which thus occurs twice—a word as to this presently. The index of pages is not fitted to Bach's MS., with which it does not agree, but to that of the copyist. The number of pages is generally greater than in Bach's MS., since Bach's writing is proportionately smaller and closer than that of the copyist. Consequently we must suppose the owner originally to have possessed the whole first part of the Wohltemperirte Clavier, only in the copyist's writing, and to have added "und Fugen" on the title-page and the table of contents in accordance with this. Subsequently he inserted what had been written by Bach himself. As regards this I firmly believe that Bach, for some unknown reason, did not write any more than we here find, but intended to add what was wanting. For instance, while the autograph everywhere shows a very economical use of space, the first leaf, before the D minor Fugue, is ruled but not written on. Here the Prelude belonging to it was intended to come; since, however, this is too long for a single page we may conclude that he intended to write in the whole of the remainder, and had calculated that this was space enough.

Like all Bach's fair copies this is clearly, and in some parts beautifully, written. Over each prelude the name is written in a large bold Roman hand with its number, and in the fugues the number of parts is added, excepting in the case of the twelfth and twentieth: thus "*Praeludium 7.*" "*Fuga 7, à 3.*" After the 10th, 20th, 21st, 23rd, and 24th preludes is written "*Fuga seq.*" once, after the E flat major Fugue the number of bars is marked, 37. It is evident from its whole aspect that the MS. was written during the Leipzig period; quite clear from the character of the writing, less so from the watermark—a shield with crossed swords on the sinister field—(the cover is of different paper and has the double eagle on one side). But if we compare the contents with those of the other autographs, which are so carefully described in the introduction by Fr. Kroll to the edition of the Bach-Gesellschaft, the different readings afford conclusive evidence that this Zürich autograph (Nägeli's) is the latest and best of all. It may possibly have been Ph. Em. Bach's private copy which he took with him when, in 1735, he quitted his father's house; granting the correctness of this hypothesis, we may assert with the more confidence that this copy had been made by Seb. Bach not long before. The most important of the variorum readings are not to be found in any of the other MSS., nor in any printed edition so far as Kroll's admirable edition enables us to judge, although they are conspicuous improvements, as I shall proceed to show. This can only be accounted for by supposing that this autograph was removed from the reach of Bach's pupils—who were chiefly instrumental in multiplying and copying his works—soon after it was written. It is, besides, reasonable to imagine that Bach should have given to his two elder and most distinguished sons, between whom his musical papers were subsequently divided, a copy of the Clavier work to which he attached so much importance. That belonging to Friedemann Bach passed first into the hands of Müller, organist to the Cathedral at Brunswick, who died there in 1835, and by him was bequeathed to Prof. Griepenkerl, after whose death it was acquired by the Royal Berlin Library.

A Second Autograph now in the possession of Prof. Wagener, of Marburg, was probably the composer's private copy. This was written in 1732. The Zürich autograph must, at any rate, have been written later than this. According to Griepenkerl, Forkel had seen a MS. bearing at the end the note "Scripsit 1734." This cannot have been Friedemann's autograph, since it has no such note; nor can it be objected that it is imperfect towards the end, for it is quite clear that nothing has been lost and afterwards supplied by a strange hand; on the contrary, Bach himself, with visibly increasing impatience, has written as far as bar 68, inclusive of the A minor Fugue, and the rest has at once been added by another hand, no doubt Friedemann's.

Then the volume, or rather Cahier, has lain by for some time and afterwards been completed by the same hand. This is plainly discernible by the arrangement of the MS. and the various kinds of paper used in it; nor does the Zürich autograph contain this note, nor, again, that of Fischhoff, of which the genuineness is accepted. Hence, unless an autograph written in 1734 is lost, that which Forkel saw either was not genuine or it was Wagener's, and he mistook 1734 for 1732; this seems to me the most probable idea.

However, the Zürich autograph had not been without influence on the present text of the Well-tempered Clavier though to be sure it is infinitesimal. I am much deceived if the two earliest German editors had not seen it—used it I really cannot say—namely, Ch. F. G. Schwenke, the musical director and successor to Ph. Em. Bach at Hamburg, who in 1800 supervised Simrock's edition, and Nägeli, who soon after brought out an edition of his own. That Nägeli should not have done so till the autograph was in his possession is extremely probable, its influence betrays itself in both, in a number of small alterations, common, and at the same time peculiar, to them both; I shall here indicate a few of these. We are forced to reflect that diplomatistic accuracy was at that time a thing unknown in the province of music, and that mere guess-work was the only guide to comprehend how these editors could have taken up such trifles and passed over the more important alterations.

Since in this place my first object is to prove that the Zürich autograph is preferable to all the others, I have not attempted to give a complete list of all the variorum readings. I shall not reproduce the places where the embellishments differ or are lacking; such minutiæ and the omission here and there of tie-marks, are evident oversights, and are besides comparatively rare. Another opportunity will doubtless offer for mentioning these. Everything else I have set down in its order, the readings from the Zürich autograph being quoted first which occur only in the Zürich autograph (and here and there in the editions by Schwenke and Nägeli) and then a number of those whose genuineness, though authenticated by several authorities, still admits of a doubt which may now be diminished or dispelled. I have used as a basis Kroll's edition for the B.-G.

D minor Fugue, bar 31, second crotchet, in the right hand

(N.B.—The notes of the upper part are always supposed to be in the treble clef.) This reading, which contradicts the sequence of the episodical treatment, is perhaps only an oversight.

E flat major Prelude, bar 9, third crotchet, right hand,

APPENDIX.   669

Nägeli's reading no doubt was derived from this (see the variants given by Kroll). Bar 56, tenor last crotchet, G is a whole crotchet, matching bar 65 where the E flat is always given as a whole crotchet.

E flat major Fugue, bar 21, right hand, [music] bars 6 and 29 show why this is an improvement.

E. flat minor Prelude, bar 37, left hand, there is no seventh in the three chords, which seems to me grander and better considered; Nägeli has adopted this, bar 38, right hand, [music] a closing chord without a fermata; so in Nägeli.

E major Prelude, bar 3, left hand, [music] E major Fugue, bar 6, left hand, twelve semiquavers on d'; bar 13, left hand, the seventh quaver f sharp, both perhaps mere slips of the pen.

E minor Prelude, bar 16, right hand, [music] This is an evident improvement, as corresponding with the earlier bars: bar 37, left hand, [music] a closing chord with a minor third; thus in Nägeli and others.

F major Prelude, bar 13, left hand, [music] bar 17, left hand, in the tenth quaver F as analogous to bar 7; it is probably on this ground that Czerny has already altered it.

F major Fugue, bar 45, right hand, [music]

F minor Prelude, bar 5, left hand, first half-bar, [music] Bar 14, right hand, first beat [music] The closing bars of this piece have been hastily written and many notes are missing; the D flat in the eighth semiquaver in the bass is an error likewise.

F minor Fugue, bar 13, in the third beat of the tenor, a flat is a full crotchet. Bar 36, the two lower parts are as follows:—

so that the *a* flat in the tenor goes to *b* flat, a progression which might have been carried out by an *e* flat in the bass instead of *e*. In bar 44, the uppermost part in the second crotchet has *d″* written as a minim, which is certainly only a slip of the pen, and the same in bar 46, where the first note in the bass *c* is written as a crotchet.

F sharp major Prelude, the time is marked 12-8, in contradiction to the real time of the piece, an oversight which may perhaps have arisen from the recollection of the measure originally intended; and it may be in connection with this that the last bass note is here, as in the other autographs, written ℘. Bar 28, left hand, the second note *c* sharp to avoid the octave with the upper part. Bar 29, right hand, the second semiquaver *e″* sharp seems to be an error.

F sharp minor Fugue, bar 29, second half right hand,

is at any rate finer and more logical. Compare bars 32 and 33, the last note in the alto *a′*.

G major Prelude, bar 7, left hand, the last six semiquavers are only the repetition of the previous six to avoid the two octaves; one instead is introduced, but only one.

G major Fugue, bar 8. There is B in the bass in all the other autographs, but here there is *b″* in the treble, which truly suits the whole feeling better.

G minor Prelude, bar 5, in the right hand:—

is a charming effect of rhythm. And similarly in bar 6, right hand, in the last beat, and in bar 8, left hand, last beat, which deviations are both given by Nägeli (see Kroll, p. 230). Bar 15, right hand, last beat:

G minor Fugue, bar 21, the first note in the alto is *a″* flat, instead of the universally accepted A flat.

A flat major Prelude, bar 36, in the alto, has no *f′*, apparently an oversight when compared with bar 38.

A flat major Fugue, bar 6, the last note of the tenor is *a* flat, crotchet; this is better, for the theme is more prominent. The slur also over the *a″* in the upper part is found here, as in some MSS. and editions, and serves the same end. Bar 13, right hand, last beat:

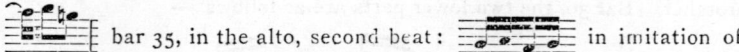

the bass figures in the last beat of the previous bar.

APPENDIX.

G sharp minor Prelude, bar 2, right hand, the first note in the alto is only a quaver to avoid hidden octaves with the bass.

G sharp minor Fugue, bar 7, second half in the tenor, [music] is better as being a more melodious progression. Bar 15, the first two quavers in the tenor are *c* sharp, possibly, however, by a mistake, although it sounds well. Bar 32, left hand, [music] is by far the least harsh of all the versions of this remarkable passage.

A major Prelude, bar 9, left hand, the *e'* is sounded on the fourth quaver after the previous semiquaver in the middle part. See also Nägeli and Simrock. The effect in performance is exactly the same as that of the more correct version given by Kroll after the other autographs. Bach often allowed himself such liberties, in order to exhibit his musical intentions to the eye of the player. So in this Zürich autograph he writes the 29th bar of the F minor Fugue, right hand:—

[music]

A major Fugue, bar 50, right hand, the sixth quaver is *c″* sharp to make it correspond to the progression of the middle part in bar 53. On the other hand, in bar 43, fifth quaver, *g′* sharp, is certainly an error.

A minor Prelude. The nine quavers from the seventh of bar 22 to the seventh of bar 23 are struck out, which indeed hinder the course of the final development, and besides that sound very harshly, for the seventh on F is either not at all or only partially resolved. For this reason the F is changed into A in Fr. Chrysander's edition (Wolfenbüttel, L. Holle), but I know not on what authority. There is a fermata on the last chord, as in Nägeli and Schwenke's MS.

A minor Fugue, bar 41, in the third beat, the *e* in the bass is a quaver. Bar 59, in the tenor, [music] is harsh, but not impossible. Bar 63, in the alto, the eight semiquavers in *e″*; the seventh, as it seems, has to be prepared. Bar 64 in the third beat of the tenor part *d′*, a crotchet; the skip to *a* is, as a matter of fact, superfluous. Bar 69 is the tenor, [music] Bar 81 in the upper part, [music] the strict stretto is discarded, perhaps because we have heard enough strettos before.

B flat minor Prelude, bar 1, left hand, second quaver, no d' flat; in the following bar the progression of parts is somewhat differently managed. Bar 24, the ties to the notes $\genfrac{}{}{0pt}{}{g'\text{ flat}}{e'\text{ flat}}$ are wanting, and there is a fermata over the last chord.

B flat minor Fugue, bar 20, left hand, [music] is to be preferred, because it carries on the crotchet movement in thirds longer. Bar 36, right hand, the last quaver but two, f'', not f'' flat, is better as preparing for A flat major. In bar 74—75 there is no tie between f and f.

B major Fugue, bar 4, in the tenor, [music] is better, since the chord of the six-four is avoided.

B minor Prelude. The direction *Andante* is lacking. B minor Fugue, bar 63, second half in the upper part, a'' alone as a crotchet, coming in anew in the next bar to keep the imitation more strictly.

D minor Fugue, bar 35, the reading of the Zürich and the Fischhoff autograph agree and confirm each other; a little harshness more or less is not of great account in this fugue.

E flat major Prelude, bar 34. The reading agrees with that of the Wagener and the Fischhoff autograph. Kroll pronounces it (p. xxiv) quite correct, and has put it into his text as the chief reading.

E flat minor Prelude, bar 10, right hand, the last chord is a complete E flat minor chord with b' flat as the lowest note, but of not much weight because it has been written in later with different ink. These traces of a foreign hand occur also in the F sharp major Prelude and in the Fugue, and are evidently corrections of Bach's errors from too quick writing, but of course I attach no further value to them.

E flat minor Fugue, bars 20 and 21, 41 and 48, all the autographs agree against the reading adopted by Kroll.

E major Fugue, bars 16, 26, and 27, are alike except that in the last of these the fifth quaver in the bass is dotted, to correspond with the foregoing bass. E minor Prelude, bar 5, right hand:—

and with a little variation in the autographs, bars 7, 9, and 11 agree with all the the autographs, except in bar 9, before the opening B in the right hand there is an accent marked. Forkel says (p. 63) that the E minor Prelude was at first overloaded with running passages, and afterwards simplified by Bach. In this there is, as we see, a grain of truth, but Forkel took the first sketch in Friedemann's book for the simplified form, and we do not know whether the ornamentation in the extended form really proceeded from Bach himself. Forkel must have heard something of the story from Bach's sons, but either understood or applied it wrongly, as was frequently the case with him.

E minor Fugue, bar 21, right hand, g' sharp as the eighth semiquaver, which is adopted by Kroll as his chief reading. The same with bar 40, right hand, g' the third crotchet.

F major Fugue, bar 42, agrees with all the autographs against Kroll's reading.

F minor Prelude, bar 22, the final chord is minor against the majority of the MSS.

F minor Fugue, bar 32, the bass has g flat in the second beat. Bar 41, alto, has e' in the third beat.

F sharp major Prelude, in bars 5, 17, and 29, there are no slurs between the ninth and tenth semiquavers.

G major Fugue, bar 82, agrees with all the autographs against Kroll's reading.

G minor Prelude, bars 13 and 14, the c in the bass remains; this is also in Simrock's edition. Bar 19, a fermata on the last b, agreeing with the Fischhoff autograph.

G sharp minor Fugue. The last bar is major, so is the Fischhoff's autograph, and a few MSS. of Nägeli and Simrock.

A major Fugue, bar 53, the fourth crotchet in the middle part is g sharp ; so that the erasure in Friedemann Bach's autograph can hardly be the work of the composer.

A minor Fugue, bar 69, in the last beat to the upper part as is in Kroll's chief reading.

B flat major Fugue. The final chord has no fermata. B flat minor Fugue, bars 50 and 51, without a stretto are thus:—

In bar 59 the second crotchet of the alto is d'', agreeing with all the autographs, as in Wagener's, which Kroll follows. A sharp can be clearly seen. I believe that this most diligent editor has been misled into an erroneous judgment in respect to the corrections inserted by a second hand in the Wagener autograph. As the reader will have remarked, the Zürich autograph always confirms the first reading. If the corrections are later improvements by the composer, it is inconceivable why none of those in the Zürich autograph should be indentical.

14 (p. 225). **Leipzig Musical Unions.** In the Leipzig Directory (Das jetzt lebende und florirende Leipzig) for 1723 we find p. 59 "The ordinary Collegia Musica in Leipzig are two: I. Under the direction of Herr George Balthasar Schott, organist of the New Church ; in the summer on Wednesdays, from four to six, in the garden of Herr Gottfried Zimmermann, in the *Wind-Mühl-Gasse*, and in the winter on

Fridays, from eight to ten, in the Coffee house in the Cather Strasse. II. Under the direction of Herr Joh. Gottl. Görner, organist of the Church of St. Niclas; on Wednesdays, from eight to ten, in Herr Schellhafer's house in the Closter Gasse." The notice is almost identical in the edition of 1732, excepting that under No. I. Bach and not Schott is named as the director; and again in that of 1736, only that under No. II. it is added "also during the winter months on Mondays, from eight to ten, at Herr Enoch Richter's Coffee house, in Herr D. Altner's house in the Market Place."

The first part of Mizler's Musikalishe Bibliothek contains on p. 63 the following: " Report of the Musical Unions of Leipzig. The two Musical Unions or Meetings, as here held every week, continue to flourish. One is directed by Herr Johann Sebastian Bach, Capellmeister to the principality of Weissenfels and musical director in the churches of St. Thomas and St. Nicholas; and it is held, out of fair-time, once every week at Zimmermann's coffee house in the Cather Strasse on Friday evenings from eight to ten, and in fair-time twice a week, on Tuesdays and Fridays, at the same hour. The other is directed by Herr Johann Gottlieb Görner, director of music in St. Paul's church and organist of St. Thomas'. This is also held once a week at Schellhafer's hall in the Closter Gasse, on Thursday evenings, from eight to ten, and in fair-time twice a week, namely, Mondays and Thursdays, at the same hour. The members who constitute these unions are for the most part students of the place, and there are always good musicians among them; so that often, as is well known, skilled performers appear from among them, and every musician is permitted to perform in public in these meetings, and for the most part there are hearers present who can judge of the merits of a skilled musician." In 1746 this register tells us (p. 69): " The ordinary *Collegia Musica* are three: I. Will for the future meet under the direction of Herr Gerlach, organist to the New Church, at Herr Enoch Richter's in the Catharinen Strasse, in the summer-time on Wednesdays, in his garden in the back Street, from four to six, and in winter, on Fridays, in the Coffee house, from eight to ten. II. Is held on Thursdays, from eight to ten, under the direction of Herr J. G. Görner, organist of St. Thomas', at Schellhafer's house in the Closter Gasse. III. On Thursday also a meeting will be held from five to eight, under the direction of the Merchants' Guild and others, at the Three Swans in the Brühl, where the greatest masters when they come here may be heard, and where the attendance is considerable and they are listened to with great attention." The directory for 1747 mentions the same three *Collegia Musica*, but instead of Gerlach "Herr Trier" is named as the director of the first union. This was Thomas Trier, of Themar, who, after Bach's death, came forward as candidate for the post of Cantor to St. Thomas', and afterwards was organist at Zittau. For several years the Leipzig Directory fails us. But from the notices here collected it is clear that it was Telemann's union which passed out of Schott's hands into those of Bach, from Bach to Gerlach, and finally to Trier. These two young musicians were pupils of Bach, and this seems to prove that Bach maintained a certain influence over the society, and had only withdrawn from the personal management because he must have observed that it could no longer hold its own against the new ones which were then becoming effective.

In the performances at the New Church after Schott's time, five instrumentalists were employed, appointed by the Council. In 1736 a sixth was engaged in consequence of Gerlach's representations. After 1738, four students, who composed the choir, received a honorarium of twelve thlrs. ; in 1741, two more violins were appointed, with four thlrs. a year, but they were required occasionally to perform as singers, and in 1744 eight thlrs. a year were granted to two musicians who were to play the stringed bass and the organ. Thus the musicians in the New Church, appointed by the Council, now consisted of fourteen persons, four singers and ten players. These facts are extracted from the accounts of the New Church.

15 (p. 282). **The Leipzig Organs.** Rust has very acutely inferred that some alterations of the same kind must have been made about this time in the organ of St. Thomas' (see his Preface to B.-G. XXII., p. xiv). In the oldest score of the St. Matthew Passion, which exists in Kirnberger's MS., we find as yet no double organ part, while in the Rathswahl-Cantata, on the contrary, "Wir danken dir Gott," two organs, one *obbligato*, the other for accompaniment, are employed. The former work was first performed in 1729, the latter in 1731. And in the accounts of St. Thomas' for Candlemas, 1730 and 1731, we find the following item : " 50 thlrs. to the organ builder, Johann Scheibe, for repairs, according to the vouchers of Zettel and Görner, the organists." Thus the *Rückpositiv* is not named here, and what was done to it was not in fact repairs. However, we must not take the accountant's style so literally. He writes concerning the great restoration of the St. Thomas' organ finished in 1721 : " 217 fl. 3 ggr. or 190 thlrs., were furthermore paid to Joh. Scheibe—in agreement with the new contract made with him, altogether 390 thlrs —for the entire and complete repairing of the great organ, with four new bellows, and 400 new pipes for the mixture." These " 400 new pipes for the mixture " include, however, an independent Sesqualtera for the *Rückpositiv*, which—as Rust was the first to remark—had no Sesqualtera in the year 1670, while Bach ordered this stop to be used on the *Rückpositiv* in the Matthew Passion. An experienced organ builder, whom I consulted, told me that, at that time, the alterations here spoken of could quite easily have been made for the sum of 50 thlrs. Only, indeed, with one keyboard, and that for the manual: for, if a separate pedal-stop had been put into the *Rückpositiv*, such as Scheibe put into the organ of the Neuekirche, which had, however, an independent *Rückpositiv*, the

thing would have cost more. This agrees entirely with the use which Bach subsequently made of the *Rückpositiv*. But this must refer to the Thomaskirche organ, for the organ in the Nicolaikirche had no *Rückpositiv* which could be played alone, and in the accounts no trace can be found of its *Rückpositiv* having been made fit for such a treatment in Bach's time.

16 (p. 286). **Organ-pitch in Leipzig.** In the Preface to B.-G., I., p. xiv., Moritz Hauptmann goes upon the assumption that the Nicolaikirche organ stood in chamber pitch, and concludes from this that of the double parts for the figured bass, which are frequently found in Bach's cantatas, the transposed part was used in the Nicolaikirche, and the other in the Thomaskirche. He may have founded his opinion upon the fact that the organ in the Nicolaikirche, which was replaced in 1862 by a new one, stood in chamber pitch. But this organ was not the same one that stood in the Nicolaikirche in Bach's time. On the contrary, the old organ, the specification of which is given in the text, was replaced in the year 1793 by a new one, built by the brothers Trampeli, of Adorf, for 7,000 thlrs. It must have been badly built to require renewing in less than seventy years. But there is nothing to show that the old organ stood in chamber pitch. In the seventeenth century the "chorus pitch" was still in ordinary use, so that the organ must have been altered to chamber pitch during the repairs of 1725. But at that time this pitch was still very little used, and we may be sure that if it had been introduced here in 1725, the fact would have been mentioned somewhere or other. Probably the first organ in chamber pitch in Saxony was one built by Silbermann, at Zittau, 1741, and its pitch was considered a curiosity. (See Sammlung-einiger Nachrichten von berühmten Orgel-Wercken in Teutschland. Breslau, 1757, p. 103.) There was also no reason whatever for changing the pitch of the organ in the Nicolaikirche, and thus making it different from the Thomaskirche organ, for at times the same church music had to be accompanied on each organ by turns; and considering the difficulty, especially in old organs, of lowering the pitch, which involved lengthening the pipes, it could only have been called a foolish waste of money and trouble. If in any way the chorus pitch were found to be a difficulty in the case of concerted church music, we may be sure that it would not have been allowed to remain in the Neuekirche. But the organ there, which was built in 1704, had this pitch, and kept it even after the thorough repair which Scheibe undertook for the sum of 500 thlrs. in 1722 (see acounts of the Neuekirche from Candlemas, 1721 to 1722, and the inserted documents that contain the contracts and advices). The specification of this organ is given in Niedt, Musikalische Handleitung Andere Theil. 2nd edition. Hamburg, 1721, p. 189.

17 (p. 325). **Clef and Pitch of Instrumental Parts.** Kuhnau's Christmas Cantata "Nicht nur allein am frohen Morgen," for chorus, two violins, viola, bass, two oboes, two horns, drums, and

continuo, is in A major; the score and parts are in the Town Library of Leipzig. All the stringed instruments as well as the figured continuo are set in this key. The violins and oboes are to play from the same parts, as is indicated by the signature; That is to say, the oboes play in the key of G without any signature—*i.e.*, in C major, which in the low chamber pitch sounded like B major; the strings, on the contrary, and the organ, in A major at chorus pitch, which in the same way sounded B major. In the cantatas "Erschrick mein Herz vor dir," "Ich hebe meine Augen auf," "Und ob die Feinde Tag und Nacht," which also exist in the Town Library at Leipzig, the organ parts are likewise in the original key, so that the violins had to be tuned to chorus pitch. This is not the case in the cantata "Welt ade, ich bin dein müde"; here the continuo is in G, the strings in A, the flutes and oboes in B flat, but from this disposition we perceive that the wood wind instruments were in the low chamber pitch. Bach's cantata "Höchst erwünschtes Freudenfest," which was originally written for the dedication of an organ at Störmthal, is in B flat major. When Bach conducted it on the feast of Trinity at Leipzig, he had only oboes at chamber pitch at his disposal, as is proved by the notes "*tief Cammerthon*" added in his own hand. Thus the B flat sounded as A, and the part for the organ which was at chorus pitch had to be written in G major, as is in fact the case. That this G major copy was prepared for a Leipzig performance is shown by this note inserted above the second portion, "*Parte 2da sub Communione*." Hence all the stringed instruments, in order to be able to play from the copies which already existed in B flat major, had to be tuned a semitone lower, a process which Bach in fact prescribes by the words "*tief Cammerthon*." When, on a subsequent occasion, he once more conducted this cantata, he had oboes at his command in high chamber pitch, and so wrote out very carefully a new figured bass part in A flat major, while the strings could remain at their normal pitch. This fresh part is remarkable, moreover, for having a bass figure only in the recitatives, the figures being generally replaced by a voice or instrumental part above the bass, and two portions have neither. It may also have served as the conductor's copy, particularly as the cantata is abridged, and the different sections have been displaced. Finally, as regards the pitch of the trumpets (compare Vol. I., p. 343, note) a MS. note may here be quoted as supporting what is said in the text. It is on a Whit Sunday cantata by Kuhnau, now in the Berlin Library ("Daran erkennen wir"), and runs as follows: "1. N.B. This piece is set for the violins (and all the strings) at chorus pitch, and for the voices and bass, in the key of B flat. 2. The trumpets are written in C natural. Thus a particular mouthpiece must be added to the trumpets so that they may sound a tone lower in the chamber pitch, and the drums must be tuned a tone lower, down to chamber pitch.

3. The *hautbois* and *bassons* must tune to chamber pitch, and in writing out the parts the music must be transposed a tone higher, so in this way all may agree."

18 (p. 331). **Accompaniment on the Harpsichord.** Gesner says that while he conducted Bach played either the organ or the clavier, which agrees very well with the circumstances, if by the *clavier* we understand the keyboard of the separate *Rückpositiv*. But the mention of a pedal does not agree with this, as the *Rückpositiv* had no pedal. Still the picture of Bach sitting at the great organ and thus playing the figured bass part cannot be reconciled with the rest of Gesner's description, and is hardly conceivable in itself; for, if he was to conduct thirty or forty performers, and keep them in order by movements of his hands and head, beating time and indicating emphasis, he cannot have kept his back turned upon them. Gesner's account is more compatible with solo organ-playing; it is quite possible that he should think of Bach's grand performances on the organ when wanting to give an instance of how various functions might be fulfilled at once, and that he overlooked the importance of accuracy in the picture as a whole.

In this place I must refer the reader to a statement already quoted from Kittel (ante p. 104, and note 6 of this Appendix) that when Seb. Bach conducted church music he always had an accompaniment played by one of his best pupils. For the purpose mentioned in the text this statement is useless, since, as I have already said, Kittel's remark can only refer to rehearsals, for this reason—irrespective of any others—that when Kittel was studying under Bach the harpsichord had been removed from St. Thomas. He wrote the above more than fifty years afterwards, so it is not surprising that his memory should have failed him. Still the notice is interesting within its limits, as proving that Bach stood to beat time at the rehearsals and entrusted the figured bass playing to a pupil. The scene of these rehearsals must have been the school hall; and yet there was no harpsichord there in Bach's time. An inventory of the instruments in the Thomasschule was added to the accounts of the school every year, and from 1723 till 1750 it is always as follows:—

Musical Instruments.

1 *Regal*, old and quite done for.

1 *dito*, bought āō. 1696.

1 *Violon* āō. 1711.

1 *Violon* āō. 1735, bought at the auction (not included in previous years, of course).

2 *Violons de Braz* (? iron fiddles).

2 *Violins*, repaired āō 1706.

1 *Positiv*, upright, with 4 stops and tremulant, yellow striped with gold, bought āō 1685.

1 *Positiv*, in the form of a *Thresores* (? in trapeze form), with four handles, which gives a stopped 8 ft. tone.

1 Another with a 4 ft. tone.

1 *Principal*, 2 ft. tone, bought āō 1720, to be used at family weddings.

Now since, irrespective of Kittel's evidence, the rehearsals—excepting the chief rehearsal on Saturdays in the church—would hardly have been carried on with only the accompaniment of a *Positiv*, particularly when Bach's intricate compositions were being practised, it follows that the scene of them must have been Bach's own residence, in which several harpsichords were at command for all the purposes of practice.

19 (p. 353). **Bach's Probationary Cantata, and Cantatas written between 1723 and 1727.** The autograph score of the cantata "Du wahrer Gott und Davidssohn," is written on paper with the watermark of the wild man with the fir tree. This mark occurs in no paper used by Bach at Weimar or at Leipzig. On the other hand, we find it in the autograph of the Cantata "Durchlauchtger Leopold" (see note 1 of this Appendix), and in that of another piece of occasional music written at Cöthen, and not long since come to light again, "Mit Gnaden bekröne der Himmel die Zeiten"; thus it is a token of the Cöthen period. Now the following observations may also be made note of: 1. In Cöthen itself Bach had no opening for the performance of church cantatas; what he wrote there were composed for other places. 2. Bach must have passed his tests at Leipzig on Estomihi (or Quinquagesima) Sunday, 1723. 3. He wrote the score of "Du wahrer Gott" with unusual care, with a view to some special occasion, but he did not quite complete it; the final chorus is wanting. It is clear that it was not Bach's original intention to close with the chorus in E flat major, and that the final chorus owes it existence to a later idea, from the fact that the closing initials, *S. D. G.*, were not put at the end of the chorus in E flat major. Besides, the chorale brought into the recitative by the instruments would be quite vague in intention if the introduction of this chorale in the final chorus had not been planned from the beginning; indeed, the cycle of keys presented by the different subjects of the cantata is only completed by the chorale in chorus closing in C. 4. From the character of the original parts we see that some of them only were written at Cöthen, while the rest were finished during his early years at Leipzig. 5. The texts of the cantata under discussion, and of that finally used as his test of proficiency, "Jesus nahm zu sich die Zwölfe," are evidently by the same poet, and we know the date of this latter cantata on the best authority.

The conclusion is obvious. When Bach had made the clean copy of "Du wahrer Gott," as far as the closing chorus, he changed his mind, wrote another piece, and put the first aside for a more fitting opportunity; when this offered, in Leipzig, he completed the parts and performed it no doubt on Estomihi Sunday, 1724. I observe that W. Rust has already come to the same conclusion (see preface to B.-G., V.,[1] p. xxi).

Two original scores exist of the cantata "Jesus nahm zu sich die Zwölfe," both in the Berlin Library. One is autograph throughout, and the almost total absence of corrections, as well as the ruling of the bars in the first chorus, show that it is a fair copy. It is not possible to determine exactly when it was made; we can hardly base a definite opinion on the watermark—a shield with crossed swords—though it is not improbable that it indicated the Cöthen period (compare notes 9, 41, and 57 of this Appendix). The second copy is written by Anna Magdalena Bach, and Sebastian has only written in one or two passages—*e.g.*, a line of the text on p. 7, and the careful figuring of the bass which extends as far as bar 42 of the first aria. From the watermark this score was written at an early period in Leipzig; and, at the beginning, in Anna Magdalena's writing, are the words: "N.B. This is the probationary work for Leipzig."

The watermarks in the paper of the original MSS. afford, perhaps, the most valuable data for the chronology of Bach's cantatas; a thorough investigation and comparison of the MSS. reveal the fact that a few marks, very distinct in character, constantly recur. From this we may infer that at different periods Bach used quantities of certain sorts of paper, and consequently that those MSS. which are written on paper with the same watermark belong to the same period. This sign of course is not absolutely infallible in every case, but, in the state in which the chronology of Bach's works is at present, much is gained even when we have succeeded in grouping them within different periods. When once this is done other means may be more successfully adopted for the more exact determination of the date of composition, when any such exist. I hope that I have succeeded in thus dividing and grouping them, and I will now deal more fully with the first section of the Leipzig period. I must begin by saying that to identify these marks I have had to look hundreds of MSS. several times over, because many which at first seemed undecipherable or worthless after fresh observations became clear and important.

The first period extends from 1723 till October, 1727, and the latest example is the score, written in this month, of the mourning ode for Queen Christiana Eberhardine. The watermarks of the MSS. within these limits are, on one half of the sheet, this sign:—

and on the other a half moon. These marks occur in the original MSS. of the following:—

    1. Aergre dich, o Seele, nicht
    2. Christen ätzet diesen Tag
    3. Christ lag in Todesbanden
    4. Dazu ist erschienen der Sohn Gottes
    5. Die Himmel erzählen die Ehre Gottes
    6. Du Hirte Israels
    7. Du sollst Gott deinen Herren Lieben

8. Du wahrer Gott und Davidssohn
9. Ein ungefärbt Gemüthe
10. Erforsche mich Gott
11. Erfreute Zeit im neuen Bunde
12. Erwünschtes Freudenlicht
13. Halt im Gedächtniss Jesum Christ
14. Herr gehe nicht ins Gericht
15. Herr wie du willt
16. Herz und Mund und That und Leben (Remodelled for Leipzig)
17. Himmelskönig sei willkommen (Remodelled for Leipzig)
18. Höchst erwünschtes Freudenfest
19. Jesus nahm zu sich die Zwölfe
20. Jesus schläft, was darf ich hoffen?
21. Ihr Menschen rühmet
22. Leichtgesinnte Flattergeister
23. Lobe den Herrn, meine Seele (XII. Sunday after Trinity)
24. *Magnificat* in E flat major (D major)
25. Mein liebster Jesus ist verloren
26. Nimm was dein ist und gehe hin
27. O Ewigkeit, o Donnerwort (F major)
28. O heilges Geist- und Wasserbad
29. und 30. *Sanctus* in C major and D major
31. Schauet doch und sehet
32. Schau lieber Gott, wie meine Feind
33. Sehet, welch eine Liebe
34. Siehe zu, dass deine Gottesfurcht nicht Heuchelei ist
35. Sie werden aus Saba alle kommen
36. Sie werden euch in den Bann thun (G minor)
37. Singet dem Herrn ein neues Lied
38. Wachet, betet (Remodelled for Leipzig)
39. Weinen, Klagen
40. Wahrlich, wahrlich, ich sage euch
41. Wo gehst du hin?

To prove the practical use of this critical text it can be shown that other marks exist on these compositions which assign them to the period between 1723—1727. Of five of the number we know the date precisely. The original score of "Aergre dich, o Seele, nicht," and "Die Himmel erzählen die Ehre Gottes," bear the date 1723, that of "Mein liebster Jesus" 1724. "Höchst erwünschtes Freudenfest" was composed for the dedication of the organ at Strömthal, near Leipzig, as we learn from a note in the composer's hand in the score, and copy of the words for the voices; and the church accounts of that place inform us that this took place November 2, 1723. The cantata "Jesus nahm zu sich die Zwölfe" was performed at Leipzig on Estomihi Sunday, 1723, as is noted by Anna Magdalena Bach. It has been shown that it is probable that some of the original parts of "Du wahrer Gott" date from 1724. The words of "Ein ungefärbt Gemüthe" were by Neumeister, written in 1714, and are to be found in the Fünffachen Kirchenandachten (p. 334). The text of "O heilges Geist- und Wasserbad" is by Franck, and was written in 1715; it exists in his "Evangelischem Andachts-Opffers." It is natural that in the early Leipzig period, when Bach was not yet in intimate relations with Picander, and could not at once find in the place a writer whose texts might suit his needs, he should return to poems already known to him. This can be proved to have happened in two other cases. The words of "Aergre dich, o Seele, nicht," which bear the date 1723, are also by Franck, as also those of "Ihr die ihr euch nach Christo nennet," and I shall presently show that this last must have been composed in 1723 or 1724 at the latest. Consequently it seems probable that the cantatas "Ein ungefärbt Gemüthe" and "O heilges Geist" must belong to the early Leipzig period. Thus, of the forty-one compositions in the list, eight may, from other marks, be assigned to a time before 1727, and the same can be proved with a number of others. The marks are of an external character; but internal evidence also exists to a considerable extent. As this, however, is derived from the characteristics of the separate works, I will discuss them in the body of the work. What has been said will suffice to show that we are justified in assigning MSS. which are on paper with the same marks to approximately the same period.

20 (p. 354). **Oboe d'Amore in two Cantatas.** W. Rust has pointed out, and with justice, the resemblance in the complicated notation for the oboe d'amore as indicating the connection of these two cantatas (B.-G., XVIII., Preface, p. xiii). For this instrument both the G clef on the second line and the C clef on the first line were used.

The oboe d'amore was pitched a minor third lower than the ordinary oboe, thus C minor sounded as A minor, G minor as E minor. Bach here uses this instrument for the first time, and, to make it easier to read the part, he added the C clef as well, a plan he afterwards discarded. The intention is perfectly evident from the middle portion of the Cantata "Die Himmel erzählen," where, if the oboe parts had been correctly assigned to the C clef, a sharp would have to be added. This has not been done; Bach was satisfied with a reminder merely of the unusual pitch. However, such double notation was not unusual, particularly for the oboes (see No. 17 of this App., and Vol. I, p. 628), and, as in the present instance, could only be intended to facilitate the performance of the part on an ordinary oboe or violin when no oboe d'amore was at hand. Whether this was Bach's purpose cannot be decided, for the orchestral parts are wanting. If the different parts of the Cantata "Die Elenden sollen essen" had not been lost, we could perhaps determine what is meant by the words "*col accomp.*," which occur at the beginning of this score, under the first line of the instrumental bass. Was this a hint to the copyist of the bass part, as the score has no figures—that he should add the figures corresponding to the instrumental harmony; or does it rather indicate something like *colla parte*, and did Bach thereby intend a freer performance of the Ritornel, as is quite conceivable from the passages given to the first oboe? The watermarks on the autograph MS. are on one page a W, on the other a horse.

21 (p. 358). **The Cantata "Ein Ungefärbt Gemüthe,"** if it was not composed in 1723, must have been written in 1726 or 1727—since we are here dealing only with the period between 1723—1727 (see No. 19 of this Appendix). For, in 1724 the fourth Sunday after Trinity fell on the Feast of the Visitation, and in 1725, on St. John Baptist's Day, so that no ordinary church music would be performed. Still, it is improbable that Bach, who was by this time acquainted with Picander, should have selected this insignificant text by Neumeister. In Neumeister's "Neue Geistliche Gedichte" (V. I.), we find on p. 105 a text for Whitsunday, "Gott der Hoffnung erfülle euch," and a setting of this, bearing Bach's name, exists in the Amalien Library at Berlin (oblong folio volume, 43). This composition is not at all like Bach, and certainly not his; in style it greatly resembles a cantata in the same volume, "Herr Christ der einge Gottssohn," which was discussed in Vol. I., p. 633. It would seem that out of the Fünffache Kirchenandachte, Bach also composed that for the seventh Sunday after Trinity; but at present we only know of the existence of the music from its being mentioned in Breitkopf's list for Michaelmas, 1770, according to which it was set for four voices, two flutes, two violins, viola, and bass.

22 (p. 361). **"Ihr, die ihr euch von Christo nennet."** The date of this cantata is derived from a particular mark in the part for the second oboe; this part being extant in the Berlin Library among the autograph score and original parts. At the end of this part we find: *Il Fine*, and this monogram  which supplies the initials W. F. B., and is a boyish triviality on the part of the copyist, William Friedemann Bach. The title and the words "*Aria tacet*," "*Recit aria tac:*" "*Recit tacet*," with the violin or G clef, in what is now the first section; the two flats, the common time mark ₵, the indication "*Aria all unis:*" and, below, the *volti* and the larger *Fine* under the monogram, are all in the writing of Anna Magdalena Bach; the rest is by Friedemann, and in a very stiff and childish hand. In the summer of 1723, Friedemann was in his thirteenth year; we may therefore infer that his mother, who was a good musician, set him the part as he was to write it. With regard to the watermark, see No. 43 of this Appendix.

23 (p. 362). **Wachet, betet.** The first performance of this remodelled piece must have taken place either in 1723 or 1725, since the

twenty-sixth Sunday after Trinity did not again come into the Church year till 1728. I incline to the year 1723, because I believe that Bach, during the early years at Leipzig, never let this Sunday, which occurs comparatively seldom, pass without performing some composition of his own, and all the less because the character of the services of the day must have strongly affected his imagination. The evidence of a second performance of the remodelled work lies in the early obbligato violoncello part, and one of the figured organ bass parts in B flat. These parts are on paper with the watermark M A which is the distinguishing sign of a group of cantatas about 1730 (see No. 44 of this Appendix). Between 1725 and 1736, the only years in which there were twenty-six Sundays after Trinity were 1728 and 1731, so the second performance must have been on one of these days.

24 (p. 369). **Christmas Day Cantatas.** It may be assumed as certain that during the early years of his official residence at Leipzig, Bach composed the cantatas and other part music at any rate, for all the more important festivals, since it was universally the custom for the Cantor or Capellmeister himself to supply the greater part of the music needed for the church year. And when a man like Fasch could compose during the first year of office a double series of cantatas, that is to say, above 100 (see Gerber, N. L. II., col. 92), Bach was certainly not the man to be behindhand in the matter. Besides, he himself says in a *Promemoria* dated August 15, 1736, and to which his squabble with Ernesti gave rise, that the cantatas which he performed with the first choir were mostly of his own composition. All then that is required is to select the right ones from among the festival cantatas. Bach wrote five annual series, at least five therefore for each festival. There are indeed six extant for Christmas Day, as follows:—

1. Christen ätzet diesen Tag.
2. Ehre sei Gott in der Höhe.
3. Gelobet seist du, Jesu Christ.
4. Jauchzet, frohlocket, auf preiset die Tage.
5. Unser Mund sei voll Lachens.
6. Uns ist ein Kind geboren.

No. 6 is an early work written at Weimar (see Vol. I., p. 487), and cannot come under consideration here, unless we assume, what is most improbable, that Bach, in his new office, could find nothing better for the first celebration of one of the greatest Church festivals than a *rechauffé* of a not very important early work. No. 2 is the setting of a text written by Picander for Christmas Day, 1728. No. 3 bears the marks of having been composed between 1735 and 1750 (see note 4 of Appendix A to Vol. III.). No. 4, which belongs to the Christmas oratorio, was composed in 1734, according to Ph. Em. Bach. No. 5 contains, as the fifth movement, a duet, "Ehre sei Gott in der Höhe," which is an extended re-arrangement of the setting of *Virga Jesse floruit* belonging to Bach's great Latin *Magnificat*. This was written for vespers at Christmas, at Leipzig; hence the Cantate No. 5 cannot have been written for Christmas, 1723, as this would make the remodelled work older than the original.

Only No. 1, therefore, can possibly have been written for this first Christmas in Leipzig, and this inference is borne out by the watermark of the original parts (see No. 19 of this Appendix), and by the circumstance that the part of the oboe in the A minor duet was subsequently given by Bach to an organ *obbligato*, and this would not have been possible before 1730, when the *Rückpositiv* in St. Thomas' was altered; so the cantata must have been written before that year.

We only have one other setting of the *Magnificat* by Bach, and it is doubtful whether there ever were more than these two. The marks on the score assign it to a period between 1723 and 1727 (see No. 19 of this Appendix), and the probability that it must have been written for Christmas, 1723, is increased by its being connected with a cantata by Kuhnau.

25 (p. 385). **St. Stephen's Day Cantatas.** There are but four cantatas for the second day of Christmas, St. Stephen's Day:—
1. Christum wir sollen loben schon.
2. Dazu ist erschienen der Sohn Gottes.
3. Selig ist der Mann, der die Anfechtung erduldet.
4. Und es waren Hirten in derselbigen Gegend.

Of these No. 1 has the watermark of a half moon without the corresponding mark on the other half sheet, which refers it to 1735—1750 (see note 3 of Appendix A to Vol. III.). No. 4 belongs to the Christmas oratorio,[3] and so to 1734. In No. 3 the watermark is the shield with crossed swords, and if any date may be founded on this it would at any rate not be the first half-year in Leipzig, as will be shown in No. 41 of this Appendix. No. 2, therefore, remains, and the watermark confirms this.

26 (p. 386). **St. John's Day Cantatas.** There are four Cantatas for St. John's Day:—
1. Herrscher des Himmels erhöre das Lallen.
2. Ich freue mich in dir.
3. Sehet, welch eine Liebe.
4. Süsser Trost, mein Jesu kommt.

No. 1 belongs to the Christmas oratorio. (No. 24. Hear, King of Angels. (Ed. Novello.) No. 2 was composed at the same time as the *Sanctus* in six parts, which Bach afterwards used in the B minor Mass; this *Sanctus* was written between 1735—1737. No. 4 is set aside by the watermark—the shield with crossed swords. No. 3 has been included in the list in note 9, Picander probably wrote the text, which displays a certain resemblance with one known to be by him for 1729.

---

[3] No. 11. "And there were shepherds." (Ed. Novello.)

686   APPENDIX.

27 (p. 387). **New Year's Day Cantatas.** There are five complete New Year's Cantatas :—
  1. Gott, wie dein Name so ist auch dein Ruhm.
  2. Herr Gott, dich loben wir.
  3. Jesu, nun sei gepreiset.
  4. Lobe den Herrn, meine Seele.
  5. Singet dem Herrn ein neues Lied.

And besides these the chorus "Fallt mit Danken," out of the Christmas oratorio (No. 36, "Come and thank Him." Ed. Novello.) Of No. 5 Rust thinks that only the four voice parts and two violins ever existed (see Preface to B.-G. XII.,[2] p. v.). However a considerable portion of the autograph score remains in the form of a separate cantata "Lobe Zion deinen Gott," in the Berlin Library. But the absence of any title, and also of the initials *J. J.* at the beginning, shows that this is not perfect, as well as the fact that the cantata begins with an aria in A major and ends with a chorale in D major, while three oboes, three trumpets, and a drum are introduced into this chorus, the remainder being accompanied only by the strings. The connection of this score with the imperfect parts of "Singet dem Herrn" seems to have been recognised by L. Erk (Joh. Seb. Bach's "Mehrstimmige Choralgesänge und geistlichen Arien," Th. II., p. 124). It is very evident, if only from the identity of keys, and of the watermark which are alike; and again from a comparison with the text of the cantata written by Picander for the first day of the Jubilee of the Augsburg Confession, 1730, which is throughout merely a recasting of the verses in this New Year's Cantata. It is clear not merely that the two sets of MS. belong to the same composition, but also that the cantata "Singet dem Herrn" is not lost, as has hitherto been supposed. Even the closing chorale has been preserved in the Trinity cantata "Lobe den Herrn, meine Seele (see No. 52 of this Appendix)"; nothing therefore is lacking but the recitative.

It is probable that the cantata "Singet dem Herrn" was first performed in January, 1724, in the first place from the watermark which differs from those of Nos. 1, 2, and 3. No. 1 is out of the question, because the text was not written by Picander till the New Year of 1729. No. 3, from the watermark, belongs to 1736 (see No. 3, Appendix A to Vol. III.). No. 2 has an eagle for the watermark, but no date can be inferred from this, as it occurs in MSS. of various periods. But a part subsequently added for the violetta has the initial M A, the watermark for 1727—1736. Hence it is not impossible that the work should have been written before 1727 at the same period as No. 5. If so, No. 2 may have been performed so early as 1724, and then No. 5 would belong to 1725—26 or 1727. Still it is more probable that No. 5 belongs to the same brief period as Nos. 2, 3, 4, 11, 25, 29, 30, and 35 of the list in Note 19, as all bear the same watermark. As to No. 4, in the total absence of any original MS. it is impossible to decide to what period it belongs, excepting on internal evidence. The circumstance that in the verse set to a tenor aria allusion is made to a great war then being carried on, in which Saxony itself was not involved, indicates a period after 1723—30, as during those years all Europe was at peace.

28 (p. 389). **Epiphany Cantatas.** Besides "Sie werden aus Saba alle kommen" we have two others for the Epiphany, "Liebster Emanuel" and "Herr, wenn die stolzen Feinde schnauben." This last is in fact the sixth section of the Christmas oratorio, "Lord, when our haughty foes assail us" (Ed. Novello), and the former must be assigned, from the watermark, to a period after 1735 (see No. 3 of Appendix A. of Vol. III.). We here and there find mention (as in Mosewius, Op. cit., p. 21) of a cantata for Epiphany, beginning "Die Könge aus Saba kamen dar," but this is the same cantata, the second movement having been wrongly placed before the first in some copies. The peculiarities noted in this work, as well as the watermark, assign this cantata to the same period as "Christen ätzet," and if the latter was written for Christmas Day, 1723, the former no doubt was composed for Epiphany, 1724.

29 (p. 392). **Cantatas for the Purification.**
1. Der Friede sei mit dir.
2. Erfreute Zeit im neuen Bunde.
3. Ich habe genug.
4. Ich habe Lust.
5. Ich lasse dich nicht.
6. Mit Fried und Freud ich fahr dahin.

No. 5 belongs to the year 1727 (see ante, p. 412). No. 3, as will presently be proved, must have been written in about 1730. No. 6 is of the period subsequent to 1735. It cannot be absolutely proved that, of the first two, it was the second that was performed in 1724. No. 4 is lost, and the only trace of its existence is in Breitkopf's list for Michaelmas, 1761, p. 19, where we find "Bach's, Joh. Seb., Capellm. und Musikdirectors in Leipzig cantate: *In Fest. Purificat. Mariae.* Ich habe Lust zu, etc., à 2 *Oboi*, 2 *Violini*, *Viola*, 4 *Voci*, *Basso ed Organo*, à 1 thl. 4 grs." This statement does not deserve absolute confidence, for a few lines farther on two cantatas are entered under Bach's name, which are certainly not by him, and the publisher, J. G. J. Breitkopf, in his preface, apologises for possible mistakes. Still there is no means of proving that "Ich habe Lust" is not genuine. No. 1 was used for the Purification and for Easter Tuesday. In its present form it is not exactly suited to either day. The first movement, a bass recitative, and the closing chorale "Hier ist das rechte Osterlamm," are connected only with Easter, and more particularly with the Gospel for Easter Tuesday. The aria, on the other hand, and the recitative that follows deal exclusively with the feelings aroused by the Gospel for the Purification. As the aria is the only independent composition in the cantata that is cast in any definite form, it must have been adapted to the earlier date, for the Purification, and only remodelled later for the Easter festival. Still we can hardly suppose that the aria, the recitative, and perhaps some other chorale better suited to the day were at first all it consisted of. This is a question that could only be solved by the discovery of the original MS., which, at present, is not known. The text I take to be Franck's; the music I also ascribe to the Weimar period, and any one who compares it with the first aria in the the cantata, "Komm, du süsse Todesstunde," will certainly share my opinion. In each we find a chorale in one part, allotted to a solo voice, and allied to the emotional character of the piece; in the first case, "Wenn ich einmal soll scheiden," and here "Welt ade! ich bin dein müde"—words which freely reproduce the sentiment of the text of the aria. The combination with the instrumental parts is also identical, many passages are almost exactly alike, and so is the fundamental feeling of the two pieces. If the cantata was written in Weimar we must still hesitate to assume that it could have been performed again before 1724, since Bach would surely have prepared some new piece for the first occurrence of the Purification rather than have repeated a composition so small and unpretending, notwithstanding its depth of feeling. Nor is there anything to hinder our regarding No. 2 as the composition in question; the watermark in fact indicates this very period.

30 (p. 392). **The Cantatas for Easter Day** are five in number:—
1. Christ lag in Todesbanden.
2. Denn du wirst meine Seele.
3. Der Himmel lacht.
4. Ich weiss, dass mein Erlöser lebt.
5. Kommt eilet und laufet.

No. 2 is a work of Bach's youth (see Vol. I., p. 229). He remodelled it, however, for use at Leipzig, but even this was when he was but thirty years old. No. 3 was composed at Weimar, but certainly not for the first Easter of his residence there. No. 4 was also written at Weimar (see Vol. I., p. 501). No. 5 stands in connection with the cantata for the second day of Easter, "Bleib bei uns," and this was also written, as I shall presently prove, in his thirtieth year. Only No. 1 remains, and the watermark assigns this also to the period between 1724—27.

31 (pp. 399 and 404). Of **Whitsuntide Cantatas**, four remain:—
1. Erschallet ihr Lieder.
2. O ewiges Feuer.
3. Wer mich liebet.
4. Wer mich liebet (a longer setting).

No. 2 is re-arranged from a wedding cantata. Even if it were not improbable that Bach should have produced a mere re-arrangement for his first Whitsuntide, it can be shown that No. 2 was written at the same time as Kirnberger's copy of the St. Matthew Passion. It can

APPENDIX.

be determined with tolerable accuracy when Kirnberger's copy was made; here it suffices to note that the St. Matthew Passion was composed in 1729. As regards Nos. 3 and 4, I would refer the reader to Vol. I., pp. 512 and 631. It will there be seen that No. 3 is assigned to 1716 and No. 4 to 1735.

Notwithstanding a date on an old copy of the score of No. 3, it is certain that this cantata was not written in 1731, because the original parts are on paper bearing the watermark of 1723—27, while the autograph score has a watermark which does not recur on any Leipzig MS. Bach must therefore have had new copies made of the parts early in Leipzig, and of course with a view to a performance. It cannot be proved with any certainty that this was not in 1724. But here, again, it is probably safe to assume that Bach would not come forward with an old work. If, however, this was the case, No. 1 must be assigned to Whitsuntide, 1725; it was certainly composed before 1730. In that year an independent and separate *Rückpositiv* was added to the organ in St. Thomas' (see No. 15 of this Appendix). This was not brought into use in the duet, as we may see from the original parts, till a second performance of the cantata; originally the *cantus firmus* in the duet was given to an instrument, and Bach could certainly not have done this if he had had the *Rückpositiv* at his command when he first composed the work. We farther gather that this mus have been before 1728, since the parts prepared for the revival of the work have the watermark M A, and the paper with this mark was used by Bach after the autumn of 1727. The watermark of the older parts is given in this woodcut, and it is in paper of remarkably

fine and thick quality. The same paper was used in some of the original parts of the Christmas Cantata "Christus ätzet diesen Tag," and in some of the autograph parts of "Wachet betet." This, which was written in Weimar, was often performed in Leipzig between 1723 and 1727, either 1723 or 1725 (see No. 23 of this Appendix). Moreover, the same paper is found in the autograph score, and some of

II.

the original parts of the Cantata "Himmelskönig, sei willkommen," which was also composed in Weimar for Palm Sunday, and performed at least twice in Leipzig at the festival of the Annunciation, since there no music was performed on Palm Sunday, as it was in Lent. One performance, to judge from the watermark M A, must have been about 1730, the other between 1723—27 (see No. 19 of this Appendix). But this cantata, of which the text underwent no alterations, could only be used for the Annunciation when this festival fell on Palm Sunday (see ante, p. 271). This happened in 1723 and 1725; Bach, however, was not yet in Leipzig on Palm Sunday, 1723, and thus the first performance of this cantata, "Himmelskönig," must have taken place March 25, 1725. We may therefore limit the possible period of the composition of "Erschallet ihr Lieder" to 1723—25. In 1723 Bach was not yet in a position to perform the Whitsuntide music in the churches of St. Nicholas and St. Thomas, so this cantata, if it was composed then, can only have been intended for the University Church. This hypothesis is contradicted by the fact that it seems to have been composed at the same time as the cantata "Weinen Klagen," for Jubilate Sunday, which falls four weeks before Whitsuntide.

The paper on which they are written is the same; the structure of the text is very similar, and both are undoubtedly by Franck; and in both these cantatas we find double viola parts, as also in the Easter cantata, "Christ lag in Todesbanden," and these would have been of no use at Leipzig. If "Weinen Klagen" and "Erschallet ihr Lieder" were composed in the same year, this cannot have been 1723, since in that year Bach's official work did not begin at the University Church till Whitsuntide, and not till the first Sunday after Trinity as Cantor of St. Thomas'. Hence we must decide on 1724 and 1725.

32 (p. 400). **In the Cantata "O heilges Geist,"** the watermark is the only evidence of the date (see No. 19 of this Appendix), besides the fact of the text being by Franck. Bach made use of his writings in the early time of his residence in Leipzig, but he had not yet entered on his office of Cantor by Trinity Sunday, 1723. The MS. is in the Amalien Bibliothek at Berlin, and in the handwriting of Anna Magdalena Bach.

33 (p. 401). **Cantatas for the Second Sunday after Christmas.** A Sunday after New Year's Day (the second Sunday after Christmas) does not occur every year. During Bach's residence it fell in 1724, 1727, 1728, 1729, 1733, 1734, 1735, 1738, 1739, 1740, 1744, 1745, 1746, 1749, and 1750. We may also throw out of this list 1728, since, in consequence of the death of the Queen Christiana Eberhardine, from September 7, 1727, till Epiphany 1728, "all organ playing, and all other stringed or jovial music, part singing in all the churches, at weddings, baptisms, funerals, in the streets, or by scholars at the doors," &c., were forbidden.

We have a second cantata for the second Sunday after Christmas,

"Ach Gott, wie manches Herzelied" (B.-G., XII.,[2] No. 58). The
original parts have the watermark M A, indicating that it was composed
either in 1729 or 1733—35. It is not very probable that Bach should
have composed a second work in these years for a Sunday which does
not regularly recur, and is of no particular importance, and the
extensive use of the simple four-part chorale makes it almost certain that
"Schau lieber Gott" was an early work, since this mode of treatment
is frequent in the early Leipzig period, for instance in "Dazu ist
erschienen der Sohn Gottes" (see No. 25 of this Appendix), and in the
cantata written for the first Sunday after Epiphany, 1724, "Mein
liebster Jesus ist verloren." Irrespective of these considerations, the
watermark assigns this cantata to the period between 1723—1727 (see
No. 19 of this Appendix), that is either 1724 or 1727. The treatment
of the chorale seems akin to that in "Dazu ist erschienen" and
"Mein liebster Jesus," and the text has an unmistakable resemblance
to this last. In each we find a Bible verse set *Arioso*, with a figured
bass in imitation, and a text for an aria, consisting of four trochaic,
and two dactylic lines, of four feet each, in a form which is rarely met
with in Bach's compositions, though it occurs in the cantata by Franck,
"Der Friede sei mit dir" (see No. 29 of this Appendix).

34 (p. 403). "**Mein liebster Jesus ist verloren.**" The
original score and parts are in the Berlin Library. The score is imperfect: it contains the whole of the first number, the second only in a
fragment, the third minus the text, the fourth in a fragment; the
remainder is wanting. In each of the figured bass parts are the words
senza basso, but there is an autograph figured part for the harpsichord
in A major. The harpsichord bass in this works throughout with the
continuo of the score, but an octave lower, and is figured even in the
closing cadence, and wherever the oboes are silent. It is easy here to
see the use made of the harpsichord as an instrument for conducting
from (see ante, p. 328). It cannot be meant to be essential to the whole
effect, since its natural tendency to give a light and brilliant character
to the piece is counteracted by the added weight in the bass; but it
was very useful in keeping the performers together, being clearly
audible to those who sat near, and yet not so strong as to intrude
on the audience. The chords in the figured bass during the rests
for the oboes serve the purpose of rendering the progression of the
harmony more solid and intelligible to the other players, but they
are not essential to the effect on the hearer, because the harmonic
progression is perfectly accounted for by the bass. Consequently,
indeed, the harpsichord is not mentioned in Breitkopf's list for
Michaelmas, 1761, and it is simply styled "Cantata: In Dom. I. Epiph.
Mein liebster Jesus ist, etc., à 2 Oboi, 2 Violini, Viola, 4 Voci, Basso
ed Organo." It may be assumed that the same method was followed
by Bach in similar solo pieces, as for instance the soprano aria in the
Ascension Oratorio (B.-G., II., p. 35).

35 (p. 403). "**Jesus schläft.**" B.-G., XX.,[1] No. 81. The fourth Sunday after Epiphany occurred during Bach's life at Leipzig only in the years 1725, 1728, 1731, 1733, 1736, 1738, 1739, 1741, 1742, 1744, 1747, 1749, and 1750. Hence, as the watermark assigns the cantata to a period between 1723 and 1727 (see No. 19 of this Appendix), it must have been composed in 1726 or 1727, if not indeed in 1724. But 1727 is out of the question, because the Feast of the Purification fell on the fourth Sunday after Epiphany, and we must therefore decide between 1724 and 1726. On the last page of the autograph score there is a fragmentary sketch of the first chorus of the Epiphany cantata "Sie werden aus Saba alle kommen." From this it appears that both the cantatas were composed at about the same time, for, when we consider Bach's free use of music paper, it is incredible that after the lapse of at least two years he would have used a sheet which had been previously used for the first sketch of "Sie Werden aus Saba." Thus, if this work was composed in 1724, "Jesus schläft" must have been also.

36 (p. 407) "**Siehe zu, dass deine Gottesfurcht.**" The chronology is based in the first instance on the watermark (see No. 19 of this Appendix), and on the resemblance of the first subject of this cantata with that for Trinity Sunday, "O heilges Geist." No further data are available, but no one can fail to attribute great importance to this resemblance who has duly observed, again and again, the way in which Bach was never satisfied with one attempt in any new or peculiar form, but could not rest till he had worked out its powers of utterance on all sides, and in a certain sense exhausted it.

37 (p. 407). "**Lobe den Herrn meine Seele.**" The Gospel for the twelfth Sunday after Trinity relates the miraculous healing of one born deaf and dumb; of course it is not impossible to connect with this event a cantata of thanksgiving for the numberless blessings bestowed by God on man throughout life, but, apart from any other motive, it seems remote enough. When we carefully consider the details of the text we must be convinced that the first purpose of this cantata was not a performance on the twelfth Sunday after Trinity. Nor need we seek far to discover what this main purpose was, for about the same time as this Sunday fell the annual election of the Town Council; and if we compare it with the words of other "Rathswahl cantatas," as "Preise Jerusalem den Herrn" (B.-G., XXIV., No. 119) and "Wir danken dir Gott" (B.-G., V.,[1] No. 29), their general resemblance shows that the Rathswahl must have determined the choice of the text. Moreover Bach, in order to emphasise its purpose, subsequently had the text remodelled, in part giving it a distinct allusion to the municipal authorities. This alteration, which involved a change in the music, was made about 1730, for the sheets with the new parts inserted among the old ones have the watermark M A (see No. 43 of this Appendix), while the older ones have the watermark of 1723—27 (see the B.-G.

edition for a comparison of the two forms). In bar 16 of the original
composition there is a choice of two readings in the text, "Sprich
dein kräftig Hephata"—"Speak, the almighty Epphatha." This
refers directly to the Gospel for the day, and could be sung so at the
Sunday service; but over the word "Hephata" the more general "gnädig
Ja" ("Gracious Yea") is inserted. It is easy to detect that this has
been written in by Bach himself, after the part had been finished by
Anna Magdalena Bach. This variant in the text can have nothing to
do with the alterations made in 1730, for those involved the removal of
the original recitative and the insertion of another. We may then
certainly infer: (1) That the first performance of this cantata in
the service on the occasion of the Council election must have been
between 1723 and 1727. (2) That the performance for the Sunday
must have preceded that for the election; and since, from the first, the
cantata was evidently better fitted for the latter than for the former
purpose, both performances must be ascribed to the same year, and
this must be a year in which the twelfth Sunday after Trinity preceded
the election. In 1723 Bach composed another cantata for the election
(see ante, p. 362); in 1726 and 1727 the Sunday came after the election.
We have to choose between 1724 and 1725. I must decide on 1724,
and the reason is at hand; in 1724 the election was on Monday, August
28, the next day after the twelfth Sunday after Trinity, and this would
probably suggest to Bach the possibility of composing a piece which
might serve for both occasions (compare No. 40 of this Appendix).

38 (p. 411). "**Bringet her dem Herrn Ehre seines Namens.**"
(The reader who is interested in comparing Bach's verses with Picander's
will find both poems at length in the German edition of this biography,
Vol. II., p. 993. Translators). On comparing the two texts it is easy
to detect that, irrespective of the Bible text and chorale, the lines by
Bach are largely derived from Picander. Though no whole stanza
has been borrowed from it many entire lines are repeated almost
word for word, and special stress must be laid on the resemblance of
their general purport. In both a feeling of rejoicing in the House and
Word of God is the ruling sentiment, and the Gospel for the seventeenth
Sunday after Trinity strictly speaking does not give rise to this. It
is only in one place that the sanctity of the Sabbath is mentioned at all,
and there Jesus acts in an opposite sense. It is only by a reference to
the original source of the poem that we can understand how it should
have occurred to Bach to compose this cantata for this particular
Sunday. Picander's verses form the sequel to a satire in rhymed
Alexandrines reflecting on the worldly and unworthy use of the Sabbath
by the multitude; the poet there exhorts us to keep holy the Sabbath
day, and closes with the hymn. Thus, without the antecedent lines by
Picander, the cantata text, as used for this Sunday, is unintelligible.
There can be no doubt therefore that Bach's and Picander's poems
must have been written about the same time (see too the remarks as to

the Michaelmas cantata, p. 408). Accordingly this cantata must have been composed for September 23, 1725, and is an inseparable companion piece to the Michaelmas cantata first performed in this year. No autograph copy exists, but a MS. by Harrer, Bach's successor as Cantor, is a tolerable substitute; it is in the Berlin Library. The text of the chorale is not given, and Erk (Choralgesänge No. 13) thinks it was verse vi. of "Auf meinen lieben Gott." It seems to me that the last verse of "Wo soll ich fliehen hin" is more suited to the idea of the composer.

39 (p. 412). "**Herz und Mund und That.**" The first four sheets have the Weimar watermark, the last two M A; the parts on the other hand have the mark of the period between 1723—27 (see No. 19 of this Appendix). We find the M A paper used throughout for the first time in the autograph parts of the mourning ode for Queen Christiana Eberhardine, which was finished October 15, 1727. Hence the new MS. of "Herz und Mund" must have been written at a transition time, the summer probably of 1727.

40 (p. 413). **Telemann's Influence in certain Cantatas** of the early Leipzig period. In No. 27 of this Appendix I have already spoken of the cantata "Herr Gott dich loben wir," and shown that it is possible, but not probable, that it should have been composed by January 1, 1724. Strictly speaking nothing can be inferred from the watermark beyond the fact that it cannot have been written later than 1736. For the period during which Bach used paper marked M A extended from 1727 to 1736; but during this time he used various other kinds of paper, consequently there is nothing to hinder our supposing that this New Year's cantata was composed later than 1727, and then only completed in 1736 by the addition of the well-known Violetta part. My reasons for attributing it to the earlier date are founded on internal evidence; the imitation of Telemann's style is so conspicuous that it cannot have been unconscious and accidental, but clearly betrays a purpose; and this purpose is accounted for by the position, as to musical taste, in which Bach stood during his first years at Leipzig. These made a certain reference to the ruling taste desirable, nor was Bach unwilling, within the limits set by his own character and powers. By 1730 he himself had become the ruling authority in matters of musical taste in Leipzig, and I know no example of his having subsequently set so deliberately to work to write in a style not his own.

Even Winterfeld expressed an opinion that Bach had been deeply influenced by the art of the greatest masters of his time (Ev. Kir. III., p. 385). As to the fact I agree with him, but I cannot regard the instances he adduces as satisfactory evidence. I can find nothing in the Cantatas "Jesus schläft" and "Halt im Gedächtniss" which points to the direct influence of the Hamburg masters. Nor is Winterfeld's suggestion a happy one, that Bach first became intimate with their music at the time when he wrote these compositions;

it will be seen, when speaking of the Passion music, that even in Weimar Bach had a Passion set by Keiser performed. As regards the cantata "Gedenke Herr," in which Hasse's influence is said to be perceptible (Ev. Kir., III., p. 386), until more reliable evidence is forthcoming I am absolutely incredulous as to its genuineness. Breitkopf's list for the Michaelmas book fair, 1761, in which it figures under Bach's name, is insufficient warranty, as has been said in No. 29 of this Appendix, and so too is the MS. copy (44 in the Joachimsthal Gymnasium at Berlin). The first chorus of the cantata shows a few points of resemblance to the final chorus of Part I. of the St. Matthew Passion; but these are mere similarity, the true spirit of Bach is not in this, still less in either of the other pieces. In the whole work there are scarcely any polyphonous passages, and not one striking combination. Single passages avail not; the whole effect is to me conclusive.

41 (p. 415). **The Shield and Crossed Swords watermark.** The autograph score and the original parts, in the Berlin Library, have the watermark of a shield with crossed swords; this is very frequent in paper of that time, and it occurs in paper used by Bach at the most various periods of his life. The form of the outline varies considerably, however, and where the well-known peculiar Saxon shield is not recognisable no importance is to be attached to the circumstance. The sign occurs in Cöthen MSS., as in the orchestra Partita in B minor (see No. 10 of this Appendix), in portions of the St. John Passion; again in the cantata "Gott, wie dein Name," written for the New Year, 1729, and in the autograph alterations of "Vergnügte Pleissenstadt" written in 1728; in portions of the St. Luke Passion, in the cantatas "Ihr werdet weinen" and "Gott fähret auf" of 1735, and in a report to the Council made by Bach with reference to his dispute with Ernesti, August 12, 1736. This, however, does not exclude the possibility of Bach having used paper with a similar watermark early in his residence at Leipzig, but the evidence at hand, so far, does not allow us to assume that he did so during the first year and a half. The cantatas which have this watermark and no other data as to the time of their composition are:—

1. Gottlob, nun geht das Jahr zu Ende, Sunday after Christmas.
2. Unser Mund sei voll Lachens (see No. 24 of this Appendix).
3. Liebster Jesu, mein Verlangen, first Sunday after Epiphany.

That all three belong to the Leipzig period is beyond a doubt from the character of the writing and of the compositions themselves. No. 1 may, at soonest, have been first performed on December 31, 1724, since in 1723 there was no Sunday after Christmas; No. 2 on December 25, 1724, for reasons given above; No. 3 on January 7, 1725, since "Mein liebster Jesus ist verloren" was written for 1724 (see No. 27 of this Appendix). The fact that the text of "Gottlob, nun geht das Jahr," is by Neumeister forbids our placing it later than 1726, because Bach evidently did not have recourse to his "Fünffach Kirchenandachten" after he had become acquainted with Picander. The same may be said concerning Franck's text "Alles nur nach Gottes Willen," and this was certainly composed in Leipzig at an early date, but not before 1725.

42. (Incorporated in note 49, p. 703).

43 (p. 431). **The original parts of "Liebster Gott, wann werd ich sterben?"** display in some of the sheets a small shield-shaped outline, in some a C, and in some a device with a scroll above and two hanging sack-shaped tips, which is not in the middle of the page but nearer to the fold down the sheet. This mark only occurs again in the original MS. of the cantata "Ihr, die ihr euch von Christo nennet," of the year 1723 or 1724 (see No. 22 of this Appendix); we must therefore assign "Liebster Gott" to the same date. The original parts in the Thomasschule are in D major instead of E major, and in the first chorus the parts of the two oboi d'amore are given to two violins *concertante*. A comparison with copies of the score—for the original score is lost—proves that E major was the original key. As the introduction of the oboi d'amore precludes the possibility of its having been written earlier than in Leipzig, while the parts must have been copied out by 1723 or 1724, it is clear that Bach must have made the arrangement—which greatly facilitates the labours of the oboe players—soon after composing the cantata, and probably before it was performed in public.

44. **The watermark M A as marking a period.** I here give a list of cantatas, with secular and occasional music, which constitute this group of MSS., including a motett and a mass. Many have, besides the M A, other watermarks which shall be presently discussed.

1. Ach Gott, wie manches Herzeleid (C major).
2. Der Geist hilft unsrer Schwachheit auf (Motett, 1729).
3. Der Herr ist mein getreuer Hirt.
4. Erhöhtes Fleisch und Blut.
5. Es ist das Heil uns kommen her.
6. Es ist nichts Gesundes an meinem Leibe.
7. Gelobet sei der Herr, mein Gott.
8. Geschwinde, geschwinde (the contest between Phœbus and Pan).
9. Herr Gott, Beherrscher aller Dinge.
10. Ich glaube, lieber Herr.
11. Ich liebe den Höchsten.
12. Ich ruf zu dir, Herr Jesu Christ (1732).
13. In allen meinen Thaten (1734).
14. Jauchzet, frohlocket (Christmas Oratorio, 1734).
15. Jauchzet Gott in allen Landen.
16. Jesu nun sei gepreiset.
17. *Kyrie* and *Gloria* of the B minor Mass (1733). B.-G., VI., p. xv.

APPENDIX.

18. Lass, Fürstin, lass noch einen Strahl (Mourning ode for Queen Christiana Eberhardine, 1727).
19. Lasst uns sorgen (cantata in honour of the birth of the Crown Prince-Elector, 1733).
20. Nun danket alle Gott.
21. Preise dein Glücke, gesegnetes Sachsen (*Cantata gratulatoria in adventum Regis*, 1734).
22. Schweight stille, plaudert nicht (Coffee Cantata).
23. Schwingt freudig euch empor.
24. Sei Lob und Ehr dem höchsten Gut.
25. Tönet, ihr Pauken (cantata in honour of the Queen, 1733).
26. Wachet auf, ruft uns die Stimme (1731 or 1742).
27. Was frag ich nach der Welt.
28. Was Gott thut, das ist wohlgethan (in G major, a setting of the Hymn).
29. Was soll ich aus dir machen, Ephraim.
30. Wer da glaubet und getauft wird.
31. Wer nur den lieben Gott lässt walten.
32. Wir danken dir Gott (1731).

The oldest of these works is the mourning ode. Bach finished it, as stands attested in his own handwriting, on the autograph score, October 15, 1727. We find the dates noted on others, as they are given above. No. 2 was written for the funeral of Ernesti, who died October 16, 1729. The date of No. 20 is doubtful, because it was written for the twenty-seventh Sunday after Trinity, which occurred only twice during Bach's residence at Leipzig (see ante, p. 459). With regard to No. 8, the text is by Picander, and occurs in Part III. of his poems (Gedichte, p. 501). The preface to the volume is dated February, 1732, and the title to the poem itself states that this *Drama per musica* had already been performed; at p. 564 of the same volume we find the *Caffee cantate*.

It must finally be mentioned that in the Berlin Library a MS. copy of the Partita out of Part II. of the Clavierübung exists which has hitherto been ascribed to Bach himself, but which appears to me to be in the writing of his wife, and which has the watermark M A. According to a MS. note by J. G. Walther, in his own copy of the Lexicon (now in the collection of the Gesellschaft der Musikfreunde of Vienna), Part II. of the Clavierübung came out for the Easter fair, 1735. After the work was printed Bach could of course have had no farther occasion to copy it or to have it copied; this MS. must therefore date before 1735; but since Part II. was certainly not composed till Part I. was finished, not earlier than 1731.

The period when Bach ceased using the paper marked M A was about 1736. On August 13, 15, and 19, 1736, Bach wrote three documents in the matter of his quarrel with Ernesti, and the watermark in the paper is a man on horseback blowing a post horn; but in Bach's musical MS. this watermark occurs but twice so far as is known: in the cantata "Schleicht, spielende Wellen," and in the autograph score of the C minor concerto for two claviers and strings, now in the Berlin Library (B.-G., XXI.,[2] No. 3, and preface, p. IX.). And of the latter MS. only the first eight sheets have this watermark ; the last has the letters M A, and it can be seen that this had shortly before been intended by Bach for another purpose, and then laid aside, for it has upon it the sketch of the commencement of a cantata in D major. The trumpet begins :—

This autograph belongs to the period when Bach was ceasing to use the M A paper. No. 28 is to be assigned to this period, and it must at the same time be dated 1736, because it has the same mode of notation for the oboe d'amore as the cantata "Schleicht, spielende Wellen" (see No. 50 below). The most frequently recurring watermark in the last period, beginning in 1736, is a half moon, somewhat differing from that of the period between 1723—27, but the paper is more easily distinguished by the absence of the corresponding watermark on the other half sheet (see Note 3 of Vol. III., on this watermark). By 1737 the M A mark seems to have disappeared ; in a few autographs, as in "Wo soll ich fliehen" and in "Jesu nun sei gepreiset," both marks occur.

45 (p. 438). **The date of the Cantata " Wer nur den lieben Gott lässt walten "** is approximately known from the watermark M A (see the foregoing note 44). It is farther restricted by the fact that in 1731 the fifth Sunday after Trinity coincided with St. John the Baptist's day, while another cantata "Siehe ich will viel Fischer aussenden" was written for either 1732 or 1733; this however does not help us greatly. But a close study of the text reveals the probability of its being by Picander. It includes a paraphrase of the fifth verse of a hymn by Neumark, which much resembles a poem by Picander (Part II. of the Gedichte, p. 89, "Bey dem Grabe Herrn C. K. Freyberg, July 6, 1728"), and certain references to verses of the Bible such he was wont to display his knowledge of the Scriptures in. And by comparing the dates of this funeral poem with that of the performance of the cantata we may infer that this was probably June 27, 1728. Nor can it be regarded as seriously adverse to this conclusion that, of the three continuo parts for this contata, the original MS. of the organ part is on paper with the half moon, which is the watermark of the latest period, beginning in 1735, for this part may very likely have been added at a later date. (The student who is interested in a detailed comparison of the cantata texts is referred to the German original of this work, Vol. II., App. A., Note 34, Page 798.)

APPENDIX. 699

46 (pp. 444, 445, 447, 455, and 473). The watermarks here figured occur

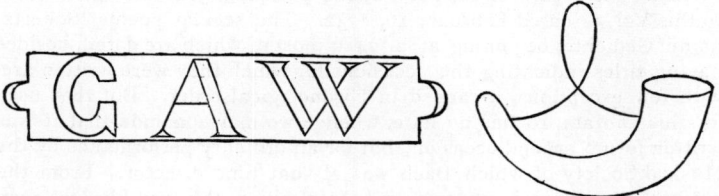

in the autograph score, and most of the original parts of the following compositions, ten in all :—

1. Der Herr ist mein Getreuer Hirt.
2. Ehre sei Gott in der Höhe.
3. Geist und Seele sind verwirret.
4. Gott, wie dein Name.
5. Herr Gott Beherrscher aller Dinge.
6. Ich bin vergnügt.
7. Ich habe genug.
8. Ich liebe den Höchsten.
9. Lobe den Herrn, den Mächtigen König.
10. Der Streit zwischen Phöbus und Pan, a secular cantata.

In the original MSS. of Nos. 1, 5, 7, 8, and 10, we find mixed with these watermarks the M A mark (see No. 44 of this Appendix), which of itself indicates certain limits to their date of composition. No. 4 has a text from Picander's cantatas for the year, and was written for January 1, 1729. But in No. 3 an obbligato organ part is introduced, and this cannot have been earlier than 1730 (see No. 15 of this Appendix). With regard to No. 5, the original parts, all marked M A, are written by Ph. Emanuel Bach, and have only a few additions in Seb. Bach's hand, so it must have been composed before 1733, when Emanuel Bach left his father's roof for Frankfort on the Oder. No. 7 underwent a subsequent revisal, and the new parts prepared in consequence have for the most part the watermark of an eagle and for counterpart HR (H I R). These marks, which rarely occur together in Bach's MSS., are found combined in one earlier, in the text of the cantata " Preise dein Glücke, gesegnetes Sachsen," in Bach's own hand, which is added to the autograph score. This was performed October 5, 1734, so No. 7 must have been written earlier. No. 8 has in the autograph score a long passage written by W. Friedemann Bach in the staves for violins and violas, from bar 24; and farther on again other portions are in his hand. Now in 1733 Friedemann Bach, who till then had lived with his father, became organist to the Sophienkirche at Dresden. In this year, moreover, all church music was suspended from Estomihi Sunday till the fourth Sunday after Trinity, in consequence of the public mourning, and No. 8 is a Whitsuntide cantata. It must therefore have been written in 1732 at latest. No. 10 is set to a text written by Picander; it occurs in Part III. of his poems, p. 501, and the Preface to this Vol. is dated February 18, 1732. The secular poems (Schertzhaffte Gedichte, beginning at p. 241), most of which are dated, besides having titles indicating the occasions for which they were written are, with few exceptions, arranged in chronological order. But that used for this cantata, 10, has no date, whence we may conclude that it was written for no special occasion, but for an ordinary performance by the Musical Society of which Bach was at that time director. From the order of the remaining poems and the position this one fills, we may ascribe it with certainty to the year 1731; and since above it we find the words "performed in a *Dramate*," the music must have been composed when Picander's volume was printed, and we may therefore assign the composition to the summer of that year. Thus we have determined the church year 1730—31 as being most probably the period when these ten cantatas were composed, and 1731—32 as its latest term. It is to be remarked, however, that there are among them two (Nos. 3 and 9) for the twelfth Sunday after Trinity; they must therefore extend through two church years; but to which each cantata belongs can only be guessed with more or less probability—in many cases not even guessed.

With regard to No. 2, the preface to B.-G., XIII.,[1] has an inaccurate statement. The autograph fragment contains, before the bass recitative "Das Kind ist mein," the last nineteen bars of the alto aria "O du angenehmer Schatz." This fragment consists of only one folio sheet, which is marked 7 in the top right hand corner. Judging from this, the opening chorus must have been of considerable length, and perhaps introduced by an instrumental symphony; the closing chorale "Wohlan so will ich mich" is included in Erk's collection, No. 278, but the text is inaccurate, and he does not mention the source whence he derived it.

The autograph score of No. 4 has the watermark figured above in the first two sheets only; the two last have the shield with crossed swords. They are evidently of later date, and contain the two borrowed movements of the cantata "Jesus soll mein erstes Wort" (out of "Der zufriedengestellte Aeolus," and "Lass uns dass Jahr vollbringen" (out of "Jesu nun sei gepreiset"). The first is lowered and the second raised a tone, and both are fair copies, while the recitative that divides them is, on the other hand, a rough copy. Bach probably began this cantata towards the end of 1730, left it unfinished, and did not complete it till after 1736.

47 (p. 446). "**Ich habe meine Zuversicht.**" The Berlin Library possesses a copy of this cantata, formerly belonging to Professor Fischhoff, of Vienna, transcribed from the original autograph, but of which the chief part has been lost; some fragments, including bars 24—67 of the first aria, and the second aria from the last crotchet of

bar 24, with the recitative that follows, and the closing chorale, were in the possession of Professor F. W. Jähn, of Berlin. He, after taking a careful copy, gave them to Petter, a collector of autographs in Vienna. The earlier half of the cantata, which included the beginning of the first aria, Petter parted with to Herr Otto Usteri, of Zurich, who was good enough to send it to me to look at in January, 1870. It is numbered 7 in the top right hand corner. This number cannot refer to the paging, since in that case there would be no more than five pages for the introductory organ concerto, a space too small for the first movement only. If the number refers to the sheets, we must suppose that it was not the first movement only, but the whole concerto that served as the introduction; the number of sheets would suffice for this, but would be much too large for the first movement by itself. This suggestion is confirmed by the note on the title-page of Fischhoff's copy, that the organ concerto served as the *Introitus* to the cantata (the chief subject is added in notes). The first two subjects were used again by Bach for the cantata "Wir müssen durch viel Trübsal," the first forming the instrumental introduction, while the principal chorus is worked into the *Adagio*. The upper part is taken down an octave, but is not otherwise altered from the original form of the clavier concerto as it is given B.-G., XVII., p. 275. We may suppose that this older form lay at the root of the re-arrangement for the cantata "Ich habe meine Zuversicht," since there is too much in it which is properly suited only to the clavier. The aria "Unerforschlich ist die Weise" is set in E minor for the voice and violoncello, and in D minor for the organ obbligato; according to a note on the cover of the Fischhoff copy, in the original autograph, the voice part in the violin clef was written an octave too high; and since in this, as well as in the Jähn fragment, the aria is in the alto clef, this plainly indicates that in the autograph score both the alto and violin clefs were given. I believe the reason to have been that E minor in the alto clef being in the same notes as D minor in the violin clef, the singer had only to sing his part an octave lower. We often meet with such two-fold notation in Bach (see No. 20 of this Appendix). It naturally follows that Bach must have had this aria sung sometimes in E minor and sometimes in D minor, and in the latter case would perform the organ part on an instrument tuned to chamber pitch. It cannot be denied that the key of E minor is foreign to the scheme of modulation of the work as a whole; still it flows more smoothly with the movements which immediately precede and follow it than that of D minor; when we consider that D minor has already largely predominated, the composer may have thought a digression into a remote key a desirable variety.

48 (pp. 448, 451, 453, 455, 461, 463). **The Watermarks of 1731 —1733.** To determine the date of "Gott soll allein" certain marks must be collated. A small number of Bach's original MSS. have a

702   APPENDIX.

shield-shaped watermark supported on each side by palm branches, and bearing a chevron. The cantatas which have this watermark are the following:—

1. Ach Gott, wie manches Herzeleid, C major.
2. Es wartet alles auf dich.
3. Gelobet sie der Herr.
4. Gott soll allein mein Herze haben.
5. Ich armer Mensch, ich Sündenkneckt.
6. Ich will den Kreuzstab gerne tragen.
7. Siehe ich will viel Fischer aussenden.
8. Vereinigte Zwietracht der wechselnden Saiten.
9. Vergnügte Ruh, beliebte Seelenlust.
10. Was Gott thut, das ist wohl gethan, B major.
11. Wer weiss, wie nahe mir mein Ende.

Of these, No. 8 may be set aside; it is an occasional secular cantata, written for December 11, 1726, but that this cannot be the date at which this group generally was composed is easily proved. Bach must have accidentally used for it a kind of paper which, though the watermark is the same, is itself quite different.

Nos. 1 and 3 have in some of the parts the M A watermark, which assigns them to a date between 1727 and 1736. In Nos. 4 and 11 we find an organ obbligato; this limits them to the years between 1730—1736, and we are still farther limited as to the organ part for No. 11, which is in Bach's own writing. It has a different watermark from the score:—

These marks occur in only two other cantatas, whence we may refer them to the same period, and that a short one. These are:—

1. Herr deine Augen sehen nach dem Glauben.
2. Wir danken dir Gott, wir danken dir.

And in the first of these they only occur in the paper forming the original wrapper, in which they are very distinct. In the second they occur only in the autograph score, while the M A mark distinguishes the parts; this cantata has also an organ obbligato, and finally from a note in Bach's own hand we learn that it was written for the Rathswahl service on Monday, August 27, 1731. We may therefore assume that this year is the earliest limit of the period during which these cantatas were composed, and this is confirmed by our finding the same paper used for the Credo of the B minor mass. The next point is to ascertain the latest limit. We shall here find that the cantatas in question, with the exception of No. 8, must have originated within a by no means lengthy period. No. 7 was intended for the fifth Sunday after Trinity, which in 1731 coincided with the festival of St. John, and was merged in it. No. 3 is for Trinity Sunday, and so cannot have been written in 1733, the year of the public mourning. No. 5, again, composed for the twenty-second Sunday after Trinity, cannot have been written in 1733, since in that year it fell on the same day as the Reformation festival. No. 1, on the other hand, being composed for the second Sunday after Christmas cannot have been written in 1732, nor in 1730 or 1731, since there was no second Sunday after Christmas in those years. We must therefore date 1733, and this determines a period of about a year and a half, beginning with the tenth Sunday after Trinity, September 9, 1731, and ending with the second Sunday after Christmas, January 4, 1733.

49 (p. 456). **The autograph score of "In allen Meinen Thaten"** (which is published B.-G., XXII., No. 97) is in the possession of Professor Rudorff of Lichterfelde, near Berlin. It consists of six sheets, and the title on the cover is in the hand of Anna Magdalena Bach. The watermark of the wild man and fir tree is to be seen only in the wrapper and the first two sheets, the others have no watermark. These four contain all the cantata after the opening. Not only is the paper different, the staves have been marked with a narrower ruler, and different ink has been used; but whether these are superficial accidents, or the greater part of the work was written later than the beginning, cannot be determined. It is quite possible that a different composition originally followed the opening chorus; this is too short ever to have formed an independent cantata. An investigation of the original parts preserved in the Thomasschule makes a remodelling of the cantata—the first chorus being retained—seem highly probable. These parts were copied in 1735 at the earliest, the watermark of 1723—27 is visible only in the bass and tenor parts. Bach no doubt used for these some blank sheets of paper remaining from the original MS.; that he was careful in such matters has already been mentioned (Vol. I., p. 643). Before the autograph score of this cantata was known, W. Rust had been of opinion that it was an early work, perhaps of the Weimar period; and he thought himself justified in adhering to this view, although the score bears at the end, in Bach's own hand, the words "*Fine. S. D. Gl.* 1734." He founded it on certain discrepancies between the figuring of the organ part and the harmony of the upper parts. There are two original parts written out for the organ, the older in A flat and the later in G. This last is not here of any consequence, for if in one or two places the figured harmony does not agree with that of the other parts, Bach may have intended to alter it subsequently. But the variations in the other organ part certainly

point to the conclusion that it was prepared for an earlier score than that we now possess, and it suggests some quite different original readings. Those adduced by Rust are but three; and of these, that in bar 8, of p. 218 of the B.-G. edition, does not, in my opinion, argue a different progression in the *obbligato* parts; the collision between the b' in the second violins and b' flat in the organ part is a passing discord, such as we find in innumerable cases in Bach's works. In bar 5 of the same page there is a real discrepancy, the question being as to the presence or absence of a natural before the e flat; the older organ part would require e' flat, the latter e', in accordance with the score; but there is nothing to hinder us from supposing that in the subsequent performance, for which he used the second organ part, he added the natural. The third deviation is in bar 1 of p. 193, and in the same passage again, bar 2, p. 201, and here, to render the figured bass at all possible, the first half of the bar must be entirely altered. I consider the passage in the organ bass to have been simply incorrectly copied. It not unfrequently occurs that Bach's organ parts do not accurately correspond to the other parts, for in hastily writing them out he often neglected to refer to the score, trusting to his memory for the harmony, and so making mistakes. Thus, in the first chorus of the Cantata "Halt im Gedächtniss," in bar 72 in the continuo, which is not transposed, we find on F sharp in the bass, the chord $\smash{\begin{smallmatrix}9\\7\\4\end{smallmatrix}}$, while in the organ part, which is transposed, we have $\smash{\begin{smallmatrix}7\\4\\3\end{smallmatrix}}$. Rust has accepted the former, but in my opinion the latter is the correct form; comparing bars 51 and 102 it is evident that, when figuring the continuo, Bach had in his head a different scheme of harmony from that which was required by the strings. Or if we assume that the chord for the strings was wrongly copied, bar 2 ought to have an analogous harmony, and then it is the organ part that Bach wrote wrong. There is an error in either case. As to the curious clerical mistakes of which Bach was capable even in notation, we have an instance in the Cantata "Angenehmes Wiederau" (B.-G., V.,[1] p. 404, bar 3), where, in five MS. parts written out by himself for violins and violas, he has set down a wholly inaccurate and ill-sounding phrase, and it is clear that at the moment he had in his mind a bass quite unlike what he had previously written. Still, these discrepancies between the organ part and the original score are few and trifling; it seems to me very bold to derive from them so serious an inference as that the date at the end of the score counts for nothing as regards the date of composition, and refers only to that of the copy or of some special performance; it is too contrary to our experience of Bach's practice.

Nor is the evidence as to style, adduced by Rust, convincing. It is precisely in the first movement that I detect that mature and artistic development which, from all we know of Bach's work in Weimar, he had not as yet attained. As regards the final chorale in some parts and its relationship to the Cantata "Wachet, betet," I must remind the reader that this cantata is known to us only through the remodelled Leipzig form (see No. 23 of this Appendix), and that a composition similar in every respect occurs at the end of the Cantata "Lobe den Herren." The date in the master's own hand, the internal resemblance of the work with other cantatas of the same period, and the watermark in the paper (the letters M A, see No. 44 of this Appendix) all afford such sufficient grounds for assigning the composition of "In allen meinen Thaten" to the year 1734 that Rust's arguments to the contrary are quite inadequate to disprove it.

50 (p. 463). **In the original parts of the Cantata, "Christus der ist mein Leben"** there is no watermark by which to judge of the date of the composition. But the method of notation used for the *oboi d'amore* leads us to assign it to the year 1732. At three different periods Bach adopted three different and unusual methods of noting music for this instrument, and I entirely agree with W. Rust in regarding these as unerring chronological data; see B.-G., XXIII., p. xvi. It may also be added that during the whole of the Leipzig period Bach used for this instrument the same notation as for the ordinary oboe; as in "Es erhob sich ein Streit," 1725; Trauerode, 1727; St. Matthew Passion, 1729; "Es ist das Heil uns kommen her," about 1731; Christmas Oratorio, 1734; "Freue dich, erlöste Schaar," after 1737; "Du Friedefürst," 1744. From this it is plain that it was only occasionally that Bach was compelled to write for *oboi d'amore* which were a minor third lower than chamber pitch; and we have all the more reason, therefore, for assigning all the compositions which exhibit this peculiarity to a narrow period of time, unless other circumstances prove the contrary. The *Kyrie* and *Gloria* of the B minor Mass were presented to the Elector of Saxony, July 27, 1733; in the *Kyrie* we find the *oboi d'amore* in this low pitch, and noted in the same way as in the Cantata "Christus der ist mein Leben." Both these portions of the Mass were at any rate composed by the spring of 1733. (See the next note.) I have, however, one objection to raise against the views taken by Rust (in the preface just referred to). I cannot understand why, at p. xvii., he refers the re-arrangement only of "Lobe den Herrn meine Seele" to a period between 1733 and 1737, on the ground of the notation of the *oboe d'amore*, since it is identically the same in the first state of the work, and it will be seen, on reference to p. 692, that my researches have led me to a different conclusion. Still, the notation of the *oboe d'amore* supports me in assigning it to 1724, for it is the same as that in the Cantatas "Die Elenden sollen essen" and "Die Himmel erzählen," written in 1723. The use of the C-clef on the first line is intended to facilitate the reading of the score (see No. 20 of this Appendix). It is not possible to determine whether the same

method was followed in the score of " Lobe den Herrn," since it is lost, while of the other cantatas it is the parts that are missing.

51 (p. 463). In this cantata, or **Dialogus between Hope and Fear**, there is no watermark. It must, however, be observed that if it is to be assigned to the early Leipzig period, though the form of it is opposed to this inference, it can hardly have been written before 1730—33, since in 1726—29 there was no twenty-fourth Sunday after Trinity, and in 1727 the general mourning put a stop to all music after the thirteenth Sunday after Trinity; and in 1728, when the twenty-fourth Sunday after Trinity fell on November 7, Bach was so busy with other work that it is not likely that he should have found time for this. The twenty-fourth Sunday after Trinity was again wanting in 1734, 1737, 1740, and 1745. The parts for the *oboi d'amore* have the same pitch and notation as in the Cantata " Christus der ist mein Leben " (see No. 50 of this Appendix), so that we may assign them to nearly the same date ; nor is there any reason to prevent our referring them to the same year, 1733. On the same grounds we may fix the performance of the re-modelled form of the grand Magnificat at Christmas, 1732, and the watermark M A agrees with this (see No. 19 of this Appendix).

52 (p. 469). **The Wedding and Jubilee Cantatas.** " Herr Gott, Beherrscher," cannot have been composed later than 1733, for the reasons given in No. 46 of this Appendix. The Jubilee Cantata " Gott man lobet dich in der Stille" was performed June 26, 1730. The text is like those of the first and third Jubilee Cantatas by Picander. These were first reprinted in Sicul's " Des Leipziger Jahr-Buchs zu dessen Vierten Bande Dreyzehente Fortsetzung." Leip. 1731, p. 1126, with a note that they were composed (that is, the music for them) by the *Cantor oppidano*, Herr Johann Sebastian Bach. In 1732 Picander himself published them in the third part of his " Gedichte," and on comparison it is apparent that the text of the Rathswahl Cantata is adapted to the same music. This fact has already been pointed out by A. Dörffel, the editor of the music, as regards the second aria (" Heil und Seegen " in the Rathswahl Cantata—" Treu im Gläuben " in the Jubilee Cantata). The designation of *Aria* given to the lines in the printed edition by no means excludes the possibility of their being set by Bach as a chorus ; the word in this use only indicates a text of the *da capo* character. On comparing the two texts with the music belonging to them it is evident, from the adaptation of the music to certain words, that it was intended only for the Rathswahl Cantata ; hence this must be the earlier of the two. The text of the Jubilee Cantata suits the music so badly in some places that we might doubt whether they were meant for each other; but the text of the Wedding Cantata " Herr Gott, Beherrscher," suits equally ill, and yet it was certainly sung to the music (B.-G., XIII.,[1] p. xiv). Indeed, the Rathswahl Cantata, in the form in which it exists for us, cannot be in its first state, but must be a later copy and probably

remodelled; on this point I would refer the reader to Dörffel's careful disquisition in the preface to B.-G., XXIV., p. xxxiv.

Though the text of the Rathswahl Cantata is not to be found in Picander's poems he no doubt wrote it, and in the collected edition of his works, p. 192, we find a Rathswahl text, which, apart from the recitatives and closing chorale, corresponds in every part of its metrical structure with the Jubilee text, proving that the third of the Jubilee cantatas, " Wünschet Jerusalem Glück !" is nothing else than a Rathswahl Cantata with fresh words. Only an *arioso* corresponding to " Der Höchste steh uns ferner bey " is wanting. But Bach does not seem to have set this *arioso* at all; for in " Nützlichen Nachrichten " for 1741, where the text of this Rathswahl Cantata was first printed, it is also wanting. In such investigations as these the recurrence of similar cases strengthens the argument. Thus, at p. 686, it has been seen that the first Jubilee Cantata, " Singet dem Herrn," is founded on an older composition. Thus, not one of these three works was, strictly speaking, new; the first having been written as a New Year's cantata and the two others as Rathswahl cantatas. And with due reserve we may also trace out something as to the history of the closing chorale of the first Jubilee Cantata, " Es danke Gott," in place of which we find another in the Christmas music. " Lobe den Herrn meine Seele " was also a Rathswahl Cantata for 1724, and we find at the end of it the chorale in question. It is probable that it was transferred hither from the end of the first Jubilee Cantata; and all the more so because we find it again in " Herr Gott, Beherrscher," and there is no doubt as to the relationship between this and the Jubilee Festival, though it is an indirect one.

53 (p. 474). **Cantata for the Purification,** " Ich habe genug." The remarks in the preface to the B.-G. edition as to the original MS. of this cantata I cannot but regard as inaccurate. The part for mezzo-soprano which is there spoken of must certainly, and indeed evidently, have been originally in the key of E minor, and so, properly, for a soprano voice. Afterwards the signature and the accidentals were carefully altered; the signature being ♭♭. This does not indicate E flat minor, as this preface asserts, but C minor. The autograph score, in which the cantata is set in C minor, was begun by Bach with the intention of giving the voice part to a mezzo-soprano or an alto, and the voice part of the first aria was even written in the alto clef. Then the composer changed his mind; he wrote the rest of the cantata for a bass voice, and wrote at the bottom of the first page this note with reference to the aria :—" The voice part must be transposed into the bass." But the whole cantata is written for a mezzo-soprano voice. From this it is plain that the copy in E minor, for a soprano, must have been the first, and there even exists a complete part for the

flute in E minor (subsequently given to the oboe in C minor). It was then set in C minor for an alto or mezzo-soprano, and finally a bass voice was substituted for the alto. A part of the cantata, worked out for the soprano, occurs in Anna Magdalena's book, which thus betrays its original plan and purpose. In the preface to the B.-G. edition it is not mentioned that the score has the autograph note "*Festo Purificationis Mariae. Cantata.*" This explicit designation is not without its bearing on the character of the work.

54 (p. 508). **The Score of the St. Luke Passion** consists of fourteen sheets of closely written paper, and one sheet which serves as a wrapper; on the first page of the wrapper is the title, and the back half of the sheet forms the last two pages of the score. The title is as follows: "*J. J. Passio D. J. C. secundum Lucam à 4 Voci, 2 Hautb., 2 Violini, Viola e Cont.*," but does not enumerate all the instruments, since there are also flutes and a bassoon. The score has the appearance of having been written at intervals, and it is on three different kinds of paper. The sheet which serves as the cover and sheets 1, 11, 12, and 14 have the watermarks shown on p. 699. Sheets 2 to 10 have the shield with crossed swords; and sheet 13 has the eagle with the corresponding H I R. This mark indicates the year 1734 (see No. 46 of this Appendix); the post-horn and G A W, the years 1731—32. Hence, the score was written between 1731—34, and, as it was in 1731 that the St. Mark Passion was first performed, it may have been produced in Leipzig in 1732—33 or 1734. It probably was not in 1732, since in that year the Passion music had to be performed in the Church of St. Nicholas, in which Bach must have been much straitened for room. The real history of the score appears to me to have been this: Bach would have begun it towards the end of 1732, with a view to performing the work on Good Friday, 1733. The King-Elector died, however, unexpectedly on February 1, 1733, and there was a general mourning in consequence until the fourth Sunday after Trinity. Bach, therefore, laid the work aside for the time, but took it up again in the course of the year and finished it in the early part of 1734, when it was first performed at St. Thomas' on Good Friday. It is evident, both from the style of the work and from the aspect of the MS., that the score, as we possess it, was a revised copy from some quite early work.

55 (p. 509). **Keiser's St. Mark Passion.** That this copy was made at Weimar is proved by the fine and elegant writing which was characteristic of Bach in his youth, and by the watermarks figured in Vol. I., p. 638, which are an unfailing indication of the Weimar period. These parts are preserved in the Berlin Library, and consist of soprano, alto, tenor, and bass, with 1st and 2nd violins, and 1st and 2nd violas—all single copies—and harpsichord. There are also belonging to them duplicates, and an organ part, which were copied at Leipzig. Bach must, therefore, have performed Keiser's St. Mark

Passion again at Leipzig, and (if we may judge from the watermarks of the organ-part, which are the same as those of the cantatas "Denn du wirst meine Seele," "Gott fähret auf," and " Gott de rHerr ist Sonn," to be discussed in No. 2 of Appendix A. to Vol. III.) this must have been about 1735. In the same library there is a copy of this Passion, which must have been in Bach's possession, for the whole of the words are in his writing, as Chrysander pointed out in his life of Handel. It is dated 1720, which has been subsequently altered to 1729. It is clear that this cannot refer to the time of its composition, which must have been before 1725 at the latest.

56 (p. 522). **The St. John Passion** in the B.-G. edition, Vol. XII.,[1] was edited by W. Rust after a most careful critical study of all the MS. authorities, and his conclusions may be taken as final so far as regards the assigning of the different MSS. to the several phases of remodelling it underwent, and the evidence of at least four performances of the work during Bach's lifetime. With regard to the chronology, however, much remains to be desired, and I will here give the results of my own investigations.

The oldest copy of the score of the St. John Passion no longer exists. The one we possess is a much later copy, and only the first twenty pages are in Bach's writing. The parts copied out for the first performance, on the other hand, still exist, and, with them, some that were added for a second performance, and others for a fourth. It is chiefly from this collection of parts that we are able to judge of the alterations which Bach made by degrees in this work. The question now is when these three different batches of parts were prepared. It will be best to begin with the middle set, which have the same watermarks as the cantatas enumerated at p. 680, and thus belong to the period between 1723—27. But he did not set the first copies of parts altogether aside at the performance for which the second set was made. This we learn, 1st, from finding the "concerto," soprano, tenor, and first violin parts of the old set stitched up with a new leaf on which the opening chorus, composed for the second performance, is written; 2nd, from the old continuo part, in which a sheet has been inserted for the arioso " Betrachte meine Seele " and the aria which follows " Erwäge "—also new compositions. Both the early sheets, the inserted portion, have the watermark M A, and, as has been said, this distinguishes another group of cantatas written between the years 1723—1727 (see ante, p. 696). This second set of parts must, therefore, have been written just about this time, that is to say, in 1727.

As has been said, the chorus " Herr unser Herrscher " is already substituted for the original chorus " O Mensch, bewein dein Sünde." Hence, Rust's hypothesis cannot be correct, that this chorus was expunged from the St. John Passion, and used instead at the end of the first part of the St. Matthew Passion (which was not written till 1729) in order to give it a more dignified close when it was remodelled later—thus assigning the composition to a date after 1729. Why should its position have been changed? It is impossible to suppose that Bach was no longer satisfied with it. If this had been so with some of the arias the reason might have been obvious; but Bach never wrote a more perfect composition than "O Mensch, bewein," nor a piece more overflowing with the genuine sentiment of Passion music. Thus, he must have used it previously for some other purpose, and from its character this can only have been in some other Passion. But then, during Lent and Passion week, concerted music was performed only on one occasion, namely, Vespers on Good Friday; not to omit a single possibility, however, it may be mentioned that on Quinquagesima Sunday, though not properly in Lent, it was usual to introduce some reference to Christ's Passion into the music performed. Still, it is almost impossible that Bach should have used this chorus for the service on that Sunday; 1st, because it was quite contrary to his practice to despoil a great work for a small one; 2nd, because the chorus is constructed on a scale which does not fit into the limits of an ordinary cantata; 3rd, because it is highly improbable that, of five Quinquagesima cantatas which Bach wrote, he should have composed no less than four between 1723 and 1730. We must conclude from all this that, between the writing of the first and second sets of parts of the St. John Passion, Bach must have composed another Passion, a theory which is confirmed by the fact that a text for a Passion exists, written in 1725, by Picander, who had already worked for Bach. Having got so far we can fix the date of the first performance of the St. John Passion with tolerable exactitude.

Bach wrote in all five Passions; the St. Matthew Passion was composed in 1729, the St. Mark in 1731, the St. Luke in Bach's very earliest time, probably during his first years at Weimar. Of course the first Passion music conducted by Bach at Leipzig was his own composition. It cannot have been the St. Luke Passion, since Bach would not have made his first appearance in Leipzig with an immature and youthful work, nor would he under any circumstances have repeated it without retouching it, and it did not in fact undergo this process till after the performance of the St. Mark Passion in 1731 (see ante, p. 708). There remain then only the St. John and the lost Passion music. If this lost work was that performed in 1724—setting aside for the moment its probable identity with that of which Picander wrote the text in 1725—we must assign the St. John Passion to a date prior to the Leipzig period. It cannot have been written for Cöthen, because Bach had not to provide any church music there, and Passion music was not usual in the Protestant churches. It bears at the first glance a stamp of maturity and distinction beyond those of the Weimar time.

We must also give due weight to the connection between the St. John Passion and the cantata "Du wahrer Gott." Now, a multitude of clear indications render it almost certain that this cantata was originally composed for performance at Leipzig on Quinquagesima Sunday, 1723 (see ante, p. 679); it closes with the chorale chorus "Christe, du Lamm Gottes." But when Bach determined to compose another cantata for Quinquagesima Sunday, and set this one aside for a time, he transferred this chorus to the St. John Passion, and I have already endeavoured to prove (see ante, p. 679) that this was in fact what happened, and not as Rust assumes (B.-G., V.,[1] p. xix) that the chorale was adopted into the cantata from the Passion, which therefore must have been finished before the cantata. It was in May, 1723, that Bach moved to Leipzig, and if we insist on holding the vague theory that Bach composed the St. John Passion for some other town—for an installation or some other occasion—no alternative remains but to conclude that the St. John Passion was first performed April 7, 1724. For it is improbable that Bach should have written it for any other place when all his energies were claimed by the worthy performance of his new duties in Leipzig, and we have only two settings of the Passion to choose from—of which the fifth, now lost, was the later—while it is certain that, on the first Good Friday of his residence in Leipzig, Bach would have performed a composition of his own.

Of course, it does not necessarily follow that it must have been composed immediately before its performance; and an investigation of the oldest copies of the parts proves that it was not. The paper on which they are written has the shield and crossed swords watermark, and though he used such paper in Leipzig he does not seem to have done so within the first year and a half (see ante, p. 695) while there is evidence of his having used such paper in Cöthen (see No. 41 of this Appendix). I consider from this that the St. John Passion was sketched when in Cöthen at the same time as "Du wahrer Gott," and, like that, with a view to his removal to Leipzig. The original form of the text also points to a want of time, and it was one main cause of the subsequent remodelling of the whole work. I have already discussed this in the text, p. 520.

Thus, after the first performance in 1724, there was a second in 1727, this is farther confirmed by the circumstance that, as the Passion music was given in the two churches in alternate years, and had taken place in St. Nicholas', in 1724, the performance in 1727 must have been in St. Thomas'. Between these dates came the lost Passion music which, on Picander's account, I have dated 1725. In 1729 came the St. Matthew Passion, and 1731 the St. Mark. We here see the determined effort made by Bach to secure the performance of his works, which required ample space, in the larger church of St. Thomas. The St. Luke Passion must have been performed in 1734 by reason of the general mourning in 1733, when it was the turn for St. Thomas' again, and we may date the third performance of the St. John Passion 1736. Thus Bach would have performed his four Passions in the order of the four Gospels. It is not possible to determine the date of the fourth performance of the St. John Passion; no certain data can be derived from the watermarks in the paper on which the parts were newly written out; and even the time when the earliest score we possess was written out can only be approximately given—it was after the latest copies of the later parts, and so at any rate towards the end of the master's life.

57 (p. 537; this note will also be referred to in Vol. III.). **The Text of the St. Matthew Passion** is to be found in Part II. of Picander's " Ernst-Scherzhaffte und Satyrische Gedichte," p. 101. The year of its production is not given here, nor is there any date on the autograph score; but the printed text is accompanied by a note to the effect that it was written for music for the church of St. Thomas. Still, as this Part II. of the poems came out at Easter, 1729, and the Passion music in 1728 and 1730 was performed in St. Nicholas', while in 1731 Bach brought out his St. Mark Passion (also at St. Thomas'— the text is in Part III. of the poems, 1732); this is undoubtedly the year of the performance. It is certain that it was not already before the public in 1727, because in that year Picander published Part I. of his poems, and would of course have included it.

The date of the second performance of the work in its extended form can only be approximately discovered. The first state of the score is not an autograph, but in Kirnberger's writing, a copy preserved in the Library of the Joachimsthal Gymnasium at Berlin, and the date of this copy can be approximately ascertained.

Kirnberger, who was born at Saalfeld, in Thuringia, in 1721, left his native town to study music under J. Peter Kellner, of Gräfenrode and in 1738 went to advance his studies first to Sondershausen and then— by the advice of Heinrich Nikolaus Gerber—to Leipzig to learn from Bach. He was his most assiduous and devoted pupil from 1739 to 1741; he then went abroad and only returned in 1751, after Bach's death. Hence 1741 must be the latest date for his copy, and probably 1739 is the earliest. For, even if the St. Matthew score should have come under his ken through the intervention of Kellner or Gerber, is it likely that, as their pupil, he should have been competent to understand it? Is it not more likely that his copying out so extensive a work is evidence of his enthusiasm, and of the close intimacy between him and his master? Certain circumstances contribute an element of certainty to these views. The paper on which Kirnberger wrote is the same as that which Bach often used for his first copies, and is of so unusual a character that we cannot suppose Kirnberger to have met with it accidentally elsewhere. The watermarks here given—

APPENDIX.

are also found in Bach's score of the cantata "O ewiges Feuer," in the chorus "O Jesu Christ mein's Lebens Licht," and in the autograph of a Latin Christmas hymn constructed on the Gloria of the B minor Mass. Kirnberger may even have written out the score in Bach's own house, if, like some other pupils, he resided there. The St. Matthew Passion could only be performed in St. Thomas' on account of the large scale of the work, and it came to St. Thomas' turn in 1740, so that the revised form of this work cannot have been performed before 1740 at earliest; this might suggest that Kirnberger had made his copy in the previous year. Still there are certain facts which make it probable that it was not made before 1741. Kirnberger, who went through a complete course of composition, cannot have been introduced to such a work till the end of the course; moreover, there exists a St. John cantata by J. G. Goldberg, who was also Bach's pupil just at this time ("Durch die herzliche Barmherzigkeit unsers Gottes," score and parts in the Berlin. Library), and the autograph parts are on similar paper to that of Kirnberger's copy. From this the earliest date of the revision would be 1742. On examining the paper and writing of the autograph with other autographs by Bach, we find a resemblance between the original parts and the autograph clean copy of the *Ricercar* in six parts out of the "Musikalisches Opfer," preserved in the Berlin Library. This work was written in 1747, and according to this the revision of the St. Matthew Passion was not made till 1746 or even 1748.

There is a discrepancy in time between these results and a date given by Marpurg (in Legende einiger Musikheiligen, Cöln, 1786, p. 62). Friedemann Bach at Halle was commissioned to compose a serenata in 1749 (see Bach, F., in the index to Marpurg's Legende), and had made use of some airs out of "a certain highly artistic Passion oratorio." When the serenata was performed "there was among the audience a cantor of Saxony, not far from Leipzig, to whom the parodied airs were well known." Through him it came out that the airs were not new at all—" far from it, being at least thirty years old, and taken from the Passion of a certain great master of double counterpoint." Though Sebastian Bach is not named, the whole story leaves no doubt that it was a Passion music by him. But which? This we cannot infer from Marpurg's narrative, but if the date is seriously meant it must have been a work composed before 1719, and this date does not correspond with either of the known five Passions. My opinion is that no importance attaches to the number of "thirty years." It would seem as though Marpurg himself was not clear as to which of the Passions Friedemann had taken such a liberty with, and his saying "at least thirty years" shows that he did not intend to give an exact date.

58 (p. 591). **The Easter Oratorio** exists in three forms. The first we know from the original parts; the second in the autograph score and a few parts belonging to it; the third, again, only from some autograph parts copied out later. The score supplies the most exact data for the chronology. The watermark in the front half sheet is this:—

and in the back half sheet an M surrounded by arabesques. Both these marks recur in the autograph score of the St. John's Cantata "Freue dich, erlöste Schaar." This is a remodelled form of a secular composition "Angenehmes Wiederau," which was performed September 28, 1737. But the watermark in the autograph of the secular work is different, we have there—

But from various experience I find that such a variation, in detail, does not imply a radical difference in the character and manufacture of the paper, and is less important here, as the other watermark, shown above, is exactly the same in both. Another MS. helps to prove that this paper was in use in 1737; namely, a petition addressed by Bach to the King-Elector, October 18, 1737, on the occasion of his quarrel with Ernesti. Hence, we may argue that the St. John's Cantata was written immediately after the secular cantata, and so for June 24, 1738; and consequently, that the score of the Easter Oratorio was written for Easter Day, April 6, 1738. Among the parts which represent the oldest state of the Oratorio, and which have the shield and crossed swords watermark, we find a continuo part, and on the last page of this is the beginning—subsequently crossed out—of the Cantata " Bleib bei uns." This was written for the second day of Easter; thus it is clear that the copyists were employed on it and on the Easter Oratorio at the same time, and perhaps at the same table; whence we may fairly conclude that the two works were composed for the same Easter-tide. The watermark of the Cantata "Bleib bei uns" is a half moon; Bach began using paper of this description in 1735. If, then, the revision of the work, as it stands in the score, was made in 1738, the copying of the older parts and the antecedent composition of the Oratorio must have taken place within the two or three years preceding. It cannot be assigned to 1737, because the revision was no doubt undertaken with a view to a performance, and it can hardly be supposed that Bach would repeat the same music at an interval of only a year. Thus probability points to the year 1736 as that of its composition.

The latest set of parts, finally, in which we have the third state of the Oratorio, are on paper with the watermarks of an eagle and HR (H I R); this paper has, till now, never been found to have been used by Bach excepting in 1734 (see ante, p. 699), and, if this is to be considered as final, any contradictory evidence is, of course, useless. We have here an irreconcilable discrepancy. Still, it may be observed that the same watermark also occurs in one single part-sheet of the Cantata " Gelobet sie der Herr," which must be attributed to the year 1732 (see ante, p. 702), so that it is not an immutable landmark. Thus, as all the strongest probabilities are in favour of the autograph score having been written after 1737, since the paper on which that is written can, on various grounds, be shown to belong to that date, and as it is also probable that the oldest part-copies were not made before 1735—and these two probabilities harmonise as to date—we must found the chronology of the work on them, and assume that, by chance, some old paper was used for the latest part-copies.

59 (p. 595). **On Bach's Motetts.** In the Necrology, p. 168, we find, as No. 4 of the list of Bach's unprinted works, "Einige zweychörige Moteten." Forkel, on the other hand, writes, p. 61, " *Viele* ein - und zweychörige Motetten," and farther on informs us that, of the motetts for double chorus, eight or ten still remained in existence, but in the hands of different possessors; and, at p. 36, he says that

Bach wrote motetts for single, double, and more choirs ("ein- zwey- und mehrchörige"). We must suppose that Forkel was not talking at random, and, therefore, that several more motetts have been lost, or have not come to light. We certainly, at present, know of none for three choirs, for instance; and even if we accept all of those for two, which are known under Bach's name, they fall short of the highest number given by Forkel. But some of these are certainly spurious, or at least doubtful. At any rate, the score published in 1819, by Breitkopf and Härtel, of the Motett "Lob und Ehre" is not genuine, though it afterwards was republished as No. 3 in the new edition of Schicht's collection, after the Motett " "Ich lasse dich nicht" had been rejected as being by Joh. Christoph Bach. The fact that it bears Bach's name in a MS. copy in the Gotthold Library at Königsberg (No. 13569, 2) proves little, since the copies of Bach's motetts in this collection appear to have been made by Schicht.

The Motett "Lob und Ehre" is full of the grossest musical blunders, and it is difficult to imagine how it can so long have passed for Bach's work. In the collection of Herr Hauser (Kammersänger), of Carlsruhe, the same motett is to be found in score and two parts, as a composition by Georg Gottfried Wagner, whom we may more easily suppose to be the author. (Wagner, born in 1698, was a St. Thomas' scholar from 1712—1719; studied theology till 1726, and was still a performer in Bach's choir from 1723—26, then he became Cantor at Plauen. It is easy to understand from this that the work should frequently remind us of Bach, whom Wagner evidently took as his model).

The double chorus motett "Jauchzet dem Herrn alle Welt" is also in part certainly spurious; but it is mentioned in the catalogue of Emanuel Bach's property under the heading of Seb. Bach's compositions, p. 73, as "*Motetto:* Jauchzet dem Herrn alle Welt. Für 8 Singstimme und Fundament, in 2 Chören. In Partitur." This MS. is now in the Berlin Library; in the right hand bottom corner of the first page is the name J. G. Parlaw (Parlaw seems to have been a musician of Hamburg; the name occurs on MSS. in the Library of the Academy for Singing at Berlin which formerly belonged to Pölchau. Hilgenfeldt says, p. 112, that this motett was edited by J. S. Döring, and published by Kollmann, Leipzig, but I have not seen a copy). The motett is in three subjects, of which the middle one is the chorale chorus out of the cantata "Gottlob, nun geht das Jahr," but with the fifth verse of the hymn instead of the first, so this is in fact Sebastian's work. The third movement, an "Amen, Lob und Ehre und Weisheit und Dank," for double choir, is to be found in Röchlitz's publication "Sammlung vorzüglicher Gesangstücke," Mainz, Schotts Söhne, 1835, Vol. III., p. 66, where it is given as a separate piece by Telemann; and that it is actually by Telemann, and certainly not by Seb. Bach, any one can see who knows the style of the two men. The genuineness of the first must remain doubtful, for though it bears unmistakable traces of Telemann's hand, it has a certain breadth of outline, and that richness of harmony and certainty of part treatment which are peculiar to Bach (see Vol. I., p. 633). I think it probable that it was patched together by Emanuel Bach out of heterogeneous elements; but the title in the catalogue is not decisive, since it was not compiled till after Em. Bach's death.

Sebastian's Bach's compositions have often been tampered with in this way. At the Academy for Singing, in Berlin, there is a Christmas motett in D major for four voices, " Kündlich gross ist das gottselige Geheimniss," with figured bass, and a long instrumental interlude which was probably played originally by violins and bass. Professor A. W. Bach, of Berlin, died 1869, also had a MS. copy of this work. Zelter arranged this for a double chorus, and gave the interlude to a soprano and alto soli above a figured bass, and in a note at the end, dated December 31, 1805, he says this piece is by the great Sebastian Bach, and originally set with instruments. This however is certainly not the case; the smooth flowing style and lack of richness of harmony point to Graun, of whose chorus "Christus hat uns ein Vorbild gelassen" it frequently reminds us; indeed, the venerable musician, Professor Grell, informs me that it was formerly ascribed to Graun. But appended to it are two interludes really by J. Seb. Bach, one from the great Magnificat, "Vom Himmel hoch," transposed into D major, and "Freut euch und jubilirt," set in A major; in the second the continuo has been given to a bass voice.

The motett "Ich lasse dich Nicht" was also lengthened by the addition of a chorale set by J. Seb. Bach (see Vol. I., Appendix A., No. 6).

60 (p. 600). I found the five-part motett "Nun danket alle Gott" in a very inaccurate, but—so far as I know—unique, MS. in a volume in the Gotthold Library at Königsberg (No. 13569), containing also the motetts "Komm Jesu komm" and "Lob und Ehre," all written by the same hand. It may have been copied in about 1800, and probably belonged to Schicht. The title is "3 Motetten | von | J. S. Bach." It is expressly added with regard to the last two that they are by J. Seb. Bach (which I have shown to be incorrect as regards the last, see ante, p. 716), but this is simply called "Motett | Nun danket alle Gott." But the title of the volume proves that it was regarded as Bach's work by the writer of this copy, and from internal evidence it is the one whose genuineness we have least reason to doubt. The final chorale has the same bass, two notes excepted, as Bach's chorale-subject, reprinted by Erk as No. 270 of his collection; but the middle parts move rather more freely in places, though without affecting the harmony, so that they are essentially identical, though the copy according to Erk seems rather more ornate, as if for a different purpose. The MS. no doubt came from Leipzig, as shown by this among other proofs; over the chorale "Gott Vater, dir sei Preis," in the motett "Lob und Ehre," it is noted: "Aus No. 584 Altes Gesb. O Gott du frommer Gott," and this hymn is to be found in the Leipzig Hymn-book, No. 584.

61 (p. 620). The Cöthen Secular Cantata is an autograph MS. in upright folio, and on the cover is written, in another but antique writing: "Sechs Bogen starckes Fragment einer eigenhändigen Cantate von Joh. Sebastian Bach auf das Geburtsfest Leopolds, Fürsten zu Anhalt-Cöthen" (six sheets, being part of an autograph cantata by Joh. Seb. Bach for the birthday of Leopold Prince of Anhalt-Cöthen). Below this in a different hand ("Durch Tausch anderer Manuscripte von D. Pölchau, in Berlin erhalten. D. Fst.") [Obtained by exchange of other MSS. from D. Pölchau.]

D. Fst. is Dr. Feurstein, formerly an autograph collector at Pirna, and after his death it was acquired by the father of the present owner. The watermark is the wild man and fir tree (see ante, p. 679), and the writing is evidently that of a mere sketch, hasty, abridged, small, and full of corrections. For the repeats the bars only are marked off and left empty, to be filled up subsequently. The last page is blank, and on the foregoing page is a little addition sum in money, which no doubt relates to some of Bach's pecuniary affairs. The MS. begins with the second half of a tenor aria which closes with an indication of *da capo*. The principal key—as may be gathered from the sacred cantata which was altered from it, "Ein Herz, das seinen Jesum"—was B flat major, and the cantata closes with a four-part air in the same key. Since the fragment which remains is of considerable length, almost exceeding that of an ordinary cantata, the tenor aria, with perhaps an introductory recitative, was probably the first movement and no more than a sheet of the MS. may have been lost. The text (which the curious reader will find at full length in the German edition of this work) is in honour and praise of the Prince and his family, and the second recitative appeals to the "happy land" to contemplate the prudence and enlightenment of one princess and the "crown of virtue" of the other.

The second number was a recitative, and must have begun on the lower staves of the lost sheets, being continued on the first of those remaining. It is a dialogue between alto and tenor soli, no doubt as allegorical personages. The third number is a duet between alto and tenor in E flat major common time. Number 4 is the second recitative, and the next is an alto aria in G minor common time; then follows another recitative for alto and tenor, and then the finale, which is in the Italian aria form, and consists of 418 bars in 3-8 time. In the second portion of this finale the alto and tenor stand out from the general four-part background in various places in duet, and then retire and mingle with it again. These were in fact the two solo parts of the cantata, as is made clear by their not having precisely the same words; the alto sings *Gnade*—Mercy—where the tenor has *Segen*—Blessing. The alto and tenor being intended for solo voices, it is evident that the whole work was meant to be sung in single parts, not in chorus. The first words of the text, as well as those of the foregoing recitative, are written under the corresponding words of the altered form.

It is, however, perfectly clear from the character of the text that the statement on the cover, as to the work having been composed for Prince Leopold's birthday, is founded in error. The text, "Bedencke nur, beglücktes Land" (Reflect, O Happy Land!), is a guide to the chronology, for it contains allusions to all the members of the royal family then living. Leopold himself, the reigning prince; his brother, Prince August Ludwig; their youngest sister, Princess Christiana Charlotte (Eleonore Wilhelmine, the elder sister, had married in 1716, Duke Ernst August of Saxe-Weimar); and, finally, the princess. The question is, whom are we to understand by *the princess*? If Prince Leopold's wife is intended, the cantata can only have been written between December 11, 1721, and January 13, 1722, for Leopold was married on the former date, and on the latter, Prince Ernst August, whose wife, however, is not mentioned at all; moreover, it must be the first wife of Prince Leopold who is alluded to, since on September 21, 1722, she had a daughter, who certainly would not have been omitted. From all this we should be inclined to date this cantata somewhere about the beginning of 1722. But, on the other hand, the attributes of "prudence and enlightenment" are far more proper to Gisela Agnes, the prince's mother, than to Leopold's first wife, who died very young, and was eulogised for her gentleness and patience. And if Gisela Agnes is meant by "the princess," as I am convinced is the case, the cantata can only have been written between the Autumn of 1717 and the end of 1721.

62 (p. 631). **The date of the performances of "Schleicht Spielende Wellen"** is founded, 1st, on the allusions in the recitatives; 2nd, on the subsequent alterations in these portions of the text of the recitative, and in the title to the score; 3rd, on the connection between the score and the parts; 4th, on the watermarks in the original MSS. In the text of the bass recitative (see p. 27, &c., of the B.-G. score) we find the following allusions and inferences, which help to fix the chronology: 1st, the words on p. 56, Sop. Rec., "To-day we see the longed-for day, on which our heart's desire, the gracious Augustus was bestowed on us and the world," assure us that the cantata was composed, as the title-page tells us, for the King's birthday; this fixes the day as October 7; 2nd, on p. 27, the Bass Rec. says :—"*My stream, so lately like Cocytus flowing and bearing corpses,*" &c., referring to some recent carnage in Poland, or somewhere on the banks of the Vistula, where King Augustus had remained the master of the field. This only agrees with the year 1734, when the bloody battle of Cracow was fought in April, and the siege of Dantzig was carried on from April till June. In 1733 the enemy had not yet been reduced, and after 1735 peace reigned on the shores of the Vistula; 3rd time, p. 33., Ten. Rec., it appears that the king was in Saxony when the cantata was written—"*Blessed be thou, O Vistula, if only thou respect my wish, and take not my king from me again;*" but on p. 55 we find that his

departure was imminent. "*I (the Elbe) feel the loss bitterly, but thy (the Vistula's) advantage over-rules my will.*" In fact, the king started for Poland, November 3, 1734, and did not return till August 7, 1736 (see Mittag, op. cit., pp. 490 and 612); 4th, at p. 47, the Sop. Rec.:—"*I am filled with delight * * my nymphs rejoice, and at our hero's coming we will dance and play,*" implies that the cantata was performed soon after the king's return to Leipzig, and we know that this was on October 2, 1734.

The alterations made in the text later all refer to the passages here quoted, and in them we can clearly trace the attempt to fit the allusions to somewhat different circumstances. The cantata was no longer intended for a birthday, but for a *fête* or name-day, August 3, and the recitative, on p. 56, had to be altered accordingly; the king, too, was in Poland, involving a change in the text at p. 33—"*My king is taken from me again.*" In 1736 the king had not returned from Poland—as has been said—by August 3, and later in Bach's life, in 1748, he again spent his name-day in Poland (see Fürstenau, Zur Geschichte der musik und des Theaters am Hofe zu Dresden, II., p. 260), but this year, for reasons presently to be given, is quite out of the question. The title to the autograph score is exactly as follows, and I give it to correct and supplement the preface to the B.-G. edition, XX.[2]: "Drama auf Hohem Geburths —("Nahmens" written above) Fest Augusti 3 Regis Poloniarum unterthänigst aufgeführet von J. S. Bach." (Misprinted? *Poloniorum* in the B.-G. ed.)

Rust has already pointed out in this preface that the score is a clean copy and gives the best readings for the parts. I agree in this, and even believe that the progression of the continuo on p. 18 of the score is the better of the two, and ought to have been included in the B.-G. edition. At the same time, the idea that the work is a remodelled form of older compositions with a different text seems to me very doubtful from the highly characteristic style of the first chorus. The watermark in the autograph score is a man on horseback, blowing a horn; this mark, which we never meet with again in Bach's MSS. of music, recurs in three documents addressed by Bach to the Leipzig Town Council, dated August 13, 15, and 19, 1736 (see ante, p. 697); and as the name-day of the king was August 3, and the year 1736 is the only one besides 1748 which agrees with the alterations made in the text, this score must have been written in the summer of 1736. Since the first performance must have been on October 7, 1734, it is probable that Bach copied out the older score, and took the opportunity of making such alterations as he thought desirable, while the parts still existing are transcribed directly from the older score. Nor does the fact that the Cantata still bears the original title ascribing its use to the king's birth-day contradict this view; he may have wished to preserve the remembrance of the first purpose of the work, or it may have been a mere oversight. Several corrections and improvements and the omission, from the score of whole sections of the B minor aria (whereas the parts, on the other hand, have always contained those readings which were ultimately fixed upon) can be accounted for by an attempt, subsequently abandoned, to substitute new readings for the older ones when writing out a fresh score. Still, the part copies lead us to a singular issue. To all appearance, although they undoubtedly contain the older reading, they have been written later than the score we now possess. The watermarks, for instance, are the same as in the autograph score of " Angenehmes Wiederau," which was performed September 28, 1737, and in a petition addressed to the king, October 18, 1737 (in the town archives at Leipzig). Judging from this Bach must have had the parts copied out afresh in 1737, and, for some unknown reason, from the old score. This brings us to the question whether Bach repeated the performance of the Cantata both in 1736 and 1737, and whether for the name-day each time—on which it may be noted that the king was staying in Leipzig from September 29, 1736, for a few days. Bach himself reveals a certain vagueness as to the purpose of the work, since the alterations in the text stand side by side with the older words, so that either one or the other could be used according to circumstances; and the same holds good with regard to the title. The upshot is that the Cantata "Schleicht spielende Wellen" was composed October 7, 1734, and that Bach intended to repeat it August 3, 1736, and that the repetition must have taken place in 1737 at latest—either August 3 or 7—but probably it was in 1736.

63 (p. 473). **The autograph score of the Coffee Cantata** is in the Royal Library at Berlin; the autograph parts in the Royal Court Library (Hofbibliothek) in Berlin. The watermark is M A (see ante, p. 661). The text first occurs in Part III. of Picander's poems, p. 564, the last piece in the volume; so that, as the preface is dated February 18, 1732, it must have been written at latest at the beginning of that year, and as the "schertzhafften Gedichte" in this Vol. are arranged in chronological order we can hardly date it much before. It does not, of course, follow that the music was written in the same year, but the watermark shows that it cannot well be of later date than 1736. If Bach intended it for use in his own home circle, the composition of the music probably followed hard on that of the words, for this was the most flourishing period of Bach's family band, while his sons, Wilhelm Friedemann and Philipp Emanuel, were still living at home. It is certain that the words of the last recitative and finale are not by Picander, for if he had added them in obedience to Bach's wish they would have appeared in the collected edition of his works, 1748; but here also the text ends with Lieschen's second aria (p. 1244).

PHILIPP SPITTA

JOHANN SEBASTIAN BACH

HIS WORK AND INFLUENCE ON THE
MUSIC OF GERMANY, 1685-1750.

TRANSLATED FROM THE GERMAN BY
CLARA BELL
AND
J. A. FULLER-MAITLAND.

IN THREE VOLUMES
VOL. III

LONDON
NOVELLO & CO., LTD.

NEW YORK
DOVER PUBLICATIONS, INC

*Copyright 1951 by
Dover Publications, Inc.
180 Varick St., New York 14, N. Y.*

*This edition has been produced
under special arrangement with
Novello & Company, Ltd.,
London.*

Printed and bound in the United States of America

CONTENTS.

BOOK VI.

THE FINAL PERIOD OF BACH'S LIFE AND WORK.

I.—DISPUTES WITH ERNESTI. THE *COLLEGIUM MUSICUM*.
THE ORIGIN OF THE CONCERT 3

II.—BACH'S MASSES. THE MASS IN B MINOR... 25

III.—THE LATER CHORALE CANTATAS 64

IV.—THE CHORALE COLLECTIONS. THE ECCLESIASTICAL MODES, AND BACH'S RELATIONS TOWARDS THEM 108

V.—WORKS OF THE LAST (LEIPZIG) PERIOD. CONCERTOS. THE LATER WORKS FOR CLAVIER AND ORGAN 135

VI.—BACH'S PRIVATE FRIENDSHIPS AND PUPILS; HIS GENERAL CULTURE. BLINDNESS AND DEATH 222

APPENDIX (A, TO VOL. III.) 279

APPENDIX B 296

MUSICAL SUPPLEMENT 363

INDEX 407

ERRATA 421

BOOK VI.

I.

DISPUTES WITH ERNESTI.—THE *COLLEGIUM MUSICUM*,— THE ORIGIN OF THE *CONCERT*.

GESNER'S successor as rector of the Thomasschule was Johann August Ernesti,[1] who in 1732 had been appointed Conrector. Ernesti, born in 1707, was still very young when he was placed at the head of the school; but he was qualified for the post by his learning and accomplishments, grounded on a thorough knowledge of the authors of antiquity, and by a conspicuous talent for methodical teaching. Under his guidance the school made wonderful progress, and in this respect he was worthy to succeed Gesner, whom he even excelled in his writings by their lucidity, accuracy, and a high and pure standard of Latinity. On the other hand, he lacked the geniality, the sympathetic kindliness, and the breadth of culture by which Gesner's highly successful work had been carried out, as well as his judgment and delicate tact. Ernesti held the place of Rector till 1759, when he accepted a professorial Chair in the University.

Bach, who was now nearly fifty, was at first on excellent terms with his superior, whose father he might very well have been in point of age. His family continued to increase; in 1733 he had already requested Ernesti to be godfather to one of his sons, and he did so again on the birth of his last son, Johann Christian, in September, 1735. But these friendly relations were not destined to be

---

[1] See Vol. II., p. 261.

of long continuance. In the choir composed of the foundation scholars the prefects filled the office of deputy to the cantor, and among them the head prefect held a particularly important place. In 1736 a certain Gottfried Theodor Krause of Herzberg filled this post. He had been expressly enjoined by Bach to keep a strict watch over the smaller boys, and when, in Bach's absence, any disorderly conduct should arise in church, to meet it with due punishment. When Krause found that he could no longer control the ill behaviour of the troop of boys by admonition, and when, on a certain occasion of a wedding, their misconduct had gone too far, he proposed to flog some of the worst. They resisted, and finally had a severer dose of the cane than had been intended. A complaint was laid before the rector, who was furious with the prefect. Krause's previous character was blameless; he was on the point of going to the University, and had taken part in the school speeches of April 20.[2] In spite of this Ernesti condemned him to the ignominious punishment of a public flogging in the presence of the whole school. Bach interposed and took Krause's fault entirely upon himself, but without success. When a second attempt to obtain remission was angrily rejected by the rector, Krause, to elude the disgrace that threatened him, took upon himself to quit the school. His little possessions and singing money, which had gradually accumulated to thirty thalers and which was in the rector's hands, Ernesti withheld, but he had to restore them by order of the Council (dated July 31) whom Krause petitioned for redress.

Bach felt himself aggrieved in the person of his prefect, and the proceeding implanted in his heart a dislike to Ernesti from which evil results were to follow. The school regulations of 1723 describe it as the duty and right of the Cantor to compose the four choirs out of the pupils fitted for each, and select the choir prefects. In this last choice he was to obtain the consent of the Superintendent,

---

[2] See Ernesti's school report for April 20, 1736, p. 15, in the Library of the Thomasschule.

and the rector had to give his approval to the composition of the choirs, though he had no right to any initiative. Traditional custom allowed the cantor even wider control, and left him all but unlimited mastery in all matters relating to the choir; here, as in many other cases, the old practice survived in spite of the new regulations. When Gottfried Krause had refused to submit to the ignominious punishment to which he was condemned, the rector had suspended him from his office, and at the same time had, on his own authority, promoted the second prefect, Johann Gottlob Krause of Grossdeuben[3] to fill his place *pro tem*. But the cantor had for a long time past disapproved of this individual, and had not concealed the fact. On the twenty-second Sunday after Trinity (Nov. 6) of the previous year, Magister Abraham Krügel, of Collmen near Colditz, had been married to the daughter of pastor Wendt of that place. Ernesti and Bach had been invited, and as they were returning home together in the evening the conversation fell on the new appointments to the post of prefect which were always made before the Christmas perambulations. Johann Krause had a right to one from his age and place in the school, but Bach hesitated and said he had always been "a disreputable dog." Ernesti admitted this, but still opined that he could not well be passed over, since he was distinguished for his talents, and seemed to have improved in morals, if only he possessed sufficient musical knowledge. Certain pecuniary benefits were attached to the place of prefect; Krause might be able in this way to release himself from his debts, and thus the school would be spared a discreditable report. In musical matters Bach considered him sufficiently competent, at any rate as a lower prefect, so he was made fourth, third, and finally second prefect; and even his temporary promotion to be first prefect Bach had agreed to, though the rector's independent action annoyed him. However, at the end of a few weeks, he was convinced that Krause was not equal to this responsible and onerous position; he therefore set him down again to be

---

[3] See Vol. II., p. 240

second prefect and gave the more competent Samuel Küttler (or Kittler) of Bellgern[4] the first place, and communicated both nominations in writing to the rector. Ernesti did not like it, but he yielded; not so Krause himself. He complained to the rector, and was referred by him to the cantor. Now Bach's vexation and wrath blazed out. He let himself be provoked into answering that the rector had shoved him into the place of first prefect on his own authority, and he, the cantor, turned him out again to show the rector who was master here. And he repeated this in the rector's room, to his face. Ernesti thought he ought not to submit to such a mode of proceeding, and, fortified by the Superintendent, he required Bach, in writing, to reinstate Krause. Bach must have seen that he had gone too far; he showed himself disposed to an amiable accommodation, and even promised to yield to Ernesti's demands. But, at the very next practice, Krause behaved so badly that this was impossible. On July 20 he went on a journey, and did not return till August 1. Ernesti, who was expecting Krause's reappointment—and it would seem that Bach's conduct justified him in this—became impatient, and when Bach still took no action he wrote to him on Saturday, August 11, a letter categorically stating that if Bach did not at once do what was required of him he himself would re-instate the prefect on Sunday morning early. Bach remained silent. Ernesti carried out his threat and allowed Krause himself to inform Bach of the fact.

It was before Matins; Bach went at once to Deyling, the Superintendent, and laid the matter before him, and Deyling promised, after making enquiries, to do his best to settle the dispute. Meanwhile the service had begun. Bach fetched Küttler, who, as second prefect, had gone by the rector's order to the Church of St. Nicholas, took him into St. Thomas' and turned out Krause in the middle of the hymn, stating, without any grounds, that he was authorised by the Superintendent. Ernesti saw the proceeding, and after church he also told his story to the Superintendent, whom

---

[4] See Vol. II., p. 240.

he won over to his side. This he reported to Bach, who retorted with growing indignation that he would not retract a word in the matter, let it cost what it might, and that he had besides laid a complaint in writing before the Council. Before the beginning of Vespers the rector made his appearance in the organ choir and publicly prohibited the boys, under penalty of the severest punishment, to carry out Bach's orders regarding the prefects. When Bach came and found Krause again in the first prefect's place he once more turned him out with much vehemence; but now, the foundation boys having been intimidated by the rector's threats, there was no one to lead the motett, and Bach's pupil Krebs, who had been at the University since 1735, and who happened to be present, at his master's request undertook to direct it. The result was a second appeal from the Cantor, who felt himself deeply aggrieved with regard both to his authority and his self-respect. When that same evening Küttler came to table, Bach angrily sent him away, because he had obeyed the rector and not the cantor.[5]

On the following Sunday, Aug. 19, these irritating proceedings were repeated; Bach would not allow the prefect nominated by the rector to direct and lead the singing, and not one of the other scholars dared to take his place. Bach had to make up his mind to conduct the motett himself, contrary to custom; and a University student once more led it. On this Bach addressed a third appeal to the Council. He represented that if things went on thus, public scandal and disorder must continue to increase, and that if the Council did not at once look into the matter he would hardly be able to maintain his authority over the scholars. Meanwhile Ernesti was also required to give his account of the affair, in which he ingeniously tried to justify himself and cast all the blame on the cantor. But no interference of the Council followed, notwithstanding the pressing need of it.

Bach now tried other means. His application for a Court title, forwarded to Dresden under the date of July 27, 1733,

---

[5] Bach must have been the School Inspector for these weeks, and must have had to eat with the foundation scholars.

had remained unanswered, the reason no doubt being the political confusion at the beginning of the reign of August III., and his absence from Saxony for nearly two years (Nov. 3, 1734—Aug. 7, 1736). Since then Bach had endeavoured to please the Court by several festival performances, the times were quieter and a repetition of his suit seemed to promise a better issue. His talents as an artist had not been sufficient to secure him from an unworthy, nay, an ignominious position; but certain dignity conferred by the Court might avail to release him from it. On Sept. 27, 1736, he renewed his petition, and since the King was to be in Leipzig for a short time after Sept. 29, it might be conferred on him then and there. But Bach was not even yet immediately successful, and greatly annoyed by this unendurable state of affairs, in November he had drawn up a fourth appeal addressed to the consistory of Leipzig, when the appointment he coveted reached him:—

"*Decret* | Vor Johann Sebastian Bach, as Composer to the King's Court band.

"Whereas His Kingly Majesty of Poland and Serene Highness the Electoral Prince of Saxony has been graciously pleased to grant to Johann Sebastian Bach—at his humble petition presented to His Majesty, and by reason of his good skill—the *Predicate* of Composer to the Court band; this present decree is issued under his Kingly Majesty's most gracious personal signature and Royal Seal. Prepared and given at Dresden, Nov. 19, 1736."

The transmission of the patent was undertaken by the Russian Ambassador to the Royal and Electoral Court, who received it on Nov. 28. His name was Baron von Kayserling, and we shall meet with him again presently.

However, the confidence which Bach had felt in the happy results of his title was disappointed. The contest with Ernesti did not cease, and the Council still showed no signs of intervening. Thus the appeal drawn up in November, but not forwarded, had to be presented on Feb. 12, 1737. Six days previously the Council had indeed made up its mind to issue a letter to accommodate matters, but it lay by for two months, only reaching Ernesti on April 6, Bach on the 10th, and Deyling on the 20th. Nor had the Council taken any particular pains to get at the kernel of the matter in dispute. It selected the simplest issue and pronounced both parties in the wrong; Johann Krause was left first prefect, "his term at the school ending at Easter." Easter fell on April 21, thus the decision had now no practical importance; still Ernesti had conquered so far as Krause was concerned, since Bach was no longer in a position to deprive the incompetent prefect of his post.

He was not disposed, however, to rest content with the decision of the Council. Immediately after Bach's statement of Feb. 12, the Consistory had enjoined the Council and the Superintendent to investigate the matter and to adjust it without delay, as also to arrange that divine service should be performed without hindrance or interruptions. Deyling, however, took Ernesti's part and he certainly would do nothing to give the order in Council a different character to that which it ultimately possessed. Bach's relations to Deyling at that time were of course highly strained, as we may see from the following circumstance. On a certain Wednesday, April 10, in the Church of St. Nicholas, the scholar from St. Thomas' whose duty it was to lead the singing pitched the communion-hymn after the usual sermon by the Superintendent so low that the congregation could not sing with it; but instead of speaking privately to Bach on the subject, as would have befitted its trifling character, Deyling laid a complaint before the Council through the sacristan. The Council at once summoned the cantor, charged him to enquire into the matter, reprove the leader, and for the future to appoint a fit person to do the duty. Thus Bach was once more set down.

But, on Aug. 21, Bach laid the following facts before the Consistory: "The rector had publicly, and in the presence of the assembled first class, threatened with suspension and the loss of all the singing money due to them, each and all of the scholars who should obey Bach's orders." This he regarded as an injury to the position he held and as a personal indignity, for which he was entitled to demand equally personal satisfaction. Moreover, he disputed the authority of the School Regulations of 1723, on which the decision of the Council was based. These lacked the necessary ratification of the Consistory; and for this reason their validity had already been denied by the elder Ernesti, Gesner's predecessor, while, in point of fact, in all that related to the rights of the cantor, not they, but the traditions of the place, had always been acted on. More particularly, he disputed the conclusion that it was not within the cantor's power to suspend or to dismiss a scholar from a post he had once held, "seeing that cases must arise when, *in continenti*, some change must be made, without waiting for a long investigation into trifling details of discipline and school management; and such changes in matters relating to music stood within the cantor's province, since otherwise the lads, if they know that nothing can be done to them, will make it impossible for him to govern, and fulfil his office satisfactorily." In the document of Feb. 12, Bach had contested even the "concurrence" of the rector in the appointment of the prefects, under the new School Regulations. This was so far justifiable that the rector was not allowed any positive share in the selection, but only a veto. Still, Bach must clearly have seen that he could not, under these regulations, attain what he desired—perfect independence as cantor in all matters musical.

On Aug. 28 the Consistory announced to Deyling and the Council that Bach had again applied to them and required them to furnish a report of the affair within fourteen days; but though the report was not sent in, they did not trouble themselves any farther, and Bach saw only one way open to him to secure his rights. This was by a petition addressed directly to the King, to whom he appealed Oct. 18; and on Dec. 17 the King called upon the Consistory to settle the complaint of Bach, after enquiring what was due to him. On Feb. 1, 1738, the document was handed to the Consistory, and on Feb. 5 they once more emphatically demanded of Deyling and the Council a report within fourteen days. At the time of the Easter Fair, the King himself came to Leipzig and, as has already been told, Bach on this occasion performed in his honour an "Abend-Musik," which was received with universal approval. Now, at last, his suit was brought to an issue, which, under the circumstances, we may infer was in every way favourable to him. The sudden lack of all documentary evidence points to the conclusion that this was brought about by the personal intervention of his Majesty.

The struggle had continued for nearly two years—almost as long as the dispute over his uncle's matrimonial difficulty in Arnstadt sixty years previously.[6] Both had shown themselves to be genuine sons of their race, and had fought with the utmost determination for their rights. It is not, however, on this account alone that these unsatisfactory proceedings have here been narrated at full length.[7] They were, in fact, of decisive importance to a whole section of Bach's existence. I regard it as a specially happy circumstance that I can point this out and treat it as the background of my picture of the last twelve years of his life. Johann Friedrich Köhler, Pastor at Taucha, near Leipzig, in 1776 planned a history of the Schools of Leipzig,[8] in which he gave due importance to Bach. His knowledge of that period he derived for the most part from the oral statements of past scholars of St. Thomas'. He speaks as follows:—
"Bach altogether fell out with Ernesti. The occasion was this: Ernesti turned off the head prefect Krause, who had too severely punished a lower scholar. He dismissed him from the school, as he had chosen to leave it, and put in his place another scholar to be chief prefect—a right properly pertaining to the cantor, whose deputy the chief prefect must be. The person chosen being unfit to lead the church music, Bach selected another. Thus differences arose between him and Ernesti, and from that time they were enemies. Bach now began to hate the scholars who devoted themselves to the *humanities* and pursued music

---

[6] See Vol. I., p. 162.
[7] The original documents are inserted at full length in App. B. to Vol. II. of the German edition.
[8] The MS., which was never printed, is in the Public Royal Library at Dresden.

only as a secondary study; and Ernesti was a foe to music. If he came across a scholar who was practising on an instrument, he would say: 'Do you want to be a beer-fiddler?' By his importance in the eyes of Burgomaster Stieglitz, he succeeded in procuring that he, like his predecessor Gesner, should be excused from the duty of inspection, which was handed over to the four under-masters. Now, when it came to cantor Bach's turn, he referred to the example of Ernesti, and came neither to table nor to prayers; and this omission had the most adverse influence on the moral training of the scholars. Since that time, though many persons have held both places, there has been but little harmony between the rector and the cantor."

This quotation gives a clear hint as to the side on which the public opinion was ranged in this contest between the two officials. Bach's vehement nature carried him into various indiscretions; still, on the main point, he was in the right. He would have been still in the right, even if he had had less sound ground for appealing to the old tradition, which had never yet been disputed or abrogated. He could not possibly fulfil his office as he himself wished and as he was required to do, unless he was allowed to manage all the affairs of the choir as he judged proper. This Ernesti must have seen, even if the greatness of his antagonist had never dawned upon his mind, nor the absurdity of employing a musician like Bach to teach a parcel of schoolboys. But though he gave himself an air of treating the question from the standpoint of discipline, he nevertheless presumed to try to prove what none but a musician could decide—that Krause was not incompetent to be head prefect; and when Bach asserted that he could not direct the music, he appealed to the opinion of the scholars. Judgment must be pronounced altogether against Ernesti if we only compare the tone and feeling of the documents on either side. Bach's language is stern and sharp, and strictly to the point; and in his numerous utterances not a word is to be found personal to his opponent. Ernesti proceeds very differently. He not only takes the opportunity of denouncing Bach to the Council in general terms as a negligent official, who was, properly speaking, alone guilty of the misfortunes of the "unfortunate" Gottfried Krause, and as a haughty musician who thought it "beneath him" to direct a simple chorale; he accused him of never having given the prefect a lesson or rehearsal in conducting, so as to lead him into a snare; and he charges him with "a lie," because Bach only mentions the appointment of Johann Krause as prefect in the New Church, and does not allude to his former place as fourth prefect in the New Year's singing. He does not hesitate even to represent Bach as corruptible by bribery, and says "he could adduce yet other evidence that Bach's testimony is not always to be depended on, and that he, for his part, would sooner make a discantist (treble singer) out of an old specie thaler than out of this boy, who was no more fit for the place than he himself was." Such accusations as these no man should utter without proof on the spot, and in default of this he is a slanderer.

The evil results of the contest must therefore be laid, for the most part, to Ernesti. All that the Thomasschule owes to him, as an educational institution, must be remembered to his honour, but he destroyed the harmonious co-operation of the different teachers, which had been Gesner's happy achievement. The most nationally German of all the arts was looked upon no longer as a means of culture, but as a ground of contention. And not by him alone; for the rest of the professors, some of whom had been Ernesti's pupils, followed their rector's lead, and thus Bach fell more and more into an isolated and doubtful position. This he bequeathed to several of his successors, and from Ernesti in the same way a dull and pretentious aversion to music was handed down to his; thus a reciprocal dislike became hereditary on both sides. Bach, to a certain extent excluded from the school, now fixed his attention more resolutely than ever on his independent musical occupations. Even before this he had avoided appearing publicly as cantor, certainly not out of petty conceit, and now he thought of himself chiefly as composer to the King and director of music to the princely courts of Weissenfels and Cöthen[9]—appointments which gave him occupation as a musician at a distance, and without interfering with the amount of freedom he desired. His fellow officials at Leipzig must have felt this very strongly, and that they should not have been pleased at it is quite intelligible. A certain ill-feeling towards him survived among them even after his death, for at a meeting of the Council of August 7, 1750, it was announced, with a touch of irony, that the " Cantor—or rather the Capell-director—Bach was dead"; and as regards a new appointment to the post, one of the Council was of opinion that "the school needed a cantor, and not a Capellmeister, though, of course, he must understand music."

However, it is but justice to point out that more occult influences were at work in the division between the scholar and the artist. By the beginning of the eighteenth century music had reached a stage of development which made its close connection with the school an impossibility; the new vitality in which even the Protestant schools of the time were beginning to expand was unequal to the task of reconciling this difference; it stood in direct opposition to the vigorous growth which in music was struggling towards daylight. For just as music is, in history, the youngest of the arts, so in the course of each separate period it comes last in the train. The musical art of the eighteenth century has its foundations in the sixteenth—the era of the renaissance; but it did not flourish till all the other grand and splendid phenomena of the period had bloomed and faded—like a last ethereal and glorified reflection of their dying light. Indeed, it is only thus that it seems possible—even with the widest allowance for the happy characteristics of the German nation—that music should have survived uninjured the ruin of the thirty years' war. It had remained in a slowly rising, yet irrepressible, condition of development, which was destined to swell to a triumph only in the following century; while

---

[9] Königlicher Hofcomponist, Capellmeister der Fürstenhöfe Weissenfels und Cöthen, are his titles.

science and poetry—more nearly allied to it—began in this very century a course towards new and various ends. Different objects were desired, theories gradually ceased to be taken for granted, and at last our very greatest poets and most learned men regarded music with indifference or contempt. It was the beginning of this divergence that we may detect in the quarrel between Ernesti and Bach.

But, irrespective of this, any permanent reciprocal co-operation between music and learning was no longer to be hoped for; music was growing so mightily both in height and breadth that it must necessarily break through the narrow bounds of the schools if not transplanted into freer soil. Bach was by no means the only artist who had run hard at the fence. As much as thirty years before the relations between a rector and cantor had been the subject of a violent dispute and of extensive discussion. About the year 1703 the rector of the school at Halberstadt, on the occasion of the interment of a distinguished Prussian official, had forbidden the whole body of choral scholars, in the open street, and under threats of expulsion from both the choir and the school, to obey the orders of the cantor, although he had previously desired them to co-operate with him, as was customary. On this, Johann Philipp Bendeler, cantor at Quedlinburg, took occasion to define in a searching discussion the limits of the interdependence of the rector and cantor.[10] In matters musical the pre-eminence rests with the cantor; the rector is no doubt the head of the school, but what has the church music to do with the school? What can it matter to the cantor to whom his assistants are subordinate, except as regards music? As to the appointment of the prefects, the rector and cantor must come to an understanding, but the rector must not be able to appoint any prefect against the will of the cantor. "When I am required to perform a

---

[10] "*Directorium | musicum, | oder | Gründl. Erörterung | Dererjenigen | Streit-Fragen, | Welche bisshero hin und wieder zwischen | denen Schul-Rectoribus* und *Cantoribus* über dem *Directorio Musico movir*et | worden, &c. | Von | Joh. Phil. Bendeler, *Cant.* | zu Quedlinb. | Gedruckt im Jahr 1706. Royal Library at Berlin.

musical *Actus* I announce the fact to the rector and order my class, and then enjoin the scholars each to excuse himself properly for his absence to his master." The cantor is to have the power of excluding any scholar who is unavailable for musical purposes from the church music, from the choir, and from the emolument attached to it; nor shall he be bound previously to consult with the superiors or with the rector. "Any one who deprives the cantor of this power will become the cause of many sins, and treats the cantor not much better than if he turned him out, with his hands tied, among a swarm of bees, humble-bees, and hornets." "When the scholars ought to come they stay away, when they ought to stay they run away, and so on. And in order that they may be the safer they insinuate themselves in all sorts of ways into the good graces of the rector, slander the cantor, and so forth; whereby the rector is the more zealous to support them, but the cantor the more moved to think of his own safety." The close of the treatise consists of three judgments on the subject, which all pronounce in favour of Bendeler's views; they are derived from the Universities of Halle and Helmstädt, and from the Court of Assessors of the Electorate of Saxony at Leipzig. It would be easy, as we read this dissertation, to fancy that it proceeds neither from Halberstadt nor from Quedlinburg, and that it was Bach—not Bendeler—who wrote it.

In fact, there was only too good reason for such squabbles, and a similar one took place at Freiburg during Bach's lifetime. Music was forcing its way to freedom—the freedom of the concert. But, in Germany, the time was not yet come for allowing it such freedom. Bach still stands on the old ground, but already a breath of air from the new land is wafted towards him, and shows its influence on his art. How it pervades his compositions has been shown again and again. Now we shall be able to detect it even in the circumstances of his outward life.

What I have here said is not to be understood as meaning that Bach's art lacked in anything requisite to secure its perennial vitality, or that its full thriving and bloom were checked by any narrowness of its surroundings, particularly in his later years. Once more must it be repeated, the whole course of his life was throughout and thoroughly favourable to him; the difficulties were never greater than his genius could surmount without injury. When the turn came in Bach's affairs, which is marked by his quarrel with the school authorities, he had already reached the highest mark. As we glance over the whole of his accomplished work we never have the sense of anything having been left unfinished. We cannot conceive of Bach in circumstances such as those by which Handel was surrounded, and his creations promulgated, in London. Bach had developed early and quickly, and naturally came early to a standstill; at the period when he was writing his Passions, the Christmas Oratorio, and the first two numbers of the B minor Mass, Handel was still far from his goal; and when Handel was in fact just beginning Bach was ceasing to write. Above all, however, it must be explained that Bach took no direct part in the revolution which gradually made its way in the social life of Leipzig during the fifth decade of the century; although this very revolution was destined to bring about a change which is indicated in many ways in Bach's works—*i.e.*, the neglect of the Church as the chief centre of music, and the gradual evolution of the independent public *Concert*.

The natural and obvious centre for this was offered by the well-known *Collegia Musica*. At an earlier period these had consisted merely of weekly meetings of musicians of repute, and in this form were tolerably general, at any rate in Saxony. Their objects were, as Kuhnau had once said, "constant practice and improvement in a noble art, and to establish side by side with a pleasing harmony of sounds an equal harmony and agreement of minds, which among people of this stamp was too often lacking."[11] Kuhnau himself, in 1688, was member of such a *Collegium Musicum* in Leipzig.

Then, in the beginning of the eighteenth century, we find the first *Musik-vereine*, musical associations of students,

---

[11] Kuhnau, Der Musicalische Quack-Salber. 1700, p. 12.

III.

notably that founded in 1704 by Telemann, which was subsequently under the direction of Bach himself, and which became a very important institution. The endeavour to extend the breadth and freedom of musical practice by their means is conspicuous, particularly as these societies included not performers only but listeners as well; still they were for the most part confined to the students of the Universities, and it was not till after 1740 that we find the citizens caught in the current. In 1741 Zehmisch, a member of the worshipful class of merchants, undertook to form a new society for giving concerts.[12] But the preparatory arrangements were long in hand, possibly in consequence of the first war with Schleswig then in progress; it was not till two years later that the society was formally called into existence. "On March 11, 1743, the grand *Concert* was founded by sixteen persons, both nobles and citizens; each person being required to pay annually for its support the sum of twenty thalers—that is to say, one Louis d'or per quarter; the number of performers was likewise sixteen selected persons, and the Concert was given at first at the house of Herr Bergrath Schwaben in the Grimmische Gasse, but four weeks later, as these quarters proved to be too small, at that of Herr Gleditzsch, the bookseller." The enterprise at once achieved such a brilliant success that on March 9, 1744, its first anniversary was celebrated with a solemn festal cantata.[13] In the Leipzig address book or directory of 1746 and 1747 it is thus mentioned among the permanent musical institutions: "(3) on Thursdays a *Collegium musicum*, under the direction of the worshipful company of merchants and other persons, is held from five to eight o'clock at the Three Swans in

---

[12] I am unable to give any more exact information concerning this person who was so important to the advance of Concert music in Leipzig; I find, at least, three of this name at that time, two merchants and one *Doctor juris utriusque*.

[13] I derive this information from the programme of the Gewandhaus concert of March 9, 1843, a centenary celebration of this foundation. The authority there given is the "*Continuatio Annalium Lips.* VOGELII, *anno* 1743." This work has not come to light again, however, there is no doubt of the accuracy of the programme.

the Brühl, where the greatest masters, when they come hither, are wont to perform; they are fashionably frequented, and admired with much attention."

This transfer from a private house to a public situation is sufficient evidence of the rapid progress made by the new society. It is, in fact, the same which still exists as the "Gewandhaus Concert," and the very day of the performance remains unaltered, though the hours are slightly changed. An interruption occurred during the seven years' war, but the concerts were resumed in 1763 under the conduct of Joh. Adam Hiller, and again given at the Three Swans. Zehmisch was still at the head of the society; in 1768 it was said of him "Who will not, even in time to come, laud that zeal for the advancement of music which, for the last seven and twenty years we owe to our friend, Herr Zehmisch? His indefatigable efforts have given to the Leipzig Concert a character which has not merely earned for it the unflattering praises of many foreign professors and connoisseurs, but has even won it the honour, three times already, of the most gracious and illustrious Presence (of the king or princes as audience) in which honour every member of the society has his share."[14]

The interest in music thus suddenly aroused in the citizens of Leipzig found expression in other ways. For a long period fifty gülden (= 43 thlrs. 18 gr.) a year had been paid as a *beneficium* to the foundation scholars of St. Thomas', out of the funds of the church of St. Nicholas. From 1746 the Council caused a similar sum to be paid out of the treasury of St. Thomas' and twenty-five thlrs. out of that of the New Church for the maintenance of the Thomasschule, and the advancement of church music.[15] This source of public support, which had hitherto only yielded driblets in the cause of music, suddenly began to flow more freely; in 1745 Bach could venture on his own account to appoint his pupil Altnikol as bass in the church choir. When he presented himself to receive his salary, May 19,

---

[14] From a historical explanation of pictures in the collection of Gottfried Winkler at Leipzig (Hist. Erklärung der Gemälde, &c., Leipzig, Breitkopf, 1768).
[15] Accounts of these churches from 1747.

1747, one of the members of the Council expressed his opinion that for the future Bach ought to give notice of such a step; however, the salary was paid—six thlrs., for assisting as bass singer in the two principal churches, from Michaelmas, 1745, till May 19, 1747.

Under this new aspect of affairs the old students' musical unions lost their importance. Görner, to be sure, held his society together with characteristic tenacity, but Telemann's was visibly dwindling. Bach gave up the direction of it; in 1746 it was conducted by Gerlach; in 1747 by Johann Trier, a student of theology, who must have kept up the society till Bach's death, for he was a good musician, and did not quit Leipzig till 1754. Thus the direction, after having remained for a long period in the hands of famous artists, once more fell into those of the students, as at its beginning, and such a relapse is always a sign of decay. Why Bach withdrew is not known, only that he did so at some time between 1736 and 1746. We may safely guess that he felt no satisfaction in remaining at the head of a musical union which had now fallen into the second rank; however, we find no trace of his having taken any part in the newly-formed concert union of the citizens. His supreme superiority and great fame would have entitled him to no less a position than that of conductor if he had been concerned in it at all; and, as he did not fill this post, we may safely assume that he had—or desired to have—no influence in its formation. It was announced that the greatest "foreign" (to Leipzig) masters would perform at the concerts of the new society, but the greatest master of all, a resident in the town, was not mentioned. In 1744, on the first anniversary of its formation, a grand cantata was performed, but it was not Bach who composed it, but a young student of twenty-eight, Johann Friedrich Doles, who had been living in Leipzig since 1738, and who must have known Bach well, but who followed a quite different path of art, more pleasing to the modern taste. Five years after Bach's death Doles was appointed to his place, and held it till 1789. He, and with him Joh. Adam Hiller, forsook the lines of Bach; and, so far, it is a striking circumstance that his name should have been connected from the first with the "great Concert Union." The aims for which this society strove had found their forecast here and there in Bach's life and works, but they had nothing in common with the essential character of the master.

Bach's illustrious position was, however, firmly rooted in the mind of the inhabitants, and nothing could now shake it. He was the glory of their city; no musician of repute ever visited it without paying his respects to Bach. Pupils streamed to and fro, and to be received by him was a coveted honour. He was always regarded as the first authority on organ building; in 1744 he was required to test the new organ built by Joh. Scheibe in the church of St. John. Although Scheibe's son, the author of the "Kritischer Musicus," had incurred Bach's displeasure, the Cantor was thought impartial enough to try the organ, and he pronounced it faultless; though Agricola frankly says that he put it to the severest tests that any organ had perhaps ever undergone.[16] Two years later Scheibe completed another organ, under a contract, for 500 thlrs., at Zschortau, near Delitzsch, and here Bach again was called in to test it, in the beginning of August, 1746.[17]

Some of Bach's secular compositions seem to have become popular and to have long kept a hold upon the people. In a description of the *Kirmess*[18] at Eutritzsch, near Leipzig, in 1783, we are told "the band of musicians strikes up bravely, beginning with sonatas by Bach and ending with ballads."[19] By this must be meant portions out of orchestra-partitas, for in the parlance of the town and tower musicians, even in the nineteenth century, the word *sonata* retained its original meaning of a single instrumental piece in several parts.[20] At any rate, this notice proves that Bach's name

---

[16] Adlung *Mus. Mech.*, p. 251.

[17] Bach's certificate was, in 1872, in the possession of Herr Clauss (General consul) in Leipzig. Bach received 5 thlrs. 12 gr. for his pains.

[18] The Saint's day of the parish church; a festival known as the "wake" in some parts of England.

[19] *Tableau* von Leipzig im Jahre, 1783. 1784.

[20] "In the performances on wind instruments on the tower, besides the simple chorales, *Sonatas* composed expressly were given on the widely

survived among the people. When the concert-room of the Gewandhaus was finished, in 1781, the painted ceiling (by Oeser) showed the older music expelled by modern music, which took its place, and the art triumphant was represented by a genius with a scroll, on which was inscribed only the name of Bach; Forkel observes this was the highest panegyric he could receive,[21] and, at any rate, this was the intention; even that generation, which understood the master's mind less than any before or after it, would not deny him the highest rank and praise, and across a blank gulf of thirty years the mighty name sounded out to command veneration and stir a worthy pride.

Still, even before his death, he had ceased to be the heart of musical vitality in Leipzig; duly contemplating all the facts, we are forced to conclude that, though still admired, he had ceased to be understood or loved. His fate has this in common with that of Beethoven, who, in his later years, in spite of the undiminished respect of the Vienna public, gradually lost his popularity under the foreign influence of Rossini. But Bach could look down calmly enough on all the bustle of the young world at his feet.

A society for musical science (Societät der Musikalischen Wissenschaften), founded in Leipzig in 1738, made itself a good deal talked about. Its promoter and centre was Lorenz Christoph Mizler, born July 25, 1711, at Wurtemberg; he had been educated at the Gymnasium at Anspach, and subsequently, with a brief interruption, was a student at Leipzig from April 30, 1731, till 1734. He had been a diligent musician from his boyhood, and at Leipzig he was under Bach's personal teaching—probably through Gesner's good offices—in clavier playing and composition. In 1734 he took the degree of *Magister*, and on June 30 disputed

---

resounding instruments, which all the inhabitants of a village or of a whole town might hear." Ch. C. Rolle, Neue Wahrnehmungen zur Aufnahme der Musik, Berlin, 1784. Friedrich Schneider, a native of Saxony (1786-1853) in his youth composed twelve so-called *Thurmsonaten*, for two trumpets and three trombones. See Kempe, Friedrich Schneider, als Mensch und Künstler. Dessau, 1859.

[21] Musikalische Almanach für Deutschland, 1783. Leipzig: Schwickert.

publicly on a dissertation, "*Quod musica ars sit pars eruditionis philosophicæ.*" He dedicated his essay to four musicians: Mattheson, Bach, Bümler, and Ehrmann—the last, a native of Anspach, had grounded him in the elements of musical knowledge. In his address, dated June 28, he says: "I have derived great profit, most famous Bach, from your instructions in the practice of music, and lament that I can no longer enjoy them."[22] He was in fact leaving Leipzig, and soon addressed himself to other studies in Wittenberg. His dissertation met with a favourable reception from those to whom he had, in the first instance, dedicated it, and this must have been one of his reasons for returning to Leipzig, where, from 1736, he gave a series of lectures on Mathematics, Philosophy, and Music, and began to bring out a critical monthly journal entitled "Neu eröffnete musikalische Bibliothek." Mizler had a friend and patron in the person of Count Lucchesini, a man of musical culture, who was captain of the Sehr regiment of Cuirassiers in service of the Emperor Charles VI., and died a soldier's death in 1739. When, during the Polish war of succession in 1735, the Emperor was hardly pressed by France and her allies, Mizler dedicated to the Count a little facetious Latin treatise, in which he represents the course of the war under the figure of the concord and discord of various musical notes.[23] Lucchesini was the first to assist him in founding the Society, while of the four musical patrons to whom he had dedicated his treatise only one supported him, old Bümler of Anspach. Mattheson was already on bad terms with Mizler, since he found himself "scoffed at in a covert manner" in the latter's Musikalische Bibliothek, and he soon became his implacable opponent. To the real aim of the

---

[22] A second edition of his essay appeared in 1736, with a new preface and under the title, "*Dissertatio quod musica scientia sit et pars eruditionis philosophicæ.*"

[23] "*Lusus ingenii de præsenti bello augustissimi atque invictissimi imperatoris Caroli VI. cum fœderatis hostibus ope tonorum musicorum illustrato.*" The key note C stands for France, the fifth, G, for Spain, the third, E, for Sardinia, the octave C for the Emperor Charles. The first three endeavour by deviations from the original key to lower the octave from C to B; they cannot succeed however; by the help of England, A, they are forced to return to their allegiance.

Society, which was to advance the science of music and reduce it to a system, Mattheson was quite indifferent; all his life through he had taught and acted on his own principles only, and had got on very well. What should he care if the Society should propound the question as to why consecutive fifths or octaves were incorrect? In his treatise on figured bass playing, he says: "Two fifths or two octaves must not occur in succession, for not only is this a fault but it sounds badly," and says no more. Or if Mizler tries to connect thorough-bass with mathematics? "In figured basses the left hand plays the prescribed notes, the right adds concords and discords in such wise that a well sounding harmony may be produced to the honour of God, and the permissible diversion of the mind. Where due heed is not paid to this there is no true music, only a diabolical clang and clatter." So Mattheson dismisses the subject. It was a well known fact that Bach never let the connection between mathematics and music worry him for an instant; Mattheson observes: "He (Bach) certainly and positively no more showed him (Mizler) these hypothetical mathematical principles of composition than the other master (meaning Mattheson himself); that I can warrant."[24] Even the writers of the Necrology state plainly: "Bach never went into a deep theoretical study of music."

Unions such as this Society was intended to be, might have their uses, but their possible results are of value only to the average mind. To the soul of genius they say too much or too little. Besides this, Bach must have been averse to Mizler's personality and projects; Mizler had had a wide education and even, for a man of his position, had sound practical views of music; his industry was untiring, and he knew how to put a good face on things in the eyes of the world. But he was vain and a swaggerer, and, at the best, but a barren soul; the compositions which he was so rash as to publish could only excite a compassionate smile in any true musician. Thus it is easy to understand that Bach should have held aloof from this musical society.

---

[24] Ehrenpforte, p. 231, note.

Mizler keenly felt the implied slur thus cast upon his undertaking. In the bye-laws of the Society, to be sure, it was stated that mere practical musicians[25] could find no place in it, since they were in no position to do anything towards the advancement or extension of scientific music. But men of Bach's stamp were not included in this category; for musicians who were strictly speaking practical, as Telemann and Stölzel, were members of it within two years of its being founded, and Handel—practical if anything—was unanimously elected an honorary member in 1745. Besides, Bach, though he had never written a treatise, was well known as a learned composer. Mizler, who by 1743 was no longer at Leipzig but living with Count Malachowski in Poland, took, in fact, great pains to induce Bach to join; in 1746 he proudly announces that it is possible the Society may soon be increased by the addition of three illustrious members; these were Graun, Bach, and Sorge. Graun in fact joined it in July of that year; Bach was not in such haste, he waited till June, 1747. When once he was a member, however, he fulfilled his duties; he composed for the Society a triple canon in six parts, and the variations upon " Vom Himmel hoch." After his death it was asserted that he would undoubtedly have done much more if the short period during which he was a member—only three years—had not prevented it. This may have been more than mere conjecture, for in later years Bach's proclivity for profound and introspective musical problems grew stronger than ever, and in this Society he would have found for these a small but very competent public.

II.

BACH'S MASSES.—THE MASS IN B MINOR.

In the foreground of a picture of Bach's later labours as a composer stand his Latin masses. Of these he wrote five,

---

[25] In the Musikalischer Staarstecher, Leipzig, 1740 (*The Musical Oculist!* to cure the blind in matters musical), Mizler designates as " practical musicians " those who only sing or play and do not compose; this is probably how it was understood in the laws of the Society.

and the first question must be whether, and in what way, these stood in any connection with the Protestant form of divine service.

The form of worship in the principal churches of Leipzig had remained nearly allied to that of the Catholics, as was also the case in various other places, and I have already fully discussed the subject (Vol. II., p. 263). With the Latin hymns and responses and Latin motetts, the *Magnificat* was also sung in Latin at Vespers on the three great festivals, and we have seen how Bach thus found the opportunity for composing one of his most important works (Vol. II. p. 369). In this way the principal portions of the choral Latin mass maintained their existence and actually the very same places in the service that they had always held: partly, indeed, the German texts that replaced them did the same. However, the *Kyrie, Gloria*, and *Credo* only were retained for all Sundays and Holy days, and even these were not always sung all through. The *Sanctus* was introduced only on the great festivals, and we have no evidence as to the *Agnus Dei;* it does not seem to have been used in the choral form.

The case is different as regards the figured treatment of the sentences of the mass, which was never used in a connected form in the divine service as performed at Leipzig. The introduction of the Protestant Church Cantata had partly displaced and throughout restricted all other part music. Indeed, there was not even space left to perform the whole of the shorter mass, *Kyrie,* and *Gloria*. From the account previously given of the forms of worship at Leipzig we can only find certain evidence of the use of the *Kyrie* on the first day of Advent and the Reformation festival, and of the *Sanctus* on the three great festivals. If any other portion of the mass was sung, it must have been an occasional performance. Christmas offered an obvious opportunity for the use of the *Gloria*; it then took the place of the cantata, and Bach did actually make use, for a Christmas performance, of the *Gloria* from his B minor mass in a somewhat abridged form. If we desire to acquire any further information as to occasional use of parts of the mass, we must derive it from a time later than Bach's; this can be done with considerable certainty. During the whole eighteenth century the tendency was towards the limitation or even elimination of the Latin portions of the liturgy. What the Town Council began in 1702 was still being carried on by Johann Adam Hiller in 1791, although by that time the existence of the Latin hymns or canticles had left no trace behind.[26] It is very certain that everything that remained in use at a later date must have been usual in Bach's time, and probably much more. Thus, even at a late period, we find the *Agnus* sung at the communion on the feast of the Visitation, and the *Gloria* after the motett at evening service on the same day. In the church of the University—which may be cited on this occasion because its service grew after and out of those at St. Nicholas and St. Thomas—during the ecclesiastical year from 1779-1780—the following parts of the mass were sung: on Christmas-day the whole *Gloria* from a mass by Gassmann; on the feast of Epiphany the *Sanctus, Benedictus*, and *Agnus* from a mass by Haydn; on Jubilate Sunday a *Gloria* by Hasse; on Trinity Sunday a *Credo* by Haydn; on the second Sunday during the fair a *Gloria* by Graun. Thus every part of the mass has its representative excepting the *Kyrie*.[27]

Though in the liturgy only a few portions of the mass were prescribed to be performed in the ornate style, still even this kept up a certain sense of connection with the mass as a whole, particularly as it still appeared almost entire in its original chanted form in public worship. Both artistic and religious reasons tended to allow its survival, at any rate, as the *Missa brevis*. Nor could the composition of a solitary *Kyrie* satisfy the soul of the Christian composer; for its melancholy sentiment craved the relief of a happy contrast to follow, the feeling of guilt demanded absolution.[28] Since on the first Sunday in Advent a *Kyrie* in the elaborate

---

[26] In the preface to Hiller's Four-part Latin and German choral hymns. Part I., Leipzig, 1791.

[27] Collection of texts for church music in Leipzig in the Royal Library at Berlin.

[28] They were always spoken of as *Kyrie cum Gloria*, not *et;* this indicated that the *Gloria* was regarded as supplementary to the *Kyrie*.

style was directed to be sung, and afterwards at Christmas, the *Gloria in excelsis* supplied and expressed the leading sentiment of the festival, it was very natural that the composer should refer the one to the other and unite them in a single composition. And accordingly, we find that Bach's predecessors, Knüpffer and Kuhnau, composed the " short mass."[29] So also did Görner and Hoffmann, who for a time was the director of Telemann's musical union; as Hoffman instituted performances in the New Church, though they took place only on the high festivals and during the fair time, we may assume that the *Kyrie* and the *Gloria* were sung in succession during one and the same service. However, something else must have contributed in Bach's case to urge him to the composition of a mass, and to give rise in the first instance to several works of his of this class. This was, in fact, his connection with Dresden, and the interest he thus acquired in Italian and in Catholic church music generally; finally, too, his duty, as composer to the Royal and electoral court, of producing some works for the King from time to time. Bach's liking for Lotti's music has already been mentioned. Besides a mass in G minor by this master which Bach copied, at about the middle of the Leipzig period, almost wholly with his own hand,[30] there is also extant a mass in G major, in the Italian style for two choirs (and another choir to supplement them) and with instruments, of which Bach also wrote out the whole of the first twelve pages at some time about 1738. In this we can perhaps recognise the hand of Lotti, though it is unusually diatonic in style for this master's work: at any rate, it is the composition of an Italian, or of a German who has wholly adopted the Italian style.[31] But going farther

---

[29] Two masses by Knüpffer are mentioned in Breitkopf's list for the New Year 1764, and one by Kuhnau in his list for Easter, 1769.

[30] The watermark in the paper, M A, indicates a period between 1727 and 1736.

[31] This MS. is in the Library at Berlin. The title is much worn and has almost disappeared, still *d L* can be traced (*di Lotti ?*). The watermarks agree with those of the MS. of the Easter oratorio, which gives an approximate date for the writing of it; see App. A. to Vol. II., No. 58.—Schicht erroneously calls the work a mass by Joh. Seb. Bach, published by Breitkopf and Härtel.

back than Lotti, Bach also devoted much attention to
Palestrina, and copied out a grand mass of his in parts
for the singers and the supporting instruments, with
a figured bass.[32] He also wrote out a short mass in
C minor, in full score, by an unknown Italian composer of
his time.[33] A *Magnificat* by Caldara, in C major, exists in
Bach's handwriting[34] and one by Zelenka, in D major,
in that of his son Wilhelm Friedemann.[35] A number of
other settings of the *Magnificat* by anonymous but certainly
Italian composers exist, partly in Bach's writing and in
score, partly in separate parts, which are most of them
written by Anna Magdalena Bach, and supplemented here
and there by additions in her husband's hand.[36]

From this we may conclude that he not merely studied
these works, but had them performed. It even appears
that he introduced into them subjects of his own composition. We have already seen that he extended his
own grand *Magnificat* by inserting four Christmas hymns.
He did the same with a *Magnificat* in D major by another
hand; but, with the exception of "Freut euch und jubilirt,"
the hymns are inserted in different places. In the C minor
mass just alluded to, the *Christe eleison*—a short but very
artistically-written duet with *Basso quasi ostinato* — is his
work. He also collected Latin masses or portions of the
mass by other composers; for instance, a mass by Wilderer
(Capellmeister to the Elector Palatine), and various others
which were long erroneously supposed to be his own compositions.[87]

---

[32] MS. in the Berlin Library.
[33] MS. in the possession of Messrs. Breitkopf and Härtel, Leipzig.
[34] Berlin Library.
[35] Thomasschule Library.
[36] Berlin Library.
[87] Rust has given a list of these in the preface to B.-G. XI.[1] We must,
however, eliminate from this catalogue the mass in E minor (No. 3)—a work
by Nikolaus Bach which is not in Bach's writing. (See Vol. I. p. 132 and note
293, on p. 574). Breitkopf's List for Easter, 1769, like that for Michaelmas 1761,
mentions, on pp. 12 and 13, six masses by Seb. Bach, of which four are
certainly not genuine. The first must be No. 5 of Rust's Syllabus (C minor
mass, with the interpolated *Christe eleison*); the fourth No. 4 of Rust; the
fifth No. 10 (G major mass for three choruses); the sixth No. 3.

From all this, we can plainly see that Bach directed his attention to Catholic, and more particularly Italian, Church music; and this is all the more noteworthy because it was not till the period of his own ripest maturity, when he no longer had any imperative and practical necessity for studying such pieces. From the time and style of their composition, it seems probable that his own masses were for the Court of Dresden, even when we do not know it for certain, as we do with regard to the two first numbers of the B minor mass. Of the four shorter masses, two were written about 1737, when Bach had just been appointed Court composer; and we must examine them all in detail.

Bach's shorter masses are in G major, G minor, A major, and F major. No chronology of them can be given, since all that is certain is that all four were written after 1730, and the first and third about 1737. I may, however, express my own conviction that Bach composed them all within a short time; and I arrange them in accordance with certain internal evidence. The masses in G major and G minor were not new compositions—they consist entirely of portions of cantatas written previously.[38] For this, however, various re-arrangements were necessary, which—as always with Bach—are highly instructive, and in many respects admirable. Even without any direct reason arising from the fresh text and purpose, he has in many cases given the compositions a richer and freer form; but equally unmistakable is the violence he has often done his own creations by converting them into portions of the mass. There are among these remodelled pieces some which are elevated by the process and severed from a connection with some less dignified theme; and this commonly occurs when Bach transfers a composition from a secular to a sacred purpose. There are also re-arrangements which work back to the original germ of the idea, and under

---

[38] This has already been pointed out by Mosewius (J. S. Bach in seinen Kirchen-Cantaten und Choralgesängen, p. 11), and after him by M Hauptmann in his preface to Vol. VIII. of the B.-G., which contains the masses, as far as the final chorus of the G major mass, which is a re-arrangement of the opening chorus of the cantata "Wer Dank opfert, der preiset mich."

the new conditions give it quite a new form. Finally, there are some which are only a vivid reproduction of a piece; and just as a finished composition may differ each time it is repeated, varying with the character of the performers and the feeling, time, place, and surroundings at the moment, so it has happened that Bach makes a composition serve with different effect, though with but slight alteration, under different conditions of feeling. All these modes of treatment have artistic justification, but none of them have been used in the masses under discussion, which, so far as was possible to Bach, are mere mechanical arrangements. To see with what relentless objectiveness Bach could sacrifice the noble proportions of his compositions, we need only compare the *Gloria* of the G major mass with its prototype. The tremendous opening chorus of the cantata "Herr deine Augen sehen nach dem Glauben" is also gravely injured, though only in details, when we find it forced into the mould of a *Kyrie* for the G minor mass. Other pieces have suffered less reckless treatment, but no artistic purpose in their transformation is anywhere to be detected; and even a superficial comparison must result in favour of the cantata forms. There each piece seems to have sprung from a living inspiration. It corresponds to the poetical purpose, and adequately fills its place as part of a whole; but here each gorgeous blossom is severed from the stem and bound in an ill-assorted nosegay. In the G minor mass Bach has not even regarded that necessary contrast between the *Kyrie* and the *Gloria* which, being based on the nature of the words, had already become typical. The *Gloria* does not stand out in radiant contrast of Christmas glory after the passionate and agitated *Kyrie*, but, on the contrary, continues the same strain of sad and unfulfilled longing. Even the closing chorus, though impressive, retains the same gloomy solemnity.

It is at once evident that Bach cannot have written the G major and G minor masses for his churches at Leipzig. As the chanted mass as a whole had no place in the Leipzig Liturgy, it is impossible to imagine any reason which could have prompted him to make up two such questionable pieces out of some of his finest cantatas, and to set them before the congregation on some special occasion in this fragmentary and ineffective form. These masses must have been intended for some other place, and Dresden at once occurs to the mind. If we may assign the G minor mass to about the same period as the other, Bach may have intended to make his mark as Court composer by thus enriching it, and, at the same time, on account of his immediate difficulties at Leipzig, to keep himself in mind at Court. The work, which was evidently written in haste, indicates lack of time and of the humour for original production.

We trace this also in parts of the A major mass written in 1737; in this, with the exception of the Aria in F sharp minor, the *Gloria* can be shown to be put together from portions of cantatas,[39] and I have no doubt that even this air might be found to have its original home elsewhere. Our judgment of this work can be no more favourable than of the two former masses. That no task was too severe for Bach is sufficiently proved by the first section, where, to make the original subject serviceable, a four-part chorus had to be inserted into the instrumental portions, while a solo for bass voice had to be amplified into a full chorus. This is accomplished in the most facile manner, but the glorious poetry of the original composition, to the words "Friede sei mit euch"—"Peace be with you" (from the Cantata "Halt im Gedächtniss Jesum Christ")—is almost completely destroyed. The voice solos are rounded off and extended, often illuminated by masterstrokes, and on the whole by no means ill-fitted to their purpose; the final chorus has brilliancy enough, but the characteristic ardour which gives the fundamental feeling of the original has been effaced by the accommodation to a new text. This *Gloria* is cast into still deeper shade from the *Kyrie*

---

[39] Hitherto this has only been established with regard to the Arias for the soprano and the alto (see Mosewius, *loc. cit.*). The final chorus, however, is also a re-arrangement of the opening chorus of the cantata "Erforsche mich Gott und erfahre mein Herz."

which precedes it. This is not, as in the other two masses, composed as a single number, but divided into the three sections indicated by the sentences; the first gives us the image of a simple and timid soul in fervent supplication; the second section, *Christe eleison*, displays an amalgamation of the freest with the strictest form, achieved with the daring of genius—it is a chorus in canon, but with the character of a recitative.[40] The last section also is in canon, but more strictly worked out as to form. Both are stamped with the sentiment of helpless weakness and a passionate desire for redemption, kept, however, within the limits prescribed by the first section. The style of the canon treatment contributes greatly to this result; since the parts always follow each other at equally wide intervals —of fourths or fifths—the modulation deviates more and more from its starting point, presently to return by an unexpected phrase into the original path. The feeling of a fundamental key is thus entirely eliminated, conspicuously in the last section.[41] There is no piece by Bach in which depth of purpose and sweetness of sound have more closely joined hands. As a whole this mass will be only fully understood when the last of the four, in F major, is discussed.

The conditions are almost the same in the *Gloria*; the final chorus and the arias for alto and soprano may be pointed out as borrowed from other works;[42] and the opening chorus is unquestionably not written for it; its aria-like structure of itself betrays this, the repeated portion having a different text—quite contrary to all tradition and sense of form. The recurrence of the principal theme (the first sixteen bars) no less than five times in the three sections, and of the middle theme (bars 101—118) no less than three times, is also not in Bach's usual manner. We

[40] An analogous, but far less artistic, example is the four-part recitative at the close of the Christmas Oratorio "O'er us no more shall fears of Hell" (Novello's 8vo ed., p. 168).

[41] The *Christe* is quoted by Kirnberger—Kunst des reinen Satzes, II., 3, p. 63—as a masterpiece of canon writing, and he observes that "it is quite unlike all church-music previously written because that was generally in the so-called heavy style in which hardly any variety in the dissonances was admitted."

[42] Hauptmann, preface to B.-G. VIII.

III.

seem actually to see the joins in this chorus, though it is externally compact, and is carried rapidly onwards by its animated flow.[43] The only bass air which remains could not, of course, be an exception among all these borrowed pieces. It is otherwise, however, with the *Kyrie*. Expressive and appropriate fugal movements develop the text in three sections; a leading theme runs through them, and in the second section comes forward in a form which is the outcome of a free inversion; in the third section only the second half is thus modified, while the total subject thus obtained is once more answered in perfect inversion. While this is going on in the three upper parts, the bass voice as *Cantus firmus* sings the *Kyrie eleison! Christe eleison! Kyrie eleison!* of the Litany. As a second *Cantus firmus* the chorale "Christe du Lamm Gottes" is given out by horns and oboes, only the *Amen* is somewhat altered, and in another position, so as to close in the original key.

It was not unusual at that period to introduce a Protestant sacred melody into a setting of the mass. Ernst Bach had made the attempt with the chorale " Es woll uns Gott genädig sein " (see App. B., II.) ; Zachau, again, with the Easter hymn, " Christ lag in Todesbanden,"[44] Kuhnau with the hymn for Whitsuntide *Veni sancte Spiritus;* an unknown composer with the Advent hymn, *Veni redemptor gentium;* while Telemann adapted the *Kyrie* no less than five times to Protestant hymn tunes, both cheerful and mournful.[45] A capital piece of music is Nikolaus Bach's E minor mass, in which the *Gloria* is combined with the chorale "Allein Gott in der Höh" (Vol. I., p. 133); here the words of the mass and of the chorale have just such a correspondence of feeling as in Sebastian Bach's mass, but it was only Sebastian himself who could weld them into an organic unity both of form and purpose, because his conception,

---

[43] The original form may be approximately traced in bars 1—28+65—83 as the first and third sections, and 84—118 as the second; only, of course, as to the main material of the music. Compare this with the construction of the opening chorus of the cantata " Es erhob sich ein Streit," B.-G. II., No. 19.

[44] Chrysander, Händel I., p. 25.

[45] See Breitkopf's Easter list, 1769, and New Year's list, 1764.

even of the Latin words, was strictly Protestant in character.[46] The *Kyrie* of the F major mass is one of his profoundest and most impressive pieces, and transcends even that of the A major mass by what I may designate as a monumental character, which suggests to our minds that Protestantism is not the reaction from Catholic church feeling, but rather the outcome of its development and continuity; this work, with its Protestant chorale, could not, of course, have been written for any Catholic congregation. But its abstruse affinities yet remain to be indicated, and by them the clue to its full comprehension. The *Cantus firmus* given to the bass voice is not the ordinary *Kyrie Dominicale* but the *Kyrie* from the Litany, and of that the closing and not the opening phrase :—[47]

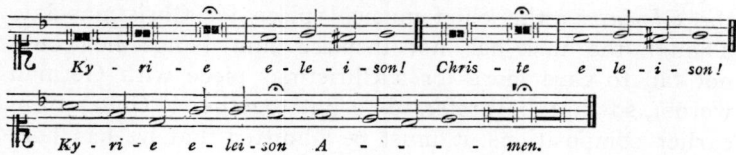

In the Litany this is immediately preceded by the appeal to the Redeemer, "O Lamb of God, Thou that takest away the sins of the world," &c., the very words of the *Cantus firmus;* thus Bach has entwined and supported his setting of the mass with the last two sentences of the Litany.

There were only two church seasons during which the Litany was used at Leipzig: Advent and Lent. Since during Lent, and on the three last Sundays in Advent, no concerted music was admissible, this composition, from internal evidence, must have been written for the first Sunday in Advent; the congregation is expressly prepared for that humble supplication for redemption which was

---

[46] One of the two MSS.—not in Bach's hand—in which the work exists gives the chorale to a soprano voice. This was certainly not Bach's intention; the characteristic effect here, as in so many of his similar compositions, depends essentially on the inarticulate delivery of a familiar melody quite apart from the confusing mixture of German and Latin words.

[47] Comparison shows how Bach has added character to the phrase by the addition of chromatic and other passing notes.

offered up after the reading of the Epistle, in which they all took part, and which was to be repeated every Sunday, till Christmas-tide brought fulfilment to the prayer.

If the *Kyrie* of the F major mass was written for a Protestant service the *Gloria* must have been also; and here, undoubtedly, that religious motive of his art which has been mentioned strongly influenced Bach. The spirit that has once so thoroughly imbibed the Advent feeling embodied in the liturgy, as Bach has done in this composition, must needs allow the fulfilment to crown the anticipation; he must rise from this depth of dejection to joy, must gladden the contrite sinner by the gospel of salvation. He has worked out this line of feeling and has supplemented the *Kyrie* of the first Sunday in Advent by the *Gloria*, as the chief musical piece for Christmas-day. Though the task has not inspired him (since he could not fail to care more for a Christmas piece with German words), so that he preferred to put together a *Gloria* from earlier compositions, it must be admitted that he has here produced a *Gloria* which is greatly superior to those of the other short masses, setting aside the diffuseness of the first chorus. The forced effects of the other three do not occur in this one; the modified fragments, so far as we can recognise them, are more carefully selected, particularly the last chorus, which in its original state is a Christmas chorus: "For this purpose the Son of God was manifested, that He might destroy the works of the devil" (I. John iii. 8). This selection of a text is still farther evidence that the *Gloria* was intended as a Christmas piece. With the exception of the abridged opening and a few other small changes, the Latin chorus is a faithful reproduction of the original under somewhat different conditions, and the feeling of the situation also remains entirely the same. This was not possible with the aria *Quoniam tu solus*, which, after the elimination of the most picturesque features in the original state, remains a neutral composition in no respect suited to its purpose. The *Kyrie* and *Gloria* of the F major mass, as they now exist, were conceived of as a complete whole; this is clear from the last word of the *Cantus firmus* delivered by the bass, which is not *Amen*, as prescribed by church custom, but once more *Eleison*. This could only occur when the artist's conception found its fitting form in closing with the *Kyrie*. At the performance on the first Advent Sunday the *Amen* was indeed sung,[48] but when the idea was continued and carried out by the *Gloria*, the *Amen* was set aside and found its fit place at the end of the whole piece.

If we now glance once more at the A major mass it is evident that it holds an intermediate position between the F major mass, on the one hand, and the G major and G minor masses on the other. It displays no essential affiliation to the Protestant service, but the well considered and loving treatment of the original *Kyrie*, in contrast to the compiled *Gloria*, makes it probable that this *Kyrie* also was originally intended for the Protestant liturgy; it could, certainly, be equally well used in Catholic worship. But whether this *Gloria* arose under the same demand as that of the F major mass, or was merely added in order to complete it for Catholic use, remains an open question; still, its having been produced as a whole, at the same time as the masses in G major and G minor, renders it probable that its purpose was the same.[49]

But all these works—even the F major mass—are but feeble offshoots of a *Kyrie* and *Gloria* which subsequently formed part of the B minor mass—the only grand and complete mass that Bach ever wrote. The earliest trace of this *Kyrie* and *Gloria* we detect in a passage of the *Domine Deus*, which, in the Leipzig form of prayer, deviated from the canonical text of the Catholic liturgy. In the Leipzig service it was sung thus: *Domine Deus rex cœlestis*,

---

[48] As is proved by a MS. copy of the *Kyrie* as a single piece which belonged to Joh. Adam Hiller, and is now in the Berlin Library.

[49] The autograph scores of the G major and A major masses indicate 1737—1738 as the year when they were written, the paper having the same watermark—in part at any rate—as the score of the Easter Oratorio. The original parts of the A major mass, which have lately come into the possession of the Berlin Library have, it is true, the watermark figured in note 48 of App. A. to Vol. II. Still, as the cantata " Herr deine Augen " has been borrowed from, and as all four masses seem to have been written at about the same period, it is easy to imagine that some remains of paper of a former date may have fallen under Bach's hand.

Deus Pater omnipotens. Domine Fili unigenite, Jesu Christe altissime.[50] The word *altissime* is inserted; it was not used in the Catholic service, and, so far as I can discover, it was so sung nowhere but in Leipzig. In the B minor mass Bach has followed the Leipzig custom; but when he became more familiar with the Catholic mass he left the word out in other works of this class.

When Friedrich August II., the King and Elector, died, Feb. 1, 1733, Bach resolved to show his devotion to his successor and to raise himself in the estimation of the Leipzig functionaries by connecting himself more closely with the Court. He, therefore, composed these two subjects from the mass, and presented them himself in Dresden, July 27, 1733. The dedication that accompanied them is well-known, but must not be omitted here:—

To the most illustrious Prince and Lord, the Lord Friedrich August, King and Prince of Poland and Lithuania, Duke of Saxony, &c., &c., my most gracious Sovereign,
Most illustrious Elector,
Most gracious Lord.

I lay before your Kingly Majesty this trifling work (or proof) of the science which I have been able to attain in music, with the very humble petition that you will be pleased to regard it, not according to the measure of the meanness of the composition, but with a gracious eye, as befits your Majesty's world-famed clemency, and condescend to take me under your Majesty's most mighty protection. For some years, and up to the present time, I have had the direction of the music in the two principal churches in Leipzig; but I have had to suffer, though in all innocence, from one and another vexatious cause—at different times a diminution of the fees connected with this function, and which might be withheld altogether unless your Kingly Majesty will show me grace and confer upon me a *Prædicate* of your Majesty's Court *Capelle*, and will issue your high command to the proper persons for the granting of a patent to that effect. And such a gracious acceding to my most humble petition will bind me by infinite obligations; and I hereby offer myself in most dutiful obedience to prove my indefatigible dilligence in composing church music, as well as in your orchestra, whenever it is your Kingly Majesty's most gracious desire, and to devote my whole powers to your Majesty's service, remaining with constant fidelity your Kingly Majesty's most humble and obedient servant,

Dresden, July 27, 1733.[51]　　　　　　JOHANN SEBASTIAN BACH.

---

[50] Vopelius, *ob. cit.*, p. 422.
[51] See the preface to the B.-G. edition of the B minor mass, VI., p. xv.

His desire to prove himself serviceable to the Court was the inducement which led Bach to undertake the composition of a complete full mass. The *Kyrie* and *Gloria* were, in this instance, conceived of as a whole from the first, flowed from the same fount, and were cast in the same mould. This is evident even from the scheme of key by which the last subject of the Kyrie is set—not in B minor, but in F sharp minor, to obviate the effect of a full close, and also, as the *Gloria* was to consist of two movements in B minor, to avoid monotony. The fact, too, that the first *Kyrie* is in five parts, and the last only for four, is significant from this point of view.[52] The remaining portions — the *Credo*, *Sanctus*, and *Osanna* to the *Dona*—were written separately and by degrees. It is not quite certain that the *Credo* was written later than the first two portions; if we may trust certain tokens, it may be assigned to as early a date as 1731-32. As, however, the composition of a *Credo* was not obviously necessary under the conditions of the Leipzig liturgy, it is more probable that Bach did not write this portion till the idea of writing a full mass was suggested to him by the magnificent success of the *Kyrie* and *Gloria*. The *Sanctus* was probably written in 1735; certainly not sooner, but not later than 1737. Since, then, for the remainder, which consists almost entirely of re-arranged pieces, no great trouble was involved, and as Bach seems to have been anxious to get the work finished, we may consider 1738 as the latest date of Bach's labours on the B minor mass.[53]

There is not the slightest indication that Bach ever presented the last three portions of the B minor mass to the King; and the two first even were never performed in Dresden, if we may derive any inference from the state of the parts as they remain in the Berlin Library. Their

---

[52] The autograph score is a clean copy, which, to judge from the initials M. A. must have been finished, if not actually in 1733, at any rate soon after. The *Kyrie* and *Gloria* are closely connected, for on p. 20, where the *Kyrie* ends, the *Gloria* is at once subjoined; and in the original score they were included together under No. 1.

[53] See App. A., No. 1.

unusual length unfitted them for use in the Catholic church; and Bach was no doubt well aware of this, and had not reckoned on any such performance. Still, so thoroughly practical a musician as he was never wrote anything—and least of all such a mighty work as this—simply to leave it buried unheard. He had intended it for the churches of St. Nicholas and St. Thomas in Leipzig, and, at any rate, performed it there—not as a whole, to be sure, but in detached portions. This can be proved with regard to the *Gloria* and *Sanctus* and each of the following numbers. Of the first there is still extant a score copied in 1740, with the significant note: " On the Feast of the Nativity."[54] Bach must, however, have discerned that the whole *Gloria* was not fitted for the intended Christmas performance; and he prepared an arrangement which was limited to the first and last choruses and the duets between. The text of the first chorus remained unchanged; but he set the words of the Doxology *Gloria patri et filio et spiritui sancto* to the duet which ends at bar 74 of the original work. The rest of the Doxology—*Sicut erat in principio et nunc et semper et in sæcula sæculorum. Amen*—he set to the final chorus; and to this end he had to alter the opening bars. This re-arrangement, however, does not preclude the possibility that Bach should have performed the whole *Gloria* on occasion, as a piece of ceremonial church music. The *Sanctus*, too, was used as a Christmas piece, and was, indeed, originally composed for that purpose, though, at the same time, its grandiose proportions were, no doubt, determined by the general character of the mass of which it was destined to form a part.

The *Sanctus* in ornate style, as has already been said, had a fixed position in the Leipzig liturgy. At the three great festivals it was sung at the close of the preface before the Communion. Bach wrote many other settings of the *Sanctus* for this purpose, one of which has already been

---

[54] "*Festo Nativitatis Christi. Gloria in excelsis Deo.*" The autograph belongs to Herr Kammersänger Hauser, of Carlsruhe, who was good enough to introduce it to my notice. See App. A. of Vol. II., No. 57.

mentioned when speaking of the Christmas festival of 1723 (see Vol. II. p. 369). The most important of them, however, which was also written during the early years at Leipzig, is in D major; it is a piece full of solemn inspiration constructed on attractive and beautiful themes, while an accompaniment of violins hovers seraphically above it.[55] All these compositions are somewhat meagre and only intended as a finish to the preface; the *Osanna* and *Benedictus* are wanting in all. Certainly, the mighty proportions of the *Sanctus* of the B minor mass somewhat outstep the bounds of the traditional liturgy; still it is quite clear that it was originally composed to follow the proper preface at one of the great festivals, from the fact that the *Osanna* and *Benedictus* do not form part of it.[56] Bach had combined this with the *Agnus*, a circumstance which affords an unmistakable hint as to the application of the last four numbers of the mass. It must be remembered that during the Communion service at the great festivals elaborate music was always performed. Here the *Agnus* finds its most natural place, and it can be shown to have held it even at a later date. It only was appropriate to begin the Communion music with the *Osanna* when the *Sanctus* had been previously sung. Thus Bach performed both sections in the course of one service, so that the preface ended with the *Sanctus*, then the Sacramental words were recited, and afterwards, during the distribution of the Lord's Supper, the *Osanna, Benedictus, Agnus,* and *Dona* were sung as a consecutive whole. The *Kyrie* of the B minor mass was

---

[55] B.-G., XI.,[1] p. 81. The *Sanctus* in D minor and that in G major, published in the same volume, are of inferior worth, particularly the latter; that in D minor, notwithstanding its simplicity and brevity, cannot be said to have any characteristic feeling. Two *Sanctus*, in F major and in B major which appear under Bach's name—and seem to have been set down as his original compositions even in Breitkopf's list for Michaelmas 1761—are obviously spurious; on the other hand, there is an eight-part *Sanctus* in D major which I cannot repudiate without further evidence: see Rust's Preface to B.-G., XI.,[1] p. xvii. No. 7 and 8, p. xvi. No. 6.

[56] It is a noteworthy detail that Bach always in the *Sanctus* wrote *Gloria ejus* instead of *Gloria tua*; following the text of the Bible and not that of the Canonical Mass.

too long for the customary service at the festivals of the Reformation, and the first Sunday in Advent; by itself it attains the dimensions of a moderately long cantata. But as the principal anthem it may have been performed on the Sunday next before Lent, for instance. The feast of the Trinity, with its dogmatic character and purpose must have seemed especially suited to the performance of the *Credo*, and this, as I have before shown, subsequently became the established custom. But the Saints' days were also to be considered, because, on those days, after the Gospel was read, the entire Nicene Creed was sung by the choir, and the ornate *Credo* was very fitly connected with this. There were, however, no independent services kept up for the Saints' days at that time; they were merged into the nearest Sunday.[57]

Though no portion of the B minor mass may ever have been performed at Dresden, even in Bach's lifetime it was not unknown to circles beyond Leipzig. We know that he sent the *Sanctus* to Count Sporck. Franz Anton Count von Sporck, born at Lissa, in Bohemia, in 1662, and at one time Stattholder of that Province, was a man of superior culture, many-sided interests, and great wealth. His services to music were conspicuous. He sent native German artists to be educated in Italy, and was the first person to introduce the Italian opera into Bohemia. When the "French horn" was invented in France, he made two of his servants learn the new instrument, and so introduce it into Germany. His noble Christian spirit was always occupied in works of benevolence and undertakings for the benefit of the world at large. He was a Catholic, but of a breadth and independence in his religious views far in advance of his time. He suffered at the hands of the spiritual authorities, since he would not confess the sole saving efficacy of the Catholic faith, but admitted the equal value of the different forms of religion, deeming it sufficient to seek salvation through Christ, to love God and one's

---

[57] The merging of the Saints' days into the nearest Sundays seems to have become general in Leipzig by the end of the seventeenth century. See Vopelius, *op. cit.* p. 20.

neighbour according to His laws, and that he who did this would be saved, whatever creed he might profess. He founded a printing press at Lissa, by means of which he spread abroad his religious works, some of which were translated into French by his daughters.[58] He had long been connected with the artists and learned professors of Leipzig; Picander dedicated to him, in 1725, the first fruits of his sacred poetry: the Collection of Edifying Thoughts ("Sammlung erbaulicher Gedanken"), and in the dedicatory poem he sings the praise of "pious Count Sporck." Count Sporck died at an advanced age, March 30, 1738, at Lissa.[59] His introduction to Bach's masses must therefore have been one of the latest occurrences of his life.

The mention of this Mecænas of the eighteenth century has led us away from the study of the B minor mass, its essential character, and the spirit in which it is worked out. Though the external suggestion was afforded by the Catholic form of service, Bach wrote the work for the Protestant worship of the churches of St. Thomas and St. Nicholas. He accepted the form established by the Catholic church just as it stood, and adhered to the vein of sentiment which had already become typical for each section of the mass: the absorbed gravity of the *Kyrie*, the jubilant animation of the *Gloria*, the strong confidence of the *Credo*, and solemn grandeur of the *Sanctus;* he even intensified these feelings. In the choruses of the *Credo* a kind of polyphony appears, to which we are unused in Bach's Cantatas; an effect produced by broad and simple phrases of melody, a highly artistic extension of the theme, and elaborate stretto of the *Cantus firmus*, which is borrowed, not from any congregational hymn, but from the priest's chant. We also find in this mass the subdivision of the greater sections into several independent smaller ones, the utilisation of the aria and the duet—all of which had made its way under the

---

[58] Zedler, Universal Lexicon, Vol. 39. Leipzig and Halle, 1744. Also, G. B. Hancken's Weltliche Gedichte. Dresden and Leipzig, 1727, p. 30 and p. 123.—J. Ch. Günther's Gedichte. Breslau and Leipzig, 1735, p. 137.
[59] Gerber, N. L. IV., col. 243.

influence of the Italian opera during the seventeenth century. Still, Bach could not be false to his own Protestant style. This, which was founded on the firm basis of German organ music, had, with careful eclecticism, grasped every other form worthy to survive, and now proved itself capable of absorbing that element which distinguishes the B minor mass from all the rest of Bach's church music. Wherever the Protestant liturgy required it, Bach has deviated from the main lines of the Catholic mass. Thus the B minor mass is scarcely less essentially Protestant than the rest of Bach's church music, but its roots strike deeper. Luther's purer creed was born in the lap of the Catholic church, and it was only the ill-founded pretensions of the Mother Church, which had nothing in common with her original constitution, which forced Protestantism to fight for an independent position. The political exigencies of Princes, and the antagonism of nations and races, roused a hostile fury which led to the most terrible religious war ever waged, and left an enduring bitterness, even late in the eighteenth century. Nowhere was this bitter spirit stronger or more stubborn, on the Protestant side, than in Saxony, and it was precisely there that the great work of art was destined to be created which showed Protestantism no longer as the antagonist and foe of Catholicism, but as an inevitable outcome and development from it, grown from the same soil. The B minor mass plainly reveals how immeasurably deeper and broader Bach's church feeling was than that of his age. In him dwelt the true spirit of the Reformation-epoch, with all its assertiveness and its personal meditative sentiment, but also with its comprehensive and assimilative power. When Luther arose, all the most cultured and honest minds were agreed as to the necessity for the self-examination and reconstruction of the church; and all the nobler souls, even though they might not go over to Protestantism, were of one mind with Luther in this. Almost all Germany was at once devoted to the new doctrine, and in the enlightened classes throughout Europe it found numerous adherents. The reformers themselves were far from purposing a breach with the Catholic church; they accepted the Nicene Creed in the books containing their profession of faith, as well as the *Credo unam sanctam catholicam et apostolicam ecclesiam*, in token of their community of belief with true Catholics. But this lofty conception of the work of reformation had totally vanished within two hundred years. At Bach's time the old warlike spirit still lived in the orthodox, though under restraint, it is true, and the fervent religious sentiment survived in the Pietists; but scarcely any one preserved the sense of the historical continuity and internal connection of Protestantism with the Catholic church. It remained for Music to re-unite all these different currents of thought, and to show them to the world in an immortal work, and that in the same part of the German empire whence the most powerful impetus was first given to the Reformation—a solemnly suggestive fact, but scarcely understood.

At that time the art of music had not yet been fully adapted to mirror all the new ideas of the period. All the Protestant music that attained any importance displays, under merely superficial variations, forms of art which sprang not from contemporary life, but from the later middle ages. It was once more to be proved that, of all the arts, music requires the longest time to become available for the utterance of a new type of culture. Just like Handel's music—nay, like all the music of the eighteenth century—Bach's music is based on the period of the renaissance. It was his vocation to produce the most thoroughly objective—because the latest—the purest and most glorified image of the spirit of the reformed church of that great epoch in his B minor mass.

It is only in certain portions that he shows himself subject to the conditions of the Protestant liturgy; indeed, it is only in the *Sanctus* and *Agnus* that this necessitates any particular form. The *Gloria*, *Kyrie*, and, above all, the *Credo* are only slightly and more arbitrarily connected with it. The structure of the whole work rested solely on the personal will of Bach, who found in the Protestant form of worship only the ruins of a magnificent liturgical work, which was both capable and worthy to be reconstructed in the spirit of the Reformation. This, more than anything else, is the free expression of his own powerful individuality—of an individuality which has drawn all its nourishment from the life of the church, down to the rock of its foundation. This mass is more absolutely inseparable from the Protestant Church of his time than even the cantatas and Passion music. Though the Passion according to St. Matthew extended to a length which made the Vesper service of Good Friday seem almost a secondary object, the connection was real, and, if only for the sake of the chorales, quite indispensable. In the B minor mass Bach has refrained from any use of the congregational hymn, although there were examples at hand for each portion of the mass. He adopted no Sunday nor holy day, no church solemnity as its background; and nevertheless there is no work which more amply satisfies the true spirit of Protestantism. But when Bach purposed to work down to the very core of the liturgy, the height of the structure had to correspond to the depth of the foundations; he could not let the edifice run up like a spire in the Protestantism of the time; it must over-arch it. In this work the artist addressed himself with independent Protestant feeling to the "one holy and universal Christian church"; any one who yet recognised that, and had cherished its spirit, could understand his work.

Though, even in the B minor mass, certain portions are recognisable as remodelled from cantatas, still the nature and purpose of the whole work at once dismiss any idea of this having been done for convenience sake, or from pressing haste; a comparison of the re-arrangements with the originals shows, too, that Bach carefully selected only such pieces as agreed in poetic feeling with the words to which they were to be adapted. In the *Gloria*, the sentence *Gratias agimus tibi propter magnam gloriam tuam* was set to a chorus, of which the original words were "Wir danken dir Gott"—"We thank Thee, O Lord, we thank Thee, and proclaim Thy wonders"—(see Vol. II., p. 450). In the same portion of the mass, the words *Qui tollis peccata mundi*, &c., are based on the first portion of the opening chorus of a

cantata, of which the words are "Schauet und sehet"—
"Behold and see, was ever sorrow like unto my sorrow"—
(Vol. II., p. 427). The text of the second chorus of the *Credo*
is: *Patrem omnipotentem, factorem cœli et terræ;* the text of the
original "Gott wie dein Name so ist auch dein Ruhm"—
"God, as Thy Name is, so is Thy glory, even unto the world's
end"—(Vol. II., p. 441). The *Crucifixus* is a revival of the
cantata chorus "Weinen, Klagen, Sorgen"—"Weeping,
anguish, terror, pain, and grief are the Christian's bread of
tears"—(Vol. II., p. 404). All these subjects are precious
gems which, in their new setting, not only sparkle more
brightly in themselves, but add to the magnificence of a
splendid whole. Bach has left nothing wholly unaltered,
though the pieces have not been reconstructed from the
foundations. In many cases some small detail adds to their
characteristic fitness; in the *Crucifixus*, the *tremolo* bass and
the closing modulations; in *Qui tollis*, the muffling of the
sound by the cessation of the supporting wind instrumennts.
The chorus "Gott wie dein Name" really seems to have been
awaiting its conversion into the *Patrem omnipotentem*; the
slight modification in the theme, which was necessitated by
the Latin text, first fully brought out its sinewy structure,
and the rhythm of the words from the mass fit the melody
better than the Bible text.

Besides those already mentioned, there are but two that
are not perfectly new compositions, and these must be
judged somewhat differently. The *Agnus* is founded on the
alto aria in the Ascension oratorio "Ach bleibe doch, mein
liebstes Leben" (Vol. II., p. 593 f.)—but only one long phrase
of it is used, and the remainder is quite a new composition.
The *Osanna* occurs at the beginning of the secular cantata,
"Preise dein Glücke" (Vol. II., p. 631). But even there it
is not in its original place; on the contrary, it bears conspicuous marks of arrangement, and must be more unlike
the true original than the *Osanna*, so that in this instance
nothing can be said as to the connection between the
original and the reproduction.

Among the twenty-six numbers into which the B minor
mass is divided there are six arias and three duets. There

are no recitatives, as they were not admitted into Catholic church-music, and must have seemed unsuited to the grand generalisation of the text in the Protestant service. Five-part writing predominates; it is absent only in the re-arranged choruses and the *Sanctus*, which was originally independent, and in the second *Kyrie*, which may therefore, perhaps, be regarded as a remodelled piece. Bach has rarely written five-part music, and here the influence of Italian church music is unmistakable.

The preference for the chorus-form was required by the nature of the great undertaking, and was also the outcome of the structure of the text, which is nowhere open to subjective treatment; and though solos could not be wholly avoided in so colossal a work, as they were indispensable for the sake of contrast, it is very intelligible that they should assume a less personal character than is usual even with Bach. But the intrinsic contradiction which is inherent in the very nature and idea of *impersonal solo* singing can be removed by the whole work which this subserves. And this is the case in the B minor mass; these arias and duets would have less charm apart from their connection than those even of the cantatas and Passion music, but in the course of the work they adequately fill their place. The duet, *Christe eleison*, conveys something of the trustful and tender feeling of the sinner towards the Divine Mediator, and the introduction of the sub-dominant in the first bar of the symphony suggests it at once. The fervent devotion to the Saviour which impresses us in the *Agnus Dei* is not yet attained; this is prohibited by the juxtaposition of the two choruses. In the course of ideas presented by the text of the *Gloria*—which suggests the proceeding of Christ from God, His deeds and sufferings on earth, and His return to the Father—the aria, *Laudamus te*, is intended as a transition from the lofty rejoicing over the incarnation of the Son of God to the solemn thanksgiving for God's glory. The duet, *Domine*, then enlarges on the mysterious Unity of the Father and Son, on which the possibility of the atonement depends, and ends with the vocation of Christ on earth. This doctrinal aspect is the source whence Bach derived the tone-picture, which cannot be understood but by a reference to it. The violins and violas playing *con sordini* the *pizzicato* in the basses, and the fantastically wandering passages for the flutes have a very mystical effect.

The musical germ which diffuses its life through every portion of the piece is a motive of four notes :—

It is, as it were, the musical symbol of the Unity which this dogma inculcates, and is thus put forward at the very beginning of the piece. The phrases *Domine Deus rex cœlestis, Deus Pater omnipotens—Domine Fili unigenite, Jesu Christe altissime*, are not sung straight through as the mass text gives them, but the tenor addresses himself to God the Father, and the soprano, beginning a bar later, to God the Son; each develops the melody, which proceeds in imitation, by extensions of the motive quoted above, and presently both sing it together in its original form. The way in which the motive constantly recurs, not prolonged to any fuller melody, but isolated, distinct, and stern as a dogma, is unique among Bach's compositions. In bar 42, the descending passage of octaves for the whole body of violins and violas can have none but a symbolical meaning, coming in as it does without any organic sequence, and quite unexpectedly, in a way which is not usual with Bach. Comparing it with the duet, *Et in unum* (in the *Credo*), which resembles it in many respects, we may fancy it intended to suggest the descent of God to assume the form of men.[60]

---

[60] An observation as to the performance of this may find a place here. The first bar for the flute part is thus written by Bach:—

Later on, where the theme recurs, we find, in the second half of the bar, simple semiquavers, phrased in pairs; thus the dotted mode of notation only indicates that the first is closely joined to the second, and to be accented, and not that it is of less value than the second. A manual of music by J. G. Walther, of 1708, of which I possess the original autograph, says on

No human emotion anywhere finds utterance as yet, for the words do not give rise to it till later, when the contemplative mind is directed to the atoning death of Christ; and if Bach desired to work out the sections of the mass in a variety of subjects, and not merely as music for music's sake, this was the only course that could result in a profound and impressive composition.

Here his theological learning—which the discovery of the catalogue of his theological library proves to have been considerable—stood him in good stead. Doctrinal theology assigns to Christ a three-fold office—as Prophet, High Priest, and King. The text offered no opening for treating the prophetic aspect—only the priestly and the kingly. As, in considering Christ as a priest, there is again a distinction between Atonement and Mediation (*munus satisfactionis* and *intercessionis*), Bach has figured the former by the chorus *Qui tollis*, and the latter by the alto aria *Qui sedes*, but in close connection, for the key is the same in both. The chorus itself at the end delivers the words *Suscipe deprecationem*, preparing for the aria by a half-close; thus the function of intercession, in accordance with the orthodox dogma, appears as a personal outcome of the work of Atonement—an application of it to the individual soul. The bass aria which follows, *Quoniam tu solus sanctus*, thus refers to the kingly office, which is broadly indicated by the dignified form of the principal subject, and by the solemn blast of two bassoons and a horn added to the organ and bass solo. The purport of the *Credo* is the presentment of the doctrine of the Trinity. Here it was indispensable that the Unity of the Father and the Son should be more strongly insisted on than in the *Gloria*. The duet *Et in unum* does this by the canonic treatment, which is employed

---

this subject: "*Punctus serpens* indicates that notes written as follows should be slurred," *e.g.*:—

Here it is evident that Bach differs as to this mode of performance, and a note in B.-G. XIII.,[1] p. xvi., might be made more exact.

for the instruments as well as the voices. But, to represent the essential Unity as clearly as possible, Bach treats the parts in canon on the unison at the beginning of the principal subject each time, not using the canon on the fourth below till the second bar; thus both the Unity and the separate existence of the two Persons are brought out. The intention is unmistakable, since the musical scheme allows of the canonic imitation on the fourth below from the very beginning.[61]

Indeed, much more may be said without over-straining the idea. Wherever the chief subject is given to the instruments, Bach makes the last quaver of the first bar in the leading part *staccato*, and in the second part *legato*, thus:—

Now, the object of this effect, which is consistently carried out all through, can only be to distinguish the parts in imitation and already in unison by a somewhat different expression; and so, even here, to suggest a certain distinction of Persons within the Unity. There are yet other highly significant features. In bars 21, 22, and 66, we again meet with the passage of hovering descending octaves, which is not worked episodically; and as it accompanies the words *et ex patre natum* and *et incarnatus*, and in the second case is followed by the voices, its purpose is easy to be understood. To express *descendit de cœlis*, the instruments sink through three octaves on the chord of the dominant seventh. It was not until later that Bach cast the words *Et incarnatus est de Spiritu sancto* in a separate chorus and, as may still be seen, inserted the score of it on a sheet by itself. Originally these words were included in the duet, and the division of the text was, consequently, different. This, however, must not be set aside in judging

---

[61] Mosewius detected the symbolical meaning. See, in Lindner, Zur Tonkunst, p. 165.

the composition: the startling modulations at the close, which seem to reveal another world, are to be accounted for by supposing that they are intended to embody the miracle of the passage from the Divine into the human state of existence. By the other and subsequent arrangement of the text Bach greatly obscured this subtle reference; but, by leaving the original distribution of the text standing in the score side by side with the inserted chorus, he, no doubt, meant to indicate that they could be thus sung, even when the choral subject was used.

The last three solo pieces are more full of warmth and sentiment. The confession of faith in the Holy Ghost, through whose instrumentality the new and holy life is shed upon mankind, is given in a bass aria. The sentiment of the melody, which flows softly, like a breath of spring, is only fully understood when we find it again forming the basis of certain Whitsuntide cantatas (as " Erschallet ihr Lieder," "Also hat Gott die Welt geliebt "). The *Benedictus* is delivered by the tenor; graceful intricacies on a solo violin[62] mingle with the sweet and solemn song, which makes such an impressive effect between the twice-sung *Osanna*—a grandly massive jubilant chorus. But the human sentiment is uttered in the most fervent manner by the alto in the *Agnus Dei*. The character of the feeling Bach here intended to express is clearly indicated by his having borrowed the music from the fervid farewell passage in the Ascension oratorio. But the *Agnus* as a whole was required to be something quite different, because the text demanded that the music should be in two sections. Only a faint resemblance remains to the original form in three portions; of the two subjects, each forming a section by itself, the first is newly-invented, and out of the long-drawn lamentation the song is worked out to a pitch of passionate supplication.

The solo songs stand among the choruses like isolated valleys between gigantic heights, serving to relieve the eye that tries to take in the whole composition. The choruses,

---

[62] Or flute; the autograph score gives no directions on this point.

indeed, are of a calibre and grandeur which almost crush the small and restless generation of the present day. As throughout the whole work the most essential portions are given to them, a general consideration of the whole is the best way to understand them. The liturgical elements in the mass are four—the consciousness of sin in man (the *Kyrie*), the Atonement through Christ (the *Gloria*), the Christian Church as proceeding from Him (the *Credo*), the memorial supper in which the Church celebrates its union with and in the Founder (the *Sanctus* and subsequent parts). That which in this mass gives artistic connection to the five sections into which the materials are worked out is not the under-current of congregational feeling which is derived from the performance of a solemn function, and which finds its highest union in the Catholic mass. The predominant sentiment in Bach's work is, of course, absolutely free from any such theatrical element. The inherent continuity of the liturgical theme is alone insisted on; it is an ideal and concentrated presentment of the principal factor in the development alike of Christianity and of the individual Christian up to the solemn realisation of the Holy Sacrament. And even this is but half realised, inasmuch as the music belonging to it is conceived of as inseparable from the other portions of the mass, though it is not, and never can be, performed as a part of Divine Service. The communion music, however, marks the culminating point, at which the essential difference from a mere historical picture of Christianity is defined. It is in the intrinsic connection of the various parts, from the religious point of view, and in the profound contemplation of the special bearings of certain portions of the text which this has induced, that we find the source of that deviation from the typical forms of utterance which has already been alluded to when speaking of the prototype offered to Bach by the Catholic mass. A vein of serious meditation was not lacking, even in the Catholic *Kyrie*; but it rather lent itself to the character of an introduction to a solemn ceremony, and as such, under the increasing frivolity of Catholic Church music, it grew more and more vapid. Bach's *Kyrie* goes at once to the heart of the matter, without second thoughts of any kind. Man, convicted of sin, cries in his need to God for mercy; and the unusual proportions of the first chorus remove every doubt as to the composer's purpose of representing in it the common supplications of all Christendom. A fugue, which lasts from twelve to thirteen minutes, is worked out in 126 bars of slow *tempo*, in extremely simple passages and modulations upon a marvellously bold theme steeped in sorrow. It may be safely asserted that a purely personal emotional idea has never been worked out so persistently and with such unflagging strength of feeling, while the subordination of the expression of pain, so acute as to be almost physical, to the powerful governing will of the artist, is incomparably sublime. This gives us the key-note of feeling for the whole work; but, even within the limits of the *Kyrie*, it has its value. The condition of mankind as craving redemption—of which the three-fold *Kyrie* is the symbol—is attributed by the Church to all the generations before Christ. As it is expressed in the first *Kyrie*, the elect people of God are crying to the Redeemer from the very first introduction of sin into the world. As the time of fulfilment draws nearer, their longing is more urgent and passionate; and to depict this is the aim of the short, agitated closing cry of *Kyrie*, almost desperate in some places (see the last nine bars). The beginning is epic, the close dramatic—if I may be allowed the terms. A distinct reference in the separate subjects to the Three-fold Person of God, which is, of course, intended by the three cries of the text, is not to be imagined—the position which this portion of the work was to occupy precludes this. I regard the *Christe* rather as a lighter musical subject to give relief, which need not exclude the idea that its softer character was induced by the image of the loving Saviour, especially when we remember Bach's way of letting himself be led by incidental suggestions.

At the beginning of the *Gloria* stands the *Hymnus Angelicus;* the Bible text of the song of the angels on the night of Christ's birth: Bach has treated it as a chorus, which was not the custom in the Catholic mass. In the settings of the *Gloria* in his shorter masses, the first chorus always has some sentences of the doxology which come after, besides the words of the angels' hymn. Here Bach has severed the Bible words from the liturgical amplification which follows them; even if we did not know that in later years he made use of this chorus for Christmas music, we could not fail to recognise the Christmas feeling that pervades it, and of which there is no trace in the Catholic Masses. The treatment of the words *Et in terra pax hominibus bonæ voluntatis* displays a certain resemblance with the angels' chorus in the Christmas oratorio; even the 3-8 time is characteristic of a festival of which Paul Gerhardt could sing:—

> Dance my heart with triumph springing,
> On this day
> When for joy
> Angels all are singing.

In fact, this measure recurs in several choruses of the Christmas oratorio and in the cantata "Christen, ätzet diesen Tag" (Vol. II., pp. 367—369). But the general impression is definitive; this is less a hymn of rejoicing mankind, on whom the day of redemption has risen, than an innocent jubilation, strung, it is true, to the highest conceivable pitch possible to this type of feeling. We must accustom ourselves to the colossal proportions of the B minor mass before we can accurately discriminate between the different characters that stamp each chorus; but then we cannot fail to recognise in this chorus the old blissful Christmas feeling which we have met with so often and so touchingly in Bach, not least in the happy tranquil middle subject, from which the development of a flowing fugue is as natural as it is characteristic, while the joyful voices combine in the greeting of "Peace." It is not till we come to the following air that we are led to the serious presentation of the dogma of the work of Atonement initiated by the birth of Christ. Here the splendid and solemn chorus, which originally stood in the Rathswahl cantata as "Wir danken dir Gott" has found a worthy place as *Gratias agimus tibi*. The farther course of this section is to a great extent given to solo singers; only the climax, the atoning death of Christ, is expressly emphasised by the deeply pathetic chorus *Qui tollis peccata mundi*; and at the close we have a triumphant hymn to Christ, Who, having finished His earthly course, sits on the right hand of the Father. The bold onward march of the theme of the fugue, the victorious *Amen* that bursts into the middle of it, and the surge and roll, so characteristic of Bach, bear the impress of Protestantism. Still, the long-drawn harmonies of the chorus, through which it breaks like a flash of light, eclipsing for the moment all the other individual forms, reveal another and more general sphere of ecclesiastical feeling.

In the *Credo* the Church founded on Christ declares its faith in the words of the Nicene Creed. The opening chorus, *Credo in unum Deum*, stands up like an over-arching portal, by which the precincts of the Church are thrown open to us. As the theme for the fugue Bach has chosen the church tone:—

Cre - do   in   u - num   De - um.

Above the five voices the two violins come in, piling up the structure of the fugue, while the *continuo* wanders up and down in a constant movement of crotchets. At the close the bass delivers the theme in augmentation, and at the same time the second soprano and alto give it out in the proper measure, and the first soprano in syncopation; agreeing with this, syncopation occurs also in the violins. As such complicated arrangements occur rarely elsewhere with Bach, the connection here indicated with the polyphonous church music of the sixteenth century is pretty obvious, and Bach's study of Palestrina thus acquires a peculiar significance. The utilisation of the priests' chant and of the mixolydian mode remove every doubt that the master had intentionally reverted to that period, since there was as yet no question of a severance, but only of a reconstruction of the whole Church. The symbolism of the augmented theme in the bass and the intricacy of the voice parts above—the immovably rooted unity of the faith—is thus made clear at once. In the course of the creed the image of God the Omnipotent is indicated in broad outline, and a chorus full of brilliancy and of nervous vigour lauds Him as the Creator of heaven and earth. It falls to the part of an intermediate movement for solo voices to announce the mysteries of His Unity with the Son, and then Christ Himself appears, the incarnate God. The broad descending intervals of the opening theme represent His descent to mortality. A maiden fervency breathes through this quite simple chorus, which is chiefly homophonous, but it fills us with a mysterious thrill, and the accompaniment—chiefly by means of the bold passing notes—affects us like a foreboding of deep grief.

At the close the clouds of sorrow gather; then we have a new scene—Christ crucified. Bach had gradually so ennobled and inspired the old forms that he could venture in this place to introduce a *passecaille*. Nothing more characteristic can be imagined; the theme, which recurs thirteen times, holds the fancy spellbound in contemplation of the stupendous scene that is being enacted. The subject, which is taken from an earlier cantata, seems even there to be referable to some still more remote inspiration; the bass theme itself had haunted the musician from his earliest youth, and here is cast in its final mould as a *passecaille*. Indeed, this chorus and the cantata "Jesu, der du meine Seele," which treats the same subject in the form of a *chaconne*, indicate the sum total of Bach's development in a certain direction. It is an aid to a keener comprehension of the predominant characteristic of this subject in the mass to consider the two settings in connection; here we have not a mere infused colour inspired by a general sentiment of sacred solemnity, still less a histrionic illustration of a thrilling event. Beneath the words of the narrative the inner ear may detect a fervent prayer to Jesus—Who once, through His death redeemed the world—that He will vouchsafe evermore to fulfil the work of redemption in all who seek Him. All is pathetic and piteous, but purified from every trace of egotism. And what the parts have to say above the bass theme in their excess of chromatic diminished intervals, either alone or in harmony, is as stupendous as the event they are intended to shadow forth. When at last the thematic bass is released from its rigid progression, and the chorus sinks into the deep cool repose of the shadow of the grave, the hearer is left under the sense of a tone-picture by the side of which anything that has ever been written for this portion of the mass is a pale phantom. Even the chorus *Qui tollis* is cast into the shade by this. And so it should be; there the sufferings of Christ were only a factor in the whole work of redemption, which forms the subject of the section in which it occurs; while here we have the poetical image of the very nature of the Son as contrasted with the Father and the Holy Ghost. Originally the chorus *Crucifixus* was intended to suffice for this purpose; afterwards Bach thought this conception inadequately emphasised and made a chorus of the *Incarnatus est* also. Thus he intended to balance the different sections of the mass against each other.

Out of the silence of the tomb, to which we are led by the closing bars of the *Crucifixus*, the chorus triumphantly starts afresh and raises the standard of the Resurrection; a long instrumental symphony is introduced to accustom the mind to the return of the light. Then with renewed vitality the chorus soars up again and rejoices, not in prolonged phrases, but with constant interruption from the instruments; this gives the subject a character which, in spite of all its vigour and of the defiant boldness of the basses, which declare the promise of Christ's coming again, tempers the movement and keeps it down, for yet another climax remains.

The third Person of the Trinity, the Holy Ghost, is revealed through the Church, and the symbol of admission and fellowship in the Church is baptism. Hence a confession of faith in baptism is at the same time a confession of faith in the Holy Ghost. The elaborate choral treatment of the *Confiteor unum baptisma* is founded on this conception, after the belief in the Holy Ghost Himself has been declared in an aria which is musically indispensable. Here again, as in the first *Credo*, and with the same allusion to the universal Christian Church, we find the Gregorian chant as follows :—

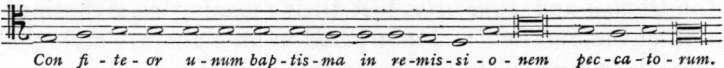

But this tune is not fitted to be the theme of a fugal subject, and one had to be invented. It is not till bar 73 that the chant first appears in diminution and close imitations between the bass and alto, and after this in full time by the tenors alone. This leads up to the full close; the Church lives on beyond the grave in the life eternal, where it attains to perfection. Through a slow succession of marvellous harmonies, wherein the old world sinks and fades, we are conducted to the conception of "a new Heaven and a new earth." Hope in that future life is poured forth in a chorus full of solemn breadth in spite of its eager confidence.

The fourth portion of the mass, which belongs to the Lord's Supper, is in two divisions. In the Catholic mass the *Sanctus, Osanna,* and *Benedictus* form the first, the *Agnus* and *Dona* the second. In the absence of any information as to the Leipzig usages, it has hitherto been customary to adhere simply to the Catholic custom in all the editions and performances of the B minor mass, thus ignoring Bach's express indications.[68] But the peculiar arrangement of the B minor mass in this place is important and significant in more respects than one. In the earliest times the Sanctus, with the introductory preface, was regarded as a thanksgiving for the beneficence manifested in the creation, of which the first fruits, generally in the form of bread and wine, had been previously offered by the members of the congregation in the *Offertorium.* The addition of the *Osanna* and *Benedictus* was made when this symbolical thank-offering sank into the back-ground by the side of the later conception of a symbolic sacrifice of the Body and Blood of Christ by the hands of the ministering Priest; for the *Osanna* and *Benedictus* point to the coming of the Saviour, and in this place to

---

[68] The oldest edition, by Nägeli and Simrock, is guiltless of this in so far as it omits all indication of the distribution of the mass in the liturgy.

His presence in the Bread and Wine. At the Reformation, however, this conception of the sacrifice was rejected, and thus the *Osanna* and *Benedictus* lost their meaning as a continuation of the *Sanctus;* and as treated in an elaborate style they were even omitted in the great churches of Leipzig as early as in the seventeenth century.[64] Bach, who intended this *Sanctus,* like his others, to be used in the service, restored the usage of the primitive church; whether consciously or unconsciously cannot be known, the fact remains that here the *Sanctus* is restored to its original form as a portion of the mass. To realise the effect of this *Hymnus seraphicus* (Is. vi. 3) we must connect it with the words of the preface, which varied, and still varies, according to the festival. But the main paragraph was always the same and very similar to that now used in the Anglican Communion service—"It is very meet, right, due, and of saving power that we should at all times and in all places give thanks unto Thee, Holy Lord, Almighty Father, Everlasting God; through Christ our Lord, through Whom the Angels laud Thy Majesty, the Dominions adore and the Powers fear it. The Heavens and the Powers of Heaven and the blessed Seraphim praise Thee with one shout of triumph; with them we beseech Thee let our voices reach Thee and say, entreating and acknowledging Thee: Holy, Holy, Holy," &c. The overwhelming idea of a hymn of praise in which the Powers of Heaven and the Angels unite with man may have prompted the composer not only to replace the *Sanctus* in the B minor mass but to extend the harmony to six parts. In fact, we find that the words of Isaiah have determined even the details of the composition—" I saw the Lord sitting upon a throne high and lifted up and His train filled the temple. Above it stood the Seraphims; each one had six wings," &c. The majestic soaring passages in which the upper and lower voices seem to respond to each other are certainly suggested by the last words "and they cried one to another." In the bars where the five upper parts hold out in reverberating harmony against the broad pinion strokes of the violins and

---

[64] As may be gathered from Vopelius, p. 1086.

wooden wind instruments, the blare of trumpets and thunder of drums, while the bass marches solemnly downwards in grand octaves, we feel with the prophet that "the posts of the door moved at the voice of him that cried, and the house was filled with smoke." After this majestic *Sanctus* follows an animated setting of *Pleni sunt cœli*, which so far exceeds any similar movement in the mass in ecstatic jubilation that we cannot help feeling that till this moment Bach has only given us the hymns of praise and joy of mortal Christians, but that here "the morning stars are singing together and the sons of God shouting for joy" (Job xxxviii. 7).

The second section of the fourth portion of the mass Bach has begun with the *Osanna* and it closes with the *Dona*. We might say, as the *Sanctus* can only give expression to the most universal form of thanksgiving for the mercies of God, that it constitutes by itself a fourth division, and that the *Osanna* and what follows form a fifth. But it has been shown above that they must have been written to be performed in connection, and indeed an internal relationship can be traced. It is evident that by a division which should treat the *Osanna* as introductory and the *Dona* as final in an independent section of the mass this would have a very different common character from that which would stamp only the *Agnus* and *Dona* taken together. In this latter case— as the observant student has long since detected—the impression cannot be other than unsatisfactory, not only as regards each of these numbers separately but as to their connection and their position as finishing the whole mass. But in point of fact the *Agnus* is only intended to supply a very effective transition; it lies like a deep and gloomy lake between lofty heights; the character of the closing section is not penitent entreaty, not an overwhelming sympathy with a tragical event, not even the mystical exaltation of the Lord's Supper—it is joy and thankfulness and, so far, a reiteration of the feeling of the *Sanctus*, but brought down to the level of humanity. The double chorus *Osanna* has also much more of the character of an introductory chorus than of a finale, and the critic who objects to the absence of a ritornel may remember that no concerted church music was ever performed without an organ prelude. Now, whatever might have been the original purpose of the subject employed for the *Osanna*, it cannot be disputed that these crowding and competing strains of rejoicing are admirably well suited to the words of the multitude who accompanied Christ in His entry into Jerusalem. Indeed, from the point of view adopted for this section as a whole, all that seems strange in the *Dona* even, which simply repeats the music of the *Gratias agimus tibi*, disappears. It is not, nor ought it to be, a prayer for peace. As the grouping of the sentences stood, the close could be nothing else than a solemn hymn of thanksgiving. It can hardly, however, be asserted that this mode of treatment has not given rise to a contradiction between the music and the words—though it is but superficial and, in the whole work, unimportant—nor that a finale with a different setting might be quite conceivable. But at any rate the reproach can no longer be raised that the B minor mass has no close of due importance, based on the main line of feeling that pervades the whole.

The B minor mass exhibits in the most absolute manner, and on the grandest scale, the deep and intimate feeling of its creator as a Christian and a member of the Church. The student who desires to enter thoroughly into this chamber of his soul must use the B minor mass as the key; without this we can only guess at the vital powers which Bach brought to bear on all his sacred compositions. When we hear this mass performed under the conditions indispensable to our full comprehension of it, we feel as though the genius of the last two thousand years were soaring above our heads. There is something almost unearthly in the solitary eminence which the B minor mass occupies in history. Even when every available means have been brought to bear on the investigation of the bases of Bach's views of art, and of the processes of his culture and development; on the elements he assimilated from without; on the inspirations he derived from within and from his personal circumstances; when, finally, the universal nature of music comes to our aid in the matter, there still remains a last wonder—the lightning flash of the idea of a mass of such vast proportions—the resuscitation of the spirit of the reformers, as of waters that have been long gathering to a head, nay, the actual resurrection of the genius of primitive Christianity, and all concentrated in the mind of this one artist—as inscrutable as the very secret of life itself. A feeble quiver from this movement is to be noted, indeed, in the following generations of Protestants; down to quite modern times the idea of a musical setting of the mass has had a mysterious power to tempt composers. But even with the best of them —Spohr and Schumann—it was in great part merely an antiquarian and romantic whim; although Schumann's saying that "it must always be the musician's highest aim to address his powers to sacred music" betrays an evident feeling for the realm whence the fountain head of art must flow. No comprehensive treatment of the abstract conception of a universal Church, in the form of an ideal liturgy, could proceed from the Catholics, since they have not the requisite freedom within the limitations imposed by the Church; indeed, it has never been attempted.

No one can set Beethoven's Second Mass side by side with Bach's—as it is just now the fashion to do— who does not wilfully shut his eyes to the unmistakable gulf that yawns between what the idea of such a work demands and the spirit in which the execution of the former work is undertaken. In Beethoven's work we cannot but admire the grand individuality of its creator, and the Mass will be understood and loved as long as a hundred other works exist which reveal his genius more purely and fully. But though all of Bach's compositions might be lost, still the B minor mass, even to the remotest future, would bear witness to the artist's greatness with the weight of a divine revelation. There is only one other work that can really be set by the side of it. Handel's "Messiah" has often been compared to Bach's St. Matthew Passion, but this must inevitably lead to an unfair judgment of both these works, which, in reality, have hardly anything in common. The real companion work to the "Messiah" can only be the

B minor mass. The aim fulfilled by both works is the artistic presentment of the essence of Christianity. But the two men apprehended the subject differently; Handel viewed it from the independent and historical standpoint; Bach from the more limited doctrinal side. Though the latter was beyond a doubt the most suggestive as regards the depth of the world of feeling to be expressed, still the former afforded an opening for a more intelligible dramatic treatment, which is no less pure in art. As all the musical inspiration of that period was embodied in these two equally sound and gifted artists, and consequently each can only be perfectly understood through the other, in any honest historical review we must refrain from elevating one at the expense of the other. But the German nation may rejoice in boasting that both these incomparable geniuses were her sons.

III.

THE LATER CHORALE CANTATAS.

BACH devoted himself, even in the later years of his life, to the composition of church cantatas. It is only at first indeed that we observe that royal profusion which created such an unlimited wealth of musical forms during the middle of the Leipzig period.[65] He falls gradually back on one particular form of chorale cantata, he becomes more silent, and, when he speaks, it is in the regular typical form. He gives the finishing touches and the final shape to two of his greatest sacred works, the St. John and St. Matthew Passions, "setting his house in order," as it were, till at last he seems to become quite silent as a composer of vocal church music. His life-work is done, and he prepares himself for death.

We can point with certainty to the cantata with which Bach welcomed the New Year, 1735. This year found Europe plunged in war. In Italy the French, Sardinians,

---

[65] Compare Vol. II., pp. 477 and 434 ff.

and Spaniards were fighting against the Austrians, and the French were attacking the Austrian possessions on the Rhine. The petty rulers of Germany were seized with panic, and in the province of Reuss special weekly hours were set apart for prayer for God's mercy "in these fearful and dangerous circumstances of war." Meanwhile, however, peace had been restored in the empire of August III., after the subjugation and amnesty of Poland, and on his arrival in Warsaw the king was able to publish a pacificatory proclamation dated December 16, 1734. We have seen that on October 7, 1734, Bach composed a birthday cantata for the king,[66] in which he was celebrated as the peacemaker. The writer of the text of the New Year Cantata is inspired by the same idea. He views Saxony and Poland as a secure island around which may be seen the troubled waves of strife. While praising the king in this strain, he prays to Christ the Prince of Peace to perform His office. During all the time Bach was at Leipzig, there is only one occasion which will exactly suit the idea of these words, and that is the beginning of the year 1735. In the Silesian war Saxony was directly and essentially implicated; so that the date of the cantata is fixed beyond all doubt.

The work itself contains much that is remarkable. It is founded on vv. 1, 5, 10 of Psalm cxlvi., vv. 1 and 3 of Ebert's hymn "Du Friedefürst, Herr Jesu Christ"—"Lord Jesu Christ, Thou Prince of Peace," and only two sets of verses in madrigal form. Of these last, the second "Jesu, Retter deiner Heerde" can only be considered as partly in madrigal form, for the tenor solo with the bassoon and bass serves as the counterpoint to the chorale melody played by the violins and violas. So that the chorale comes in three times—first, as the second number, sung alone by the soprano with lovely interwoven accompaniments of the violins and basses; secondly, against the tenor solo in sombre colouring, the dominant idea of which must be contained in the second verse of the hymn;

---

[66] See Vol. II., p. 631.

and lastly, as a chorale fantasia for the chorus and all the instruments. The words of the psalm are used with no less frequency. The cantata begins with the words "Praise thou the Lord, O my soul" ("Lobe den Herrn," &c.). The tenth verse of the psalm is set to one of those bass ariosos which approach so nearly to the form of the aria, and which we have already pointed out as being a characteristic innovation of Bach's, in the years that immediately preceded this period.[67] Verse five is set to a simple tenor recitative; this kind of treatment is one which we have not hitherto met with in Bach, except in the "mysteries" which do not come into comparison with these works, and even there it is but of rare occurrence. In the cantata the words of the Bible and of the chorale strive, as it were, for the mastery. In this sense it is significant that the first chorus is of very limited extent (only thirty-five bars), and has no thematic development; so that it is not to be regarded as giving the emotional key-note of the whole; and in the final chorus the chorale is not wholly triumphant. While in most cases the voices that have the counterpoint sing the same words as those of the *Cantus firmus*, they here give out the last word of the Psalm, "Hallelujah"; both elements being thus united. In the relation thus established lies the individuality of this remarkable work.[68]

To the effect produced by the wars on the Rhine and in Italy is due the composition of the cantata "Wär Gott nicht mit uns diese Zeit"—"If God were not on our side"—which was performed about four weeks later, on the fourth Sunday after Epiphany (January 30, 1735). It is well known that Luther wrote a paraphrase of Psalm cxxiv. in three verses. This forms the germ of the text, but the second verse is paraphrased in the madrigal style.[69] The first verse is in the Pachelbel organ

---

[67] See Vol. II., p. 471.
[68] I know the cantata only by a copy in the Royal Library at Berlin. I know nothing of where the autograph may be.
[69] In B.-G. II., where this cantata is published, the words start on p. 126, "Ja hätt es Gott nicht zugegeben," apparently by a clerical error (for it is

chorale form. The chorus begins in a fugal style with
very ingenious answering in contrary motion, the themes
thus introduced being then used as counterpoint against
the *Cantus firmus*, which is entrusted to the horn and the
two oboes. To work out this new form—not technically,
for it is clear at the first glance, but figuratively, as it were,
realising in it the representation of an inward experience,
must have been immensely difficult. When the chorale is
only played, it leaves the greatest room for subjective and
subsidiary fancies with regard to its meaning; but when a
solo or chorus is added, with words and a melody of its own,
we find two contrasting elements, one subjective and fleeting,
the other objective and permanent, of which, however,
the first, as representing fully the Church element, must
predominate, elevating and sanctifying the effect of the
objective feeling. This crossing and alternation of emotions
is in the truest and most characteristic spirit of Bach's
romanticism. In the present case, however, as no essential contrast subsists between the instrumental *Cantus
firmus* and its vocal counterpoint, Bach allows an objective element to enter into the subjective character of
the melody that is played, but not so as to usurp its
place. This would hold good, even though the *Cantus
firmus* were sung. Bach has taken care to keep it
subordinate. Foreshadowings of the form here ventured
upon are found as early as the first chorus of the
cantata "Es ist nichts gesundes an meinem Leibe,"
so far, that is to say, that the themes of the vocal fugue
are derived from two lines of the chorale which is played.[70]
The two arias are very important, the first with its quaint
rhythm being most characteristic, while the second, as full
of sentiment as of ingenuity, is most elevating, especially

---

nonsense) for "Ja hätte Gott es zugegeben"; the cantata is also published P.
No. 1297. By way of exception to his general rule, Bach himself gives the date
of this cantata as 1735. The paper on which the parts are written is the same
as that containing those of the cantata "Vereinigte Zwietracht," in its form
as altered for the king's birthday. The year fits very well for the performance
of this music; see Vol. II., p. 628, note 766.

[70] Vol. II., p. 466.

in the second part, by the expression of a firm and valiant faith.

In March, 1735, Bach completed his fiftieth year. That his creative activity remained in undiminished strength is shown by the fact that to no year can so great a number of church cantatas be ascribed, either with absolute certainty or with reasonable probability. No fewer than twenty cantatas seem to have been produced by him in this year; among them, it is true, are several re-modellings of works written at Arnstadt, Weimar, and Cöthen, and the cantata "Komm du süsse Todesstunde"[71] was left entirely unaltered, excepting that it was now to serve, not for the sixteenth Sunday after Trinity, but for the festival of the Purification (February 2). For Easter day (April 10) he had recourse to a work of his earliest youth ("Denn du wirst meine Seele nicht in der Hölle lassen")[72] and, for the Tuesday in Easter week, to an occasional cantata written at Cöthen ("Ein Herz, das seinen Jesum lebend weiss").[73]

The music for the Monday in Easter week[74] owes its pleasing character to the circumstance of its coming between those two last mentioned. Bach had the gift of throwing himself, up to a certain point, into various kinds of styles, whether those of other persons or his own in his earlier phases. Careful comparison will at once show that there is a relation between the occasional cantata "Erfreut euch ihr Herzen" and the same in its remodelled form. A pleasing character, aiming rather at breadth than at depth, is not the only characteristic that is common to both. The first chorus of the earlier composition agrees exactly in its plan with the last chorus of the later work, and even the passages set as duets, especially those of the middle movement, which in the occasional compositions were necessitated by the text, were copied in their setting in the Easter cantata. Both are full of genius and elegance, although they cannot lay claim to a prominent place among

---

[71] See Vol. I., p. 549 ff.
[72] See Vol. I., p. 229 ff.
[73] See Vol. II., p. 619 ff.
[74] B. G. XVI., No. 66. P. 2145.

Bach's Easter compositions. It will not escape the attentive observer that the last bar but one of the bass recitative is referred to at the beginning of the second part of the aria which follows it. There was possibly some accidental reason for this, and for the somewhat uncalled-for introduction of the same figure in the recitative itself. A leaf, on which is written the first idea of the beginning of the cantata for the Sunday after Ascension day in the same year, also contains the following sketch:—

This may have been the subject he first intended for the bass aria in the second Easter cantata, and when he altered his mind as to the chief subject he may have introduced this as a subsidiary.

Two cantatas for Whitsuntide—"Wer mich liebt, der wird mein Wort halten," and "Also hat Gott die welt geliebt"— were in part only made up from old compositions. The first is built upon the cantata with the same beginning which Bach wrote in Weimar, to a text by Neumeister.[75] Two numbers from this work are introduced; but the whole cannot be considered as finer than the older work. The arias for tenor and alto have a remarkably undevotional character, being showy and almost secular; there is no relation between the length of the cantata and its importance.[76] For the arias in "Also hat Gott die welt geliebt" the material was supplied from the much used occasional composition "Was mir behagt, ist nur die muntre Jagd." The cleverness of this recension, which contains the lovely soprano aria "Mein gläubiges Herze"—"My heart ever faithful"— has been pointed out in another place.[77] Each of the arias

---

[75] Vol. I., 511 ff.
[76] B.-G. XVIII., No. 74.
[77] Vol. I., 568 ff. The instrumental working-out of the bass theme, which in the autograph score of the secular cantata is found on the back of the final chorus, does not belong to the soprano aria "Wenn die wollenreichen Heerden," and there is no evidence for supposing that they have any connection. It would seem that Bach at first threw off this working-out on the blank page that remained, when he undertook to alter the cantata for Whitsuntide, intending it for a coda or appendix to the new recension of the work.

is supplemented by a newly composed chorus, the character of which is given by the aria; the soprano aria closes with the opening chorus, a four-part choral aria, the accompaniment to which makes the whole work into a charming sort of sacred Siciliano. The feeling of the bass aria, on the other hand, is carried out in a powerful concluding fugue, surrounded with passages for the trombones.[78] This cantata is as original throughout as it is important. In all five of these remodelled cantatas, however, the chorale plays now no part at all, and now but a subordinate one.

Bach seems to have written a new cantata for every Sunday and holy day which fell between Easter and Whitsuntide in the year 1735, with the single exception, perhaps, of the first Sunday after Easter For the Ascension he wrote two cantatas; a rare occurrence, but then for this festival it was necessary to perform two pieces of concerted music. The disposition of the text is noteworthy. The poet frequently turns back from the well-worn track of madrigal poetry to the simpler hymn verse, thus originating some agreeable forms; and the words of Scripture are used more frequently than usual. The feeling of the text is deeper and purer than the average of those in the earlier cantatas in the madrigal style, and it often rises to real devotional strength. We should like to know whether this signifies that a new poet was employed, or whether Bach, after plainly showing his distaste for Picander's meaningless doggerel, by composing music to the verses of hymns exclusively, succeeded in inspiring the poet with some of his own earnest spirit. The chief part of the Ascension cantata (Himmelfahrts-Cantate) "Gott fähret auf mit Jauchzen"—"God is gone up"—consists, from the soprano aria onwards, of a poem in six verses, neat in form and not devoid of a certain fervour.[79] The first aria of the third Whitsuntide cantata, "Er rufet seine Schafe mit Namen," is set to graceful words cast in the metre of the hymn "Ach Gott und Herr."[80]

---

[78] B.-G. XVI., No. 68. P. 1287.
[79] B.-G., X., No. 43 p. P. 1658. In English, published by Novello.
[80] The autograph score and original parts of the cantata are in the Royal Library at Berlin.

The similar arrangement of the text of the cantatas for the fourth Sunday after Easter (" Es ist euch gut, dass ich hingehe ")[81]—" It is good for you that I should leave you"—for the fifth Sunday after Easter (" Bisher habt ihr nichts gebeten in meinem Namen ")[82]—" Hitherto ye have asked nothing in my name "—for Ascension (" Gott fähret auf"), and for the third day of Whitsuntide ("Er rufet seine Schafe mit Namen")—" He calleth His sheep by His name "—in which a text of Scripture is introduced both at the beginning and in the middle, shows that they are by the same hand.[83] Bach, for his part, who, in remodelling his earlier compositions for these cantatas was somewhat fettered by their original form, has left us a set of new compositions of the rarest beauty. He has here developed a form which he had previously attempted in single instances —namely, a new style of vocal solo setting of Scriptural words—a form in which the addition of concerted instrumental parts to the ordinary arioso with contrapuntal figured bass should bring it nearer to the richer form of aria, while remaining quite distinct from it. We can conceive of nothing which could more perfectly point out both the meaning of the sacred words and the necessity of their personal application than the lovely bass solos with which the compositions for the second Sunday after Easter (" Ich bin ein guter Hirt "),[84] for the fourth Sunday after Easter (" Es ist euch gut, dass ich hingehe "), and for the fifth Sunday after Easter (" Bisher habt ihr nichts gebeten ") begins. The arioso with the mere *basso continuo* is not, however, discarded in these cantatas. Several passages of Scripture are also treated in the pure recitative style, contrary to Bach's general usage in the church cantatas properly so-called. Two of the cantatas begin thus modestly with a recitative of this kind; as also does the cantata for the Sunday after Ascension day (" Sie werden euch in den

---

[81] B.-G., XXIII., No. 108. P. 2148.
[82] B.-G., XX.,[1] No. 87.
[83] The words of the Whitsuntide cantata " Wer mich liebet," in which three Bible texts are introduced, seem to be by the same author.
[84] B.-G., XX.,[1] No. 85. P. 2140.

Bann thun")[85]—"They will cast you out." On the whole, a great freedom of formation, both in outline and detail, characterises these cantatas. A good part of this is due, of course, to the exceptional forms of the texts. As the words were not written with constant regard to the scheme of the Italian aria, the composer had to accommodate himself to them. This gave rise to various ingenious departures from the usual form. The shorter cantata for Ascension ("Auf Christi Himmelfahrt allein")[86]—in which the text is incoherent and generally inferior—offers the original phenomenon of an aria which loses itself in recitative, but yet so far verifies its character that at the conclusion the opening ritornel of the aria is brought in as a close. Many of the solos, for the splendid wealth of melody, the tender expressiveness, the blending of colours, the swing and majestic pathos which they display, are worthy to rank among the very highest of Bach's productions in this kind. The choruses are comparatively few. Chorales are for the most part introduced only as simple closing movements; the shorter cantata for the Ascension alone begins with a chorale fantasia, and in the music for the second Sunday after Easter such a fantasia occurs in the middle, but the soprano alone takes part in it. And yet where free choruses do occur they are instinct with character and life, and replete with that freedom and boldness of form which characterise true genius.

For the third Sunday after Easter he begins with a chorus, "Ihr werdet weinen und heulen"[87]—"Ye shall weep but the world shall rejoice"—in which the contrasts of tears and joy are depicted with marvellous power, and are at length united in a double fugue; after this there comes a bass recitative, followed by the opening movement developed at greater length and with other words.

The great cantata for Ascension ("Gott fähret auf"), the character of which is in the first three movements rather that of the oratorio, while the rest is more devotional,

---

[85] Autograph score and original parts in the Royal Library at Berlin.
[86] B.-G. XXVI., No. 128.
[87] B.-G., XXIII., No. 103. P. 1697.

exhibits a tone-picture of Christ's Ascension full of majestic and graphic movement, both in outline and in detail; everything presses upward, the form of the different themes as well as the construction of the fugue, and one important subject comes crashing in like the blast of a trumpet. The opening of the choral movement is thoroughly original; it is a short and solemn instrumental adagio which foreshadows the first fugal theme by way of a prelude. When the allegro is reached, the theme and the counter subject are developed at first on the instruments, and after the most complete musical preparation the chorus bursts in with the clash of trumpets and drums.

In the cantata "Es ist euch gut, dass ich hingehe," all the middle of the work is usurped by one grand chorus on the words "Wenn aber jener der Geist der Wahrheit kommen wird"—"Howbeit when He, the Spirit of truth, is come, He will guide you into all truth." It comes in immediately after a *recitativo secco*, and speaks to us in inspired and overpowering tones, like those of the Apostles at Pentecost. In the fugal theme which breaks through every barrier:—

Wenn a - ber je - ner der Geist der Wahr - heit kom - men wird

we seem to feel that "the Lord's word is like a hammer that breaketh the rock."[88]

There are but few cantatas having their chief choruses founded on freely invented themes, besides those which have been mentioned, for during the last period the chorale cantata is decidedly in the ascendant. In works which can be confidently assigned to the same date as those just described, his predilection for new forms, ingeniously wrought out and sharply characterised, is very evident. The text ordered for the sermon for the Reformation festival of 1735—which was kept on October 30, the

---

[88] In the parts Bach changed the *d* of the first entry into *e*, so that the skip of a seventh becomes a sixth, probably only in order to facilitate the attack. We may be allowed to retain the original reading. For the date and watermark, see Appendix A., No. 2.

twenty-first Sunday after Trinity (the proper day, October 31, falling on a Monday)—was Ps. lxxx. 14—19. The words "Look upon it . . . it is burned with fire, it is cut down," were evidently chosen with application to the results of the war in 1735. And in the text of the cantata "Gott der Herr ist Sonn und Schild" the lines—

> Denn er will uns ferner schützen,   His protecting hand shall save us,
> Ob die Feinde Pfeile schnitzen   Though the foe exult and brave us
> Und ein Lästerhund gleich billt—   Raging and blaspheming still—

suggest a vein of feeling that cannot be adequately accounted for by the ordinary character of the Reformation festival. The music too is by no means devoid of martial character. In the duet "Gott, ach Gott," the violins do not content themselves with a contrapuntal accompaniment in Bach's usual manner, but paw the ground like impatient chargers, and burst in upon the voice parts with unrestrained energy. A similarity to the second movement of the cantata "Ein feste Burg" must not be overlooked; and it is worth mentioning that this movement, when it is transferred to the G major Mass, has a quite different accompaniment. One of the chief subjects of the opening chorus too is instinct with martial vigour and the roar of battle. In the construction of this chorus Bach was imbued with the idea of the concerto form, for two strongly contrasted subjects are introduced side by side and alternately. But the most striking and typical feature is the way in which the chorus is crystallised, as it were, on the instrumental form. At first it only affords a background of wide-spread harmony to the second subject, to the words "Gott der Herr ist Sonn und Schild," clearly symbolising that every battle is to be fought in the name of God. This subject is subsequently turned into the theme of a fugue. The elaborate combinations of instruments which are found throughout this great work, down to the rhythm of the drums, cannot be gone into in detail. The poetic meaning of the first subject becomes perfectly clear in the third movement, where it serves as an accompaniment to the chorale "Nun danket alle Gott," sung by the chorus in simple and beautiful harmony.[89]

At the time of the first performance of the Easter oratorio—*i.e.*, 1736—Bach wrote a cantata for the second day of Easter (April 2, 1736); "Bleib bei uns, denn es will Abend werden"—"Bide with us," &c.[90] The Gospel for the day narrates the touching story of the two disciples on the way to Emmaus. In the principal chorus a movement full of deep feeling and yet of mild expression, composed of several subjects most skilfully interwoven, is enclosed between two slow movements, the longing and yet simple character of which is deeply affecting. This kind of triple division, in which the emotion passes from quiet stillness to strongly moved passion, and then back again, is a new phase in Bach's cantata choruses; for the choruses in overture form have nothing in common with these, owing to the totally different character of that form.[91] The chorus is also remarkable for an intensely vivid picture of nature, which has already been noticed.[92] In the middle movement the long-drawn notes "Bleib bei uns" are heard through and above the tangled web of parts, as though distant voices were heard calling across a plain through the twilight. Long dark shadows fall across the landscape at the beginning and end; Bach himself took care that there could be no misunderstanding of his meaning here, for the same effect of tone is used by him in the tenor aria of a secular cantata, where it serves to illustrate the words "Frische Schatten, meine Freude, sehet wie ich schmerzlich scheide"—"Cooling shades, my great delight, see with what regret I quit you."[93] The downward motion of most of the parts suggests the approach of night; and in various passages of the alto aria, which is steeped

---

[89] B.-G., XVIII., No. 79. P. 1013. See Appendix A., No. 2.
[90] B.-G., I., No. 6. P. 1015. See Appendix A. of Vol. II., No. 58; and Appendix No. 2 of this Vol. Published in English by Novello.
[91] Among Bach's solos the aria "Seligster Erquickungstag," from the cantata "Wachet, betet," has this kind of form. See B.-G., XVI., p. 364 ff.
[92] See Vol. II., p. 561.
[93] B.-G., XI.,² p. 181.

in noble yearning, the gloom of darkness is represented in a strangely weird manner by means of sequences of tone and harmony. Whether Bach thought that enough had been done in this direction, or from some other cause, the beautiful evening hymn "Ach bleib bei uns, Herr Jesu Christ," which was such an especial favourite for evening services, and which forms the middle movement of the cantata, is rather wanting in depth for a composition of Bach. As in an organ trio, the soprano sings the *Cantus firmus* while the figured bass and a *Violoncello piccolo* have counterpoint upon it. The last-named instrument derives its subject from the first line of the tune, but soon goes off into wonderful leaps, arpeggios, and runs. At the close, where we should expect the return of the same chorale, we find the second verse of "Erhalt uns Herr bei deinem Wort," which was probably suggested by the doctrinal turn now taken by the words of the cantata. At all events, the remainder of the work cannot be compared with the impressive beauty of the first two movements.

The cantatas "Es wartet alles auf dich" (seventh Sunday after Trinity)[94] and "Wer Dank opfert, der preiset mich" (fourteenth Sunday after Trinity)[95] were used for the Masses in G minor and G major, so that they must have been written before the date of these works, and that is about 1737. The Gospel tells of the feeding of the four thousand, and the chief chorus of the first cantata extols the goodness of God, Who nourishes all His creatures with inexhaustible gifts, in the words of the Psalmist (Ps. civ., 27 and 28). The chorus is grandly conceived and splendidly worked out, excepting that, considering the subject, it is curiously gloomy. In the course of the work, which goes on to a general glorifying of God's mercies, this character is not however maintained. On words taken

---

[94] The autograph score and original parts in the Royal Library at Berlin; the original instrumental parts are in the possession of Herr Professor Rudorff of Lichterfelde, near Berlin. I have lately observed that the watermark seems to assign it to July 7, 1732. It is that described in note 48, Appendix A., of Vol. II.

[95] B.-G., II., No. 17. P. 1282.

from the Sermon on the Mount (Matt. vi., 31 and 32) there is developed one of those bass ariosos which so nearly resemble the aria, full of that doctrinal zeal which Bach knows so well how to clothe with beauty. The arias for alto and soprano have a character of waving fulness like ripe cornfields with their golden blessings. There is in them something of that warm and lovely sensuousness which Mozart was the first to bring to perfection. There is a most captivating passage in the soprano aria *Un poco allegro* in 3-8 time, in which the chief subject of the Adagio appears again with altered rhythm. The old-fashioned style of the hymn of praise, "Singen wir aus Herzensgrund," leads us back into the feeling of the principal chorus. The cantata " Wer Dank opfert," also extols the goodness and might of God, but throughout in brighter colouring. The joyful strain of the chief chorus is brought in in a splendid fugue. The middle of the cantata is built on a short passage of Biblical narrative, which suggests a new starting point. The same method is pursued in the Ascension cantata " Gott fähret auf mit Jauchzen."

In the period between 1738—1741 there are three more cantatas with choruses in a free style, which consist more or less of remodelled forms of other works. A composition for St. John's day "Freue dich, erlöste Schaar,"[95*] is to be assigned to June 24, 1738. It is founded on the secular cantata " Angenehmes Wiederau,"[96] and expresses a calm and happy feeling which is less suited to the dogmatic character of the festival than to the time of year at which it is held.

To about the same time, perhaps to the late autumn of 1737, belongs an important work for a marriage ceremony, " Gott ist unsre Zuversicht." All the chief numbers of its second part are taken from the Christmas cantata " Ehre sei Gott in der Höhe."[97] The great opening chorus is very effective and intelligible, and yet full of purpose. The broadly treated melodies of the alto aria are full of a certain enchanting sweetness of a kind which

---

[95*] B.-G. V., No. 30. P. 1017.
[96] See Vol. II., p. 635, and Appendix A., to Vol. II., No. 58.
[97] See Vol. II., p. 440.

occurs only in Bach's wedding cantatas. The title which the work bears, *In diebus nuptiarum*, indicates a great and solemn ceremonious occasion, very likely the marriage of some exalted personages. One wedding cantata, which, however, has only come down to us in an incomplete state, but which seems to have been a very important and interesting work, was used by Bach about 1740 or 1741 for a Whitsuntide cantata "O ewiges Feuer, o Ursprung der Liebe."[98] It contains two choruses and an alto aria "Wohl euch ihr auserwählten Seelen"—"Rejoice, ye souls elect and holy"—between two short recitatives. The re-arrangement is perceptible in the brevity of the final chorus, which had not the same position in the wedding cantata; it is also conspicuous in the aria, in no part of which can the original bridal feeling pass unnoticed. For its pure and ardent atmosphere, its magic charm of tone, its lovely melodies, it is indisputably fitted to rank at the head of all Bach's works of this kind, and to be considered an unapproachable model. In the lavish adornment of the first chorus there is a touch of ardent human love, which, from Bach's pure ideal nature, is in no way incompatible with the emotional aspect of Whitsuntide, but which only receives its full explanation when we understand the original purpose and object of the chorus.

It has already been pointed out that Bach adapted instrumental works for his church cantatas, not merely as symphonies but also as solos.[99] There are indeed two cantatas in which instrumental movements are turned into choruses—a Christmas cantata, "Unser Mund sei voll Lachens,"[100] and a work for the third Sunday after Easter, "Wir müssen durch viel Trübsal in das Reich Gottes eingehen."[101] As in the *Gloria* of the Mass in A major, the

---

[98] B.-G., VII., No. 34. P. 1291. In English, "O light everlasting" (Novello). See Appendix A. to Vol. II., No. 57. I may here remark that, in 1742, owing to the fortnight's mourning for the death of the widowed Empress Maria Amalia, no church music was performed on Whitsunday or Trinity Sunday, so that the cantata "O ewiges Feuer" cannot have been performed in 1742.

[99] See Vol. II., p. 448.

[100] B.-G., XXIII., No. 110. P. 1681.

[101] In a later MS. in the Royal Library at Berlin.

chorus is worked into the fabric of the instrumental piece, without altering it in any essential particular. For the Christmas cantata Bach used the overture of an orchestral suite in D major.[102] We cannot be certain when this suite was written; but a comparison between it and the suites in C major and in B minor makes it probable that Bach wrote it, not in Cöthen, but in Leipzig, where works of this kind were naturally suggested by his post of director of the Musical Society. As we already know three cantatas for Christmas Day written by him between 1723 and 1734, and as he did not enter upon the conductorship of the Musical Society until 1729, it is highly improbable that the cantata "Unser Mund sei voll Lachens" can have been written before 1734.[103] Only the fugal allegro of the overture is transformed, and that with astonishing power, into a chorus, the slow movements forming the symphonies before and after it. Bach adopted a similar method in the cantata "Preise Jerusalem den Herrn," and in the chorale-cantata "In allen meinen Thaten,"[104] while elsewhere ("Nun komm der Heiden Heiland," "Höchsterwünschtes Freudenfest," "O Ewigkeit, du Donnerwort") he makes the chorus take part in the grave movement.[105] The pompous character of the French overture made it very suitable to a Christmas composition, although it cannot be denied that, by confining the chorus to the allegro movement, he gives that movement a predominance which is foreign to the form from which it is taken. In this cantata we also find the *Virga Jesse floruit* from the *Magnificat*[106] reset to the words "Ehre sei Gott in der Höhe"—"Glory to God in the highest." The rest of the solo numbers, which seem to be entirely new, are of great merit, and, especially in the alto aria, have an earnest and manly character, which we do not find in the earlier Leipzig cantatas. The same may be said of the cantata "Wir müssen durch viel Trübsal in das Reich

---

[102] See Vol. II., p. 141.
[103] See Appendix A. to Vol. II., No. 24.
[104] See Vol. II., pp. 362 f. and 457.
[105] See Vol. I., 507, Vol. II., pp. 365 and 421.
[106] See Vol. II., p. 269.

Gottes eingehen." A comparison of the alto aria with organ *obbligato* " Ich will nach dem Himmel zu " with the aria " Willkommen will ich sagen," from the cantata " Wer weiss, wie nahe mir mein Ende," [107] will make this clear. The arias have many points of resemblance, even in the disposition and accompaniment of the vocal parts; but the ecstatic longing for death is more intense in the later work. As an instrumental section Bach uses for this cantata the same D minor concerto that he had already used for the introduction to the cantata "Ich habe meine Zuversicht" after the Rückpositiv of the organ in St. Thomas's Church was made to be played upon independently.[108] Here, however, only the first movement is used for the symphony; in the adagio the principal chorus is brought in in such a manner that the part for the solo instrument goes on simultaneously and independently, a masterpiece of ingenuity and skill.

Of the four cantatas for Trinity which have been preserved, one still remains to be mentioned, as to the date of which nothing can be asserted with any certainty. As, however, no fewer than three of these cantatas date from the year 1723—1732 ("Höchsterwünschtes Freudenfest," " O heilges Geist- und Wasserbad," " Gelobet sei der Herr "), it may be presumed to have been written at a later period; and the style of the music favours this presumption. In particular, this is seen in the chief chorus, the characteristic energy of which makes it a worthy companion piece to the choruses " Herr deine Augen sehen nach dem Glauben," " Ihr werdet weinen und heulen," " Gott fähret auf mit Jauchzen," and others. But, with all its musical beauty, the cantata, as a whole, is an inexplicable enigma. The Gospel contains the story of how Nicodemus came to Jesus by night for fear of the Jews. In this the writer of the words sees an instance of weak cowardice, and, accordingly, he begins his poem with the text, "Es ist ein trotzig und verzagt Ding"—"There is a

---

[107] See Vol. II., p. 451.
[108] See Vol. II., p. 446 f.

perverse and cowardly thing in the heart of man." He goes on to say that when the Christian dares not seek Jesus openly a feeling of shame for his own unworthiness takes possession of him. But he may take courage in the hope of salvation through faith. The sequence of ideas must have been as clear to Bach as it is to us; for that reason it is inexplicable why he set himself in the most distinct opposition to it in his music. He lays the emphasis in the first chorus not on the faint-heartedness but on the resistance of the heart. And it is a Titanic resistance that storms out against heaven in the fearfully energetic theme of the fugue. The number by itself is a masterpiece of the highest order, but its character deprives it of all inner connection with what follows. In strong contrast to this is the first aria, which is in the style of a gavotte. It is charming as a piece of music, but quite unsuited to its text, which treats of the timidity of the Christian in approaching his divine and wondrous Master. Although, as the work goes on, there is more connection between the words and the music, yet a harmonious co-operation of effect is rendered impossible. If the condition of the autograph did not prove it otherwise, we should suspect that the first two numbers were transferred from elsewhere. But as it is, we can do no more than simply point out the discrepancy that exists.[109]

In each of the two cantatas, having their chief choruses in free style, which still remain to be mentioned, as in the compositions for Trinity Sunday and for Christmas-day, just described, we find a text of Scripture in the middle as well as at the beginning: a form of which the frequent recurrence in the cantatas written at this time will have been noticed. "Brich dem Hungrigen dein Brod" —"Give the hungry man thy bread"—is for the first Sunday after Trinity, and "Es ist dir gesagt, Mensch, was gut ist" for the eighth.[110] The resemblance between the two, as regards the musical conception and execution, lies on the surface. The chorus of the first, set to two

---

[109] The autograph score is in the Royal Library at Berlin.
[110] B.-G., VII., No. 39. P. 1295; and B.-G., X., No. 45. P. 1016.

beautiful verses of Isaiah (lviii. v. 7, 8), brings out the meaning of that text in the Sermon on the Mount, "Blessed are the merciful, for they shall obtain mercy," and the cantata is fitly concluded with the sixth verse of the paraphrase of the beatitudes.[111] It is an affecting picture of Christian love, softening with tender hand and pitying sympathy the sorrow of the brethren, and obtaining the highest reward. The peculiar accompaniment, allotted to flutes, oboes, and strings, was very likely suggested to Bach by the idea of the breaking of bread. But how little he cared for such trivial realism is seen, as the number goes on, in a passage where the accompaniment is continued to entirely different words. It gives the piece a tender, dreamy tinge, and this was what Bach chiefly wanted. The second part of the cantata opens with the text in Hebrews, xiii. 16: "Wohlzuthun und mitzutheilen," &c.—"To do good and to communicate forget not, for with such sacrifices God is well pleased." This is sung by the bass in the usual way. The two arias have an expression of loving activity and kindliness. The Gospel for the day treats of the rich man and Lazarus.

The second cantata, which is wholly different from this, is a grand Protestant sermon on the duties required of the Christian by God, in order that he may stand before His judgment seat. The whole has a character of severity, and the chief subject of the first chorus is not without a trace of orthodox harshness. But its source is no dead and barren consent to received dogmas; it is a living enthusiasm for a lofty purpose which imbues this elaborate chorus with earnest vigour.

A mighty *torso* of a church cantata is preserved in a double chorus with the richest orchestral accompaniment, on the words from Rev. xii., 10, "Nun ist das Heil," &c.—"Now shall the grace," &c.[112] It must be called a torso, for it seems that in its present form it falls in with no church use whatever. It is too short to lay claim to the title of a

---

[111] "Kommt lasst euch den Herren lehren," ascribed to David Denicke.
[112] B.-G., X., No. 50. P. 1657. In the Bach Choir Magazine with English words.

church cantata in regular form; it cannot serve as a motett by reason of its concerted accompaniments, and its subject of course would forbid all thought of its being performed during the communion service; and there is no other occasion for which it would be suitable. It must certainly have formed the opening of a complete cantata for Michaelmas, and it may be assumed to have had an orchestral prelude. The colossal piece, however, by itself, with its ponderous march and its wild cries of victory is an imperishable monument of German art.[113]

There are extremely few solo cantatas remaining which can be assigned to the later Leipzig period. A *Dialogus* between Christ and the Soul, for the second day of Christmas ("Selig ist der Mann, der die Anfechtung erduldet"),[114] belongs to the class of those compositions which represent domestic sacred music rather than church music. It contains no trace whatever of special fitness, and if the intention of the work were not expressly given no one would conceive that it was meant for Christmas. The case is different with a work for the third day of Christmas, "Süsser Trost, mein Jesus kommt."[115] Deeply and thoroughly as Bach had given expression to the ecclesiastical intention of the festival, both in the Christmas Oratorio and in separate arias in earlier cantatas,[116] this work testifies that he had not yet exhausted the subject. The pure happiness of Christmas is here transfigured and glorified. The bright and silvery soprano voice with its simple, sweet, and longdrawn melodies, accompanied by tender waving passages on the oboe, seems to present to us the idea of an angel of peace hovering over the dark city. The middle movement of the first aria differs

---

[113] A cantata is preserved, though only in incomplete parts, which begins with the chorus "Ihr Pforten zu Zion, ihr Wohnungen Jacobs freuet euch"—"Rejoice, ye gates of Zion, and ye dwellings of Jacob." Zion here represents Leipzig, and the occasion for which the music was composed was the election of the Council. No more definite evidence than this of its date can be given. The fragments of this cantata are preserved in the Royal Library at Berlin.

[114] B.-G., XII.,² No. 57. P. 1661.

[115] The original parts are in the Royal Library at Berlin.

[116] "Nun komm der Heiden Heiland," see Vol. I., p. 507; "Tritt auf die Glaubensbahn," Vol. I., p. 560.

in time and pace from the rest, a form not unfrequently employed by Bach in his Leipzig period.[117] The second aria has a gentle rocking motion, and though wholly terrestrial is very lovely; the Christmas chorale "Lobt Got ihr Christen allzugleich" closes the beautiful little work.

Another *Dialogus* composed for the first Sunday after Epiphany ("Liebster Jesu, mein Verlangen")[118] has a particular interest for us in the fact that, in the duet "Nun verschwinden alle Plagen," an idea which he had evolved in an earlier cantata for the same Sunday after Epiphany is developed at greater length.[119]

Side by side with this *Dialogus* we must consider a cantata for the second Sunday after Epiphany, "Meine Seufzer, meine Thränen."[120] It betokens the poverty of thought and one-sidedness of view in the sacred poetry of the time that no lesson could be deduced from the Gospel story of the marriage in Cana, in which a tone of noble cheerfulness is evidently predominant, than this, so endlessly reiterated, that Jesus helps the despondent sinner in his need. With more or less variation all the chief cantata librettists of the period harp on this thought alone. The three cantatas composed by Bach for this Sunday[121] are one and all founded on this idea. In the one now under consideration the feeling of trouble, of "groans and piteous crying for salvation," is most persistently brought forward. Until the chorale at the close, hardly a single ray of sunlight breaks in upon this overclouded world. Bach never wove a more compact and consistent fabric of tones of mourning than the arias for bass and tenor. It seems impossible that this combination of deep and spontaneous emotion with the display

---

[117] For instance, in the cantatas " Ihr Menschen, rühmet Gottes Liebe," " Es wartet alles auf dich," " Gott ist unsre Zuversicht," "Am Abend aber desselbigen Sabbaths," " Ach lieben Christen seid getrost," " Ich freue mich in dir."
[118] B.-G., VII., No. 32. P. 1663.
[119] See Vol. II., 402.
[120] B.-G., II., No. 13.
[121] The others are " Mein Gott, wie lang ach lange" and "Ach Gott, wie manches Herzeleid" (in A major).

of infinite ingenuity should ever be outdone. Led by his natural artistic feeling, the master gave to the chorale fantasia which separates the two arias a somewhat milder and simpler character. The semiquaver figures on the violins have a breath of coolness, although this conception is not suggested by the words of the chorale, and therefore no real repose is attained. Bach revelled in sorrow, and to him the setting of cantata poems of this kind was a welcome task. But from this work we again see that music was gradually gaining in independence and beginning to free itself from the Church. A work of this kind is certainly not suitable to a service which ought, in accordance with the Gospel, to point the lesson that Christ holds communion with His elect, showering on them His gifts, and rejoicing in the joy of men.

Still, it can seldom be said that Bach's church music does not fulfil its liturgical purpose. Highly suitable, for instance, are the cantatas for the twenty-fifth Sunday after Trinity ("Es reifet euch ein schrecklich Ende") and for the first Sunday after Easter ("Am Abend aber desselbigen Sabbaths"). The first, in spite of the scanty materials employed, is a powerful and affecting work.[122] The second opens with a lovely symphony, which exhibits the form of the first movement of a concerto in ingenious combination with that of the aria in three sections, and must have been adapted from some secular instrumental composition;[123] the beginning of the alto aria seems also to be an adaptation. The duet between the soprano and the tenor is set to a verse of a chorale "Verzage nicht, o Häuflein klein," but only using the words; the music is freely invented. Here Bach again treads the same path which he struck out in the cantatas "Gelobet sei der Herr," "In allen meinen Thaten," and others.[124] Every one of these solo

---

[122] B.-G., XX.,[1] No. 90. The instruments to be used are not given in the autograph score. They can be supplied, however, from Breitkopf's catalogue for Michaelmas, 1761. There it stands in so many words: *à Tromba*, 2 *Violini*, *Viola*, 4 *Voci, Basso ed Organo*.

[123] B.-G., X., No. 42. P. 2144. In the autograph score, the symphony and the first part of the aria are written out fair.

[124] See Vol. II., p. 457.

cantatas is divided among different voices, and they all conclude with chorales in four parts. Only in the cantata "Meine Seele rühmt und preiset" does the tenor sing alone, and there is no chorale. It thus belongs to the same class as the vocal compositions "Ich habe genug," "Widerstehe doch der Sünde," and others.[125] I mention it here at the end because, although there is no doubt of its genuineness, as yet there is no documentary evidence for it whatever.[126]

We come now to that form of cantata in which Bach's creative power as applied to sacred music finally expired. It must not, it is true, be forgotten that a portion of his church cantatas has been lost. As among these some would surely be found which were composed at Leipzig, the existing relations between the groups wherein the different forms are represented, would probably be altered were the works to come to light again. But yet it is clear, from the large number of works in the same form and written in the same period, that during the last part of his life Bach adhered to one and the same form with a persistency not elsewhere to be observed even in him. This form is the chorale cantata.

The reader must be again reminded that about 1732 the idea occurred to Bach of setting complete hymns to music; that is, that for certain verses the proper melody is retained, while other verses are made to serve as texts for recitations and arias without the help of any words but those of the hymn.[127] I called these cantatas works of transition or of digression, lying on or beside the way to higher developments of form. Works of this kind, treated in a different way, also occur, a chorale being taken as the central idea;

---

[125] See Vol. II., p. 473.

[126] A late MS. copy from the collection of Professor Fischhoff is in the Royal Library at Berlin. I would expressly draw attention to the general fact that all the solo cantatas described above are only assigned to the later Leipzig period on an assumption of probability. For the correctness of this assumption there is one piece of negative evidence—namely, that they bear no outward marks whatever which would oblige them to be assigned to an earlier date— and positive evidence is borne by their ripeness and depth of feeling.

[127] See Vol. II., p. 454 ff.

but its character, as a hymn, is not fully brought out, since sometimes the words and sometimes the tune only is made use of. In a cantata for the nineteenth Sunday after Trinity[128] the chorale occurs only in an instrumental form in the principal chorus. The chorus sings passages of imitation to the words (Rom. vii., 24) " Ich elender Mensch," &c.— " O wretched man that I am," &c. To the instruments is allotted an independent subject, twelve bars long, which is played again and again from the beginning, according to the method which we first became acquainted with in the chorale "Was Gott thut, das ist wohlgethan," from the cantata " Die Elenden sollen essen."[129] Against this subject, the trumpets and oboes play the chorale in canon on the fifth ; at the close the last two lines are repeated by the trumpet alone, so that the form becomes, as it were, a musical interrogation. In order to understand the complete poetic meaning of the chorus it is necessary to know the words which belong to this chorale melody. They run thus :—

> Herr Jesu Christ, ich schrei zu dir
> Aus hochbetrübter Seele,
> Dein Allmacht lass erscheinen mir
> Und mich nicht also quäle.
> Viel grösser ist die Angst und Schmerz,
> So anficht und turbirt mein Herz,
> Als dass ichs kann erzählen.
>
> Lord Jesu Christ, I cry to Thee,
> When heart and flesh are failing,
> Show Thine Almighty power to me,
> To keep my soul from quailing.
> So great my pain and deadly smart,
> So troubled and distressed my heart,
> That words are unavailing.

The cantata is brought to a close with the 12th verse of the hymn simply set in four parts. In the middle, however, we meet with a second chorale melody ; the fourth verse of "Ach Gott und Herr" is set in four parts, with modulations

---

[128] B.-G., X., No. 48. P. 1699.
[129] See Vol. II., p. 355.

of incredible boldness. The middle movements, in a madrigal style, stand in no sort of relation to the chorale. Among them the tenor aria is especially interesting in respect of rhythm and modulation.

Another cantata, for the fifteenth Sunday after Trinity, is set to the first verses of Hans Sachs's hymn, "Warum betrübst du dich, mein Herz."[180] The first two verses, which follow one another almost immediately, are separated by recitatives on words in the free madrigal style, of the same kind as we have noticed in the cantatas "Wer weiss, wie nahe mir mein Ende," "Herr wie du willst, so schicks mit mir," and elsewhere. Each of the first three lines of the first verse is prefaced by a short tenor arioso on the words of the hymn, and in this method we see the development of the plan begun in the cantatas above-mentioned, which are set to hymns throughout. The phrases of the arioso portions are formed upon the principal subject on which the instrumental movement is built, which makes it seem as though the whole movement together were to be in the form of a chorale fantasia. But it is not so, for whenever the chorus enters the instruments are restricted to the simplest accompaniment. It is also contrary to the nature of the chorale fantasia that each of the first three lines should be played over on the oboe before the voices begin. The second verse of the chorale is characterised by a certain want of unity and of repose. It appears first in the simplest four-part setting, while the last two lines have rich thematic counterpoint, and—which is most remarkable—after an inserted recitative they are repeated almost exactly as before. The third verse forms the close of the work; it is a real chorale fantasia, although the instrumental movement is of a simplicity not very often met with in such cases in Bach. This, as well as the unpretending character of the chorale movements which precede this movement, is probably accounted for by the

---

[180] The autograph score is in the Royal Library at Berlin. It is also published in Winterfeld, Ev. Kirchengesang III., Musical Appendix, p. 145 ff.

nature of the Gospel for the day, in which the faithful Christian is exhorted to leave all anxiety for bodily well-being, in childlike confidence, to his heavenly Father, like the birds of the air and the lilies of the field. That which chiefly gives the cantata the character of an imperfect and transitional production, is the subjective and arbitrary treatment of the first verse of the chorale, and the strange arrangement of the whole. A chorale movement at the beginning and at the end, with middle movements in the madrigal style, or three different treatments of chorales at the beginning, in the middle, and at the end respectively, with madrigal interludes, is intelligible; but not a form in which two chorale movements treated in various ways follow consecutively, after which comes a bass aria in madrigal style, the whole closing with a chorale fantasia. The date of this, and of the cantata described before it, cannot be fixed with certainty. The first of the two must, however, have been written before the Mass in G major, for the bass aria appears again in the last-mentioned work as a transferred movement.

Of the chorale cantata, strictly speaking, Bach has only left isolated specimens of the first Leipzig period. The cantatas "O Ewigkeit, du Donnerwort" (F major), "Liebster Gott, wann werd ich sterben," "Wer nur den lieben Gott lässt walten," "Was Gott thut, das ist wohlgethan" (second composition), and "Es ist das Heil uns kommen her," display the form in its completeness. Nearly trenching on these are "Wachet auf, ruft uns die Stimme," and "Wär Gott nicht mit uns diese Zeit," which dates from the beginning of the last period. The whole body of chorale cantatas may now be presented in a tabulated form:—

1. Ach Gott vom Himmel sieh darein (second Sunday after Trinity).
2. Ach Gott, wie manches Herzeleid (second Sunday after Epiphany).
3. Ach Herr, mich armen Sünder (third Sunday after Trinity).
4. Ach lieben Christen seid getrost (seventeenth Sunday after Trinity).
5. Ach wie flüchtig, ach wie nichtig (twenty-fourth Sunday after Trinity).
6. Allein zu dir, Herr Jesu Christ (thirteenth Sunday after Trinity).
7. Aus tiefer Noth schrei ich zu dir (twenty-first Sunday after Trinity).

8. Christum wir sollen loben schon (second day of Christmas).
9. Christ unser Herr zum Jordan kam (St. John's Day).
10. Das neugeborne Kindelein (Sunday after Christmas).
11. Du Friedefürst, Herr Jesu Christ (twenty-fifth Sunday after Trinity).
12. Erhalt uns Herr bei deinem Wort (sixth Sunday after Trinity).
13. Gelobet seist du, Jesu Christ (Christmas Day).
14. Herr Christ der einig Gottssohn (eighteenth Sunday after Trinity).
15. Herr Gott, dich loben alle wir (Michaelmas Day).
16. Herr Jesu Christ, du höchstes Gut (eleventh Sunday after Trinity).
17. Herr Jesu Christ, wahr' Mensch und Gott (Quinquagesima).
18. Ich freue mich in dir (third day of Christmas).[131]
19. Ich hab in Gottes Herz und Sinn (Septuagesima).
20. Jesu, der du meine Seele (fourteenth Sunday after Trinity).
21. Jesu nun sei gepreiset (New Year's Day).
22. Liebster Immanuel, Herzog der Frommen (Epiphany).
23. Mache dich, mein Geist, bereit (twenty-second Sunday after Trinity).
24. Meinen Jesum lass ich nicht (first Sunday after Epiphany).
25. Meine Seele erhebet den Herren (Visitation of B. V. M.).
26. Mit Fried und Freud ich fahr dahin (Purification of B. V. M.).
27. Nimm von uns Herr, du treuer Gott (tenth Sunday after Trinity).
28. Nun komm, der Heiden Heiland, B minor (first Sunday in Advent).
29. Schmücke dich, o liebe Seele (second Sunday after Trinity).[132]
30. Was frag ich nach der Welt (ninth Sunday after Trinity).
31. Was mein Gott will, das gscheh allzeit (third Sunday after Epiphany).
32. Wie schön leuchtet der Morgenstern (Annunciation of the B. V. M.).
33. Wo Gott der Herr nicht bei uns hält (eighth Sunday after Trinity).
34. Wohl dem, der sich auf seinen Gott (twenty-third Sunday after Trinity).
35. Wo soll ich fliehen hin (nineteenth Sunday after Trinity).

Putting aside numbers 3, 6, 16, 20, 27, 32, 33 and 34, the twenty-seven cantatas which remain belong, as is sufficiently proved by the evidence of the manuscripts, to one and the same limited space of time. The most that can be asserted

---

[131] Compare, on the melody, Appendix A., No. 1. The form there given differs only in unimportant details from that in the cantata. There are differences of greater importance in the form given in König's Harmonische Liederschatz (Frankfort à M. 1738), p. 280 (O stilles Gotteslamm)."

[132] Entered in Breitkopf's catalogue for Michaelmas, 1761, p. 23, as a communion cantata.

as to their dates is that "Du Friedefürst, Herr Jesu Christ," is the latest, falling as it does on November 15, 1744. It would seem, on the other hand, that among the earliest were the following : "Was frag ich nach der Welt," "Wo soll ich fliehen hin," "Ich freue mich in dir," and "Jesu nun sei gepreiset." Of these the first was probably written for August 7, 1735, the second for October 16, 1735, the third for December 27, 1735, and the fourth for January 1, 1736.[133]

In these thirty-five cantatas a series of the most beautiful and the best known Protestant chorales of the sixteenth and seventeenth centuries is subjected to elaborate musical treatment. As the subject has hitherto been only lightly touched upon, it may be well here and now to analyse completely the way in which the chorales were used, and the nature and character of Bach's chorale cantatas properly so called.

The particular hymn in its entirety forms the groundwork of each cantata. In the case of long hymns several verses are sometimes omitted ; but this neither alters nor impairs the course of thought, and is in accordance with the ordinary custom, even in congregational singing, of leaving out some stanzas for convenience ; but the first and last stanzas always appear in their exact original form, and combined with the original melody. In rarer cases one or more of the other verses appear, as the work goes on, with the original words, to which the church melody is always retained as an inseparable adjunct. For the most part the verses of the

---

[133] See Appendix A., No. 3. The autographs of 3, 33, and 34 are not known, they only exist in later copies. The following have been published after the originals: 1 B.-G., I., 2, P. 1194; 2 B.-G., I., 3, P. 1195; 4 B.-G., XXIV., 114; 5 B.-G., V.,[1] 26, P. 43; 6 B.-G., VII., 33; 7 B.-G., VII., 38, P. 1694; 8 B.-G., XXVI., 121; 9 B.-G., I., 7, P. 1198; 10 B.-G., XXVI., 122; 11 B.-G., XXIV., 116; 12 B.-G., XXVI., 126; 13 B.-G., XXII., 91, P. 2147; 14 B.-G., XXII., 96, P. 2142; 15 B.-G., XXVI., 130; 16 B.-G., XXIV. 113; 17 B.-G., XXVI., 127; 19 B.-G., XXII., 92, P. 2143; 20 B.-G., XVIII., 78, P. 1294; 21 B.-G., X., 41, P. 1656; 22 B.-G., XXVI., 123; 23 B.-G., XXIV., 115, P. 1687; 24 B.-G., XXVI., 124; 25 B.-G., I., 10, P. 1278; 26 B.-G., XXVI., 125; 27 B.-G., XXIII., 101, P. 1678; 28 B.-G., XVI., 62; 30 B.-G., XXII., 94, P. 2146; 31 B. G., XXIV., 111; 32 B.-G., I., 1, P. 1193; 35 B.-G., I., 5, P. 1197. The autograph score of No. 18 is in the hands of Herr Ernst Mendelssohn-Bartholdy of Berlin ; that of 29 belongs to Madame Pauline Viardot-Garcia of Paris.

hymn are transformed into poems in the madrigal style, and in this form they serve as the poetic material for the free solos and concerted numbers. The altered text adheres closely to the original words, with the exception of a few abridgments in some places and prolongations in others.[134] The thoughts, and in part even the turns of expression, and individual words, remain the same. The hymn appears, not essentially changed, but as it were beneath a veil. This rule may be applied with more or less strictness to all the cantata texts. For instance, in the cantatas "Ach Gott vom Himmel sieh darein," "Ach Herr, mich armen Sünder," "Erhalt uns Herr bei deinem Wort," "Liebster Immanuel, Herzog der Frommen," "Schmücke dich, o liebe Seele," "Wo soll ich fliehen hin," the recension follows the original text with remarkable exactitude. Elsewhere, as in "Christum wir sollen loben schon," "Mache dich, mein Geist, bereit," "Nimm von uns Herr, du treuer Gott," "Nun komm, der Heiden Heiland," "Wie schön leuchtet der Morgenstern," "Herr Christ der einig Gottssohn," the poet takes greater liberties, sometimes deviating widely from his original.; but by retaining certain striking words he preserves the feeling that it is still a paraphrase. Only exceptionally and in unimportant passages do independent and original insertions occur, as for example, in the bass recitatives in "Ach lieben Christen seid getrost" and "Das neugeborne Kindelein"; and in the alto recitatives in "Mit Fried und Freud" and "Wohl dem, der sich auf seinen Gott." The alto recitative of the cantata "Du Friedefürst, Herr Jesu Christ" is also set to independent words; this is probably accounted for by the particular character given to the work by the war of 1744.

If the hymn to be used happened to be too short, more words had to be inserted, in order to fill up the requisite number of arias and recitatives. This is done, for instance, in the case of "Meine Seele erhebet den Herren." Frequently, too, in cases of this kind the verses were divided

---

[134] An example of the process as applied to the hymn "Christ unser Herr zum Jordan kam" is given in the German original of the present work, Vol. II., p. 570.

into two portions; the alto aria of the cantata "Ich freue mich in dir" is founded upon the first half of the second verse, and the soprano aria upon the first half of the third, both the second halves being employed for the recitatives which follow the arias. In the cantata "Jesu nun sei gepreiset" all that is in madrigal style—viz., two arias and two recitatives—is founded upon the middle verse. In the cantata "Mit Fried und Freud" the text of the first aria seems to be suggested by the first verse of the hymn, although this has previously been sung through in its original form. And, on the other hand, in the case of hymns consisting of many verses, we meet with combinations of two or more in one number. In "Ach Gott, wie manches Herzeleid," verses 7—10 are taken for the tenor recitative, and in "Mache dich, mein Geist, bereit," verses 3—6 form the groundwork of the bass recitative. The extreme limit in this respect is reached in the cantata "Ach wie flüchtig," where no fewer than seven verses (3—9), each compressed into a single line, are employed for the recitative. It has been already mentioned that omissions are sometimes made; an example occurs in "Ach Gott, wie manches Herzeleid," where after the tenor recitative, verses 11—14, and then verse 17 are omitted, verses 4 and 5 having previously been left out.

A special kind of paraphrase occurs, also, in which the verse of the chorale is interwoven with recitatives. Either the chorale is sung set for one or more voices, and recitatives are inserted between the lines, or else it is played on the instruments, and a solo voice sings a recitative against it. This never occurs either at the beginning or end of a cantata, but is not unfrequently found in the course of it. I would mention, for example, the cantatas "Ach Gott, wie manches Herzeleid," "Aus tiefer noth schrei ich zu dir," "Das neugeborne Kindelein," "Gelobet seist du, Jesu Christ," and "Ich hab in Gottes Herz und Sinn." Here the words of the recitatives stand in no close connection with the chorale verses, but are merely generally suggested by them. Most of these paraphrases are a kind of explanation of the words, but here this is the case in a quite remarkable degree. In the unrestricted madrigal form certain applications which it was thought desirable to introduce were easily brought in. Nicolai's hymn, "Wie schön leuchtet der Morgenstern," was not originally written for the festival of the Annunciation. When used for this occasion, it was fitting that reference should be made to the event; hence the mention of the angel Gabriel in the tenor recitative, which is built upon the second verse of the hymn. Besides this, several allusions to the first verse run through the words, a method which is not peculiar to this cantata. The words of the tenor aria "Jesus nimmt die Sünder an," from the cantata "Herr Jesu Christ du höchstes Gut," enlarge upon the ideas of the fourth verse which has been previously sung through word for word, and at the same time refer to the Gospel for the day—the parable of the pharisee and the publican. Any one who carefully studies the texts of the chorale cantatas, comparing them with the original hymns, will notice that the poet frequently transposes the subjects, ideas, and words in a way which is sometimes ingenious and interesting in its effect, but which often produces the impression of laboured and mechanical work.

The style of the texts of all these chorale cantatas is so very much the same that we may assume them all to have been the work of one poet. Picander had a great facility for this kind of work. It appears frequently in his own poetry, and the "Erbauliche Gedanken" show that he could put together very good hymns. The hymns divided into regular verses in this collection are for the most part mere compilations; he was almost entirely devoid of originality in this line, but he compiled with such success that some of his poems have passed into use in the church.[135] We can with safety assert him to be the author

---

[135] Koch. (Geschichte des Kirchenlied), V., p. 500 f. (3rd edition) states that the hymns, "Bedenke, Mensch, die Ewigkeit," "Das ist meine Freude, dass, indem ich leide," and "Wer weiss, wie nahe mir mein Ende, ob heute nicht mein jüngster Tag," are still in ordinary use. The opening lines, which are almost identical with well-known hymns, sufficiently prove Picander's plagiarism.

of at least one of the cantata texts. It must be remembered that he had in his time adapted the Michaelmas hymn from "Erbauliche Gedanken" for Bach, for the cantata "Es erhub sich ein Streit."[186] He again made use of the same materials for the Michaelmas cantata, "Herr Gott, dich loben alle wir," by amalgamating the second verse of the hymn with the closing chorale of the cantata poem for the text of the tenor aria.

We cannot be wrong in assuming Bach to have originated the idea of this new style of text-writing, since he could not find a poet after his own heart. It is easily intelligible that Bach would feel a growing aversion to the dull poetry of the ordinary cantata texts of the time. He must soon have discovered, however, that it was impossible to adapt the verses of hymns to recitatives and arias. The resistance which the form of the verse would show to free musical treatment was not the greatest difficulty he had to encounter. To a refined perception, the contrast between the modern form of verse for musical setting and the hymn-verse was very nearly as strong as that between the modern form of verse and the words of Scripture. Whatever could be done from the side of music to bridge over the chasm had certainly been done by the style in which Bach's church music was cast; and now the work had to be completed from the poetical side of the gulf. In the paraphrase in madrigal style he found a form of poetry which solved the problem, and which must be owned to be unsurpassable for its purpose. By its means the connection with the hymn-form was kept up, and a noble and fitting style of thought was rendered possible; at the same time, it lent itself readily to the existing forms of musical expression. In consequence of the rule that the first and last verses, at least, and sometimes some of the middle verses, must remain untouched in words or melody, so as to constitute the groundwork of the whole, the numbers in madrigal form stand towards the whole in the relation of members, of a lighter structure, but of the same substance.

---

[186] Vol. II., p. 408 ff.

In the cantatas written by Bach in early life, on hymns unaltered in form, he had frequently indulged his fancy with playful imitations of the melody. In this respect he now returned to the strict rules of sacred composition. The chorale melody is something sacred and unalterable. It serves as a central point round which the music crystallises, without itself being drawn into the movement and action consequent upon the process of formation. The most that is allowable is to surround and adorn it. The most effective treatment which can be applied is the combination of the melody with a poetic effusion of a more subjective character. The middle parts more especially of the chorale cantatas present to us interesting examples of Bach's adherence to this fundamental principle. In these we not seldom find fragments of poems in madrigal style built upon detached lines of the chorale melody. They are always perfectly clear and definite, and are inserted with admirable art into the free flow of the piece. In only one case that I know of is such a line of a chorale set to new words. In the tenor aria in the cantata "Herr Jesu Christ, du höchstes Gut," in bars 41—43 and 52—55, the last line of the melody comes prominently forward in both cases in an adorned form to the words "dein Sünd ist dir vergeben." It may have been because the text of the aria stands only in a very loose connection with one verse of the hymn—viz., the fourth; and so this reference in the music may be in order to unite it more closely to the hymn. It is, however, observable that the words are not in a free madrigal form, but are taken from the Bible.[137]

As a general rule, the line of the melody is set to the proper words in the particular verse which forms the basis of the poet's madrigal paraphrase. Examples of this occur in the duet from the same cantata. It is founded on the seventh verse of the hymn, and of this the first, third, fifth, and seventh lines are introduced word for word;[138] these lines are united to the proper fragments

---

[137] Spoken by Christ on various occasions: Matt. ix., 2; Luke vii., 48.

[138] The third line, which is not very happy in its original form, receives a slight alteration.

of the melody, which are afterwards prolonged and developed by freely inverted passages. The bass recitative in "Schmücke dich, o liebe Seele" closes in the same way, with the last line of the eighth verse of the hymn brought in almost in its original form; and the last line of the melody is developed in a florid arioso. As models of ingenuity in expressively adorning a given musical phrase, these melodic fragments may take very high rank. The two opening bars of the alto recitative in the cantata "Ach Herr mich armen Sünder," which begins with the words of the fourth strophe " Ich bin von Seufzen müde," testify to a wonderful genius for manipulating and transforming musical phrases, as well as to an incomparable depth of feeling. Nor are these qualities less prominent in bars 44—47 of the tenor aria in the same cantata, where the voice, instead of going down by the interval of a second, suddenly rises a seventh. The bass recitative in the cantata "Jesu, der du meine Seele" culminates in the latter half of the tenth verse, of which the actual words are retained. The strings have an independent accompaniment throughout, to which the voice sings the last four lines of the melody, varying them so that they become full of fervour and deep devotion.

Often, too, the fragments appear in their natural simplicity. In the duet of the cantata " Nimm von uns Herr " this occurs at the beginning, and again at the repetition of the same portion; and in the alto aria of the cantata " Ach Gott vom Himmel sieh darein" it occurs in bars 55—59. On those who understand it aright it has a peculiarly thrilling effect, when in the midst of strange phrases and passages the well-known tune of the chorale unexpectedly falls upon the ear; it is like the sun breaking through clouds, and flooding the world with light, to give assurance that although the glory was obscured for a while its life-giving influence was yet present. In two instances fragments of this kind are not merely introduced in passing, but as determining the meaning and character of a whole number. In the cantata " Herr Jesu Christ, wahr' Mensch und Gott," there is a bass solo entitled by Bach " Recitativ und Arie." The " Arie " is to be conceived of as beginning at bar 13. The name was probably chosen because of the more concise character of the text, and because a musical phrase from the beginning recurs at the end, though in part, indeed, in another key. In the text, which is in this instance really poetical, the sixth and seventh verses are paraphrased, the first, third, and fourth lines being taken bodily from the sixth verse. The sections of the melody are treated analogously; for the fourth line the melody appears in florid extension (33—37), for the first and third in a simple form, but frequently repeated and imitated on the instrumental bass. This becomes the leading idea in the whole "Arie," and it is used alternately with phrases of a more animated kind in 6-8 time and freely invented. There is also an "Arie" for bass in the cantata "Nimm von uns Herr, du treuer Gott," set to the fourth verse "Warum willst du so zornig sein" ("Wherefore so wrathful still, O Lord?"). A prelude of an agitated character seems to betoken anger. Then the bass gives out, *Andante*, the opening line with its proper melody. At the third bar, however, the more animated instrumental movement breaks in, carrying the voice part onward with it in its course. It is again interrupted by the measured movement of the chorale melody, and it then changes into the relative major. As it must, however, in accordance with the aria form, return to the original key, the instruments take up the chorale melody. The first line is not enough; the tune in its entirety soars above the agitation of the voices and of the bass part, as though it had been roused by the vocal opening at the beginning, and had waited the opportunity to assert itself as the chief constituent part of the movement. A form which is of frequent occurrence in these cantatas— viz., the combination of the chorale melody on the instruments with vocal recitative on freely invented words, is here united with the aria, with Bach's inexhaustible power, in order to form an entirely new musical entity.

Thus the hearer perceives in the madrigal paraphrases numerous musical allusions to the source of all these apparently free productions. Still, Bach was not in such subjection to this rule as that whenever the poet inserted a verbal allusion to a well-known hymn he should always second it by the use of the proper tune. In the cantatas "Allein zu dir, Herr Jesu Christ," "Herr Gott dich loben alle wir," "Ich freue mich in dir," "Jesu, der du meine Seele," "Mache dich, mein Geist, bereit," and "Wo Gott der Herr nicht bei uns hält," many verbal reminiscences occur which are allowed to pass without their associated melodies being introduced. Just as these transformed poems themselves were an intermediate and transitional form, so Bach remained at liberty to be guided by his judgment either to the purely musical or to the devotional view of the work.

The two chief pillars of the chorale cantata—viz., the first and the last verses—have so far the same general form that the latter always appears in a simple setting, while the former is always worked up into a composition of considerable length. I have before pointed out what must have been the origin of this arrangement, by which the more ornate musical factor is placed first and the simplest last;[139] it bears the plainest testimony to the religious spirit which pervaded Bach's imagination. The circumstance that the instruments follow the course of the vocal parts simply and without adornment shows also how the whole fulness of the individual life flows at the end piously and humbly back to the source from whence it rose. It is significant, too, that the upper part in which the melody occurs is often strengthened by as many instruments as possible, giving it undue prominence from the purely musical point of view; a sense of its high symbolic meaning has led the composer to emphasize it strongly. The cases in which any of the instruments have an independent part in the accompaniment of the closing chorale are quite exceptional. In the cantatas "Gelobet seist du, Jesu Christ," "Herr Gott dich loben alle wir," and "Wie schön leuchtet der Morgenstern" the passages on the horns and trumpets serve only to give colouring to the orchestration,[140]

---

[139] See Vol. I., p. 500.
[140] The ascending trumpet passage in the cantata "Herr Gott dich loben alle wir" may possibly be intended to convey the same poetic meaning which

the short and fanfare-like ritornels in "Jesu nun sei gepreiset" refer back to the opening chorus and have no independent importance.

With regard to the great chorale choruses at the beginning of the cantatas, a general survey shows clearly that Bach considered that he had found once for all the musical form best adapted to this purpose. The proclivity to new varieties of form, which is so evident elsewhere, here has nearly disappeared. By far the greater number of these choruses are chorale fantasias. The motett style is only met with in the cantatas "Ach Gott vom Himmel sieh darein," "Aus tiefer Noth schrei ich zu dir," and "Christum wir sollen loben schön." These grandiose movements show to what a height the motett form could rise when transplanted into the new and fertile soil of the Bach cantata.[141] The so-called Pachelbel chorale form is found in its purity only in the cantata "Ach Herr, mich armen Sünder," and there as a matter of course, at this period of Bach's development, with simple counterpoint. He solved the difficulty of keeping to the same motive, and at the same time preluding before each separate line, by working in the contrapuntal motive and the phrase that is to serve as prelude together. The great organ chorale "Jesus Christus unser Heiland" offers a beautiful example of this.[142] Here he rendered the task still more difficult by using the first line in diminution for the counterpoint throughout, in direct and inverse motion. But his prolific fancy was not satisfied; for besides this he introduces the fragments of the melody which are to serve as preludes to each line in constantly different and distant keys.

---

is clearly brought out in the aria which precedes it and in a similarly treated closing chorale in "Es erhub sich ein Streit."

[141] See Vol. II., p. 598, f.—To the list of chorale choruses in motett form should be added the splendid movement published in B.-G., XXIV., No. 118, "O Jesu Christ meins Lebens Licht," which, however, exhibits a more independent treatment of the wind instruments in the accompaniment. It was probably performed in the open air at a funeral ceremony, and afterwards adapted for indoor performance. Its date is about 1737. See Vol. II., App. A., No. 57.

[142] See Vol. I., pp. 607 and 613.

This method of adorning the Pachelbel chorale form with one and the same subject used as counterpoint, makes it a transitional form towards the chorale fantasia.[143] Accordingly we shall not be surprised, in the cantatas "Nun komm der Heiden Heiland," "Wohl dem, der sich auf seinen Gott," and "Wie schön leuchtet der Morgenstern," to meet with forms that are half one thing and half another. In the first case only the opening line, which is identical with the last line, is introduced by a prelude; in the second, an almost independent movement is developed from the first line, only showing its relationship to the Pachelbel form by the fact that before the fifth line a reference is made to the first line; in the last case the relationship is shown by the union of the melody with the pervading motive in the prelude to the *Cantus firmus* of the second and fifth lines.[144] Another hybrid form occurs in "Jesu nun sei gepreiset"; the opening chorus is at the beginning and at the end a chorale fantasia, but in the middle it is partly in free fantasia form, and partly in that of the motett.

Several longer choruses reveal their origin in the fully-developed Pachelbel form only by the fact that the music by which the *Cantus firmus* is surrounded and upborne takes its subject from the first line of the chorale. In all other respects they belong to the class of free chorale fantasia. In the cantata "Erhalt uns Herr," only the first three notes are thus used as a theme, while in "Was frag ich nach der Welt" and "Wo soll ich fliehen hin," the whole line is employed. This method is employed also in the cantata "Herr Jesu Christ, wahr' Mensch und Gott." Here, however, the ear is attracted by another combination of great subtlety. The cantata is written for Quinquagesima Sunday, which immediately precedes the church

---

[143] See Vol. I., p. 609.

[144] This motive is also connected with the first line of the melody by its two first notes being identical. It is moreover remarkable that the very same motive is also employed in the chief chorus of "Ach Gott, wie manches Herzeleid." I believe the cantata "Wie schön leuchtet der Morgenstern" to have been written immediately after this, so that Bach still had the motive in his head.

season commemorating the Passion. For this reason, while choir and instruments perform the chief subject, the chorale "Christe, du Lamm Gottes" is introduced in a fragmentary way, so that the idea of the Passion dawns on the hearer. This reminds us somewhat of the tenor recitative in the cantata, also for Quinquagesima Sunday, "Du wahrer Gott und Davidssohn."[145] Bach's careful attention to minor details, even in the firmly-fixed form of chorale-fantasia, is no less clearly illustrated in the cantata "Liebster Immanuel." The hymn—written, probably, by Ahasverus Fritsch—and the melody belong to the second half of the seventeenth century. Notwithstanding its beautiful feeling and expressiveness, it has an almost secular and playful character, and the melody is in the form of a sarabande.[146] Bach does not disregard this character, and, accordingly, the music which takes its subject from the first line of the melody is conceived rather in a melodious and homophonic than in an imitative and polyphonic style.

The opening choruses of all the other chorale cantatas are free chorale fantasias. In the course of its transference from the sphere of organ music to that of concerted church music, the style underwent only those changes which were demanded by the alteration in the tone material. The independent musical creation which was to bring out all the sentiment of the chorale was no longer worked out by means of organ-stops, but by Bach's orchestra on the background of the organ. The *Cantus firmus*, which, when played, ran the risk of being drowned by the important part taken by the rest of the instruments, fills, when sung, a higher and more dominating position. The vocal parts of the chorus fill an intermediate place, connecting the chorale and the instrumental portion. Sometimes they support the voice that has the melody (generally the soprano) in simple harmonies, but more often they appear as a factor in the instrumental portion of the work; and it is obvious

---

[145] See Vol. II., pp. 350—353.
[146] This was remarked upon with displeasure, even in Bach's time. See Witt, Neues *Cantional* mit dem *General-Bass*. Gotha and Leipzig (1715). Preface.

that that portion must hold its place in the work as a whole, not merely by richer musical treatment, but by a picturesque use of the means at hand. As a model of an organ chorale treated in this poetic way, the arrangement of the melody "Ach wie flüchtig, ach wie nichtig," from the "Orgelbüchlein," which numbers only ten bars, has been already mentioned."[147] It may be compared with the first movement (sixty-five bars long) of the cantata on the same chorale. The mode of construction is the same in both, as is also the fundamental form, although, by degrees, the longer work rises to a height which renders it scarcely possible to recognise their common origin. In outline these choruses have almost a stereotyped form. The composer is revealed in them rather as having attained a high plain and walking serenely over it than as striving upwards—the marks of progress are to be seen in the details.

Still, the impulse towards the development of new forms was not yet exhausted. The splendid opening of the cantata "Jesu, der du meine Seele," which is built on the same theme as the *Crucifixus* of the B minor mass, exhibits a chorale-fantasia in the form of a chaconne. In the cantata "Meinen Jesum lass ich nicht," the chorale-fantasia form is combined with that of the instrumental concerto. This last plan is evident, too, in the cantatas "Herr Jesu Christ, du höchstes Gut" and "Was mein Gott will, das g'scheh allzeit," although, in accordance with Bach's refining method, no solo instrument is here opposed to the *tutti*. On the other hand, the cantata "Christ unser Herr zum Jordan kam" has the regular concerto form; nay, more, it is a *concerto grosso*, the *concertino* consisting of a solo violin and two *oboi d'amore*, while the *tutti* is formed by the violins, tenors, and bass, with the organ. Meanwhile, the chorale chorus, with the *cantus firmus* in the tenor part, keeps on its way, never heeding this mass of sound; and yet these two elements are homogeneous. It must, however, be admitted, that in this case the forms

---

[147] See Vol. I., p. 602.

are intertwined in a rather perplexing way. It is not enough to analyse how this or that element originated and grew, absorbing others in its development, or how some other has met it half-way till a grand unity is the result. A work of art must be estimated as a whole, and followed with lively sympathy. Every form contains, as its soul, a certain general emotion; and this must be understood before the form itself can be properly appreciated. Unless, by the study of the concertos of that time, we make ourselves familiar with the realm of feeling in which these works have their organic being, we can never understand Bach's intention in these combinations. They will appear strange, and even absurd. The characteristic and vigorous spirit of the first movement of a concerto, as understood at that time, appeared to Bach suitable for giving a particular colouring to the chorale-fantasias in these works. But even a perfect comprehension of this will not bring us to the root of the matter. The chorale fantasia, as it appears in the chorale cantatas, is derived from a higher and by no means a simple artistic idea.

If we examine the nature of the chorale cantata, we find it to be nothing less than the perfect poetic and musical development of some particular hymn by means of all the artistic material which Bach had assimilated by a thorough study of the art of his own and former times. The course of his formation shows that he took organ-music, and especially the organ chorale, as his starting point, and that he mastered and appropriated the various forms of art by welding them into the style of the organ—the only sacred style of the time. This was the case both with the instrumental and the vocal forms of musical art; for, closely as the latter forms were connected with the chorale, they remained latent in Bach's organ music. The poetic element, which is essential to the organ chorale, naturally forced its way into the sphere of vocal music as the form was developed. That form aims at expressing the sentiment and feeling of a hymn, as that sentiment appeals to individuals in the congregation; this is attempted with ever-increasing materials of sound, and in ever larger proportions. The voice is found necessary for the melody, in order to keep the overflowing current of individual emotion within the bounds of sacred feeling. Thus we have a vocal and instrumental composition which, however, exhausts only one verse of the chorale; the instrumental portion is based on the chorale as a whole. One last and crowning step remained: to change the essentially musical idea of combination and concord for that of a succession of parts, according to the sequence of thought suggested by the poetry, which now took an active share in the whole; and to subject the chorale, which had been paraphrased, so to speak, as a whole in the purely instrumental composition, to a detailed treatment of its separate verses. Thus the chorale fantasia at the beginning became once more merely a part of the whole; only, as it were, a single petal, the perfect flower being the chorale cantata.

Still, the chorale fantasia determines the style of the whole cantata. In such a work as the Easter music, "Christ lag in Todesbanden," which retains the chorale melody for all the verses—its model in instrumental music being the chorale "Christ ist erstanden," in the "Orgelbüchlein"—the highest stage of development is not yet attained. The strong individual character, peculiar alike to Protestantism and to Bach's style, demanded a wider space in order to attain its full expansion. This it obtained in the chorale cantata. In the recitatives and arias it appears to have perfect liberty; even the limits assigned to it by the sequence of thought and expression in the words of the hymn are enlarged by the transformation into the madrigal form, and at times seem to vanish altogether. Often all that remains is the general emotional character left by the remembrance of the hymn. Then, it may be, the feeling of connection is again aroused by the introduction of a line of the hymn word for word; now a fragment of the melody appears, and now the whole, on some instrument or, perhaps, in the voice part, although surrounded and intertwined with recitatives; and the limits to the emotional character are once more emphasised. If now the hearer, having thoroughly taken in the great chorale chorus which preceded these freer middle portions of the cantata, is brought up to the same chorale as the end and aim of the work—and the more unadorned its form the more impressive will be its effect—he will feel that not a moment has passed during which he has not been hearing the chorale, either in its material form or in its inner spirit. In the chorale cantata as a whole—as in the chorale fantasia—the parts are treated in the most independent way possible, so that they seem to build up a fabric on their own account, and yet at the right moment they are referred again to their origin by means of the *cantus firmus*.

In the chorale cantata we meet with the last and highest possible development of the organ chorale. That little piece of ten bars on the melody " Ach wie flüchtig, ach wie nichtig " was the germ of which the cantata with that title, with its great chorale fantasia, its simple closing chorale, its recitatives, and its elaborate arias, is the ultimate fruit. The chorale cantata can only be properly understood by one who has made himself perfectly familiar with the nature of the organ chorale; and both demand that the hymn and its melody should form part of the vital *experience* of the hearer. While, however, in the former case it is only needful to feel the delicate aroma of the general feeling of the hymn, and the recognition of special feelings and details is a secondary requisite, an exact knowledge of the import of each separate verse is indispensable for the cantata. Without this even the words and general arrangement of the "madrigal" text cannot always be understood. We do not see that the consuming flames and foaming billows referred to in the bass aria of the cantata "Ach wie flüchtig" are not to be understood metaphorically, but actually of perils by fire and water, until we compare it with the tenth verse of the hymn. The words in the bass aria of the cantata " Ich hab in Gottes Herz und Sinn," which run thus : "The roaring of the mighty wind leads to the ripening of the grain," point out that the fruits of the field are not brought to perfection without storms and tempests; this is not made clear unless we know the ninth verse of the hymn. And, which is of greater importance, the character of the music is often suggested by images or ideas not expressed in the inserted words, but only in the original text. How did Bach come to give that tender idyllic character to the duet in the cantata "Jesu, der du meine Seele," which consists of a prayer for help in sickness and weakness? The second verse of the hymn runs thus:—

Treulich hast du ja gesuchet	Faithfully Thou, Lord, hast sought
Die verlornen Schäfelein,	All Thy lost and trembling sheep,
Als sie liefen ganz verfluchet	As they ran with fear distraught
In den Höllenpfuhl hinein.	Headlong into Hell so deep.

We have frequently noticed Bach's way of giving force and point to a tedious or digressive cantata text by seizing upon the emotional character of the Sunday or festival. This example shows how, in the chorale cantatas, he composed the music, not so much with regard to the madrigal version or paraphrase as to the original text. All these works are founded on the assumption that the hearer will have constantly before his mind the hymn in its original form. The church-goer of those days could compare the printed text of the cantata with the version in his hymn book; or he could even dispense with this material aid, since those hymns were in every heart as a possession common to all. He had sung them times without number in church, had taken them as his guide in daily life, and had drawn consolation and edification from isolated verses under various experiences. This was the audience to which Bach addressed himself, and such an audience do these compositions still require, for to such alone will they reveal all their meaning and fulness, both in outline and detail.

As we glance backwards from this point over Bach's life, we are struck by the completeness and rounding-off of his artistic development. His starting-point in early youth was the sacred song of the people, and to it he returns at the end of his career. He felt that all he could create in the sphere of Church music must have an inherent connection with the chorale and the forms of art conditional to it. He must have deemed it the noblest goal of his ambition to give his genius that direction which should create a form that displays the chorale in its highest possible stage of artistic development. The chorale cantatas lack, it may be, that profuse variety of form which during the earlier and middle periods of his life calls forth our highest wonder. But the serene mastery over the technical materials of his art, the deep mature earnestness which pervades them, can only be regarded as the fruit of such a superabundant art-life. In considering these works in their unalterable and characteristic grandeur, we seem to be wandering through some still, lofty, Alpine forest in the peaceful evening that closes a brilliant summer day.

IV.

THE CHORALE COLLECTIONS.—THE ECCLESIASTICAL MODES, AND BACH'S RELATIONS TOWARDS THEM.

BACH, while in Leipzig, had made a collection of chorale melodies with figured basses. It comprises all the melodies in ordinary use there, in number about two hundred and forty. In the year 1764 the manuscript was in the possession of the musicseller, Bernhard Christoph Breitkopf, of Leipzig, who offered copies of it for sale at ten thalers each.[148] This important collection is lost.[149] A few fragments of it, however, seems to have been saved. Pupils of Bach who took down copies of his organ chorales appended to them the two-part figured settings from Bach's chorale book, when they could get access to them. Thus, when they played the organ chorale as a prelude, they could afterwards use the melody, as harmonised by their revered master, for accompanying the congregational singing. In this way the figured setting of the melodies " Christ lag in Todesbanden," " Herr Christ der einge Gottssohn," " Jesu meine Freude,"

---

[148] " Bachs, J. S. Vollständiges Choralbuch mit in Noten aufgesetzten Generalbasse an 240 in Leipzig gewöhnlichen Melodien. 10 thl." Breitkopf's catalogue for New Year, 1764, p. 29.

[149] Herr W. Kraukling, of Dresden, possesses a chorale book with figured bass, in small quarto; on the pig-skin cover stand the words: "Sebastian Bachs Choral-Buch." The volume, however, exhibits neither Bach's writing nor, in the writing to the chorales, a single trace of Bach's style or spirit.

"Wer nur den lieben Gott lässt walten," have come down to us.[150] Johann Ludwig Krebs has also handed down to us Bach's figured setting of four Christmas hymns—viz., "Gelobet seist du, Jesu Christ," "In dulci jubilo," "Lobt Gott, ihr Christen allzugleich," and "Vom Himmel hoch." The interludes introduced in them show that they were written for the very purpose of accompanying the congregation. The harmonising, which is of rare originality and power, makes us feel how much we have to regret in the loss of the whole chorale book.[151]

A third source, which may possibly contain some fragments of the chorale book, is a collection printed and published in Bach's lifetime. In May, 1735, a student from Zeitz, Christian Friedrich Schemelli, entered the Leipzig University.[152] His father, George Christian Schemelli, "Schloss-Cantor" at Zeitz, occupied himself with the editing of a hymn book with tunes, on the model of Freylinghausen's popular and widely known "Geistreiches Gesangbuch"; but it was to be free from pietistic colouring and sectarian spirit, bringing together whatever was good from any quarter, and containing also the recognised treasures from among the old hymns. At that time Bach was still intimately connected with the students through the medium of the Musical Society. The fact that Bach's name appears as the arranger of the musical portion of the work, which was published in 1736, was probably due to the persuasion of Schemelli's son.[153] Hitherto, so far as we know, Bach had not done any work of this kind. If, however, the

---

[150] These will be found in P. S. V., C. 5 (244), No. 53; in the same volume in the Appendix to No. 7; also in C. 6 (245), No. 16; and same volume, No. 29. What is known as to their origin is given in the preface.

[151] These Christmas chorales are in an organ book by Krebs in the possession of Herr F. A. Roitzsch, of Leipzig (p. 241 f.). Probably the setting of "Jesu der du meine Seele" on p. 253 is also by Bach. The chorale "Gelobet seist du" is given in the Musical Appendix to this volume (4, A.). A comparison of this with the one spoken of in Vol. I., p. 594, will show that it has the same harmonies, and that the interludes are identical.

[152] See the books of the University.

[153] Schemelli, the father, b. 1676, was sixty years old when the book appeared; Schemelli, the son, b. 1712, succeeded him, when he was superannuated, in his office (Archives of the Schlosskirche at Zeitz).

book was to have a success equal to that of Freylinghausen's, a well-known name was indispensable. The preface therefore duly announced that "the melodies to be found in this musical hymn book are in part newly composed, and in part improved by the addition of figured bass, by Herr Johann Sebastian Bach, Capellmeister to the Grand Duke of Saxony, and *Directore Chori Musici* in Leipzig." In spite of this the book did not meet with great success. A second enlarged edition, which was advertised to appear after the sale of the first, did not appear, although the first edition was only a small one.[154]

It is doubtful whether Bach cared much what became of the book. It was published in Leipzig by Breitkopf, and the name of the engraver, who, by the way, did his work remarkably well both in the title-vignette and in the melodies, is a Leipzig name. But it bears traces of carelessness in the copies for the plates, and in the supervision of the work. Otherwise, putting aside engravers' errors, it would never have happened that over every hymn that has a melody appended to it the same number appears twice; that above No. 627 there is an engraved title: "*Di S. Bach, D. M. Lips.*," while many of the other melodies—as, for example, 397 and others—were composed by Bach; and that in the case of the hymn "Jesu, meines Glaubens Zier," not only are the two numbers, one engraved and the other printed, set over it, and the melody is in its right place before the hymn, but it is notified again: "N.B.—This melody belongs to No. 119." It is also evident that many hands took part in the preparation of the copies for engraving. Thus it seems very probable that Bach had the greater part of the melodies simply written out from his chorale book. The number of the

---

[154] " Musicalisches | Gesang-Buch, | Darinnen | 954 geistreiche, sowohl alte als neue | Lieder und Arien, mit wohlgesetzten | Melodien, in Discant und Bass | befindlich sind | Vornemlich denen Evangelischen Gemeinen | im Stifte Naumburg-Zeitz gewidmet, | .... von | George Christian Schemelli,| Schloss-Cantore daselbst. | . . . Leipzig, 1736. | Verlegts Bernhard Christoph Breitkopf, Buchdr." | Copies of the work, which has now become scarce, are in Count Stolberg's Library at Wernigerode and in the Royal Library at Berlin.

melodies is sixty-nine. Among these are forty melodies by various composers, dating from the sixteenth, seventeenth, and eighteenth centuries. In the preface we are informed that about two hundred more melodies were ready to be engraved, and that they would be given in a second edition if that were called for. This makes altogether two hundred and forty melodies, exactly the number contained in Bach's chorale book.

In those chorales which still remain to us we have a valuable possession. These melodies, with their slight but animated figured basses, reveal to us, as in a sketch, the art which the master carried out in the elaborate four-part chorale movements. To their publication, also, we are indebted for our knowledge of a series of short original compositions set to hymns which should have a still higher value for us. Since Bach himself wrote a part of the tunes contained in Schemelli's Hymn book, and since a thorough examination of the melodies in use up to that time in the evangelical church shows that forty out of the sixty-nine melodies are by other composers, we have a perfect right to assign the remaining twenty-nine to Bach. We know by documentary evidence that two are by him ("Dir, dir Jehovah will ich singen" and "Vergiss mein nicht, vergiss mein nicht, mein allerliebster Gott"). But the others, at least the majority of them, bear upon them the unmistakable impress of Bach's style.[155] In the case of seven out of the twenty-nine, we have ground for assuming that they were written expressly for Schemelli's Hymn book.[156] Bach can hardly have intended them as

---

[155] It is to Winterfeld (Ev. Kirchenges III., p. 270 ff.) that we are lastingly indebted for the first thorough investigation of this subject, although he was partly led to wrong results. Since, as I have endeavoured to prove (see Vol. I., p. 369), Bach's having taken any share in Freylinghausen's Hymn book is out of the question, the number of the compositions in Schemelli's Hymn book which may be ascribed to him is considerably diminished. All the melodies contained in that hymn book have been brought out by Breitkopf and Härtel, edited by C. F. Becker. According to this edition the tunes by Bach are Nos. 4, 7, 8, 10, 11, 14, 19, 21, 24, 26, 30, 31, 32, 42, 44, 46, 47, 51, 52, 53, 56, 57, 59, 62, 63, 64, 66, 67, and 68.

[156] Nos. 31, 32, 47, 59, 64, 67, and 68. In preparing the plates for engraving, two different copyists were employed. The above Nos., together with

melodies suitable for public worship. Schemelli's collection, like that of Freylinghausen and others, was intended particularly for domestic devotion, and for this purpose, towards which Bach showed a leaning even in many of his greater sacred compositions, these works written for Schemelli were also intended. After devoting his whole life to the work of chorale arrangement, with an energy which fathomed every hidden quality of the chorale, he must have known what was wanted for a popular hymn for public worship; and he must also have known that his own manner of expressing himself, animated, melodious, and refined by the highest art, lent itself readily to the most subjective expression, but ran counter to the character of congregational use. We must not apply the standard of the church chorale to these melodies of Bach's. They are sacred arias, and the fact of their not having passed into church use—not more than five of them have even been included in any of the later chorale collections— in no way derogates from their value.[157] Their charm, which is all their own, is like that of a pious family circle, musically cultured, and we may delight to fancy that these touching hymns, so delicately worked out in their small limits, were sung, at the master's household devotions, by one or other of the members of his family.

The music-book of Anna Magdalena Bach, made in 1725, which was intended purely for domestic use, actually contains, on p. 115 ff, one of the hymns, Crasselius's "Dir, dir Jehovah will ich singen," and Bach's name is expressly

---

40, 194, and 281, are in the same handwriting. They differ from the rest in the form of the clef and in the use of the brace. The soprano clef exhibits pretty clearly the form most usually employed by Bach at that time. Besides this we may notice a larger and more hasty style of writing the notes, although it is much equalised in the engraving. It is probable that Bach himself wrote out the tunes which he composed specially for Schemelli's Hymn book.

[157] Winterfeld (*loc. cit.*, p. 278 f.) says that in J. B. König's "Harmonische Liederschatz" (Frankfort a. M. 1738), only Nos. 4 and 63 are given without alteration, while to the hymns Nos. 24, 56, and 57 are found mere popular tunes. Here, however, he is mistaken. No. 8 is also found in König unchanged; No. 57 is found in an altered form; and No. 24 with a newly written first part to the tune (p. 490 in König); No. 56 has quite a different tune.

given as the composer. There are besides several other pieces of the same kind in it. A composition, also entered in Bach's name, to Gerhardt's "Gieb dich zufrieden und sei stille," surpasses all the others by its lofty and individual beauty. Another melody to the same hymn leaves us in doubt as to its composer; it is strikingly simple for a composition of Bach's; but at all events it is new.[158] A melody to the hymn "Wie wohl ist mir, o Freund der Seelen" has a similar character. Though we must hesitate to assign these two to Bach without further evidence, the hymn "Schaffs mit mir Gott nach deinem Willen" shows plainer tokens, and the compositions "Warum betrübst du dich und beugest dich zur Erden" and "Gedenke doch, mein Geist zurücke" show unmistakable marks of his style.[159] It seems probable that Bach wrote several hymn-tunes expressly for Schemelli's book; his pupil Krebs, who was very intimate with him just about this time, being his deputy in the Musical Society and also in the Church,[160] has handed down to us five such hymns, which there is ground for regarding as Bach's compositions.[161]

After Schemelli's hymn-book had gone forth to the world to meet with so poor a reception, Bach continued his work upon it for his own pleasure in a remarkable way. He added to his copy no fewer than eighty-eight chorales, written fully out. At his death the volume came into the hands of Philipp Emanuel Bach, but has since been lost.[162] Doubtless, among these, Bach had set many, if not all, of his own melodies, in four parts. Of these we are sure of four—"Dir, dir Jehovah," "Jesu, Jesu, du bist mein,"

---

[158] König gives it at p. 340 of his "Harmonische Liederschatz."
[159] Comp. Vol. II., p. 148 f. The hymn "Gedenke doch, mein Geist" is published in Bitter (Vol. I., Musical Supplement).
[160] See p. 7 of this volume. In the Musical Society he accompanied on the harpsichord; see Gerber, Lex. col. 756.
[161] These five hymns with the melodies alone, as well as the hymn "Warum betrübst du dich" are given in the Musical Appendix to this volume (4, B.). Further remarks as to their genuineness will be found at App. A., No. 4.
[162] In the printed catalogue of Emanuel Bach's effects (Hamburg, 1790), on p. 73, it is mentioned among the compositions by Sebastian Bach: "The Naumburg Hymn-book, containing printed chorales, and also eighty-eight chorales written out in parts."

"Meines Lebens letzte Zeit," and "So giebst du nun, mein Jesu gute Nacht."[163]

Besides those written for Schemelli, Bach left several other original melodies to hymns, set in four parts, and among them, fortunately, is the splendid one "Gieb dich zufrieden." What was said above of the character of the tunes in one part with a figured bass holds good of these, which we only know in their four-part form, so far as we are justified in assuming them to be of Bach's composition. They are not so much chorales as devotional songs for household use, and as such are, for the most part, full of character and beauty. The fact that two original melodies such as these are introduced in their entirety in the fourth part of the Christmas Oratorio was before adduced as bearing significantly on the character of that work.[164] A similar chorale opens the cantata "Also hat Gott die Welt geliebt,"[165] and another is found at the end of the motett "Komm, Jesu, komm."[166] Among the nine four-part compositions which, besides these, we are more or less justified in attributing to Bach, are found some which approach more nearly to the chorale character. Two of these ("Da der Herr Christ zu Tische sass" and "Herr Jesu Christ, du hast bereit") were included by Johann Balthasar Reimann in his Hirschberg Chorale Book in 1747.[167] Reimann was with Bach in Leipzig from 1729 to 1740, and, as he himself tells us, was "affectionately received by him and charmed with him."[168] The melody with which Bach prefaces the cantata "Ach ich sehe," written in Weimar, has the character of a chorale. It is,

---

[163] The first is in Anna Magdalena Bach's book, the others in Philipp Emanuel Bach's editions of his father's four-part chorales. (Leipzig, Breitkopf, 1784-1787). In "So giebst du nun, mein Jesu" (No. 206 there), a whole bar is omitted after bar 7.

[164] See Vol. II., p. 586. One of them is set as a chorus, and forms the closing movement of the part, and the other appears with a four-part instrumental accompaniment and with counterpoint in a recitative bass part.

[165] See Vol. II., p. 695.

[166] See Vol. II., p. 605.

[167] As has been remarked by L. Erk, Choralgesänge II., No. 178 and 222.

[168] Mattheson, Ehrenpforte, p. 292.

however, nothing more than a compound produced by the fusion of the melodies "Herr ich habe missgehandelt" and "Jesu, der du meine Seele."[169]

On the whole, Bach can hardly have cared to exhibit his greatness as a church composer in this particular branch of work. When we consider the grand tasks it was his fate to fulfil, his hymn tunes can only be regarded as subsidiary work.

We have evidence that Bach's style of harmonising a chorale melody in four parts had, even in early times, excited the admiration of a large circle. After his death the melodies so treated by him were collected, and, at the New Year of 1764, Breitkopf, of Leipzig, was in possession of a manuscript containing one hundred and fifty chorales in score, copies of which were sold by him.[170] This was, probably, the collection of which Birnstiel, of Berlin, undertook to publish a printed edition in 1765. At the last moment he very wisely thought of getting Emanuel Bach to revise the manuscript. He did this so far that the collection, containing one hundred chorales, received correction from his hand;

---

[169] See Vol. I., p. 554. I am now thoroughly convinced of Bach being the author of this melody, which occurs nowhere else. Erk (see Choralgesänge II., No. 159) holds it to be merely another version of "Jesu, der du meine Seele." But, to my mind, it betrays an unmistakable connection with "Herr, ich habe missgehandelt" (see Choralgesänge, 1784, No. 35). The remaining hymn-tunes, in four parts, which may be attributed to Bach are "Nicht so traurig, nicht so sehr" (C minor), "Ich bin ja Herr in deiner Macht," "Was betrübst du dich, mein Herze," "Für Freuden lasst uns springen," "Gottlob es geht nunmehr zu Ende," and "O Herzensangst, o Bangigkeit." Winterfeld (op. cit., p. 282 ff) thinks that there are still a number of other hymns which may be assigned to Bach. With regard to the hymns "Ist Gott mein Schild," "Schwing dich auf zu deinem Gott," "O Mensch, schau Jesum Christum an," and "Auf, auf mein Herz und du mein ganzer Sinn," the incorrectness of the assumption that they are by Bach is demonstrated by Erk in his admirable edition of Bach's chorale melodies. In the case of the hymns "Meinen Jesum lass ich nicht," "Das walt Gott Vater und Gott Sohn," and "Herr nun lass in Friede," Winterfeld himself has half recanted his opinion. But the hymns "Dank sei Gott in der Höhe," "O Jesu du mein Bräutigam," and "Alles ist an Gottes Segen," appear in König in forms which show plainly that the reading of both Bach and König are nothing but variations of older original forms.

[170] Breitkopf's Catalogue, New Year, 1764, p. 7: "Bach, J. S. Capellmeister and Musikdirector in Leipzig, 150 chorales, in 4 parts. 6 Thlr."

but he had nothing to do with the second collection, which appeared in 1769. Not long before his death he arranged a second, a better and more complete, edition, which was published by Breitkopf. It is in four divisions, and contains three hundred and seventy chorales.

These chorales were taken for the most part from Sebastian Bach's concerted church compositions, so that they were intended for voices with instrumental accompaniment. For the convenience of the organ or clavier player, the editor arranged them on two staves. The announcement that the collection was to constitute a complete chorale book was only made in order to increase the number of purchasers. Under the existing circumstances it could not be a chorale book, comprising all the melodies in ordinary use in any particular town or neighbourhood; and if it had been, the various arrangements of one and the same melody would be purposeless. The real purpose of the book was to delight those who knew and loved the art of music, and to provide models for the study of the rising generation of composers. The last object may have been the cause of the son's treading so much as he did in his father's steps; for the chorale played an important part, not only in his own compositions, but also in his course of instruction in composition.

Johann Philipp Kirnberger says with regard to Bach's method of teaching composition, that it led up step by step from the easiest subjects to the most difficult; that among all those known to him it was by far the best; that he always referred back to fundamental principles, and sought to lay bare all the secrets of the world of music.[171] From this it follows that we may regard the scheme of study in Kirnberger's theoretical writings as having been invented and approved by Bach himself, if not the foundation of the method and special points of teaching. Kirnberger's chief work, "Die Kunst des reinen Satzes in der Musik,"[172] stands high as an instruction book for composition. It treats in

---

[171] Kirnberger, Gedanken über die verschiedenen Lehrarten in der Komposition. Berlin, 1782, p. 4 f.
[172] Part I., 1774; Part II. (up to third section), not before 1776—1779.

succession of the scales, the temperament, of intervals, of chords, with their combinations, and of modulations; also of the formation of melodies, and of simple and double counterpoint. The greater part of the work deals with counterpoint; Kirnberger intended to append to this an additional section of instruction in vocal composition, and the character of the dance-forms, and then to conclude with instruction in fugue. His plan, however, was not fully executed. The title ("The Art of pure composition") does not exactly correspond to the contents of the work, which is chiefly intended to cultivate the composer's taste as well as his technical knowledge, by giving especial prominence to the works of Bach. Though in matters of detail it leaves much to be desired as an instruction book, it acquires a special importance as a reflex of Bach's practical teaching, even in its scientific defects, since from these we may infer what the life, variety, and wealth of Bach's personal influence must have been. As a necessary introduction to his school of composition, Kirnberger subsequently published a book of instruction in thorough-bass.[173]

From his twenty-second year, through a space of forty-three years, Bach had had a very large number of pupils in composition. Putting aside the fact that a method of instruction takes a considerable time to form, it is not to be supposed that his teaching was always exactly the same even in his mature life. Friedrich Wilhelm Marpurg, when engaged in a controversy with Kirnberger on some questions concerning the theory of music, seeing that his adversary always quoted Bach, said: "Good God! Why should old Bach be dragged into a discussion in which he would have taken no part if he had been alive? No one will ever be persuaded that he would have expounded the principles of harmony according to the views of Herr Kirnberger. I believe that this great man had more than one method of instruction, and that he always adapted his style to the capacity of each pupil, according as he was more or less gifted by nature, or as he turned out to

---

[173] Grundsätze des Generalbasses als erste Linien zur Composition. 1781.

be pliable or stupid, clever or a mere blockhead. But
I am perfectly well assured that if there still exist any
introductions to harmony in manuscript by this master,
they will nowhere be found to contain certain things which
Herr Kirnberger wants to palm off upon us as Bach's way
of teaching. His celebrated son in Hamburg ought to
know something about it."[174] There is no doubt some truth
in this. And yet Kirnberger was sufficiently intimate with
the elder sons and other pupils of Bach to escape suspicion
of his having built up his system solely on the basis of
what he himself had learnt from Bach. The appeal to
Emanuel Bach turned out ill for Marpurg; for he distinctly
took Kirnberger's side, with the most decisive disapproval
of the polemic tone adopted by Marpurg, and he authorised
Kirnberger to say that Sebastian Bach had in no way shared
Marpurg's opinions of Rameau.[175] And as regards Bach's
writings on the art of composition, they only serve to prove
that the method ascribed to Bach by Kirnberger was really
that of the master.

The short rules of thorough-bass which Bach noted down
for his wife, Anna Magdalena, in her later clavier book are
all that has hitherto been known of his method.[176] There
exists, however, a more elaborate work on thorough-bass
by Bach, which Kirnberger seems to have known nothing
about, for he was of opinion that Bach had written
nothing on the theory of music.[177] It was preserved to
posterity by Johann Peter Kellner; its date is 1738, and
its title is as follows: " Des Königlichen Hoff-*Compositeurs*
und Capellmeisters ingleichen *Directoris Musices* wie auch
Cantoris der Thomas-Schule Herrn *Johann Sebastian Bach*
zu *Leipzig* Vorschriften und Grundsätze zum vierstimmigen

---

[174] Marpurg, Versuch über die musikalische Temperatur, nebst einem Anhang über den Rameau- und Kirnbergerschen Grundbass (Essay on musical temperament, with an appendix on the thorough-bass of Rameau and Kirnberger. 1776, p. 239.)

[175] Kirnberger, Kunst des reinen Satzes. II. 3, p. 188: "You are at perfect liberty to say that my late father's principles and my own were opposed to Rameau."

[176] They are given in Appendix B., XIII.

[177] Gedanken über die verschiedenen Lehrarten, p. 4.

Spielen des *General-Bass* oder *Accompagnement,* für seine *Scholaren* in der Musik." (Principles of thorough-bass, and directions for performing it in four-parts in accompanying, for his scholars in music, by Herr J. S. B., royal court composer, capellmeister, director of the music and cantor of the Thomasschule in Leipzig.) The copy, corrected here and there by Kellner, seems to have been made by a person of only moderate musical education, and from the manuscript made by a pupil of Bach's. The first assumption rests upon the numerous and silly blunders of writing, and the second upon frequent inaccuracies in the four-part writing. Details, such as that of the often recurring word *modus* for *motus,* seem to imply that Bach dictated the text of the manuscript. He may have prepared the work for use in class instruction, which would account for the fact that the basses set in four parts contained in the manuscript have not been corrected by him, or at least not throughout, and thus there remain mistakes.[178]

The instruction in thorough-bass is divided into two portions, a "short instruction in Thorough-bass so-called," and a "complete instruction in Thorough-bass"; the first apparently intended for mere beginners, the second for more advanced pupils. Even this little work testifies to the deep and powerful moral earnestness which pervaded all Bach's artistic activity. "The ultimate end and aim of thorough-bass should only be the glory of God and the recreation of the mind. Where these are not kept in view there can be no real music, only an infernal jingling and bellowing." The book is remarkable for clearness, conciseness, and an admirable methodic progress. In these respects it reveals a considerable gift for tuition, and confirms Kirnberger's statement that Bach used to go step by step from the easiest things to the most difficult. It is important to note that Bach calls thorough-bass the beginning of composition, adding that if any one who is willing to learn can take in thorough-bass and imprint it on his memory, he may be

---

[178] The MS., printed in its entirety in Appendix B., No. XII., is now in the possession of Herr Professor Wagener of Marburg, to whom I hereby offer sincere thanks for his kind permission to publish it.

assured that he has already grasped a great part of the whole art.[179] This shows that when Kirnberger insists on beginning his instructions in composition with thorough-bass, calling it "the first lines of composition," it is quite in accordance with Bach's views. The great importance attached by Bach to a knowledge of thorough-bass, not only for the purpose of accompanying, is confirmed by other evidence. In the instruction book now before us the pupil is led up to the accompaniment of short fugal movements. There can scarcely be any doubt that a collection of sixty-two preludes and fugues which remains to us, written throughout on one stave, with figured basses, and bearing the name of Bach as the composer, served as the continuation of his thorough-bass instructions,[180] and that Bach was accustomed to lead his advanced pupils up to the point of making an *ex tempore* accompaniment, even to independent pieces of music, by means of a figured bass and a few other indications.

However, this work of Bach's on figured bass is not altogether original. In chapters 1 to 9 of the section which is entitled "Gründlicher Unterricht des General Basses," Bach has relied largely on Part I. of Friedrich Erhardt Niedt's Musicalische Handleitung, and in parts the original arguments of Niedt are merely abridged and compressed.

In other places, it is true, the development is differently worked out, particularly in chap. 8, where not a trace of Niedt's hand remains; the examples of musical notation also are some of them new, and some of Niedt's are rendered more instructive and concise. The whole work is so treated and altered that Bach might well regard it as his own. At the same time, it is interesting to know that Bach was familiar with Niedt's instruction book and found material in it of which he could make use for his own purposes.

In a preparatory instruction book for composition such as this is, it seems evident that Bach, like Kirnberger, should in actual teaching have preferred the method which leads straight on, after treating of intervals, to chords, chord-

---

[179] In Cap. 5 ("Of the Harmonic Triad").
[180] See Vol. II., p. 98, note 138.

combinations, and modulations, and after that too, not to begin with two-part counterpoint, but with simple counterpoint in four parts. Quite in accordance with Bach's opinion is Kirnberger's statement " It is best to begin with four-part counterpoint, because it is impossible to write good two or three-part counterpoint until one is familiar with that in four parts. For as the harmony must necessarily be incomplete, one who is not thoroughly acquainted with four-part writing cannot decide with certainty what should be left out of the harmony in any given case."[181] This rule is based on the principle that all combinations of notes which can be placed in juxtaposition are to be referred to, or connected with, certain fundamental harmonies— the only principle adopted in musical practice even so early as the seventeenth century. Bach's compositions take it for granted. The boldness and freedom of his part writing, polyphony, and modulation, his way of resolving discords by the interchange of parts, and even his occasional overstepping of all generally held rules of composition, are always limited by this " harmonic " theory, which he developed to such an astounding degree of certainty that he could dare even the boldest flights. At the same time his ear was so delicately trained to follow the parts, even of the most elaborate and complicated pieces of music, that not only would he immediately notice the slightest theoretical mistake while the piece was being performed, but in his own works he would give himself endless trouble in order to obtain the greatest purity of writing. In his thorough-bass instruction book he explains that consecutive fifths and octaves are the greatest errors in composition; they had indeed always been considered as such, but the composers of the beginning of the seventeenth century, and, in part also, the contemporaries of Bach, are much more lax in practice than himself. His ear was exceedingly sensitive with regard to hidden fifths and octaves, even in the inner parts.[182] As to

---

[181] Kunst des reinen Satzes I., p. 142.
[182] See Vol. II., p. 173 f. (on the G sharp minor prelude of the Wohltemperirte Clavier) and Kirnberger, Kunst des reinen Satzes I., p. 159. Bach is less intolerant of fifths and octaves when they occur in passing notes or in

the reduplication of discords and of the leading note, he imposed the strictest rules both upon himself and upon his pupils. Kirnberger says that, so far as his own observation goes, Bach only doubled the major third in the chord of the dominant in one single case in a four-part composition.[183] We know, also through Kirnberger, that in five-part writing he forbids the reduplication of the superfluous second, the fourth, the diminished fifth, the superfluous sixth, the seventh, and the ninth.[184] In the chord of the sixth with the diminished triad Bach characterised the doubling of the sixth as a mistake, because it sounds badly.[185] And yet all these carefully stated rules of part-writing were only a secondary consideration with him. He could even allow them to be disregarded without his sensitive ear being offended, so long only as the logical sequence of harmonies remained intelligible and not to be mistaken. His infallible certainty of feeling with regard to these points allowed him frequently to venture upon things of such a kind that even Kirnberger is forced to admit that Bach's works demanded a quite peculiar style of performance, exactly suited to his style of writing. The player should know the harmonies perfectly, otherwise many of them can scarcely be listened to.[186]

We cannot suppose that simultaneously with the practical music of the seventeenth century the theoretical branch of the art developed in proportion, for in all times artistic

---

ornaments. But here he availed himself of one single license which had long ago been taken by his predecessors. This subject is thoroughly gone into by J. G. Walther in his instruction book, fol. 116 f.

[183] Grundsätze des Generalbasses, p. 83. The passage referred to occurs in B.-G., III., p, 194, bar 5. Compare Kunst des reinen Satzes, I., addendum to p. 37.

[184] Kunst des reinen Satzes, II. 3, p. 41: "*Regula Joh. Seb. Bachii: In Compositione quinque partibus instructa non sunt duplicandæ* 2, 4, 5^b, 6, 7, *et* 9."

[185] Notwithstanding, this is done in the Thorough-bass instruction book, Cap. 8, Reg. 4, example, bar 1, last crotchet, nor can it be regarded as an uncorrected error on the part of the pupil who made the copy. Kirnberger (Grundsätze des Generalbasses, p. 57, note) interprets this rule, which is too strict if universally applied in this manner, that the reduplication is only to be avoided when the sixth appears as the leading note.

[186] Kunst des reinen Satzes, I., p. 216 f.

instruction has always been considerably in arrear of artistic practice. So much, however, is certain that there were practical musicians even at the beginning of that century who knew that the easiest introduction to the art of composition was the knowledge of thorough-bass. In the year 1624 the Berlin Cantor, Johann Crüger, calls this method, which he himself employed, a well-known one;[187] and although for a time it held less prominence than various other methods which he describes, it is certain that it never died out again, but that, like practical music itself, it waxed stronger and stronger, and, in Germany at least, gradually became the prevailing method.[188] Thus in employing this method Bach did nothing new, for it had been long in use. Of course it had its weak points, and was liable to misuse and superficial treatment in unskilful hands. On this account it was assailed in the year 1725 by Joseph Fux, Capellmeister in Vienna, with his *Gradus ad Parnassum;* in this he begins the course of composition with simple two-part counterpoint, note against note, and, after a thorough working out of the five kinds of simple two, three, and four-part counterpoint, he proceeds gradually to imitation, to fugue in two, three, and four parts; he next treats of double counterpoint, applying the same again to fugue, and concludes with some chapters on the church style and recitative, thorough-bass and harmony remaining unnoticed. This method was really new at that time in certain circles, and Fux designated it as such, nor does he attempt to conceal the reactionary spirit which led him to oppose the increasing arbitrariness and law-

---

[187] Crüger, *Synopsis Musices.* Berlin, 1624, *page* 57: "*Nos incipientibus gratificaturi compendiosissimam illam et facillimam ingrediamur componendi viam, qua nimirum ad Fundamentum prius substratum et positum reliquæ superiores modulationes adjici possint. Hoc enim qui poterit, facillime postmodum melodiæ regali Tenoris et Cantus reliquas adjunget voces.*"

[188] J. G. Walther, in his MS. instruction book of 1708, says, with evident reference to Crüger's *Synopsis*: "And this is the most compendious and the easiest way to compose, by building the other parts up from the bass, taking that as the foundation. . . . Therefore we will keep to this said easy method, and make a beginning of composition with four parts (as that whereon so much depends)."

lessness in music. In fact, it was only a revival and completion of the musical teaching of the sixteenth century, and refers only to unaccompanied vocal music in the polyphonic style, and Fux wished this to be regarded as the starting-point of all musical education.

Almost at the same time, in France, Rameau had made a first attempt to justify on scientific grounds the practice of the time, which rested upon the "harmonic" theory, and to reduce it to a system. His "Treatise on harmony," which appeared in 1722, quickly attracted notice even in Germany; Nikolaus Bach told Schröter about it in 1724,[189] and, from Emanuel Bach's statement to Kirnberger, we must conclude that Sebastian Bach was also acquainted with it. The chief points of Rameau's system — the determining of the chord by means of the bass, as well as the inversions arising from the alteration of the bass note — had of course been long known and practised, and so far Bach and Rameau were agreed. But some of the conclusions arrived at by the Frenchman in the course of his system were highly disapproved of by Bach. We cannot with certainty point out the passages which met with Bach's disapprobation, but we may assume them to have treated of those opinions with which Kirnberger waged war.[190] On the other hand, it

---

[189] Schröter, Deutliche Anweisung zum General-Bass, 1772, p. x.

[190] These were chiefly, perhaps, the chord of the 6-5 constructed on the subdominant with the addition of the major sixth, and stated by Rameau to be a root-chord, and also the distinction between essential and non-essential dissonances. Rameau's method of drawing out the fundamental bass of a connected piece of music—*i.e.*, of pointing out the series of notes to which the harmonies of a piece are to be referred as to their proper ground-tones—was applied by Kirnberger to two of Bach's compositions, but not indeed in an irreproachable manner (Die wahren Grundsätze zum Gebrauch der Harmonie, p. 55 ff. and 107 ff.). The same thing has been done with the chief autograph of the French Suites in the case of the Sarabande and the two minuets in the D minor suite, as well as in Fischhoff's autograph of the Wohltemperirte Clavier, in the C minor fugue and the D minor prelude. I cannot confidently affirm that the numbers and letters inserted in the MS. for this purpose are in Bach's own writing. It is not, however, impossible that Bach himself may have demonstrated the nature of Rameau's ground bass theory on his own compositions.

is allowable to conclude that Bach paid full recognition to the method of Fux. For no other than a pupil of his own, Mizler, translated, under Bach's very eye, as it were, the *Gradus ad Parnassum* into German, and when Mizler, referring to the value of the work, says that it had been well received by those who really knew what a good composition was, he must doubtless refer most directly to Bach.[191] As a matter of fact, there was much more sympathy between Bach and Fux than might appear at first. With regard to the development of the art in Germany, it must not be thought that up to Bach's time a strictly contrapuntal method of teaching prevailed and that Bach introduced a freer style. A more exact examination of the German composers, particularly those of the latter half of the seventeenth century, and among them notably the organ and clavier masters who were the glory of that period, will show plainly that the contrary is the case. Awkwardness in polyphonic vocal writing had much increased during the century, and even the authorised freedom of instrumental composers threatened to degenerate into arbitrary laxity. Only to a very limited degree can Bach be said to have inherited from his predecessors his astounding contrapuntal skill, and the strictness and purity of his style. He re-introduced these qualities into the art, with a leaning, indeed, towards the old classical models, but following rather the leading of his own genius. Through him the old approved rules of part-writing again came to be duly honoured, with such modifications, it is true, as were rendered necessary by the alteration in the tone material, and it was Bach who once more taught the organ and clavier composers to write as a rule in real parts, and to keep the same number of parts throughout a whole piece. Hence, notwithstanding his approval of Fux's method, it was only natural for him to prefer another style of instruction. That was very well fitted for vocal compositions; for that alone could it afford a safe groundwork. With certain limitations it was also useful for violin players and for writers for

---

[191] See the Introduction to Mizler's translation of the *Gradus*. Leipzig, 1742.

the violin; it could not be used for the instruction of writers for the organ or clavier, since it placed the learner in direct opposition to the demands of his instrument. Whether, as a general rule, it is better to begin with vocal or with instrumental composition is not the question here; we are stating historical facts. It cannot be denied that the only method of instruction that can succeed is one which from the first, let the pupil begin where he may, will awaken his individual feeling for art. In true art there is nothing mechanical, there exists no essential antagonism between reproduction and production; the first phrase sung, or the easiest little clavier piece played, is a starting-point in the art of composition, or it may become so. Thus it was quite natural that in Italy, and in those districts of Germany that remained under Italian influence, the old contrapuntal method, revived by Fux, should predominate, since the musical culture of the Italians was mainly based on vocal music, while in Germany, where instrumental music was always more and better performed than vocal music, the other method was of course preferred. From this point of view we can understand the different positions held by fugues in the method of Fux and Bach (or Kirnberger) respectively. Vocal fugue moreover rested on other conditions than those of instrumental fugue, and what Bach made out of the latter might well be regarded as the highest pinnacle of art, to which none but the most thoroughly and diversely cultured of his disciples could possibly attain.[192]

In the case of two pupils of Bach's, Heinrich Nikolaus Gerber and Agricola, we have sufficiently precise information with regard to the method pursued by the master in their instruction to enable us to see the way in which Bach used the instrument with which the pupils were already

---

[192] Bach himself once was heard to call the fugues of an "old and laborious contrapuntist" dry and wooden, and some fugues by a "modern but no less great contrapuntist" pedantic, at least in the form in which they were arranged for the clavier, because the former kept persistently to his chief subject without change, and the latter did not show enough invention in enlivening his theme by means of interludes (Marpurg, Kritische Briefe über die Tonkunst, I., p. 266). It is not known to what composers he alluded.

familiar as the vehicle of instruction in composition suitable in each case. Gerber had, when receiving instruction from Bach, to " study thoroughly " (" durchstudiren ") his (Bach's) Inventions, a set of suites, and the Wohltemperirte Clavier. Then came the practice of thorough-bass, but not extempore; Gerber had to write out a four-part accompaniment to Albinoni's violin solos, from the bass part.[193] Similarly, Agricola was first instructed in clavier and organ-playing, and after that " in the harmonic art."[194] In neither case was the pupil a beginner nor the instruction elementary; these young men were led on by Bach to real composition by means of harmony and the four-part accompaniment to be written on Albinoni's figured basses. For this purpose he also availed himself of chorales. His method of beginning to teach simple counterpoint was to give a chorale melody to be harmonised in four parts. This is rendered certain by what Kirnberger and Emanuel Bach say. The latter asks, alluding to the chorales by his father that he edited, who can dispute the advantage of an education in composition which begins, not with a stiff and pedantic contrapuntal exercise, but with chorales?[195] Kirnberger recommended beginning with four-part counterpoint, and extols the chorales of Bach as models of four-part writing which it would be impossible to surpass, and in which not only all the parts had their own flowing sequence, but one kind of character was preserved in all.[196] He also holds a diligent practice in chorales to be of the highest value, or indeed indispensable, and that

---

[193] Gerber, L. I., col. 492.
[194] Ch. C. Rolle, Neue Wahrnehmungen, p. 93.
[195] Preface to " J. S. Bach's vierstimmige Choralgesänge." The idea in the sentence above quoted is not expressed with sufficient clearness, for it admits of the interpretation that the method alluded to merely substituted the more animated and interesting chorale melodies for the old-fashioned *cantus firmi* which were of the greatest possible simplicity, and which were used for contrapuntal studies, which does not exclude the possibility of this method having begun with two-part counterpoint. And the sentence was in fact understood thus by Vogler (Choralsystem, p. 61). There would have been nothing essentially different in this system, however, and yet Emanuel Bach doubtless intended to indicate something different to Fux's method.
[196] Kunst des reinen Satzes, I., p. 157.

it is prejudice to consider this kind of exercise as superfluous or pedantic, since it forms the true foundation, not only for good writing, but for the art of composing well and expressively for the voice.[197] Thus this method was not to do away with those difficult yet indispensable first studies at the cost of thoroughness, nor was it ever supposed to do so, though there were persons who considered it pedantic, while, on the other hand, those who knew about such things extolled it for its thoroughness.[198] When Kirnberger regards Bach's chorale movements as models of vocal composition, it must not be forgotten that he meant accompanied vocal music, and that he considered the chief aim in such work to be the invention of simple and expressive melodies. They were not intended as models of an *a cappella* movement in several parts, and nothing was farther from Bach's thoughts than that they should be treated so. These chorale movements are treated for the most part with a rich instrumental accompaniment, never without any accompaniment; they are almost all integral parts of grand and ingeniously elaborate church compositions, and always strictly adhere to the particular feeling prompted by the words; this is the manner in which they were conceived, and in this way only are they to be judged. Carl Friedrich Fasch, whose opinion (according to Gerber) was that in Bach each separate part was very vocal, but the combinations between the parts were utterly unvocal, that they were beautiful parts, and yet not welded together to a beautiful whole,[199] had not found this standpoint on which to form his opinion. It cannot be denied that the publication of the chorales as separate pieces in four parts, with the announcement that

---

[197] Kunst des reinen Satzes, I., p. 215. Forkel's account of Bach's method of instruction agrees with this, but he drew his information mainly from Kirnberger; he had opportunities, however, for learning about it from Bach's sons.

[198] Lingke, Die Sitze der musikalischen Haupt-Sätze. Leipzig, 1766. Introduction. He says, alluding to Emanuel Bach's statement quoted above, "It would be difficult to find a more thorough method than this." It was afterwards related of Kirnberger that he made his pupils work at chorales for three years; see Vogler, Choralsystem, p. 24.

[199] Gerber, N. L., II., col. 86.

they were to form a complete hymn-book, may have given rise to false impressions, and have done harm. Bach's book was soon taken as a pattern to an undue extent, and in 1790 Abraham Peter Schulz, in discussing the influence of music and of its introduction into schools upon the national culture, was obliged to admit "that in arranging a simple chorale the greatest harmonists of the Bach school sought rather to display their erudition by multiplying unexpected and dissonant progressions—often rendering the melody quite unrecognisable—than to regard that simplicity, which is necessary to render the chorale intelligible to the common people."[200]

In connection with the four-part chorales of Bach there is still another question, equally important as regards the master's attitude towards the music of the past, and the way in which he taught his disciples. A great part of the chorale melodies set by him date from centuries when the formation of melodies followed other laws than those of our own, or even of Bach's day. The formation of a passage in accordance with one of the six, or, counting the plagal modes separately, the twelve kinds of octaves, gave rise to very characteristic modulations, and, when set in several parts, to a harmonic accompaniment of corresponding individuality. Until the middle of the seventeenth century, the feeling for the different characters of the various modes was still kept up to a moderate extent among Protestant musicians. It then began to die out, and eventually the multiplicity of the church modes gave place to the duality of major and minor. Though the former continued to exist in name for a time, no definite idea attached to most of them. Johann Schelle once asked Rosenmüller, whom he so highly admired, what was his opinion with regard to the old musical modes. Rosenmüller laughingly replied that he only knew the Ionian and Dorian modes.[201] The amalgamation of those scales which contained the greater

---

[200] Schulz, J. A. P., Gedanken über den Einfluss der Musik auf die Bildung eines Volks, und über deren Einführung in den Schulen der Königlich Dänischen Staaten. Kopenhagen, 1790.
[201] (Fuhrmann), Musicalischer Trichter. 1706. P. 40 f.

third into the C clef was more easy than the concentration of the Dorian, Phrygian, and Æolic modes, whose triads contained the lesser third, into the minor key. Werkmeister at first agreed with Rosenmüller in thinking that the Dorian mode was the best representative of the minor,[202] but was afterwards more in favour of the Æolian.[203] Johann Gottfried Walther took a middle course, for, in the year 1708, he taught that three modes were in ordinary use at that time, Dorian, Æolian, and Ionian.[204] Lastly, Bach recognised only two modes, the Ionian with the greater and the Æolian with the lesser third.[205] This gradual process of simplification is very interesting. The Dorian and Æolian modes were adapted each to its special purpose. The Dorian allowed of a perfect cadence on the key-note, on the fourth, and on the fifth, without overstepping the limits of the diatonic system (putting aside the raising of the leading note). In the Æolian a diatonic cadence on the fifth was impossible. On the other hand, it had the more distinct minor character, inasmuch as the triads on the key-note, on the fourth, and on the fifth, all had the minor third, while in the Dorian mode the triad on the fourth has the major third. Our minor mode appears to be more strictly a combination of the Dorian and Æolian. Although the modern two-mode system is firmly established in Bach, he yet keeps up a close connection with the system of six modes, by simply taking the Æolian mode for the minors. Theoretically there only existed for him one and the same scale for ascending and descending passages alike, while Rameau, and Kirnberger, used the scale a b c d e f♯ g♯ a, in ascending, and Lingke, in order to get over the

---

[202] *Harmonologia musica.* 1702. P. 59.
[203] Musikalische Paradoxal-Discourse. 1707. P. 86.
[204] MS. Musical Instruction Book, Fol. 152*b*.
[205] Clavier Book of Anna M. Bach. 1725. P. 123. "The scale with the minor third is: 1st, a tone; 2nd, a whole tone; 3rd, a half; 4th, a whole; 5th, a whole; 6th, a half; 7th, a whole; 8th, a whole tone; from this comes the following rule: (for the intervals) the 2nd is in both scales major; the 4th remains the same in both; the 5th and 8th are perfect; and as the 3rd is so are the 6th and 7th."

anomaly of having two different scales, made up the scale a b c d e f g♯ a, for both kinds of movement.[206]

Bach's minor scale has no distinctive bearing on his practice, but it serves to explain his attitude towards the two opposing systems. When, as organist at Weimar, he first displayed his full mastery over the chorale, Mattheson and Buttstedt were disputing about the *raison d'être* of the ecclesiastical modes, and his success and achievements served as the tragi-comic death-knell to their quarrel. When Fux, in his *Gradus*, several years later, attributed a fundamental importance to the church modes, the opposition of so eminent a man to the modern reform could not fail to have a certain effect; but Fux was thinking chiefly of Catholic church music. In the Protestant church, the old system of modes could only find the protection it deserved through a Protestant musician, and this it found, in part, through Bach. In the principles of composition, as in chorale treatment, he had achieved a position where all contrasts were reconciled; he kept to the historically developed and fundamental principle of major and minor scales, and used the church modes as a kind of subsidiary keys. He obtained from them the full wealth of modulation which they afford, but always kept them subordinate to the simpler radical feeling of major and minor.[207] The consciousness that the beauty of many of the old chorale melodies would be much impaired by forcing them to submit to the laws of modulation prescribed by the "harmonic" system was to Bach only a secondary consideration. He felt that the art ideas which had taken form in these hymns carried in them, by reason of their having been nurtured for centuries in the bosom of the church, an

---

[206] Lingke, Die Sitze der Musicalischen Haupt-Sätze, p. 16 ff.—Lingke laid this scale, invented by him, before the Society for Musical Sciences in Leipzig in 1744, and all the members approved of it (see Mizler, Musikalische Bibliothek, Vol. III., p 360). Bach had not become a member at that time.

[207] In consequence of this, Kirnberger (Kunst des reinen Sätzes, I., p. 103) says: "In the music of the present day, we not only have twenty-four different keys, each with a definite character of its own, but we can retain beside them the modes of the ancients. Hence arises an immense variety of harmony and modulation."

inalienable wealth of genuine religious feeling. This he neither could nor would dispense with in forming his own church style. The system of church modes appears in Bach not as one ingeniously employed for certain subjects; it came to a new birth in his genius, and finds its place not merely in this or that chorale, but in all his music. When it seemed suitable he would arrange a chorale strictly in accordance with the rules of its mode; for example, in the Mixolydian melody "Komm, Gott Schöpfer, heiliger Geist."[208] Generally, however, he used what his pupil Kittel calls the "mixed" style of harmonising,[209] giving now more and now less prominence to the characteristic modulations of a particular mode. Instances of this are the Dorian chorales "Das alte Jahr vergangen ist," "Erschienen ist der herrlich Tag";[210] the Mixolydian "Gelobet seist du, Jesu Christ," "Gott sei gelobet und gebenedeiet," "Nun preiset alle Gottes Barmherzigkeit";[211] the Phrygian "Christum wir sollen loben schon," "Erbarm dich mein, o Herre Gott."[212] It also occurs that melodies belonging to one of the ecclesiastical modes appear quite in modern harmonising; and, on the other hand, Mixolydian modulations (it may be) are introduced into chorales that do not naturally belong to this mode. The chorales "Jesu nun sei gepreiset," "Es ist das Heil uns kommen her," and "Vom Himmel hoch da komm ich her," in the cantatas of the same name, and especially the chorale in the middle of the second part of the Christmas oratorio, are examples of this treatment. For all three are, strictly speaking, Ionian, even the second; it serves, at any rate, to prove that Bach thought that the closing chorale of that cantata was already more than a century old. From all this it is evident that Bach had evolved from the church modes a means of expression which he used freely wherever the poetic meaning and the musical sequence seemed to him

---

[208] In the chorales published 1785, No. 187; in Erk's edition, No. 255. Compare Kirnberger, *loc. cit.*, II., 63.
[209] Der angehende praktische Organist. Section III., p. 37 ff.
[210] Nos. 180, 29, and 30 in Erk.
[211] Nos 41 and 213 in Erk, and No. 222 in the chorales of 1786.
[212] No. 175 in Erk, and No. 33 in the chorales of 1784.

THE DEVELOPMENT OF *MAJOR* AND *MINOR*.

to require it; and for the same reason he harmonises one of his favourite melodies, "O Haupt, voll Blut und Wunden," now in the Ionian, now in the Phrygian mode. The inexhaustible wealth of harmony, which he exhibits not only in the chorales but in all his compositions, and generally without any far-fetched modulations, arises from these two sources; a thorough familiarity with the ecclesiastical modes, and an unfailingly keen and certain appreciation of the harmonic relations subsisting in the systems of major and minor.[213]

That Kirnberger understood his great teacher's principles is shown by his not only having felt moved to comment[214] upon the harmonic nature of a set of organ chorales in the third part of the "Clavierübung" which had made their appearance at the time when he was studying with Bach in Leipzig, but also by his having recognised a certain form of modulation in the style of the ecclesiastical modes, even in fugues in the "Wohltemperirte Clavier" and in the free portions of the cantatas. In his manuscript copy of the first part of that collection of fugues he designates those in C major and C sharp major as Ionian, those in C minor, E flat minor, and G sharp minor as Æolian, and those in C sharp minor and F sharp minor as Dorian. The terzett in the cantata "Aus tiefer Noth" he assigns to the Æolian mode.[215] This nomenclature can, of course, only have very limited weight, inasmuch as it expresses nothing more than a preponderance of those modulations which are peculiar to this or that mode; and more than this Kirnberger certainly did not mean to say. Strictly speaking, scarcely any connected groups of bars could be found in these pieces in which the diatonic laws are not somewhere set aside. But such a depth and variety of harmony was only attainable by the most comprehensive use of the means of modulation afforded by the ecclesiastical modes.

---

[213] Compare with this the beautiful passages devoted to this subject in Winterfeld, Ev. K., III., p. 299.
[214] MS. in the Amalienbibliothek at Berlin. Inserted in App. B., XII.
[215] His MSS. of both are in the Amalienbibliothek, Nos. 57 and 58.

Bach's connection with the system of church modes is also recognisable in his style of writing key signatures. Thus he signs E in the Dorian mode with two sharps, and, consequently, F in the Dorian mode with three flats, and G in the Dorian mode with one flat; using, of course, no signature in the case of the Dorian, Phrygian, and Mixolydian when they are used in their original positions. To show how very much the character of the scales had been lost at that time, it may be stated here that Mizler calls the Dorian mode D minor, the Phrygian mode E minor, the Lydian mode F major, &c., without more ado.[216] Bach, however, only selected those signatures when he wanted to give to his work the stamp and character of the particular mode. Where this is not the case, he keeps to the simple major and minor. It often seems, indeed, as though his method of procedure was dictated only by custom, and that Agricola's remark was partially true of him, that many composers of the first half of the eighteenth century indicated the Dorian mode when they probably intended the Æolian.[217] This is more particularly to be understood of Bach's free compositions. In the alto aria of the "Homage" cantata, written at Cöthen,[218] Bach gives the first line the signature of G minor and all the rest that of G in the Dorian mode, to explain which there is no inner reason whatever. But from such things as this we see how, in the search after a comprehensive minor key, musical feeling wavered for a long time between the Æolian and the Dorian modes.[219]

---

[216] Musikalische Bibliothek, I., pp. 30, 31, 34, Notes.
[217] Remarks to Tosi, p. 5.
[218] See Vol. II., p. 619 f.
[219] I cannot close this chapter without mentioning Vogler, who, partly in his own "Choral-System," p. 53 ff., and partly through his pupil, C. M. von Weber (Zwölf Choräle von Sebastian Bach umgearbeitet von Vogler, zergliedert von C. M. von Weber. Leipzig, C. F. Peters), attempted to point out the supposed incorrectness and want of beauty in Bach's four-part chorales. That such a ridiculous procedure should have been possible is partly the fault of the editor, Emanuel Bach. As Vogler knew neither the object, position, arrangement, nor, in many cases, even the words of these chorales, and since they were supposed to constitute a complete chorale book, he must have started with utterly false impressions. He is not to be blamed for not having known Bach's relations towards the ecclesiastical modes, since it required a much more com-

V.

WORKS OF THE LAST (LEIPZIG) PERIOD.—CONCERTOS.—THE LATER WORKS FOR CLAVIER AND ORGAN.

WHILE Bach, during the first part of his life—that is, down to the close of the Cöthen period—shows himself chiefly as an instrumental composer, during the Leipzig time the sphere of his chief activity is seen to be concerted church music. A natural progress and growth from the one period to the other is perceptible; and, putting aside, as far as may be, the relative intrinsic importance of the different provinces of art, the earlier period may be said to stand to the later in the relation of preparation to fulfilment. The art which had been applied to instrumental music was not given up, but only found expression through a higher medium; since the two-fold importance of song as a means of expression inevitably introduced a new and independent element. But since instrumental music was the true source of Bach's art, it was only natural that he should continue to draw from it, even in the second period of his life. The number of his later instrumental compositions is by no means small, and their character is that of hardly-won fruits ripened in the prosperous harvest of his life.

The true artist is in the happiest position when all his works are "occasional" compositions; and, conversely, he will be instinctively prompted to apply his art as often as possible to the events that are nearest to his own life. Both these statements hold good in the case of Bach. He was never greater than when he had to make his music conform to the requirements of his position and office; and he always embraced the opportunity of composing music for a definite object. From 1729 until 1736 he directed Telemann's Musical

---

prehensive view of the work and historical importance of the master than was possessed by any one at that time. As regards the supposed want of beauty, Vogler's remarks reveal that his inability to enter into the genius of Bach's harmonic and melodic sequences was as great as his so-called "improvements" are lame and tasteless.

Society. Mention has already been made of the chamber vocal compositions with accompaniment that he wrote for it. There is no doubt that his own instrumental works had been performed there. These, no doubt, were orchestral suites. Those "Bach Sonatas," which the musicians at Eutritzsch performed at the opening of the "Kirmess" (fair), in 1783, were probably works of this kind.[220] It is impossible to say with certainty which of his orchestral suites were composed expressly for the Musical Society, since it is very probable that he was engaged in this form of composition even at Cöthen. In the case of the better known of the two D major suites, it is certain that the original parts prepared for its performance were written between the years 1727 and 1736—just about the time when Bach was director of the Musical Society. The other D major suite seems, judging by its contents, to belong to the Leipzig period.[221]

The music performed by the Society was of various kinds; hence we may assume that violin and clavier concertos by Bach were also performed, though more frequently, perhaps, at Bach's house. As no fewer than five claviers, two violins, three violas, two violoncellos, a viol da gamba, and other stringed instruments were left at his death, it is evident that he was well prepared for concerts at home. Nor was there any lack of talented, or, at least, available pupils for these performances. The most flourishing time in Bach's domestic band was, no doubt, from about 1730 until 1733, since the grown-up sons, Friedemann and Emanuel, were still living in their father's house, Bernhard was already grown up, and Krebs, who had been Sebastian's pupil since 1726, was beginning to display his great talents, not to mention the vocal performances of Anna Magdalena and her stepdaughter, Katharina. And we know from Bach's own words what a pleasure it was to him at that time "to get up a vocal and instrumental concert."[222] Whether Bach ever wrote violin concertos expressly for them must remain un-

---

[220] Compare ante, p. 21.
[221] See Vol. II., pp. 141—143, and 661. Vol. III., p. 78.
[222] Compare Vol. II., p. 254.

decided; but it is certain that about this time he had works of this kind performed.[223]

In this branch of art he devoted himself chiefly at Leipzig to the clavier concerto. The sonatas and suites for the violin were subsequently arranged by Bach, either wholly or in part, for the clavier or organ; and the comparison of the arrangement with the original shows that the idea of many of these pieces had its root in the clavier style, rather than in that of the violin.[224] The same thing was done with some of his violin concertos, and the comparison of the two versions gives a precisely similar result. But not only did he arrange the violin concertos in A minor, E major, and D minor for clavier and orchestra, transposing them into G minor, D major, and C minor; he also has left three concertos (D minor, F minor, and C minor) which are evidently re-arrangements of violin concertos, the originals of which are unfortunately lost.[225] A fourth, in E major, bears no undoubted signs which point to a violin concerto as its original, so that we must, for the present, assume it to have been originally written for the clavier. After receiving its first form, the whole of it was used for two church cantatas, and then it was re-arranged for a clavier concerto.[226] A fifth, in D minor, also became part of a church cantata; but the original has been lost, with the exception of a small fragment.[227] The A major

---

[223] The time when he was most engaged in the composition of violin concertos was when he was in Cöthen. See Vol. II., p. 125 f. The fact that the original parts of the A minor concerto and two autograph parts of the D minor concerto bear the water-mark M A only proves with certainty that their performance took place between 1727 and 1736—not that they were written during that time.

[224] See Vol. II., p. 80 ff. and 98.

[225] See Rust's dissertations in the preface to B.-G., XVII. Rust (preface to B.-G., XXI.,[1] p. xiii.) has also completely convinced me that the other concerto of the two in C minor for two claviers originated in a concerto for two violins.

[226] See Vol. II., p. 447 ff.

[227] See Vol. II., p. 446 f. It is very probable that among the cantata-symphonies may be embodied parts of lost violin and clavier concertos. One such symphony, its cantata no longer existing, has been published by Rust among the violin concertos in B.-G., XX.,[1] No. 4. I have alluded, on p. 85 of

concerto stands alone, without any direct connection with other works. The two concertos originally composed for two violins and tutti were also arranged for two claviers.[228] The number of simple clavier concertos is seven, including one which only exists in the church cantata.[229]

In the case of several of these concertos the date of composition can be more definitely asserted. The first of those in D minor was turned to account in two different church cantatas, first for "Ich habe meine Zuversicht," which must have been written for the twenty-first Sunday after Trinity, in 1730 or 1731, and afterwards for the music for the third Sunday after Easter, " Wir müssen durch viel Trübsal in das Reich Gottes eingehen."[230] The other D minor concerto is found in the cantata written apparently in 1731, for the twelfth Sunday after Trinity, "Geist und Seele wird verwirret."[231] The first two movements of the E major concerto are contained in the church composition "Gott soll allein mein Herze haben," and the last in the cantata "Ich geh und suche mit Verlangen"; these two are intended for the eighteenth and twentieth Sundays after Trinity, apparently in the years 1731 or 1732.[232] The original compositions must therefore have been written before these dates. The re-arrangement for two solo instruments of one of the two C minor concertos was made in

---

this volume, to another case of this kind. Forkel (p. 60) says that Bach wrote instrumental pieces to be played during the Communion, and that they were always so arranged as to be instructive to the player, but that most of them have been lost. These may also partly have been arranged movements from instrumental concertos, and partly pieces of the same kind as those which are found at the beginning of the second part of the church cantatas in two sections—compare "Die Elenden sollen essen" and "Die Himmel erzählen die Ehre Gottes."

[228] B.-G., XXI.,[1] Nos. 1 and 3.—P. S. II., C. 10 (257). Both in C minor.
[229] G minor {B.-G. XVII., No. 7; P. S. II., C. 2(249).
  D major   „   „   3;   „   4(251).
  D minor   „   „   1;   „   7(254).
  F minor   „   „   5;   „   3(250).
  E major   „   „   2;   „   6(253).
  A major   „   „   4;   „   5(252).
  D minor (see B.-G. XVII., p. xx. and vii., No. 35).
[230] See Vol. II., p. 446. Vol. III., p. 79.
[231] See Vol. II., p. 449.   [232] See Vol. II., p. 448 f.

the year 1736.[233] In Bach's later years, when he undertook the final revision of the most important of his organ chorales, and gave the final form to the Passions according to St. John and St. Matthew, he also collected his clavier concertos and put the finishing touches to them.[234]

In Weimar, as we know, Bach had worked diligently at the arrangement of violin concertos for the clavier; and in Leipzig he transcribed a great number of Vivaldi's compositions. The work was identical, excepting that in the former case the tutti parts are included in the clavier arrangement, while here the clavier part simply takes the place of the violin. Beside the alteration of those passages and melodic phrases which were too exclusively fitted for the violin, and their extension to deeper registers inaccessible to the violin, he had to add a part for the left hand. Merely to allot the figured bass part to the clavier was a makeshift, of which indeed he availed himself frequently, especially in the D major concerto and in the middle movement of a concerto in C minor. When he undertook a more thorough remodelling he generally surrounded the figured bass with more animated passages in the clavier bass, and sometimes introduced an independent third part between the upper part and the figured bass, or, when the figured bass stopped, turned the one part of the solo instrument into a trio. In this respect the G minor, and in its last recension the D minor, concerto underwent especially careful treatment, as also the first movement of that C minor concerto which still exists in the original. As contrasted with stringed instruments, the peculiarity of the clavier is its power of playing in two parts, or even in three or more parts. By this, as well as by quality of tone, the clavier can be brought into sharper contrast with the tutti than the violin. Since the time of Mozart the clavier has come into more and more prominence as the solo instrument in instrumental concertos; and nowhere is its style more purely and perfectly displayed than

---

[233] B.-G., XXI.,² No. 3. See Appendix A. of Vol. II., No. 44.

[234] The autograph, in the Royal Library in Berlin, contains seven of these and also the fragment of the second D minor concerto. One of them, however (in F major), is a *Concerto grosso* with clavier, of which more anon.

in the hostile position, so to speak, which it is made to assume towards the orchestra in works of this class by Mozart and Beethoven. The same thing cannot be asserted of Bach's clavier concertos, even when due allowance is made for the conditions of development afforded by the difference between the harpsichord and the pianoforte. It must be remembered that at that time the clavier formed a part of every concerto, taking the figured bass part. As such it had not only to support the solo instrument, but, in Bach especially, to bind together the different instruments which took part in the tutti to an unity to which it gave the general stamp.[235] It has been stated before that even in Bach's clavier concertos the accompanying harpsichord was employed. Thus there could be no sort of distinct opposition between the tutti and the solo instrument; either externally or internally. Bach of course knew this, and he now attempted rather to let the clavier appear openly in the character which had hitherto been as it were latent in it when treated as the exponent of the figured bass. The musical form which arose from the antagonism of two equally matched forces is preserved, as in the Brandenburg concerto, but the clavier is always predominant. These works are, we may say, clavier compositions cast in concerto forms, which have gained, through the co-operation of the stringed instruments, in tone, parts and colour. Accordingly Bach allows the solo clavier to play during all the tutti portions, or to surround them with figures and passages. He thus deprives himself of even the simplest effect of contrast; sometimes to a surprising extent, as, for instance, in the Andante of the G minor concerto. But his great object was to obtain a predominance of the clavier tone, in which he must have succeeded, if we consider that another harpsichord was added to play the figured bass part, and that the tutti portions were generally very thinly orchestrated. The part filled by this second harpsichord in the way of supporting the harmonies is generally so slight that it could easily have been undertaken by the solo instrument, and, indeed, it

---

[235] Compare Vol. II., p. 108.

seems to me that in the last recensions of the D minor concerto and the G minor concerto it was intended that this alteration should be made. We shall see that Bach followed out to its extreme consequences the idea of making the clavier the predominant part in a clavier concerto. The germ of this idea may be traced in these compositions, even in their original state as violin concertos. For that Bach undertook their re-arrangement merely because he did not care to write new clavier concertos is an assumption utterly contrary to his character, and is disproved even by the large number of these re-arrangements. No doubt he felt that the style of his violin concertos was so much moulded by his clavier style that their true nature could only be fully brought out in the shape of clavier concertos. It cannot be denied that many details, and notably *cantabile* passages, lose in effect in the clavier arrangement; but as a whole we must regard them as new and higher developments, rather than arrangements.

It is clear that in respect to the form, especially of the first movement—to the relation between the solo instrument and the tutti—to the characteristic qualities of tone in each and in the whole, the Bach clavier concerto and the newer form which originated with Mozart must be judged from different points of view. When we have discovered the right one our enjoyment is perfect.[236] Omitting the three concertos which still exist in the shape of violin concertos, and have already been spoken of,[237] and that D minor concerto which only survives in the cantata "Geist und Seele wird verwirret," those for one clavier in F minor

---

[236] Forkel (p. 57) briefly calls Bach's concertos for a clavier with accompaniment antiquated. That they can never be, so long as they are approached with the right kind of preparation. An artist has a right to demand that his productions be viewed as he intended them to be. Hilgenfeldt (p. 127) was led by Forkel's opinion to ascribe the four clavier concertos, which he knew and called "somewhat in the old French style," to the first part of the Cöthen period, and he is followed by Bitter (II., 292). How utterly without foundation this assumption is, is proved by the fact that among them are included one of the Brandenburg concertos, and one remodelled from the Brandenburg concertos which are known to have been written in 1721.

[237] Vol. II., p. 125 f.

and A major demand attention for their clear and compact form. They are particularly well fitted to elucidate the structure of the older form of concerto. The middle movement of the F minor concerto consists only of a continuous and richly ornamented *cantilena* for the solo instrument, while in the Larghetto of the other the stringed instruments, together with the figured bass, perform a kind of free chaconne, which may be compared to the Adagio of the E major violin concerto (in D major for the clavier); only that the theme lies in the upper part, is used to form interludes, and appears also in an inverted form. The concertos in E major and D minor are of large proportions; the first is characterised by a cheerful activity, with a tender fervour in the Siciliano, and the second by a passionate and touching pathos. The latter is indisputably the most important of all, never ceasing to rivet our attention by the mighty swing and deep earnestness of its subjects, as well as by the ingenuity of their treatment. The allegro movement is rather episodical than thematic, as befits the concerto style; the Adagio is a chaconne, the theme of which remains in the bass, excepting that as the key changes short episodical interludes are continually introduced for the purposes of modulation. The wild, uncontrolled energy, which only comes to rest in the passages of deep lament, scarcely relaxing even at the close from its earnest gravity, gives to the work a character unusual in concertos; for at that time, much more than in later periods, it was usual to use this form for nothing more than an agreeably and lightly moved play of feeling. The C minor concerto for two claviers, based upon a lost composition for the violin, has the same gloomy character, and is rather more elegiac in style. The peculiarity of orchestration in this concerto, which is polyphonic and very full, especially in the first movement, is chiefly due to repeated remodelling. Bach evidently bestowed especial care on the task. In an original concerto for two claviers the relation of these to the strings would have been very different.

Besides the simple concerto there was the *Concerto grosso*, in which the contrasting element to the tutti consisted, not of one, but of several instruments in combination. Bach's concertos for two violins belong to this class, as, of course, do those for two claviers which grew out of them. We must assume a concerto for oboe and violin, which is unfortunately lost, to have been cast in this mould.[238] We know from the Brandenburg concertos how fond Bach was of setting the so-called *Concertino* in a very original way. Thus, in the second concerto, the *Concertino* consists of trumpet, flute, oboe and violin; in the fifth, of flute, violin, and harpsichord; and in the fourth, of violin and two flutes.[239] The last-mentioned concerto was re-arranged by Bach, the key being changed from G major to F major, and the clavier appearing in place of the violin; thus the *Concertino* resembles that of the fifth Brandenburg concerto,"[240] and in both the clavier predominates even over the other solo instruments. A third concerto of the kind also employs flute, violin, and harpsichord as the solo instruments. Its allegro movements are grandly developed from a prelude and fugue for clavier alone, which even in that form were seen to be designed for concerto movements; they have already been spoken of.[241] The middle movement is taken by Bach from an organ sonata in three parts in D minor. There is no tutti part in the movement, for the fourth part, rendered necessary in order to weld the solo instruments together, is not *obbligato*, but serves merely to fill up the harmonies.[242]

If we could determine the chronological sequence of the clavier concertos it would be easier to recognise the different stages by which Bach developed and raised this form. Many impulses of various kinds, from without and

---

[238] Breitkopf's catalogue for New Year, 1764, p. 52: "Bach, *G. S.* I. Concerto, *a Oboe Concert. Violino Conc.* 2 *Violini, Viola, Basso.* 1 thlr." From this it would appear as though Bach had written several oboe concertos.
[239] See Vol. II., pp. 132, 134, and 133.
[240] B.-G., XVII., No. 6, and P. S. II., Cah. 1 (248).
[241] Vol. I., p. 421 f.
[242] According to the MS. sketches, which although not original are yet very trustworthy, the solo clavier in this concerto would seem to have taken the part of the figured bass in addition to its own. It is published in B.-G., XVII., No. 8.

from within, assisted him in the attainment of that excellence which is perfectly exemplified in the C major concerto for two claviers, the two concertos for three claviers, and the so-called Italian concerto. The tradition that Bach wrote the concertos for three claviers in order to play them with his elder sons is quite trustworthy.[243] This was an incitement, which well accorded with his own inclination, to make the concerto more and more into a form for clavier alone. At all events the C major concerto for two claviers preceded them, and if they were composed by 1733 at latest its date will be probably between 1727 and 1730.[244] Of the two concertos for two claviers which exist only in an altered form, one, as has been said, is to be assigned to the year 1736. If the other is to be assigned to an earlier period, we may assume that Bach was moved, by the very fact of altering this concerto, to attempt an original composition of the same kind.[245] But it is easily conceivable that the mere use of a second harpsichord for the figured bass would suggest to the composer that it might be raised from its dependent position to a more prominent one.[246] For it must be regarded as an indubitable fact that in the C major concerto no instrument is meant to play the figured bass part; and even in the older C minor concerto—to judge from the parts which date from Bach's time—it was not regarded as indispensable. Finally, it must not be forgotten that a composition for two claviers was nothing new. Bach may not have had the form suggested to him by Hieronymus Pachelbel's Toccata for two claviers, if indeed he knew of it; Couperin had written an Allemande for two claviers which Bach, his great admirer, must certainly have known.[247]

---

[243] S. F. K. Griepenkerl, in the preface to the D minor concerto for three claviers; P. S. II., **Cah. 11** (258).

[244] The autograph clavier parts have the watermark M A. Published in B.-G., XXI.,² No. 2, and P. S. II., Cah. 9 (256).

[245] Forkel (p. 58) says that the C minor concerto here referred to is "very old," as compared with that in C major. But he seems only to mean that its style was antiquated.

[246] A theory with which Rust agrees (B.-G., XXI.,² p. vi.).

[247] See Couperin's works, edited by J. Brahms (Denkmäler der Tonkunst, IV., p. 160).

Be that as it may, the C major concerto leaves not a moment's doubt as to Bach's conception of this form. There is no longer any idea of strife or opposition between the solo instruments and the tutti; the tutti has nothing to do but supply an accompaniment to the harmony, or to support the passages played on the claviers. In the Adagio it is silent, and in the other movements it could quite well be dispensed with without detriment to the construction of the work. Its use is to give fulness and colour. The few short episodes and polyphonic phrases which it has to itself are apparently accounted for by the fact that Bach could not endure the tedium of writing parts which were not *obbligato*. The working-out falls entirely to the share of the claviers, but with this exception it exactly follows the method prescribed by the concerto form. A tutti phrase (bars 1—12) and a solo-phrase (bars 12—28) come into prominence in the first movement, which is developed out of their different combinations and contrasts in different keys. Within the limits of these two chief groups, however, the solo instruments have concerted passages of a very animated kind among themselves. This movement can thus be called a concerto in a two-fold sense, both because it preserves the form of Vivaldi's concerto style, which proceeds from the contrast between the solo and the tutti, and also because it actually contains a strife or competition between two instruments, although these are of different kinds.[248] The last movement of the concerto generally has a dance character and some kind of three-time, and, as compared with the more pathetic first movement, it must always be gay, light, and brilliant. This requirement is fulfilled by Bach in the C major concerto; but the employment of the fugal form is remarkable. The fugue belongs to the sonata form, or to that of the concerto in the sonata style;[249] it has nothing in common with the strict concerto form, since that originates not in polyphony, but in homophony, and its working-out

---

[248] The opening of the tutti-phrase recalls the opening of the first movement of the cantata "Wer mich liebet." I mention this because even the cantata form is built originally upon that of the concerto. See Vol. I., p. 512.

[249] See Vol. II., p. 136, f.

is not thematic but episodic. Bach often employs a fugue for the last movement, especially where the clavier appears as a solo instrument; this is the case, for instance, in the fifth Brandenburg concerto, in the concerto in A minor for clavier, violin, and flute, and also in the fourth Brandenburg concerto, the violin part of which was re-arranged by Bach for the clavier. There, and in the C major concerto, he succeeded in a most masterly way in suiting the form to the character of the movement, by the style of invention and treatment, especially by means of longer episodes, or even interludes, quite in the free style; and he was led to introduce them by the style of the harpsichord and the organ, which always influenced his imagination.[250] Although the fugal style would appear to afford but little temptation for anything of the kind, Bach contrives in this movement to employ the two claviers in such a manner as to make them appear as two factors of equal importance. By this means, the working-out of the fugues, even putting aside the interludes, is characteristic and especially interesting. The two allegro movements, and, in no less a degree, the delicately woven and melancholy quatuor which serves as an Adagio, reveal a fresh though controlled inventiveness, a feeling of strict moderation, which, when united to the highest perfection of form—for the work corresponds absolutely to the ideal of the concerto—make the work a classic model.[251]

The two concertos for three claviers are constructed upon the same principles. The tutti (if, indeed, this name ought

---

[250] See Vol. I., p. 422.

[251] Rust (B.-G., XXI.,² pp. 6—8) holds, on the authority of an older copy which contains only the first movement, and that without orchestral accompaniment, that the first movement was written as a separate work, and that the accompaniment was not added till afterwards. This cannot be gathered from the style of the accompaniment, for I cannot see that of the first movement to be different from that of the last, except in so far as is required by the different form of the two movements. It is not organically necessary either in one movement or the other; it has been already noticed by Forkel (p. 58) that the concerto could very well dispense with the accompaniment of the strings, and that it has a quite good effect without it. It may have even been played in this way at an earlier date. Rust cleverly inferred that Bach probably made no full score of the concerto, but may have written the parts for the stringed instruments separately

to be used any longer) serves, with few exceptions, only to support and strengthen, the musical development being left to the claviers. In the working of these together, indeed, another method of treatment is to be remarked. On account of the requisite response in the parts, it is harder to work three claviers together than two. Chiefly for this reason, as I believe, Bach generally used them all together in the first movement, and without any opposition between a tutti phrase and a solo phrase. The consequent close symphonic structure in the first movement—which he, not content with clearly-articulated form and marvellous variety, succeeds in adorning with endless invention—forms an effective contrast to the last movement, and one which is thoroughly justified by the rules of the concerto. Of the two concertos, that in D minor is certainly the earlier.[252] Its character is delightful and soothing; still, it lacks that perfect workmanship which makes the strong, grave, majestic concerto in C major one of the most imposing of all Bach's instrumental compositions.[253] In the first two movements of the D minor concerto the first clavier comes into marked prominence, and in the charming Siciliano, it predominates exclusively, the other claviers being used only for accompaniment and support. The string quartet takes much the same place as that held in concerted chamber music by the harpsichord on which the figured-bass part is played. In this sphere it fulfils its task with great discretion and taste. It should also be noticed how carefully and circumspectly Bach uses it to strengthen the real parts of the claviers. But, even then, the bass lies for the most part in the clavier parts; if the string quartet were entirely omitted, the full and perfect effect of tone might be affected, but not the organism itself,

---

[252] P. S. II., Cah. 11 (258).
[253] P. S II., Cah. 12 (259). This work, of which—as also of the D minor concerto—the autograph is wanting, exists in MS. in D major as well. In opposition to Griepenkerl's view, I hold this to be the original key, because of bar 33 of the Adagio. The published form contains many errors. Thus, in the bar just mentioned, after f the violas play again in unison with the violins; in bar 48 of the last movement, first clavier, left hand, the fourth beat of the bar should be fg, instead of fe.

excepting in a few passages which are not intelligible without the string bass.[254]

The C major concerto allots an equally important part to each of the three claviers; and the string quartet, without overstepping its modest limits, has more independence than in the D minor concerto, or even in the concerto in C major for two claviers. In this case the string parts contain the true and indispensable bass; but it is frequently identical with the clavier bass, or forms a central line for the adornments with which it is surrounded by the latter. In a very limited way, too, the quartet is allowed an independent share in the development of the whole. In the first movement it opposes themes of considerable importance to the subject, which enters in unison on all the claviers. On two occasions the string bass gives out, quite alone, the ponderous, hammer-like subject of the last movement; and in the Adagio we meet once more with a veritable *tutti* contrast, so far as such a device can have place in the Adagio of a concerto. This movement is built, as is frequently done by Bach, upon a *Basso quasi ostinato*, which is seven times repeated in its entirety, but in different keys, and is also dissected and used episodically. To this the strings oppose, four times over, their own contrapuntal passage, while the claviers take up the figured bass part; after which the string quartet resumes the accompaniment. Thus this concerto fulfils all the demands which can be made for independence in each co-operating part, according to its importance to the whole effect; and, in respect to the general feeling of the music, the concerto character is kept up throughout. But it is evident that such rich materials treated in such a polyphonic style must necessarily result in grandeur and gravity. It imparts weight to the stern vigour of the first movement; to the melancholy Adagio a slight feeling of austerity; and the last movement soars far above the cheerful vigour of an ordinary finale. It rises with a broad steady flight above the ponderous theme :—

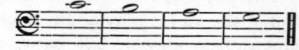

---

[254] Bach must have meant the string bass part for the violoncello. Passages like bars 22 and 116 of the first movement preclude the use of the double-bass.

and spreads to completeness in six parts. Although richly
adorned with brilliant passages thoroughly befitting the
concerto style, it is almost solemn in its broad, majestic flow.
There also exists a concerto for four claviers, with accompaniment for string quartet (A minor).[255] Forkel[256] considered
it an original composition. We now know it to be only an
arrangement of a concerto by Vivaldi for four violins. The
original is in B minor, and is accompanied by two violas,
violoncello, and bass. As in his other arrangements of
Vivaldi's concertos, Bach has given the basses greater
independence, and worked out the middle parts more richly
and fully. He gives the solo instruments more work in
counterpoint, and here and there the violin passages
display a character more suited to the clavier, he also
often adds a fuller accompaniment not unfrequently varied
with episodes. But even here the accompaniment is
generally used only to fill up and support the harmonies.
And yet the work affords new evidence of the master's
ingenuity in writing four *obbligato* parts, even in the lightest
style. We must assume the arrangement to be about
contemporary with the original concertos for three claviers.[257]

At Easter, 1735, Bach published, in the second part of
the "Clavierübung," a concerto which, in respect of style,
must be admitted to be the maturest of his labours in this
form. It is for clavier alone, and composed "in the Italian
taste" (nach italiänischem *Gusto*").[258] This description
summarises the whole history and character of the Bach
instrumental concertos. The concerto was a form of violin
music invented by the Italians. Ever since Bach had first
taken an active interest in the form, his constant endeavour
had been to employ it in the most widely different provinces
of music, and we have already seen how he brought it even

---

[255] P. No. 260.
[256] See p. 58 of his work on Bach.
[257] Vivaldi's concerto is to be found in Walsh's *Vivaldi's most celebrated Concertos, opera terza, No. X*. In Roitzsch's edition of Bach's arrangements (P. S. II., Cah. 13) there is a mistake at the beginning. The second violin begins four bars too soon; and in bar 8, which must then be bar 12, of the second violin part, the third crotchet should be c'', instead of b'.
[258] B.-G., III., p. 139 ff. P. S. I., Cah. 6 (207) No. 1.

into the chief choruses of several chorale cantatas. But above all he wished, from the first, to turn it to account for solo music for the organ and clavier. We found that under the name of Toccata he had written, even in the first period of his full maturity, pieces for the organ and clavier which exhibit the concerto form in perfection; there is a clavier composition of the very earliest period, with the title of Concerto,[259] and the arrangement of Vivaldi's concertos ought to count as free compositions for the clavier. Subsequently, when he wrote real concertos, he clung more or less to the idea that the prominent part in the work must always be a single instrument. We have already seen how, in the case of the clavier concertos with accompaniment, the clavier gradually gained prominence, while the tutti sank down into a mere accompaniment and the string bass to a *basso continuo*. This method is carried furthest in the concerto in the "Clavierübung." While it is a masterly composition for the harpsichord, it is, at the same time, a vivid reflex of a form which was really invented for the violin and a band of instruments in contrast with it. The influence of the violin is most easily perceived in the *Andante*. This influence is what is meant by the words "in the Italian taste," as also by the phrase "*Alla maniera Italiana*," used by Bach to describe his earlier clavier variations, in which he approached the style of treatment most characteristic of the violin.[260] He was not the only composer of his time to write a concerto for only one instrument; others had made various attempts, for the clavier and for the lute. But they were nothing more than attempts, for only a genius of the first rank could succeed, and there was but one German musician capable of uniting two opposing styles with such harmonious results. This was felt even by Scheibe, who was by no means an unreserved admirer of Bach. But every one was obliged at once to allow this clavier concerto

---

[259] See Vol. I., p. 417. A Concerto in G major, for clavier alone, which bears Sebastian Bach's name in Zelter's handwriting, has lately come to light in an old MS. in Grasnick's bequest. I cannot, however, accept this dry, stiff composition for Bach's.

[260] See Vol. I., p. 431.

to be a perfect model of its kind; very few, or indeed hardly any, concertos could be mentioned having such splendid qualities and sound working-out. "It was no less a master of music than Herr Bach, who has made the clavier his especial study, and with whom we can safely venture to compete with any foreign nation, who was to bequeath to us a piece in this form, which should provoke the envy and emulation of all our own great composers, and be vainly imitated by foreigners."[261] The sound working-out consists of clearly grouped and sharply contrasted subjects, which scarcely need the aid of different effects of tone to make them intelligible. Still, a form which is developed on the principle of subjects of different kinds relieving one another in succession, involves a predominance of the homophonic style, and an extension by means of episodes rather than a thematic treatment in many parts. In this way the concerto style resembles that of the modern pianoforte sonata, and Bach's Italian concerto was undoubtedly the classical predecessor of this later form, and may even be regarded as in many respects its prototype. The modern sonata not only took from the concerto the division into three parts, but it found there the Adagio and the last movement fully developed. The first movement is however quite different in the two forms. The sonata movement being the result of a combination of the dance-form with that of the aria in three sections, it could derive nothing from the concerto but the episodic development, and even this had reached its full growth in the aria. The last step towards the attainment of the modern sonata form was not destined to be made by Sebastian Bach, although he was well acquainted with that combined form in two sections, and employed it himself in isolated cases;[262] for this step led downward at first from freedom to narrow and petty limitations; and the master can have felt little impulse in this

---

[261] Critischer Musikus, p. 637 f. Not the great, but the little, German composers strove to emulate Bach, for example, Michael Scheuenstuhl, Stadt-Organist in Hof, who in 1738 published a G minor concerto for clavier alone (Balthasar Schmidt, of Nuremberg).

[262] See Vol. II., p. 60.

direction, particularly at the time of his highest maturity. He left it for his son, Emanuel.

We know that Bach also wrote real clavier sonatas. They are not in the three-movement form, nor are they built upon that of the concerto; their style is that of the Italian violin sonata, and they are merely arrangements for the clavier of violin compositions, some his own and some by other composers.[263] It would have been very surprising if Bach had not transplanted the sonata into the realm of harpsichord music. But his fancy for this form was not of long endurance; in writing in the free style for the clavier he preferred the *Suite*. Bach is not only the last great composer of suites; he is also the greatest, the perfecter of this form in every way; after him there was nothing more to be said in the form of clavier suites, and this accounts for the rapid disappearance of that form from the practical history of the art after 1750. No fewer than twenty-three such works, comprising many movements, have come down to us entire. Of these, six form the collection known as the French suites. Three more appear, so to speak, as the supplement of these; they are very similar to them in character, and in a few MSS. two actually figure among them. One, again, full of unpretentious grace, belongs to the earliest period of Bach's maturity.[264] Here we have only to deal with the remaining thirteen, twelve of which, again, are comprised in two collections of six each.

One of these collections is known by the name of the "English suites." The name refers, even less than in the case of the French suites, to the character of the music. It would be hard to say what material for suites Bach could have found among the English. According to trustworthy tradition they were written for an Englishman of high standing.[265] As to the date of their composition, which

---

[263] See Vol. II., p. 77 f. and 81 ff.—I am not speaking now of the sonata in Kuhnau's style (see Vol. I., p. 243).

[264] See Vol. I., p. 432.

[265] So says Forkel (p. 56), who must have got it from Bach's sons. In Johann Christian Bach's manuscript of the English suites, above the title of the A major suite, stand the words "*fait pour les Anglois.*"

has hitherto been quite uncertain, so much may now be definitely said, that five of them must have been written, at the latest, by 1727. But apparently they were all written before 1726—*i.e.*, in the first years at Leipzig or the last at Cöthen, for the French suites are certainly older.[266] We have already considered the fundamental form of the clavier suite, its component parts, their meaning and value, and their connection with one another.[267] That Bach thought it needed no further development is evident from the fact that he adheres to it without the least change, not only in the earlier French suites, but now in the English suites, and also in the last six partitas. The four chief parts are the Allemande, Courante, Sarabande, and Gigue. Before the Gigue, came in Bourrées, Minuets, Passepieds, and Gavotte by way of *intermezzi*. The English suites are distinguished from the fanciful and beautiful French ones by their strong, grave, and masculine character. They are in the keys of A major, A minor, G minor, F major, E minor, and D minor; so that the minors preponderate. The richer style of the music demands forms of greater extension. The character of the separate pieces is sharply and distinctly marked, and their feeling intensified by richness of harmony: Bach never wrote sarabandes of such breadth and beauty, or gigues of such wild boldness.[268] In the A major suite he introduces two courantes, and, moreover, furnishes the latter, in Couperin's manner, with two *Doubles;* the sarabande of the D minor suite also has a *Double* of the same kind. These are perfect and complete variations, intended to bring out the character of the pieces they belong to; thus they have their own special place in the development of the whole. On the other hand, the pieces that follow the sarabandes of the A minor and G minor suites are not to be regarded as variations, because they do not, as it were,

---

[266] See App. A., No. 5.
[267] See Vol. II., pp. 84—92.
[268] The gigue in the D minor suite has its prototype in Buxtehude (see my edition of Buxtehude's organ compositions, I., p. 94 f.). Forkel (p. 28) quotes it as an instance of bold harmonic sequences.

deck out the piece that went before in a new musical dress, but merely adorn separate portions of the melody with embellishments. It was the fashion at that time to embellish simple melodies in performance. Since, however, for the most part, more was lost than was gained by this practice, Bach, following Couperin's example, wrote out the ornaments in full. Thus it was not intended that the simple and the adorned sarabandes were to be played in succession, but it was left open to the performer to choose between the two.[269] The *intermezzi* are in each case two in number; they belong together, as principal subject and trio. And lastly, the character of the English suites, which strives after what is rich and grand in effect, is revealed in the preludes which are affixed to them, while in the French suites they are entirely dispensed with. These, which at once lift the hearer into a higher and graver atmosphere, are, one and all, masterpieces of Bach's writing for the clavier. With the exception of the prelude in A major they are planned on the grandest scale and elaborated with great variety. The perfect aria-form in three sections is seen in the A minor prelude; that in G minor is developed on the plan of the first movement of a concerto, and its form is also similar to that of the concerto, but is more fantastic, as is also that of the prelude in F major. The E minor prelude may be described as a rapid and powerful fugue combined with the aria form; the same combination is seen in the D minor prelude, but it is preceded by broad passages of broken chords, eminently characteristic of the prelude form. The whole comprises no fewer than one hundred and ninety-five bars.[270]

The second collection forms the first part of the "Clavierübung," which Bach himself published in 1731. The name had been invented by Kuhnau, who in 1689 and 1695 published two works, each consisting of seven clavier suites, under the title of "Clavierübung."[271] We drew

---

[269] In agreement with this is the fact that Johann Christian Bach left out the ornamented sarabande in his MS. of the G minor suite. The MS., No. 50, in the library of the Princess Amalia of Prussia is also without it.
[270] P. S. I., Cah. 8. (203). B.-G., XIII.,² p. 3—86.
[271] See Vol. I., p. 237.

attention in another place to the fact that in his church compositions, and also in his earlier clavier works, Bach was influenced in several ways by Kuhnau.[272] In choosing for a collection of clavier suites the same title which had been used by his predecessor for the works that first made him famous as a composer for the clavier—in publishing works of this kind as *Opus I.*, and calling them, in contradistinction to his earlier practice, not suites, but Partitas (*Partien*) like Kuhnau—it is evident that he wished to appear before the world as Kuhnau's successor. He indeed gradually extended the scope of this modest title, which was subsequently made use of by others, such as Vicentius Lübeck, Georg Andreas Sorge, Balthasar Schmidt, Friedrich Gottlob Fleischer, and his pupil Ludwig Krebs. There appeared at Easter, 1735, a second part, containing the Italian concerto already spoken of and another partita. And about 1739 a third part appeared, with a great organ prelude and fugue, a number of organ chorales, and four clavier duets; and finally, in 1742, the fourth part appeared, with a grand set of variations.[273] The first part, however, was published piecemeal by Bach, so that from 1726 onwards one partita appeared every year until the completion of the work in 1731. As he published it himself, and the engraver's name occurs nowhere, it is possible that Bach may have managed, or at least superintended, the engraving. This supposition receives support from the fact that his son, Emanuel, who at that time resided in his father's house, was occupied with engraving; his first original composition, a minuet, introducing much crossing of the hands, was engraved by himself, and appeared in 1731.[274] The third part was also published by the author himself; the other two appeared in Nuremberg, the second published by Christoph Weigel, and the fourth by Balthasar Schmidt.

---

[272] See Vol. II., pp. 369 ff. and 392 f.; and Vol. I., pp. 236 ff., 243 ff., and 320.
[273] See App. A., No. 6.
[274] See Burney's Musical Tour, III., p. 203. (*Sic* in original; but ?) We know about the prices from Breitkopf's Catalogue for the new year, 1760; Breitkopf received for the whole six partitas (the No. 5 in the Catalogue must be an error) five thalers; and for the second partita alone, 8 gr.

At that time it was the custom to put opus-numbers to instrumental works only. The "Clavierübung" is, however, universally regarded as the first work published by Bach; the Mühlhaus "Rathswechsel-Cantate," of 1708, was printed but had not yet come out.[275] The fact that Bach only began to publish his compositions at the age of forty-one years, does not mean, of course, that he had hitherto been in any way averse to letting them become known. At that time German music was most widely disseminated by means of transcribed copies. Long before 1726 Bach had become widely known as a composer; even in 1716, in Hamburg, Mattheson spoke with admiration of his compositions for the church and also for the clavier.[276] Even later, and indeed until the beginning of the nineteenth century, the chief part of his works were only known to the musical world in a manuscript form. Besides the four parts of the "Clavierübung," only three other works by Bach appeared in his lifetime; he died as he was preparing a fourth for publication.

The name "German suites" has been given to the six suites of the "Clavierübung" in order to distinguish them from the French and English suites.[277] And not without good reason. Bach's employment of the name "Partita" implies more than a mere imitation of Kuhnau; for the form bore this name in Germany at the end of the seventeenth century, before the influence of the French had become authoritative, and Bach's returning to it showed that although he fully recognised the services rendered by French and Italian musicians to German art the groundwork was still German. The suite is a German form of art, although foreign nations have done much

---

[275] See Vol. I., p. 344.
[276] See Vol. I., p. 393.
[277] Verzeichniss aller . . . Musikalien . . . welche zu Berlin beim . . . J. C. F. Rellstab zu haben sind (1790), p. 67: "Bach, J. S., 6 Partite ou Suites françoises, 4 Thlr.
— — 6 dito Tedesche   6 Thlr. 12 gr.
— — 6 dito Anglaises   5 Thlr. 12 gr."

(For this information I am indebted to the kindness of Herr G. Nottebohm of Vienna.)

in assisting its development.[278] It was precisely in the use made of the foreign elements that the German spirit asserted itself most distinctly, and the fact that Bach had a full consciousness of his position with regard to his predecessors is proved by other things besides the choice of the name Partita. The six partitas in the first part of the "Clavierübung" constitute a comprehensive and self-contained work, in which all the elements that bear an important relation to the suite form receive careful and earnest attention. The wealth of structure which they exhibit is quite extraordinary; but yet the outlines of the form as a whole are strictly preserved, a fact which serves to distinguish them from the suites published by Handel in 1720, which are true to their title of Suites only in being free clavier-pieces played one after another.[279]

While in the case of the English suites the external form and arrangement is in each the same, and the invention is almost entirely confined to the treatment of the separate subjects, each partita lays before us a new series, we can hardly say of forms, but rather of artistic types. The Partita in B flat major begins with a Prelude of the kind developed by Bach, which stands as it were on the threshold of the fugal style; there is a real theme, but it is more like a passage than a melody, and its working-out has only time to begin before a section of light episodical work comes in and bears all before it. On the other hand, the partita in C minor opens with a *Sinfonia*. As the name implies, the piece is influenced by Italian art-elements. An Andante in common time, with a richly ornamented upper part, followed by a fugue in 3-4 time, was the usual opening of the Italian violin sonatas. In order, however, to prevent the resemblance to this form being too strong, so as to disturb the unity of the suite, Bach prefaces it with a broad full *Grave* movement. This is in the style of the opening of the French overtures, and thus the prelude-like character of the piece is defined and established. The A minor Partita opens with a *Fantasia*, a two-part piece in the style

---

[278] See Vol. II., p. 72 f..   [279] "Suites de Pièces pour le Clavecin."

of Bach's *Inventions*, only more elaborate. The D major Partita begins with an *Ouverture* in the French style. That in G major is introduced by a *Præambulum*, which is distinguished from the Prelude of the first partita by not being thematically worked out, but consisting of passages and broken chords; as regards the grouping of its sections, it resembles a concerto movement. Lastly, the E minor Partita opens with a *Toccata;* this is considerably simpler even than the toccatas in F sharp minor and C minor;[280] it goes on in one movement without change of time; fantastic passages at the beginning and end form as it were the light husk, the kernel of which is a noble and grave fugue.

It has been remarked farther back that in the hands of the Italians the Courante acquired a special form of its own, so that at the beginning of the eighteenth century the name *Corrente* and *Courante* represented two different types.[281] From the partitas in the "Clavierübung" it appears that Bach regarded them as distinct and of equal importance, as the expression of two nationalities. The first, third, fifth, and sixth partitas have *Correntes*, the second and fourth *Courantes*.[282] The former are in 3-4 and 3-8 time, and in a smooth, rapid style, while the Courantes are impassioned, and yet solid and grave. They alone have the agitated change between two and three time. Usually in Bach the three time predominates, and the displacement of accent occurs only at the close of each section. Cases like that in the French suite in B minor and the isolated suite in E flat major, where the 6-4 time is the principal rhythm, the pure 3-2 time occurring only now and then, are only exceptional; but there the capriciousness with regard to rhythm is carried to such an extent that often the two hands are playing simultaneously in different kinds of time. In this particular the two Courantes in the partitas are in contrast to one another; in the C minor

---

[280] See Vol. II., p. 31 ff.   [281] See Vol. II., p. 85.
[282] In the later editions, and unfortunately even in the B.-G. edition, the title Courante has been given to all, against Bach's express injunction. And in other dances the important distinction which is implied by the use of Italian or French titles has not been regarded by the editors.

Courante 3-2 time predominates; and in the D major 6-4. Yet even there it is marked 3-2 in order to prevent the feeling of the two beats from becoming too strong. The time of the courante was properly neither one nor the other, but a mixture of both. It was a rule that even passages in 6-4 time were to be played in 3-2 time, that is to say that in the case of phrases which were evidently, both by their formations and natural character, in 6-4 time, the remembrance of the 3-2 time was to be kept up by means of occasional 3-2 accentuation.[283] At any rate, it was possible to give to the courante a special character of its own, according to the prominence given to this rhythm or to that, and Bach availed himself of this privilege, showing at the same time that he wished to introduce into the set of partitas all the different forms which this dance was capable of taking.

The *Gigue,* too, underwent a two-fold and divergent development, although not so decisively as in the case of the courante. In the French and German suites the fugal form generally predominated, the theme being treated in inversion in the second section. The Italian *Giga* was homophonous and consequently much lighter in character; it was always in triple time, whether simple or compound.[284] Gigues in fugal style are found in the third, fourth, fifth, and sixth partitas. That in the third (A minor) has no peculiarity in its form; in the fourth (D major) the second part contains, instead of the inversion, a new theme, which serves as counter-subject to the first; yet here the result is not an ordinary double fugue, as in the gigue in the fifth partita, which in other respects resembles it closely.[285]

---

[283] Marpurg, Kritische Briefe über die Tonkunst II., p. 26: "The proper rhythm of the Courantes in the French style, which indeed are strictly to 3-2 time, yet approaches in various places very near to the 6-4 time, in respect of the external form of the measure; the difference is generally to be marked by playing the 6-4 passages with a 3-2 accent. The late Herr Capellmeister Bach has left us a sufficiency of genuine models of this proper Courante time."

[284] See Vol. II., p. 89 ff.

[285] In this gigue it is remarkable that in bars 17—19 of the second part he would seem to have formed the idea of a higher compass than was contained in any clavier of that time. The chief subject ought properly to begin on e''',

The gigue in the sixth partita (in E minor) is in direct contrast to the other three, inasmuch as it dispenses with triple time. Bach had already ventured to employ the two-part *alla breve* time for the gigue, in the French Suite in D minor. The dance-form proper is thus abjured; there remains only a characteristic piece, full of energy and passion. This more general form, however, distilled, as it were, from the gigue, seemed to the master worthy of being immortalised in this place as a special type. Undoubtedly, too, the gigue served as a model for the last movement of the second partita (in C minor). It is in two sections; the development is fugal, and in the second part the theme is inverted. But both in time (2-4) and general character it departs so far from the gigue character that Bach does not use the name, but calls the piece *Capriccio,* as a token that the piece was created by his own artistic volition.[286]

At the end of the first partita (in B flat major) there is not a Gigue, but an Italian *Giga*. The different character of this piece, which is in a graceful rocking style, is very obvious. Bach even draws attention to the relation which it bears to Italian art by means of the kind of clavier *technique* which he employs in it. The crossing of the hands, which Bach first makes use of in the *Giga* of the B flat major partita, was a specialty of Domenico Scarlatti's.[287] That devices of this kind were beginning to become popular in Germany about this time is proved by Emanuel Bach's first work, which appeared in 1731, and was written in his father's house. Sebastian Bach also availed himself of this technical device in a part of the G major partita, and, again, in the C minor fantasia, which we shall notice

---

but Bach was obliged to begin it an octave lower down. Such cases are of rare occurrence in his works, for, as a general rule, he accustomed himself to keep strictly and steadily within the limits of his instrument. Two passages of the same kind occur in the Wohltemperirte Clavier, see Vol. II., p. 163.

[286] The arbitrary skips of tenths in the theme bear a strong resemblance to the first movement of the Concerto in D minor for two violins.

[287] We do not indeed know that D. Scarlatti had published clavier compositions in print before 1726; but there is no doubt that the works of this master, who was born in 1683, had become widely known in the musical world by means of manuscript copies.

presently; and in the B flat major prelude, in the second part of the Wohltemperirte Clavier; it then became the fashion for a time among composers, and fell again into disuse about the middle of the century. Subsequently, Emanuel Bach was of opinion that, in many pieces of the kind, the natural position of the hands was preferable to such tricks, but that the crossed position might sometimes be a means of bringing out good and new subjects on the clavier.[288] The latter is the case with the *Giga;* the device of crossing the right hand over the left [? left over right] is not employed only here and there, but the whole piece is to be performed in the same way, and has a peculiar charm of its own.[289]

In other passages in the partitas, also, Bach paid due attention to the different characteristics of the Italian and French styles. The Italians had a tendency to obliterate the distinguishing rhythmic marks of the dance-forms. In order to do this the more effectually, Corelli often merely intimates that a certain piece is to be played in the tempo of this or that dance, and leaves himself free in all other respects. So in the E minor Partita, we find a *Tempo di Gavotta,* while in other respects the piece is more like a *Giga,* or, except for the common time, a *Corrente.* In the partitas in B flat major and D major there are *Menuets;* in that in G major a *Tempo di Minuetto*,[290] a lovely, playful piece with a charming displacement of accent, in which crossing of the hands is employed here and there; but it has nothing more in common with the *Menuet* proper than the 3-4 time. Corelli's free style of treating the *Allemande* seems also to have been noticed by Bach. While, for the most part, his Allemandes flow with extreme smoothness, that of the E minor Partita has a rugged character, which is considerably

---

[288] Versuch über die wahre Art das Clavier zu spielen, I., p. 35 f.

[289] J. A. P. Schulz remarked that this *Giga* must have been haunting Gluck when he wrote the aria *Je t'implore*, in "Iphigenia in Tauris" (see Jahn, Mozart, IV., p. 715). As Marx discovered (Gluck und die Oper I., p. 201), this aria, in essential points, is the same as one in Gluck's opera *Telemacco*, which dated from 1750. It cannot be doubted that the resemblance is no accidental one.

[290] *Minuetta* in the original edition must be a printer's error.

intensified by means of dotted semiquavers. This figure is also found, however, in Corelli's Allemandes,[291] and Bach gave the piece, not the French title, as usual, but the Italian form, *Allemanda*. The partitas contain three other short pieces with Italian titles, which bear no relation to any dance-types; these are called *Burlesca, Scherzo*, and *Aria*. The first two are in the A minor and the third in the D major Partita; they are free characteristic pieces in dance-form. To balance these there are some in the French style; in the C minor Partita there is a *Rondeau*, and in the E minor an *Air*. A comparison of the *Air* with the *Aria* shows a distinct difference in style, the latter being more *cantabile*, while the latter is rather in instrumental style.

But even in the pieces which Bach developed independently he exhibits, within the limits of one and the same type, a variety, which perfectly corresponds to the gradual development and modifications carried out by time and nationality. In order to convince ourselves of this we have only to compare the Allemandes of the French and English suites, and afterwards those of the first five partitas. It would seem that Bach wished to test the elasticity of the form to the uttermost in these compositions. It would not be easy to find any pieces more different than the Allemandes in B flat major and D major; and yet full justice is done in each to the requirements of the type. A similar power of variation is displayed in the Sarabandes; but in this case Bach several times does really overstep the limits of the type (in the A minor, G major, and E minor Partitas).[292] Finally, regarding the six partitas as a whole, we see that the differences between them are as marked as the unity which prevails in each by itself. The surest evidence of this is afforded by the introductory pieces, which establish the character of each partita beforehand, while they stand in

---

[291] See the Allemandes on pp. 90 f., 99, 102 f., 110 f., 219 f., of J. Joachim's edition (Denkmäler der Tonkunst III.).

[292] Sarabandes beginning on the up-beat of the bar are found elsewhere only in the orchestral Partie in B minor, the Clavier Sonata in A minor, and the Violin Suite in A major.

complete contrast to one another, and a different title is intentionally given to each.

The contemporary musical world at once appreciated this precious gift. The novelty of the music itself and of the clavier *technique* excited wonder and admiration. To know them perfectly soon became the highest aim of the clavier player, and "he who had learnt to play a few of them could make his fortune."[293] That Bach's contemporaries should recognise the full value of this masterpiece of art was hardly to be expected, and, indeed, it is never confidently asserted that they did.[294] Even Sorge—although in a "dedication" he calls Bach the prince of clavier and organ-players,[295] and recommends his clavier works as models to show how to bring out a good inner part in the left hand, how to treat discords correctly in rapid passages, and how to elaborate the art of improvising—always classes him with Kuhnau, Handel, Mattheson, Walther, and others.[296] Without undervaluing these others, and especially Kuhnau, we now see clearly that the art of harpsichord composition—and that they were designed for that instrument and not for the clavichord is plain from their extensive compass—had attained to the highest conceivable point in these six partitas—a point which no one but Handel, in the suites of 1720, ever reached again. A simple comparison of the works will show that Handel here holds his own. The

---

[293] Forkel, p. 50. Mizler, in an advertisement of the "Wegweiser zu der Kunst, die Orgel recht zu schlagen" (Musikalische Bibliothek I., 5, p. 75), says: "He who cannot finger better than this, will scarcely be able to learn to play the Partitas (Partien) for the clavier by our famous Herr Bach, of Leipzig." The Partita in C minor is considered the easiest (Breitkopf's Catalogue for the New Year, 1760, p. 18: "the easiest of all").

[294] Forkel, who considered that "such excellent clavier compositions had never been seen or heard," is no impartial witness to the statements made by Bach's sons, even if they did really tell him that this was the general opinion of Bach's contemporaries.

[295] See his Drittes halbes Dutzend Sonatinas vors Clavier. Nuremberg; Balthasar Schmidt. The dedication has been reprinted by Schlattener in the Monatsschr. für Musikges., 1079, p. 65.

[296] Sorge, Vorgemach der musikalischen Composition. Pt. III. Lobenstein (1747), pp. 338 ff., 404, 416, 425.

difference between them is only a difference of personality; each hearer, according to his own character, will feel greater sympathy with the one or with the other. Handel's suites, in which a great artist's soul pours out its wealth in free abundance, are more brilliant and attractive; but Bach's partitas leave a deeper and more lasting impression, because he adheres more strictly to the prescribed form. Viewing them from the high historical standpoint, Bach must bear the palm, for he stands out, not only as an individual of equal greatness with Handel, but as the representative of the typical form. Handel, with his Suites, flashed like a comet through the heaven of art; if they had never been written or had been lost we should have been deprived of a brilliant apparition, but there would have been no gap in the system of co-operating influences. Bach's partitas may be called a light-focus, in which alone we see the brilliancy that might be attained by concentrating the rays of all the smaller luminaries.[297]

To the cursory observer it might appear surprising that after Bach had, in the six partitas, said all that was to be expressed in the suite form, he should return to it in the second part of the "Clavierübung." This contains, besides the before-mentioned Italian concerto, another partita (in B minor).[298] But here his purpose was quite different. He meant to write, not a new clavier suite, but a piece which should be, as it were, a transference of the orchestral *Partie* to the clavier. The difference between the two forms has already been discussed.[299] Bach's intention betrays itself even in the title "*Ouverture*"; for it was usual to give the name to the orchestral *Partie* from its opening piece. The arrangement of the movements leaves no doubt of his intention. There is no Allemande, because, being exclusively a form of clavier music, it had no place in the orchestral

---

[297] Of the editions of the first part of the "Clavierübung," I will only mention two in this place—viz., P. S. I., Cah. 5 (205, 206), and B.-G., III., p. 46—136.

[298] P. S. I., Cah. 6, No. 2 (208). B.-G., III., p. 154—170. A MS. made by Anna Magdalena Bach, which is preserved in the Royal Library at Berlin, gives the piece in C minor. See App. A. to Vol. II., No. 44.

[299] See Vol. II., p. 140 ff.

suite. The Courante, the Sarabande, and the Gigue are all there, but before the Sarabande are inserted two Gavottes and two Passepieds, and after it two Bourrées—a series so numerous that they cannot be called *Intermezzi*, but must claim to be of equal importance with the other movements. It was the special privilege of the orchestral suite that it might have an irregular and arbitrary number of dances, and this treatment is carried on even beyond the Gigue. Bach has finished his orchestral suite in B minor with a *Badinerie*, and one of the two in D major with a *Réjouissance*. In this case an *Echo* closes the series; it is in a dance form, without exhibiting any definite type. It takes its name from certain imitations of the effect of an echo, which are especially charming from the fact that the phrases are not repeated exactly, but with soft, supplementary passing notes, as if the sound became indistinct in the distant repetition, as, for instance:—

Bach, of course, had no intention of imitating the orchestral style on the clavier. This would indeed have been superfluous, for there was so much of his clavier style in his compositions for the orchestra that when the orchestra form was transferred to the clavier it prospered quite naturally. He only requires the hearer to approach this work in the same frame of mind as that in which he would listen to an orchestral suite; just as the Italian concerto can only have its full effect upon one who knows what a real concerto of that time was like. Both works have some intrinsic connection, which may explain why Bach chose to unite them into a part of the Clavierübung by themselves. They are reflections on the surface, as it were, of clavier music, of forms which were invented for a number of different instruments. An outward bond of union between them is that, for the proper performance of both, a harpsichord with two manuals was requisite. As regards the character of the separate sections of this partita, the nature of the original model is plainly perceptible. It is easiest to recognise in the Gigue—for so it must be called, and not a *Giga*—yet how entirely does it differ from the Gigue of the clavier partitas proper and of the English suites. Even those of the French suites are more complicated in their form. The orchestral *partie* adhered more closely than the clavier *partita* to a simple and popular style of Gigue. This is very marked in the Gigue now under notice; it is simpler and more intelligible than even the Gigue of the second French suite, to which it has a strong resemblance in other respects. And the first Bourrée and the second Gavotte have a blunt simplicity about them, which is evidently intentional. Where richer materials are employed there yet remains in the melodies a popular character which is not found in the real clavier partitas. The beautiful and impassioned Sarabande alone seems to look down from inspired heights at the artless gambols of the other pieces.[300]

Bach also wrote three partitas for the lute, which may be briefly mentioned in this place.[301] Under the express title of a composition for the lute we have only one work, in three movements (in E flat major), by him, which however we may regard as a part of the same collection.[302] Strictly speaking, indeed, it is not a partita, but a sonata without a second Adagio and with a prelude in the place of the first. Yet it might pass under this name among other more genuine partitas, provided that Bach really gave it its title. Possibly too the E minor suite, which we have already mentioned in connection with the French suites, may have belonged to the

---

[300] In this place, while referring to Bach's treatment of rhythms and the views of art revealed in it, I might go farther and fulfil a promise half made in Vol. II., p. 175, note 251. But since writing that note, Rudolf Westphal has made renewed researches into the subject, which are shortly to be given to the public, and I believe that their appearance will exempt me from the fulfilment of my promise.

[301] Breitkopf's Michaelmas catalogue, 1761, p. 56: "*Bach, J. S., Direttore della Musica in Lipsia, III. Partite à Liuto solo. Raccolta I.* 2 thlr."

[302] "*Prelude pour le Luth ou Cembal.*" P. S. I., Cah. 3, No. 4. The autograph is in the possession of Henry Huth, Esq., London. There is an independent prelude for lute or clavier transmitted to us by J. P. Kellner, see P. S. I., Cah. 9, No. 16 (200), III.

collection;[303] the low compass within which it is confined makes this supposition very probable.[304] Bach frequently employed the lute in concerted vocal composition, as, for instance, in the St. John Passion and the Trauerode, and he himself possessed one among his numerous instruments. It does not follow from this that he played it himself. The fact of his possessing one is explained when we remember that Bach had an idea of combining the lute with the harpsichord; the result of this was the "Lautenclavicymbel," which Hildebrand was commissioned to build on Bach's own plan in 1740, and of which Bach possessed two specimens.[305] It is possible that the lute-partitas may have been written expressly for this instrument, just as he tested the powers of his *Viola pomposa* by means of a special suite;[306] but he may have been incited to composition for the lute by his intimacy with the Dresden band, which numbered among its members Sylvius Leopold Weiss, the first lute player of his time.[307] He may also have been personally acquainted with Ernst Gottlieb Baron, who speaks of Bach with admiration.[308]

In the two-part inventions Bach had created a form altogether new to the time. He proved that he laid great stress on this form by placing four clavier pieces in the form of inventions at the close of the third part of the "Clavierübung," which was intended to contain only organ music.[309] In this case he calls them duets, thus drawing attention to their two-part writing. All that has previously been said of the Inventions[310] holds good of these, and they are written with the obvious intention of showing that it was possible to display the greatest harmonic richness with perfect distinctness even within the limits of a

---

[303] See Vol. II., p. 160.
[304] Observed before by F. A. Roitzsch, see the preface, P. S. I., Cah. 3, No. 8.
[305] See Vol. II., p. 46 f., and Appendix B, xiv., cap. vi. ?
[306] See Vol. II., p. 100.
[307] Fürstenau, Zur Geschichte der Musik am Sächsischen Hofe, II., p. 126 f.
[308] Baron, Untersuchung des Instruments der Lauten. Nuremburg, 1727, p. 126 f.
[309] P. S. I., Cah. 4, No. 11 (208). B.-G. III., p. 242—253.
[310] See Vol. II., p. 55 ff.

clavier piece in two parts. In this respect they are really wonderful; but they are not to be regarded as models of strict writing. On the contrary, Bach has made free use of all the license which the "harmonic" system afforded, especially in the use of the interval of the fourth, although Kirnberger was probably justified in asserting that Bach considered the fourth not to be a correct interval in two-part writing.[311] The remark that it is characteristic of these duets that they admit of no third part[312] is obviously beside the mark. For if they are works of art at all they must be capable of expressing all that the composer wished to express, without further filling up. The duets have always been received with surprise rather than with enthusiasm, and this is partly to be understood. The capricious duet in E minor and the more cheerful one in G major are intelligible as continuations of the Inventions; the other two, however, leave behind them an impression that the enormous wealth of harmony, the severity of the ideas and the length of the working-out, are not quite proportionate to the poverty of the material which represents them. They are compositions for those who are connoisseurs of such things, who delight in realising the full harmonic sequences in the bare outlines; they must always be "caviare to the general." The A minor duet looks like a two-part fugue, but the freely-treated sections preponderate to such an extent that it can only be called an elaborate Invention. The theme is capable of an extraordinary variety of harmonies, and notwithstanding all the limitation of means we feel in this duet the free movement of a creative spirit. On the other hand, in the middle section of the F major duet, it is very hard to get rid of a strong scholastic character which pervades it.

Just as the suite-form attains its ultimate development in the partitas, the violin concerto in the Italian concerto, and the invention in the duets, so the ultimate development of the variation form is reached in the fourth part of Bach's

---

[311] See Marpurg, Kirtische Briefe, I., p. 183.
[312] Forkel, p. 51.

"Clavierübung." This part contains an aria with thirty variations for a harpsichord with two manuals.[813] The circumstances which led to the composition of this work will be related presently; here we have only to do with the music itself. The *Variation*, as the oldest form of independent instrumental music, exercised the most diverse influences upon the musical organisms which arose in the sixteenth and seventeenth centuries; on the clavier suite and the chorale arrangement, on the sonata and on music for the violin, but its principal sphere was that of the keyed instruments. The singular manner in which it grew for a long period simultaneously with the suite has been shown before.[814] The chorale variations, or chorale partitas properly so called, flourished however when the difference between clavier and organ music was not sufficiently appreciated. The variation form—*i.e.*, an artistic entity consisting of a theme and a set of alterations or variations upon it—did not belong to the church; and this not only because there was no place for it in divine service, but because it would impair the solemn effect of the chorale melody. Hence it became gradually defined as a form of clavier music, but as such, in spite of its great historical significance, it had few attractions for a man like Bach. Inventive power could only be displayed either in the general outline, in the order of the different variations, or in the finest details of ornamentation. No thorough working out of the theme lay within the scope of a single variation. As regards the extent of the variation, it was strictly confined to the limits of the theme, while the alteration of the theme must consist exclusively of surrounding ornamentation, so that the sequence of harmonies had to remain essentially the same throughout. All this gone, the form had a superficial character, antagonistic to depth and scientific elaboration. It is a significant fact that the variations of Sweelinck, Frescobaldi, or Cornet have the strongest family likeness to those of Mozart, written two hundred years afterwards.

---

[813] P. S. I., Cah. 6, No. 3 (209). B.-G. III., p. 263 ff. The copies of the original edition were sold by Breitkopf in the year 1763 for 2 thlr. 8 gr.
[814] See Vol. II., pp. 74 and 77.

Setting aside Bach's earliest chorale partitas, in which he imitated Böhm, he wrote only one set of variations in the usual style, the variations *alla maniera Italiana*, which we believed ourselves warranted in assigning to the Weimar period.

But he perceived that there existed a kindred form in which the limited and monotonous style of the variation could be avoided. This was the Passacaglio or the Chaconne. The quality which it has in common with the variation is that it consists of a certain number of bars which must remain the same at every repetition, and that the sequence of harmonies must be essentially the same. In practice the Passacaglio or Chaconne is very often confounded with the Variation, the one form passing into the other. The rule that the bass theme is to remain unchanged throughout was not always strictly adhered to; its position and even its notes were altered, and it was resolved into figures; and frequently there was no bass theme at all, but only a number of phrases of four bars each, strung together in the same rhythm, and in triple time.[315] After this Handel tried to combine the Chaconne and the Variation by retaining the upper part throughout as the theme of the variations, supported at first by the bass.[316] Now if a theme in song-form of two sections be given out in the regular way, adhering exactly in the variations, not to the melody but to its bass, then on the one hand the strictly enclosed form of the Chaconne (which consists of either four or eight bars) is enlarged in the best possible way, while on the other hand opportunity is left for the most various combinations in the sphere of the variation form. This is exactly what Bach did.

For the theme he employs a sarabande, which is to be found in the larger clavier book belonging to his wife; it is in her handwriting.[317] As its place in that book is before

---

[315] See the Chaconne and Passacaglio in George Muffat's *Apparatus Musico-Organisticus*.

[316] In the G major Chaconne. Handel Society's edition II., pp. 110—122; Peters, 4 c.

[317] Pp. 76 and 77. It comes between the two divisions of the hymn tune "Bistu bey mir"; she turned over two pages together by mistake, and then

THE THIRTY VARIATIONS. 171

that of the aria " Schlummert ein, ihr matten Augen," from the cantata " Ich habe genug," it is older than that, and must have been written ten years when Bach made it the central point of his great set of variations.[318] It must certainly have been originally written for Anna Magdalena. In giving it this new form he was probably influenced by special motives of a personal kind, with which we are not acquainted; something of this kind may be surmised in the Quodlibet which forms the last variation. If Bach had written the sarabande especially for the variations he would have allowed the bass to be played through quite simply, since the chief part falls to its share, instead of decorating it with passing notes, subsidiary notes, and ornaments, which tend to obscure its outlines. It is most plainly heard in the thirtieth variation, where it runs thus, in the original time:—[319]

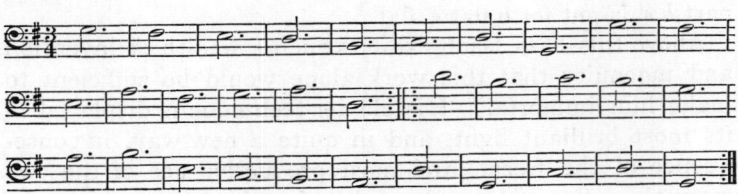

When speaking of a set of variations by Johann Christoph Bach, it was remarked that Sebastian may have known of it, because his own variations remind one of it here and there.[320] It is worth noticing that the theme of that set is also a sarabande in G major.

Bach treats the bass of the theme for the most part in Passacaglio style, making it the basis of the harmony even in the variations. As a rule, it comes in to mark the first note of the whole or the half bar. In certain cases it springs up or down an octave, according to the

---

filled up the space with the clavier piece. There is no title; the name *Aria* was not given until Bach used it for variations.

[318] See Vol. II., p. 473 f.
[319] The fundamental form of the first half is given by Kirnberger. Kunst des reinen Satzes II., 2, p. 172 f.
[320] See Vol. I., 128.

requirements of the movement of the piece; but it seldom departs from its course, although here and there a chromatic alteration is found. Sometimes it goes into the inner or upper part in the style of a chaconne, for instance, in variations 18 and 25, but in such a manner that it is always perceptible as the prominent part, and does not remain long in its altered position. The plan of working-out is so consistently adhered to that even in the minor variation 25 the complete bass appears in the major. This is effected by chromatic treatment, which involves very bold and strange harmonic sequences. He had indeed to break through the rule that the significant note of the bass part should always come in at the beginning of the bar. Once, in bar 4 of the second part, he uses c' sharp instead of b; in bars 9—11 of the first part the bass appears in the tenor register; and in bars 7 and 8 of the second part $b\ e'$ is put for b flat e' flat.

Above this bass Bach exhibits such a wealth of invention and ingenuity that this work alone would be sufficient to make him immortal. His clavier technique is displayed in its most brilliant light, and in quite a new way, in consequence of the freest and most productive use of the two manuals. Free ideas alternate with those in strict style, clear and almost homophonous sections with polyphonic forms of the greatest possible ingenuity. Thus variation 7 exhibits a gigue, variation 16 a complete overture, variation 25 a richly ornamented Adagio in the style of a violin sonata, variation 26 a sarabande alternately in the left hand and the right, while the other hand plays rapid passages of sextolets of semiquavers. Variation 10 is a fughetta, the theme of which is derived from the ground bass, and which follows the course of the bass up to the end, notwithstanding that its development is perfectly regular. Canons are introduced on every interval, from that of the second to that of the ninth; that on the fifth is in inversion, and moreover its commencement, like that of the canon on the second, is derived from the bass of the theme by diminution. The last variation is a Quodlibet in which two popular songs are combined and worked out in imitation above the bass. And all this is without a trace of constraint. The hardest chains become wreaths of flowers; the whole result is gay and easy, the few variations that are in the minor serving only to intensify the effect of the pervading ground-tone of contentedness.

The question that now presents itself is what part is left to the melody in this elaborate extension of the variation-form. The musical motive of the whole is not the melody but its bass, and the harmonies evolved from it form a framework in which new forms are generated. And yet the effect of the melody is not lost. Setting aside the fact that it fills the hearer with a certain characteristic feeling which serves as a clue to all the variations and accounts for the various changes, it remains as the general design, so to speak, invisible yet everywhere present. The course of harmony which remains in all essential features the same in all the variations, together with separate reminiscences of the melody scattered throughout the work, assists the mental ear in calling up again the outlines of the melody. And indeed the full enjoyment of the variations is dependent upon this. A stronger and more conscious effort is demanded of the hearer in these than in ordinary variations; he must, as it were, project the outlines of the melody into each new musical form; this work is written for cultured hearers or connoisseurs.

The reader who recalls the nature and character of the chorale cantata, as we have described it,[321] will not overlook the near affinity which exists between that form and this set of variations. In both the full melody in its simple form constitutes the end and ultimate goal of the work, in both there are suggestive reminiscences of it in the intermediate sections, and, just as here the bass and the harmony remain essentially the same throughout, so there unceasing reference is made to the hymn which is the basis of the work, though lightly veiled under the madrigal form. This is Bach's contribution to the variation-form, which was an element of the first rank in the development of instrumental music. He

---

[321] See above, pp. 105-6.

lifted it into a high ideal region, giving it an intellectual and inspired form, which, on the incomparably higher level to which Bach had raised instrumental music, was once more worthy of the attention of the most earnest musicians. Although in later times the older variation form has continued to flourish, just because its relations with the origin of instrumental music were of so close a kind, and has borne new fruit in the shape of many delightful creations, yet the best musicians, even down to the latest times, have recognised in Bach's thirty variations the highest model of the form.

We must now return to the Quodlibet which forms the last variation. Musical diversions of this kind, in which even words in different languages were sometimes mixed up, might be performed in two ways. Either a number of well known melodies were sung or played one after another, so that those tunes which had the least inward connection were brought into juxtaposition; if each could be strung on to some unimportant phrase of the words or the music of the preceding one, the joke was all the better. The other method consisted of performing simultaneously several well known melodies with the most incongruous words. This last and more ingenious method was the one chosen by Bach. The words of the two popular songs used are handed down by his pupil Kittel.[322] They run thus:—

Ich bin so lang nicht bei dir gewest,	I long have been away from thee,
Ruck her, ruck her, ruck her;	I'm here, I'm here, I'm here,
Mit einem tumpfen[323] Flederwisch	With such a dull and dowdy prude
Drüber her, drüber her, drüber her.	Out there, out there, out there.

and

Kraut und Rüben	Kail and turnips
Haben mich vertrieben;	Don't suit my digestion;
Hätt mein' Mutter Fleisch gekocht,	If my mother cooked some meat
So wär ich länger blieben.	I'd stay here without question.

---

[322] The information comes to us through Pölchau, who inserted a note to this effect in his copy of the original edition of the variations. This copy is in the Royal Library at Berlin.

[323] Dialect for "Stumpfen."

THE QUODLIBET.

The melody of the second song, which is sung by the people to this day, is used by Bach in its entirety:—

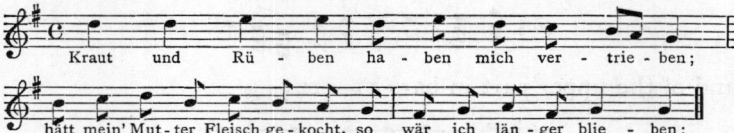

The lines of the tunes in that Quodlibet were frequently torn from their proper connection and used in fragments, just as it suited. So here the first two bars of the tune are found in the upper part in bars 3 and 4, the last two in the tenor part, bars 7 and 8, and in the upper part at the end; besides this the two first bars are worked out through the whole piece in imitation in the three upper parts. Only the first half of the first tune can be reconstructed from this Quodlibet. It runs thus:—

So far as I have been able to find, this tune is now extinct; the well known tune "Es ist nicht lang, dass g'regnet hat" is identical in metrical structure and almost so in the beginning of the melody. Nor do I find it mentioned in any of the numerous collections of popular songs which have appeared since Herder's time. Only the first line is carefully worked through in its exact form. The second is seen, but not clearly, in bar 2 of the tenor part, but seems not to occur again unless it may be alluded to in bar 10 of the bass part. Of the remaining lines the rhythm alone is employed, especially that of line 3:—

The fourth line is alluded to in the rhythm of the bass in bars 4—5 :—

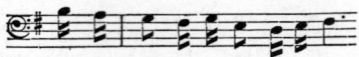

and of the upper part in bars 11—12 :—

The Quodlibet formed from these elements is an extraordinary performance, on account of its ingenious construction, but it has a higher interest as throwing light on Bach's artistic personality. In the realm of church music he drew his chief strength from the sacred popular song; and here he shows his intimate sympathy with the secular popular song. For this is the only supposition which can explain his adding such a close to a work on which he had expended his highest ability. This is clearly a trace of that old family spirit of the Bachs, which displayed itself in somewhat rude merriment on the family birthdays, by singing Quodlibets of this kind.[324]

We possess, however, yet another piece of documentary evidence of Bach's interest in popular vocal music, which we will consider, for this reason, here, although as a vocal composition it ought to have been mentioned in another place. On the 30th of August, 1742, the chamberlain (Kammerherr) Carl Heinrich von Dieskau received allegiance as Lord of the Manor (Gutsherr) of Kleinzschocher.[325] As Provost of the district he was inspector of the land, liquor, and income taxes, and of the quarterly tax.[326] We have already seen that in 1743 Picander held the post of receiver of the land and liquor taxes.[327] Whether he was anxious to retain the situation, or to express his thanks to the authorities for allowing him to retain it, his making a *Cantate en burlesque* for the occasion evidently has some

---

[324] See Vol. I., p. 153.
[325] Schwartze, Historische Nachlese zu denen Geschichten der Stadt Leipzig. 1744, p. 231.
[326] See "Das jetzt lebende und *flori*rende Leipzig." Years 1736 and 1746.
[327] See Vol. II., p. 340.

connection with the circumstances. The subject of his poetry was taken from peasant life at Kleinzschocher. He begins with a chorus sung in the dialect of Upper Saxony, by peasants who are rejoicing in the holiday which has been granted them.[328] After this a couple of peasant sweethearts are brought in, who praise the new lord of the manor and his wife, and then make sundry allusions to the receiver of taxes, to the conscription of recruits, to " Herr Ludwig and the tax reviser," and end by going into the inn to dance and make merry. As the upper classes at that time were surfeited with French and Italian theatrical representations, they were naturally pleased, if only for the sake of the contrast, with a German peasant *Divertissement*. Not, however, the representation of healthy, uncultured humanity, such as was afterwards made the groundwork of Christian Felix Weisse's opera librettos; they preferred to regard themselves as cultured and fashionable rulers, governing an oppressed and unmannerly peasantry. *Divertissements* of this kind were got up, not only at the luxurious court of Dresden, but also in the more simple and sober Weimar, and in other places.[329] Picander's burlesque cantata is a work of this kind.

We need not be surprised that Bach should have felt impelled to set it to music. He probably left the ethical side of the affair quite out of the question; it gave him evident pleasure, for once, to compose a secular work in a popular style almost throughout. The aristocratic element appears only in two arias (" Klein Zschocher müsse" and " Dein Wachsthum sei feste"), the second of which is taken, not without some violence, from the " Streit zwischen Phöbus und Pan." All the rest is confined within the narrow limits of popular forms. There is no chorus; the first number and the last are only duets, between which the peasant and his sweetheart alternately sing recitatives and arias. The instrumental accompaniment is in the style of a

---

[328] The rest of the poetry is free from dialect, but several provincial expressions occur, such as Guschel for Mund, Dahlen for Kosen, &c.

[329] See Vol. I., p. 228, note 69. Fürstenau, Geschichte der Musik am Hofe zu Dresden. II., p. 158 f.

village orchestra, consisting only of violin, viola, and bass; a horn is introduced in one passage, but for a special reason. The first number is preceded by an instrumental movement. This is a Quodlibet consisting of seven tunes put together. It begins with a waltz which, however, is not heard in its entirety until the end; on its first appearance it breaks off after seven bars, to make way for a variety of short movements in different rhythms and rates of speed. Many of these sound like genuine popular melodies, especially in this case:—

In the middle occur sixteen bars in sarabande rhythm, the solemnity of which is rendered doubly effective by the use of the unison. Dance tunes predominate even in the vocal portions of the work. The opening and closing numbers are bourrées, and the arias "Ach es schmeckt doch gar zu gut" and "Ach Herr Schösser" are polonaises. On Kirnberger's authority, who lived long in Poland and studied the dance music of the country, the aria "Fünfzig Thaler baares Geld" must be regarded as a mazurka.[330] The aria "Unser trefflicher" is a sarabande, and the song "Und dass ihrs alle wisst" is a Rüpeltanz (or *Paysanne*).[331] No distinct type can be recognised in the aria "Das ist galant," but the general character is that of a dance. All these pieces are short, rough, and merry, but ingeniously worked out;' witness the delicate way in which in the sarabande the chief subject recurs again and again, yet always in the bass, or in the "Rüpeltanz," the representation of intoxication in the displacements of accent. The melodic forms are so

---

[330] See Marpurg's Kritische Briefe, II., p. 45. The words of this song, which in Bach's work vary slightly from those given in Picander (see his Poems, Part V., p. 285), seem to refer to editions made on the occasion of paying allegiance to the lord of the manor.

[331] See Gregorio Lambranzi, Neue und Curieuse Theatralische Tantz-Schul. Nuremberg, 1716, Pt. I., fol. 4. The words of the *Paysanne* have a genuine rollicking humour about them, and are decidedly too good for Picander. It seems to me that he has made use of some student song or other, for he was very apt to turn the ideas of other poets to account.

thoroughly popular in style that we may suppose that here and there certain well known melodies must have been running in the composer's head. The supposition receives support from the fact that three popular melodies are used consciously and confessedly in the cantata.

After that, the sweetheart having sung the praises of the new governor in a " genteel " aria, the peasant says, " That is too fine for me, and more fit for the town, we peasants don't sing so softly as that. Listen to the sort of thing that suits me :—

> Ten thousand golden ducats
> Go to the chamberlain every day,
> He drinks his glass of good old wine,
> And on it he flourishes fair and fine.
> Ten thousand golden ducats
> Go to the chamberlain every day.

The melody to which he sings these words is still in popular use in Germany, to the words " Frisch auf zum fröhlichen Jagen." At the time when Bach composed this peasant cantata both the words and the tune had been popular for about eighteen years. The words were written by Gottfried Benjamin Hanke, a Silesian, who lived in Dresden as Clerk of Excise. He wrote them in 1724 for the feast of St. Hubert, for Count Sporck, who was very fond of hunting. The tune is of French origin and belongs to the hunting song " Pour aller à la chasse faut être matineux," which seems also to have served Hanke as a model for his song.[332] About 1730 it was very popular in the Count's dominions in Bohemia, and must have become widely and quickly known in Saxony, if in 1742 it could be called a peasant's tune. It is interesting to notice another

---

[332] See Gottfried Benjamin Hanke's Weltliche Gedichte. Dresden and Leipzig, 1727, p. 144. Poetischer Staar-Stecker, In welchem sowohl Die schlesische Poesie überhaupt, als auch Der Herr v. Lohenstein . . . verthaydiget . . . wird. Breslau and Leipzig, 1730, p. 30. For the information as to its origin we are indebted to Hoffmann von Fallersleben; see his Horæ Belgicæ. Pars secunda. Vratislaviæ. MDCCCXXXIII., p. 100. The words originally began "Auf, auf auf, auf zum Jagen! Auf in die grüne Heyd !" and numbered twelve verses. In newer collections it is reduced to six verses, and altered in other details. The tune is found in Erk and Irmer, Die deutschen Volkslieder mit ihren Singweisen. Leipzig, 1843. Book I., p. 47.

appearance of Count Sporck in the history of Bach's life. It will be remembered that he was connected with the B minor mass, and that in 1725 Picander dedicated to him the first fruits of his religious poetry.[333] Both the composer and the librettist had a special purpose in introducing the melody; we can now understand why, in this piece alone, a horn should be introduced, there being no opportunity for its use in Picander's words.

The girl replies in a parody of her lover's song as being too "tuneful," and saying that it would make the gentlefolks laugh as much as if she were to sing an old tune—then follows the second popular melody:—

She sings this to words which express hope of a numerous male offspring of the house of Dieskau, which hitherto has been blessed only by daughters. From the character of the tune it may be assumed that it belongs properly to some lullaby or child's song.[334]

A third popular melody is used as an interlude in the first recitative. Its first half comes at the end, and its second half in the middle. When put together and set in the same key, the result is as follows:—

It is evident that the first half is essentially the same as the first half of the tune. "Ich bin so lang nicht bei dir gewest," from the Quodlibet in the variations. In that place Bach employed only the rhythm of the second half; we may be certain that this is its original form, for if we compare the words of the recitative in the passage where the interludes (consisting of parts of this tune) occur, with the words in

---

[333] See ante, p. 43.

[334] I have not succeeded in finding any older source of the melody and its original words. Possibly the researches of Herr L. Erk or Herr F. M. Böhme have been more successful.

the popular song which correspond to those parts, a strict connection displays itself in the poetic idea. The ideas suggested in the recitative are expressed in each interlude, and especially in the passage where the first interlude occurs, the wishes of the lovers are explained very plainly. If we assume the composition of the variations and of the peasant cantata to have taken place at the same period, it is easy to understand that it might occur to Bach to bring the same melody into the peasant cantata, as it is of rather later date than the variations.[335]

After this digression, into which we were led by the subject of the Quodlibet in the thirty variations, we return to Bach's clavier compositions, and to the consideration of the fugal works and canons of his later life. And first the Chromatic Fantasia and Fugue. This celebrated work must have been written, at the latest, in 1730,[336] and internal evidence convinces us that it must be assigned to a considerably earlier date; the Fantasia bears a perceptible resemblance to the piece of the same name which precedes the great organ fugue in G minor.[377] This was written before 1725, and we found reason for connecting it with Bach's journey to Hamburg in 1720. Thus it is quite possible, and even probable, when we take into consideration another and an older form of the work which still exists, that the Chromatic Fantasia and Fugue may also date from before the Leipzig period. The effervescent character which pervades both pieces is not in the spirit of the Leipzig productions; and even in his fantasias Bach is not wont to dispense with strict forms, developed either thematically or episodically. Here all is uncontrolled "storm and stress."

---

[335] Compare Appendix A., No. 6. The peasant cantata was first published by Gustav Crantz of Berlin, edited by S. W. Dehn. A second edition appeared published by C. A. Klemm of Leipzig. In 1879 it appeared in the B.-G. edition, xxix., p. 175. There is a tradition, which may be mentioned here, that Bach was the composer of a comic song which was very widely known about the middle of the century, entitled " Ihr Schönen, höret an." I myself do not believe the tune to be by Bach, but I cannot do more than simply state my opinion in this place, leaving the subject, which would branch off into all kinds of different questions, for discussion elsewhere.

[336] P. S. I., ch. 4, No. 1 (207). See App. A., No. 7.
[337] See Vol. II., p. 23 ff.

The bold idea of transplanting the recitative into a clavier-piece had already been embodied in Bach's earlier fantasia in D major;[338] its germ is found in the works of the northern organists, and its development was assisted by Bach's intimacy with Vivaldi's violin concertos.[339] In the Chromatic Fantasia this idea attained a grand perfection. The piece, in which the boldest feats of modulation are crowded together, has the effect of an emotional *scena*. The fugue worthily carries out the chromatic character and the startling modulations of the piece, and the treatment of the fugue is full of genius, with a mighty demoniacal rush.

Bach wrote another fantasia with a fugue (in C minor) about the year 1738. In composing this work he seems to have been guided by the idea of bringing the technical resource of crossing the hands to bear upon larger forms than those to which it had been applied in the B flat major Partita. As the fantasia is in two sections with repeats—a form nowhere else employed by Bach in pieces of this name—it may be supposed that he was influenced in adopting it by Domenico Scarlatti; and it is interesting to observe the character which the Italian composer's style acquires in Bach's hands. This energetic and brilliant piece is, moreover, the forerunner of Emanuel Bach's sonata form. Bach has only left us the first forty-seven bars of the fugue. We may assume that he finished it, for the autograph is not a sketch, but a fair copy. He must have been prevented by some mischance from writing it out in its entirety. This is doubly to be lamented, since, to judge from the fragment that remains, the fugue must have been a work of especial boldness, and designed on a grand scale.[340]

---

[338] See Vol. I., p. 437.

[339] One of Vivaldi's concertos, which Bach arranged for the organ, has an Adagio in the style of a recitative. See P. S. V., Cah. 8 (247), No. 3; and Vol. I., p. 414.

[340] The autograph was discovered in 1876 by Moritz Fürstenau, of Dresden, and it belongs to the musical collection of the King of Saxony. It bears the same watermark as the Easter oratorio: see App. A., of Vol. II., No. 58. Forkel and Griepenkerl have been unfortunate in judging of the work, for they consider that the fugue did not originally belong to the fantasia and that only the first twenty-nine or thirty bars are authentic. Both these opinions are proved false

The irregular proportions which are apparent between the fugue and the fantasia, which in many cases has the character of a prelude, are found in many preludes and fugues of this period. We shall return to this presently.

A third fantasia, with a (double) fugue, is in A minor. By way of fantasia, we have a piece in imitation in strict style. The work has an elevated and steady glow of feeling, and is in every respect so pure and mature in style that it may well have been written in the beginning of the Leipzig period.[341]

Of the many characteristic pieces in fugue form which yet remain to be considered, the most important is a collection which forms a counterpart to the Wohltemperirte Clavier, and, indeed, is known as the second division of that work. Taken in its strict sense, the title conveys a false impression, and it is not at all certain that Bach gave it that title. The loving interest taken by Bach in the first part, and shown in his having transcribed it at least three times, was not bestowed upon its younger brother. We do not possess a single transcript of the second part made by the composer; indeed, there is hardly more than one complete copy in existence.[342] It was certainly finished in 1744; nay, if we may trust one piece of evidence, as early as 1740.[343] I called it a collection. The long-cherished and popular notion that it was only in the first part that Bach incorporated older compositions is certainly an error.

---

by the autograph. Griepenkerl, moreover, believes that the fantasia was written in 1725. See Forkel, p. 56, and Griepenkerl, in the preface to P. S. I., Cah. 9, where both are published—the fantasia as No. 7 (207), and the fugue as No. 18 (212).

[341] The autograph is wanting. Published in P. S. I., Cah. 4, No. 6 (208).

[342] Only one autograph, and that of the A flat major fugue, is at present known. It is in the Royal Library at Berlin. Fragments of a valuable MS. of equal date with this are in the possession of Professor Wagener, of Marburg. They have until now been mistaken for autographs; Kroll even considered them as such (see B.-G. XIV., P. xviii., No. 14A). The rest of this MS. was discovered by Fürstenau in 1876, in the musical collection of the King of Saxony. This makes the MS. complete, with the exception of the prelude and fugue in G major and the fugue in B major.

[343] Schwencke's MS., in the Royal Library at Berlin, gives the date as 1744 (see B.-G. XIV., p. xvi., No. 11). Hilgenfeldt (p. 123) must have had in his hands an autograph, with the date 1740, which was part of Emanuel Bach's bequest.

The C major prelude existed long before 1740. In its original form it contained only seventeen bars, and the fugue was entitled *fughetta*. The prelude was afterwards extended to thirty-four bars; but it did not receive its ultimate form until its third revision.[344]

The G major fugue has also undergone several changes. It was originally in a shorter form, coupled with another prelude, and its counterpoint was of so simple a kind that it must be assigned to the very earliest years of the Weimar period. The next step was to change the prelude, but the new one was not yet the same as that in the "Wohltemperirte Clavier." The fugue itself remained for a time unchanged. Finally, it was revised for the collection, and associated with a third prelude.[345]

The A flat fugue was originally in F major, and only half its present length; it also had another prelude.[346]

The C sharp prelude also bears evident traces of having been an older work, only used again in the Wohltemperirte Clavier. Apparently it was a piece by itself, for the first half alone is in the manner of a prelude, the second being a fughetta of a light and transitory character, it is true, but yet perfectly worked out. It was originally in C major; and its prelude-like portion was so nearly allied to the C major prelude of the first part of the book that Bach could not use it in its original key.[347] But all of the second

---

[344] The prelude appears as a piece of seventeen bars in a clavier book of J. P. Kellner's, bearing the date "3 Juli, 1726," belonging to Herr Roitzsch, of Leipzig, of which Kroll did not know the existence. Fürstenau's MS. contains it in its second form, together with those numbered 2, 3, 9, 16, and 18 in Kroll, which appear to be connected with it.

[345] The fugue in its original form and the two accompanying preludes are edited by Roitzsch in P. S. I., Cah. 3, Nos. 10 and 11 (214). Of the preludes the second is particularly fine, but it must have appeared too important for the fugue. The first of the two is also given, together with the fugue, by Kellner in the book above mentioned.

[346] P. S. I., Cah. 3, No. 9 (214), and introduction to it.

[347] A MS. of Johann Christoph Bach's still exhibits it in C major. This son of Sebastian was born in 1732; the piece must, therefore, have continued to exist in its old key after the completion of the second part of the Wohltemperirte Clavier, for before its completion the prelude cannot possibly have been written out by Johann Christoph. See B.-G., XIV., p. 243.

part of the Wohltemperirte Clavier that belongs to a later period must be regarded as having been composed, not so much with a view to writing a new work to consist of twenty-four preludes and fugues, as separately and by degrees. In the last ten years of his life we find Bach occupied in collecting his most important works, revising them for the last time, and in many other ways " setting his house in order." In this way the second part of the Wohltemperirte Clavier is chiefly to be regarded. As might be imagined, here and there pieces were wanting which would make the cycle of the work complete and proportionate to the first part. In these cases, whenever he did not feel inclined to compose new pieces he availed himself of older works.

These facts leave the value of the whole undiminished. Bach admitted nothing that was not of equal merit with the newly composed pieces, or that could not be made so by revision. In the first part of the " Wohltemperirte Clavier " some few pieces are somewhat inferior to the rest. This cannot be said of the second part. But, regarded as a whole, it is not a greater or less degree of genius that makes the difference between the two parts. It lies in the periods of the master's life during which the majority of the pieces in each were composed. In the second part, as compared with the first, are revealed an imagination more richly saturated with music, a greater grasp and more developed formative power, and the endeavour to give the fugue-form more sharply cut outlines and more characteristic features. As a whole, the second part, like the first, is distinguished by a certain general community of character between the separate pieces. Compositions like the great fugue in A minor,[848] which is older, or the chromatic fantasia and fugue, would never have been included by Bach in this collection, which was intended to bear that particular stamp of intensity of expression and contemplativeness which the clavichord was so well fitted to interpret.[849]

---

[848] See Vol. II., p. 33 f.

[849] Bach, as a rule, adheres to the limited compass of c to c''' in the second part also; only in the A flat Prelude the note d''' flat occurs once; in the B major Prelude, B occurs twice; and at the close of the A minor Fugue, A occurs. The latter piece has indeed more of the harpsichord character.

Even in the first part of the " Wohltemperirte Clavier " it was noticeable that several of the preludes were not coupled to their fugues with such homogeneity as we should have expected. In Bach the prelude soon grew to a greater independence than was warranted by its name. Not a few of the preludes in the first part were independent pieces as well as the C sharp prelude in the second part; and indeed after the completion of both parts Bach made, or intended to make, a collection consisting of preludes alone.[350] The preludes of the second part, even more than those of the first, bear the stamp of independent organisms. In the first part the greater number are nothing more than an animated development of harmonic sequences. In the second part, however, it is remarkable that no fewer than ten preludes have the dance-form in two sections—a form which occurs only once in the first part—and are short sonata-movements in Emanuel Bach's meaning of the word, only for the most part much more polyphonic than his movements.[351] Others make good their independent character, if in no other way than by a careful working-out of an important theme. It has already been remarked that the B flat minor prelude belongs to the class of three-part clavier Sinfonien (inventions).[352] It differs in no essential respect from a fugue; though the theme is accompanied by counterpoint on its entry, and though later on it is transferred from one part to another (bars 43, 44 and 49, 50), yet the former instance is hardly different from what happens in Bach's vocal fugues, when the first entry of the theme is accompanied with full harmony, and the latter case occurs also in the G minor fugue (bar 12, in the counter-subject).[353] The C sharp minor prelude is developed from three suggestive themes most ingeniously worked out. That which ought to be the strict purpose of the form—viz., to prepare for the

---

[350] See Vol. II., pp. 165 f., and 666.
[351] Kiruberger has attempted to resolve the first part of the A minor Prelude, with its characteristic wealth of harmony, into its proper fundamental harmonies; see Die wahren Grundsätze zum Gebrauch der Harmonie. P. 107 ff.
[352] See Vol. II., p. 59.
[353] And also in the F minor Fugue of the first part, bar 19.

most important part of the music which is to follow, and to keep up the interest in that part—is not fulfilled by most of them. They move us too much by their own importance, and each of them exhausts the particular emotion which it expresses. They not only assert themselves as equal to the fugues, but often stand in direct contrast to them. Every one must feel this who will compare the easy grace of the E flat major prelude with the stiff gravity of the fugue, or the devotional aspirations of the E major fugue with the cheerful activity of its prelude.[354]

Of course the relations between them are not fortuitous, nor did they arise simply in arranging the pieces in the collection, but they are Bach's design. From the prelude and fugue a new form in two movements has been developed. It is not entirely strange to us. Not to speak of the fantasia and fugue in C minor, lately spoken of, since that work dates from the last period of Bach's life, an example of the form occurs even in the first part of the "Wohltemperirte Clavier" in the E flat major prelude and fugue; and another example is offered by that prelude and fugue in A minor which was afterwards developed into a concerto.[355] This last instance is particularly instructive, because by his re-arrangement of it Bach has left evidence as to his ideas of the relation between prelude and fugue, and also because the gigue-like style in which the fugue is written is found again in five fugues in the second part of the "Wohltemperirte Clavier" (C sharp minor, F major, G major, G sharp minor, and B minor).[356] Remembering how the fugal gigue was used in the Suites; we may suppose that a similar idea suggested its use in this place. In three cases this is evident. The preludes in C sharp minor, F major, and G sharp minor express a grave, subdued, and in parts a

---

[354] It may be mentioned, by the way, in this place, that S. W. Dehn has discovered a Fugue by Froberger in the Phrygian mode, built upon almost the same theme and with the same counterpoint (see his Analysen dreier Fugen Joh. Seb. Bachs. Leipzig, Peters. 1858. P. 31). As Bach was familiar with Froberger's compositions (see Vol. I., p. 323), the agreement cannot be an accidental one.

[355] See Vol. I., p. 420, and ante, p. 143.

[356] In the first part the G major Fugue is the only one in the gigue style.

resigned mood; the fugues break in upon this repose with a different and lighter feeling.[857] This emotional contrast is also felt in the pairs of pieces in C major, C sharp major, and F minor, where more of the fugues are in gigue form. In the C sharp major set, the allegro end of the dreamy prelude is intended as a connecting link between that and the vivacity of the fugue. As far as the difference of style will allow, we may find analogies to these forms in the two-movement clavier sonatas of Emanuel Bach and of Haydn. Other pairs also stand in opposite contrast; animation is subdued to firmness and repose; sometimes the contrasts are so sharp that no common ground of feeling is to be found for the two. Cases also occur in which the two movements are distinguished only by delicate shades of difference (D major, F sharp major, F sharp minor, B flat minor, &c.), remaining essentially the same in feeling and character. But in all cases—and this must once more be insisted on—the prelude and the fugue are regarded as factors of equal dignity and importance. Instances where the prelude is left to its old function of merely ushering in the fugue are rare, and can scarcely be proved to exist elsewhere than in the pairs in D minor, G minor, and B major.

The artistic skill displayed in the fugues is astounding; and after careful comparison it must be admitted that in this respect the first part is not so finely conceived as the second. The devices of inversion, stretto, and augmentation are used equally often in both. But, on the other hand, in the first part Bach makes very sparing and only occasional use of double counterpoint on the tenth and twelfth,[858] while in the second part it is used with great effect. Although in the B flat major fugue (bars 41 ff. and 80 ff.) it is only introduced incidentally, in the B major fugue counterpoint on the twelfth forms a chief element in the working out (see

---

[857] Kirnberger (Kunst des reinen Satzes II., 1, p. 120) says of the F major Fugue: "In order to perform this Fugue correctly on the clavier, the notes must be struck lightly and with a rapid motion, and without the least pressure."

[858] See the counterpoint on the twelfth in the F sharp major fugue, bars 15, 16 and 32, 33.

bars 36, 43, 54, 94). And an instructive example of the musical wealth which is to be obtained by these artifices, without exaggeration, and without detriment to the sharpest and most characteristic expression, is afforded in the G minor fugue—a piece full of "swing" and determination, as if beaten out with hammers, in which free use is made of double counterpoint on the octave, the twelfth, and the tenth.

Still, if we expect to find in the second part of the "Wohltemperirte Clavier" a model of fugal work in the technical sense we mistake its purpose. That it was never so intended may be concluded from the fact that an equal and not rarely a greater degree of importance is given to the preludes than to the fugues. There is also this difference between the two parts, that in the second the technical treatment is nowhere made prominent. The fugues in C major, D minor, E flat minor, A minor, and B flat minor of the first part, in which the art tends to artifice, find no counterparts in the second part. A comparison between the A minor fugue (of the first part) and the similarly constructed fugue in B flat minor in the second part shows how entirely Bach could eliminate that savour of display which we feel in the older work. The augmentations of the theme which in the C minor and C sharp major fugue of the second part are brought in simultaneously with the themes in their original and inverted forms have, when compared with the same device in the E flat minor (D sharp minor) fugue of the first part, rather a light and purposeless effect. Such combinations are difficult and unusual in instrumental fugues, and are not intelligible unless the theme is quite a short one; the theme of the E flat minor fugue is almost too long, and the effect produced is somewhat over elaborate and erudite. The opportunities offered for ingenious complications increase, of course, in proportion to the number of parts employed; and it is significant that in the first part there are ten fugues in four parts and two in five, while the second contains only nine in four parts, all the rest being in three.[359] Among these last

---

[59] In the C minor fugue the fourth part does not enter until near the end, and then comes in with the augmented subject Bach introduces the same

there is a wonderful triple fugue (F sharp minor), unapproached in ingenuity by the three-part fugues of the first part; and yet this ingenuity is perfectly unobtrusive. Various liberties, too, are taken which would ill befit the work if they were not intended to prove how gracefully the strictest writing might be treated. More than once we find alterations in the theme in the course of development (F major, bars 86 and 87; E major, bar 23 ff.; F sharp minor, bars 54, 55 and 60), or an undue increase in the number of parts, even in other places than at the close (F major, bars 86 and 87; G major, bar 60). These, like the division of a passage between two parts (G minor, bar 12), or the daring entry of the answer to the theme on the augmented fourth, and the long, almost homophonous digressions in the F sharp major fugue [360]—these are not the remains of that youthful audacity which characterised Bach's early fugal compositions. They prove that his aim now was to create music which should be full of life and energy and with a meaning of its own.

One of the highest triumphs of art is so to modify a form, which, by its limitations, seems adapted only to convey the most general ideas, as to render it capable of expressing individual and personal feeling. It was this which, even in Bach's lifetime, more than all else, surprised the musical world in his fugues. Soon after Bach's death one writer spoke of their "strange character that is so different from the ordinary run of fugues."[361] And another says: "Just at the time when the world (of music) began to take a new course, when the lighter kind of melodies became popular, when people were tired of stiff and harsh harmonies, the late Herr Capellmeister Bach hit upon a clever idea, and taught us how to unite a pleasant and flowing melody with the richest

---

effect in the great organ fugue in C major (B.-G., XV., p. 234), the construction of which in other respects also has a great similarity to that of the C minor fugue. Apparently, therefore, the dates of both must be near together; and very possibly the C minor fugue may have been originally written for the organ.

[360] Compare Vol. II., p. 402.
[361] In Marpurg, Kritische Briefe I., p. 192.

harmonies.[362] The fugues in the second part of the "Wohltemperirte Clavier" are some of the most affecting and characteristic pieces in the whole range of music, and in their own class they are unapproached. The first part has nothing to show that can be compared in this respect with the fugues in D minor, E minor, F minor, G minor, and A minor. But even in the second part of the "Wohltemperirte Clavier" the great master of fugue has not solved—nor did he wish to solve—exhaustively and once for all the problem of what art in its highest development could do in the fugue form. This task was left till the evening of his life.

There are two works which, from this point of view, present themselves to our notice—the so-called "Musical Offering" (Musikalische Opfer) and "The Art of Fugue" (Kunst der Fuge). The former work is, as it were, the vestibule through which Bach passed to the latter. Without depreciating the value of its several component parts, the Musical Offering, as a whole, must be regarded only as a study by which the master prepared himself for the second and greater work. A part of the work appeared in July, 1747 —two months after Bach played before King Frederick II. at Potsdam. Bach resolved to use the theme set him by the King on that occasion as the basis of a number of thoroughly developed and artistic compositions, since, as he said, his improvisation had not done justice to so expressive a theme. The name "Musical Offering" is derived from its having been dedicated to the king.[363] The

---

[362] Marpurg, Abhandlung von der Fuge, Part II. Berlin, 1754 (in the Dedication). A kind of imitation of the Wohltemperirte Clavier by Gottfried Kirchhoff has come down to us (see Vol., I., p. 521) in his *A.B.C. Musical*, four-and-twenty fugues comprising all the keys. It was a work of similar kind with his " Clavir Ubung, in sich haltend das I. und II. halbe Dutzend Von 24. melodieusen, vollstimmigen und nach modernen Gustu durch den gantzen Circulum Modorum Musicorum gesetzten Præludiis." (Clavier Practice, containing the first and second half-dozen of twenty-four melodious preludes, fully harmonised, and written in accordance with modern taste in the complete circle of musical modes.") Published about 1738 by Balthasar Schmidt, of Nuremberg.

[363] P. S. I., Cah. 12 (219)—an edition in oblong folio by Breitkopf and Härtel. An autograph of the six-part Ricercar is in the Royal Library at

work, which was both composed and engraved piecemeal, contains one fugue in three parts and one in six, eight canons, a fugue with answer on the fifth in canon form, a sonata in four movements, and a two-part canon over a free *basso continuo*. All these are worked out more or less upon the same theme.

Bach gave to both the three-part and the six-part fugue the name of *Ricercar*. In the case of the second, the name seems to be accounted for by the fact that it was a hitherto unheard-of undertaking to write a strict six-part fugue for clavier alone without pedals. The idea would, probably, not have occurred to Bach had not King Frederick, on the occasion of Bach's playing at Potsdam, expressed a wish to hear an *extempore* six-part fugue played by him. At the time Bach complied with the king's request, improvising the fugue upon a fitting theme chosen by himself. In this work he wished to show that he could also treat the given theme, although it was not so well adapted for the purpose, in a six-part fugue. Such a composition, of course, implies a close interweaving of the parts, more especially when it is played by two hands alone. If we compare it with the C sharp minor fugue in the first part of the "Wohltemperirte Clavier," which contains, not indeed six, but five *obbligato* parts, and if we notice how much more intelligible it is than the fugue in the "Musical Offering," it will appear that Bach has, intentionally, made all the six parts work together. The mechanical difficulty of the piece is a second reason for calling it a Ricercar. There are no other special feats of ingenuity, in the use of counterpoint, or clever episodical developments. An old-fashioned style, the effect of which is heightened by the use of the *Tempus imperfectum*, and which agrees very well with the harmonic richness, and the bold chromatic treatment running through the whole work, are the contrasting elements from which the individual character of this mighty work is developed.[864]

Berlin; it is from Emanuel Bach's effects. As to the original edition, see App. A., No. 8.

[864] In the original edition Bach had the six-part Ricercar printed in score, in order to present the intricate interweaving of the parts in a clearer form to the eye. In the autograph it is brought together on two staves.

The three-part fugue corresponds less to the idea of a Ricercar. It is a simple fugue with moderately extensive digressions, in which, however, only the chromatic portion of the theme undergoes thorough episodical development, and the counterpoint is proportionately less rich. The countersubjects which are associated with the theme in bars 61—66 are repeated together with it no less than three times; twice, it is true, with double counterpoint in the octave. The passage in bars 38—52 is repeated, almost note for note but in another key, in bars 87—101, and in the same way bars 66—71 are repeated as bars 81—86. Some of the digressions are in themselves rather strange. That sudden triplet figure, which begins at bar 38, giving place after four bars to another digression strongly contrasted with it, ceasing entirely, never to be used in such a way as to account for its strange appearance, must, when viewed from the standpoint of Bach's fugal writing, be regarded as almost anomalous. The case is the same with the digression in bars 42—45, with the passage from bar 108 to the entry of the chromatic quaver-figure, and finally with the close. Without venturing to criticise the master, we believe ourselves fully justified in referring these strange phenomena to the influence of some external circumstances. If we remark how the idea in bar 42 f.—

is repeated again in the first Allegro of the sonata (bars 48 ff. in the flute part, 69 ff. in the violin part, &c.), how the passage in bar 108 onwards is used afterwards to prepare for the Andante, then we must come to the conclusion that Bach intentionally gave to this fugue a light and prelude-like character. There is no evidence to show that while writing it he had any idea of writing the sonata; for at first he offered to the king nothing more than this three-part fugue, the six canons, and the fugue in canon-style. It is probably, however, to be explained by supposing that when he was writing out the work he bethought himself of the fugue which he had improvised at Potsdam, and which had so pleased the king, and that he preserved in it more of the ideas that had occurred to him then than he would have thought permissible under other circumstances. At all events, it seems clear that Bach thought this "Offering" an insufficient one, and so followed up his first gift with a second of more importance, containing the six-part Ricercar, the Sonata, and three more canons.

Bach had five of the canons together with the fugue in canon-style printed on a sheet by themselves and designated by the general title *Canones diversi super Thema Regium.* Besides this he gave them the title of *Regis Jussu Cantio Et Reliqua Canonica Arte Resoluta* (*i.e.*, The theme given by the king's command, together with additions, resolved in canon-style), the initials of the words of the title giving the word *Ricercar*. Above the fourth canon (seventh in Peters), which is in inverted augmentation, Bach inscribed the words (in the dedicatory copy which is still preserved): *Notulis crescentibus crescat Fortuna Regis* (As the notes increase in value, so may the fortune of the king increase). At the fifth (eighth in Peters) an infinite canon, which modulates by ascending a whole tone at every repetition, we find the words *Ascendenteque Modulatione ascendat Gloria Regis* (And with the rising modulation may the glory of the king rise). In this pretty and playful symbolism, which undoubtedly proceeded from the composer himself, may be traced the spirit of the contrapuntists of the Low Countries. The solutions of the canons are indicated by Bach himself. The first (fourth in Peters) is in two parts, and by retrogression (*cancrizans*), the theme lying in the parts that are in canon. The rest are also in two parts, but at the same time they form counterpoint to the theme which is brought in as a *Cantus firmus*, so that the result is a three-part piece. Without any of that dry and unimaginative ingenuity which generally intrudes itself into the domain of canon, these little pieces reveal not only real ingenuity but also true genius and characteristic feeling. Notwithstanding their astounding elaboration, there are very few harsh notes, and those are only transitory.[365] A more delightful or interesting little work of its kind than the canon *per augmentationem, contrario motu*, written in the French style, certainly does not exist. The three-part fugue in canon form, too, is a masterpiece of artifice worked out with playful facility. Bach has omitted to say what instrument is to play the uppermost part with the two clavier parts, but it is apparently a flute.

The other three canons stand, perhaps from typographical reasons, one after the three-part fugue and the other two after that in six parts. (In Peters they are placed before the other five and numbered 1, 2, and 3.) The solution of the last two—one in two parts and the other in four, without a *Cantus firmus*—is not given by Bach, but are left with the words "Seek, and ye shall find" (*quærendo invenietis*), to the ingenuity of the player or reader. In the four-part canon the entrances follow each other at the distance of seven bars; but the effect of this daring chromatic treatment verges on the abstruse.[366] In the two-part canon the bass as the second part enters, in the strictest contrary motion, on the second crotchet of the fourth bar. Moreover, the canon is so constructed that a correct solution is obtained by precisely the reverse method—letting the bass begin with the contrary motion, and the alto follow in direct motion on the second crotchet of the fourth bar. And again the bass may be made to follow on the second crotchet of the fourteenth bar. But the first of these three solutions is the one intended by Bach.[367]

---

[365] Kirnberger solved three of these canons in the Kunst des reinen Satzes II. 3, p. 47 ff.). In the canon *per motum contrarium* (No. 6 in Peters) bar 3 cannot possibly be right. In Bach's dedicatory copy the fourth semiquaver *b* is corrected in red ink to *b* flat, probably by Kirnberger when the copy came into the possession of Princess Amalie. Then, however, the *a* before it would have to be changed into *a* flat; this would be the most beautiful arrangement, and thoroughly satisfactory, were it not for the fact that the two naturals of the original plate receive strong confirmation by the express insertion of the flat before the last *a* of the bar.

[366] The solution is given by Hilgenfeldt, p. 122.

[367] A MS. of a part of the Musikalisches Opfer, in the Amalienbibliothek, contains this canon and the first two solutions. Above the solution which begins with the bass Kirnberger has written, "This solution is not the one intended by the author," and above the other "The true solution." The third was given by an anonymous writer in the Allgemeine Musikalische Zeitung for 1806, p. 496, note.

The Sonata, in ordinary four-movement form, displays no less genius, but has, in addition, a warmth and richness of emotional character which the other pieces have not. The Largo is not strictly built upon the theme, but preludes, as it were, upon its characteristic interval of the seventh (see bar 4 in the bass, bars 13 and 14 in the flute and violin, &c.). In the fugal Allegro, however, the theme makes its way gradually into all the parts as a *Cantus firmus*, with very beautiful effect; the proper theme of the fugue itself forms the first countersubject to it, and the second consists of a passage taken from the three-part Ricercar. The expressive Andante is, as it were, a fantasia built chiefly upon ideas taken from the three-part Ricercar, but the beginning of the theme also makes itself distinctly heard in several places. In the final Allegro the theme appears cleverly transformed into 6-8 time, and is developed into an animated fugue.[868] And, as though the composer's powers of combination were quite inexhaustible, he inserts after the sonata yet another canon, this time written out in full. It is in two parts with a figured bass and worked out in strict inversion, the second part coming in first on the fifth above, and afterwards on the fifth below. Quite at the end Bach, by way of a joke, brings in the theme in the figured bass part.

The infinite capacity for combination, the ingenuity, penetrating even to the deepest source of harmony, and the powerful imagination which preserves its full vitality even within the narrowest limitations, make of the "Musikalisches Opfer" a monument of strict writing which will endure for all time. It is not, however, an organic whole, since different instruments are required for the different parts. The first two fugues are clavier music, and some of the canons are intended for the clavier, or at least can be performed upon it. For the other pieces stringed instruments are necessary, and in some their presence is expressly required; the fugue

---

[868] The figured bass accompaniment to the Sonata has been written out in full, in four parts, by Kirnberger. The work is published in P. S. III., Cah. 8, No. 3.

in canon style requires a clavier and flute (or violin); the sonata and the final canon are written for clavier, flute, and violin, probably with reference to the fact that King Frederick himself played the flute. The work is simply a group of various pieces composed at different times, and thrown together quite arbitrarily, intended to exhibit one and the same idea under the greatest possible variety of aspects. In this instance, Bach set aside the higher idea of uniting these ingenious component parts into an artistic whole. On that account the Musikalisches Opfer cannot be called anything but a study for a greater work, a half abstract creation, in which the technical points must be chiefly insisted on.

The case is different with the second work, the "Art of Fugue." The idea of creating a great work of art in many parts, but as a perfectly organic whole, from a single theme, by employing all the devices of strict counterpoint, is here fully worked out. By mere chance the master's work has not come down to posterity as he had completed and perfected it. We must assume the main portion of the "Kunst der Fuge" to have been composed in the year 1749. When Bach was intending to have it engraved on copper he revised and completed his manuscript. The greater part had been engraved under his personal supervision, when death overtook him, in the middle of the year 1750. His family were not sufficiently informed of his intention with respect to the manuscript, as it was found amongst the master's effects; his grown-up sons were at a distance, and the task of editing fell into the hands of ignorant people, who put everything on the plates as it came—sketches beside completed movements, original settings beside arrangements, parts that had a connection beside those which had none —in dreadful disorder; and in this state the work was published. But the parts which are to be separated from the mass, as not intended for the complete work, are plainly discernible, although the arrangement designed by Bach for the last part of the work must remain in uncertainty. By mistake or ignorance an older sketch for the tenth piece has crept into the work as published, besides a fugue with three subjects, on which Bach had been working a short time before his death, but which has nothing whatever to do with this work, and lastly, two fugues for two claviers. Both these last are arrangements of those two three-part fugues, of which the second is the inversion of the first in all parts; portions of this are very difficult, nay, impossible, to play, for the parts often lie so far apart that two simultaneous notes cannot be reached except by a spring. In order to make this "eye music" appreciable by the ear, Bach arranged the fugues in such a manner that one clavier plays two parts, while the other takes the third and a free part added in besides. A few free additions are also made at the beginning of each fugue. The newly added parts give fresh evidence of Bach's enormous talent in contrivances of this kind. The result, however, has but a doubtful value as a work of art, and was never intended for insertion in the complete work, since it is a distortion of the idea of writing a fugue in only real parts, all capable of inversion; while the introduction of a second clavier radically alters its style and character.[369]

After the excision of these inorganic elements, we have a work consisting of fifteen fugues and four canons, on one and the same theme. In it we meet with simple, double, and triple fugues, fugues built upon the theme altered either in melody or rhythm, fugues with strettos, with the answer in contrary motion, both in notes of the same value and in diminution and augmentation, fugues in double counterpoint, in the octave, tenth, and twelfth, and lastly fugues in which

---

[369] In the present century the "Kunst der Fuge" was first republished by Nägeli of Zurich, then edited by Czerny, by Peters (P. S. I., Cah. 11—218), and of late edited by W. Rust (B.-G., XXV.[1]); the last an excellent production, rich in valuable results of critical labour. Rust has not ventured to omit the portions that do not belong to the work. The relations between the three-part fugue with its inversion and the fugues for two claviers struck, strange to say, both him and M. Hauptmann, to whom we owe a fine analysis of the work (Leipzig, C. F. Peters). From this relation, however, it is easy to account for the mistakes to be found in the fugues for two claviers, which Rust attempts in part to explain away by very bold conjectures; they were overlooked and left standing when the fourth part was added in, and can indeed be only due to this.

all three or four parts are in contrary motion to each other, and that in different positions, besides two-part canons in augmented contrary motions, and in the three practicable kinds of double counterpoint. There are forms from the very simplest up to the hardest that is conceivable, such as even Bach himself never produced before in his life. Although it is not certain whether the title "Kunst der Fuge" proceeded from Bach himself, yet the work bears sufficiently plain traces of having been intended to serve an educational purpose. This is proved by the circumstances that the work is written in score, and that the pieces are not called fugues, but "counterpoints," with direct reference to the school. Still, to regard it merely as an instruction book in fugue, with examples, and not as a genuine work of art, would be to misapprehend its nature. The mere idea of imparting instruction awoke Bach's artistic inspiration. Never had he done more excellent work than when he undertook to aid, by his example, in the advancement of the disciples of art. In the "Orgelbüchlein," in the Inventions and Sinfonias, and in the first part of the "Wohltemperirte Clavier," this purpose is avowed with dignified simplicity. The practical and educational purpose and the free artistic ideal are so inseparable here, at the highest point of their development, that there is scarcely a trace of any compromise, or sacrifice of the highest demands of the one factor, in favour of the other. From both points of view we have before us a perfect work.

Whoever seeks instruction in it will confine himself to studying the separate portions. He will perceive the incomparable capability for combining parts, and that almost miraculous wealth and variety of harmony, which will make him feel that Bach has exhausted all the possibilities of harmony, and that after him there is no more to be said in that province of art. Our view of the work as an artistic creation must include it as a whole and rest upon the impressions which it produces as a whole. It differs essentially from both parts of the "Wohltemperirte Clavier,' and from the Inventions and Sinfonias, by the earnestness of its emotional character. There is a greater uniformity in the characters of the parts, and a more contemplative nature suggests itself to the hearer even in the first fugue, which seems like the solemn repose of a winter's night. Even the theme, which has been unjustly designated as practicable for musical purposes, but insignificant in itself, is steeped in this feeling.

The inner development of the work is in a sequence of grand, majestic groups. The first group consists of the four fugues already mentioned. Even in this group we can clearly trace a progressive development. The second fugue contrasts with the first by its somewhat more animated counterpoint; the third has the theme altered by inversion, and here the theme reveals a deep and yearning character of its own. It is continued in the fourth fugue on broader lines, rising from bar 61 onwards to a striking degree of power.

The second group consists of the fifth, sixth, and seventh fugues. In contrast to the first group the second is worked out on a modified form of the theme, the effect depending chiefly on the rhythm, and it exhibits the theme in different kinds of time, combined with itself in double counterpoint. The animation thus produced is increased by Bach in the sixth fugue by the use of the rhythms of the French overture style, and in the seventh fugue, which introduces the theme in its natural time, and diminished and augmented simultaneously, the animation verges on restlessness. The progressive development of the second group lies chiefly in these external qualities.

In the third group (fugues 8—11) external and internal elements combine to attain the climax. The principal subject is now associated with independent and contrasting themes. The eighth fugue begins at once with a subject of this kind, which glides in with stealthy, snake-like windings, and is full of peculiar individuality both in rhythm and in melody. After it has been thoroughly worked out, a second and very agitated theme, not less important in rhythm and melody, accompanies it. A double fugue is thus produced, in which the strange little taps, as it were, in the second theme, increase to hammer beats, and the animated movement to violence of unrest· and it is not till the original quiet movement has been restored that the chief subject enters and is treated fugally. But dissevered as it is by crotchet rests, it also conveys an impression of inward agitation. The other themes now join in with it, and they work together to produce a progressive intensity of effect in a triple fugue in three parts and containing 188 bars. The four-part fugue which follows has only one counter-subject which, owing to its being treated in counterpoint at the twelfth, does duty for two, although it cannot of course appear in two places simultaneously. Setting out with a mighty spring the counter-subject goes on without stop or stay, while against it the chief subject goes grandly and solemnly on in augmentation. In the tenth fugue, which is written in counterpoint at the tenth, there is again only one counter-subject. Its mild and flowing character is felt as a repose after what has gone before, and it prepares us for the full appreciation of the last fugue, in which the three themes of the eighth fugue are worked out again, but in four parts, which intensifies their expression to the last degree and entirely exhausts their harmonic capabilities. This concise and cyclic group reveals the master in that gloomy grandeur which we seek in vain in any other composer. It also shows, in the plainest way, how firm his purpose was to create this work on the truest principles of art.

The fourth group consists of the last two pairs of fugues. From a technical point of view they exhibit Bach on a dizzy pinnacle of eminence. At a height where existence would be an impossibility to others, he breathes with ease and freedom. The formal limits to which he subjects himself, in his case serve only to give to the pieces their true artistic character. The solemn repose of the opening returns upon us, but sublimated to a feeling which is best expressed by the words " a cold grandeur."[370] Bach's last intention with respect to the arrangement and order of the pieces, we only know up to the eleventh fugue. But it may be confidently assumed that of the last two pairs, those in three parts were intended to precede those in

---

[370] Hauptmann, Erläuterungen zur Kunst der Fuge, p. 10.

four.[871] How he intended the four canons to be associated with the fugues—whether they were to be inserted after the third group, or only added on to the whole as an important appendix—must be left in impenetrable mystery, owing to the master's premature death. To us it appears scarcely credible that Bach could have intended to interrupt the sequence of the grand structure revealed to us in these fugues by inserting a number of stricter forms, which, though very clever and ingenious, suffer from the same disproportion of idea and material that we have already noticed in the case of two of the clavier duets, and which are therefore not well fitted to intensify the harmonic effect of the fugue-sequence.

The two parts of the "Wohltemperirte Clavier" must be considered as a whole, although each separate fugue, either with or without its prelude, gives us a feeling of pleasurable satisfaction. Bach himself regarded them as one, for he would sometimes feel impelled to play them from beginning to end at one sitting.[872] In a much higher degree then must the "Kunst der Fuge" be regarded as a self-contained unity; and not only because the same theme is the subject of all the fugues. The result of this need only have been a method of treatment which would have shown off each fugue by itself; no one would look for any inner connection between Schumann's fugues on the name "Bach." But in the "Kunst der Fuge" the separate parts bear a relation to one another, and only through one another can they be perfectly understood. The effect of a piece written in double counterpoint depends upon the hearer's realizing the two different registers in which the contrasting subjects are placed, as a unity, and, at the same time, a duality; and the meaning of the separate fugues in this work can only be seen in the right light when the different modifications undergone by the theme, the various countersubjects, and the organic entities resulting from their association are compared, not only with the original form of the theme and the simplest fugal form built upon that,

---

[871] This opinion is shared by Rust, see B.-G., XXV.,[1] p. 28.
[872] Gerber, N. L., col. 492.

but also with one another. By this is meant an emotional, not an intellectual comparison; indeed, to apply this to the work before us requires special musical culture. But —to take an instance—in the seventh fugue, where the theme is worked in direct and contrary motion, and in the natural, diminished, and augmented forms, it will scarcely be possible so much as to understand the relations of the parts to each other, unless the ear has been prepared for the task by the combinations that have gone before, and has become thoroughly familiar with the chief subjects. This preparation alone will enable us to see the full justification of such ingeniously intricate forms. In reality this last work of Bach's is a single gigantic fugue in fifteen sections. Hence it may be traced to a single idea, in a much stricter sense than was the case with the two parts of the "Wohltemperirte Clavier." Hence, too, each section necessarily leaves an impression of incompleteness, and this in great part accounts for the fact that no single fugue of this wonderful composition has taken such firm hold on posterity as even the lightest of the more pleasing fugues of the "Wohltemperirte Clavier." Few, perhaps, have the ability and the inclination to understand it as a whole. The obscure state in which it has hitherto lain has rendered this task all the harder, and it has thus come about that a composition of incomparable perfection and depth of feeling, although it has always been mentioned with especial reverence as being Bach's last great work, has never yet formed part of the life of the German nation.

The engraving of the original edition was clumsy and full of mistakes.[373] Emanuel Bach undertook the publica-

---

[373] Rust supposes J. G. Schübler, of Zella, near Suhl, to have been the engraver. The monogram on p. 25, in which A is the principal letter, seems to contradict this supposition. Rust's researches, however, have made it certain that it was not done by any of Bach's sons, as has hitherto been generally supposed. And there is nothing to show that Bach's sons took part, even indirectly, in any supervision of the engraving. The note in the first supplement of the Berlin autograph (see B.-G., XXV.,[1] p. 115), which may have given rise to such a supposition, is, I am convinced, not by Emanuel Bach's hand.

tion, but his expectations of a large and rapid sale were disappointed. He therefore lowered the price from five thalers to four, and commissioned F. W. Marpurg, of Berlin, to write a longer preface, in place of the short notice which had preceded the work in explanation of its supposed incompleteness. Thus provided, the "Kunst der Fuge" was brought out at the Leipzig Book Fair, at Easter, 1752. But even then its success was very moderate. In the autumn of 1756 Emanuel Bach, having sold no more than about thirty copies, gave up the attempt, and offered the copperplates for sale "at a low price."[374] In the year 1760, Breitkopf was again selling it for a Louis d'or (=5 thalers) a copy.[375] In spite of the slight reception of the work, the few who became acquainted with it were competent judges, and they appreciated it at its real value. Among the first purchasers was Mattheson, who, to the end of his life, took an eager interest in all new works of importance. "Joh. Sebast. Bach's 'Art of Fugue,'" he writes, "a practical and splendid (praktisches und prächtiges) work of seventy plates in folio, will at once amaze all the French and Italian fugue-makers; provided only that they can understand them rightly—I will not say play them. Let every one, whether a German or a foreigner, lay out his Louis d'or on this treasure! Germany is and remains without doubt the true land of organ music and of fugues!"[376]

Bach's creative activity did not rest even after the completion of the "Kunst der Fuge." He now took in hand the composition of a clavier fugue on the grandest scale, in

---

[374] The conclusions drawn by Rust, from the supposed preparation of two editions of the "Kunst der Fuge," as to its reception by the public are untenable. For he has overlooked the fact that Forkel, whose opinion he attacks, is here speaking, not on the authority of oral tradition, but from a printed source of information of a trustworthy kind—namely, Emanuel Bach's own statement, made September 14, 1756, given in Marpurg's Historisch-Kritische Beyträge II., p. 575 f. The title and the preface are the only new portions of the so-called second edition.

[375] Breitkopf's New Year Catalogue, 1760, p. 7.

[376] Mattheson's Philologisches Tresespiel. Hamburg, 1752, p. 98. The book was published, as appears from its dedication and preface, at Easter, 1752, and thus must have been written in 1751.

which the last of the three themes is on the letters of his own name. He only wrote 239 bars of this work; at least no more have come down to us, and Emanuel Bach states that his father died while engaged upon it. It is the same fugue that has, by misunderstanding, crept into the original edition of the "Kunst der Fuge."[377] As the fragment only reaches down to the beginning of the combination of all three themes, it probably contains only about three-quarters of the work as originally conceived; from this we can see on what colossal dimensions the fugue was planned. The emotional character, as in the "Kunst der Fuge," is devotional and grave; showing that this must have been Bach's general range of feeling at the time.

Notwithstanding the long line of musical ancestors on which Bach could pride himself, it remained for him to discover that the name of the family could be expressed in musical notes.[378] Subsequently the musical phrase thus generated, which has great melodic individuality and harmonic value, has been used, times without number, for fugues and canons by Bach's sons and pupils, and by later musicians, down to our own time. A well known prelude and fugue in B flat major on the name Bach, ascribed for a long time, as a matter of course, to Bach himself, is now generally considered spurious.[379] We have indeed no manuscript evidence in the case, and there are several other fugues upon the same theme, which have sometimes been considered as Sebastian Bach's. Forkel once asked Friedemann Bach what was the real truth of the case. He answered that his father was not a buffoon, but had only used his name as the theme of a fugue in the "Kunst der Fuge."[380] This sounds very decisive, but it is inaccurate in a double sense. As regards the "Kunst der Fuge" we now know that the piece mentioned by Friedemann has no place in it; and we know from Walther that Sebastian

---

[377] It is found on p. 93 of the B.-G. edition.
[378] It will be remembered that B flat is called B by the Germans, and B natural H (Tr.).
[379] P. S. II., Cah. 4, Appendix.
[380] The story was told by Forkel to Griepenkerl and by him to Roitzsch.

must have written a composition on his name long before this. In the short article upon Sebastian Bach in his Lexicon, he says: "The Bach family must have come from Hungary, and all who have borne the name, so far as can be known, have been attached to music; which may perhaps arise from the fact that even the letters b′a′c″h′ (b′♭ a′c″b′♮) are melodic in their arrangement. (This was first remarked by Herr Bach of Leipzig)." No one can believe that Bach could rest satisfied with the mere observation and not at once use so practicable a phrase as a theme. Walther's lexicon appeared in 1732; but his knowledge of Bach's composition was almost entirely confined to the old times of intimacy at Weimar, and more especially to the first half of that period.[381] From internal evidence the fugue in question must date from the first ten years or so of the eighteenth century. If Bach were not its composer, we are confronted with the strange fact that we possess a fugue, and that a very fine one, by an unknown master, while the authentic work of the most celebrated master of fugue has been lost. In my opinion there is no substantial internal evidence against its genuineness, as long as we admit that the fugue is a youthful work. The prelude is formed on the model of the French overture-form, which Bach used frequently as early as the Weimar period. Passages analogous to that in bars 8 ff, &c., occur in the prelude to the great organ fugue in D major.[382] The working-out of the fugal theme is throughout in Bach's style; the phrases in the second bar occur again in the theme of the B minor fugue in the first part of the "Wohltemperirte Clavier"; and in the "Kleine harmonische Labyrinth," they are even found with the same counterpoint.[383] Passages in thirds, ascending at each repetition, are of quite common use in the compositions of that time, for which Kuhnau's clavier music was the standard; the

---

[381] See Vol. I., p. 396.
[382] P. S. V., Cah. 4 (243), No. 3.
[383] See Vol. II., p. 43. I do not forget that this amount of evidence is not sufficient to decide the authenticity of the work. But there is not enough evidence on the other side to allow of its rejection.

interruption, for the sake of effect, near the end, is a peculiarity of the style of the Northern masters, from whose influence Bach had not yet freed himself. The whole piece with its youthful freshness and happy playfulness agrees well with the character of Bach's earlier Weimar compositions. Of the other anonymous fugues on the name "Bach," the greater part bear evident marks of their not being by Bach himself. One alone :—

has an old-fashioned air about it, reminding us of Buxtehude's great C major fugue.[884] On this account it may possibly be by Bach, and would in that case have been written before the other, about 1707. If the one mentioned by Walther be one of these, as I assume, it must certainly have been the former, not the latter; as may be seen from his writing down the notes as b′ a′ c″ h′, and not bac′h.

The reader will now expect that Bach's organ compositions should again come under our notice. If in Leipzig Bach's labours in the other branches of instrumental music did not diminish, how much less would they cease in the sphere of organ music, his oldest and most particular domain. Moreover, organ music stands in the closest relation to cantatas, Passions, motetts, and masses, since, like them, it is devoted to the service of the church, and a faculty chiefly exerted in the domain of church music must inevitably recur to the organ. In fact, the concerted works, and the organ works written in Leipzig, have much in common. In the first period, when a great freedom and variety of form prevailed in the cantatas, independent forms also preponderated in the organ compositions. In the second period the chorale cantata came distinctly into the foreground,

---

[884] No. XVII. of the first Vol. of my edition of Buxtehude. The anonymous fugue above cited is from Schelble's bequest; I owe my acquaintance with it to Herr Roitzsch.

and at the same time in instrumental composition the master devoted himself principally to the organ chorale.

On the whole, however, the number of organ works written at Leipzig is not large as compared with those written at Weimar, and they are less remarkable for number and variety than for fulness of import. Of the organ works completed by Bach in Leipzig, many must have originated at an earlier period. Two preludes and fugues in C minor and F major give evidence, in their second portions, of dating from the period of his earlier maturity; probably, too, these are not the original preludes.[385] The preludes which Bach added later are so grand as almost to force the fugues into the background. No exact date can be given for these or for the gigantic F major toccata—for this belongs to the same category—or for the C minor prelude, consisting of harmonic and fugal sections worked out in the form of *tutti* and *solo*. And we must also be content to assign the noble toccata and fugue in D minor (called Doric) in general terms to Bach's central period.[386] A prelude and fugue in G major was completed in 1724 or 1725, and a similar work in C major about 1730.[387] The schemes of these, however, point back to a time before the Leipzig period. When, in Weimar and Cöthen, Bach was working into his art the forms of the Italian chamber music, the idea occurred to him of creating an organ-form in three movements analogous to the Italian concerto. He carried this idea into execution only once, however, as we must assume, in Weimar;[388] he does not seem to have found the form of the first movement of a concerto productive enough for an organ piece. It appears once again in the above-mentioned C minor prelude, although in a very much modified form; but the form

---

[385] The fugues have already been noticed in Vol. I., p. 590 f.
[386] P. S. V., Cah. 3 (242), No. 3. B.-G., XV., p. 136 ff.
[387] P. S. V., Cah. 2 (241), Nos. 2 and 1. B.-G., XV., p. 169 ff. and 212 ff. The date is inferred from the condition of the autographs. That of the work in G major has the same authoritative signs of date as the older portions of the Whitsuntide Cantata, " Erschallet ihr Lieder," see App. A. to Vol. II., No. 31. That of the work in C major has M A. See App. A. to Vol. II., No. 44. As to the gradual alterations which the works have undergone, see Rust's preface to B.-G., XV. [388] See Vol I., p. 421 f.

which chiefly asserts itself is that of the prelude, worked thematically. Nevertheless, Bach attempted, in the works quoted in G major and C major, to adhere at least to the triple form of the concerto by inserting, between the prelude and the fugue, middle sections in three parts and of a quieter character. This experiment leads us to assume that both works may have been conceived at the same time; finally, however, he abolished the middle sections, perceiving that his idea could be realised in another way. Thus two pairs of movements remain, which may be said to hold the happy medium in the highest sense, like the toccata and fugue in D minor; they are mature in their form, and nowhere overstep a delightful moderation. A festal character, common to both, develops in the C major work into a feeling of devotion, and in the other into one of rejoicing. The theme of the G major fugue, which was used in the minor in the Cantata "Ich hatte viel Bekümmerniss," has a rhythm which generates a free motive in the prelude.

The composition of the famous A minor fugue, in which science and effect are united in the most perfect manner, also extends apparently over two periods of his life.[389] The first conception of the prelude, which, as compared with Bach's later style, seems to consist only of passages, may, from certain characteristic reminiscences of the school of Buxtehude (see bars 22 ff. and 33 ff.), be referred to a moderately early time. Only four great preludes and fugues are to be regarded as the immediate fruits of the Leipzig period; they are in C major, B minor, E minor, and E flat major, four stupendous creations, in which are embodied the highest qualities that Bach could put into this branch of art.[390] The C major fugue, with its lovely

---

[389] P. S. V., Cah. 2 (241), No. 8. B.-G., XV., p. 189 ff. See also the preface.

[390] P. S. V., Cah. 2 (241), Nos. 7, 10, 9, and Cah. 3 (242), No. 1. B.-G., XV., pp. 228, 199, 236 ff., and III., pp. 173 and 254 ff. The first three, together with the fugues in A minor (P. S. V., Cah. 2 (241), No. 8), C major (Ibid., Cah. 2 (241), No. 1), and C minor (Ibid., Cah. 2 (241), No. 6), are known as the six great preludes and fugues. As they exist united together in one MS. it is possible that Bach may have collected them into one work during the last years of his life.

III.

structure in five parts, rising from the broad foundations of the prelude—like Bach's own artistic greatness from the great middle class of the German people — has a fellow in the C minor fugue in the second part of the "Wohltemperirte Clavier." The latter work is, indeed, in a more unpretending style; but the ingenious interweavings of the theme in direct and contrary motion, and more especially the late entry of the lowest part in augmentation, have the same grand and majestic character as the C major fugue effect.

In the prelude and fugue in B minor he strikes a chord of deep elegiac feeling such as we find nowhere else in the organ works. The prelude, with its firm and close texture, leads us into a labyrinth of romantic harmony, such as has never been constructed by any more modern composer. The fugue is in a vein of quiet melancholy. Bach's power of embodying this feeling in an organ piece in the strictest style, and of keeping it up throughout a work of the longest proportions, would alone secure him imperishable fame. In contrast to this work, in the prelude and fugue in E minor, the whole energy and vitality of the master are displayed. It is a composition not sufficiently described by its present title; it should be called an organ symphony in two movements to give an adequate idea of its grandeur and power. The prelude numbers 137 bars, while the fugue is extended to 231. It is the longest of Bach's organ fugues. The theme is of the greatest possible boldness, and yet, like the rest of the work, in the highest degree dignified. Its date must be between 1727 and 1736.[391]

The prelude and fugue in E flat major, which were published in the third part of the Clavierübung about 1739, belong to the first years of the last period of his life. They form the beginning and the end respectively of that collection which was properly intended to contain only organ chorales. Although so widely separated in position

---

[391] According to the mark M A in the original MS., which is autograph as far as bar 20 (inclusive) of the fugue. (The splendid autograph of the prelude and fugue in B minor, in the possession of Sir Herbert Oakeley, bears the watermark M A, thus showing that it belongs to the same period as the work in E minor.—Tr.)

they have an inward connection, which may be seen in the quiet stateliness common to both, and also in the fact that both are in five parts, besides which Forkel expressly testifies to their connection on the authority of Bach's sons.[392] The elaborately-constructed prelude belongs to the toccata class. And yet, in the portions which are not fugal, there is an individual character which reminds us of the clavier music of Emanuel Bach and of Haydn.[393] The fugue, on the other hand, exhibits in a remarkable degree the Buxtehude form in several sections;[394] in the second and third of these the theme undergoes hardly more than a change in rhythm alone, but its new counterpoints give it quite another aspect. This work may be most fitly called central or intermediate, for it points towards a period yet to come, and at the same time refers to one already past.[395]

That form in three movements which Bach vainly attempted to produce as long as he tried to insert a contrasting middle movement between the prelude and the fugue, he discovered in the six so-called organ sonatas. These are compositions in which the forms of the Italian chamber sonata, as developed by Bach, and of the instrumental concerto, appear united. They form a companion work to the six violin sonatas with clavier obbligato, adhering to the three-part form with even greater strictness, for in the violin sonatas we here and there find chords in the figured bass. In the organ sonatas two manuals take each a part, the third being allotted to the pedal. Bach produced this form under the influence of his chamber music, and it holds a central position between the style of organ and chamber music; it is accordingly best suited to an instrument which

---

[392] See Griepenkerl's preface to P. S. V., Cah. 3 (242), No. 1.
[393] Compare Vol. II., p. 121.
[394] Compare Vol. I., p. 324 f.
[395] A beautiful five-part fantasia with fugue in C minor is unfortunately preserved only in a fragmentary condition, inasmuch as the fugue only reaches to bar 27. Griepenkerl's supposition, which I followed in Vol. I., p. 590, that the fantasia belonged originally to the fugue in P. S. V., Cah. 2 (241), No. 6, can scarcely be held, now that the autograph has come to light. The autograph, in a large powerful style of writing, on fine strong paper, was formerly in the possession of Professor Wagener, of Marburg, and is now in the Royal Library at Berlin.

would give expression to this medium character. That instrument is the pedal clavier with two manuals. The original MSS. distinctly state them to be for that instrument, and the title of "Organ Sonatas" now in use is, strictly speaking, incorrect. In his organ music proper, Bach turned to account much of his chamber music. But he took care not to transfer the forms without alteration, and in their entirety. We possess neither genuine organ sonatas by him nor organ concertos. In contrast to Handel, he never ceased to regard the organ as devoted to the service of the church.

These six sonatas were intended to complete the education of his eldest son, Wilhelm Friedemann, as an organist.[896] They were produced gradually; the first movement of the D minor sonata dates from about 1722; the Adagio and Vivace of the E minor sonata belonged originally to the church cantata "Die Himmel erzählen," of the year 1723; the last movement of the same sonata and the Largo of the C major sonata must have been middle movements inserted between the preludes and fugues in G major and C major respectively, and therefore date from the Cöthen or the Weimar period. The whole collection must have been completed between 1727 and 1733, since in the latter year Friedemann Bach was appointed organist in Dresden; we may with more exactitude put the completion of the work soon after 1727.[897] It is easiest to understand these sonatas if we take them up after the six violin sonatas already mentioned. Fully equal to them in wealth of ideas, in interesting working-out, in masterly treatment of the three-part writing, and in sharpness of contrast between each other, they have a limited individuality consequent on the

---

[896] Forkel, p. 60.

[397] Both the existing original MSS. bear the sign M A. See Appendix A., No. 34. Only one of these is autograph; the other was written as far as to page 48 inclusive, by Friedemann Bach, and from that point by Anna Magdalena, Sebastian having only put in a few additions. For the rest see the preface to B.-G., XV., in which Vol. the sonatas are published, and P. S. V., Cah 1 (240), with Griepenkerl's preface.

more limited powers of expression of the organ tone-material.[398]

The first, second, and fourth parts of the "Clavierübung" contained works which, each in its way, must be regarded as the highest of its kind, and (for the most part) complete in itself. This is equally true of the third part, which contains the greatest number and the most important of the chorale arrangements made by Bach in the last period of his life.[399] We are forced to regard the twenty-one choral pieces of the third part as a whole, to which unity is lent by one poetic idea. The root of the work consists of twelve arrangements of the so-called "Catechism Hymns." For each of the five chief divisions of the Lutheran catechism, and also for confession, one of the finest and best known chorales is chosen; the first is "Dies sind die heilgen zehn Gebot" ("These ten are God's most holy laws"); the second, "Wir glauben all an einen Gott" ("We all believe in one Lord God"); the third, "Vater unser im Himmelreich" ("Our Father which art in Heaven"); the fourth, "Christ unser Herr zum Jordan kam" ("When Christ our Lord to Jordan came"); for confession we have "Aus tiefer Noth schrei ich zu dir" ("From deep distress to Thee I cry"); and for the Holy Communion, "Jesus Christus unser Heiland" ("Jesus Christ our Lord and Saviour"). Each is twice treated, first with pedals and then *manualiter* alone. These are preceded by a two-fold

---

[398] Forkel, p. 60, speaks of yet more sonatas written by Bach, in addition to these six. There are, however, no more extant, except two separate three-part movements in D minor and C minor, which are not complete. The first is found in P. S. V., Cah. 4 (243), No. 14, and the second is in a MS. in the Royal Library at Berlin, containing an Adagio and a fragment of an Allegro. Forkel was probably thinking of the *Pastorale* for two manuals and pedals, which, together with three short clavier pieces which follow, but apparently do not belong to it, has been published by Griepenkerl, on Forkel's authority, as one continuous whole (P. S. V., Cah. 1 (240), No. 3). The *Pastorale* itself, which, as it now stands, begins in F major and ends very unsatisfactorily in A minor, is certainly only a fragment.

[399] B.-G., III., pp. 184-241. The original edition, which appeared in 1739, or at latest, at Easter, 1740, cost 3 thalers; see Mizler, Musikalische Bibliothek, II., p. 156. Kirnberger's remarks on these chorale arrangements (see ante p. 133) will be found in Appendix B., XIV.

arrangement of "Kyrie Gott Vater in Ewigkeit" ("Father of Heaven, have mercy"); and a threefold one of "Allein Gott in der Höh sei Ehr" ("To God alone be glory"). These two chorales—German versions of the *Kyrie* and *Gloria* of the mass—have here a peculiar importance as being substituted in the Lutheran church for the two first numbers of the mass, and sung at the beginning of the service in Leipzig.

The task of glorifying in music the doctrines of Lutheran christianity which Bach undertook in this set of chorales, he regarded as an act of worship, at the beginning of which he addressed himself to the Triune God in the same hymns of prayer and praise as those sung every Sunday by the congregation. And the fact that the chorale "Allein Gott in der Höh sei Ehr" is treated three times has an ecclesiastical and dogmatic reference. For this hymn is sung in praise of the Trinity, to whom also the prayer of the *Kyrie* is addressed; but in the *Kyrie* the three different melodies require as many different treatments, while in "Allein Gott in der Höh sei Ehr" all the verses were to be sung to the same melody. I have said before that the organ chorale of Pachelbel appears as a kind of ideal act of service in a purely instrumental form.[400] It is the same artistic idea which in the organ chorale, as treated by Bach, became so grand and inspiring. It led on to the chorale fantasia, which, by the introduction of the *Cantus firmus*, is transferred to vocal music; and its ultimate outcome is the chorale cantata, the last and highest development of the Pachelbel organ chorale. It has evidently influenced the character of the set of chorales in the third part of the "Clavierübung," in so far as its character may be viewed as a whole. The different parts of this work have indeed no connection whatever in a musical sense, and their poetical connection is only indirect, depending upon an idea derived from without. Bach had long recognised the fact that the ideal of that form could not be completely carried out in instrumental music, and he proved most conclusively that he recognised it, at

---

[400] See Vol. I., p. 112 f.

the very time when he wrote those organ chorales. But he could not abandon the plan once formed, and all the less so as his intention was to sum up here the whole of his life's work.

Bach has here employed the pure form of the chorale fantasia only for the first arrangements of the melodies "Dies sind die heilgen zehn Gebot," "Christ unser Herr zum Jordan kam," and "Jesus Christus unser Heiland." These grand pieces are at the same time eloquent witnesses to his depth of nature, both as a poet and as a composer. Bach always deduced the emotional character of his organ chorales from the whole hymn, and not from its first verse alone. In this way he generally obtained from the poem some leading thought, which seemed to him of particular importance, and in accordance with which he gave to the composition a poetic and musical character of its own. We must follow out his method in detail in order to be sure that we have grasped his meaning. In the hymn for the Holy Communion, "Jesus Christus unser Heiland," the counterpoint, with its broad, ponderous progressions, may, to the superficial observer, seem unsuitable to the character of the hymn. The attentive reader of the words will, however, soon find the passage which gave rise to this characteristic musical phrase. The fifth verse runs thus :—

Du sollt gläuben und nicht wanken,	Hold it true, nor ever waver,
Dass ein Speise sei der Kranken,	That a feast of sweetest savour
Den'n ihr Herz von Sünden schwer,	Granted is to hearts distressed,
Und für Angst ist betrübet sehr.	With their load of guilt oppressed.

Faith, lively and immovable, together with the solemnity of a consciousness of sin, are the two elements which constitute the emotional groundwork of the piece. A comparison with the no less grand earlier arrangement of the same chorale, or with the mystic composition of "Schmücke dich, o liebe Seele,"[401] gives new insight not only into Bach's wealth of imagination, but into his personal character, by revealing the subjective element in the organ chorales. Again, when in the arrangement of the chorale "Christ

---

[401] See Vol. I., pp. 613 ff. and 617.

unser Herr zum Jordan kam" an unceasing figure of flowing semiquavers makes itself heard, it needs no skilled critic of Bach's works to find in this an image of the river Jordan. Bach's real meaning, however, will not reveal itself thoroughly to him until he has read the whole poem to the last verse, in which the water of baptism is brought before the believing christian as a symbol of the atoning Blood of Christ. In the five-part organ chorale "Dies sind die heilgen zehn Gebot," the *Cantus firmus* brought in in canon on the octave sheds a light on the poetic meaning. The idea of bondage to the law, which is thus suggested, is one which we have met with before in the cantata "Du sollst Gott deinen Herrn lieben," although there it was not so fully elaborated. In the composition of the organ chorale Bach certainly had the cantata chorus in his mind, as I have already argued.[402]

Of more frequent occurrence are those less free forms —intermediate between the chorale fantasia and a form approaching nearly to the Pachelbel type—which arise when a motive or theme, derived only from the first line of the melody, pervades the whole piece. I will first mention the longer arrangement of "Vater unser im Himmelreich." Here the melody appears against three parts in counterpoint in canon on the octave. Bach may have intended by this to symbolize the believing, childlike obedience with which the Christian appropriates the prayer prescribed by Christ Himself. The counterpoints with their peculiar movement seem to have a specially importunate character.[403] To the same category belong the two three-part arrangements of "Allein Gott in der Höh," in the second of which, however, the *Cantus firmus* is brought out alternately in different parts, and here and there in canon or with a repetition of the same line. And lastly, the first arrangement of the *Kyrie* in three sections, the majestic structure of which culminates in a grand five-part composition.

---

[402] See Vol. II., p. 431.
[403] In the way in which certain similar phrases are to be played compare ante, p. 49, note 60.

As may be imagined, the pure Pachelbel type, to which Bach was so much indebted in his organ chorales, could not be absent from so exhaustive a work as this. It appears in the penitential hymn "Aus tiefer Noth schrei ich zu dir," both in the four-part arrangement for manuals alone and in the six-part arrangement for manuals and double pedals. It is significant of Bach's manner of feeling that he should choose this particular chorale for the crowning point of his work. For it cannot be questioned that this chorale is its crowning point, from the ingenuity of the part-writing, the wealth and nobility of the harmonies, and the executive power which it requires. Even the northern masters had never ventured to write two parts for the pedals throughout, though they had first introduced the two-part treatment of the pedals, and Bach did both Pachelbel and them full justice in this piece. And not in this piece only; in the fughetta on "Dies sind die heilgen zehn Gebot" we recognise the spirit of Buxtehude, although it has received new birth from Bach's genius. The manner in which the ideals of Bach's youth appear again in this last great work for the organ is characteristic of his art and of his nature. We also feel the influence of Georg Böhm, Bach's favourite model during his "apprenticeship" at Lüneburg; the *Basso quasi ostinato* in the organ chorale "Wir glauben all an einen Gott" points clearly to his style. The piece is a fugue on the first line of the melody, and has nothing whatever to do with the bass. And this makes the reference to Böhm quite clear; for it was not Bach's usual method to write fugues in which the pedal-part neither took any share in the working-out of the fugue nor brought out any *Cantus firmus*, but only repeated from time to time a short, independent phrase.[404] Lastly, that the organ chorale should not be absent in that simplest form, which Bach had elaborated at Weimar and subsequently established by the most splendid examples in the "Orgelbüchlein," he incorporated in the collection the beautiful second arrangement of the "Vater unser."

---

[404] Compare Vol. I., p. 213.

Still he wished to increase the scope of the collection. Not content with producing organ chorales in the most widely different forms, he went back to their original germ and found room even for real chorale preludes, a form that he had never had anything to do with, except, perhaps, in his earliest youth.[405] The organ chorale differs from the chorale prelude in this, that it is an independent work of art, requiring only a previous acquaintance with the melody, while the chorale prelude is nothing more than a preparatory piece, having its central point of interest not in itself, but in the congregational hymn that is to follow it. Accordingly, the treatment of the melody is quite different; the hymn-tune is only referred to, and must never appear entire in the organ piece. Now the second arrangement of "Kyrie Gott Vater in Ewigkeit" in this collection is pre-eminently of this kind; only the first three notes of the chorale are treated in all three parts. In the second arrangement of the baptismal hymn the whole of the first line of the tune is introduced, but no more. But in order to appreciate the almost immeasurable difference between the old chorale prelude and that of Bach, compare the works of this kind composed by Johann Christoph Bach with those just mentioned.[406]

An independent species of chorale prelude is the chorale fugue. And this form, too, had been scarcely touched by Bach during his riper years until now.[407] The third part of the "Clavierübung" contains five such pieces; they are upon the melodies "Allein Gott in der Höh," "Dies sind die heilgen zehn Gebot," "Wir glauben all," and "Jesus Christus unser Heiland." With the exception of the first the fugue theme in each is derived from the first line only. Three of them are fughettas, thrown off with masterly ease, and full of sharply defined character; the smaller work on "Wir glauben all" is in the French style, and the first and most important one is in Buxtehude's manner. The fugue on "Jesus Christus unser Heiland," with which the collec-

---

[405] See Vol. I., p. 219. [406] See Vol. I., p. 100 ff.
[407] Compare Vol. I., p. 605 f.

tion closes, has very brilliant strettos, and, finally, a combination of the theme in its natural form and in augmentation. The great fugue on "Wir glauben all" has been already spoken of. In it two form-ideas are united, that of Böhm's chorale with that of the prelude.

Let it now be decided whether the chorale collection in the "Clavierübung" may rightly be called a comprehensive and self-contained work. With its completion Bach considered his life-work practically over as regards the treatment of the organ chorale. From that time until his death he collected and arranged his older works,[408] but created very little that was new. One of the works collected and arranged by him at this time is that set of six chorales for two manuals and pedals, which he had published between 1746 and 1750 by Johann Georg Schübler, of Zella St. Blasii, near Suhl. They are taken from cantatas written at Leipzig, as can be proved in the case of all but one, which can form no exception to the rest; and, to be rightly appreciated, they must be studied from this point of view.[409]

On the other hand, we have an original composition in the "canonischen Veränderungen" (variations in canon), upon the Christmas hymn "Vom Himmel hoch da komm ich her," which appeared, published by Balthasar Schmidt, of Nuremberg, elegantly engraved.[410] According to a tradition, which we have no reason to doubt, Bach wrote them for the Leipzig Musical Society, which he joined in June, 1747.[411] They were, however, written and even engraved a year before he joined the Society.[412] From the title, "Variations,' it would be easy to draw a false conclusion as to their form. It is, however, valuable, as it teaches us the purpose that Bach had in view in the work. About the year 1700 the name "Variation" was also used for "Partita," which last word originally betokened the separate portions of a collective composition, consisting of pieces in the same key,

---

[408] See Vol. I., p. 655 f.
[409] B.-G., XXV.², II. See App. A., No. 9.
[410] P. S. V., Cah. 5 (244). Abth. II., No. 4.
[411] See ante, p. 25.
[412] See App. A., No. 10.

and was afterwards used for the collective composition itself. So carelessly were the two names used that not unfrequently the first "Partita" of a set of variations—*i.e.*, the theme itself—was called the first variation. Now we know the chorale partita to have been a form adopted by Bach in his youth from Böhm's model.[413] Its superficial and secular nature could no longer satisfy him, and in the chorale collection of the "Clavierübung" there is no reference to it. But the feeling that he must complete his task in the most perfect way possible seems to have given him no rest. The "Variations" upon the Christmas hymn are *partitas*. This is evident from the fact that the work begins at once with the so-called first variation; this is proper to the partita form, and is of frequent occurrence, in Pachelbel, for example, whereas a true variation without a given theme is an impossibility.[414] Thus, this work is to be compared with the youthful works "Christ, der du bist der helle Tag," "O Gott du frommer Gott," and "Sei gegrüsset, Jesu gütig."[415] The same immense advance will then be perceived as in the case of the chorale preludes.

The strict variation-form, which keeps exactly to the length and construction of the theme, was not adhered to in the chorale partitas, though it played an important part in them. In the partitas upon the Christmas hymn Bach quite abandoned it, and with it the adornment of the melody by means of figures more suitable to the clavier style. He also abjured the too secular disruption of the theme by episodes, after the confused method of Böhm, who brings in the theme now here and now there, setting in motion a varying number of parts. Bach's are true organ chorales, of the type which predominates in the "Orgelbüchlein," and what unites them into a whole is the style of canon treatment which is common to all. In the first four partitas

---

[413] See Vol. I., p. 210 ff.

[414] Kirnberger (Kunst des reinen Satzes, II., 2, p. 173) seems to have quite misunderstood the form; he was misled by the term "Variation," and by the fact that in the majority of the partitas the *Cantus firmus* lies in the bass.

[415] See Vol. I., p. 604.

the canon lies in the parts that have the counterpoint—a form with which we became acquainted in the "Musikalisches Opfer." The imitation follows in the octave, the fifth and the seventh, and in the octave with augmentation. But Bach set himself a more complicated task, not merely deriving the counterpoint from the melody, but in the third partita interweaving between the *Cantus firmus* and the two parts in canon a very *Cantabile* melody, freely invented. In the last partita, which properly consists of four complete workings-out connected together, the melody itself appears in canon in contrary motion, and in the sixth, third, second, and ninth. The three concluding bars are of marvellous ingenuity, for in them all four lines of the melody are heard simultaneously in the different parts.

In freedom of movement the partitas on "Vom Himmel hoch" are in no way inferior either to the "Musikalisches Opfer," the "Kunst der Fuge," or even the thirty Clavier Variations. Bach was assured of perfect success, even in the most difficult problems; and he here gives fresh proof that it was not the mere fascination of technical difficulties to overcome that led him to adopt these elaborate forms in his later works, but that his musical sense grew deeper, and imperatively demanded new modes of utterance. These partitas are full of passionate vitality and poetical feeling. The heavenly hosts soar up and down, their lovely song sounding out over the cradle of the Infant Christ, while the multitude of the redeemed "join the sweet song with joyful hearts." But the experiences of a fruitful life of sixty years have interwoven themselves with the emotions which possessed him in earlier years at the Christmas festival, and which he embodied in undying creations in the organ chorales, in the Christmas Oratorio, the Magnificat, and other works, and mingle with the feeling which in his early youth had inspired him to produce those forms of art which he was destined to ennoble for ever. The work has an element of solemn thankfulness, like the gaze of an old man who watches his grandchildren standing round their Christmas tree, and is reminded of his own childhood.

If we found, as a characteristic feature of Bach's vocal

church music, that his genius, after traversing the whole range of art, should at last find its goal in the chorale cantata, it is no less important to notice the way in which he cultivated the organ chorale until the end of his life. His art described a mighty orbit, returning to the point from which it originally started. Handel's course of development is that of a mighty river, resting only when it reaches the great world-ocean. Bach, after traversing all the heights and depths of life, finds his ultimate repose in the peace of home. He does not throw himself into the world, but absorbs it into his own individuality. This subjective character may be traced in his music in all his periods, but it is even stronger in the last period of his life than during the years of his fullest prime. The organ chorale is, of all the forms employed by Bach, the most subjective, and it is that which he used most freely. His devotion to it was proved a few days before his death. The last composition which he undertook, with the help of Altnikol—for his sight was gone—was an elaboration of an early organ chorale "Wenn wir in höchsten Nöthen sein." To the end he consecrated the highest powers of his life to a form, of which the very essence is the joy of praising and praying to God in the congregation. Bach felt, like Augustine, that "Thou hast created us for Thyself, and our heart is unquiet till it finds repose in Thee."

VI.

BACH'S PRIVATE FRIENDSHIPS AND PUPILS; HIS GENERAL CULTURE.—BLINDNESS AND DEATH.

ALL that has been written as to Bach's life, from the year 1723, bears principally on the various aspects of his official position and his relations to the public generally at Leipzig. It has not been possible hitherto to discuss his position during all this time with regard to the musical world at large, nor his domestic life and his personal characteristics as developed in it.

Bach never saw Italy, that Eldorado of the German

musician; indeed, he never crossed the German frontier. Still, he inherited a taste for wandering, and he travelled from Leipzig a good deal, considering the times and circumstances, and thus contributed himself to extend his fame as an artist.[416] He was still Capellmeister at Cöthen when he settled at Leipzig, and in the same year was appointed "outdoor" Capellmeister at Weissenfels. These offices required the person who held them to provide compositions for the use of the respective Courts when called upon to do so, and to present himself from time to time in person. A curious fate presided over Bach's connection with Weissenfels; it has left scarcely a trace to posterity. Not a single composition can be pointed to as written by Bach for the Court of Duke Christian, after his first dedicating a cantata to him from Weimar in 1716; all we know is that this same cantata was subsequently used again for a Court festival at Weissenfels.[417] From this we may be allowed to suppose that after Duke Christian's death in 1736 Bach's services were scarcely ever again claimed for any musical productions. His successor, Johann Adolph II., introduced a very economical régime, with a view to retrieving the fortunes of his family which was deeply in debt. He died in 1746, and with him this Saxon branch became extinct. Bach enjoyed the title of "*Hochfürstlich Weissenfelsische wirkliche Capellmeister*" till his death,[418] so it does not follow that he should have fulfilled any duties even during Johann Adolph's lifetime. However, he frequently visited the Court there between 1723 and 1736. He himself mentions incidentally that he had to quit Leipzig once or twice between 1723 and 1725, *ob impedimenta legitima;* on one of these occasions he had gone to Dresden,[419] and on the second we may suppose he went

---

[416] Forkel is to a certain extent in error when he says (p. 48): "If he had only chosen to travel he might have concentrated in himself the admiration of the world, as indeed even his enemies allowed." The "enemy" must have been Scheibe, but he makes the remark in a different sense. Crit. Mus., p. 62.
[417] See Vol. I., p. 566 ff.
[418] See "Nützliche Nachrichten von denen Bemühungen derer Gelehrten ... in Leipzig." 1750, p. 680.
[419] See Vol. II., p. 220.

to Weissenfels, where he had lately been appointed. In the band under his direction there, and which was not unknown to fame, were his father-in-law, Joh. Caspar Wülcken, and Adam Emanuel Weltig, godfather to his son Ph. Emanuel. Rather more is known regarding Bach's doings as Capellmeister at Cöthen after his removal to Leipzig. It is evident that after the death of his first wife Prince Leopold recovered his strong interest in music, and his second wife, the Princess Charlotte, seems herself to have had a love for it since Bach ventured to dedicate to her a birthday cantata, November 30, 1726.[420] On the 12th September of that year she had presented the Prince with an heir, just as Bach had brought out the first partita of the Clavierübung as *Opus* 1. This coincidence prompted him to copy out the partita with particular care, and offer it on the Crown Prince's cradle with a dedicatory poem; and this poem, which is undoubtedly original, is strong evidence of the friendly and informal relations subsisting between Bach and the royal family. The dedication runs as follows:

"To the Most Serene Prince and Lord | the Lord *Emanuel Ludwig* | Crown Prince of Anhalt, Duke of Saxony, Enger and Westphalia, Count of Ascania, Lord of Bernburg and Zerbst, &c., &c., &c. | these small musical first fruits are dedicated with the humblest devotion | by Johann Sebastian Bach."

The verses are to this effect:

Serene
       and Infant Prince,
              whom swaddling bands encumber
Although thy princely glance argues maturer age,
Forgive me if I dare to wake thee from thy slumber,
   And humbly crave thy grace for this my playful page.
These first fruits of my lyre to thee I dare to bring,
   Thou Prince first born to feel thy royal mother's kiss,
Hoping that she, to thee, the lay may some time sing,
   Since, in the world, thou art a first fruit too, like this.
The wiseheads fain would scare new-comers with a warning,
   Because into the world we come with cries and tears,

---

[420] See Vol. II., p. 158.

As though they could foretell the evening from the morning,
And see thy future destiny beyond the veil of years.
But I will answer them—and say that as these chords
That round thy cradle swell are sweet and clear and pure;
So shall thy life flow brightly on, through all that earth affords
Of joy and harmony—calm, happy, and secure.
May I, most hopeful Prince, play for thy delectation
When all thy baby powers are increased a thousandfold ;[421]
For my own part I only pray for constant inspiration,
 And remain,
   most noble Prince,
     Thy
      humble servant,
       Bach.[422]

The Crown Prince Emanuel Ludwig, however, gave the master no chance of "playing for his delectation," for he died August 17, 1728, and Prince Leopold followed his only son to the grave in a very few months. I have already spoken of the grand funeral ode composed by Bach for his patron and friend; I have been unable to find any further records of his connection with the Court of Cöthen.

Bach had enjoyed a great reputation at Dresden from 1717, and had been well remembered there. He certainly visited it frequently from Leipzig, and often took his eldest son with him.[423] When, in 1733, Friedemann was appointed organist to the Sophienkirche at Dresden, and Bach himself, in 1736, had an appointment at Court there, his visits became more frequent; however, we have direct evidence of only four between 1723 and 1750. The first of these was between 1723 and 1725, and from what Bach says of it incidentally we may infer that he went by a special command from the King.[424]

---

[421] In German *tausendfach*, supplying the rhyme to *Bach*.

[422] The existence of this autograph was not known to the world till February, 1879, when the possessor of it sent a description of it and a copy of the dedication and verses to the Magdeburg *Zeitung*. It was subsequently copied into the *Berliner Fremdenblatt* for February 20, 1879. The steps I then took to obtain a sight of the autograph failed of any result; I had no answer to my application, and I cannot, therefore, be responsible for the accuracy of the communication. Any doubt, however—though under the circumstances justifiable—may, in my opinion, be allowed to vanish when we consider the form and purport of the documents, which seem to me a guarantee of their genuineness.

[423] Forkel, p. 48.   [424] See Vol. II., p. 220.

In 1731 the first performance of *Cleofide* was given, and this tempted him to Dresden; on July 7 of that year Johann Adolph Hasse and his wife Faustina arrived from Venice with an introduction to the Capellmeister, and his opera, in which Faustina played the principal part, was given for the first time September 13. This was a great event in Dresden, and indeed of supreme importance to the Italian opera in Germany. The enthusiasm for Hasse's music and Faustina's singing knew no limit; one reporter says that to praise this pair in worthy terms is as vain as to try to light a candle at the sun. Bach proposed to give what would in these days be called an organ recital in the Sophienkirche on the following day at three o'clock. It does not appear that any member of the Court honoured him by attending it; but the whole band—including Hasse no doubt —were present, and Bach's success among his fellow artists was great and complete. Even the public prints took notice of it, which is certainly significant considering the impression made by *Cleofide*; and a rhymester of the day spoke of the marvellous power of his "nimble fingers" as transcending that of Orpheus.

After Bach, in 1733, had dedicated the *Kyrie* and *Gloria* of the B minor mass to King August III., and presented it personally in Dresden, and then, three years later, after repeated applications, had obtained the style and title of *Hofcomponist*, he again performed there for a few days, December 1, 1736, from two till four in the afternoon on the new organ by Silbermann in the Frauenkirche. On this occasion the company did not consist exclusively of his band; many distinguished persons met to admire the master.[425]

Among these was Baron Hermann Carl von Kayserling, who, a few years later, came into closer contact with Bach, and who must, even at this time, have known him, and have been favourably disposed towards him. The patent conferring on Bach the coveted title was ready by

---

[425] Fürstenau, Zur Geschichte der Musik am Hofe zu Dresden. II., pp. 171 and 222.

November 19, and given to the Baron to hand on to Bach on November 28. Bach, however, had known beforehand of the contents of the document, and immediately prepared to set out for Dresden and express his thanks in person, otherwise he could not have given his performance on December 1, and Kayserling's position as intermediary shows that his interest in Bach was already an understood thing. In him, indeed, Bach found a wealthy patron of high rank and wide interests. Before coming to Dresden as ambassador from St. Petersburg, December 13, 1733, he had held the president's chair at the Imperial Academy of Sciences at St. Petersburg. King August III. raised him to the rank of Count, October 30, 1741, and he remained at Dresden till 1745. As a "great lover and connoisseur of music" he was fond of collecting round him all the most remarkable artists of the city. Pisendel, Weiss, and Friedemann Bach were admitted to his house, and musicians on their travels esteemed it an honour to be introduced there.[426] All these circumstances must have contributed to make Bach's connection with the artist world of Dresden both pleasant and intimate; and that it was so was a well-known fact, even in the years from 1727 to 1731, so that trustworthy and interesting reports as to the doings of the Dresden band reached Leipzig almost every day.[427] Pisendel, who, so early as 1709, had visited Bach in Weimar in the course of a journey from Anspach to Leipzig, devoted much attention to the *Viola pomposa* invented by Bach. He liked this instrument for accompaniments. When Franz Benda came once on a visit to Dresden, in 1738, music was carried on at Weiss's house from early in the afternoon till midnight; Benda and Pisendel themselves played no less than twenty-four violin soli, while Weiss, between these, played eight or ten sonatas on the lute.[428]

One of Pisendel's best scholars was John Gottlieb Graun, and even Bach had such confidence in him that when, in 1726, he was called away from Dresden to be Concert-

---

[426] Hiller, Lebensbeschreibungen, p. 45.—Fürstenau, op. cit. p. 222, note.
[427] Scheibe, Uber die musikalische Composition, prof. p. lix.
[428] Hiller, op. cit., p. 45.

meister to Duke Moritz Wilhelm, of Sax-Merseburg, he entrusted his son Wilhelm Friedemann to his teaching for a time.[429] He was attracted to Zelenka, the worthy pupil of Fux, by his serious taste chiefly for church-music, and in this respect he no doubt set him above Hasse. Friedemann had to copy out a *Magnificat* by Zelenka, for the use of the St. Thomas's singers.[430] We know no details as to Bach's intercourse with other Dresden celebrities in art, such as Buffardin, Heinichen, Hebenstreit and Quantz, but his acquaintance with Hasse and his wife was founded on sincere and reciprocal admiration; they visited Bach several times in Leipzig.[431]

Hamburg, the goal of Bach's musical pilgrimage from Lüneburg, and of which he retained a keen recollection from the year 1720, he visited in 1727, so far as we know for the last time. Jacob Wilhelm Lustig, the son of an organist in Hamburg, a pupil of Mattheson and of Telemann, and himself subsequently an organist of mark at Gröningen, at that time heard him play. The words in which he has recorded the fact are eloquent in their simplicity: "There he heard great musicians, nay, Herr Bach himself." Everything that even great players could do appeared weak and inferior when compared with that one supreme ideal.

How Bach and Mattheson got on together at this time cannot be known; Mattheson's feeling against him has already been spoken of elsewhere. In 1731 he once more asked him to contribute an autobiographical notice to the "Ehrenpforte," and as Bach did not comply he omitted him altogether from that work, because "he had had an unreasonable objection to giving any accurate and systematic account of the incidents of his life." Now, Handel and Keiser did just the same, but they were, nevertheless, admitted to that temple of honour. Telemann, the friend of his youth—who, if he had not changed his mind, might four years before have been comfortably installed as cantor of St. Thomas's—did not pass Bach over in this way. In

---

[429] Marpurg, Hist. Krit. Beyträge, I., p. 430.
[430] Still extant, score and parts, in the Library of the Thomasschule.
[431] Forkel, p. 48.

his musical periodical, "Der Getreue Musik-Meister," published in 1728, he included a beautiful canon in four parts, of the most elaborate character, composed by Bach at Hamburg in 1727, and dedicated to Dr. Hudemann. Lustig also made acquaintance with this canon, mastered it and solved it, and twelve years later Mattheson included this solution in his "Vollkommene Capellmeister." We saw, when speaking of the cantata "Phœbus and Pan," how Hudemann showed his appreciation of the honour done him.[432]

Bach also revisited his native province of Thuringia from Leipzig. His connection with the Court of Weimar was at an end after Duke Wilhelm Ernst, in a pedantic whim, had appointed Samuel Drese's son, a quite second-rate musician, to be Capellmeister, while Bach had every right to the place. It is not even probable that any relations existed between Bach and the Court of Weimar so long as that duke remained alive; however, on the accession of Duke Ernst August (1728-1748), they were renewed. This sovereign prince, who in many respects was singularly and radically unlike his uncle, seems, like his half-brother, Joh. Ernst, to have inherited his love of music from his father, in whose service Bach had been for a short time in 1703. He was a persevering violin player, often beginning in the morning before he was up, and seems to have taken much pleasure in Italian opera music, like Duke Christian of Weissenfels.[433] One of the many purposes which the cantata

[432] At the time when I wrote Vol. I., p. 632 (of the original German. A passage omitted by the author's desire from Vol. II., p. 21, of this translation), Bach's visit to Hamburg in 1727 had escaped my notice; but it is to be inferred from the passage in Marpurg, Krit. Briefe, II., p. 470, where we have an account by Lustig of his own life, and this autobiography is chronologically arranged. The younger Kunze was born in 1720, and Lustig was teaching him in 1724-5; then followed lessons in composition, which he learnt from his father and Telemann. In 1728 Lustig went to Gröningen, but he had previously heard Bach play in Hamburg, and, as it was in 1727 that Bach dedicated the above-mentioned canon to Hudemann, he must have been in Hamburg in that year, and have written the canon there. This is evident also from Mattheson, pp. 412-13; from Mizler, Mus. Bib. III., p. 482; and Marpurg, Abhandlung von der Fuge, Part II., p. 99 (Tab. XXXIII., fig. 2). Hilgenfeldt also printed the canon without the solution, as the last musical supplement to his work.

[433] See the notices borrowed from the memoirs of Baron von Pöllnitz in Beaulieu-Marconnay, Ernst August, Herzog von Sachsen-Weimar-Eisenach, Hirzel, Leipzig, 1872, pp. 143-4 and p. 104.

"Was mir behagt" must have served was the keeping of Ernst August's birthday. Bach's eldest son has recorded for us that this Duke was "attached by sincere affection to the artist," no less than Leopold of Cöthen and Christian of Weissenfels.

Erfurt also, one of the old gathering-places of the Bach family had great attractions for Bach as he advanced in years. A cousin of his, Joh. Christoph, son of Ægidius Bach, was the leading member of the town musicians of Erfurt when Sebastian once more sought him out in the time-honoured town.[434] This must have been after 1727. Adlung was at that time organist at the Prediger Kirche, an office which Johann Bach had held in his time. Adlung on this occasion made great friends with Sebastian Bach; he made many inquiries as to his early life, and begged him to play to him on the clavier, in all of which the great master showed the polite amiability which characterised him on such occasions.[435]

Bach must have made some other journeys about 1730 and in July, 1736. It was on the former occasion that he was so severely called to task by the Leipzig Council[436] for having neglected the proper formalities in asking leave. The latter journey immediately preceded his great quarrel with Ernesti.[437] We do not know whither he went on either of these occasions, but in the course of the quarrel we learn from Ernesti's memoranda that Bach must have been absent from Leipzig tolerably often, since Ernesti mentions how he was accustomed to supply his place in such cases.

In later years Bach, as might be expected, moved about less, and his liking for "a quiet and domestic life, and a constant and undisturbed devotion to his art,"[438] preponderated. He could scarcely make up his mind, perhaps, to his last great expedition, which, however, had the most

---

[434] See Vol. I., p. 27.

[435] Adlung, Anl. zur Mus. Gel., p. 691. He does not mention any date, but it must certainly have been after 1727, since it was not till the end of that year that Adlung came to Erfurt from Jena. Adlung says "Herr Bach came to see *us*," whence we may safely infer that he was intimate already with the Bachs of Erfurt.

[436] See Vol. II., p. 243.   [437] See ante, p. 6.   [438] Forkel, p. 48.

important results of any. In 1740 Emanuel Bach had been made Capellmusiker and accompanist to Frederick the Great, and as the most important members of his band—the two Grauns, Franz Benda, Quantz, Nichelmann, and probably Baron—were all personally acquainted with Bach, and some of them had been his pupils, the king must of course have frequently heard him spoken of. The tone in which this was done excited his wish to see and hear the great artist himself. Emanuel sent word of this to Leipzig, but his father did not feel inclined to take the hint. It was not till the King became more and more urgent that he could make up his mind to set out, in May, 1747. He took Friedemann with him, and so must have taken the route viâ Halle, and on Sunday, May 7, he reached Potsdam. It was the custom to give a State Concert every evening from seven to nine, in which the king himself performed as soloist on the flute.[439] Of all that now followed we have a detailed account from Emanuel and from Friedemann.[440] Just as the king was about to perform his flute solo, a list was brought to him of the various strangers who had that day arrived. With his flute still in one hand, he glanced through the paper; he turned to the assembled band saying, with some excitement: "Gentlemen, old Bach is come!" His flute was laid aside, and Bach sent for at once to come to the château. He had put up at Emanuel's house, and he was not even allowed time to put on his black Court-dress; he had to appear at once in his travelling costume, just as he was. Friedemann tells us that his father having apologised somewhat at length for the deficiencies in his dress, the king bade him make no excuses, and that then a conversation began between the king and the artist.[441] Frederick had a high opinion of

---

[439] Marpurg, Hist. Krit. Beyträg. I., p. 76. [440] Forkel, p. 9.

[441] It may be considered an unsettled question whether in this narrative Friedemann has not given the reins to his imagination. Spener's Zeitung—which was the first to direct attention—reports the matter as follows:— "May 11, 1747. His Majesty was informed that Capellmeister Bach had arrived in Potsdam, and that he was in the king's antechamber, waiting his Majesty's gracious permission to enter, and hear the music. His Majesty at once commanded that he should be admitted."

Silbermann's pianofortes, and Bach himself had had some share in perfecting these instruments.[442] The king had several, and Bach was required to try them, and improvise upon them. On the following day Bach played the organ in the Church of the Holy Ghost, at Potsdam, before a crowded audience; but the king does not seem to have been present. However, he commanded his presence at the château again in the evening, and desired to hear him play a six-part fugue, that he might learn to what a pitch the art of polyphonic treatment could be carried; Bach was to choose his own theme, since every subject is not fitted for treatment in so many parts, and he won the king's complete approval by his performance.

From Potsdam he also went to Berlin, and visited the opera-house built there in 1741-3 by Knobelsdorf. Nothing was then being played there, however. For regular performances only Mondays and Fridays in December and January were appointed, and besides these March 27, as being the Queen Dowager's birthday.[443] But Bach was interested in any place dedicated to musical purposes. It seems almost fabulous when the historian of this wonderful man is obliged to mention one more of his many gifts; but, it is nevertheless true that Bach's keen judgment had penetrated the mysteries of the conditions of building which were favourable to acoustics; he detected everything that was advantageous or detrimental to musical effects in the opera-house at Berlin, and, without hearing a note of music in it, he saw at a glance all that others had learned by experience. He also pointed out to his companions, in the dining-room attached to the opera-house, an acoustic phenomenon which, as he supposed, the architect had probably not intended to produce. The form of the arches betrayed the secret to him. When a speaker stood in one

---

[442] See Vol. II., p. 46.
[443] See Marpurg, Hist. Krit. Beyträge. I., p. 75. Brachvogel relates, in his Geschichte des Königl. Theaters zu Berlin (Berlin, Janke, 1877), Vol. I., p. 129, that Bach played at a State Concert with Signora Astrua; but, in point of fact, Astrua did not appear till August, 1747, in a *pastoral* performed at Charlottenburg. See Marpurg, op. cit., p. 82

corner of the gallery of the hall—which was longer than square—and whispered against the wall, another person, standing in the corner diagonally opposite, with his face to the wall, could hear what was said though no one else could. Bach detected this at a glance, and experiment proved him to be right. The authority who records this adds that Bach could calculate accurately how a great composition would sound in a given space.[444] It is not superfluous to insist on this, since many might suppose that the master was so wholly absorbed in his own world of music as to disregard the sensible and physical effect of his works.

Bach had won the admiration of the Court by his extempore performance, but he had not been fully satisfied with himself. The thema for the fugue proposed to him by the king pleased him so well that he pledged himself to use it for a more extended composition, and do himself the honour of having it engraved to offer to his majesty. The result of this project was the "Musikalisches Opfer," of which the merit, as a work of art, has been duly pointed out.[445] How eager he was to fulfil his promise and how highly, therefore, he must have valued the great king's approval, is proved by the haste with which the "Musikalisches Opfer" was completed. The dedication which he prefixed to it is worthy of consideration.

MOST GRACIOUS KING,
I herewith dedicate to your Majesty, with the deepest submission, a musical offering, of which the noblest portion is the work of your Majesty's illustrious hand. It is with reverential satisfaction that I now remember your Majesty's very special royal favour, when some time since, during my stay in Potsdam, your Majesty condescended to play the theme for a fugue to me on the clavier, and at the same time graciously commanded me to work it out then and there in the royal presence. It was my humble duty to obey your Majesty's command. But I immediately perceived that, for lack of due preparation, the performance was not so successful as so excellent a theme required. I accordingly determined, and at once set to work, to treat this really royal theme more perfectly, and then to make it known to the world. This undertaking I have now carried out to the best of my ability, and

---

[444] Forkel, p. 20.   [445] See ante, p. 191 ff.

it has no end in view but this very blameless one—to exalt—though in only a trifling matter, the fame of a Monarch whose greatness and power must be admired and respected by all, and particularly in music, as in all the other sciences of war and peace. I make so bold therefore, as to add this most humble petition—That your Majesty will condescend to grant this present little work a gracious reception and to continue to vouchsafe your gracious favours to
your Majesty's
most obedient humble servant,
LEIPZIG, *July* 7, 1747. THE AUTHOR.

Through all the conventional phrases of devotion, we can see in this dedication a dignified self-reliance, arising from the feeling, not merely of having enjoyed the favour of a great king, but of having been appreciated by him as a truly great artist.

In the first instance, as was but natural, it was always Bach's eminence as an organ player which compelled the admiration of the world; according to Telemann, even during his lifetime it won him the epithet of "The Great."[446] But after about 1720 his compositions seem to have become rapidly and widely known; principally, it is true, his instrumental works, and chiefly those for the clavier and organ. His vocal compositions are but rarely mentioned. Mattheson alludes to some in 1716, and in 1725 he criticises the cantata "Ich hatte viel Bekümmerniss." The adjuvant or deputy in Voigt's "Gespräch von der Musik" boasts, in 1742, that he "already possesses three collections of church cantatas" by the most celebrated composers, as Telemann, Stölzel, Bach, Kegel, and several others.[447] The secular "Coffee" cantata seems to have reached Frankfort in 1739.[448] But, as a general rule, Bach's concerted vocal compositions were partly, no doubt, too difficult; partly, too, he had failed to hit the sentiment of his time, which had found its ideal in Telemann. It is worth noting that Samuel Petri, who was a capital musician,

---

[446] See a Sonnet by Telemann on J. S. Bach in the "Neueröffnetes Historisches *Curiosit*äten-*Cabinet*. Dresden 1751, p. 13; reprinted by Marpurg, Historisch-Kritische Beyträge, I, p. 561.
[447] Gespräch von der *Musik*, zwischen einem Organisten und Adjuvanten. (Talk about music, between an organist and his deputy.) Erfurt, 1742, p. 2.
[448] See Vol. II., p. 641.

besides being a pupil of Friedemann Bach's and a great admirer of Sebastian's instrumental works, never mentions him as a writer of cantatas. "In Telemann's time," he says, "several composers sought to distinguish themselves in the church style, as Stölzel, Kramer, and Römhild, who was Capellmeister to the last Duke of Sax-Merseburg, besides others whose talents were too moderate for them to compare with Telemann."[449]

It was, of course, inevitable that Bach's constantly growing fame in the outside world must react on his position in the town where he was settled; artists and lovers of music alike, illustrious and unknown, sought to establish relations with him from abroad, and made pilgrimages in person to the house in St. Thomas's Church-yard. Emanuel Bach speaks from his own knowledge when he writes in his autobiography: "No master of music would willingly pass through this town (Leipzig) without making my father's acquaintance and obtaining permission to play to him. My father's greatness in composition, and in organ and clavier playing, which was quite remarkable, was too well known for any musician of importance to neglect the opportunity of making that great man's acquaintance when it was in any way possible."[450]

Among the more illustrious lovers of music, and next to Count Sporck and Baron von Kayserling, Georg Bertuch holds a prominent place. Bertuch was born at Helmershausen, in Franconia; he studied in Jena, and there became acquainted with Nikolaus Bach, with whom he purposed making a journey into Italy. In 1693 he held a disputation in Kiel on a legal and musical treatise entitled *De eo quod justum est circa ludos scenicos operasque modernas* (Of the law concerning stage plays and modern operas), which was printed at Nuremberg in 1696. He subsequently entered on a military career and became a General and Governor of the fort of Aggershuys, in Norway.[451] He was so zealous in the pursuit of music that in 1738 he came before the

---

[449] Petri, Anleitung zur praktischen Musik, p. 99.
[450] Burney, Diary III., p. 201.   [451] Walther, Mus. Lex., p. 90.

public with twenty-four sonatas in all the major and minor keys. He seems to have sent a copy to Bach; at any rate, he wrote him a letter in which he extolled the German composers as superior to all others, appealing to Lotti for evidence since he, though esteeming his own countrymen for their talents, did not value them as composers, but was of opinion that the real composers were Germans.[452] We also learn—though not, it is true, from a perfectly trustworthy authority—that Bach was on intimate terms with a noble and wealthy family of Livland, with whose eldest son (who had studied in Leipzig) Emanuel Bach was to make a journey through France, Italy, and England; this plan, however, came to nothing, in consequence of Emanuel's appointment by the Crown Prince of Prussia.[453] In Bach's later life he was also on friendly terms with a Count von Würben, whose son seems to have studied in Leipzig in 1747.[454]

We must mention a few of the musicians who visited Leipzig to see Bach, Franz Benda, a Bohemian by birth, made acquaintance with him in 1734, on his way from Berlin to Bayreuth. Johann Christian Hertel, of Swabia, who at Weimar, in 1726, had taken part in the mourning ceremonial music which Duke Ernst August had had performed on the death of his wife, took the route through Leipzig to Dresden, visited Bach and persuaded him to play to him.[455] Johann Francisci, of Neusohl, in Upper Hungary, came to Leipzig, 1725, at the time of the Easter Fair, "and was so happy as to make the acquaintance of the famous Capellmeister Bach, and to derive benefit from his skill.[456] Balthasar Reimann, an admirable organist of Hirschberg in Schleswig, whose praises were often sung by his fellow

---

[452] Mizler, Mus. Bib., Vol. I., Part IV., p. 83.
[453] Rochlitz, Für Freunde der Tonkunst, Vol. IV., p. 185 (third edition). Rochlitz obtained his information from Doles, Emanuel Bach's friend. Was not this noble family that of Baron von Kayserling?
[454] In the collection of autographs belonging to Herr Ott-Usteri, of Zürich, there was in 1869 a receipt, dated Leipzig, Dec. 5, 1747, and signed *Joh. Sebast. Bach*, for the sum of 1 rthlr., 8 groschen, paid by the "Herr Grav von Werben" (*sic*) for some work for the clavier, the title of which is illegible.
[455] Hiller, Lebensbeschreibungen, pp. 44 and 156.
[456] Mattheson, Ehrenpforte, p. 79.

countryman, Daniel Stoppe, tells us that between 1729 and 1740 he travelled to Leipzig, at the expense of a noble patron, to hear the celebrated Joh. Sebastian Bach play. "This great artist received me amiably, and so enchanted me by his uncommon skill that I have never regretted the journey."[457] Georg Heinrich Ludwig Schwanenberger, violinist in the band of the Duke of Brunswick, seems to have stood in the most confidential intimacy with Bach's family.[458] He was in Leipzig in October, 1728, just when one of Bach's daughters was baptised; one of her godfathers was Johann Caspar Wülcken, and he not being able to be present, Schwanenberger stood as his proxy.

It is to the visit of an illustrious musician that we owe a fine canon written by Bach in the last year but one before his death. The master has surrounded it with ingenious Latin sentences, which remind us of the inscriptions in the "Musikalisches Opfer." The whole is as follows:—

"*Fa Mi, et Mi Fa est tota Musica*

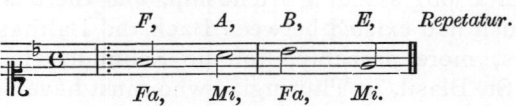

Canon super Fa Mi, a 7. post Tempus Musicum.

Lipsiæ d. 1 *Martii*
1749.

Domin Possessor
Fidelis Amici Beatum Esse Recordari
tibi haud ignotum : itaque
Bonæ Artis Cultorem Habeas
verum am Icum Tuum."

---

[457] Mattheson, op. cit., p. 292.
[458] See Chrysander, Jahrb. für Mus. Wissensch., Vol I., p. 285.

It is a canon in seven parts on a *Basso ostinato* f' a' b' e'. As these notes by the rules of solmisation represent the syllables *fa mi* twice repeated, inasmuch as the two middle notes belong to the sixth hexachord and the others to the fifth, it was possible for Bach to say that the canon was written on *fa mi* (or *mi fa*). Each of the parts of the canon enters a double bar (*tempus musicum*) after the foregoing part. We also observe in the inscription written over the *Basso ostinato*, and again in the second line of the dedicatory subscription—as an acrostic—the name of Faber (Schmidt), while the last line but one yields the name of Bach. The prominent letters I T in the last line indicate *Isenaco-Thuringum* (Eisenach in Thuringia). Who the individual may have been to whom Bach dedicated this work with its graceful arabesques can only be guessed. Balthasar Schmidt, of Nuremberg, may have been the man, the publisher of Sebastian and of Emanuel Bach's works, and himself a skilled musician. Still, the words of the dedication lead us to infer a long standing friendship, and there is no evidence that such had existed between Bach and Balthasar Schmidt. It was, more probably, Johann Schmidt, the organist at Zella St. Blasii, in Thuringia, who must have been of about the same age as Bach. In 1720 he had been the teacher of Johann Peter Kellner, who has recorded that he was justly famous for his remarkable skill.[459] He probably is the same person as the organist Johann Ch. Schmidt who copied out a clavier prelude by Seb. Bach for his own use, November 9, 1713. If so, the connection between him and Bach was a very old one. It is true that this Johann Schmidt resigned his place as organist to his son Christian Jakob in 1746; but there is nothing to show that he died at that time, and when we reflect that it was in the last years of his life that Bach had two of his works engraved and published by J. G. Schübler, of Zella,[460] we have good ground for

---

[459] Marpurg, Hist. Krit. Beyträge, I., p. 441. A composition by him exists in the Library of the R. Inst. for Church Music at Berlin.

[460] This was probably Johann Wolfgang Georg Schübler, son of a gunstock maker, of Zella. There is a composition by Schübler in a MS. Vol. of Miscellanies in the Royal Library at Berlin (see Buxtehude, Orgel-Comp.,

assuming that these dealings with an unknown man in a small and out-of-the-way place (otherwise quite inexplicable) may have been due to the instrumentality of Johann Schmidt. Of far more importance as to Bach's private life than these incidental meetings, however frequent, with musicians from outside, were the pupils who crowded round him from far and near. Of the Mühlhausen, Weimar, and Cöthen periods we have only mentioned Schubart, Vogler, Tobias Krebs, Ziegler, Schneider, and Bernhard Bach as among those who enjoyed Bach's instruction. It may be partly accidental that we do not know of a greater number; still, it is certain that it was not till he was in Leipzig that he was busiest as a teacher. That strong feeling of family coherence which always existed in the members of the great Bach clan prompted the younger men of the different branches to adopt Leipzig by preference not merely for musical but for general study, so soon as Sebastian was settled in that town. On May 7, 1732, Samuel Anton, the eldest son of Joh. Ludwig Bach of Meiningen, was entered on the books of the University there. He lived in Sebastian's house, and formed a friendship with Emanuel, who had already been a student for six months. Samuel Anton was a man of great and various talents; he pursued art with some success, and studied music with Sebastian's assistance to such good purpose that he afterwards rose to be Court organist at Meiningen.[461] A few years later Johann Ernst, the son of Bernhard Bach of Eisenach, came to the Thomasschule, and afterwards studied law at the University. His relations with Sebastian as his pupil are attested by the copy made by him of twelve concertos by Vivaldi and adapted by S. Bach to the clavier. In the spring of 1739,

---

Vol. I., pref., p. IV., No. 10). Marpurg has given Bach's canon without the Latin text in his Abhandlung von der Fuge, Pt. II., Tab. XXXVII., Figs. 6 and 7, see, too, p. 67. I discovered it entire and perfect in a beautiful copy of the last century among the papers left by Grasnick, from which it was obtained for the Royal Library at Berlin. The analysis is given in the musical supplement to this Vol., No. 4.

[461] See Vol. I., p. 10, and the lists of the Leipzig University for 1732.

came also Johann Elias Bach, already at that time the well-paid Cantor of Schweinfurth, to matriculate as a student of theology, and to advance himself in music under Sebastian. I shall presently have occasion to show that he retained a grateful remembrance of his illustrious relative.[462]

In the list of those of his disciples who were not related to him by blood, Heinrich Nikolaus Gerber claims the first place. He was born in 1702 at Wenigen-Ehrich in Schwarzburg; had been sent to the Gymnasium at Mühlhausen, and there had often heard the gifted but degenerate Friedrich Bach.[463] In 1721 he left Mühlhausen and came to Sondershausen, and in May, 1724, entered as *studiosus juris* at the University of Leipzig. He purposed at the same time to pursue the art of music, and was wont to sign himself in these years: *litterarum liberalium studiosus ac musicæ cultor*. His veneration for Bach was so great that for a year he could not find courage enough to ask him to teach him. But there was at the time a musician named Wilde in Leipzig—the same, probably, who in 1741 was Kammermusiker to the Emperor at St. Petersburg—and who had made a name by the improvements he had effected in the construction of various instruments; he constituted himself the intermediary, and took Gerber to Bach. "Bach received him, as a native of Schwarzburg, with particular kindness, and ever after called him his fellow-countryman. He promised to give him the instruction he craved, and at once asked him if he had been diligent in playing fugues. At the first lesson he set his *Inventiones* before him; after he had studied these to Bach's satisfaction he gave him a series of suites, and then the *Temperirte Clavier*. This Bach played through to him three times with his inimitable skill, and he accounted those the happiest hours of his life when Bach, under pretence of not being in the humour to teach, would sit down to his excellent instrument, and the hours seemed to be but minutes." Gerber left Leipzig in 1727, and only once returned there, about 1737, to see his beloved master. In 1731 he was

---

[462] See Vol. I., p. 156, and the lists of 1734.   [463] See Vol. I., p. 141.

appointed Court organist at Sondershausen, and did his great teacher honour as an artist at once gifted and modest.[464]

Johann Tobias Krebs, who had been Bach's diligent and gifted pupil during the Weimar period, sent no less than three sons, one after another, to the Thomasschule. The second, also Johann Tobias, came there in 1729, aged thirteen. Bach's verdict, when he had examined him, was that "he had a fine strong voice and good method" (see ante, p. 66); but with him music seems gradually to have sunk into the background. In 1740 he left the school with honours, and by 1743 was *Magister philosophiæ* of the University, and subsequently rector or warden of the school at Grimma. The third son, Johann Carl, seems also to have devoted himself to the learned sciences, and quitted the school with honours in 1747.[465] The most distinguished musician of the three was Johann Ludwig, the eldest. He was born at Buttestädt, February 10, 1713; was at the Thomasschule from 1726 to 1735; and then studied for two years at the University. Bach was on particularly confidential terms with this favourite pupil, whose musical talents he admired while he esteemed his learning.[466] He is said to have remarked in joke, "He is the one only crab (Krebs) in this brook (Bach)."[467] He made him cembalist in the musical society,[468] and recommended him to Professor Gottsched as teacher to his wife,[469] and even took pleasure in his compositions.[470] When Krebs left the school, Bach gave him the following testimonial:

The bearer of this, Herr Johann Ludwig Krebs, has asked me, the undersigned, to assist him with a testimonial as to his performances on our foundation. As I have no reason to refuse him, and can say this much, that I am persuaded that I have brought him to be a musician

---

[464] Gerber, Lex. I., col. 490.
[465] Nützliche Nachrichten, &c. Leipzig, 1740, p. 67; 1746, p. 164; 1747, p. 289.
[466] Köhler, Historia scholarum Lipsiensium.
[467] J. F. Reichardt, Mus. Almanach. Berlin, 1796.
[468] Gerber, Lex. I., col. 756.
[469] See the biography prefixed to her poems, "Der Frau Luise A. V. Gottschedinn sämmtliche kleinere Gedichte." Leipzig, 1763.
[470] See Tonhalle, Organ für Musikfreunde. Scr. 1869, p. 831.

who has distinguished himself among us, in so far as that he is skilled on the clavier, violin, and lute, and not less in composition, so that he need not be ashamed to be heard, as will be found by experience. I therefore wish him God's help to gain him advancement, and give him, to this end, my best recommendation.

Leipzig, August 24, 1735.
JOHANN SEBASTIAN BACH.
Capellmeister and *Director Musicæ*.[471]

In April, 1737, Krebs became organist at Zwickau, and a few months later Linke, the organist of Schneeberg, wrote to a friend: "A short time since I had the honour to see and to hear Monsieur Krebs, the new organist of Zwickau, a very good organ and clavier player; I must confess that what this man does better than others as an organist is something remarkable, and he is the creation of Bach."[472] In April, 1744, Krebs was appointed organist to the Castle at Zeitz, where he and the younger Schemelli worked together,[473] and in 1750 he became Court organist at Altenburg, where he died in 1780. He was, beyond a doubt, Bach's most distinguished pupil on the organ, and in every way one of the greatest of those who survived the master. Vogler may have been his equal in playing, but as a composer was far below him.

A younger artist who had received his first training from Bach in Cöthen, and who, after an interval, once more became his disciple in Leipzig, was Johann Schneider. He competed with Vogler in 1730 for the post of organist to the church of St. Nicholas, and succeeded in winning the prize, as I have already mentioned.[474] Here I need only mention his name in order not to omit all reference to an excellent musician of whom a witness says, that his organ playing was in such good taste that, excepting only his master, Bach, there was no better to be heard in Leipzig.[475]

---

[471] From the archives of Zwickau. Text quoted by Dr. Herzog in the *Zwickauer Wochenblatt*, March 26, 1875.
[472] Voigt Gespräch von der Musik, p. 103.
[473] See ante, p. 109.
[474] See Vol. II., p. 262.
[475] Mizler, Mus. Bib. III., p. 532.

In 1732 Georg Friedrich Einicke came to Leipzig; born in 1710 at Hohlstedt in Thüringia, where his father was cantor and organist. He followed the academic course till 1737; in music he formed himself on Bach, though at the same time he made further progress by means of his intercourse with Scheibe. He won a name by his compositions, and seems to have been kindly remembered by his master. We find them still corresponding at the time of Bach's death, when Einicke was cantor at Frankenhausen.[476]

A crowd of talent had gathered round Bach between 1735 and 1745. Among them, however, there was not one pupil from the Thomasschule, any more than among those who came to him at any later period, and ultimately distinguished themselves. After all the differences between Bach and Ernesti this is easily intelligible, and is rendered quite clear by the unfavourable position which, after these squabbles, Bach held as cantor. His best pupils thenceforth were the University students, or such men as had devoted themselves exclusively to music as a profession. Johann Friedrich Agricola of Dobitsch, born in 1720, matriculated May 29, 1738, studied jurisprudence, history, and philosophy, and at the same time obtained from Bach a course of thorough instruction in playing and composition. Bach thought him advanced enough to play the cembalo in the Musical Union, of which he (Bach) must therefore still have been conductor in 1738, and he also employed him as accompanist in church music. Agricola's mother was related to Handel, and is said to have corresponded with him; Agricola, indeed, bestowed much attention on Handel's works, even in Leipzig, and this again reveals to us the sincere recognition and sympathy with which Bach always regarded his equally great contemporary. In the autumn of 1741 Agricola went to Berlin, where he was soon acknowledged to be the best organ player, but he there eagerly turned his attention to the opera and to the Italian method of singing. In 1751 he became Court composer, and after Graun's death, in 1759,

---

[476] Marpurg, Kritische Briefe. II., p. 461.—Mattheson, Sieben Gespräche der Weisheit und Musik. Hamburg. 1751, p. 189 f.

Capellmeister to the king. He wrote both church and instrumental music, but particularly operas; and his excellent general culture also enabled him to be a successful author. The world owes to him and to Emanuel Bach the Necrology of Sebastian Bach, which was published in 1754 by Mizler, in his Musikalische Bibliothek; and in his notes to Adlung's *Musica mechanica organœdi* he has preserved for us many valuable details concerning Bach. His translation of Tosi's "Introduction to the Art of Singing," with notes and explanations, may still be considered a classical authority. He died in 1774.[477]

In 1738 Johann Friedrich Doles also came to the University; he was born in 1716, at Steinbach, in Henneberg, and had been educated at Schmalkalden and at Schleusingen.[478] He succeeded Bach as cantor in 1755, but his bias was towards the sentimental, the operatic and the vacuously popular, and he was not the true son of his master. During his tenure of the place the use of Bach's works was gradually lost in Leipzig; still, he attached some importance to the fact of his having been Bach's pupil,[479] and even during his master's lifetime he won friends and admirers by his pleasing talent. It was he, and not Bach, who had to write the occasional music, in 1744, for the anniversary of the newly founded Musical Union, and in the same year he became cantor at Freiburg. I shall have more to say presently of the events of his life.

A man of more serious bent, and of far more fertile talent, was Gottfried August Homilius, born 1712, at Rosenthal, on the frontier of Bohemia. He had finished his studies under Bach by 1742, for he was then organist to the

---

[477] Marpurg, Historisch-Kritische Beyträge, I., p. 148.—Burney, Diary III., p. 58. There is a list of Agricola's works in Gerber, Lex. I., col 17.

[478] These data I have derived from his Latin autobiography, which I found among the documents of the Consistory at Leipzig. The amusing scenes which are said to have occurred between Friedemann Bach and Doles, who lodged in Bach's house (see Bitter, Bach's Söhne II., p. 156), are founded in error, since at that time Friedemann had left Leipzig. He had been living n Dresden since 1733.

[479] See a prefatory note to his cantata, "Ich komme vor dein Angesicht." Leipzig, 1790.

Frauenkirche at Dresden; he was subsequently cantor of the Kreuzschule and director of church music in that capital, and he died in 1785. He was admired as an organist, but his chief fame rests on his vocal church music, and his are undoubtedly the most important works that exist in this style of the second half of the eighteenth century. Though hardly on a par with their aim and purpose, they include many passages in a really lofty and grand sacred style.[480]

A musician of quite a different type, again, was Johann Philipp Kirnberger, born in 1721, at Saalfeld, in Thüringia, and who was persuaded by Gerber to go to Bach, whose pupil he was from 1739 till 1741. He then went to Poland, whence he returned in 1751 to become Court musician to the Princess Amalia of Prussia, at Berlin, where he died in 1783. He was of no great mark as a composer, but an excellent teacher of composition; if he had had a wider course of general culture and a thoroughly disciplined mind, he would unquestionably have been the greatest theoretical musician of his time. His works, which down to the present century were almost exclusively used as the final authority in teaching composition, have been fully discussed in this book.

Rudolph Straube, of Trebnitz, on the Elster, was principally known as a performer on the clavier; he entered the University February 27, 1740, and in 1750 made a journey through Germany and to England.[481] Another player was Christoph Transchel of Braunsdorf, born 1721; he matriculated as student of theology and philosophy June 21, 1742, and soon became known as Bach's pupil and friend. He was in great repute as a teacher in Leipzig till 1755, and then, till his death in 1800, at Dresden; he was distinguished alike for the finish of his playing and his elegant general culture.[482] Finally, we must mention

---

[480] Forkel, p. 42. Gerber, Lex. I., col. 665.
[481] Lists of the University.—Adlung Anl., p. 722.—Gerber, Lex. II., col. 599.
[482] Lists of the University.—Gerber, Lex. II., col. 671, and N. Lex. IV., col. 382. Transchel wrote six Polonaises for the clavier, which Forkel could regard as the best in the world next to those by Friedemann Bach. A MS. Vol. of miscellanies in the Royal Library at Berlin contains a composition by him (see Buxtehude's organ works, Vol. I., preface, p. iv., No. 10).

Johann Theophilus Goldberg, who had been brought to Dresden while still very young by Baron von Kayserling; he is said to have been born at Königsberg.[483] Friedemann Bach was his first master;[484] subsequently, about 1741, his patron frequently took him with him to Leipzig to obtain instruction from Sebastian. His facility and skill on the clavier, to which he devoted himself with untiring zeal, soon became quite astonishing; indeed, we have a standard by which to judge him in the thirty variations composed for him by J. S. Bach at Kayserling's request. Kayserling's health was feeble and he suffered from sleepless nights, and at such times he liked to have his melancholy dispelled by soft and somewhat cheerful music. These variations seem perfectly adapted to this end, and Forkel tells us that he was never weary of hearing them, and recompensed Bach for his artistic offering with a snuff-box containing a hundred Louis d'or.[485] Goldberg afterwards entered the service of Count Brühl, as Kammermusicus; he died early. His was a whimsical and eccentric nature, much resembling that of Friedemann, as it would seem from concurrent testimony; there is no doubt that his talent for composition was not great; still, a grand prelude by him for the clavier, in C major, is a solid work besides being brilliant, which of course was to be expected.[486]

During the last years of his life, Altnikol, Kittel, and Müthel were the pupils to whom Bach devoted most of his teaching. Johann Christoph Altnikol, who in 1749 became Bach's son-in-law, lived in Leipzig till 1747, and assisted in the church performances.[487] An attempt made by Friedemann Bach to have him appointed in his own place when he gave it up and left Dresden, in 1746, was unsuc-

---

[483] Forkel, p. 43. Reichardt, Mus. Alm., 1796, under App. 10, says he was born at Dantzig.
[484] Fürstenau II., p. 222, note.
[485] Forkel, p. 51, also p. 43.
[486] MS. in the Royal Library at Berlin, where, too, there is a sacred cantata by him.
[487] He was paid six thalers "for having assisted in the two high churches in the *Chorus musicus* from Michaelmas, 1745, till May 19, 1747." Accounts of St. Thomas's Church for 1747—48, p. 54.

cessful;[488] in 1747 he became organist to St. Wenceslaus, at Naumburg, where he died in July, 1759, after doing good work.[489] Joh. Christian Kittel of Erfurt was one of Bach's latest pupils, for at the time of the master's death he was only eighteen, he may even then have been with him. From Leipzig he went to Langensalza, and in 1756 he was at Erfurt where, in process of time, he became organist to the Predigerkirche. Kittel was an excellent organ player and composer, and a favourite teacher; he taught a great number of the best organists of Thuringia, and, with pious reverence for his own great teacher, did his utmost to transmit the traditions of Bach's art and style. He died at an advanced age in 1809.[490]

Joh. Gottfried Müthel had enjoyed the advantage of Bach's teaching for a few weeks only when his master's weak sight and ill-health disabled him from teaching. Müthel had come from the Ducal Court at Schwerin, where, after Michaelmas, 1748, at the latest, he had held the post of Court organist as his brother's successor. The Duke was so much attached to the young artist (born in 1729 at Mölln) that, in May, 1750, he granted him a year's leave, with the continuance of his salary; that he might perfect his knowledge under Bach. An introduction from the Duke himself secured him a friendly reception, and the additional favour of a lodging in Bach's house. He also was a witness of the master's last illness and death, and then, in order to make up as far as possible for the loss to himself so far as teaching was concerned, he went to Naumburg to Altnikol, Bach's son-in-law, and he was still there on June 2, 1751. From Schwerin he went

---

[488] See a document quoted by Bitter, Bach's Söhne II., p. 356 and 171.
[489] Register of deaths of that church. See also Forkel, p. 43.
[490] I have not been able to ascertain the exact date of Kittel's move to Langensalza, but in February, 1752, he was married there to Dorothea Fröhmer; he was organist there at the church of St. Boniface and "teacher in the girls' school." His successor, who lived to be a very old man, said that he had not been content after a time to keep this post in the girls' school, for his love of composition and writing music often made him do this in school hours, and so brought him into collision with authorities. Finally, he gave up the post. (Communicated by Herr Kirchner Stein, of Langensalza.) See, too, Gerber Lex. I., col. 728, and N. Lex. III., col. 57.

in June, 1753, as organist, to Riga, where he died. Members of his family still live and flourish in Livonia. His talent for clavier and organ playing was remarkable and thoroughly developed. His compositions, which are not numerous, are a test of the highest executive skill.[491]

From all this it may be fully understood how widely, in every direction, Bach's fame extended, and how vital was the influence he exercised over art. Nor is the list of his pupils by any means exhausted; disciples crowded round him in many cases only to be able to call themselves his pupils to the world, and boys who had been at the school and learnt from him only in class with a score of others prided themselves on it. One of these was Christoph Nichelmann, of Treuenbrietzen, born 1717, afterwards Capellmeister to Frederick the Great. From 1730—33 he was first treble in the church, and learned the clavier from Friedemann Bach. Johann Peter Kellner, of Gräfenroda, born 1705, who had been introduced to Bach's works by Schmidt, the organist of Zella, and had been greatly inspired by them, publicly expressed his satisfaction at his good fortune in having enjoyed the acquaintance of this admirable man, and though he was not, strictly speaking, his pupil, he may be called his disciple.[492] Thomas Trier stood in the same relation to him; he was born in 1716, at Themar, studied theology at Leipzig, and in 1747 was also the director of the Telemann Musical Union. In 1750 he was a candidate for the post of cantor after Bach, but Harrer was preferred. In Zittau, in 1753, he beat all competitors, among them Friedemann and Emanuel Bach, Krebs and Altnikol, all four of them pupils of Bach's. He remained there as organist of the church till 1790, enjoying the reputation of being one of the best organists in all Saxony. Bach's recommendation had been the foundation of his fortunes. Gerlach, by the same influence, obtained the place of organist

---

[491] I have been able to confirm and extend the information concerning Müthel as given by Burney III., p. 268. Müthel's presence at Naumburg, June 2, 1751, is proved by an entry in the Register of baptisms of the church of St. Wenceslaus.

[492] Marpurg, Hist. Krit. Beyträge, I., pp. 431 and 439.

to the New Church at Leipzig, and Christian Gräbner the younger and Carl Hartwig referred to the instruction they had received under Bach when in 1733 they were candidates for the place of organist to the Sophienkirche at Dresden.[493] Johann Christoph Dorn seems to have obtained the post of organist at Torgau by a written testimonial from the famous master. This testimonial has been preserved, and is dictated by a gentle spirit of kindliness which reminds us of the worthy Heinrich Bach, of Arnstadt. It runs as follows :—

A representative of Mons. Johann Christoph Dorn, who is a diligent musician, has applied to me, the undersigned, to give him a testimonial as to his proficiency and knowledge of music. Since I find, from the examples he has given me, that he has acquired a considerable degree of proficiency on the clavier, as well as on other instruments, and is thus in a position to do good service to God and to the State, I have not refused his proper request, but, on the contrary, shall testify that as he grows older, with his good natural gifts, he bids fair to become a very skilful musician.

Leipzig, May 11, 1731.

Joh. Seb. Bach,
Hochf. Sächss. Weissenfl. Capellmeister und
Direct. Chori Musici Lipsiensis.

According to the old and universal custom, the pupils formed part of the family. Bach, like his forefathers, clung to the idea of a guild among musicians, and Müthel cannot have been the only scholar whom he admitted to live in his house. Speaking of his pupils has led us, as it were, into the midst of Bach's domestic life, and we may now see, as far as the means at hand admit, how this remarkable personage appeared as the head of a family. If we enquire into the ordinary circumstances of his life, irrespective of his position and doings as an artist, when we remember the seclusion in which a citizen's household lived at that time, and Bach's simple and domestic nature, it seems probable that it was quiet and unvaried enough. The persons chosen by Bach as sponsors for his numerous children, thus indicating a certain intimacy, were, for the most part,

---

[493] See Vol. II., p. 225. Bitter, Bach's Söhne II., p. 159.

members of the better class of officials or respectable merchants, though we find among them two lawyers and teachers in the University. In his later years there seems to have been a warm intimacy between the Bachs and the family of a merchant named Bose, and we may also assume a constant intercourse, founded on personal regard, with Johann Christian Hoffmann, musical instrument maker to the Court (Hof-instrumentenmacher). Hoffmann had settled at Leipzig so early as 1725 (his house was in the Grimmai Steinweg), and he was always eager to co-operate with Bach, who took pleasure and interest in the improvement of musical instruments. For instance, he prepared several specimens of the *Viola pomposa*.[494] When he died, February 1, 1750, it was found that out of respect for Bach he had bequeathed to him an instrument of his own making.

We find but very few traces of his social intercourse with any of the literary magnates of the city, excepting with his colleagues in the school. He had come into contact with Gottsched on the occasion of the mourning ceremonial in honour of Queen Christiana Eberhardine. He is not likely, however, to have had any very warm sympathy with a man who was so emphatically antagonistic to the opera,[495] and Gottsched's attempts in the way of poetry he can hardly have found interesting. However, a connection was formed between the two houses for a time by Gottsched's selecting Krebs as his wife's musicmaster. Luise Adelgunde Victoria, who came to Leipzig as a bride in 1735, had, besides other remarkable gifts, a very great talent for music. She already played well on the clavier and lute, and now wished to learn composition; she made such rapid progress as soon to be able to compose a suite and a cantata. Krebs himself was absolutely bewitched by his pupil, who was then about twenty years of age, and so late as 1740 he dedicated to her —writing from Zwickau—a cahier of six "Preambles," which he had engraved on copper by Balthasar Schmidt, of

---

[494] Hiller, Lebensbeschreibungen, p. 45, note.
[495] His friend Hudemann, of Hamburg, wrote a treatise against Gottsched in defence of the opera in 1732, see Mizler, Mus. Bib. II., 3, p. 120.

Nuremberg, and in a romantic set of verses expressed himself as the obliged person—
> Who having thought to guide and school your ear,
> When you but played, was quite content to hear.[496]

Gottsched's house became for a time a centre of musical meetings. Sylvius Leopold Weiss, who once came from Dresden to Leipzig, visited the gifted lady, heard and applauded her playing, and played to her himself; Gräfe dedicated to her the second volume of his odes, and even Mizler made himself agreeable to her in a dedication. It is probable, therefore, that Bach would have been at her house; indeed, the admiration with which T. L. Pitschel speaks of Bach—in the Belustigungen des Verstandes und Witzes[497]—in 1741, would seem to prove that at this period he must have moved within the Gottsched circle, and sometimes have charmed it by his playing.

Bach also had formed a lasting friendship with another of the professors at the University, who was also very popular, though less famous than Gottsched. Johann Abraham Birnbaum was promoted to the degree of *Magister*, February 20, 1721, at the early age of nineteen, and qualified as an instructor on October 15 of the same year.[498] Rhetoric was his principal branch of learning, and his lectures were eagerly frequented. There was in Leipzig a private debating society of which Birnbaum was a member, and in 1735 he published a volume of the discourses which he had delivered there, in pure and flowing, but somewhat pedantic German.[499] He played the clavier elegantly, and his taste for music brought him into contact with Bach. He lived in the Brühl, and died, unmarried, August 8, 1748.

Bach mentions his death in a letter to Elias Bach, dated November 2, 1748, in a way which leads us to infer that the

---

[496] A copy of this work is in the Royal Library at Berlin.
[497] (Johann Joachim Schwabe), Vol. I., Leipzig, 1741, pp. 499 and 501.
[498] Sicul, Leipziger Jahr-Geschichte, 1721. pp. 199 and 236.
[499] Leipziger Neue Zeitungen von gelehrten Sachen, 1735, p. 603. One of the speeches made by him in this society, "uber den Hohen Geist des erblassten Thomasius," was published before 1729. A copy of it exists in the Library at Wernigerode.

cousin had enquired for him, and the friendship between Bach and Birnbaum was known to a wide circle. In a periodical, published by Johann Adolph Scheibe, under the title "Critische Musikus" (beginning March 5, 1737), on May 14, 1737, an anonymous letter was printed, containing an attack upon Bach. He was not named, but the identity was unmistakable. Scheibe subsequently confessed that the letter was a got-up affair, and he himself the author. In it Bach's extraordinary skill in playing the organ and clavier were highly praised, but his compositions were found fault with for their lack of natural grace and pleasing character, for a turgid and confused style and an extravagant display of learned art. The writer alluded chiefly to his vocal part-music, and came to the conclusion that Bach was in music what Lohenstein had been in verse.[500] A year later Scheibe wrote again: "Bach's church pieces are constantly more artificial and tedious, and by no means so full of impressive conviction or of such intellectual reflection as the works of Telemann and Graun."[501] Nor was he the only one to hold this opinion, as it is easy to believe from the taste of the time and the false ideas then prevalent as to church music. Nevertheless, the whole affair has an unpleasant aspect, not only by reason of the exalted and almost unapproachable merit of the person attacked, but more particularly from the tone in which fault is found, and considering who it was that made the attack. Scheibe was a young man of knowledge and acumen, and a talented writer, but only a second rate practical musician. Every one in Leipzig knew perfectly well that his test performance, when he hoped to obtain the post of organist at St. Nicholas's, had found no favour in Bach's opinion; and very unsatisfactory—though no doubt exaggerated—rumours were rife on the subject. Scheibe was ambitious and jealous; he had "agitated" against Bach ever since, and had stirred up, or, at any rate,

---

[500] Critischer Musikus, p. 62.
[501] Mattheson, Kern melodischer Wissenschaft, Hamburg, 1738. Appendix, "Gültige Zeugnisse," &c., p. 10.

promoted, a tide of opinion, which, so early as 1731, had put Bach and his adherents on the defensive. I have, I think, made it clear in a former page that the character of Midas in the cantata " Der Streit zwischen Phöbus und Pan " was meant for Scheibe ;[502] and in this attack, made under safe cover, the world saw an act of undignified personal revenge.

Bach was more deeply offended than perhaps he ought to have been; but it occurred just at a time when he was already in a highly irritated mood, in consequence of his squabble with Ernesti. It would almost seem that he himself had at one time thought of taking up his pen against Scheibe. However, it was Birnbaum who stepped in ; he began, in January, 1738, by publishing an anonymous article, " Unparteiische Anmerkungen über eine bedenkliche Stelle in dem sechsten Stücke des critischen Musikus "—(" Impartial remarks on an important passage," &c.). When, two months after, Scheibe brought out an answer to it, he came forward in March, unmasked, with a categorical defence. He dedicated both his articles to Bach, the latter, which is by far the best, in a long introductory letter. In the first he tried to weaken the case as against Bach's method of composition, and this was a province of art with which he was not sufficiently familiar. Scheibe's retort is an impertinent pamphlet. All sorts of petty scandal, which he had procured from Leipzig, is to be found in it, mixed up with insinuations, misrepresentations, and invective, and he reaches the utmost limits of insolence when he asserts that Bach has no special breadth of view in those branches of knowledge which may be particularly required of a learned composer. " How," he asks, " can a man be faultless as a writer of music who has not sufficiently studied natural philosophy, so as to have investigated and become familiar with the forces of nature and of reason ? How can he have all the advantages which are indispensable to the cultivation of good taste who has hardly troubled himself at all with the critical study, the cultivation and the rules which are as necessary to music as they are to

---

[502] See Vol. II., p. 645.

oratory and poetry, so that without their aid it is hardly possible to write with feeling and expression." Such writing as this deserved to be thoroughly taken in hand, and Birnbaum dealt to Scheibe the measure that was due. He was himself a perfectly competent authority in all that related to rhetoric and poetry, and all that he replied to Scheibe's strictures from his own experience of Bach has every claim to our belief. But the matter has already been discussed.[503]

Scheibe was silenced for the time, and revenged himself only by a spiteful pasquinade flung at Bach under the form of a letter witten by one Cornelius in the "Critische Musikus" of April 2, 1739.[504] But in a second edition he could not refrain from reprinting his opponent's letters with annotations. He was conquered in the fight, not because the faults he attributed to Bach were in fact beyond proof, but by reason of the unbecoming manner in which he carried on the dispute. But it rained blows from all sides, and having brought the whole school of Bach down upon himself from far and wide, retribution haunted him throughout his life; even in 1779 Kirnberger had a fling at him when an opportunity happened to offer.[505] And others, who would not withhold a due recognition of his talents and industry, still took care—like Mattheson and Marpurg—not to coincide in his verdict on Bach. Indeed, at a later period he himself seems to have arrived at the view that he had not taken the right tone towards this great man; this is very evident from the preface to the second edition of the "Critische Musikus,[506] written in 1745.

Bach himself took an unusually lively personal interest in Birnbaum's different articles in this controversy. Scheibe pretended to know that the first of these papers had been distributed to his friends and acquaintances so early as January 8, 1738, "with no small pleasure." Bach's own

---

[503] See Vol. II., p. 238.
[504] See Schröter in Mizler, Mus. Bib. III., p. 235.
[505] Kunst des reinen Satzes, II., 3, p. 39.
[506] Containing all the controversy, pp. 833 to 1031.

answer was given—as he best could give it—by the publication of Part III. of the Clavierübung, of which Mizler wrote: "This work is a sufficient answer to those who venture to criticise the Herr Hofcompositeur's compositions."[507] However, when Scheibe, in this pasquinade tried to show Bach in the light of a man who had never given himself time to learn to write a letter of any length, who had never attended to general culture, and rarely read even musical essays or books, it was either heedless detraction or a lie pure and simple. Bach, it is true, to Mattheson's great vexation, had never given an autobiography to the Ehrenpforte, though repeatedly solicited; but in other ways, so far as the practice of his art allowed —and he very rightly made this the chief aim of his life—he showed more interest in the literature of music than many of his contemporaries. His library of books on music, most of which at his death fell into the hands of Emanuel, must have been by no means inconsiderable.[508] He frequently watched literary disputes with attentive interest. When Sorge of Lobenstein (1745—1747) was publishing the "Vorgemach der (antechamber to) Musikalischen Composition," some persons expressed a doubt as to the authorship; Telemann was supposed to have written it while Sorge had only lent his name. Bach took much interest in the matter, and, to Sorge's great distress, expressed the same opinion, "that he had ploughed with another man's heifer," and even after his death this unfavourable view was brought up against him.[509]

Bach's participation in an affair which heated the temper of the musical world for a time in and after 1749, also gave rise to a vehement war of words. Doles had gone to Freiberg as cantor in 1744, and the *Rector* there, since 1747, was Johann Gottlieb Biedermann, a man distinguished for his learning. In 1748 he, as qualified to decide in such

---

[507] Mus. Bib. II., p. 156.
[508] Emanuel gave several of these "rare old books and dissertations on music" to Burney. Diary III., p. 215.
[509] See Marpurg, Krit. Briefe I., p. 139.

matters, determined to have a *singspiel* performed in commemoration of the peace of Westphalia, concluded just a century before. A blind poet, named Enderlein, wrote the text, and his subject was "Germany made happy after a long war by a general peace." Act I. "The miseries of the thirty years' war." Act II. "The prospect of peace and the hindrances to it." Act III. "Peace concluded." Act IV. "Peace declared to the advantage and joy of all." The music was by Doles, who, as Biedermann remarked in a note to the printed text, had so done his work "that both skilled and delicate ears would be charmed by it." The piece was not an opera in the sense in which the word was then used, but a vocal performance mixed up with spoken dialogue ; such pieces were very commonly performed in the schools of Saxony during the first half of the eighteenth century, though their importance in the history of German opera has hitherto been by no means fully appreciated. The town council had a stage erected and encouraged the project in every possible way. The performance took place in the Kaufhaus on October 14, at four in the afternoon, and was repeated on the following days.[510] The audiences were very numerous, and flowed in from the neighbouring towns and country, but the chief applause was bestowed on the music. It may have been an annoyance to Biedermann to see the cantor's importance and influence among the scholars greatly increased by this circumstance, and to feel that the success of music threatened the interests of learning ; and it was even whispered that various unpleasant details had occurred in calculating the receipts and awarding the composer's honorarium. In short, the constantly recurring feud between rector and cantor broke out afresh, and this time Freiberg was the scene of the struggle. Biedermann, in his next school programme, May 12, 1749, expressed his irritation only too emphatically ; the main idea of this document, printed in

---

[51)] A copy of the printed text with an introductory note by Biedermann, two sheets in folio, is preserved in the Library of the Freiberg Society of Antiquaries. See also Beyträge zur Historie und Aufnahme des Theaters. Stuttgart, 1750, p. 596.

Latin, and having for its text a passage from the Mostellaria of Plautus (*Musice hercle agitis aetatem, ita ut vos decet,* Act III., Sc. 2, v. 40), is that the over-much practice of music is apt to lead the young astray into a life of dissipation, and so far he is right and unanswerable; but when, with the memory still fresh of the triumph achieved by music, though he had been annoyed by it, he goes on to name certain *mauvais sujets* who had devoted themselves to music, when he refers to Horace—who classes musicians with bayadères, quacks, and beggarly priests—and says that the Christians of old excluded them from their pious meetings, and only allowed them to take the sacrament once a year, the whole musical community was justified in feeling itself insulted; and this naturally was the issue. Those named and those ignored alike fell upon Biedermann with equal virulence; Mattheson alone wrote five articles against him. Others again tried to take his part, and a vehement literary war broke out which lasted till 1751. Biedermann had to endure many unreasonable attacks as a punishment for his want of tact. When Bach learnt what had been going on at Freiberg, and saw Biedermann's programme, his old wounds ached anew; however, he had lived through similar experiences with his rector. He sent the document to Schröter at Nordhausen, a member of the musical society, begging him to review it and reply to it, as he could find no one capable of doing it in Leipzig or the neighbourhood. Schröter consented; he sent a review to Bach and left it to him to get it printed in some periodical paper. It met Bach's wishes, and, on December 10, 1749, he wrote to Einicke in Frankenhausen, "Schröter's review is well done and quite to my taste, and will shortly appear in print. . . . Herr Mattheson's Mithridates had caused a very violent commotion as has been told me on trustworthy testimony. If yet some other Refutations should follow, as I suspect, I make no doubt the author's ears will he purged and made more apt to hear music." However, he handed over the matter of the printing to another person, who took upon himself to make certain alterations and additions, and to add causticity to the tone of the document, which was before very moderate and dignified. In the altered review Biedermann was given to understand that he was better versed in the writings of the heathen than in the word of God, and the article, quite without Schröter's consent, was entitled "Christian Judgments."

Schröter, however, was more touchy than he need have been as to the treatment his review had undergone, and he wrote a letter, April 9, 1750, which he desired Einicke to send on to Bach. On May 26, 1750, Bach wrote to Einicke "Pray make my compliments to Herr Schröter, till I am able to write to him, and I will then excuse myself with regard to the alterations in his review; though in fact I am not to blame in the matter at all; they are solely attributable to the person to whom I entrusted it to print." Schröter would not be satisfied with this and demanded among other things a public explanation from Bach. But death interfered to put an end to the matter.

The worthy Einicke, however, had got drawn into the squabble without any fault of his own. A partisan of Biedermann's supposed him to be the author of "Christian Judgments," speaking of him as a "certain Cantor of F. [rankenhausen] who had formerly been a miserable schoolmaster at H. [ohlstedt];" and this prompted him to lay the whole facts before Mattheson.[511]

The review by Schröter was not, however, the only means of retaliation on Biedermann adopted by Bach. He once more took up the cantata "Der Streit zwischen Phöbus und Pan," which eighteen years previously he had composed as a satire upon those who attacked his music (see Vol. II., p. 473), and had it performed. It would seem that the performance was by one of the musical unions of the town, though Bach was no longer in direct connection with either of them. Johann Michael Schmidt, of Meiningen, had been studying in Leipzig since March 12, 1749, and he, a few years later, published a well-considered book, which was

---

[511] Mattheson, Sieben Gespräche, &c., Hamburg, 1751, p. 181. Adlung gives a full account of the matter, Anl. Mus. Gel., p. 70. See also Marpurg, Krit. Briefe, I., p. 253. Lindner, Zur Tonkunst, p. 64. Bitter has reprinted the "Christliche Beurtheilung," J. S. Bach, II., p. 340.

received with great approbation, entitled *Musico-Theologia*.[512]
In this work Bach is frequently referred to with admiration,
and in one place it is said that the chief aim of the composer
should be to give an adequate representation of the state
of feeling he wished to express, and the actions proceeding
from them, and to hold the mirror up to nature. "The
more nature was copied in his compositions the more
pleasure he would give his hearers. From this theory of
art sprang the Calendar of the months depicted in music,
Bach's 'Gesprächspiel,' the Lyre, the Cuckoo, the Night-
ingale, the Posthorn, and others." These names evidently
designate some cantatas of a gay, light character, which were
popular at the time but are now lost. The "Gesprächspiel"
can only have been a secular cantata by Bach; from the
context it cannot possibly have been one of his dramatic
occasional pieces. The choice, therefore, remains between
the Coffee Cantata and "Phöbus und Pan." But, since the
point under consideration is definite expression and imitation
of nature, both of which are to be found in their widest sense
in this cantata, we can hardly doubt that the latter is meant;
and all the less because we actually have a text for it dated
1749; Michael Schmidt may, indeed, have sung in the performance, or, at any rate, have been present at it.

Moreover, this text contains a passage which clearly shows
the aim and end of this performance. In the last recitative
Picander had originally written:—

"And now, Apollo, strike the lyre again,
For nought is sweeter than thy soothing strain."

Here it runs as follows:—

"Now strike the lyre with redoubled power,
Storm like Hortensius, like Orbilius roar."

And again at the end:—

"Now strike the lyre with redoubled power,
Storm like Birolius, like Hortensius roar."

Now Orbilius, as is well known, is the schoolmaster in
Horace,[513] and Birolius is an anagram serving to suggest the

---

[512] *MUSICO-THEOLOGIA*, oder Erbauliche Anwendung Musikalischer
Wahrheiten; Bayreuth und Hof. 1754. See Marpurg Hist. Krit. Beil. I., p. 346.
[513] Ep. II., 1, lines 70, 71.

name of Biedermann. I think it possible also to identify Hortensius. Quintus Hortensius was Cicero's only rival in eloquence, both being regarded as models of Latin diction. The man who revived the knowledge of these two orators among the learned in Germany in the eighteenth century was Ernesti, who had edited an edition of Cicero in 1737. It was all the more obvious to dub him by the name of Hortensius, because a name had been found in classic literature to fit Biedermann. Thus the rector at Freiberg and the rector at Leipzig were bracketed as the pair who stood as the butt of Bach's satire in the cantata.[514] From this it appears that Bach's resentment against Ernesti was deeply rooted.

Biedermann meanwhile knew very well what were Bach's feelings towards him. He said of one of the polemical papers of which he wrongly suspected him to be the author, that "the stupid lies proceeded from that foul Bach"—to such amenities of language had the disputants descended. After his *programme* it was certainly impossible to attack Biedermann on the score of indifference to music; but his previous history was very well known in Leipzig through the cantor Doles, and this showed the rector in an unfavourable light.

The eager vehemence displayed by the venerable composer in upholding the honour and dignity of his beloved art has something pathetic in it when we find him singing its praises with enthusiasm in a cantata which, at this time, he had performed on three occasions. This is the cantata "O holder Tag, erwünschte Zeit," which was now refitted with a text exclusively in praise of music. Here we find the words "Alas, beloved muse of harmony, sweet as thy music is to many ears, yet

---

[514] Though Dehn asserts (op. cit. p. 479, note) that by Hortensius Bach meant to designate a certain Gärtner who gave the Thomasschule scholars too much to do with music; but I know not on what this hypothesis rests. Karl Christian Gärtner, who at a later date edited the Bremen documents was, it is true, a native of Freiberg and he had lived for a time in Leipzig, but only until 1745, nor is there anything to prove that he ever came into personal contact with Bach.

art thou sad and standest pensive there; many there be who scorn thy charms. Methinks I hear thee complaining and saying, ' Be still, ye pipes and flutes, ye fail to please; even I myself am weary of your tones; away with songs for I am left alone.' But calm thyself, fair muse; thy glory is not dead, nor altogether banished and despised." (See Vol. II., p. 635 f.)

All these occurrences bear a two-fold significance; they show us, in the first place, Bach's intelligent interest in literary matters, and they also point to a strongly marked characteristic of his nature. There was in him—it had already often shown itself (see Vol. II., p. 648), a certain pugnacity, or even contentiousness, which shows him to have had a near affinity with the orthodox Lutherans, of the type, for instance, of Eilmar, the pastor of Mühlhausen. And the reader who has followed me in the analysis of Bach's compositions must often have detected the traces of this spirit in his works. I need only remind him here of the cantatas " Christ lag in Todesbanden " and " Es erhub sich ein Streit," and of the double chorus " Nun ist das Heil und die Kraft." He stood apart from the orthodox, it is true, by the deep personal fervency of his feelings and by a touching and childlike simplicity that betrayed itself when they were appealed to. He was a typical German; at once a hero and a child, untamed and yet impressionable and tender; most nearly akin to Luther among theologians. Even the tenacity with which he fought for his rights, and of which many instances have been given, was an essential element in such a nature. He was irritable—but so is every artist—and his masterful nature could break out in ungoverned wrath. The organist of St. Thomas's, either Gräbner or Görner, was one day playing the organ at a rehearsal and made some mistake. Bach snatched off his wig in a rage, flung it at the criminal and thundered out " He had better have been a cobbler."[515] It occasionally happened that he would turn a recalcitrant scholar out of

---

[515] Hilgenfeldt, p. 172. I cannot trace the origin of the story; but Hilgenfeldt was generally cautious enough not to accept any facts for which he had not trustworthy authority.

the choir with no small commotion in the middle of service, and then in the evening dismiss him from the supper table with an equally high hand; and as by these arbitrary proceedings he often sacrificed his dignity as a teacher, he very naturally sometimes found a difficulty in keeping the rout of boys in order. But these weak points could not diminish the high opinion in which he was held by all who knew well the honest depth and soundness of his character. His personal pupils bear witness to this, for all without exception regarded him with unbounded respect and devotion.[516]

Bach had a justifiably good opinion of himself, and this the Leipzig Council, among others, must more than once have had reason to know. Still, like all noble natures, he was devoid of conceit and full of human sympathy and consideration for others. His pupils had only to follow the example he set them of industry; he never held them accountable for their lesser or greater natural gifts. Nor was praise invariably acceptable to him. Once when some one had spoken with enthusiasm of his wonderful skill on the organ he said with indifference "There is nothing very wonderful about it; you have only to hit the right notes at the right moment and the instrument does the rest."[517] In the same way he was never known to hesitate or refuse to play out of affectation, and he was always ready and willing to perform alone or with others.

It was noted as one of his peculiarities that, though a master of improvisation, he never liked to begin with anything of his own; he preferred to play at sight a composition put before him, and could then go off into an impromptu, as if he had needed the impetus from outside to stimulate the flow of his own invention. The evidence of this is interesting enough to be quoted at length. It is recorded by a certain theological student, T. L. Pitschel, who was studying at Leipzig about 1738, took the degree of Master in 1740, and died there in 1743, aged 27. He was

---

[516] Necrology, p. 173.
[517] This characteristic anecdote is from Köhler, *Historia Scholarum Lipsiensium*. Note to p. 94.

an ally of Gottsched's, and published in a paper which was the organ of the party (Belustigungen des Verstandes und Witzes) a letter to a friend "On frequenting public worship." In this he says: "You know that the famous man who in our town enjoys the greatest praise for his music and the admiration of all connoisseurs is never able to enchant his hearers with his own musical combinations until he has played something already written, and so inspired his inventiveness"; and farther on: "That gifted man, of whom I have already spoken, always needs to play from the page something worse than his own ideas can ever be. And yet these good ideas of his result from those inferior ones."[518] And it was no doubt the same vein in his nature which made Bach prefer to find some impulse of occasion for his own artistic productions, as is constantly traceable, especially in his compositions for singing.

He also played music written by others from a genuine interest in seeing what they had done and could do, and in the same way he liked to hear others play their compositions. Things that particularly pleased him he would copy out, not only in his younger days when he was working at his own development, but at the time of his ripest maturity. Vocal works exist to this day in his handwriting, by Lotti, Caldara, Ludwig and Bernhard Bach, Handel, Telemann, and Keiser, and clavier pieces by Grigny, Dieupart, and even Hurlebusch. His collection of printed works, or of copies by other hands, was not a small one. Unfortunately they were so promptly dispersed by his sons at his death that a list of them was not even included in the inventory of his property. That his knowledge and interest extended to a remote past is proved by his having possessed Elias Ammerbach's *Orgel oder Instrument Tabulatur* of 1571, and Frescobaldi's *Fiori musicali* of 1635; and, as has been already said, he collected theoretical works on music.

When he spent a Sunday away from home he always followed the music of the service with particular attention.

---

[518] (Johann Joachim Schwabe). Belustigungen, &c., Vol. I., Leipzig, 1741, pp. 499 and 501.

If a fugue was introduced, and if one of his sons happened to be with him, he would, as soon as he had heard the thema, say beforehand what the composer ought to make of it in the farther treatment, and what, if he could, he might make of it. If it was worked out as he had sketched it, he would nudge his son's elbow, and be quite delighted.[519] Cause for blame, however, as may be supposed, more often occurred than for satisfaction. Still, his sons have put it upon record how lenient his verdicts always were; that he never allowed himself to speak in severe terms of the work of a fellow artist, though to his scholars he thought it his duty to speak the severest unqualified truth. Hurlebusch, of Brunswick, a restless, conceited player, who was always going from place to place, once introduced himself to Bach at Leipzig, not to hear him play, but to be heard; and Bach obliged him in the matter with the utmost patience. On leaving, Hurlebusch presented Bach's eldest sons with a printed copy of his own collected works with an injunction to study them diligently. The father knew full well that both Friedemann and Emanuel Bach were far beyond such things, but he only laughed quietly to himself, and was perfectly friendly and polite to the donor. He was particularly averse to seeing a fellow artist humiliated by a comparison with his own superiority, and would never willingly refer to the competition with Marchand, which had attracted attention throughout Germany.

Even during his lifetime mythical anecdotes were rife about him; for instance, it was said that he would go into a church dressed as a poor village schoolmaster and request the organist to allow him to play, and then excite the astonishment of the congregation and of the organist, who declared it must be either Bach or the devil. He would never listen to such stories if by chance they came to his ears.[520]

He lived in music, in his downsitting and his uprising; a story, of which the main idea at any rate may be true, will

---

[519] Forkel, p. 46.
[520] Forkel, p. 45.

HIS OUTWARD APPEARANCE. 265

prove this. He had often met a certain troup of beggars in whose *crescendo* supplications he fancied he had detected a certain series of intervals. At first he made believe to be ready to give them something, but to find no money about him, then their cry reached a piercing pitch; then two or three times he gave them a very small alms, and the acuteness of their tones was somewhat qualified; finally he gave them a rather considerable sum which, to his great delight, resulted in a full resolution of the chord, and a complete and satisfactory close.[521]

Bach's nature was, above all, grave and earnest, and with all his politeness and consideration for his fellow men his demeanour was dignified and commanded respect. If we may trust the portraits which remain of him his appearance answered perfectly to this. To judge from them he must have been of a powerful, broad and stalwart build, with a full but vigorous and marked face, a wide brow, strongly arched eyebrows, and a stern or even sinister line between them. In the nose and mouth, on the contrary, we find an expression of easy humour; the eyes are keen and eager, but in his youth he was somewhat shortsighted.[522]

All his actions were based on a genuine piety which was not the outcome of any mental struggle, but inborn and natural; and he clung to the tenets of his fathers. He was fond of reading theological and edifying books, and his library included eighty-three volumes of that class at his death.[523] The authors represented sufficiently indicate his religious bent and views. Luther must be first mentioned,

---

[521] Reichardt, Mus. Alm., fols. L. 2 and 3. Berlin, 1796.

[522] Necrology, p. 167. There were four portraits of him in oil; two by Hausmann, the Court painter to the Duke of Saxony. One of these is at the Thomasschule at Leipzig, and it contains the canon presented to the Musical Union. (This is worked out in Hilgenfeldt, Mus. Supp., No. 3). The other portrait by Hausmann became the property of Emanuel Bach. A third oil portrait belonged to Kittel, and a fourth to the Princess Amalie; this is now in the Library at the Joachimsthal at Berlin.

[523] See, in Appendix B, XVI., the inventory of his property. I owe this interesting document to the industry and kindness of my friend Dr. Wustmann, of Leipzig, who discovered it among the legal archives of that circuit.

as Bach seems to have possessed his works in two editions.[524] The Table Talk and a volume of Sermons are also mentioned, separately. Most of the other works are by old Lutheran divines of the sixteenth and seventeenth centuries. Calovius, who appears in three folio volumes (born 1612, died at Wittenberg, 1686), was a truculent controversialist, and one of the most passionate champions of orthodoxy. Heinrich Müller (died in 1675), Professor and Superintendent at Rostock, was a man of softer mould; Bach possessed several of his works.[525] The *Schola Pietatis* of Johannes Gerhard, of Jena (died 1637), contains a treatise on the duties of the Christian, in five books; the learning, manly character and piety of its author must have recommended it strongly to Bach, who also possessed a well known work, "Die wahre Christenthum," by Gerhard's master and friend, Johann Arnd (died 1621, at Celle). Superintendent General August Pfeiffer again is well represented; he was an esteemed professor and preacher of Leipzig in the seventeenth century, and from 1681 was for several years Archdeacon of St. Thomas's. The three titles, "Evangelische Christenschule," "Anticalvinismus," and "Antimelancholicus" have a special interest, because they are also written in Anna Magdalena's "Clavier-Büchlein," of 1722. They were inserted there not by accident or for a fancy, but indicate his particular liking for the books, of which the purport was to be reflected in the contents of the little music book.

Besides these orthodox Lutheran works we also find some of a more mystical tendency. Among these it is particularly interesting to find the sermons of the Dominican Tauler, the direct product of mediæval mysticism. Spener's edition

---

[524] That in seven Vols. is the Wittenberg edition of 1539; that in eight the Jena edition of 1556. There is an error in the inventory—which, of course, I have printed exactly as I found it—as to the author of the "*Examen Concilii Tridentini*, which is attributed to Luther. It is by *Martinus Chemnitzius* (born 1522, died after 1568, at Rostock), and the two following works are no doubt by the same writer.

[525] The folio edition of "Evangelische Schluss Kett" was brought out at Frankfort, 1734, and Bach must have bought it late in life.

of 1703, quarto, can hardly be the one in Bach's possession, for, as that is valued only at four groschen, it must have been an old and much worn copy. The pietists are represented by a work of Spener and one by Franck which serves to prove that Bach, with all his old Lutheran feeling, was not a fanatical partisan of orthodoxy, but could take a lofty and impartial position (see Vol. I., p. 360). Johann Jakob Rambach, whose writings he liked greatly, was not, strictly speaking, of the pietist school.

Bach's knowledge of the Bible, as shown by his church cantatas, was evidently as extensive as his acquaintance with hymns. We see from his owning Bünting's *Itinerarium Sacræ Scripturæ* that he must have tried to realise the Bible history as vividly and as picturesquely as possible. In this Itinerary all the travels of the Patriarchs, Judges, Kings, Prophets, Princes and their peoples, of Joseph and the Virgin Mary, of the Wise Men from the east, Christ and His apostles were traced out and estimated in German miles; it also contains a full description of all countries and towns mentioned in the Bible. Judge as we may the scientific value of such a work, it is at any rate an evidence that Bach did not regard his Bible merely as a repertory of texts for lyric verses, or even for dogmatic argument, but that he tried to make himself familiar with it in every sense. Another proof is his having owned Josephus's "History of the Jews."

At the time that the inventory was taken there was no religious poetry among the books, excepting the comprehensive collection in eight volumes made by Paul Wagner (see Vol. II., p. 278). Some of his heirs must, therefore, have already appropriated them. We find two series of fifty-two sermons, each by Neumeister, for the years 1721 and 1729, and there can be no doubt he must also have had Neumeister's sacred poems as well as those of Franck, Rambach, and Picander. But what else he may have had in this kind, or in secular literature, remains unknown.

When Bach moved, in 1723, to the cantor's house in St. Thomas's Church-yard, with his second wife, there went with him four children of his first marriage (see Vol. II., p. 8). By his second wife he had seven daughters and six sons, but only three daughters and three sons survived him. Of the rest three lived only a few days, the others their father laid in the grave at ages when they were beginning to understand something of life. Legendary lore has been busy with his family as well as with himself. He was said to have had an idiot son, David, who "having learnt music well, by his wild but most expressive and melancholy improvisations on the clavier, often drew tears from his hearers"; it was added that he died at the age of fourteen or fifteen.[526] This David never existed, nor did any child of Bach's die at that age;[527] it is possible that Gottfried Heinrich, the eldest of the second family, may have given rise to this fable. In Emanuel's opinion he had a great genius but it never developed. In the papers relating to Bach's property after his death he is spoken of as imbecile, and legally incapable of acting. He died February, 1763, at Naumburg, where he had been taken, it would seem, even before his father's death, by his brother-in-law, Altnikol.

Bach himself has told us of his happy home life, and of the little concerts he delighted in conducting with his sons, his wife, and his eldest daughter (see Vol. II., p. 254). He took conscientious care of the education of his children, and one main reason which decided his removal from Cöthen to Leipzig was the prospect of the better means of education offered by the University of that town. Wilhelm Friedemann, his favourite, was entered for matriculation as early as December 22, 1723, and he, as well as his younger brother Emanuel, had, according to the good custom of the time, the benefit of a complete course of academical education. It was not originally their father's intention that Emanuel should take up music as a profession, but when his great talent led to his doing so Bach was very well content, and

---

[526] Rochlitz, Für Freunde der Tonkunst. IV., p. 182 (3rd ed.). It has been copied from him even quite lately.
[527] See App. B., XV. The Necrology also mentions thirteen children of the second marriage. The genealogies never mention a son named David.

throughout his life he watched the musical proclivities of his sons with affectionate interest. He made their compositions known as he let them make his known,[528] published theirs and his own through the same publisher—Balthasar Schmidt, of Nuremberg—and copied out with his own hand anything by them that he particularly liked.[529] Friedemann's style was more nearly like his own than Emanuel's, and in his unbounded devotion to this eldest son he perhaps overlooked the fact that even during his own lifetime Friedemann risked becoming the mere caricature of himself. Emanuel's smaller and narrower forms are equally based on his father's work. For instance, his treatment of the first sonata movement in two sections, in which he was the precursor of Haydn, is founded on those preludes of the "Wohltemperirte Clavier" which are in two sections; and a still broader outline was offered by several of Sebastian's arias in which the first section does not close in the principal key. But even before Emanuel had brought out his well-known six clavier sonatas of 1742, Krebs had selected the same form for his "Preambles" of 1740. Emanuel's leaning towards what was popular, facile, and pleasing in style still farther diverted him from his father's road.

Bach spared no pains to smoothe his children's way through life; when Bernhard, the third son of the first family, was twenty years of age, and the post of organist to the Marienkirche at Mühlhausen fell vacant, Bach exerted himself to procure for his son the place which he himself had filled twenty-eight years previously. The letter he wrote to Mühlhausen on this occasion has been preserved:—

Most Noble and Most Learned Gentlemen, and particularly Most Worshipful *Senior* (of the Council), Most Esteemed Patron,

It has come to my knowledge that Herr Hetzehenn, organist to the town of Mühlhausen, died not long since in that town, and that

---

[528] Friedemann's Clavier Sonata of 1744 was to be had "1, of the author in Dresden; 2, of his father in Leipzig; and 3, of his brother in Berlin." While Sebastian's six three-part chorales " are to be had in Leipzig of Capellmeister Bach, of his sons in Berlin and Halle, and of the publisher in Zella."

[529] As Friedemann's Concerto for the Organ in D minor, which exists in his father's hand in the Royal Library at Berlin.

his place has not yet been filled up. Now, my younger son, Johann Gottfried Bernhard Bach, has for some time made himself so skilful in music that I undoubtedly consider him perfectly competent and capable to compete for the vacant post of town organist. I therefore request you, most noble gentlemen, with all reverence and submission, that you will be pleased to vouchsafe to my son your invaluable intercession for the obtaining of the post he applies for, and so to fulfil my desires and make my son happy; so that I hereby once more, as before for former favours, now again may find ample cause to assure you that I remain with unalterable devotion,

Your Honours',
And particularly your most Worshipful Senior's,
Most Devoted Servant,
JOH. SEBAST. BACH.[530]

Formerly organist to the church *Divi Blas*, at Mülhausen.
Leipzig, May 2, 1735.

Addressed to the Most Noble and Learned Herr, Herr Tobia Rothschieren, the Illustrious *Juris Consultus* and Honourable Member as well as Most Worthy *Senior* of the Learned and Wise Council of the Imperial and Free Town of Mülhausen, at Mülhausen.

Bach was not unsuccessful in reminding the Council of their former goodwill to him. Bernhard was elected, but before long gave up the post to turn his attention to learning. He went to Jena, where Nikolaus Bach, the worthy senior of the family, was living, and in 1738 was studying law there; but in the following year he was seized with a violent fever, which carried him off (May 27, 1739).[531] There were now, besides Gottfried Heinrich, only four sons left on which the father might found his hopes. He was still living when Johann Christoph Friedrich (born 1732) was appointed, still quite young, to be Kammermusicus to Count von Lippe, at Bückeburg. Hoffmann, the instrument maker, having bequeathed to him a clavier of his own making, Bach presented it to this son to aid in fitting him out for this place.

In the now diminished home circle, Johann Christian (born 1735) seems to have been the Benjamin of the family,

---

[530] The signature alone is autograph; the seal is a rose and crown; preserved at Mülhausen.

[531] Walther, in a MS. addition to the Lexicon, gives May 30. This is an error, as it is proved by the Jena register.

and to have enjoyed his father's particular affection. His talents were precocious, and his father gave him three claviers with pedals, all at once, so conspicuous a piece of partiality that after their father's death the children of the first marriage were prepared to dispute it.

Of the daughters, none married till Elizabeth Juliane Friederike (born 1726) married Altnikol, January 20, 1749. To enable the young people to set up a house in comfort, the father's assistance was again called in. The Council of Naumburg had in 1746 appealed to Bach's skill and knowledge with reference to the repairs of the organ of that town; since then the post of organist had become vacant. When Bach learnt this he immediately asked for it for Altnikol, without letting him know that he had done so. He urgently recommended him as his "former beloved scholar, who had already had an organ under his care for some time at Niederwiesa, and had competent knowledge both to play and manage it," who also " was exceptionally skilled in composition, in singing, and on the violin." Altnikol was accordingly chosen for the place July 30, 1748.[532] Bach's first grandchild, the issue of this marriage (born October 4, 1749), was named Johann Sebastian.

The only wedding which ever took place in his house was, of course, an event of great importance; we find traces of this in the second of two letters which he wrote at the end of 1748, to his cousin Elias Bach at Schweinfurth. These letters give us a glimpse of Bach's feelings as a householder, father of a family, and host.

Leipzig, October 6, 1748.
Most worshipful and respected cousin,

I must try to say much in a few words, as time presses, though I am heartily thankful for God's grace and blessing on the abundant vintage and on the marriage now soon to take place. With the copy of the Prussian fugue that you ask for I cannot at present oblige you, for the edition is *justement* this day sold out, since only 100 were printed, most of which have been given *gratis* to good friends. But between this and the New Year's Fair a few more will be printed, and if my worthy cousin is still minded to have a copy you have only to

---

[532] Bach's letters on this occasion were published at length by Friedrich Brauer, in Euterpe (a musical paper), published by Merseburger, Leipzig, 1864, p. 41.

give me notice of an opportunity with the remittance of a thaler, and it shall be forwarded at your desire. Finally, with salutations from us all, I remain,
Your honour's devoted,
J. S. BACH.

P.S. My son in Berlin has now two male heirs; the first born about the time when we, alas! suffered the Prussian invasion,[533] the other is 14 days old.

Addressed to Monsieur.
Monsieur J. E. Bach,
Chanteur et Inspecteur du Gymnase,
p. l'occasion.
a Schweinfourth.[534]

Elias had been in personal intimacy with Sebastian during his student days in Leipzig, and was full of regard and gratitude for him. He gave expression to these feelings in a present which gave rise to another letter from Bach.

Leipzig, November 2, 1748.
Most worshipful and respected cousin,
That you and your dear wife are still well, I was assured by your gratifying letter received yesterday with the splendid little cask of new wine for which accept hereby my thanks, as due. It is, however, much to be regretted that the little cask has suffered either from some jar in the carriage of it or other accident, for after opening it in this place for the usual inspection it was found almost a third part empty, and according to the report of the inspector now contains no more than 6 kannen (quart or so); for indeed it is a pity that of so noble a gift of God the smallest drop should have been wasted. However, for the good gift I have received from my worthy cousin I am heartily obliged, though I must *pro nunc* confess my inability, not being in a position to take any worthy revenge (make a worthy return). However, *quod differtur non affertur*, I hope to have an opportunity when I may in some way repay my debt. It is much to be regretted that the distance between our towns does not allow of our visiting each other in person; else I would take the liberty of humbly inviting my respected cousin to my daughter Liessgen's wedding, which is to take place in the next month of January, 1749, to the new organist of Naumburg, Herr Altnikol. But in consequence of the above-mentioned difficulty, and also of the inconvenient season, I cannot allow myself to hope to see you with us in person; I will only beg you, in your absence, to help them with your Christian good wishes, wherewith I beg to recommend

---

[533] November 30, 1745.
[534] On a sheet quarto, only one page written on; the seal is almost broken away. It belongs to Herr Schöne (Oberregierungsrath), of Berlin.

myself to my worthy cousin's remembrance, and with warmest greetings to you from all here,
I remain your honour's most devoted and
faithful cousin and servant to command,
JOH. SEB. BACH.

P.S. M. (agister) Birnbaum has been buried now six weeks.

P.M. (on the next page) Although my good cousin kindly offers to assist me in procuring the same liquor again, I must decline on account of the excessive expense here; for the freight was 16 gr., the delivery at the house 2 gr., the inspector, 2 gr., the town excise 5 gr. 3 pf., and the general excise 3 gr., so my good cousin may calculate that it costs me nearly 5 gr. a measure, which is somewhat too much for a present.

Addressed to
Monsieur
Monsieur J. E. Bach,
Chanteur et Inspector des Gymnasiastes de la
Ville Imperialle,
à Schweinfourth.[535]

Bach, it will be seen by these letters, was economical and exact; still, not to let this trait in his character preponderate unfairly, it must be remarked on the other hand that it was known to all that he gave a hearty welcome to every one, from far or near, who came to his house, that in consequence his house was hardly ever empty of visitors,[536] and without thrifty management it would have been impossible for him with his numerous family to keep up the comfortable and respectable style of housekeeping, simple as it was, which he so long enjoyed. The inventory of his property enables us fully to understand and appreciate this.[537]

Bach allowed himself a certain luxury in the matter of instruments; of claviers alone he had five—or six if we include the little spinett (spinettgen)—not counting the four he gave to his youngest son. Besides these he had a lute, two "lautenclaviere," a viol da gamba, and violins, violas, and violoncellos in such number that he could supply enough for any of the more simple kinds of concerted

---

[535] Two quarto pages written almost all over; the seal broken away. In the possession of Herr Schöne, of Berlin.
[536] Forkel, p. 45.
[537] See Appendix B., XVI.

III.

music. The fittings of his house, too, from the moderate stock of silver plate to the black leather chairs, and the laborious master's writing table furnished with drawers, give us the idea of modest but respectable ease. He even laid by some little savings; and it is a trait of the old Bach character—always ready to help any needy member of the family—that part of this money was lent to relations.[588]

It was in such a home as this, piously submitting to the common woes of humanity, and heartily and soberly enjoying the pleasures of German family life, that Johann Sebastian Bach lived and worked and awaited death. His vigorous body enshrined, even to advanced age, a healthy, energetic, and creative spirit. The only drawback was that a congenital weakness of the eyes was gradually but seriously increased by his incessant toil, in his youth often carried on through the night, and at last became serious disease. In the winter of 1749-50 he decided on following the advice of a friend and allowing an operation to be performed by a famous English oculist then resident in Leipzig; this was probably John Taylor, who had been much resorted to in Berlin. This unfortunately failed, both on a first and on a second trial, so that Bach was henceforth totally blind. Nor was this all. The medical treatment associated with the operation had such bad effects that his health, hitherto unfailing, was severely shaken. On July 18 he suddenly found his eyesight restored, and could bear daylight; but this was life's parting greeting; a few hours after he was stricken by apoplexy followed by high fever, and he died on Tuesday, July 28, 1750, at a quarter to nine in the evening.[589]

By his deathbed stood his wife and daughters, his youngest son Christian, his son-in-law Altnikol, and his pupil Müthel. He had been working with Altnikol only a few days before his death. An organ chorale composed in a former time was floating in his soul, ready as he was to die, and he wanted to complete and perfect it. He dictated and Altnikol

---

[588] "Frau Krebsin" was his wife's sister.
[589] Necrology, p. 167. Spener's *Zeitung*, Aug. 6, 1750.

wrote. "Wenn wir in höchsten Nöthen sein" ("Lord, when we are in direst need") was the name he had originally given it; he now adapted the sentiment to another hymn and wrote above it "Vor deinen Thron tret ich hiemit" ("Before Thy throne with this I come"). Joh. Michael Schmidt, the young theologian, who admired Bach to his dying day, said afterwards that "all that the advocates of materialism could bring forward must collapse before this one example."[540]

The funeral took place on Friday, July 31, early in the morning, in the church of St. John's; as was usual when church or school officials were buried, the whole school followed him to the grave. Many bells, too, were generally tolled on such occasions. The 31st of July was the second day of public humiliation in the year. In the church, which, for twenty-seven years, Bach's mighty tones had so often filled, the preacher announced from the pulpit, "The very worthy and venerable Herr Johann Sebastian Bach, Hofcomponist to his Kingly Majesty of Poland and Electoral and Serene Highness of Saxony, Capellmeister to his highness the Prince of Anhalt, Cöthen, and Cantor to the school of St. Thomas's in town, having fallen calmly and blessedly asleep in God, in St. Thomas's Churchyard, his body has this day, according to christian usage, been consigned to the earth."[541] His grave was near the church; but when,

---

[540] *Musico-Theologia*, p. 197.

[541] The note of this announcement written across a quarto leaf I found in the library of the Historical Society at Leipzig. And below it are the words: "Announced again on the next day of humiliation after July 31, 1750," in explanation of which I may mention that at that time three days of repentance were held annually in Leipzig.

In the Register of deaths, Vol. XXVIII., fol. 292b, we find "1750, Friday, July 31, a man died aged 67, Herr Johann Sebastian Bach, Cantor of the Thomasschule, ♂ 4 K. (children)"—"fees 2 Thlr. 14 gr." The age is given wrong, and if the statement four children is correct, Heinrich must have been already removed to Naumburg. The costs which, when the whole school attended, commonly amounted to about 20 thlrs., are here reduced to 2 thlrs. 14 gr., as was always the case when church or school officials were buried; the charges were as follows: 12 gr. for alms, 6 gr. for the burying, 12 gr. to the registrar, 1 thlr. 8 gr. to the two gate keepers.

A note which exists in the Town Library at Leipzig is as follows: "A man, 67 years, Herr Johann Sebastian Bach, Capellmeister and Cantor to the school

within this century, the graveyard was removed farther from the church and the old site opened as a road-way, Bach's grave, with many others, was obliterated, and it is now no longer possible to determine the spot where his bones were laid to rest.[542]

The mourning for Bach was universal wherever there were true musicians and friends of music. The Musical Union of Leipzig did honour to his memory by performing a mourning ode as a cantata; and Telemann dedicated a feeling sonnet to his departed friend.[543] Bach's colleague, Magister Thomas Kriegel, also celebrated him in a short but warmly expressed eulogium.[544] The Town Council seem to have less appreciated the loss the city had experienced; in two sittings, on August 7 and 8, it was remarked satirically that, "the school needed a Cantor and not a Capellmeister," and that "Herr Bach had been a great musician, but not a schoolmaster."

Bach left no will. His property was legally valued, after his eldest sons had greatly impaired it by abstracting all the musical portion of it; then the widow and children agreed to a division. Anna Magdalena undertook the guardianship of those under age, and Johann Gottlieb Görner, who during Bach's lifetime had often dared to consider himself his rival, assisted her greatly in the distribution of the property. She also was allowed the payment of her husband's salary for the half year following on his death, till December 13.

The whole family then dispersed. Friedemann, who had represented Emanuel's interests as well as his own, re-

---

of St. Thomas's; died at the school and was buried, with a hearse, July 30, 1750." This also comes from the Registry Office. The hearse (*Leichenwagen*), in contradistinction to the *Leichenkutsche*, was used for grand funerals. If the date July 30 is not a slip of the pen it may perhaps imply that the body was conveyed to the mortuary of St. John's Churchyard on the previous (Thursday) evening.

[542] The Necrology gives an approximate idea of its original situation, p. 172. There is no clue in the registers of St. John's churchyard. It is possible, though not certain, that some pages of them have been lost. Herr Heinlein, however, in his work, Der Friedhof zu Leipzig, 1844, p. 202, believes that he saw a page in which Bach was mentioned, but it had become illegible.

[543] Necrology, p. 173, Neueröffnetes, Hist. *Curiositäten-Cabinet*, 1751, p. 13.
[544] Nützliche Nachrichten, &c., 1750, p. 680.

turned to Halle, and Friedrich to Bückeburg. Heinrich remained with Altnikol at Naumburg; Christian, aged 15, was taken to Berlin for a time by his brother Emanuel. These sons, all talented, some very remarkably gifted, and with their great father's fame to support them, now tried to make their way in the world without the aid of his experience. It does not lie within the scope of this book to trace their fortunes or criticise their work. But at any rate they were so far successful that for at least one generation more the name of Bach was a name of credit and glory in the world of German art.

Anna Magdalena, left with three daughters, fell into poverty. In 1752 she was receiving moneys from the town, as she was in need, and had offered some musical relics for sale. Whether the sons could not or would not help is not known, but it is certain that her circumstances became narrower, till at last she lived on public benevolence. She died February 27, 1760, as an "alms woman," in a house in the Hainstrasse. Her coffin was followed to the grave by a quarter of the school, as was usual with quite poor folks, and the place of her burial is unknown. The town left the widow of one of its greatest sons—herself, too, an artist—to perish thus.

The then surviving unmarried daughters lived to see a day when their father's music was less and less remembered. Katharina Dorothea died on January 14, 1774; Johanna Carolina on August 18, 1781. It was only his youngest child, Regina Johanna, who lived till Germany began once more to value Sebastian Bach. She, too, lived in privation and solitude, and it is to the credit of Rochlitz that, by an appeal to the public, he at last secured her ease in the evening of life. She died December 14, 1809, the last of all the family.

Having followed the course of a great man's life to a close, we will not dwell on the melancholy picture of the ruin of all that he had constructed and the dispersal of all he had held together. What under such circumstances is lost is undoubtedly the least precious portion of what he has created. It is true that Bach's creative spirit worked less actively and fruitfully in the succeeding generation than has often been the case with a great genius. And it is especially in Bach's sons that we may mark the decay of that power which had culminated after several centuries of growth, and which utterly disappeared in their posterity. But, in truth, for nearly a century the whole German nation have entered into that inheritance; it has recovered its connection with Bach, and, through him, with the almost forgotten centuries of its own musical history. The works of his creation—the highest outcome of an essentially national art, whose origin lies in the period of the Reformation—are like a precious seed which bursts the soil at last to be garnered in perennial sheaves. Henceforth it will not be possible that Bach should be forgotten so long as the German people exist. His resurrection, in the works of a later generation of artists, has already begun; but we who are not of the mystic guild have our duty too, each in his degree, to labour that the spirit of the great man may be more widely understood and loved.

APPENDIX (A, TO VOL. III.)

1 (p. 39). **The B minor mass** exists entire, or in part, in three original MS. copies: the autograph complete score in the Berlin Library; the original parts of the *Kyrie* and *Gloria* in the library of the King of Saxony, at Dresden; the autograph score of the *Sanctus* in its first state in the Berlin Library. The parts belonging to the whole score and to the *Sanctus* are lost, as is the separate score of the *Kyrie* and *Gloria*. In the collective score these two numbers show some deviations from the parts preserved at Dresden, which proves that this complete score was written later than the Dresden parts. From the watermark, M A, it would have been about 1736 (see Vol. II., p. 697). As regards the date of the *Credo*, it is surprising to discover that it is written on the same paper as the cantata "Wir Danken dir Gott," and parts of the cantatas "Herr, deine Augen," and "Wer weiss, wie nahe" (see Vol. II., p. 702). On this we may base the hypothesis that it was composed in 1732, thus earlier than the *Kyrie* and *Gloria*. Against this, however, it may be said that in the *Agnus*, too, the same watermark seems to be traceable, though too faint for certainty. From this we might infer that the *Agnus* also was older than the *Kyrie*; but from the connection of the *Agnus* with the *Sanctus* through the *Osanna* this is highly improbable; in this instance the internal evidence seems to me to outweigh that of the watermark. In the first sketch, even the *Sanctus* is distinguished as an independent composition by the *J. J.* at the beginning, and at the end, *Fine SDG.* The theme of the fugue *Pleni sunt cœli* was at first written thus:—

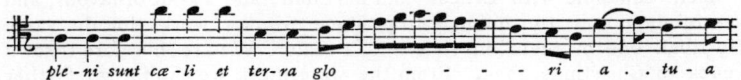

and in this form it is hastily noted down on the first page, as was usual with Bach when an idea occurred to him while he was working up another composition. But that this *Sanctus* was originally a Christmas piece is clear from the fact that the following lines are to be seen on the same page and written with the same ink as the rest of the sketch:—

Brü-der-lein zu sein, ach wie ein süs-ser Thon, wie freund-lich sieht er aus der

gro-se Got-tes-sohn.

In Thee is all my joy, of morning stars the brightest!
O Jesu, Lord of love, Thy holy word Thou plightest
My brother to be called: how gracious is the word!
How tenderly benign Thy glorious face, O Lord!

This is part of a Christmas hymn by Caspar Ziegler, with a little known air, both of which Bach made use of for the third day of Christmas (St. John's day). He has here noted down the melody, evidently in order to return to it for the cantata which he intended to perform for the first time on the same occasion as the *Sanctus*. The watermark of both is the half moon in the first half-sheet, the second being blank, and Bach began to use such paper as this in 1735. On the same page as the above quoted, staves of music are written in Bach's hand, "*NB.* Die *Parteyen* sind in Böhmen bey Graff *Sporck*" ("The parts are with Count Sporck, in Bohemia"). This Graff Sporck must undoubtedly be the same as is spoken of at p. 43 of this Vol.; he died March 30, 1738, and his family was extinct, as he left no son; so, as Bach had sent him the parts, the *Sanctus* must have been written for Christmas, 1737, at latest. We may, however, question whether in fact the score also may not have been sent to him, and he have returned that only and retained the parts. If this were the case—and it seems to me a permissible suggestion—the *Sanctus* may have been written earlier. Bach would hardly have composed it in 1736, as he was engaged in a violent squabble with Ernesti and his choir, and so out of favour, and certainly not in the humour to produce two new works for the same festival, one of them on so grand a scale. We are therefore thrown back to Christmas, 1735. From the watermark we cannot go farther back than this; but there is another point in favour of this year. If we compare the closing chorus of the Easter oratorio with the *Sanctus* we trace in them a remarkable relationship. Both are constructed in the form of the French *ouverture*, and each has, in its two sections, similar proportions; in each there is a movement in ³⁄₈ time following one in common time, with the change conducted in the same manner. Still, no one can suppose that the Easter oratorio chorus is the earlier. On the contrary, it rather seems like the outcome of a somewhat weary advance along the road opened up in the *Sanctus*. The Easter oratorio was probably composed in 1736 (see Vol. II., p. 714).

APPENDIX. 281

2 (pp. 73, 75). This new and distinct watermark

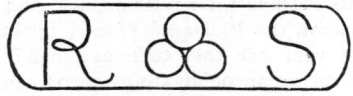

characterises the following cantatas:—
1. Ihr werdet weinen—Third Sunday after Easter.
2. Es ist euch gut, dass ich hingehe—Fourth Sunday after Easter.
3. Bisher habt ihr nichts gebeten—Fifth Sunday after Easter.
4. Auf Christi Himmelfahrt—Ascension Day.
5. Sie werden euch in den Bann thun (A minor)—Sunday after Ascension Day.
6. Wer mich liebet—Whitsunday.
7. Also hat Gott die Welt geliebt—Whitsun Monday.
8. Er rufet seine Schafe—Whitsun Tuesday.

These, as may be observed, form an unbroken series. At the top of the last page of No. 1 are the words *Dominica Quasimodogeniti. Concerto*, and below, a beginning of some music crossed out, seven bars long, of which this is the upper part:—

Whether this sketch was ever worked out is not known; if it was, the composition is undoubtedly lost. At any rate, Bach intended, with the music for the Third Sunday, also to compose a work for the First Sunday after Easter. Thus there would be missing only a cantata for the Second Sunday after Easter to complete a series from Easter to Trinity Sunday, and it will presently be shown that this cantata probably is extant. I regard these eight cantatas as composed consecutively, for one and the same year. Considering Bach's extensive use of writing materials, and the fact that paper with this watermark was never used by him, excepting in these cantatas and in a single MS., to be mentioned presently—so far as is known—considering, too, their natural connection, the contrary hypothesis hardly seems tenable. Besides, as to Nos. 6 and 7, we have yet other evidence. No. 6 is in two of its movements no more than a remodelled form of an older Whitsuntide cantata, beginning with the same words (see Vol. I., p. 512, and Vol. II., p. 688). On the back of the autograph score of that earlier composition we find the thema of the closing chorus of No. 7, and with it its first answer; thus Bach must have had recourse to the original Whitsuntide cantata to use parts of it for the new one, and while he was engaged in the work the theme of the fugue of No. 7 occurred to him, and he noted it down on the back of the sheet. These eight cantatas also exhibit a general affinity in form and structure which betrays their relationship, and especially we find one unusual instrument, the *Violoncello piccolo*, introduced into several of them. To which year are they to be assigned? The score of No. 4 directs the use of an organ obbligato as accompaniment to the duet. For this an independent Rückpositiv was indispensable, and we have seen (Vol. II., p. 282) that this was not constructed till 1730, so that No. 4, and consequently the whole series, was not composed before 1731. However, two cantatas exist, "Der Herr ist mein getreuer Hirt" for the Second Sunday after Easter, and "Ich liebe den Höchsten" for Whit Sunday, which must on good grounds be assigned to the year 1731 or 1732 (Vol. I., p. 699). Thus one of these years is thrown out of count. 1733 is equally out of the question, on account of the general mourning. Now, it is my opinion that it was only in specially exceptional cases that Bach wrote cantatas for the same Sundays in two successive years, and for this reason I believe this group cannot have been written before 1734.

But there is other evidence. Bach wrote a cantata, "Gott der Herr ist Sonn und Schild," for the Reformation Festival, and on the blank page of the fourth sheet of the autograph we find written at the top "*J. J. Doicã Exaudi* (Sunday after Ascension Day) Sie werden euch in den Bann thun," and below it a very charming and peculiar beginning for five instruments, of which the first two lines are written in the violin clef, the two middle ones in alto, and the lowest in the bass. I believe the four upper instruments to have been for oboi (*d'amore* and *da caccia*). Above the lowest line that is filled, one is left empty for a bass voice; key A minor, common time. There are but seven bars in the fragment. On comparing this with No. 5 of this group it is plain that we have here the composer's first sketch of that cantata, so that if we could determine the date of the Reformation cantata we should obtain a trustworthy landmark. We have seen that Bach and the writers of his texts always had due regard to the general character of the Sunday or festival for which they were writing, and some to the purport of the text appointed for the sermon. There was no fixed and regular text for the Reformation Festival, one was chosen each year by the authorities. I have been so fortunate as to come upon a MS. list of the texts prescribed for this festival during Bach's residence at Leipzig. It is among the "Ephoralarchiv" of that town, "*ACTA* die Feyer des Reformations-Festes betr. (concerning the Festival of the Reformation) *Superintendur,* Leipzig, 1755." In 1723 the Gospel for the Sunday on which it falls, being the Twenty-third Sunday after Trinity; 1724, Ps. xv., 1, 3; 1725, Heb. iii., 7, 14; 1726, Matt. xvi., 24, 26; 1727, Ps. lxxxv., 6, 8; 1728, Gospel, Twenty-third Sunday after Trinity; 1729, Titus ii., 14; 1730, Rev. xiv., 6, 8; 1731, Luke xiii., 6, 9; 1732, 2nd. Tim. i, 12; 1733, Is. li., 15, 16; 1734, Gospel, Nineteenth Sunday after Trinity; 1735,

Ps. lxxx., 15, 20; 1736, Rev. iii., 1, 6; 1737, Rev. xiv., 6, 8; 1738, Rev. iii., 14, 18; 1739, Ps. xxxiii., 18; 1740, 2 Thes. ii., 10, 12; 1741, Rev. xiv., 6, 8; 1742, Rom. vi., 17, 18; 1743, Rev. iii., 10, 11; 1744, Is. lv., 10, 11; 1745, Gospel, Twentieth Sunday after Trinity; 1746, omitted; 1747, Rev. xiv., 6, 8; 1748, Is. xxvi., 9, 10; 1749, Is. xlv., 22, 25. Glancing through these texts, we see that the words of the cantata "Gott der Herr ist Sonn" only suits those for 1733, 1735, and 1739. The Reformation Festival had, no doubt, a very distinct poetical sentiment of its own, and when any special occasion took the precedence, as in 1730 and 1739, the years of Jubilee, it would be misleading to seek for any close connection between the sermon and the cantata. Thus the cantata "Ein feste Burg" (see Vol. II., p. 470) may very well have been connected with the sermon in 1730; still, it is possible that it was not written till 1739. But it is hardly possible that the cantata "Gott der Herr" should have been written in 1739, since— firstly, several portions of it have been worked into the A major and G major masses, and internal and external evidence alike concur in dating these in 1737-38; and secondly, the watermark of the original parts is found again in the original MS. of the music for Easter Monday, "Erfreut euch ihr Herzen," and for Easter Tuesday, "Ein Herz, das seinen Jesum," so that we must assume that these works were composed at about the same time. But in the parts of the Easter music we also find the mark M A, which is not to be met with in any other MS. by Bach of 1739. Thus we must fall back upon the years 1733 and 1735.

The above-mentioned sketch of No. 5 is on the fourth sheet of the Reformation Cantata; but the complete score fills altogether six sheets. Of these, 1, 2, 5, and 6 are alike in the size, quality, and watermark of the paper. Sheets 3 and 4 are different in size from the rest and from each other. It is plain that Bach, when proceeding to write the Reformation Cantata, used a few sheets of old paper which he had not used before, or used but little, and which fell under his hand. The contrary idea, that he should have sketched the first idea of No. 5 in the Reformation Cantata after the other was finished, is quite untenable, since it occurs in the middle of the score, and any one can see that Bach simply skipped over the page in question, because it was already partly used (there are other notes of musical themes upon it). Besides, the watermark in this sheet, though not very distinct, is recognisable by a practised eye as the same as that in the paper of the other eight cantatas in question. From this it would follow that the cantata for the Sunday after Ascension day was composed before the Reformation Cantata, and if so it cannot possibly have been written in 1733. For in 1733 Bach could not have composed this series, since there was no concerted music at all between Quinquagesima and the fourth Sunday after Trinity. The Reformation Cantata would thus have been written in 1735, and this leads us directly to the date of this series, which must also have been 1735.

With regard to No. 4, there is still a doubt, and we must once more refer to the Reformation Cantata. Sheets 1, 2, 5 and 6 of the score have the watermarks of a stag on one half sheet, and on the other

while in most of the parts we find this mark—

These marks recur in the following cantatas :—

(*a.*) Denn du wirst meine Seele—Easter Day.
(*b.*) Erfreut euch ihr Herzen—Easter Monday.
(*c.*) Ein Herz, das seinen Jesum—Easter Tuesday.
(*d.*) Gott fähret auf—Ascension Day.

And *a* and *d* have the same watermarks as the score of the Reformation Cantata; and *b* and *c* that of the parts with the addition on the other half sheet of the figure of an eagle. It is evident that *a* and *d* must have been written nearly at the same time as the Reformation Cantata, and somewhat before it. But the paper on which the Reformation Cantata is written is of no less than four different kinds, showing that the economical composer was using up the remains of a former stock, particularly as the watermarks do not recur in any composition which can with certainty be dated later, excepting only a few scraps inserted into the parts of the St. John Passion for the third performance, probably in 1736. But if *a* and *d* do belong to the year 1735, we find that we have two cantatas by Bach for that year, a circumstance which must be considered as suspicious; and if I accept the hypothesis after mature deliberation, it is on the following grounds: Firstly, two cantatas were in fact needed for every Ascension day (see Vol. II., p. 193). Secondly, it was by no means unusual with the composers of that period to compose music for both the morning and evening services of the same day (see Vol. II., p. 270), and there can be no doubt that the impulse towards composition was very strong in Bach in 1735. Thirdly, if, in the absence of any alternative, we conclude that the cantatas *a* to *d* form a series, as 1 to 8 do, the fact that two of them, *a* and *c*, are remodellings of older works is an argument in favour of their having been written in this year, for it is impossible not to perceive that at this time Bach constantly had recourse to his earlier works; 6 and 7, in the first series, are partly derived from Weimar compositions, while we know that the St. Luke Passion had been remodelled not long before, and reproduced probably in 1734. The Weimar cantata, "Komm du süsse Todesstunde," was also taken out again about this time, and the most probable hypothesis is that it was on February 2, 1735, for the paper of the score now extant is exactly like that of October 5, 1734 (a small eagle and HR ; see Vol. II., p. 708). The title indicates first the original purpose of the cantata, the sixteenth Sunday after Trinity, and below this, its subsequent use, *Festo Purificationis Mariae*. Finally, the date of the watermark in *a* and *d* is further supported by an organ part to Keiser's St. Mark Passion, prepared by Bach for his second performance of that work in Leipzig, the first having been in Weimar. Bach, having produced his own Passions in 1729, 1731, 1734, and 1736, the year 1735 remains open for Keiser's Passion. I must not omit to mention that among the parts of No. 8 a few leaves occur, which show them to be contemporaneous with the score of the Easter oratorio (see Vol. II., p. 715). From this we can only conclude that No. 8 was performed again at Whitsuntide, 1738, and that those portions of the copy that had been lost were then replaced.

Thus we have for the year 1735 a series of cantatas from Easter day till Whit Tuesday inclusive, and two for Ascension day. Only a cantata for the first Sunday after Easter is wanting, which was no doubt included by Bach, and if performed has been lost. That for the second Sunday after Easter has, I believe, survived in the composition "Ich bin ein guter Hirt." There is no difficulty as to the watermark, since it is the same as recurs in the scores of Nos. 2 and 3 (see below, note 3) and in musical character it has a marked affinity to the others. It must be further discussed in the next note.

3 (p. 91). **The Half Moon Watermark on the first half sheet** (the other being blank) is characteristic of the greater number of the cantatas of the last period of Bach's works. It occurs in the following:—

1. Ach Gott vom Himmel sieh darein.
2. Ach Gott, wie manches Herzeleid (A major).
3. Ach lieben Christen seid getrost.
4. Ach wie flüchtig, ach wie nichtig.
5. Aus tiefer Noth.
6. Bisher habt ihr nichts gebeten.
7. Bleib bei uns, denn es will Abend werden.
8. Christ unser Herr zum Jordan kam.
9. Christum wir sollen loben schon.
10. Das neugeborne Kindelein.
11. Du Friedefürst, Herr Jesu.
12. Erhalt uns Herr bei deinem Wort.
13. Es ist euch gut, das ich hingehe.
14. Gelobet seist du, Jesu Christ.
15. Herr Christ der einig Gottssohn.
16. Herr Gott dich loben alle wir.
17. Herr Jesu Christ wahr Mensch und Gott.

18. Ich bin ein guter Hirt.
19. Ich freue mich in dir.
20. Ich hab in Gottes Herz.
21. Jesu nun sei gepreiset.
22. Liebster Immanuel.
23. Mache dich mein Geist bereit.
24. Meinen Jesum lass ich nicht.
25. Meine Seele erhebet den Herren.
26. Mit Fried und Freud ich fahr dahin.
27. Nun komm, der Heiden Heiland (B minor).
28. Schmücke dich, o liebe Seele.
29. Was frag ich nach der Welt.
30. Was mein Gott will, das g'scheh.
31. Wo soll ich fliehen hin.

These are all chorale cantatas with the exception of Nos. 6, 7, 13, and 18. We can verify the date at which this paper was first used by Nos. 6 and 13, since in the original MSS. we find, besides the half moon, the watermark (see woodcut in note 2) which distinguishes the group immediately preceding: it is 1735; and further evidence is to be found in the connection between No. 7 and the Easter Oratorio (see Vol. II., p. 715) and between No. 19 and the *Sanctus* of the B minor Mass (see note 2). In No. 11 the text bears unmistakable reference to the war in which the country was then engaged, though it must also be admitted that the immediate suggestion was given by the Epistle for the day, which prophesies the dispersion of the Jews. Still the use of the hymn " Du Friedefürst " points to some contemporary event of a similar character. The introduction of an independent verse as a recitative is, however, conclusive. The text—as is always the case with Bach's later chorale cantatas—is mainly a paraphrase of some of the verses of a well known hymn. The alto aria is founded on the second verse, the tenor recitative on the third; the fourth and fifth are omitted; verse six is adapted for the terzett; the seventh, which is the last, is simply treated as the closing chorale. But the words of the alto recitative immediately before this finale are quite independent of the hymn, and to this effect :—

> Let not Thy people bleed, O God,
> Too long beneath Thy rod!
> O Lord, who art the Lord of law and right,
> Thou know'st the adversary's wrath,
> His cruelty and lawless might.
> Put forth Thy strong and saving hand,
> Defend our terror-stricken land;
> For Thou canst bid their raging cease,
> And keep us in abiding peace.

These words indicate some stress of war in Saxony itself, and in 1774, the first year of the second war with Schleswig, hostilities commenced with an incursion of the Prussian foe through the heart of Saxony. In 1745 Saxony suffered even more severely, but there was no twenty-fifth Sunday after Trinity—for which this cantata was written—so that 1744 is the latest year in which this watermark can be proved to have been used. In Nos. 21, 29, and 31 of this list we find among the sheets with the half moon others with the M A watermark, and these must therefore be ascribed to the earlier years of the period between 1735—44; not later probably than 1736. No. 29 is for the ninth Sunday after Trinity. In 1736 Bach was absent from Leipzig on the eighth and ninth Sundays after Trinity (July 22 and 29; see ante, p. 6), so this one at any rate must be assigned to 1735. So probably must No. 31, written for the nineteenth Sunday after Trinity; since at the end of 1736 Bach's quarrel with Ernesti was going on; and for the same reason No. 21, a New Year's cantata, is no doubt to be dated 1736.

In 29 we detect a third watermark, an eagle, but this occurs in such various forms and at such various periods as to be useless in chronological enquiries. I have, however, classed together the MSS. in which it appears exactly in the same form. These are:—

Jesu der du meine Seele;
Wass Gott thut, das ist wohlgethan, G major—21st Sunday after Trinity;
Allein zu dir, Herr Jesu Christ;
Nimm von uns Herr;

so that these cantatas might be supposed to belong, like No. 29, to the year 1735, but, as it happens, the twenty-first Sunday after Trinity in 1735 was the Reformation Festival, when "Gott, der ist Sonn und Schild" was most likely given; at any rate, not a regular Sunday cantata.

4 (p. 113). **The five hymn tunes, preserved to us by Joh. Ludwig Krebs,** are to be found in a MS. collection of pieces for the organ and harpsichord which was inherited from him by his successors in office at Altenburg, until they fell into the hands of Herr F. A. Roitzsch, of Leipzig, who has always authorised me to make full use of his treasures for my immediate purpose.

The melodies are set to the hymns "Hier lieg ich nun, mein Gott, zu deinen Füssen," "Das walt mein Gott, Gott Vater, Sohn," "Gott mein Herz dir Dank," "Meine Seele, lass es gehen," "Ich gnüge mich an meinem Stande." This book includes compositions by Buxtehude, Reinken, Böhm, Leyding, Walther, Kauffmann, Bach, and others, in no sort of order, and, though it cannot be directly proved, it is most likely Krebs selected and copied them while he was Bach's pupil at Leipzig.

He was at the Thomasschule till 1735, and a student at the University till 1737, and must have been familiar with Bach at the time when he was at work on the musical portion of Schemelli's hymn-book. These five melodies are not sketches but copies, and very hasty ones, for in three the air alone is given, and the bass even not indicated. But as the transcriber thought them worthy to be included in a book with works by the most illustrious masters, he must have esteemed them highly; in fact, they bear the stamp of a master-hand, and examination will show us that that hand is Bach's.

What leads us directly to the conclusion that Bach wrote them as addenda to the melodies for Schemelli is this: The hymn "Hier lieg ich nun o Herr (mein Gott) zu deinen Füssen," though it had been made popular by Freylinghausen's hymn-book, was not included in Schemelli's, while, on the other hand, it has the five penitential hymns attributed to Johann Arndt, "Hier lieg ich nun, o Vater aller Gnaden," set to the same tune. This hymn seems never to have become more extensively known; it is not mentioned even by J. B. König. Neither it nor the fellow hymn in Freylinghausen seems to have had any original melody of its own; both are to be sung to that of "Der Tag ist hin, mein Jesu bei mir bleibe." But the words of "Hier lieg ich nun, o Vater aller Gnaden" are such as might well inspire Bach to set them to a tune of their own; and in point of fact the air preserved by Krebs not only suits Freylinghausen's charming verses less well on the whole than it does Schemelli's, but in detail it is so admirably appropriate to this latter, and its fitness is so thoroughly characteristic of Bach, that in my opinion there is not the slightest doubt that it was actually composed for it. To facilitate comparison I have given both the texts in the musical supplement.

The case seems to be somewhat different with regard to the air to "Gott mein Herz dir Dank." The verses, which are by the Countess Amélie Juliane, a thanksgiving after partaking of the communion, is not to be found in Schemelli's hymn-book, but he has a morning hymn for a communicant, which seems to have been tolerably exactly imitated from it, "Gott sei Lob, der Tag ist kommen." But there was no proper melody for either of these, and the hymns themselves seem to have been included only occasionally. "Gott sei Lob" occurs in Part II. of the Arnstadt hymn-book of 1745; "Gott mein Herz" was known in Sondershausen through Heinrich N. Gerber's "Choralbuch." If Bach wanted to supply a melody for Schemelli's hymn-book it was obvious that he should make use of the original, which he probably had known in Thuringia, rather than of the imitated version; and the air, though it suits both, is more perfectly suited to the former.

The words to the other melodies preserved by Krebs are all to be found in Schemelli's hymn-book. "Das walt mein Gott" had long had a proper tune; "Meine Seele lass es gehen" is marked to be sung to the tune of "Herr ich habe missgehandelt," but it had also proper tunes, which are given by Dretzel and by König. The tunes given by Krebs are quite unlike these, and occur nowhere else. The hymn "Ich gnüge mich an meinem Stande" has no tune of its own. Dretzel, in compiling his collection "Des Evangelischen Zions Musicalische Harmonie" (probably 1731), found only the text in the hymn-books, and himself added a tune to it; but we find no reference to this in Schemelli, who enjoins that it shall be sung to the tune of "Wer nur den lieben Gott lässt walten"; but the metre of the last two lines does not perfectly agree, so the need of a proper melody must have been felt. It may be remembered that Bach had set a cantata by Picander beginning "Ich bin vergnügt" (see Vol. II., p. 443). Picander had evidently this hymn in his mind, as we see from the resemblance in feeling, as well as in several turns of phrase. Bach, therefore, must have had a special interest in the hymn given by Schemelli, and it is probably not by mere accident that the air wrote for it, like the cantata, is in E minor.

5 (p. 153). **The English Suites** are not known in autograph. But among the literary remains of Heinrich N. Gerber there are copies of four of them, those in A major, G minor, E minor, and D minor, which, after the death of Ernst Ludwig Gerber (the writer of the "Lexicon"), became the property of Hofrath André, and then of Herr Ruhl, music director at Frankfurt-am-Main. They are now in the possession of Dr. Erich Prieger, of Berlin, who kindly allowed me to make use of them. Gerber made these copies when he was studying at the University of Leipzig, and under Bach 1724—1727. This is evident from the superscription to the first suite: "*L. [itterarum] L. [iberalium] S. [tudiosus]. a [c] M. [usicæ] C. [ultor]*; and besides this the copies are, as to writing, paper, and ink, exactly similar to a figured bass written by Gerber to a sonata by Albinoni (see Vol. II., p. 293). He also copied eight other suites, including six of the French suites, at the same period and in the same style. This was of course the result of the teaching he obtained from Bach (see ante, p. 127). Though he copied only four he must have known of a fifth, at least; for he designates that in A major as the *first*, G minor as *second*, E minor as *fourth*, and D minor as *fifth*; thus either that in F major or that in A minor must have been already written. But it is probable that all six were by this time finished, since Bach wrote them to order all at about the same time, and in 1726 we find him already at work on another grand set of suites, the six Partitas of the "Clavierübung." The most important copy of the English set, next to that by Gerber, is one by Joh. Christian Bach, Sebastian's youngest son. The MS. is in the hands of Herr Arnold Mendelssohn, in Bonn. The title page of the G minor suite, which is in the same hand as all the copies, and apparently of the same date, is as follows: *pp* (*i.e., proprium*) *J. C. Bach;* and on the title page of the A major suite the owner has added, evidently at a later date: *pp Jean Chretien Bach*, and this title is also interesting from having the note: *fait pour les Anglois*. From this we may infer that there is some truth in Forkel's statement that Bach composed these suites for an illustrious Englishman. The copy is throughout very careful, but it is not possible to determine when it was

III.

made. The writing is tolerably free and bold; if Christian Bach wrote them out at Leipzig in his father's lifetime it must have been in 1749 or 1750 at earliest, as he was born in 1735. The MS. is, however, no longer complete; the suite in F major is wanting.

6 (p. 155). **The Clavierübung (or Practice).** Only Part I. of this work has the date of publication on the title page, and the dates of the first appearance of the other parts has hitherto only been approximately known. However, as regards Part II. we find a note by J. G. Walther, written into his copy of the Lexicon, which passed from Gerber's library into that of the Gesellschaft 'der Musikfreunde' at Vienna, saying that " It was brought out for the Easter Fair of 1735 by Joh. Weigel, engraved on copper." Part III. is mentioned by Mizler (Mus. Bib., VII., Part I., p. 156) among " remarkable musical novelties." This volume of the Mus. Bib. came out in 1740; Part VI. of Vol. I. had appeared in 1738. Mizler would beyond a doubt have mentioned Bach's work as soon as he knew of it, so this assigns it to 1739 at the earliest. It is not impossible that it should have been as late as 1740, but not probable, because Mizler's book was certainly brought out at Easter 1740, and Bach's work could not have appeared earlier in that year. Mizler simply writes: "Here, too, Herr Capellmeister Bach has published Part III.," &c., and he would have noticed it more fully if he had seen it before it was given to the public. As to Part IV., which Bach himself did not designate as Part IV., but simply as "Clavierübung," while in size and shape it also differs from the other three—thus much is certain as to its first publication: It was brought out in Nuremberg by Balthasar Schmidt, who also published, in 1742, the six Sonatas by Emmanuel Bach, dedicated to Frederick the Great. Now these are numbered 20 by the publisher, and the Variations by Sebastian Bach are numbered 16; these, then, must have been printed first. But how long before? Balthasar Schmidt printed a good deal in his time, considering his position. Thus, a concerto (M. Scheuenstuhl, in G minor) published by him in 1738 is No. 9, and in 1745 we already find him at No. 27 (Em. Bach's Concerto in D major). J. G. Walther, who watched the novelties in the music market with a keen eye, says in a MS. note to the Lexicon, that Em. Bach's six sonatas came out "circa 1743." From this we may infer that the sonatas were brought out, not by Easter, 1742, but towards the end of that year. Thus Seb. Bach's Variations may quite well have appeared for the Easter Fair, 1742. This would be certain if we were justified in attributing any great weight to the fact that in the statement as to the origin of the Variations the order for them is ascribed to "Graf" (Count) Kayserling. For his original title was only Freiherr or Baron, and the archives in Dresden show that it was August III. who raised him to the dignity of "Graf," Oct. 30, 1741. But this is not very trustworthy evidence, and we must be satisfied to decide that the Variations were published not later than 1742, and probably in that year.

APPENDIX. 291

7 (p. 181). **The Chromatic Fantasia and Fugue.** Among the papers of Herr Grasnick, of Berlin, who died in 1877, there was a MS. copy of this work in an unknown hand, dated December 6, 1730. The MS. consists of fourteen sheets of small oblong quarto, and was evidently put together expressly for this work, which almost fills it, the rest is written over with little inventions, which are evidently the attempts of a beginner. Notwithstanding numerous mistakes, made apparently by an unskilled copyist, in the absence of the autograph, this is by far the most reliable copy extant. Griepenkerl has added to his edition of the Chromatic Fantasia and Fugue (P. S. I., C. 4, No. 1—207) two variorum readings, the second of which has no independent value, while the first, which is derived from a MS. of the work by F. W. Rust, Capellmeister at Dessau, 1757, shows an earlier and possibly the original form of the work. The MS. of 1730 has already lost this form, so we are justified in dating the actual composition of the work at least ten years earlier. The accuracy of the form in which—as being probably the ultimate purpose of the author—Griepenkerl has edited the work, seems in more than one respect extremely open to suspicion. For wherever we find the older (or Rust's) form agree with the later (or Grasnick's) we must agree to acknowledge Bach's unaltered purpose, as a third remodelled form from his hand is most unlikely to have existed. Indeed, we have further support of this view in a MS. by Kittel, one of Bach's latest scholars, and of the Chromatic Fantasia only, by Müthel, his very last pupil. The exact *replica* of the earlier form was in Grasnick's bequest, and the later recension has for a long time been in the Royal Library at Berlin. Both agree, in all important points, with the MS. of 1730; the most conspicuous variant in the fantasia is in bar 49, at the beginning of the recitative, where, in opposition to Griepenkerl's reading, we find the following, which is now universally accepted, and seems in itself more justified :—

The fugue also shows some remarkable differences, as in bar 72. The upper part is :—

Griepenkerl follows Forkel's MS. In 1819, one year after Forkel's death, he published, through Peters, of Leipzig, a second edition of the Chromatic Fantasia and Fugue under the title " Neue Ausgabe mit einer Beźeichnung ihres wahren Vortrags, wie derselbe von J. S. Bach, auf W. Friedemann Bach, von diesem auf Forkel und von Forkel auf seine Schüler gekommen." (A new edition, with indications of the right way of playing it, as it came from J. S. Bach, through W. F. Bach, to Forkel, and from him to his pupils.) He also added a preface that is well worth reading, on Bach's method of playing, as derived by Forkel from Friedemann Bach, and so handed down to the present time. Forkel, indeed, first obtained the Chromatic Fantasia and Fugue in MS. from Friedemann Bach (see his work on J. S. Bach, p. 56). We may therefore assume that all that we find as deviations from the most trustworthy of Forkel's other MS. copies is also due to Friedemann Bach, who, in the spasmodic vagaries of his genius, was not unfrequently deficient in reverence towards his father's great works.

8 (p. 191). The "Musikalisches Opfer." A perfect copy of the original edition exists in the Amalien Library of the Joachimsthal Gymnasium at Berlin. It has a special value as being the dedication copy sent by Bach to Frederick the Great, who must have given it to his sister. This copy proves two things. First, that under the title "Musikalisches Opfer" Bach originally included only the three-part simple fugue, six canons and the canon fugue, so that when he began the work he was not fully determined as to its extent and character. This dedication copy contains (*a*) three leaves of music and two with the title and dedication. The paper is remarkably fine and thick, in very large oblong folio, and the five leaves are bound in leather, with gold tooling. The music consists of the three-part simple fugue and one canon, in which the alto had the *Cantus firmus*, while the treble and bass have counterpoint in canon. This canon is entitled *Canon perpetuus super Thema Regium*, the fugue is called a *Ricercar*. (*b*) An upright folio sheet of the same character as to the size and quality of the paper, but only laid in, on account of its being cut the other way, and the two inside pages have printed on them five canons and a *Fuga canonica in Epidiapente*," with the title to the whole, "*Canones diversi super Thema Regium*." All this is printed, but all the sheets have written notes and additions. Besides the complimentary addresses given in the text of this volume, which are inserted to the right of canons 4 and 5 on the upright folio sheet, we find on the blank first page of this sheet the title "*Thematis Regii elaborationes canonicæ*"; on the first page, also blank, of the oblong folio, "*Regis Jussu Cantio Et Reliqua Canonica Arte Resoluta*." This is the reason why the publishers of the later editions put this sentence at the beginning of the whole work; but this was not Bach's latest intention, nor that it should be placed before the first fugue and canon. The Latin sentences—or at any rate, the second of them—evidently only occurred to him when the engraving was finished, and as he was eager to dedicate and offer this copy, he had it written in it. But the sense of it is not exact as applied to the contents of the oblong folio volume, as he himself detected, and when he afterwards had a separate slip engraved and printed with these words, he

had it gummed on to the first page of the upright folio sheet where it is more appropriate, and the copies printed for sale have it in this place.

I have said that Bach was eager to forward this dedication copy. On the 7th and 8th of May, 1747, he played before Frederick the Great at Potsdam, and the printed dedication of the "Musikalisches Opfer" is dated July 7 of the same year. Thus the composition and engraving were got through in two months, and the engraving must have taken all the more time as the engraver did not live in Leipzig. On page eight, at the bottom—not, to be sure, of these particular sheets—but of those which contain the six-part Ricercar, we find "*J. G. Schübler sc.*," but the two works are so perfectly similar in workmanship, paper, and partly in size and shape, that the engraver of both was no doubt the same person. Schübler lived at Zella St. Blasii, near Suhl, and two weeks must be allowed for delays in sending backwards and forwards with proof corrections and specimen sheets, under the conditions of communication at that time. Under these circumstances the singularities of the appearance of the work are accounted for; they are those of an over hasty production, and evidences of hurry in engraving the music may be ascribed to the same causes; for instance, in the second canon on the upright folio sheets the letters *Th* (Thema) are wanting to the lowest stave, and in the fugue in canon there is no indication as to what instrument is to play the upper part of the canon.

All the rest of the music which is now included under the title "Musikalisches Opfer" does not, as we have seen, strictly speaking belong to it, but was composed by Bach afterwards and sent to the king without any formal dedication. This also may be divided into two portions. The six-part Ricercar and two canons attached to it compose a part or cahier of four leaves in oblong folio, separately played. The presentation copy of this, too, is in the Amalien library, but in unpretending guise, not bound or even stitched into a cover; the leaves are simply held together by a pin through the fold. The sonata, on the other hand, and the canon for flute, violin and bass, are engraved in upright folio—three sheets without either title or cover. These were specially preserved in the Amalien library, and seem to have been used; at any rate, in the continuo part there are corrections added in ink. It is hard to say why Bach should have had the shape of the paper altered.

As the whole collection now exists it is a strange conglomerate of pieces, wanting not only internal connection but external uniformity. The contents are fugues, canons, and sonatas. But the fugues, as well as the canons, are dispersed in the original edition; the canons are in no less than four separate groups, and, as regards their form, have no sort of arrangement. This confusion may have been fortuitous and annoying to the author himself, but after he had once started the publication was beyond remedy. Bach had 100 copies printed off in the first instance, and gave them away for the most part to friends.

These were exhausted by October 6, 1748, and on that day he wrote to his cousin, Elias Bach, of Schweinfurt, to have a few more copies printed off at once, and that each copy was to cost one thaler. Whether the trio was included in this order is doubtful since he only mentions the "Preussiche Fuge." The work afterwards came into the market, and in 1761 it was sold by Breitkopf for 1 thlr. 12 gr., as we see by his circular for Easter, 1761. But, according to the common custom at that time, it was also much copied, and as it was not a complete or inseparable whole each one transcribed only as much as he pleased, and in whatever order was most convenient. Complete copies in MS. must have been extremely rare, and I do not know of a single one now existing. Agricola copied the three-part simple fugue and three of the canons; this MS. is in the Amalien-Bibliothek. Above the first canon he has written: "*Canone perp : sopra il soggetto dato dal Rè*," and Kirnberger has used the same superscription in his "Kunst des reinen Satzes," II., 3, p. 45.

9 (p. 219). **Six Chorales for the Organ.** Rust was the first to point out (B.-G., XXV.,[2] p. V.) that the title of the original edition affords a ground for estimating the date of its publication; his only error is that he limits it to 1747—1749. Friedemann Bach was established as organist at Halle by 1746 (see Chrysander, Jahrb. für Mus. Wissensch. Vol. II., p. 244), and it is possible the work may have been published even so late as 1750. Bach kept up business relations with Schübler till his death; in the papers relating to his affairs and bequests we find, among "other necessary payments," paid to "Herr Schübler 2 thlr. 16 gr." This small sum could not relate to the "Kunst der Fuge"; besides, it is improbable on other grounds that Schübler should have engraved that work; but it may very well relate to the six chorales, and we may infer that these either were not published till 1750, or that some new copies were at that time printed off.

10 (p. 219). **The Variations on "Vom Himmel hoch."** The Necrology says, p. 173, "Bach delivered to the Society the chorale 'Vom Himmel hoch' completely worked out, and it was afterwards engraved on copper." From this the work cannot have been engraved until after Bach had presented it to the Society. As he was busy all through the summer of 1747, and even before that, the chorales cannot very well have appeared before 1748. This result cannot be reconciled with a certain remarkable circumstance. In 1745 Emmanuel Bach brought out, through Balthasar Schmidt, of Nuremberg, a clavier concerto with accompaniments in D major (see his autobiography in Burney III., p. 203), and this concerto is numbered 27 by the publisher. Seb. Bach's setting of the Christmas chorale is numbered 28, and therefore must have been brought out immediately after his son's work. Schmidt was just then much employed (see ante, note 6); but if the chorales were not published till after Bach had joined the Society, Schmidt would have published nothing for about three years, and that is incredible. It is

my opinion that these chorale settings were already engraved by 1746 at latest, and the statement that they were composed for the Society cannot, therefore, be accurate. Mizler knew in the spring of 1746 that Bach intended to join. Possibly Bach wished to enter the Society with a perfectly finished work, which would be, to a certain extent, beyond the reach of discussion, and so, as the work was already being engraved, postponed joining the Society for one reason or another. That this was in fact the real state of things stands almost confessed by the statement in the Necrology, which in such matters as these was certainly not written by either Emanuel Bach or Agricola, or any other musician. What can the words "completely worked out" (*Vollständig gearbeitet*) mean? It was not usual to lay unfinished works before the Society, and Bach was the last man to do such a thing. Or would any musician have spoken of such a work as "the chorale"? It is clear to my apprehension that the writer has misunderstood the information given him and distorted the facts. The statement, as he received it, may have been, "Bach worked out a chorale composition on 'Vom Himmel hoch,' and had it engraved on copper, and then laid the finished work before the Society."

END OF APPENDIX A.

APPENDIX B.

[In the German edition of this work Herr Spitta has printed at length a considerable number of curious and interesting extracts from the town, school, and church archives of Leipzig. They would, however, lose much of their value by being translated, as part of their interest arises from the quaint language in which they are couched. The translators have therefore exercised their discretion in selecting and abridging, and no documents have been given entire, excepting those which have some musical interest, or which either emanated from Bach himself or bear directly on his position and proceedings. A brief summary of each note is given for the reader's guidance.]

I.

To Vol. I., p. 2.

Count Günther's Letter was discovered in September, 1868, on a yellow and half mouldy sheet in the Archives of the Principality, at Sondershausen. The oldest document extant having any bearing on the Bachs, who may have been Sebastian's ancestors, may be printed here at length. It was to the friendly help of my late lamented colleague, Professor Th. Irmisch, that I owe my success in deciphering it.

"Günther, Count of Swarzburg, and Lord of Arnstet and Sundershussen:[1]—First our greeting to you, most worthy and learned and well-beloved. You have, no doubt, not forgotten that we have written to you in manifold ways, and have also let you know by word of mouth that our subjects of Gräfenrode, Hans Bach and Hans Abendroth, who are burdened with spiritual proceedings, originating from Mentz, and because of one Hans Schuler, of Vlfinau, a subject of your spiritual jurisdiction, have craved not to be refused help and justice from us, and from our officials; or else, from your oversight as spiritual ordinary, in the place of our gracious Lord (Archbishop) of Mentz to penetrate to this place (to put themselves under our protection). And although as we have been informed you have written to Mentz about it with one and suitable offers, but still fruitlessly as regards relieving the burdens of our faithful subjects, and it is quite grievous to know that the poor folks are so undeservedly burdened. Well then—it is our gracious mind that you shall once more in regard of their destitution, and our command for the benefit of the poor folks, help them, and bring it about that they are relieved of their burdens. We will then take care that the Hans Schuler is duly dealt with. Help at once, and justice before

---

[1] Günther the XXXIX., "der Bremer," born 1455, died 1531; reigned from 1503.

us or our judges for our (men), and refuse not, but let all be done and
carried out rapidly, or presently call them to appear before you in your
court, as Ordinary presiding over justice and dues, to render and give
up to the complainant so much as (what) is acknowledged so by you.
You shall show yourself well disposed herein, and we will subscribe
ourselves and acknowledge ourselves your debtor.

"Given on Friday (? it was the 23rd Feb.) before the feast of St.
Matthias *anno Nono*.

" To the very worthy and learned Master Johan Somering, cathedral
chaplain at Erfurt, our well-beloved."[2]

II.

To Vol. I., p. 26.

Johann Ernst Bach of Eisenach. It is no part of my
plan to follow up the Bach family in all directions down to the
generation of which Sebastian was a member. Hence, I can, in this
place, mention only Bernhard Bach's only son, Johann Ernst, born
September 1, 1722, died January 28, 1777. He studied at St. Thomas's
school at Leipzig, somewhere about 1735, and left evidence in a copy of
his own making of twelve Concertos by Vivaldi, which Sebastian Bach
had arranged for the organ or cembalo that he had learned of his
illustrious relation. He studied jurisprudence at the Leipzig University,
and settled as a lawyer in his native town of Eisenach. But his
musical skill was so great that he not only became his father's colleague
as organist in 1748, and after his death held the same office for life, but
in 1756 was even appointed Capellmeister of Sax Weimar with a salary
of 400 thalers a year. That he took the greatest interest in the
improvement of the band is proved by various "*Pro Memoria*," in the
archives of Weimar. He kept his residence and appointment in
Eisenach all the same, and only went occasionally to Weimar, and so
was, according to the expression of the time "out-door" (von Haus
aus) Capellmeister. He wrote an intelligent and thoughtful preface to
Adlung's book, "Von der Musikalischen Gelahrtheit." The esteem
and respect he everywhere enjoyed were due, according to Gerber, to
his admirable character. I have, in MS., copies of the period of his
Mourning Ode on the death of his patron, the Duke Ernst August
Constantin, and a Magnificat in German. The first named work is of
smooth and tender character, but not grandiose; the Magnificat deals
with massive choral movements, and artistic, though not always happy,
contrapuntal combinations. Another Magnificat of a more pleasing
character is the property of Herr A. Dörffel of Leipzig, as it would
seem in the original autograph. Besides these the Royal Library at

---

[2] Sömmering was in 1507 Domherr or Canon of the church of St. Severus in
Erfurt, and doctor of both branches of law, civil and ecclesiastical.

Berlin has two church cantatas by him, the 18th Psalm, and a Kyrie and Gloria on the Chorale, "Es woll uns Gott genädig sein," so arranged that the two first lines of the chorale are used for the Kyrie, and the rest for the Gloria. The whole composition shows a skilful musician, and is one of the best sacred pieces of that time, which, to be sure, was not in any way on the whole important in the history of art. Of his other compositions recently published, my knowledge is limited to a Fantasia and Fugue for Clavier in F major ("Alte Claviermusik" neu herausgegeben von E. Pauer. Leipzig, B. Senff. Ser. 2, part 3). Another fantasia, with Fugue in A minor, and a Sonata in A major, exist in MS. in the Royal Library at Berlin. The dates of his birth and death I have taken from a pedigree drawn up by his great grandson, Herr Bach of Eisenach.

III.

To Vol. I., p. 156.

Valentin Bach. I have already alluded to the sons of Valentin Bach, in referring to the connection which subsequently existed between Elias Bach and Sebastian. The pedigree here referred to mentions three more of his sons: Friedrich Adam, born September 5, 1752, died March 2, 1815; Johann Michael, born 1754; Simon Friedrich, born 1755, died May 2, 1799. It is however certain that this does not complete the list of the male members of the family of that time, as is proved by a testimonial which Elias Bach granted to one of his relations, and of which the original still exists in the possession of Fräulein Emmert. As it has a particular interest from a musical point of view I here give it at full length:—

"Whereas the representative of the said Johann Valentin Bach has asked me in becoming terms for a credible testimonial as to his conduct hitherto, I am in no way averse to doing so; on the contrary, can with truth confidently assure every one, be he whom he may, before whom this testimonial is laid that the above mentioned Johann Valentin Bach during his five years residence at the *Alumneum* of this place has always been found obedient, diligent, and faithful, and in music particularly has gone so far that he can sing well in discant, tenor, and bass parts,[3] and can also play well on the clavier and other instruments. Wherefore I do not hesitate hereby to give him my best recommendation to all, be they whom they may, who are patrons, lovers, and promoters of the noble art of music for the advancement of his praiseworthy purpose. Given at Schweinfurth, August 12, 1752.

"JOHANN ELIAS BACH,
[Seal.] "*Cant.* and *Alumn. Insp.*"

But exactly who this young Valentin Bach may have been, I cannot say.

---

[3] Meaning, no doubt, he could read the different clefs.

IV.

To Vol. I., p. 258.

Specification of the Organ in the Marien-Kirche at Lübeck. Friedrich Erhardt Niedt, Musikalische Handleitung II., 189 (Hamburg, 1721), says: "The organ in the Marien-Kirche at Lübeck has 54 stops.

Werk.

(1) Principal 16 ft. (2) Quintadena 16 ft. (3) Octava 8 ft. (4) Spitz-Flöte 8 ft. (5) Octava 4 ft. (6) Hohlflöte 4 ft. (7) Nasat 3 ft. (8) Rauschpfeiffe (mixture) 4 ranks. (9) Scharff (sharp mixture) 4 ranks. (10) Mixtura 15 ranks. (11) Trommete 16 ft. (12) Trommete 8 ft. (13) Zinke 8 ft.

Brust.

(1) Principal 16 ft. (2) Gedact 8 ft. (3) Octava 4 ft. (4) Hohlflöte 4 ft. (5) Sesquialtera 2 ranks. (6) Feld-Pfeiffe 2 ft. (7) Gemshorn 2 ft. (8) Sifflet 1½. (9) Mixtura 8 ranks. (10) Cimbel 3 ranks. (11) Krumhorn 8 ft. (12) Regal 8 ft..

Rück-Positiv.

(1) Principal 8 ft. (2) Bordun 16 ft. (3) Blockflöte 8 ft. (4) Sesquialtera 2 ranks. (5) Hohlflöte 8 ft. (6) Quintadena 8 ft. (7) Octava 4 ft. (8) Spiel-flöte 2 ft. (9) Mixtura 5 ranks. (10) Dulcian 16 ft. (11) Baarpfeiffe 8 ft. Trichter-Regal 8 ft. (This was of no new invention). (13) Vox humana. (14) Scharff (sharp mixture) 4 to 5 ranks.

Pedal.

(1) Principal 32 ft. (2) Sub-bass 16 ft. (3) Octava 8 ft. (4) Bauerflöte 2 ft. (5) Mixtura 6 ranks. (6) Gross-Posaun 24 ft. (7) Posaune 16 ft. (8) Trommete 8 ft. (9) Principal 16 ft. (10) Gedact 8 ft. (11) Octava 4 ft. (12) Nachthorn 2 ft. (13) Dulcian 16 ft. (14) Krumhorn 8 ft. (15) Cornet 2 ft.

There were besides the Cimbel-Stern, two Trummeln, two Tremulants, and 16 bellows.

That No. 1 of the Brustwerk is an 8 ft. stop, and that a mistake was made in its title, was noticed by Jimmerthal in his before mentioned work, p. 6; and as to the Posaune stop on the pedal of "24 ft.," it is to be understood that the "32 ft. tone" began with the F and not with the C.

V.

To Vol. I., p. 336.

Documents from the archives of Mühlhausen relating to the post of organist in that town.

Bach's emoluments were fixed at—
  85 gulden in money and his deputy paid.
  3 malter of corn.
  2 cords of wood, 1 beech and 1 oak or aspen.
  6 loads of brushwood delivered at his door.
Dated June 15, 1707.

VI.

To Vol. I., p. 526 (referred to as V.)

A List of the Ducal Orchestra at Weimar between the years 1714 and 1716.

Private Secretary, Governor of the pages and bass singer, Gottfried Ephraim Theile (written in small) has his food at the Court.

Bassoon	Bernhard George Ulrich,
Cammer *four*ier[4] and trumpeter	Johann Christoph Heininger,
	has table allowances.
Castle Warden and Trumpeter	Johann Christian Biedermann,
	has table allowances.
Trumpeter	Johann Martin Fichtel,
	has table allowances.
Trumpeter	Johann Wendelin Eichenberg,
	has table allowances.
Trumpeter	Johann Georg Beumelburg,
	has table allowances.
Trumpeter	Conrad Landgraf.
Drummer	Andreas Nicol,
	is to have table allowances.
Court Capellmeister	Salomo Drese,
	has daily 1 loaf and 1 stoup of beer from the cellar.
Court Vice-Capellmeister	Drese.
Concertmeister and Court Organist	Johann Sebastian Bach.
Secretary and Tenor	Aiblinger.
Tenor and Court Cantor	Döbernitz.
Court Cantor and Bass (coll. quint.)	Alt.
Alto	Bernhardi.
Discantist (treble)	Weichard,
	also has free table.
Discantist	Gerrmann.
Chamber Musician (Cammer Musicus)	Johann Andreas Ehrbach.
Musician and Violinist	Eck.
Violinist and Musician	Johann Georg Hoffmann,
	lives in Jena, but when he is here he has his living at Court.
Court Secretary and Musician, also Violinist	August Gottfried Denstedt.

---

[4] Intendant of the royal apartments.

Besides these there were six boys. This list is from that of Wilhelm Ernst's suite in the grand ducal archives at Weimar. The musicians named before the Capellmeister resided in the castle, which accounts for the order of the names. Generally—as in the list for 1726—the order of precedence gave the Capellmeiste the next place to the Governor of the pages, and after him came the ministers of country parishes. The trumpeters and drummers come after the under master of the school (sub-conrector) and the Master of the Mint; then follow the rest of the musicians "in order as they are engaged;" then the writers and copyists of the Consistory, and then, after a whole list of other persons almost at the end, the town and Court cantors and organists and the rest of the school officials. Bach, however, being at the same time concertmeister had a position not far below that of the Capellmeister. The *Stadt musicus*, town musician, at the time was Valentin Balzer.

I.

To Vol. II., p. 185.

A document referring to Bach's appointment as Cantor of St. Thomas's, Leipzig.

II.

To Vol. II., p. 186 (referred to as VI.)

May 5; 1723.

Herr Johann Sebastian Bach, hitherto Capellmeister to the Princely Court of Anhalt Cöthen, appeared in the Council-chamber, and *Dominus Consul Regens D. Lange* proposed that he should be chosen Cantor to the School of St. Thomas, as being the most capable candidate, and he was unanimously elected.

Ille. Humbly thanked them and pledged himself to fidelity and diligence.

Eodem. It was accordingly announced that he would be duly notified to that effect, so that he might be pleased to arrange for the ceremony of presentation and all else.

Ille. Thanked them for the notification, and would not fail to do all that was necessary.

Eodem. Notice of the selection they had made was given also to the pastor of the Church of St. Thomas, the *Licentiate* Weisen, who thanked them and wished them every blessing.

The Cantor of the Thomasschule's counterpart agreement:—

After their worships, the Council of this town of Leipzig, had accepted me to be Cantor of the School of St. Thomas, they required of me an agreement as to certain points, namely :—

1. That I should set a bright and good example to the boys by a sober and secluded life, attend school, diligently and faithfully instruct the boys.

2. And bring the music in the two chief churches of this town into good repute to the best of my ability.

3. Show all respect and obedience to their worships the Council, and defend and promote their honour and reputation to the utmost, and in all places, also if a member of the Council requires the boys for a musical performance unhesitatingly to obey, and besides this, never allow them to travel into the country for funerals or weddings without the foreknowledge and consent of the burgomaster in office, and the governor of the school.

4. Give due obedience to the inspectors and governors of the school in all they command in the name of the Worshipful Council.

5. Admit no boys into the school who have not already the elements of music or who have no aptitude for being instructed therein, nor without the knowledge and leave of the inspectors and governor.

6. To the end that the churches may not be at unnecessary expense I should diligently instruct the boys not merely in vocal but in instrumental music.

7. To the end that good order may prevail in those churches I should so arrange the music that it may not last too long, and also in such wise as that it may not be operatic, but incite the hearers to devotion.

8. Supply good scholars to the New Church.

9. Treat the boys kindly and considerately, or, if they will not obey, punish such in moderation or report them to the authority.

10. Faithfully carry out instruction in the school and whatever else it is my duty to do.

11. And what I am unable to teach myself I am to cause to be taught by some other competent person without cost or help from their worships the Council, or from the school.

12. That I should not quit the town without leave from the burgomaster in office.

13. Should follow the funeral processions with the boys, as is customary, as often as possible.

14. And take no office under the University without the consent of their worships.

And to all this I hereby pledge myself, and faithfully to fulfil all this as is here set down, under pain of losing my place if I act against it, in witness of which I have signed this duplicate bond, and sealed it with my seal. Given in Leipzig, August 13, 1722.

Herr Johann Christian Bach signed and sealed the same. May 5, 1723.

Also Herr Gottlob Harrer (date missing).

This document, as will be seen, is merely the sketch for the bond; Bach's name is inserted wrongly.

III.

To Vol. II., p. 188.

Document relating to the dispute between the Council and Consistory.

IV.

Vol. II., p. 193, &c. (referred to as VII.)

Five memorials of Kuhnau's, addressed to the Town Council and to the University on the following dates: December 4, 1704; March 17, 1709; September 1, 1710; December 18, 1717; and May 29, 1720.

A. (Vol. II., p. 203).

The first is preserved among the archives of the Town Council of Leipzig (*Consistorialia*, Vol. X. *Varia*, 1619 till 1767, and reminds their worships that he had lately applied to them on the subject of the trumpets in the two principal churches which by long use had become worn out, bent, and useless; so that they had commissioned him to ask of Heinrich Pfeiffer, the gatekeeper of St. Thomas's, who is particularly skilled and experienced in the making of such instruments, the lowest price of a complete set of four, a *quart-posaune* (*i.e.*, a low trombone), tenor, alto and treble trombone, and how much he would allow for the old set of three.

He goes on to say that the price of such instruments has doubled within a few years; that he had laid Heinrich Pfeiffer's estimate and specification before their worships, and had humbly petitioned them to order a new set for all three churches. He then begs that to save expense a case may be made to hold six violins, which have to be carried to church, and that a Colochon (*Colascione* or Calichon) with a case may be provided as being an indispensable instrument which they are forced to borrow, not possessing one in either church, and mentions that Marcus Buchner is particularly "happy" in making them. This instrument was also known as the Italian lute, and Baron describes it as simply a bass viol lute.

He laments the diminution of numbers in the singers in the churches in consequence of their finding it more lucrative to sing in the opera, (see p. 205), and recommends to their worships' consideration certain details of arrangement, tending, as he hopes, to remedy the evil by placing more power in his hands, as he already had to appoint the hymns for all three churches (see p. 232).

The memorial is addressed to D. Joh. Alexander Christen, Governor of the School of St. Thomas, and burgomaster in office.

B. (p. 280).

In the next memorial he appeals to their "paternal care and interest in the wellbeing of the churches and schools" to consider certain points to which he proceeds to draw their attention as follows :—

1. The school violin is much broken, and so ill fitted for daily practice

that the small sum of 1 thlr. and a few groscher earned by following funerals and set aside for repairs is insufficient for its restoration.

2. A new regal is needed, the old one being constantly in need of repair; still they might do with it a little longer on the chance of such an instrument being for sale at a low price.

3. On the other hand, they are greatly in need of a good colochon both for school practice and for church use, and he again recommends Mark Buchner.

4. A sand hour-glass is much needed, and the sexton has been frequently asked to see to this, but it does not appear that he has applied for it.

5. A board with nails is needed on which to hang the violins (in the right choir) that they may not be laid on the floor.

6. Unless a new step is fitted in St. Nicholas' for the Stadtpfeiffer to stand on, one or another will chance to break his leg, or at least to sprain his foot.

7. It is much to be desired that a fund of not less than 300 thalers should be created, of which the interest should be employed to keep the two great clavicembalos (one in each church) in proper repair, since they have to be seen to each time there is a performance, and in the state they are in this costs six or seven thalers.

8 and 9 refer to the numbers of the choir, and 10 complains again of the increasing influence of the opera; this, he says, causes the greatest mischief, for the better students, as soon as they have acquired, at the cost of infinite pains to the cantor, sufficient practice, long to find themselves among the "*Operisten,*" and he suggests finally that *stipendia* shall be given to certain musicians who shall supplement the efforts of the eight Stadtpfeiffer, and particularly to two good violinists and to a good bass singer since there is a difficulty (probably an impossibility) in supplying such a voice from among the scholars.

C. (p. 214).

refers to the services to be performed in the New Church of St. Paul.

D. (p. 203).

has not much interest for the English reader. He complains of the injury done to the boys' voices by the perambulations and funerals.

E. (p. 208).

is the specification of various plans by which the church music of Leipzig may be improved. The document is in the Town Library at Leipzig. In consequence of there being no regular organist, and the music in the Neue Kirche being performed by young students, chiefly from the Thomasschule, the music had degenerated and become operatic in character, which naturally scandalised those members of the congregation who appreciated and loved the true style of church music. The organ also had suffered much injury at the hands of inexperienced players who did not understand how to remedy the effects of change of weather, &c. Kuhnau therefore suggests that a regular organist should be appointed, who should also undertake to direct the music, and to manage the choir; and that he should be made to serve three churches in rotation. He concludes by giving suggestions as to the details of payment of the organist and the choir. The document is dated Leipzig, the 29th of May, 1720.

V., VI., & VII.

are of no interest to the English reader.

VIII.

To Vol. II., p. 282.

Supplement to the account of the Thomaskirche from Candlemas, 1747, to Candlemas, 1748.

"Notice is hereby given that, since the organ in the Thomaskirche has for a long time been almost useless by reason of the great quantity of dust and dirt, and in order that it may not fall still further into decay, the following contract has been duly and carefully drawn up and concluded between Herr *D*. Gottfried Lange, Privy Councillor of War to the King of Poland and Grand Duke of Saxony, Burgomaster of this town, and appointed overseer of the church of St. Thomas, of the one part, and H. Johann Scheibe, organ builder in this town, of the other part, to this effect:—

1.

The said organ-builder, Johann Scheibe, promises to repair thoroughly and make good with glue and leather all the injuries done to the organ in the Thomaskirche through the great heat of last summer, and from other causes.

2.

As the said organ, and more especially all the pipes contained in it, are full of dust, so that most of them do not speak, the organ-builder, Scheibe, shall be bound, during their renovation, to take entirely to pieces all the pipes throughout, all the stops to which they belong, and, not only to cleanse them thoroughly from dust and dirt, but also to repair whatever injuries they may have sustained, and to put them back in their proper places.

3.

All the mouthpieces and tongues in the reed work: as the *Bass Posaune*, the *Bass Trumpet*, and in the *Rückpositiv*, the *Trumpet* and the *Krumhorn*, and also in the *Brustpositiv*, the two reed-stops, are to be cleansed and repaired with saltpetre and zinc, also anything that is damaged is to be put right.

4.

H. Scheibe promises to open all the wind-chests in the organ, so that the dust and dirt can be thoroughly removed from the valves, and also to guard against such damage for the future.

5.

He will also renew the case wherever the iron and brass-work has parted, and also

6.

Will put the two-manual couplers into repair.

7.

He promises to voice all the pipes afresh, and to tune them, as also to tune the whole organ, together in good harmony and right pitch, and to make the intonation even throughout. He is also to manage the work so that some of the more necessary stops can be used at every service during the repairs.

8.

Finally, after the work is finished, anything that may be still found defective at the trial and examination of the renovated and repaired organ, Herr Scheibe promises to correct and improve at once without further question, and without demanding payment above what has been agreed. And he hereby binds himself to be responsible according to the best of his power for good and honest work in this organ.

9.

Herr Scheibe provides all the materials required for the work above described and for the thorough repair of the organ, paying all the workmen employed by him, and promises hereby to bring the whole instrument into a state to be played upon, and to deliver it over between the present date and next Michaelmas. On the other hand, the Herr Overseer, Herr Lange, Privy Councillor of War, and Burgomaster, shall allow the necessary scaffolding to be put up in front of the organ by the carpenters, for the pipes to be laid on during the repair and renovation of the organ, and to prevent damage to the instrument, which scaffold is to be taken away by the carpenters after the work is completed.

For all and each of the points specified in this contract, the Herr overseer of the Thomaskirche shall give to Scheibe, the organ-builder, the sum of
Two hundred Thalers,
to be paid in instalments out of the funds of the church.

As the contracting parties of both parts now present are in earnest in will and intention, and they both promise to fulfil this contract in all points, for the more certain fulfilment of the contract it is made out in duplicate, and signed and sealed by both contracting parties, Leipzig, June 28, 1747.

Lange's contract above is sealed and also bears the date of June 23. The Council's order for the repair was given on June 26. Scheibe received his payment in six instalments; the last on November 4, 1747.

IX.

To Vol. II., p. 519.

The injunction commanding the first performance of the St. John Passion in the church of St. Nicholas.

X.

1. Vol. II., p. 506.

The text has no interest for the English reader.

2. Vol. II., p. 627. Title :—

Drama | *Per Musica*, | Welches | Bey dem Allerhöchsten | *Crönungs-Feste* | des | Aller-Durchlauchtigsten und Gross- | mächtigsten | *Augusti III.* | Königs in Pohlen und Churfür- | sten zu Sachsen | in unterthänigster Ehrfurcht aufgeführet wurde | in dem | *Collegio Musico* | durch | *J. S. B.* | Leipzig, dem Janr. 1734. | Gedruckt bey Bernhard Christoph Breitkopf.

The personages are *Valour, Justice, Clemency* and *Pallas*.
The text is of no interest to the English reader.

3. Vol. II., p. 630. Title:—

Drama | *Per Musica*, | Welches | Bey dem Allerhöchsten | *Geburths-Feste* | Der | Allerdurchlauchtigsten und Grossmäch- | tigsten | *Königin in Pohlen* | und | *Churfürstin zu Sachsen* | in unterthänigster Ehrfurcht | aufgeführet wurde | in dem | *Collegio Musico* | Durch | *J. S. B.* | Leipzig, dem 8 December, 1733. | Gedruckt bey Bernhard Christoph Breitkopf.

Personages : *Irene, Bellona, Pallas, Fama*.

4. Vol. II., p. 263. Title:—

"Als die von E. Hoch-Edlen und Hoch-Weisen Rath der Stadt Leipzig umgebauete und eingerichtete Schule zu S. Thomae den 5. Jun. durch etliche Reden eingeweyhet wurde, ward folgende *CANTATA* dabey verfertiget und aufgeführet von Joh. Sebastian Bach, Fürstl. Sächs. Weissenfels. Capellmeister, und besagter Schulen Cantore, und M. Johann Heinrich Winckler, *Collega* IV.

"Leipzig, gedruckt Bernhard Christoph Breitkopf."

XI.

To Vol. III., p. 11.

ACTA, concerning the appointment of the Prefects in the School of St. Thomas, in this town, 1736. (In the archives of the Town Council, Leipzig.)

Magnifici.

Most Noble, Illustrious, Learned and Worshipful Gentlemen and Patrons.[5]

August 12, 1736.

May it please your worships graciously to allow me to represent to you that, whereas according to your worships' ordering of the School of St. Thomas, it pertains (is the duty of) to the Cantor to choose from among the scholars those whom he considers fit and able to be Prefects, and in electing them to have regard, not only to the voice that it be good and clear, but also to see that the Prefects, and especially the one who leads the first Choir, shall be able to undertake the direction of the Musical Choir in the absence or illness of the Cantor; and whereas this rule has been hitherto observed by the Cantors without the concurrence of the Rectors: yet and notwithstanding, the present Rector, *M.* Johann Aug. Ernesti, has lately endeavoured, without my knowledge and approval, to assume the appointment of the Prefect in the first Choir, so that he recently appointed Krause, the Prefect of the second Choir, to be the Prefect of the first Choir, and refuses to withdraw in spite of all my civil remonstrances. Since I cannot suffer this to pass, being against the aforesaid order and traditional usage of the school, and to the prejudice of my successors and to the injury of the Musical Choir, I now present to your Worships my most dutiful petition, graciously to decide this difference between the Rector and myself in my office; and because this presumption on the part of the Rector to the appointing of the Prefects might lead to strife and to the prejudice of the scholars, I pray that in your great benevolence and care for the School of St. Thomas you will direct the Rector, *M.* Ernesti, to leave for the future, as hitherto, and according to the order and usage of the school, the appointment of Prefects to myself alone, and thus graciously protect me in my office.

Trusting to your Worships' most gracious indulgence, and abiding in the most dutiful respect,

I am yours, &c.,
Most obediently,
J. S. BACH.

August 13, 1736.

Though only yesterday I in a most respectful memorial troubled your Worships, because of the great indignity done to me by Herr Rector Ernesti, through his attempted encroachment on the function hitherto assigned to myself of Cantor and of Director of the Musical Choir in the St. Thomas School here, by his interference in the appointment of the Prefect, and prayed for your Worships' gracious protection: yet I find myself under the necessity of again most humbly bringing to

---

[5] The original is lavish in the reiteration of titles and attributes, which are here omitted.

your Worships' notice that, although I had already informed the said Herr Rector Ernesti that I had complained to your Worships, and that I expected your decided judgment on the matter, he nevertheless, regardless of the respect due to your Honourable Council, again presumed to let all the scholars know that none should under pain of relegation and castigation dare to sing or conduct the usual motett in place of Krause, who is, as I stated in my most dutiful memorial of yesterday, unfit for the direction of a musical choir, and whom he wishes to force upon me as Prefect of the first Choir. Hence it came then that in the Nicolai Church at the afternoon service yesterday to my great humiliation and dejection not a single scholar would undertake to lead the singing, much less to conduct the motett, for fear of being punished. Indeed, the service would thereby have been interrupted had not most fortunately these duties been undertaken by an old scholar of St. Thomas's, of the name of Krebs, at my request. I represented in my late most humble memorial that the appointment of the Prefect does not, according to the rules and usage of the school, pertain to the Rector; he has moreover by his mode of action greatly vexed and offended against me in my official position, and thus weakened and indeed tried to deprive me of the full authority over the scholars in all matters of church and other music which I ought to have, and which authority was also conferred on me by your most Honourable Council on my accession to office. It is hence to be feared that were such high-handed proceedings to continue the services would be interfered with, and the music of the Church greatly deteriorate, while the school itself would in a short time be so injured that it would take many years to restore it to the same degree of efficiency in which it has hitherto been. Therefore I once more submit to your Worships my most dutiful and earnest supplication, since officially I cannot pass the matter over, that you will stringently admonish the Rector, since discipline is endangered, that he will hereafter not molest me in my office, nor hinder the scholars in their obedience to me by his unjust warnings and by threats of severe punishment, but rather that he will see to it, as is his duty, that the school and Musical Choir shall be improved rather than deteriorated. I hope for your gracious indulgence and protection in my office, and abiding in the most profound respect,

I am, &c.,

J. S. BACH.

Fol. 5.   P.M.

The full and true history of the matter regarding the scholar Krause, whom the Rector wishes to force upon me as first Prefect, is as follows: Said Krause had so long as a year ago so bad a reputation on account of his disorderly life and his consequent debts, that a council meeting was held, which emphatically intimated to him that, although he had on account of his dissolute life well deserved to be forthwith expelled from the school, yet, in consideration of his needy circumstances (for he had himself owned to have contracted debts to the amount of twenty thalers), and on his promising amendment, they were willing to try him for another quarter when, according as his behaviour was improved or not, he would be informed whether he was to be retained or removed. Now, the Rector has always shown a special predilection for him, Krause, and verbally begged me to make him a Prefect, but I remonstrated that he was not fitted for it, whereupon the Rector replied that I might nevertheless do so, since it might enable Krause to free himself of his debts, and so the school be spared a constantly increasing scandal, especially as his time would soon be out, and he would thus be got rid of with a good grace. I therefore, willing to do the Rector a pleasure, made Krause a Prefect in the New Church, where the scholars have nothing else to sing but chorales and motetts, and have nothing to do with other concerted music, which is managed by the organist himself, for I considered that his school years had expired all but one, and it was not to be expected that he would ever come to conduct either the first or second Choirs. Subsequently however the Prefect of the first Choir, by name Nagel, from Nürnberg, complained at the singing at the last New Year that owing to weakness of constitution he should not be able to continue with it; it therefore became necessary to make a change out of the usual course of time in the arrangement of the Prefects, and to shift the second Prefect into the first Choir, and of necessity to receive the oft-named Krause into the second Choir. He, however, made various mistakes in beating the time, as I have been told by the Herr Con-Rector, who undertook the inspection of the second Choir, and upon enquiry concerning these mistakes all the scholars laid the blame wholly and solely on the Prefect on account of his faulty beating of the time. Moreover, I myself at a recent singing lesson took occasion to test his conducting of the time, when he acquitted himself so badly that he could not even give the accurate beat in the two chief modes of time, namely, the equal or common time, and the unequal or triple time,' but made triple into common time, and *vice versâ*, as all the scholars will readily confirm. Being, therefore, fully convinced of his incapacity, I could not possibly trust him as Prefect of the first Choir, particularly as the sacred music which is performed by the first Choir, and which is mostly of my own composition, is incomparably more difficult and intricate than that which is done by the second Choir, these sing only on festivals and in choosing their music I am mainly guided by the ability of those who are to render it. And although further circumstances might be mentioned which would still more prove the incapacity of Krause, yet I think the grounds already adduced sufficiently show my complaints to your Honourable Council to be justified, and to call for an early remedy without delay.

Leipzig, 15th August, 1736.[6]   JOH. SEB. BACH.

---

[6] Autograph throughout.

Ernesti's reply, dated August 17, 1736. He denies Bach's assertion that the appointment of the Prefects had always lain with the Cantor, who only selected such scholars as he deemed fit for the post, and then presented them to the Rector for appointment. He further asserts Krause to have been dismissed not from incapacity but out of spite to himself, since he, Ernesti, had some time previously sent Krause to the Cantor with the request that he would appoint him Prefect, and since Krause's dismissal was due to this inadequate cause, he, Ernesti, had interdicted it. After a while Bach had consented to reinstate Krause, but on various pretexts had deferred doing so. Thereupon Ernesti informs Bach that unless he forthwith reinstates Krause he shall do so himself on the next Sunday, but receives no answer. Ernesti then informs the scholars that none were to attempt, on pain of severe penalties, to undertake the post of the Prefect he had appointed. Bach, however, when he finds Krause still acting as Prefect, expels him again without ceremony. Ernesti then refers the matter to the Superintendent, who promises enquiry, and meanwhile directs that Ernesti's Prefects are to retain their places. Bach replies that he does not care, and when the Prefects had gone to their places at the afternoon service "he with much shouting and noise again drove Krause from the choir and directed the scholar Claus to take the place of Prefect, who did so, but excused himself to me while yet in the Church. And in the evening he drove the other Prefect away from the (supper) table because he had obeyed me." Ernesti declares Bach's assertions and complaints to be unfounded, and begs the Council not to entertain his complaints, but to admonish him to obedience and to attend to his duties with greater diligence. For he ascribes the unfortunate dereliction of duty on the part of Krause to Bach's neglect in not undertaking the direction of certain musical performances himself, but leaving it entirely in the hands of Krause. He concludes by begging the Council to support him in his office as Rector.

August 20, 1736.

Your Worships will call to your gracious remembrance that I was under the necessity of bringing to your notice the disorders which occurred this day week in the performance of public service in consequence of the arrangements made by the Rector of the Thomasschule here, *M.* Ernesti. To-day, both forenoon and afternoon, the same things occurred again, and I was obliged, in order to avoid a great commotion in the Church and disturbance of the service, to direct the motett myself, and to get a student to lead the singing. And as matters are likely to become worse in the course of time, and I shall scarcely be able to maintain my authority in future with the scholars placed under me without the effectual interference of your Worships, my noble Patrons, and should therefore not be responsible were further and perhaps irreparable disorders to arise, I cannot avoid respectfully representing this to your Worships with the most dutiful prayer that your Worships will be pleased to restrain the Rector without delay, and that you will, by accelerating the final resolution for which I have prayed, and according to your zeal for the common weal, prevent, as is to be apprehended, further public offences in the Church, disorders in the school, diminution of my authority with the scholars requisite in my office, and other serious consequences.

J. S. BACH.

February 12, 1737.

The Rector of the St. Thomas School here, Herr *M. J. A.* Ernesti, has lately presumed to force upon me, against my will, an unfit individual as Prefect of the first Choir which is composed of the scholars of the said school, and as I neither could nor would accept him, the said *M.* Ernesti forbade all the scholars that none, except his own arbitrarily appointed Prefect, should, under pain of relegation, either lead the singing of the motett or direct it. It was thus effected that on the following Sunday at the afternoon service not a single scholar would undertake to lead the singing nor direct the motett, out of fear of the threatened punishment, and indeed the service would have been interrupted had I not persuaded a student, who was able, to undertake these duties. By this proceeding on the part of the Rector I have not only been greatly injured and molested in my office, but have also been deprived of the respect due to me by the scholars, and thus been lowered in my position towards them. And yet, according to the orders passed by your honourable Council with regard to the St. Thomas School, cap. 14, s. 4, it pertains to me to choose the Prefecti of the Choirs without the concurrence of the H. Rector, an order which has hitherto been continuously observed both by myself and by my predecessors; and this has its reasonable grounds, since the Prefecti, according to the said school order, have to fill my, the Cantor's, place, and to conduct, as I cannot be present at the same time in all the Churches; and as I have the special care and supervision of the first Choir I must know best who is most suited to me. Therefore, secondly, the prohibition of the Rector issued to the scholars that none should sing under another Prefect, is most unjust, seeing that nothing effectual can be achieved if the scholars are prevented from obeying me in all matters pertaining to the singing. In order, therefore, that these doings may have no ill result, I have the strongest grounds for moving in the matter, and am compelled in this difficulty to apply to your Honours. My most humble petition, therefore, is that you will protect me in my office, and strictly enjoin the H. Rector, M. Ernesti, that he will no longer molest me in the same, that he will abstain in future from choosing Prefecti without my knowledge or consent, and from forbidding the boys to obey me in regard to the singing, that you will further be pleased to instruct the Superintendent or one of the clergy of St. Thomas's Church, without unbecoming restriction, to enjoin the school children to again render me the respect and obedience due to me, and so enable me for the future to fulfil the duties of my post. As I now trust by this, my not unreasonable petition, to obtain the protection and aid of your Worships, so I remain as before with continual respect,

Your most obedient, &c.,

J. S. BACH.

August 21, 1737.

Your Magnificences, &c., will graciously call to mind how I, under date February 12, of the present year, complained to your Honours of the Rector of the St. Thomas School, *M. J. A.* Ernesti, concerning his interference in my office, and also his prohibition to the scholars of obedience to me, and the consequent humiliation to me, in regard whereof I humbly craved your aid and protection. Since then, it is true, your honourable Council has sent me a decree, copy of which is enclosed under A; but, on the one hand, satisfaction is not done to me, thereby, for the humiliation inflicted on me by the said Rector, and, on the other hand, I am seriously aggrieved thereby. For as the Rector publicly and in open Church, and also in the presence of the entire first class, threatened all the scholars with relegation and loss of the caution (money?) if any should be disposed to obey my orders, wherefore I not unreasonably demand that my honour be re-established; so and in like manner the above-named decree of the Council is based upon a school order made in 1723, which differs materially from the old school orders in many points, and tends greatly to my prejudice as well in the exercise of my office and in regard to the accruing perquisites, while it has never been actually in force; for, when at one time the promulgation thereof was to be proceeded with, the late Rector, Ernesti, declared himself against, to the effect that it should be, in the first place, sent in to the honourable Consistorium, whose decision thereon was to be awaited. But the ratification has, so far as I know, not yet ensued, and I cannot therefore acknowledge a new school order so prejudicial to me, especially as the amount of my perquisites was therein to be much reduced, and the old order must still continue in force.

The aforesaid decree of the Council, based as it is on this new order, cannot therefore remedy the matter. More especially impracticable is that part of it which declares that it shall not be competent for me to suspend a scholar who has once been appointed to a function, much less to remove him therefrom. For cases occur where a change has forthwith to be made, and where a detailed enquiry in a minor matter of mere discipline or other school affair cannot be undertaken. Such changes do in all lesser schools belong to the province of the

Cantor, as it would be impossible to control the youths if they knew that one could not at once deal with them, and in other respects too, it would be hopeless to fulfil the duties of one's office satisfactorily. Your Honours have required to be informed on this matter, and I herewith again make my most humble petition: that your Worships will protect me in the exercise of my office and ensure me the needful respect; that you will prohibit all undue interference on the part of the Rector, *M*. Ernesti; also to restore my honour with the scholars, which has, through the instrumentality of the said Rector, been wounded; and that you will give the necessary instructions to defend me against the new school order so far as it militates against me, and prevents me from the due performance of my duties. For the aid thus to be granted to me I shall as ever, remain with profound respect,

Yours, &c.,

J. S. BACH.

prs. 29 Oct., 1737.
Praes. d. 13 Dec., 1737.

Most Noble, Most Mighty King and Prince,
Most Gracious Sovereign,

That your Majesty has been most graciously pleased to confer on me the title of composer to your Majesty will command my most humble gratitude through life. As therefore I claim for myself in most humble confidence the protection of your Majesty, so I now venture most respectfully to beg for the same on account of my present oppressors. My predecessors, the Cantors in the St. Thomas School here, have always, and according to the traditional usage of the school, possessed the right to appoint the Prefecti in the Musical Choirs, and that for the well-founded reason that they, more than any others, were in a position to know which individual was the most capable, and this prerogative I have enjoyed for a considerable time and without question from anyone. Nevertheless, the present Rector, *M*. Johann August Ernesti, has lately been bold enough to fill up a Prefecture without my concurrence, and that with an individual with very little knowledge of music. And when I became aware of his incompetence, and felt under the necessity of making a change on account of the consequent disorder in the music, and appointed in his place a more skilful person, the said Rector, Ernesti, not only directly opposed me, but also, to my greatest affront and humiliation, forbade all the assembled scholars, under pain of "baculation," to render me obedience in my arrangements. Now, although I have endeavoured to maintain my well founded prerogative before the magistrate here in the enclosure A, and have also implored the Royal Consistory here for satisfaction (enclosure B) for the injury done me, yet from the latter I have received nothing at all, and from the former only the instructions herewith enclosed under C. Since now, most gracious King and Sovereign, the Council here completely deprive me of the right I have hitherto enjoyed, as shown by the enclosure, and in doing so rest themselves on a new school order made in the year 1723, which I do not regard as binding on me, principally because if it is to be lawful it has never been confirmed by the Consistorium. Therefore I now, in most humble submission, appeal to your Majesty—

1. To command the Council here to see that I am not molested in my *jure quæsito ratione* of appointing the *Præfecti Chori Musici*, and therein protect me; and

2. To be pleased to direct the Consistorium of this place to require an apology from the Rector Ernesti for the indignity done to me, and also to charge without reserve the Superintendent, Dr. Deyling, to instruct the entire school (*coetus*), that all the boys of the school shall show me the customary and due respect and obedience. This most exalted Royal Favour I anticipate with undying gratitude, and remain in lowliest submission,

Your Majesty's
Most submissive and most dutiful,
Leipzig, 18 Oct., 1737.   J. S. BACH.

To the Most Noble, Most Mighty Prince and Sovereign, Friedrich August, King of Poland, &c. &c.

XII.*

(Ante, p. 118.)

Rules and Instructions for Playing
Thorough-bass or Accompaniment in Four Parts,
made for
his Scholars in music
by
Herr Johann Sebastian Bach,
of Leipzig,
Royal Court Composer and Capellmeister,
also Director of the Music and Cantor of the Thomasschule,
1738.[7]

Signatures.	6	43	76	7	98	9	6/5	65/43	♭7	4/3	♯	5♭	Without Signature.	8
Middle Parts.	3	5	3	3	3	3	3	8	3	♯	2	3		5
Additional Parts.	3 8 6	8	8 3 5 8	5	5	—	—	5 5♭	—	6	6			3

* The translators take this opportunity of gratefully acknowledging valuable assistance given by Mr. W. S. Rockstro in preparing this portion of the appendix for publication.

[7] The title is in the handwriting of Johann Peter Kellner, which is also noticeable here and there, further on, in corrections and additions.

Short Instructions for what is called Thorough-bass.

The signatures which occur in Thorough-bass are chiefly the following nine—viz., 2, 3, 4, 5♭, 5, 6, 7, 8, 9.

These are divided into consonances and dissonances. Four of them are consonances—viz., 3, 5, 6, 8.

These again are divided into perfect and imperfect.

The perfect consonances are:[8] the fifth and octave, and the imperfect are the third and sixth.

The remaining five are dissonances—viz., 2, 4, 5♭, 7, 9.

When no number or signature stands over the bass note, the chord consisting of 3, 5 and 8 is to be played.

The fifth and eighth must not, however, be taken in such a manner that either lies uppermost twice running, otherwise the result will be consecutive fifths or octaves, the greatest error in music.

N.B.—They are to be taken *vice versâ*.

Sometimes also a 3, 5, or 8 is found above a note; this generally means that the chord is to be taken in such a manner that the 3, 5, or 8 shall be the topmost note.

Here follows a list of the signatures with which the third may be taken.

With a 7
6
5 6
6 5
6
5
6
5♭
7
5
♭7
5♭
8
7
8
6
9 8
7 6

} the third is taken.

N.B.—For those who cannot remember this rule, it is only necessary to keep this much in mind; whenever 6 or 7 occur, whether alone or with other figures, the 3 may be taken.

Except in the case of $\frac{6}{4}$, resolved into $\frac{3}{5}$ when an 8 is to be played, as will be seen presently.

With $\frac{7}{4}\atop 2$ nothing else is played.

---

[8] In the original there is a note of interrogation instead of the colon.

Here follow such signatures as are resolved always into the common chord.

With 9 8, 3 and 5
„ 4 3, 5 „ 8
„ $\begin{Bmatrix}6&5\\4&3\end{Bmatrix}$ 8 „ „
„ $\begin{Bmatrix}9&8\\4&3\end{Bmatrix}$ 5 „ „

are played.

N.B.—If the rule cannot be remembered, keep the following in mind: such notes of the chord as are absent from the resolution are to be played with the preceding signature.

Here follow those signatures which must each be committed to memory by itself.

With $\frac{4}{2}$ the 6 is taken.

With $\frac{4}{3}$ the 6 is also taken.

With $\flat 7$ when a $\sharp$ stands above the bass as well, the 3 and 5 are taken.

With $5\flat$ the 6 is taken.

A $\sharp$ or $\natural$ indicates that the third is to be major, and a $\flat$ that it is to be minor.

The 4 3, 5♭, $\frac{6}{5}$, 7 and 9

(1) must in an ordinary way be already in the previous chord.
(2) must remain in the same part.
(3) must be resolved by descending one degree.

ELEMENTARY INSTRUCTION IN FIGURED-BASS.

CAP. I.

OF THE ETYMOLOGY.

The word *Bassus* is derived from the Greek βάσις, which signifies the root or foundation of a thing. Others derive it from the old Latin word *bassus*, signifying *profundus*, deep. By the word taken alone is understood the lowest part in music, or any bass part that produces a deep note, whether this note be sung or played upon a bass viol, bassoon, trombone, or the like.

By the term Figured or General-Bass, however, is understood a bass that is played on the organ or clavier with both hands in such a way that all or most of the parts in the music are played *generaliter* or simultaneously, or in general terms, together. It is also called *Bassus Continuus*, or with the Italian termination, *Basso continuo*, because it continues throughout the piece, even when the other parts pause now and then, although at the present day this Bass itself frequently pauses, especially in works of great ingenuity of construction.

Cap. II.
DEFINITION OF THE TERM.

Figured-bass is the whole foundation of the music, and is played with both hands in such a manner that the left hand plays the notes written down, while the right adds in consonances or dissonances, the result being an agreeable harmony to the glory of God and justifiable gratification of the senses; for the sole end and aim of general-bass, like that of all music, should be nothing else than God's glory and pleasant recreations. Where this object is not kept in view there can be no true music, but an infernal scraping and bawling.

Cap. III.
OF THE CLEF-SIGNATURES IN USE IN FIGURED-BASS.

All kinds of these clef-signatures are found in figured bass, such as discant, alto, tenor, &c. I will give a list of those most generally in use. (1.) The French violin clef, written on the first line thus: indicating the note g'. (2.) The German violin clef is written on the second line with the same sign: . (3.) The treble or soprano clef is the sign for that class of singers, and is written on the lowest line thus: indicating the note c'. (4.) The ordinary alto clef is written on the middle line thus: indicating the same note.

(5.) The tenor, on the fourth line indicating the same note.

(6.) The ordinary bass clef on the fourth line (sic) indicates the f.[9] (7.) More generally it may be observed that the sign indicates the once-marked g' wherever it occurs.

The sign is the once-marked c'. This

N.B.—When higher clefs than these occur in figured-bass they are called *Bassetgen*.

---

[9] In the MS. "the once-marked f'."

[10] This partially incorrect form of the bass clef is retained throughout the work by the writer. From this point onwards it has been corrected by me. It always indicates the f.

APPENDIX.

Cap. IV.
OF TIME OR MEASUREMENT.

Of this much need not here be said, for it is presupposed that a person wishing to learn figured-bass will not only have learnt the notes but also the intervals before doing so, whether by previous practice of music or from some other cause, and also the differences of time. For no one can inculcate a knowledge of time all at once. This must, however, be noticed, that in the present day one single kind of time is indicated in two ways, thus: , the second way being used by the French in pieces that are to be played quickly or briskly, and the Germans adopting it from the French. But the Germans and Italians abide for the most part by the first method, and adopt a slow time. If the piece is to be played fast the composer expressly adds *Allegro* or *Presto* to it; if slowly, the pace is indicated by the word *Adagio* or *Lento*.

Cap. V.
OF THE HARMONIC TRIAD.

The harmonic triad belongs properly to the subject of composition, but as figured-bass is a beginning of composition, and may even, because of the arrangement of consonances and dissonances, be called an extemporaneous composition, made by the person who plays the figured bass, the subject of the triad may be fitly mentioned in this place. Now if a person who desires to learn the subject can make good progress in this subject, and can impress it upon his memory, he may be sure that he has already mastered a great part of the entire art. The harmonic triad is a combination of the third and fifth which is placed above a bass note, and it may be constructed on all the notes, and both in the major and the minor—

[11]

And so on through all the notes. This is called *Radix harmonica*, because all harmony springs from it. And this harmonic triad is either simple or full (*simplex* or *aucta*). (1.) The simple root (*Radix simplex*) is the simple triad, properly so-called, and consists only of the three notes, as shown in the example. (2.) *Radix auctor* or the increased or full triad has the octaves added in, as may be seen here:—

---

[11] This set of examples exactly agrees with the MS., notwithstanding the want of connection in the case of the fifth, and the wrong signature in either the sixth or seventh triad.

APPENDIX.

Cap. VI.

SUNDRY RULES FOR PLAYING FROM FIGURED-BASS IN FOUR PARTS THROUGHOUT.

Regula I.

The written bass is to be played with the left hand alone, but it may play the other parts, too, together with the right hand, whether they are indicated or not.

Regula II.

The third may be played with most of the figured notes, when it is not prevented by a second or a fourth expressly indicated.

Regula III.

Two fifths or two octaves must not occur next one another, for this is not only a fault but it sounds wrong. To avoid this there is an old rule, that the hands must always go against one another, so that when the left goes up the right must go down, and when the right goes up the left must go down.

Regula IV.

The best way to avoid and get rid of two fifths or octaves is to take the sixth in and so to effect an alternation.

Cap. VII.

HOW TO PLAY WHEN NO FIGURES ARE WRITTEN ABOVE THE BASS.

When nothing stands written above the bass, none but consonances, especially the third, fifth, and eighth are to be played; for example, in the case of a bass of this kind:—

the consonances would be played in the right hand, as follows:—

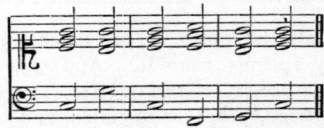

But it is not necessary to remain bound down to one kind of treatment, but the lowest part may be made the uppermost part or the treble. Then the bass just given could be also played in this way:—

The inner part when placed uppermost gives another variety, and yet the notes are of the same kind:—

But that it may be seen how, when fifths and octaves follow one another, they do not fit together, but sound wrong, although they are perfect consonances, the following example is given:—

Cap. VIII.
RULES FOR THE FIGURES WHICH STAND ABOVE THE NOTES.
Regula I.

The sign ♯ when it stands above a note betokens the major third, and those sharps which stand at the beginning of the bass part hold good for the treble also. For example, when such a ♯ stands above A the major third—*i.e.*, C♯—must be played, and so also the ♯ is indicated in the case of C where, if it were absent, it would be necessary to play C. And the case is the same with the ♭, namely, when the ♭ stands above the note the minor third is to be taken, and this rule must be strictly observed.

Regula II.

Thus, in short, all that is marked ♯ is to be in the major, and all that is marked ♭ in the minor.

Regula III.

When the number six stands above a note, it betokens the interval of the sixth, that is, I must count from the bass or from the note over which the figure stands and strike the sixth key. With this may be played either the third or the doubled sixth, and sometimes the octave, especially when a note follows immediately with the indication $^6_5$; *e.g.*—

Regula IV.

When the fifth and sixth follow one another the fifth must be prepared—*i.e.*, must be in the previous chord—the third and octave are struck together with the bass and the sixth played afterwards, but if, on the contrary, the sixth comes before the fifth, none of this holds good, but it is to be played as written—*i.e.*, the 8 3 6 are taken with the bass and the fifth afterwards, the sixth may be doubled and only the 3 taken with it—*e.g.*—

Regula V.

When 5 and 6 stand above a note, the fifth must be in the chord before, and the 3 and 6 must be added with the bass—*e.g.*—

Regula VI.

When the fourth and the third stand together, the fourth must always be prepared, and the third struck afterwards; the fifth and octave are to be played simultaneously with the bass, whether they are indicated by figures or not; the term "preparation" signifies that a note is played in the treble part by the right hand, and then remains held by the right hand over the next following figured bass note.

---

[12] The slur is wanting in the MS.
[14] The slurs wanting in the MS.
[13] In the MS. $\flat \atop 5\ 6$.
[15] The slurs wanting in the MS.

APPENDIX. 323

Regula VII.

When the second and fourth are indicated over a note ($_2^4$) the sixth is generally taken too, although not indicated. The notes in the chord of $_2^{\frac{6}{4}}$ are always struck afresh when the bass note is held from the preceding chord, and the $_2^{\frac{6}{4}}$ is resolved by means of the chord of $_5^6$ when the bass descends a semitone, as may be seen by the following example—

Regula VIII.

When the false (imperfect) fifth (♭5) is indicated, it must always be prepared; the 3 and 6, whether indicated or not, must be played together with the bass as seen in the following example—

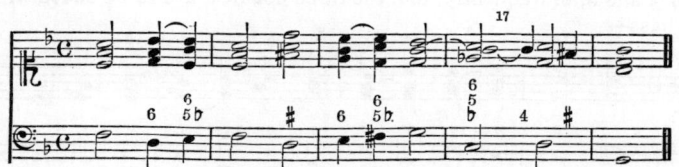

Where such closing phrases (*Cadentz Clauseln*) as $^{765}_{343}$₃ or $^{765}_{\sharp44\sharp}$ occur, they are called Syncopations, because they are, as it were, bound and intricate in their character; sometimes, too, they are indicated simply by 3-4-4-3, but yet played in full, as below :—

Where passages of rapid notes following one another occur, as they frequently do in figured-bass, it is not necessary to play chords to every note, but only in minims or crotchets; the other bass notes are

---

[16] The third chord in the right hand, and the slurs, are wanting in the MS.
[17] Figuring in the MS. $^6_{5♭}$. The slur from g′—g′ is wanting.

called passing-notes because they slide, as it were, from one principal note to the other—

Cap. IX.
Regula I.

When the seventh stands alone, it must be prepared, and the 3 and 5, or 3 and 8, or frequently, too, the three doubled, are to be added with the bass—

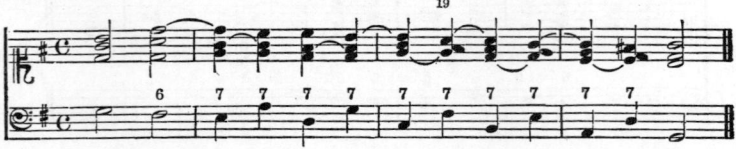

Regula II.

When the seventh and sixth come one after the other, the seventh must be prepared, and then either the 3, the 8, or the doubled 3 added with the bass, and finally the sixth, either major or minor, must be played afterwards—

---

[18] In the MS. the right hand has the chord b′ g′ d′.

[19] In the MS. the bass is

[20] The dots after the two notes in the inner parts are wanting.

APPENDIX. 325

REGULA III.

When the ninth and octave come together, the ninth must be prepared, the octave played after the bass, and the 3 and 5 played with the bass.

REGULA IV.

When the ninth and seventh are followed by the octave and sixth ($^9_7{}^8_6$) the $^9_7$ must be prepared, and the 3 taken with the bass, the $^8_6$ being played afterwards.

REGULA V.

Sometimes there also occur the figures ($^{11}_9{}^{10}_8$); in this case the $^{11}_9$ must be prepared, the 5 taken with the bass, and the $^{10}_8$ played afterwards. Those which remain cannot be well explained in words, but may be gathered from the last example.

But in order that what has hitherto been said may be impressed on the mind as completely as possible, we subjoin verbal and musical examples.

CAP. X.

EXEMPLUM I.

When no number stands over a note, nothing but the simple chord of 3, 5, and 8 is to be played, but this must be carefully observed, that always when the right hand descends, the left hand must go up, and when the left hand goes down, the right must ascend. This is called *motus*[21] *contrarius*, and in this way it is possible to avoid many consecutive fifths and octaves—

---

[21] In the MS. *modus*.
[22] B in the left hand.

APPENDIX.

Exemplum II.

When the figure 4 stands above a note it must be prepared in the preceding chord, the 5 and 8 are then added, and lastly the 4 is resolved into 3—

Exemplum III.

When the figures 7 6 stand over a note, the seventh must be prepared in the preceding chord, and the 3 and 5, or the 3 and 8, or sometimes the doubled 3, are added with the bass, and the prepared seventh is resolved into the sixth—

---

[23] The chord is ♪ in the right hand, and the figuring is $\frac{6}{5}$.

[24] Figuring ♭6. [25] Figuring ♭$\frac{6}{5}$.

[26] c″ in the upper part. [27] *Sic*, imperfect and incorrect.

APPENDIX. 327

Exemplum IV.

When 9 8 stands over a note, the ninth must be prepared in the preceding chord, the 3 and 5 taken in the bass, and then the 9 resolved into 8—

Exemplum V.

When $\frac{11}{9}\ \frac{10}{8}$ or $\frac{9}{4}\ \frac{8}{3}$ stand over a note, the $\frac{11}{9}$ or $\frac{9}{4}$ must be prepared, the 5 added in the bass, and then the discord resolved into $\frac{10}{8}$ or $\frac{8}{3}$. For the 9 is equivalent to the 2 and the 10 to the 3; the 11 to the 4 and the 12 to the 5—

[28] g′ in the alto part. [29] *Sic.* [30] *Sic.*
[31] No B flat in the bass.

Exemplum VI.

When $\frac{6}{4}$, or $\frac{4}{2}$, or 2 or 4 stand over a note, the bass must be held from the preceding chord, and the $\frac{4}{2}$ taken in the right hand, and resolved generally into $\frac{6}{5}$, if the bass descends a semitone or a whole tone—

Exemplum VII.

When 6 and 5 stand side by side over a note they are to be played one after the other, and either the 8 played with the bass or the 3 or the 6 doubled; but if one stands above the other, the 3 is to be added and played together with them.

---

[32] Figuring ♭7.   [33] Sic.

APPENDIX. 329

EXEMPLUM VIII.

When in the bass part there occurs the note that forms the third of the opening chord, the 6 must always be played to it, whether indicated or not, from which it follows that if a cadence leads into another key, the above will hold good of some other note—*e.g.*—

Starting from the C, the 6 must always be played above the E, and starting from A, the 6 must start above C—

[34] c′ in the alto.   [35] g′ in the tenor.

APPENDIX.

The following examples will throw more light on the subject—

[36] Natural wanting before the b′.   [37] *Sic.*

APPENDIX. 331

[88] g′ in the alto.
[89] The third chord is wanting in the right hand.
[40] The ♭ not before, but above the bass note.
[41] *Sic.* [42] *Sic.* [43] Only half a bar. [44] d′ in the tenor.

332 APPENDIX.

[45] *Sic.*   [46] *Sic.*   [47] *Sic.*   [48] Something is wrong in the bass part here.

APPENDIX.

[49] d' in the tenor. [50] The sharp before g' is wanting.

[51] Should be—

[52] f' ♯ in the tenor. [53] Figuring 56.

334 APPENDIX.

[54] Figuring 6 5
         5

[55] a' in the alto.

APPENDIX. 335

[56] *Sic.*

[57] This does not agree with the figuring, but might be corrected by reversing the order of the inner parts as they now stand, and making them crotchets.

[58] Inner parts

[59] *Sic.*

336 APPENDIX.

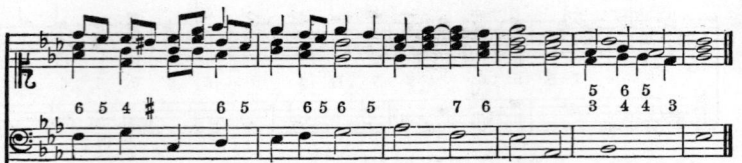

---

[60] The bass is certainly wrong. It is probably—

[61] a ♮ in the upper part is wanting.   [62] *Sic.*   [63] *Sic.*

APPENDIX.

[64] The g′ in the tenor, required by the figuring, is wanting.
[65] The f in the bass ought to be d. The sharp in the figuring thus loses its significance, but should probably be placed over the last crotchet.
[66] In the middle part, c″ alone, a crotchet.

[67] *Sic.*   [68] The a in the bass is wanting.   [69] *Sic.*
[70] The figure 6 stands incorrectly above the a.

APPENDIX.

[71] Probably intended for— [72] The d' in the tenor is wanting.
[73] f' ♯ in the tenor.   [74] The last chord is wanting in the right hand.
[75] Sic   [76] Sic.   [77] Sic.

340 APPENDIX.

RULES FOR PLAYING EN QUATRE (IN GROUPS OF FOUR NOTES).

1. Consecutive descending passages with 6 indicated.

For the first set marked with 6, the sixth may be doubled, but for every alternate group the octave is to be taken, and continued to the end.

2. Consecutive ascending passages with 6 indicated.

The octave may be taken for the first set, but the sixth must be doubled in the alternate groups.

3. Consecutive passages with 5, 6.

APPENDIX. 341

The player must begin quite high up with the right hand, and proceed *per motum*[78] *contrarium*; the octave is to be taken with the sixth.

4.

The 6 may be doubled, or the 8 may be added in.
N.B.—The 3 must always lie at the top.

5. Consecutive passages with 7, 6. N.B.—Dissonances are never to be doubled.

[78] MS., *Modum*.

342 APPENDIX.

The 5 may be taken with the 7, and the 8 with the following 6; the 6 may also be doubled. In cases where the 5 cannot be played with the 7, the 8 may be taken.

6. Passages with 7 to be resolved into 3. In playing this kind of *En quatre* the 5 or 8 may be taken with the 7.

7. The 7th which is resolved into the 3rd in such a manner that the 3rd taken with the 7th itself forms a new 7th, continuing in this way to the end.

---

[79] This example partly anticipates what is coming afterwards, probably only by an error of the transcriber.

APPENDIX. 343

With the first 7 the 5 or 8 may be taken, but if the 5 is taken with the first 7, the 8 must be taken with the next, and *vice versâ*.

8. The $\frac{6}{5}$ resolved into the 3. This kind is in four parts of itself.

The 8 must be taken with the preceding 6, for if the 6 were doubled the result would be consecutive fifths; the same holds good of the following examples.

9. The chord of 5♭ can be applied to the same passage as that in the foregoing examples.

344 APPENDIX.

10. The 4. 3.

This chord $\frac{6}{5}$ is brought in to assist in rendering many other different signatures, especially those which are consecutive, besides those with 4, 3, and they may be worked out with cognate phases (*Clausulæ cognatæ*).

11. The 9 resolved into 8 can also be well continued by means of the $\frac{6}{5}$.

APPENDIX. 345

12. The $\frac{9}{4}$, $\frac{8}{3}$ may be continued by the $\frac{6}{5}$.

[80] In bars 12-15, the sharps are wanting, both in the bass part and in the figuring.

[81] The figures on the fifth quaver are $\frac{9}{4}$ $\frac{}{5}$

[82] The figures over the first quaver are $\frac{9}{4}$ $\frac{}{5}$

[83] The figures over the fifth quaver are $\frac{9}{4}$ $\frac{}{5}$

346   APPENDIX.

Instead of the perfect 5th the 5♭ may be taken, as shown above, and in this way the $\frac{9}{4}$, $\frac{8}{3}$ can be brought in each by itself or both together.

13. The $\frac{4}{2}$ resolved into 6.

14. The $\frac{4}{2}$ may be brought in in still another way.

[84] MS. $_2$.   [85] MS. $\frac{4}{2}$.   [86] No figuring.

APPENDIX. 347

THE MOST USUAL CLAUSULÆ FINALES (CLOSING CADENCES).

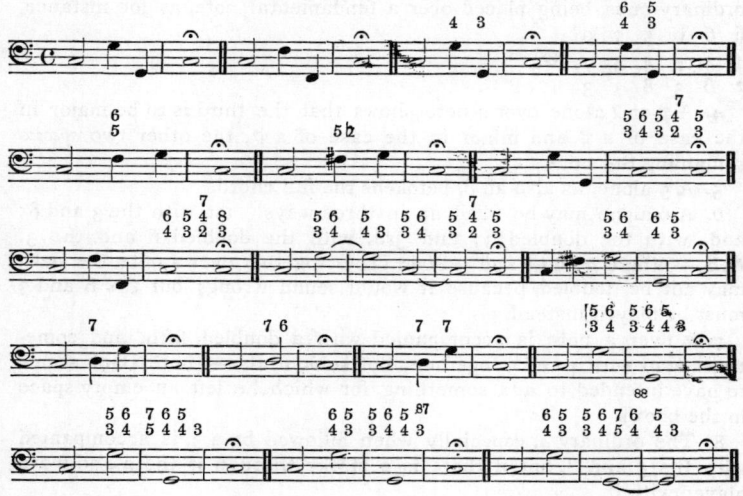

XIII.
(Ante, p. 118.)

Some most necessary rules from the *General Basso di J. S. B.*
(From Anna Magdalena Bach's Clavier book of 1725.)

The Scales. The scale with the greater third—*i.e.*, the major scale, is: (1.) *tonus*; (2.) a whole tone; (3.) a whole; (4.) a half; (5.) a whole; (6.) a half tone; (7.) a whole tone; (8.) a whole tone. The scale with the lesser third—*i.e.*, the minor scale, is: *tonus*; (2.) a whole tone; (3.) a half; (4.) a whole; (6.) a half; (7.) a whole; (8) a whole tone; from whence comes the following rule—

The 2nd is great in both scales; the 4th always smaller, the 5th and 8th are perfect, and as is the 3rd so are also the 6th and 7th.

The chord consists of 3 notes, namely, the 3rd, lesser or greater, the 5th and 8th, or, for instance, the chord of C is C E G.

On the next three pages are written the following rules in Bach' own hand:

Several rules of figured-bass.
1. Each chief note has a chord, either its own or borrowed.
2. The proper chord of a fundamental note consists of the 3, 5, and 8. N.B. Of these three *species* (sic) none can change except the 3, which may be great or small, according to whether the scale is major or minor.

---

[87] Figuring $\frac{}{4}$. [88] Figuring $\frac{7}{4}$.

3. A borrowed chord is one which consists of *species* other than the ordinary ones, being placed over a fundamental note, as for instance,

6 6 6 5 7 9
4, 3, 5, 4, 5, 7, &c.
2 6 3 8 3 3

4. A ♯ or ♭ alone over a note shows that the third is to be major in the case of a ♯ and minor in the case of a ♭, the other two *species* remaining the same.

5. A 5 alone, as also an 8, betokens the full chord.

6. A 6 alone may be filled up in three ways. 1st, with the 3 and 8; 2nd, with the doubled 3; and 3rd, with the doubled 6 and the 3. N.B.—Where 6 *major* and 3 *minor* occur together over a note, the sixth may not be doubled, because it would sound wrong; but the 8 and 3 must be played instead.

7. 2 over a note is accompanied with a doubled fifth, and sometimes also with 4 and 5 together; not seldom also—[here Bach seems to have intended to add something, for which he left an empty space in the book].

8. The ordinary 4, especially when followed by a 3, is accompanied with the 5 and 8, but if there is a stroke through it the 2 and 6 are played with it.

9. The 7 is accompanied in three ways: 1st, with the 3 and 5; 2nd, with the 3 and 8; and 3rd, with the doubled 3.

10. The 9 appears to have a similarity to the 2, and indeed by itself it is the 2 doubled, but it is accompanied in quite a different way—viz.: by the 3 and 5; instead of the 5 the 6 is put sometimes, but very rarely.

11. The $\frac{4}{2}$ takes the 6 as well, and sometimes the 5th in its place.

12. With the $\frac{5}{4}$ the 8 is taken, and the 4 resolves into the 3.

13. With the $\frac{6}{5}$ the 3 is taken, whether it be major or minor.

14. With the $\frac{7}{5}$ the 3 is taken.

15. With the $\frac{9}{7}$ the 3 is taken.

The other points which ought to be remembered are better conveyed by word of mouth than in writing.

XIV.

(Ante p. 133.)

Joh. Phil. Kirnberger's Elucidations of the third part of the "Clavierübung."[89]

Analysis

of several modulations and transpositions which occur in Herr Joh. Seb. Bach's hymns.

---

[89] Compare Kunst des reinen Satzes II., 1, p. 49.

APPENDIX.

On page 30 of Bach's collection of hymns, occurs the hymns: Dies sind die heil. 10 Gebot.

This hymn is in the Mixolydian mode, that is, G major with F natural instead of F sharp. In this mode it is possible to modulate into F, which neither suits the Lydian mode (our F) nor the Ionic (our C major) because their leading notes are only a semitone below the tonic instead of a whole tone: thus, in C major not B flat, but B is the leading note, and in F major not E flat, but E.

From the 25th to the 26th bars he modulates into F major and remains in that key until the 36th bar.

Note.—Ordinarily in a major mode, whether Ionic or Lydian, modulation is effected into the key of the fifth above with the greater third, as from C major to G major and from F major to C major.

In the Mixolydian mode (our G major) modulation cannot be effected from the chief key, G major, to its fifth above, D major, but must go into D minor, because F in the Mixolydian mode is the proper third from D.

The modulation into D minor takes place in this hymn from bars 39 and 40.

Note.—As the Mixolydian mode has no semitone below the tonic for a leading note, and consequently no major chord of the dominant, by means of which a closing cadence might be made; it has to close by going from the chord of the sub-dominant to that of the tonic, as—

The modulation to D minor on page 35 (bars 1—19), in the Mixolydian mode, occurs in other places also, and the close is made by leading from C in the bass part to the tonic G.

The hymn, *Wir gläuben all an einen Gott*, on page 37, is in the Strict Doric mode, D minor, in which the major sixth B is essential, and the minor sixth B flat is allowed.

On page 39 this hymn is found in the same mode, but transposed into E; and in order to keep to the mode there is a ♯ before C, so that just as B was the major sixth from D, so C♯ is the major sixth from E, notwithstanding that ordinarily F♯ alone is in the signature of the key of E minor.

The chorale which follows this, *Vater unser im Himmelreich* is also in the Doric mode transposed to E, and to be known by the C♯ in the signature.

On page 46 is found this same hymn in its proper place in D without B flat—*i.e.*, with the major sixth B.

The hymn on page 47, *Christ unser Herr zum Jordan kam*, is also in the Doric mode, but transposed a note lower into C, as may be seen in the signature at the beginning. Now ordinarily in C minor there is a ♭ prefixed to B, E, and A; here, however, A natural is the essential major sixth, and A with the sign ♭ written before, as ♭A, is the permissible minor sixth from the key-note.

And on page 50 this same chorale is set in the Doric mode in its proper key of D. It closes at the end not on the key-note, but in its dominant chord of A with the greater third, C♯—*i.e.*, A major.

The hymn *Aus tiefer Noth schrei ich zu dir*, etc., on p. 51, is in the Phrygian mode, in the key of E with the lesser third, or E minor.

Note.—This mode differs from the other two minor modes, the Doric and Æolic, in that its second from the key-note is a semitone, as E to F; whereas in the Doric, the interval D to E, and in the Æolic the interval A to B, are whole tones.

The last two keys, as D minor and A minor, differ from one another in its being possible in the Doric mode to modulate into E minor, because this E has the lesser third belonging to the minor triad, and the perfect fifth—*i.e.*, E G B.

From A minor, the Æolian mode, it is not possible to modulate into B, since B has no perfect fifth.

In the Doric mode it is not possible to modulate by a semitone higher into E♭ major, nor is it possible to modulate from the Æolic mode by a semitone higher into B♭ major, but it is possible in the Phrygian mode to modulate from E to F major.

This same chorale is set on page 54 in this particular mode, but it is transposed a whole tone higher, being in F♯ instead of E.

The Phrygian mode may be recognised by the signature, for usually the minor mode of this key has both F♯ and G♯ marked in the signature; here, however, the second of the scale is G, F♯ bearing the same relation to it that E bears to F.

The hymn *Jesus Christus unser Heiland*, on page 56, is also in the Doric mode. On page 60 the same chorale is transposed into F.

Note.—The Doric mode may be recognised in the signature, since there is no ♭ before D, D being the essential major sixth from F; if, however, D with the flat written before it occurs, it is the permissible minor sixth from the key-note.

XV.

(Ante, p. 268).

This note contains a complete list not only of Bach's children by his second wife, but of their Godparents. The translators have abridged it.

1. *Christiane Sophie Henriette*, born in the summer of 1723; died June 29, 1726.
2. *Gottfried Heinrich*, baptised February 27, 1724. His Godmother was Frau Regina Maria, wife of Herr Johann H. Ernesti, Rector of the Thomasschule. He was buried at Naumburg, Feb. 12, 1763.
3. *Christian Gottlieb*, baptised April 14, 1725; died September 21, 1728.
4. *Elisabeth Juliane Friederike*, baptised April 5, 1726. The date of her death is unknown.

5. *Ernestus Andreas*, baptised October 30, 1727; died November 1, 1727.
6. *Regine Johanna*, baptised October 10, 1728. Her Godmothers were Anna Magdalena, "the well-beloved daughter" of Ernesti, and another of his daughters stood proxy for the other Godmother. She died April 25, 1733.
7. *Christiane Benedicta*, baptised January 1, 1730; died January 4, 1730.
8. *Christiane Dorothea*, baptised March 18, 1731. One of her Godmothers was Frau Christiana Dorothea, wife of J. C. Hebenstreit, Conrector of the School. She died August 30 or 31, 1732.
9. *Johann Christoph Friedrich*, baptised June 23, 1732; died January 26, 1795.
10. *Johann August Abraham*, baptised November 5, 1733. The younger Ernesti was one of his Godfathers, and Abraham Krügel, *Tertius* of the school, was the other. His Godmother was Frau Elisabeth Charitas, wife of Gessner, the new Rector. He died November 6, 1733.
11. *Johann Christian*, baptised September 7, 1735. The younger Ernesti, now Rector of the Thomasschule was one of his Godfathers. He died at the beginning of January, 1782.
12. *Johanna Caroline*, baptised October 30, 1737; died August 18, 1781.
13. *Regine Susanna*, baptised February 22, 1742; died December 14, 1809.

XVI.

(Ante, p. 273.)

Specification of the property belonging to and left by Herr Johann Sebastian Bach, deceased July 28, 1750, late Cantor to the school of St. Thomas, in Leipzig. (From archives preserved at Leipzig).

CAP. I.

	Thlr.	Gr.	Pf.
A share in a mine known as Ursula Erbstolln at Little Voigtsberg . . . . Worth	60	0	0

CAP. II.
In hard cash.

(a) in Gold	112	18	0
(b) in silver money.			
(a) In specie, thalers, gulden and half gulden . . .	119	0	0
β In medals, tokens, &c.	25	20	0

CAP. III.
Assets to Credit.

A bond of Frau. Krebs	58	0	0
„ „ Unruh	4	0	0
„ „ Haase	3	0	0
Total	65	0	0

CAP. IV.

Found as cash in hand 36 0 0

Out of which some of the *Debitores passivi*, specified under †, Chap. I. and II., fols. a and b, were paid.

APPENDIX.

Cap. V. In silver plate and other objects of value.

	Thlr.	Gr.	Pf.
1 pair of candlesticks, 32 loth,[90] at 12 gr.	16	0	0
1 ditto ditto, 27 loth, at 12 gr.	13	12	0
6 cups alike, 63 loth, at 11 gr.	28	7	0
1 ditto smaller, 10 loth, at 12 gr.	5	0	0
1 ditto pierced, 12 loth, at 13 gr.	6	12	0
1 ditto still smaller, 10 loth, at 11 gr.	4	14	0
1 tankard and cover, 28 loth, at 13 gr.	15	4	0
1 large coffee pot, 32 loth, at 13 gr.	19	12	0
1 ditto smaller, 20 loth, at 13 gr.	10	20	0
1 large tea pot, 28 loth, at 13 gr.	15	4	0
1 sugar basin and spoon, 26 loth, at 12 gr.	13	0	0
1 ditto smaller, 14 loth, at 12 gr.	7	0	0
1 snuff box and spoon, 12 loth, at 16 gr.	8	0	0
1 ditto, engraved, 8 loth, at 16 gr.	5	8	0
1 ditto, inlaid	1	8	0
2 salt cellars, 11 loth, at 12 gr.	5	12	0
1 coffee waiter, 11 loth, at 12 gr.	5	12	0
½-dozen knives, forks, and spoons, in a case, 48 loth, at 12 gr.	24	0	0
1 case of knives and spoons, 9 loth, at 10 gr.	3	18	0
1 gold ring	2	0	0
1 ditto	1	12	0
1 snuff box of Agate, set in gold	40	0	0
Total	251	11	0

Cap. VI. In instruments.

	Thlr.	Gr.	Pf.
1 complete (*fournirt*) clavier, which, if possible, the family will keep	80	0	0
1 clavesin (*sic*)	50	0	0
1 ditto	50	0	0
1 ditto	50	0	0
1 ditto, smaller	20	0	0
1 lute-harpsichord	30	0	0
1 ditto	30	0	0
1 violin by Stainer	8	0	0
1 ordinary violin	2	0	0
1 ditto piccolo	1	8	0
1 viola	5	0	0
1 ditto	5	0	0
1 ditto	0	16	0
1 small bass viol	6	0	0
1 violoncello	6	0	0
1 ditto	0	16	0
1 viola da gamba	3	0	
1 lute	21	0	0
1 little spinett	3	0	0
Total	371	16	0

---

[90] A *loth* weight is about half-an-ounce.

APPENDIX. 353

Cap. VII.
In white metal.

		Thlr.	Gr.	Pf.
1 large dish	. . . .	1	8	0
1 ditto smaller	. . . .	0	16	0
1 ditto .	. . . .	0	16	0
1 ditto smaller	. . . .	0	8	0
1 ditto .	. . . .	0	8	0
1 small dish	. . . .	0	6	0
1 ditto .	. . . .	0	6	0
1 ditto still smaller	. . . .	0	4	0
1 ditto	. . . .	0	4	0
1 ditto .	. . . .	0	4	0
1 washing basin .	. . . .	0	8	0
2 dozen plates, each ¾-lb., at 4 gr.	. . . . .	3	0	0
4 jugs with metal fittings	. . . . . .	1	8	0
	Total	9	0	0

Cap. VIII.
In copper and pinchbeck.

2 dish covers with iron fittings	. . . . . .	3	0	0
3 pairs pinchbeck candlesticks	. . . . . .	2	0	0
1 pinchbeck coffee pot.	. . . .	0	16	0
1 ditto smaller .	. . . .	0	16	0
1 ditto still smaller	. . . .	0	6	0
1 pinchbeck coffee tray	. . . .	0	16	0
1 copper kettle .	. . . .	0	8	0
1 ditto smaller	. . . .	0	8	0
	Total	7	22	0

Cap. IX.
Clothes and personal sundries.

1 silver court sword	. . . .	12	0	0
1 stick, silver mount	. . . .	1	8	0
1 pair silver shoe buckles	. . . . . .	0	16	0
1 coat of *Gros du Tour* (silk), somewhat worn	. . .	8	0	0
1 mourning cloak of *Drap des Dames*	. . . .	5	0	0
1 cloth coat .	. . . .	6	0	0
	Total	33	0	0

Cap. X.
At the wash.

11 surplices . . . .

Cap. XI.
House furniture.

	Thlr.	Gr.	Pf.
1 chest of drawers	14	0	0
1 linen press	2	0	0
1 clothes press	2	0	0
1 dozen black leather chairs	2	0	0
½-dozen leather chairs	2	0	0
1 writing table with drawers	3	0	0
6 tables	2	0	0
7 wooden bedsteads	2	8	0
Total	29	8	0

Cap. XII.
In theological books.
In folio.

	Thlr.	Gr.	Pf.
Calovius, works, 3 vols.	2	0	0
Luther, Opera, 7 vols.	5	0	0
Idem liber, 8 vols.	4	0	0
Ej. Tischreden (Table talk)	0	16	0
Ej. Examen Conc. Trid.	0	16	0
Ej. Comment. über den Psalm 3ter Theil	0	16	0
Ej. Hauss-Postille (family sermons)	1	0	0
Müller, Schluss Kette	1	0	0
Tauler, Predigten (sermons)	0	4	0
Scheubler, Gold-Grube (gold diggings) 11 parts, 2 vols.	1	8	0
Pintingius, Reise Buch der Heil. Schrifft (Bünting, Itinerary of the Scriptures)	0	8	0
Olearius, Haupt Schlüssel (master key to the Scriptures) 3 vols.	2	0	0
Josephus, Geschichte der Jüden (history)	2	0	0

In quarto.

	Thlr.	Gr.	Pf.
Pfeiffer, Apostolische Christen-Schule	1	0	0
Ej. Evangelische Schatzkammer (treasury)	0	16	0
Pfeiffer, Ehe Schule	0	4	0
Ej. Evangelischer Augapffel (Apple of the Eye)	0	16	0
Ej. Kern und Safft der Heil S. (Core and Sap of Sacred Scripture)	1	0	0
Müller, Predigten über den Schaden Josephs	0	16	0
Ej. Schluss Kette	1	0	0
Ej. Atheismus	0	4	0
Ej. Judaismus	0	16	0
Stenger, Postille	1	0	0
Ej. Grundveste der Augspurg, Conf. (Ground of the Augsburg Conf.)	0	16	0
Geyer, Zeit und Ewigkeit (Time and Eternity)	0	16	0
Carried forward	29	4	0

APPENDIX.

	Thlr.	Gr.	Pf.
Brought forward	29	4	0
Rambach, Betrachtung (Reflections)	1	0	0
Ej. Betrachtung über den Rath Gottes (Reflections on the Councils of God)	0	16	0
Luther, Hauss Postille	0	16	0
Frober, Psalm	0	4	0
Unterschiedene Predigten (various sermons)	0	4	0
Adam, Güldener Augapffel (Golden Apple of the Eye)	0	4	0
Meiffart, Erinnerung (Reminiscences)	0	4	0
Heinisch, Offenbahrung Joh. (On the Revelation of St. John)	0	4	0
Jauckler, Richtschnur der Christl. Lehre (Clue to Christian Doctrine)	0	1	0

In octavo.

	Thlr.	Gr.	Pf.
Franck, Hauss Postilla	0	8	0
Pfeiffer, Evangelische Christen Schule	0	8	0
Ej. Anti-Calvin	0	8	0
Ej. Christenthum	0	8	0
Ej. Anti-Melancholicus	0	8	0
Rambach, Betrachtung über die Thränen Jesu (On the Tears of Jesus)	0	8	0
Müller, Liebes Flamme (Flame of Love)	0	8	0
Ej. Erquickstunden (Hours of Refreshment)	0	8	0
Ej. Rath Gottes (God's Councils)	0	4	0
Ej. Lutherus defensus	0	8	0
Gerhard, *Schola Pietatis*, 5 vols.	0	12	0
Neümeister, Tisch des Herrn (The Lord's Table)	0	8	0
Ej. Lehre von der Heil Tauffe (Doctrine of Holy Baptism)	0	8	0
Spener, Eyfer wider das Pabstthum (Zeal against the Papacy)	0	8	0
Hunn, Reinigkeit der Glaubens Lehre (Purity of Faith and Doctrine)	0	4	0
Kling, Warnung vor Abfall von der Luther..Relig. (Warning against the Decay of the Lutheran Religion)	0	4	0
Arnd, Wahres Christenthum	0	8	0
Wagner, Leipziger Gesangbuch, 8 vols.	1	0	0
Total	38	17	0

Recapitulation.

					Thlr.	Gr.	Pf.
Fol. 1 *a*,	Cap. I.,	one share			60	0	0
„ 1 *a*,	„	II., in cash			—	—	—
			a, in gold		112	18	0
„ 1 *a*,	„	„	*b*, in silver money		—	—	—
				a, in thalers, gulden, and half-gulden	119	0	0
„ 1 *b*,	„	„	β, in other pieces		25	20	0
„ 1 *b*,	„	III., in outstanding assets			65	0	0
		Carried forward			382	14	0

	Thlr.	Gr.	Pf.
Brought forward	382	14	0
Fol. 2 *a*, Cap. IV., in cash in hand, out of which some of the debts entered under fol. 8, *a* and *b* † Cap. I. and II. are paid	36	0	0
„ 2 *a* and *b*, Cap. V., in silver-plate and other valuables	251	11	0
„ 3 *a*, Cap. VI., in instruments	371	16	0
„ 3 *b*, „ VII., in white metal	9	0	0
„ 3 *b* and 4*a*, Cap. VIII., in copper and pinchbeck	7	22	0
„ 4 *a*, Cap. IX., in clothes and personal sundries	32	0	0
„ 4 *b*, „ X., at the wash, 11 surplices	—	—	—
„ 4 *b*, „ XI., in house furniture	29	8	0
„ 5 *a* and *b*, 6 *a* and *b*, Cap. XII., in theological books	38	17	0
Total	1158	16	0

†

Debita passiva.

According to their bills, some of which were paid out of the money specified in Cap. IV., of fol 2.

CAP. I.

	Thlr.	Gr.	Pf.
(As per bills)	143	21	6

CAP. II.

Other necessary expenses.

In necessary matters	1	8	0
To Herr Schübler	2	16	0
To the maid	4	0	0
For taxing (the property)	1	0	0
Total	9	0	0

Recapitulation.

Fol. 8, *a* and *b*, Cap. I., in payment of bills	143	21	6
Fol. 8, *b*, Cap. II., in other expenses	9	0	0
Total	152	21	6

Anna Magdalena Bach, widow.
D. Friedrich Heinrich Graff, as *curator* (representing her interests).
Catharina Dorothea Bach.
Wilhelm Friedemann Bach, for myself and for Carl Emanuel Bach, my brother, and as representing my sister above named.
Gottfried Heinrich Bach.
Gottlob Sigismund Hesemann, as representing the above, G. H. Bach.
Elisabeth Juliana Friderica Altnikol (born Bach).
Johann Christoph Altnikol, as the husband and representative of my wife Elisabeth Juliana Friderica (born Bach).

APPENDIX.

Johann Gottlieb Görner, as guardian in their father's stead, of
Johann Christoph Friedrich Bach;
Johann Christian Bach;
Johanna Carolina Bach;
Regina Susanna Bach.

THE deed of appointment of guardians sets forth that, whereas "The Worshipful Gentleman, Herr Johann Sebastian Bach, late Cantor of the School of St. Thomas, at Leipzig, had fallen asleep in the Lord, on July 28, 1750, and had left three children of his first marriage, namely:—
Herr Wilhelm Friedemann Bach;
Herr Carl Philipp Emanuel Bach; and
Jungfrau Catharina Dorothea Bach;
and no less than six children of his marriage with his present widow, Dame Anna Magdalena (born Wülcke), namely:—
Herr Gottfried Heinrich Bach.
Dame Elisabeth Juliana Friderica (married to Altnikol);
Herr Johann Christoph Friedrich Bach;
Jungfrau Johanna Carolina Bach; and
Jungfrau Regina Susanna Bach;
of which the four last are still under age, and has left his property to be divided among his said widow and heirs. Herr Johann Gottlieb Görner, *Director Musices* at the University of Leipzig, is appointed guardian of the four children under age, and with Herr Gottlob Sigismund Hesemann *L. L. Studiosus*, as representative of Herr Gottfried H. Bach, who is imbecile."

1.

All the heirs (each being named) and the guardians and representatives of those who are legally incapable, having read through the specification of the property left by their father, each and all declare it to be correct and satisfactory.

2.

With regard to the share mentioned in Cap. I. of the Specification, the heirs are all agreed that it shall remain in a common fund and put into the charge and keeping of the widow, who shall take her own third part, leaving the rest to be equally divided among the children.

3.

The cash in gold and silver moneys, specified in Cap. II., *a* and *b*, is distributed equally in kind among the heirs, the widow's portion of one-third amounting to 77 thlrs. 6 gr., and that of each child to 17 thlrs. 4 gr.

4.

The coins enumerated in Cap. II. β. were also divided *in natura* among them by lot; the widow's share amounting to 8 thlrs. 14 gr. 8 pf., and that of each child to 1 thlr. 21 gr. 11 pf. Some of the pieces were probably curious, and may have had a value above that of the bullion.

5.

Of the outstanding assets, specified in Cap. III., the widow Bach takes her sister's bond of 58 thlrs., and as her share of it amounts to 19 thlrs. 8 gr., she has paid the remaining 38 thlrs. 16 gr. to the children in cash, namely, to each 4 thlrs. 7 gr. 1 pf.; but they, with the permission of their lawfully appointed guardians, have given it back to the widow, and ceded the bond to her in such a way that she can dispose of it according to her pleasure, as her own property; and the widow has given to them a receipt for the same in proper form.

As regards the bonds of Unruh and Haase, these persons are not to be found, and the widow is, therefore, left in possession of the papers.

6.

The ready money in hand has been spent in part payment of outstanding debts, for which see Section 14.

7.

In the interest, and with the consent of all concerned, the agate snuff box, mounted in gold, is for the present withdrawn from among the valuables specified in Cap. V., and valued at 40 thlrs., partly because it is a piece of property fit only for the collector and connoisseur, and partly because it is too valuable to be assigned by lot to either of the children, and until a purchaser shall be found it is left in care of the widow. (All the rest of the things are allotted according to the valuation of Herr Berthold, the goldsmith, in such a way as that the widow, having a pair of candlesticks, the larger coffee pot, the tea pot, two of the snuff boxes, the pierced cup, and the coffee waiter, acquires the value of 72 thlrs.; and the children, each, sundry articles to the sum of 15 thlrs. 15 gr. 11 pf. Altnikol's wife became the possessor of the tankard with the lid, and the eldest son and daughter each had a gold ring.)

8.

The instruments specified under Cap. VI. (as they cannot be divided, and as no purchaser offers) are also set aside, with the hope that they may be sold before Easter. The widow is meanwhile to have the care and use of them, and when each may be sold is to keep her third of the purchase money and divide the other two-thirds among the nine children. But because Herr Joh. Christian Bach, the youngest son of the deceased, had received from his father during his

MUSICAL SUPPLEMENT

lifetime three claviers with pedal, these have not been included in the specification, since he declares them to have been given to him as a present, and has brought witnesses to that effect, the widow and Herr Altnikol and Herr Hesemann, having known of it. The guardian, however, finds something suspicious in the matter, as do also the children of the first marriage, but they refrain from urging their objections, and, on the contrary, the widow, the other heirs and their representatives acknowledge and admit the gift.

9.

The white metal, copper, and pinchbeck goods, enumerated in Caps. VII. and VIII., have by common consent been exempted from taxation, and after the deduction of one-third, amounting to 5 thlrs. 15 gr. 4 pf., each child takes a share of 1 thlr. 6 gr.

10.

The silk coat, the mourning cloak, the shoe buckles, and the stick have also been exempted from taxation, and the widow having taken her share of the 5 thlrs., children take each 1 thlr. 2 gr. 8 pf. The silver sword, belonging to the court accoutrements, was taken by the eldest son, Herr Friedemann, who paid for it; and the 6 thlrs. paid for the cloth coat, which also belonged to the court accoutrements, and which was previously left to Gottfried Heinrich Bach, was divided among the five sons, the share of each being 1 thlr. 4 gr. 9 pf.

11.

The linen of the deceased is, with the unanimous consent of the seniors, divided among the children under age.

12.

The furniture is also by common consent taken at the specified valuation, and the widow takes 9 thlrs. 18 gr. 8 pf., and each child 2 thlrs. 4 gr. 1 pf.

13.

The books are allotted according to the valuation; the widow's share is 12 thlrs. 21 gr. 8 pf., each child's 2 thlrs. 20 gr. 10 pf.

14.

Finally, the debts enumerated under † Cap. I. and II. as amounting to 152 thlrs. 21 gr. 6 pf., having been found perfectly correct after the sum of 36 thlrs., specified as cash in hand in Cap. IV., has been deducted, there still remains a balance of debt of 116 thlrs. 21 gr. 6 pf. The widow, therefore, pays for her third part 38 thlrs. 23 gr. 2 pf., and each child 8 thlrs. 15 gr. 10 pf., which is deducted from their respective shares of the inheritance. And the widow and children and their representatives agree to defray the cost and charges of their father's funeral.

15.

Mem.—Whereas Herr Görner is appointed to represent the children under age only so far as the division and distribution of the estate is concerned, and the widow is, irrespective of that, their sole and only guardian; the guardian is to be responsible for the joint division of the property in accordance with the foregoing statement.

Leipzig, Nov. 11, 1750.

Signed by all the family and representatives.

(The accounts of each child by lot, payment, and purchase are given at full length in the German edition, where the curious reader will find two petitions from Dame Anna Magdalena Bach, widow, to the Rector and Patrons of the University; the first, dated October 17, 1750, begging that, whereas her husband, J. S. Bach, had died on July 28, having four children under age, they would be pleased to appoint a *tutor* or guardian immediately for the protection of the interests of these children; and the second, dated October 21, 1750, craving that since she is fully determined not to marry again, but to take upon herself the general guardianship of the children, they will confirm her in this, and appoint Herr Görner to represent her and them in the matter of the division of her late husband's estate.

The last document quoted is a petition on the part of Johann Christoph Friedrich Bach that the Council would confirm him in the possession of the instrument left to his father by J. C. Hoffmann, for which, at the same time, he deposits a receipt).

Supplement I. (Vol. I. p.106).

ORGAN CHORALE.
"Warum betrübst du dich, mein Herz."

Joh. Christoph Bach.

*The "d" is wanting in the MS.

FUGUE.

Supplement II. (Vol. I. p. 427).

Tomaso Albinoni.

SONATA

by

JOHANN ADAM REINKEN.

372

SONATA.
For Violin and Bass
by
T. ALBINONI.

The figured bass accompaniment by Heinrich Nikol. Gerber,
Corrected throughout by Sebastian Bach.

1) Gerber; originally thus:

4) Gerber: 5) Gerber: 6) Gerber:

11) Gerber:

Supplement VII. (Vol. II. p. 551).

Final Chorale of Part I. of the Matthew in its original form.

upplement VIII^a (Vol. III. p. 109, and Vol. I. p. 594).

CHORALE
"Gelobet seist du Jesu Christ."
Arranged by Sebastian Bach for accompanying the congregation.

Supplement VIII.b (Vol. III. p.113).

SIX HYMNS.

Apparently composed by Sebastian Bach.

I.

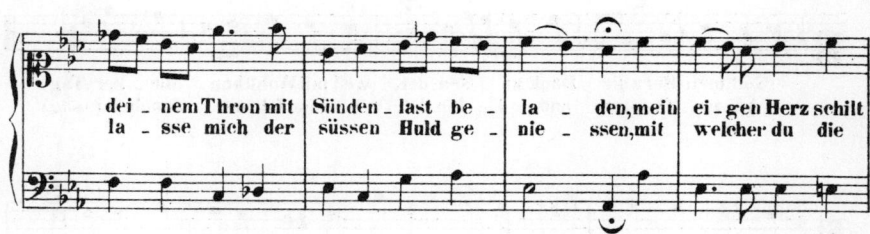

II.

III.

Gott, mein Herz dir Dank zu senden, weil mit Wohlthun dieser Tag
An-ge-fan-gen und vol-len-det, so dass ich mit Jauchzen sage:

Ich bin al-ler Sün-den los, ru-he sanft in Je-su Schooss,

ich bin Je-su Braut heut worden, steh in sei-nem Lie-bes-or-den.

IV.

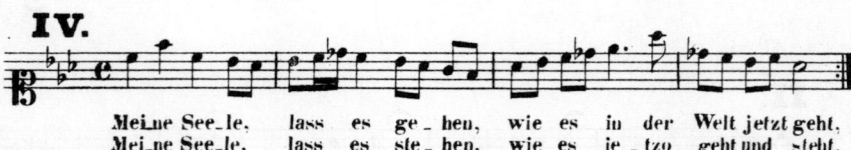

Mei-ne See-le, lass es ge-hen, wie es in der Welt jetzt geht,
Mei-ne See-le, lass es ste-hen, wie es je-tzo geht und steht.

Lieb-ste See-le, hal-te stil-le, den-ke, dass es Got-tes Wil-le.

V.

Ich gnü_ge mich an meinem Stande, in den der Höchste mich ge_setzt,
Und acht es gar für kei_ne Schande, bin ich nicht je_dem gleich ge_schätzt.
Ich darf so we_nig mei_nen Schö_pfer anklagen als der Thon den Töpfer.

VI.

Wa_rum be_trübst du dich und beu_gest dich zur Er_den, mein
Du sorgst, wie will es doch noch end_lich mit dir wer_den, und
sehr geplag_ter Geist, mein ab_ge_mat_ter Sinn.
fährest ü_ber Welt und ü_ber Him_mel hin.
Wirst du dich nicht recht fest in
Got_tes Wil_len grün_den, kannst du in E_wigkeit nicht wahre Ru_he fin_den.

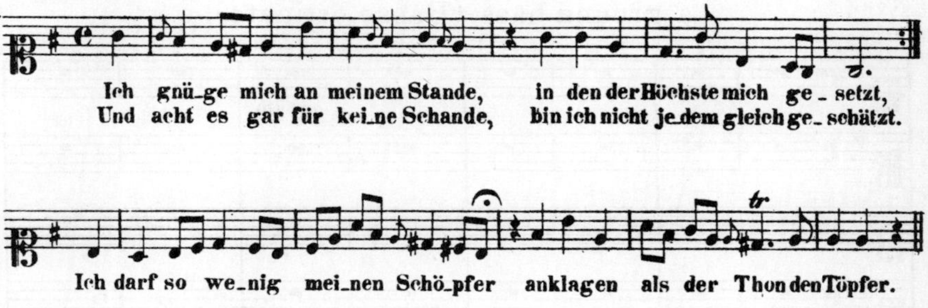

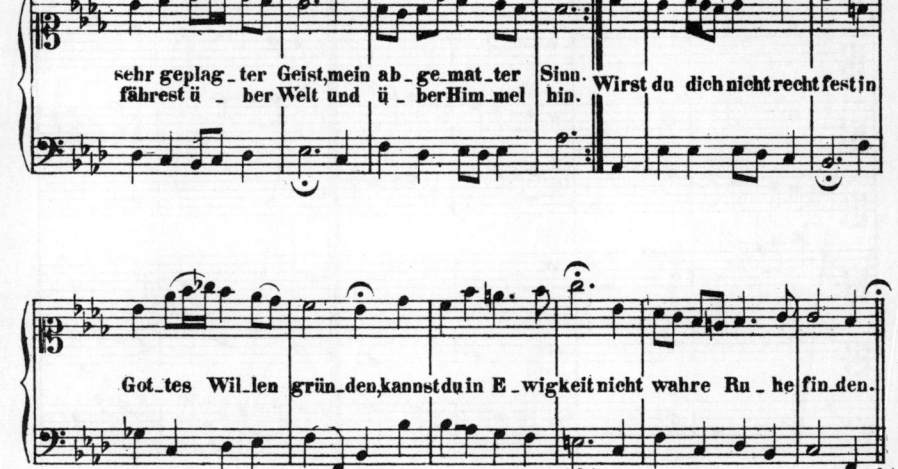

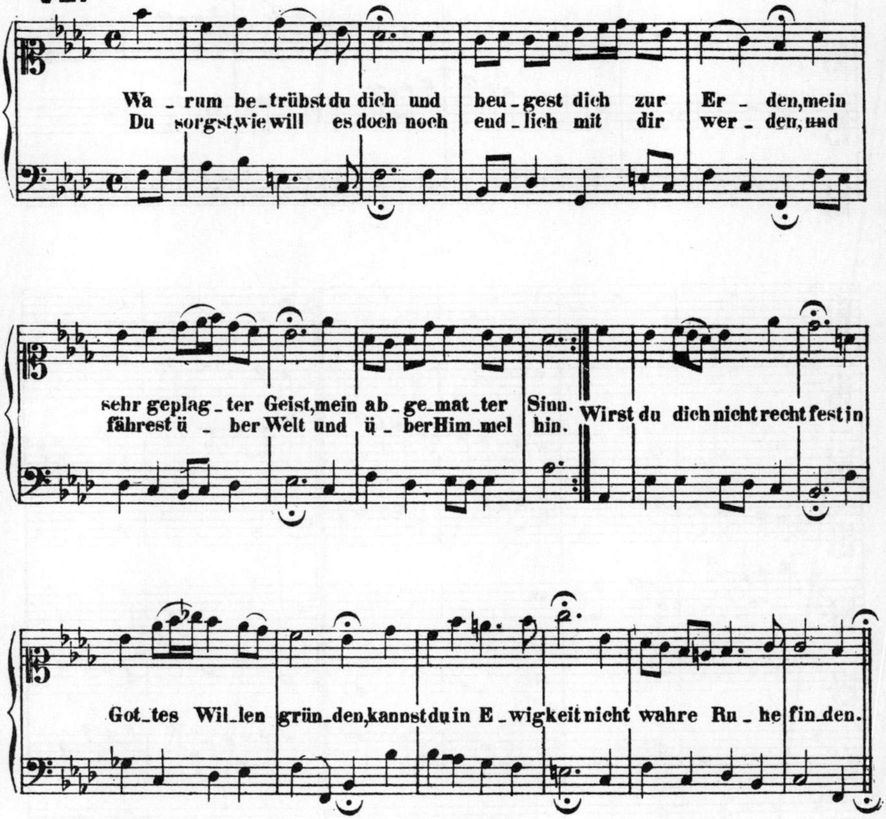

Supplement IX. (Vol. III. p. 239).

Solution of a seven-part Canon by Bach above a ground bass (Basso ostinato).

INDEX.

ABEL, C. F., Viol-da-gambist, II. 4, 100.
ABENDMUSIKEN, at Lubeck, I. 257-259.
ACCOMPANIMENTS, method of playing and conducting, II. 295-325.
"ADJUVANTEN," I. 225; II. 308.
ADLUNG, JAK., I. 137; III. 230, 244.
AGRÉMENS, II. 297, 317, 320 ff.
AGRICOLA, G. L., I. 192.
AGRICOLA, J. F., II. 237, 298, 311, 317, 329; III. 243 f.
AHLE, J. R., I. 335, 340 ff.
AHLE, J. G., I. 357.
ALBERTI, J. F., I. 99.
ALBINONI, T., I. 425-428; II. 293, 296; fugue by, III. 364 f.; sonata, III. 388-398.
ALT, W. C., I. 561; III. 300.
ALTNIKOL, J. C., II. 658; III. 19, 222, 246, 247, 248, 268, 271 f., 274, 356.
AMMERBACH, E., III. 263.
ANHALT, see Cöthen.
ANSPACH, band at, II. 640.
ARIA-FORM, I. 232, 340, 445, 469, 473.
ARIOSO, I. 230, 232, 571.
ARMSTROFF, ANDREAS, I. 118.
ARND, JOH., III. 266.
ARNSTADT, church library at, I. 37; organ at, I. 222 ff.; opera at, I. 228; accounts of church at, I. 315-317; Bach's residence at, I. 220-331.
ATTENDANCES (Aufwartungen), I. 20.
AUGUST III., II. 626; III. 226.

BACH, ANDREAS, of Ohrdruf, I. 629.
BACH, ANNA MAGDALENA (née Wülken), II. 147-153; III. 276 f., 347 f., 356, 360; her MSS., II. 149 ff.; her book, III. 112 f., 118.
BACH, CARL PHILIPP EMANUEL, I., Preface ix.; II. 8, 328, 331 f.; III. 160 f., 235 f., 244, 248, 268 f., 281; his revision of Sebastian's hymn tunes, III. 115 f., 127; his method of fingering, II. 38 f.
BACH, CHRISTOPH, son of Hans Bach, I. 19, 142, 153-157.
BACH, ERNST CHRISTIAN, I. 14.
BACH, family genealogy, I. Preface, ix.; origin of, I. 1-4; characteristics of the ancestors, I. 163, 175-178; their "family days," I. 154; their Quodlibets, I. 154; their culture, I. 153.
BACH family, of Rockhausen, I. 2, 13.
BACH family, of Molsdorf, I. 3, 14.
BACH family, of Wechmar, I. 4, 13.
BACH, GEORGE CHRISTOPH, of Schweinfurt, I. 155.

III.

BACH, GEORGE MICHAEL, of Halle, I. 12.
BACH, GOTTFRIED HEINRICH, son of Sebastian, III. 268, 270.
BACH, GOTTLOB FRIEDRICH, of Meiningen, I. 10.
BACH, HANS, "Der Spielmann," and two others of the same name, I. 6, 8, 13, 621.
BACH, HEINRICH, I. 9, 27, 32-35, 41, 169 f.; III. 277.
BACH, JAKOB, of Steinbach and Ruhla, I. 10.
BACH, JOH., of Erfurt, I. 14-20.
BACH, JOH. AEGIDIUS, of Erfurt, I. 23, 174.
BACH, JOH. AMBROSIUS, father of Sebastian, I. 21, 156-158, 173, 181 f.
BACH, JOH. BERNHARD, of Eisenach, I. 23-26, 39, 108, 118; his orchestral partie, II. 144.
BACH, JOH. BERNHARD, of Ohrdruf, I. 524 f., 629; III. 263.
BACH, JOH. CHRISTIAN (1743-1814), I. 12.
BACH, JOH. CHRISTIAN (b. 1640), I. 21.
BACH, JOH. CHRISTIAN (b. 1682), I. 22.
BACH, JOH. CHRISTIAN (1717-1733), I. 132.
BACH, JOH. CHRISTIAN, son of Sebastian, I., Preface ix.; III. 270 f., 274, 277, 289, 357 f.
BACH, JOH. CHRISTOPH of Unter-Zimmern, I. 22.
BACH, JOH. CHRISTOPH, son of Aegidius, of Erfurt, I. 27; III. 230.
BACH, JOH. CHRISTOPH, of Eisenach, I. 29, 33, 37 f., 40-51, 106-107, 119, 121, 122, 126-131, 218, 622; III. 171, 218; his compositions, "Es erhob sich ein Streit," I. 44-50, 124, 294; motetts, I. 64, 73-96, 177; "Ich lasse dich nicht," II. 716; organ chorale, "Warum betrübst du dich," III. 363; his influence on Sebastian, II. 409.
BACH, JOH. CHRISTOPH, son of the foregoing, I. 140.
BACH, JOH. CHRISTOPH, of Arnstadt, I. 156-171, 184.
BACH, JOH. CHRISTOPH, of Ohrdruf, Sebastian's elder brother, I. 183-185, 249.
BACH, JOH. CHRISTOPH FRIEDRICH, III. 270, 351, 357, 360.
BACH, JOH. ELIAS, I. Preface x., I. 156; III. 240, 271 f., 298.
BACH, JOH. ERNST, of Arnstadt, I. 171 f., 196, 222, 226, 336, 338.
BACH, JOH. ERNST, of Weimar, III. 34, 239, 297.
BACH, JOH. FRIEDRICH, I. 141 f., 375, 623; III. 240, 277.
BACH, JOH. GOTTFRIED BERNHARD, II. 8; III. 269 f.

BACH, JOH. GÜNTHER, I. 33.
BACH, JOH. JAKOB, I. 182, 235, 621; II. 154-156.
BACH, JOH. LORENZ, I. Preface ix., x., 156.
BACH, JOH. LUDWIG, of Meiningen, I. 10, 389, 574-582; III. 263; his orchestral partie, II. 144.
BACH, JOH. MICHAEL, of Gehren, I. 33, 39 f., 58-73, 94, 107, 119 f., 125; his " Ach bleib bei uns," I. 51-53.
BACH, JOH. MICHAEL, the organ builder, I. 140.
BACH, JOH. NIKOLAUS, of Erfurt, I. 27.
BACH, JOH. NIKOLAUS, of Jena, I. 131-140, 594; III. 34, 235, 270; his mass, I. 132; his operetta, I. 134-137; his organ building, I. 137-140.
BACH, JOHANN SEBASTIAN :—
His Life:—Birth, I. 181; early education, I. 182; in the scholars' choir, I. 183; he purloins the organ book, I. 186; at school at Ohrdruf, I. 186 f.; travels to Lüneburg, I. 189 f.; in the choir at Lüneburg, I. 190; studies at Lüneburg, I. 217 .f.; journey to Celle, I. 201; to Hamburg, I. 197, 200; first visit to Weimar, I. 220; appointment to Arnstadt, and residence there, I. 222-331; his salary there, I. 223; visit to Lübeck I. 256 f., 262 f.; return to Arnstadt, and difficulties with the authorities, I. 311-318, 326-328; examination by the Consistory, I. 315-317; trial performance at Mühlhausen, I. 331; residence there, I. 335-375; resignation of post, I. 370, 373 f.; second visit to Weimar, I. 370; appointment there, I. 373, 375 : his position there, I. 378, 380, III. 229, 299, 300; journey to Cassel, I. 513-515, 634 f.; invited to Halle, I. 515; tries organ there, I. 521; journey to and performances at Leipzig, I. 519; ordered to compose new works every year, I. 526; journey to Meiningen, I. 574-582; to Dresden, I. 583-586; departure from Weimar, I. 587; summoned to Cöthen, I. 586; residence there, II. 1-159, III. 224 f.; poem written there, III. 224 f.; chamber music there, II. 5 f.; relations with Prince Leopold, II. 6, 8; journey to Carlsbad, II. 7 f.; to Weimar and Halle, II. 9; examination of organ at Leipzig, II. 9; journey to Hamburg, and trial for post there, II. 13-21; writes funeral music for Prince Leopold, II. 158 f.; leaves Cöthen, II. 156-159; candidate for post at Leipzig, II. 183 f.; elected and appointed cantor of Thomasschule, II. 186-189, III. 301 f.; his position, duties, and salary there, II. 190-200, 213-215, III. 243; examination of the choir, II. 233-241; journey to Weissenfels, III. 223 f.; frequent absences from Leipzig, III. 230; at Dresden, III. 223, 225-228; at Erfurt, III. 230; application for post at Dantzig, II. 253 f.; remains in Leipzig, II.

255; conducts Telemann's society there, III. 135 f.; conducts Musical Union there, II. 674; Passion music first performed, II. 241 f.; dispute with the council, and letters to the king, II. 215-222; dispute with authorities about church music, II. 230-233; reprimanded by authorities, II. 242-251; dispute with Ernesti, III. 1-16, 280-287, 307-315; appeal to the council, III. 7; court title granted, III. 8; appointed Hofcomponist, III. 226 f.; last visit to Hamburg, III. 228; at Potsdam, III. 192, 230-232; at Berlin, III. 232-234; his intercourse with Frederick the Great, III. 231-234; his blindness and death, III. 274; his funeral, III. 275; his grave, III, 276.

His Artistic Career : His musical heritage, I. 175-178; first musical proclivities, I. 182 ff., 218; tuition by his brother, I. 184 f.; his technical skill, I. 248-252; his method of fingering and execution, I. 393, II. 35-41, 44-46, 651 f. 655, III. 240, 262 f.; his organ-playing, and use of the organ, I. 394-398, II. 29 f., 655, III. 212, 214; as a judge of organs, III. 21; his method of tuning, II. 42; method of accompanying, I. 313 ff., 592-597; II. 109 f., 293, 295-324, 678; method of conducting, II. 327-332, 678; as an acoustician, III. 232 f.; as a violinist, II. 68-70; as a lutanist, II. 46 f.; invention of the Viola pomposa, II. 69, III. 227, 250; invention of the Lautenclavicymbel, II. 46 f., III. 167; studies and copies of Italian masses, III. 28 f.; his maturity of style, I. 329-331, 504, 506; his attitude towards theoretical learning, III. 24; towards the ancient modes, III. 129-134; towards Rameau's treatise on harmony, III. 124; his treatment of figured-bass, II. 101-110, 654; III. 118-122, 315-348; relation between his preludes and fugues, III. 187; fugal writing in the second part of "Wohlt-Clav." III. 188 f.; his form of clavier concerto compared with Mozart's, III. 141; his variation form, III. 173 f.; his treatment of recitative, I. 494-498; treatment of the cantata-form, I. 485 f., II. 454; adaptation of secular music to sacred purposes, II. 447-450; comparison between his sacred and secular styles, II. 577 f.; his treatment of the chorale, II. 456-459; arrangement of his orchestra, II. 323-325; difficulties in his vocal music, II. 236 f.; love of nature in his music, II. 561; his last works, III. 275; his position in art, II. 144-146; his own view of his position, II. 225 f.; his criticisms of other players, II. 264; his contemporary fame, III. 21 f., 42, 235, 277 f.; mythical anecdotes, III. 264, 268.

As a Teacher: His style and method of instruction, II. 47-50, III. 116-122, 126-129;

INDEX. 409

his Weimar pupils, I. 522-526; list of his pupils, II. 47; his Leipzig pupils, III. 239-252; their affection for him, III. 262.

His Character: General remarks on, I. 193, 314, 329; II. 148; his piety, III. 265, 267; his Lutheran tenets, III. 44; his attitude towards Pietism, I. 362-370, 485 f.; towards the Christmas festival, III. 221; his interest in Italian music, I. 407, 432, III. 28 f.; in the opera at Dresden, II. 337; his love of travel, III. 223, 230; his relations with his intimates, III. 222, 249-252; his kindly disposition, III. 262; his fearlessness in dispute, III. 261: activity in composition in old age, III. 68; domestic habits in old age, III. 230, 249 f.; story of the musical beggars, III. 265; his hospitality, III. 273.

Personal Details: His personal appearance, III. 265; his house, III. 267; his possessions, III. 274, 351-360; his collection of musical instruments, III. 273; his library, III. 265-267, 354; music in his own house, II. 440, 443, 474-477; III. 136, 268.

His Family: His father (see Bach, Joh. Ambrosius); his mother, I. 174 ff., 623; his brothers and sisters, I. 175, 182; his first marriage, I. 327-331, 339; receives accession of fortune through first wife, II. 153 f.; her death, II. 11; her character, II. 12; children of first family, II. 8; second marriage, II. 147-153 (see also Bach, Anna Magdalena); children of second family, II. 153, III. 268, 350 f.; his daughters, III. 271, 277.

As influenced by other Musicians—by: Bach, Joh. Christoph, II. 409; Böhm, Georg, I. 210-217, 241; Buxtehude, Dietrich, I. 276, 278, 283, 285, 296, 313-326; III. 153 note, 207, 209, 211, 218; Erlebach, P. H., I. 351; Frescobaldi, I. 420; Froberger, III. 187 note; Handel, II. 426 f.; Kuhnau, I. 243-245, 320, II. 371, III. 155; Lotti, II. 638, III. 28; Pachelbel, I. 111, 255 f.; Telemann, II. 437, 694 f.

His Relations with: Ernesti, III. 1-16, 280-287, 307-315; Freylinghausen, I. 366-370; Fux., III. 124; Gesner, II. 259-263; Handel, I. 330, II. 9-11, 25-30, 144-146 f., 523 f., III. 17, 163 f. 243; Hasse, III. 226, 228; Marchand, I. 584-586, 644-647; Mattheson, III. 228 f.; Rameau, III. 124; Reinken, II.17 f.; Scheibe, III. 252-255; Schemelli, III. 109-114, 287 f.; Telemann and the musical societies, II. 242, III. 20 f., 24 f., 219; Walther, I. 381-389.

Letters, &c., by him: Application to the Mühlhausen authorities for testimonial, I. 373 f.; application to the council about church music, II. 231 f.; application to the Dresden court for a title, III. 8; correction of Gerber's accompaniments to Albinoni, III. 388-398; dedication of B minor mass, III. 38; dedication of First Partita, III.224 f.; dedication of Musikal. Opfer, III.233 f.; dedication of "Faber" canon, III. 237 f.; letter about the council to the king, II. 215-222; letter to Elias Bach, III. 271-273; letter to Erdmann, II. 253 f.; letter to Erfurt council, II. 154 f.; letter to Halle, I. 516-518, 520; letter to Mühlhausen, III. 269 f.; Memorial on the choir, II. 247-251; memorial about the prefects in Leipzig, III. 307-315; Report on the organ in the Paulinerkirche in Leipzig II. 288-290; specification for improvements to the Mühlhausen organ, I. 355 f.; testimonial to Altnikol, III. 271; testimonial to Bernhard Bach, III. 269 f.; testimonial to Dorn, III. 249; testimonial to Krebs, III. 241 f.; testimonial to Thomasschule pupils, II. 234 f., 240 f.; treatises on thorough-bass, III. 315-347, 347 f.

Documents relating to him: I. 315-317, 326, III. 8, 301 f., 351-360; autographs of, I. Preface v., 329 f.; his handwriting, I. 330; his first published work, III. 156; probable number of cantatas, II. 349 f.; editions of his works, I. Preface vii.; engraving of his works, III. 197, 203 f., 238; lives of, I. Preface ii.

His Vocal Works:

CANTATAS, SACRED—Ach Gott, vom Himmel sieh' darein, III. 285; Ach Gott, wie manches Herzeleid (in A), III. 285; *ib.* (in C), II. 463, 690 f.; Ach ich sehe, I. 554 f., III. 263 note; Ach lieben Christen, III, 285; Ach, wie flüchtig, III. 106, 285. Actus tragicus (see "Gottes Zeit"); Aergre dich, o Seele, nicht, II. 358-360; Allein zu dir, III. 287; Alles nur nach Gottes Willen, II. 415 f.; Alles was von Gott geboren, I. 563-565; Also hat Gott die Welt geliebt, III. 69 f., 281 f.; Am Abend aber, III. 85 f.; Ascension cantata, III. 70, 72, 77; Auf Christi Himmelfahrt allein, III. 72, 281, 284; Auf, mein Herz, des Herren Tag, II. 442, note 476; Auf, Menschen, ruhmet (see "Ihr Menschen," &c.); Aus der Tiefe, I. 448-456; Aus tiefer Noth, II. 298, III. 285; Barmherziges Herze der ewigen Liebe, I. 545-548, II. 314; Bereitet die Wege, I. 557-560; Bisher habt ihr nichts gebeten, III. 71 f., 281, 285 f.; Bleib bei uns, III.75 f., 285; Brich dem Hungrigen dein Brod, III. 81 f.; Bringet her dem Herren, II. 411, 693 f.; Christen ätzet diesen Tag, II. 367-369, 684; Christ lag in Todesbanden, II. 392-397, 688, III. 105; Christmas cantatas, II. 684; Christum wir sollen loben schon, II. 285; III. 285; Christ unser Herr zur Jordan kam, III. 103, 285; Christus der ist mein Leben,

II. 462 f., 705 ; Das ist je gewisslich wahr, II. 15 ; Das neugeborne Kindelein, III. 285; Dazu ist erschienen der Sohn Gottes, II. 685 ; Der Geist hilft unser Schwachheit auf (motett), II. 602, 605 ; Dem Gerechten muss das Licht, II. 468 f., 613 ; Denn du wirst meine Seele nicht in der Hölle lassen, I. 229 ff., II. 688; III. 68, 284 ; Der Friede sei mit dir, II. 687; Der Herr denket an uns, I. 370-373 ; Der Herr ist mein getreuer Hirt, II. 455, 457; Der Himmel lacht, I. 541-545; II. 314, 688; Dialogus ("Ach Gott, wie manches "), II. 463, 690 f. ; Dialogus (Hope and Fear), II. 706; Die Elenden sollen essen, II. 355 f., 682 ; Die Himmel erzählen, II. 356 f., 682; Du Friedefürst, III. 65 f., 91 f., 285 f. ; Du Hirte Israel, II. 418 ; Du sollst Gott deinen Herren lieben, II. 429, 431, III. 216; Du wahrer Gott, II. 679, 710 f. ; Easter cantatas, II. 688; Ehre sei Gott, II. 441, 684; Ein feste Burg, II. 470 f., III. 73 f., 283 ; Ein Herz das seinen Jesum lebend weiss, III. 68, 284; Ein ungefärbt Gemüthe, II. 358, 683 ; Erforsche mich Gott, II. 423; Erfreut euch ihr Herzen, III. 68 f., 284; Erfreute Zeit im neuen Bunde, II 389-392, 687; Erhalt uns Herr, III. 285; Erhöhtes Fleisch und Blut, II. 7; Er rufet seine Schäfe, III. 70 f., 281, 285 ; Erschallet ihr Lieder, II. 354, 398 f., 684; Erwünschtes Freudenlicht, II. 399 f. ; Es erhub sich ein Streit, I. 51, II. 344 ; Es ist das Heil, II. 461 ; Es ist dir gesagt, III. 81 f.; Es ist ein trotzig, III. 80 f. ; Es ist euch gut, III. 71-73, 281, 285 f. ; Es ist nichts gesundes, II. 466 f.; Es reifet auch ein schrecklich Ende, III. 85 ; Es wartet alles auf dich, III. 76 f.; Falsche Welt, II. 475 ; Freue dich, erlöste Schaar, II. 302, 322 ; III. 77 ; Funeral ode for Prince Leopold of Cöthen, III. 225 ; Fürchte dich nicht (motett), I. 93, II. 603 f. ; (Gedenke, Herr, II. 695); Geist und Seele, II. 447, III. 138 ; Gelobet sei der Herr, II. 455, 457 f., III, 80 ; Gelobet seist du, Jesu Christ, II. 684, III. 285; ib. arranged for congregation (chorale), III. 400; Gleichwie der Regen, I. 491-501 ; Gott der Herr, III, 73 f., 282 ; (Gott der Hoffnung, II. 683); Gottes Zeit, I. 456-464, 466 note; Gott fähret auf, II. 316, III. 70-72, 77, 80, 284 ; Gott ist mein König, I. 345-353 ; Gott ist unsre Zuversicht, III. 77 f.; Gottlob nun geht, II. 433, 445, 695; Gott, man lobet dich, II. 471; Gott soll allein, II. 448 ff., III. 138; Gott, wie dein Name, II. 441, 686, III. 47; Halt im Gedächtniss, II. 417. III. 32 ; Herr Christ, der einge Gottessohn, I. 633, III. 285 ; Herr, deine Augen, II. 464, III. 31, 80 ; Herr, gehe nicht, II. 424, 427 ; Herr Gott Beherrscher, II. 706; Herr Gott, dich loben alle wir, III. 285 ; Herr Gott, dich loben wir, II. 413, 686; Herr Jesu Christ, III. 97 f., 101, 285 ; Herrscher des Himmels, II. 685 ; Herr, wie du willt, II. 413-415 ; Herz und Mund, I. 573 f., II. 412, 694; Himmelskönig, I. 539 f., II. 690; Höchst erwünschtes Freudenfest, II. 324, 365-367, III. 79 f.; Ich armer Mensch, II. 472 f., 475 ; Ich bin ein guter Hirt, III. 71, 285 f. ; Ich bin vergnügt, II. 443, 475 ; Ich elender Mensch, III. 87; Ich freue mich in dir, II. 685, III. 91, 93, 286; Ich geh' und suche, II. 449, III. 138; Ich glaube, lieber Herr, II. 468; Ich habe genug, III. 473, 687, 707; Ich habe Lust, II. 687; Ich habe meine Zuversicht, II. 446 f., 700 f., III. 138; Ich hab' in Gottes Herz und Sinn, III. 106, 286, Ich hatte viel Bekümmerniss, I. 531-538; II. 319, III. 234; Ich lasse dich nicht, II. 411 f., 687; Ich lebe, mein Herze, II. 442; Ich liebe den Höchsten, II. 444 f.; Ich ruf, III. 285 ; Ich steh' mit einem Fuss, II. 441; Ich weiss, dass mein Erlöser lebt, I. 501-504, II. 688; Ich will den Kreuzstab gerne tragen, II. 472. 475 ; Ihr, die ihr euch, II. 360 f., 683 ; Ihr Menschen, rühmet, II. 422 f.; Ihr Pforten zu Zion, III. 83, note 113 ; Ihr werdet weinen und heulen, II. 72 f., 80, 281 ; In allen meinen Thaten, II. 456, 703-705, III. 79, 85 ; Jauchzet dem Herrn, II. 716; Jauchzet, frohlocket, II. 684; Jauchzet Gott, II. 473, 475 ; Jesu, der du meine Seele, I. 241 ; III. 103, 107, 287; Jesu meine Freude (motett), II. 600-602; Jesu, nun sei gepreiset, II. 686; III. 91, 93, 286; Jesus nahm zu sich, II. 350-353, 679; Jesus schläft, II. 403 f., 692 ; Komm, du süsse Todesstunde, I. 548-554, II. 315, III. 68, 284 f.; Komm, Jesu, komm, II. 604 f.; Kommt, eilet und laufet, II. 688; Lass, Fürstin (Trauerode), II. 613-618; Leichtgesinnte Flattergeister, II. 416 f.; Liebster Gott, II. 431-433, 696; Liebster Emmanuel, II. 687, III. 286 ; Liebster Jesu, II. 695, III. 84; Lobe den Herrn, den mächtigen König, II. 455-458 ; Lobe den Herrn, meine Seele, II. 407 f., 686, 692 f.; Lobet den Herrn, alle Heiden (motett), II. 599 ; Lobet Gott (Ascension oratorio), II. 315; Lob und Ehre, II. 716; Mache dich, mein Geist, bereit, III. 286; Man singet mit Freuden, II. 445 ; Meinen Jesum lass ich nicht, III. 103, 286; Meine Seele erhebt den Herrn, II. 381, III. 138, 286; Meine Seele rühmet, III. 86; Meine Seele soll Gott loben, I. 343 f.; Meine Seufzer, III. 84 f.; Mein Gott, wie lang, I. 562 f.; Mein liebster Jesus ist verloren, II. 401-403, 691 ; Mit Fried' und Freud', II. 687; Michaelmas cantata, "Es erhub sich," II. 408-410, III. 93, 286; Nach dir, Herr, II. 241, 443-448, II. 31 ; New Year cantata, II. 686; Nimm von uns, Herr, III. 97 f., 287; Nimm was dein ist, II. 416; Nun danket alle Gott, II. 456; ib. (motett), II. 599-602, 717; Nun ist das Heil, I. 51,

III. 82 f.; Nun komm der Heiden Heiland (A minor), I. 506-511, 519, II. 314, III. 79; *ib.* (B minor), III. 286; Nur Jedem das seine, I. 555-557; O ewiges Feuer, II. 688-690, III. 78; O Ewigkeit, du Donnerwort (in F), II. 421 f.; *ib.* (in D), III. 79; O heil'ges Geist, II. 4co, 690, III. 80; O Jesu Christ, mein Lebens Licht, III. 100, note 141; Preise Jerusalem, II. 362-365, III. 79; Probationary cantata, II. 679; Purification cantata, II. 687; Rathswechsel cantatas, I. 345-353; Reformation cantata, II. 470, III. 73 f., 283; Schauet doch und sehet, II. 427-429, 464 f., III. 47: Schau, lieber Gott, II. 401; Schlage doch, gewünschte Stunde, II. 476; Schmücke dich, III. 286; Schwingt freudig euch em-por, II. 471; Sehet welch' eine Liebe, II. 385 f., 685; Sehet, wir gehen hinauf, II. 442; Sei Lob und Ehr', II. 456, 458; Sei Lob und Preis (motett), II. 611; Selig ist der Mann, II. 685, III. 83; Siehe, der Hüter Israel, II. 628; Siehe, ich will viel Fischer, II. 472, 561; Siehe zu, dass deine Gottesfurcht, II. 406 f., 692; Sie werden aus Saba, II. 387-389, 687; Sie werden euch in den Bann thun (in G minor), II. 420; *ib.* (in A minor), III. 71 f., 281-283; Singet dem Herrn, II. 386 f., 469, 686, 707; *ib.* (motett), II. 602 f., 605 f., 608 f.; St. John's Day cantata, II. 685; St. Stephen's Day cantata, II. 685; Süsser Trost, II. 685, III. 83 f.; Thue Rechnung, II. 423; Trauerode, The, II. 347: Tritt auf den Glaubensbahn, I. 560-562; Und es waren Hirten, II. 685; Unser Mund sei voll Lachens, II. 684, 695; III. 78 f.; Unser Wandel ist im Himmel (motett), II. 598; Uns ist ein Kind, I. 487-491, II. 684; Vergnügte Ruh, II. 453; Wachet auf, II. 459 f.; Wachet betet, I. 569-573; *ib.* (second form), II. 362, 683; Wahrlich, wahrlich, II. 419 f.; Wär' Gott nicht mit uns, III. 66 f.; Warum betrübst du dich, III. 88; Was frag' ich nach der Welt, III. 91, 286; Was Gott thut das ist wohlgethan, II. 456, 458-460 f.; *ib.* (B. G. 100), II. 316; *ib.* (in G), III. 287; Was mein Gott will, III. 286; Was soll ich aus dir machen, II. 472; Was willst du dich betrüben, II. 456, Weinen, Klagen, II. 404-406, III. 47; Weissenfels cantata, III. 223; Wer da glaubet, II. 467 f.; Wer Dank opfert, III. 76 f.; Wer mich liebet (earlier setting), I. 511-513; *ib.* (later setting), II. 15, 688-690, III. 69, 281; Wer nur den lieben Gott, II. 437-440, 698; Wer sich selbst erhöhet, II. 12-15, 649; Wer weiss wie nahe mir mein Ende, II. 451, 453, III. 80; Whitsuntide cantata, II. 353 f., 398 f., 688-690; Widerstehe doch, II. 476; Wie schön leuchtet, III. 92, 94; Wir danken dir Gott, II. 450 f.; III. 46; Wir müssen durch viel Trübsal, III. 78 f., 138; Wo geh'st du hin? II. 418 f.; Wo Gott der Herr, III. 90; Wohl dem der sich, III. 90; Wo soll ich fliehen hin, III. 91 f., 286; Wünschet Jerusalem Glück, II. 469, 707.

CANTATAS, SECULAR—Amore traditore, II. 637; Andro dal colle al prato, II. 651; Angenehmes Wiederau, II. 636 f.; Auf, Schmetternde Töne, II. 628; Bauerncantate (see Peasants' cantata); Blast, Lärmen (August II.), II. 628; Coffee cantata, II. 641-643, 721; III. 234; Comic cantatas, II. 641-643; III. 176-181; Cöthen secular cantata, II. 619 f., 718; Die Freude reget sich (serenade for Rivinus), II. 621; Durchlaucht'ger Leopold (serenade for Prince Leopold) of Cöthen, II. 6, 620; Funeral compositions, II. 613-619; Geschwinde, geschwinde (see Streit zwischen Phœbus und Pan); Hercules in indecision, II. 629; Ich bin vergnügt, II. 640; Italian chamber cantata about Anspach, II. 639 f.; Man hat ne neue Oberkeet (see Peasants' cantata); Non sa che sia dolore, II. 639 f.; O angenehme Melodei, II. 636; occasional compositions, II. 223-225, 612-635; O holder Tag, II. 635; III. 260 f.; Peasants' cantata, III. 176-181; Pleisse und Neisse (wedding cantata), II. 634; Preise dein Glücke, II. 630, III. 47; Schleicht spielende Wellen, II. 698, 719,; Schweigt stille (see Coffee cantata); Secular cantatas, sundry, I. 566-569; Serenades, &c.:—for August III., II. 630; for Kortte, II. 627 f.; in Leipzig, II. 632; for Leopold, of Cöthen, II. 6 f.; for Müller, II. 624-626; for Ponickau, II. 411 f.; Steigt freudig in der Luft,. II. 158; Solo cantatas, II. 474, 476, 637-640; Streit zwischen Phœbus und Pan, II. 643-648, III. 253, 258; Tönet ihr Pauken, II. 224 f., 630; Trauerode, II. 613-618; Vereinigte Zwietracht (see Serenade for Kortte), Vergnügte Pleissenstadt (see Pleisse und Neisse); Was mir behagt, II. 349, 566-569, 642 f., II. 225, 632, III. 229 f.; Weichet nur, betrübten Schatten, II. 633.

THE CANTATAS AS A WHOLE—Style of the early cantatas, I. 442, 464-466; the Weimar cantatas, I. 565 f.; the Leipzig cantatas, II. 434-437; chorale cantatas, III. 89-108; chorale choruses in cantatas, III. 100-104; cantata texts, III. 92-96; the secular cantatas and their arrangement, II. 621-648; birthday cantatas, II. 631; wedding and jubilee cantatas, II. 632 f., 706 f.; music for the rebuilding of the Thomasschule, II. 632.

CHORALES—Original chorales, III. 114 f.; chorale collection published in Leipzig, III. 108 f.; five hymn tunes, III. 287-289; six hymn tunes (music), III. 401-403; work for Schemelli, III. 287-289.

MASSES, &c.—Magnificat in D, II. 369-385; the smaller Magnificat, II. 374; the masses as a whole, III. 25-64; the A

INDEX.

major mass, III. 32-35, 37; the B minor mass, I. 241, 351, III. 38-64, 226, 279 f.; analysis of, III. 48-62; "Crucifixus," III. 103; date of, III. 38 f.; dedication of first two numbers, II. 629; its historical position, III. 62-64; its origin and purpose, III. 38-40; its Protestantism, III. 44-46; its symbolism, III. 49-62; F major mass, III. 35-37; G major mass, III. 30-32; G minor mass, III. 30-32; dates of the short masses, III. 30-32; Sanctus in C, II. 369.

MOTETTS (For the separate motetts the reader is referred to the list of sacred cantatas, in which they will be found under their first lines)—The motetts, II. 459-612, 715-717; the motett form used in the cantatas, II. 598.

ORATORIOS—Ascension, II. 593 f.; III. 47; Christmas, II. 302, 313 f., 316, 318, 570-588, 684-687; its date, II. 572; its design and arrangement, II. 572-574, 582 f.; origin of many numbers in older works, II. 573 f.; Easter, II. 590-593, 714 f.; III. 280.

PASSION SETTINGS—The passions as a whole, III. 64; their style, II. 503 f.; their number and authenticity, II. 504-506, 508 f.; comparison between them, II. 518; the Matthew passion, II. 315 f., 319, 537-569, 712-714; its accompaniment, II. 543 f.; its arrangement, II. 539-544; chorales and choruses, II. 545-558; final chorale of part I. (music), III. 399; first performance, II. 568 f.; solos in, II. 557-568; the Mark Passion, II. 347, 505 f.; the Luke Passion, II. 508-518, 708; the John Passion, II. 318, 519-537, 709-712; its division and arrangement, II. 527 f.; its musical form, II. 524-537; its text, II. 519, 524; a passion, to words by Picander, II. 506 f.

SONGS—Songs in Anna Magdalena's book, II. 151 f.; "Willst du dein Herz mir schenken?" II. 661.

Instrumental works:

CANONS—Canon given to Walther, I. 387: given to Hudemann, II. 647, III. 209; for the Musical Society, III. 25; published by Telemann, III. 229; on themes by the king, III. 94 f.; for Faber, III. 237; in six parts, solved, III. 404.

CLAVIER WORKS—Capriccio, for J. C. Bach, of Ohrdruf, I. 249 f.; Capriccio on the departure of a brother, I. 236-245; Clavierbüchlein, for Anna Magdalena (smaller), II. 148, (larger), II. 149 ff.; *ib.* for W. F. Bach, II. 48, 50-53, 664, III. 266; Clavierübung, III. 154-157, 168-176, 213-222, 224, 255, 290; analysed by Kirnberger, III. 348-350; autograph of, II. 619; Concertos, MS., I. 417; "Italian concerto," I. 417; III. 149-152; Vivaldi's concertos, arranged, III. 139, 149; Duets, 167 f.; Fantasias (four),I. 436-438; II. 59; fantasias and fugues, in A minor, II. 33 f., III. 185; in A minor (double fugue), III. 183; in C minor, III. 182; "chromatic" fantasia and fugue, III. 181 f., 185, 291 f.: Fugues, in A, on a theme by Albinoni, I. 426; two in A, I. 432 f.; in A minor, l. 433; in C, II. 52; in C minor, II. 58; in D minor, II. 53 note; Kunst der Fugue (Art of Fugue), III. 191, 197-207; unfinished three-part fugue, III. 204-207 (see also under Musikal. Opfer); Inventions (Inventionen und Sinfonien), II. 58-68; Labyrinth, der kleine harmonische, II. 43; Musikalisches Opfer, III. 191-197, 221, 233 f., 292, 294; Partitas, III. 154-157, 224; in B minor, III. 164-166; Pieces, two little, III. 385 f.; Preambulen, II. 50, 59; Preludes—two early preludes, I. 433 435; in C, III. 387; Preludes and fugues—p and f., afterwards concerto for flute, violin and clavier, I. 420; short preludes and fugues, with pieces, II. 53; "Wohltemperirte Clavier," II. 6, 161-178, 664-673; various readings, II. 668-673; second part, III. 183-191, 199, 202 f., 210; the two parts compared, III. 186; use of the ancient modes in, III. 133; Ricercar, III. 192-196; Sinfonien, see inventions; Sonatas, early work, in D, I. 243 f.; in D minor, II. 80; in G, II. 81; arrangement from Reinken, I. 429 f.; later Sonatas, III. 152; in Musikal. Opfer, III. 196; Suites, in F, I. 432; French suites, II. 159-161, 663; English suites, III. 152-154, 157, 289; summary of the suites, III. 152; Toccatas, in C minor, II. 32 f.; D minor, I. 439 f.; E minor, I. 441 f.; F sharp minor, II. 31-33; F minor, II. 650; G, I. 418 f.; G minor, I. 440 f.; Variations, in A minor, I. 130, 431 f.; thirty variations, in G (Goldberg), I. 128, III. 168-176, 246; quodlibet, used in them, III. 172, 174-176; sarabande used in them, III. 170 f.

CLAVIER WITH OTHER INSTRUMENTS — Concertos for clavier (seven), III. 138-152; for two claviers (three), II. 698, III. 142, 144-146; for three claviers, III. 144, 146-149; for four claviers (Vivaldi), III. 149; for flute, violin, and clavier, II. 134; for two flutes and clavier, III. 143.

FLUTE AND CLAVIER—Sonata in B minor, I. 26; three sonatas for flute and obbligato clavier, II. 121-124; three for flute with accompanying clavier, II. 124 f.; sonata in C major (trio for two flutes and clavier), II. 105; for flute, violin and clavier (trio) in G, II. 105: trio in C minor, Musikal. Opfer, III. 196 f.; for viol-da-gamba and clavier, three sonatas, with obbligato clavier, II. 117-121; for violin and clavier, six sonatas, II. 110-117; suite in A, II. 124; sonata in E minor, II. 124; fugue in G minor, II. 124 trio for two violins and bass, II. 125; C major trio, Qu. Goldberg? II. 660; A minor sonata, for two violins and bass, II. 659.

INDEX. 413

FLUTE—for flute and clavier, see under Clavier; concertos, for flute, oboe, trumpet, and violin, II. 132 f.; for flute, violin, and clavier, in D, II. 134 f.; for the same in D minor, III. 143; for two flutes and violin, II. 133 f.
LUTE—three partitas (sonata and prelude), III. 166 f.
OBOE—concerto for oboe and violin, III. 143; concerto for oboe, trumpet, flute, and violin, II. 132 f.
ORCHESTRA—Concerti grossi, or Brandenburg concertos, II. 128-136, III. 137-143, 146; march, II. 627 f.; Suites (partien), II. 140-146, 661; III. 136.
ORGAN—Canzone, I. 421; chorale arrangements, I. 254 ff., 312 f., 594-619, III. 294; "Ein feste Burg," I. 394-397; "Gelobet seist du," III. 400; chorale fugues, I. 218 ff.; partitas, I. 211-217; partita on " Sei gegrüsset," I. 604 f.; on " Vom Himmel hoch," III. 221. Concertos arranged from Vivaldi, I. 411-416. (Fantasia)—in C (early work in four sections), I. 324; in C, I. 399; in C minor, III. 211 note, 395; in G (unpublished), I. 320; in G, I. 322. (Fugues)—in C, I. 399; in C minor, I. 252 f.; fugue (fragment), III. 211; fugue on theme of Legrenzi's, I. 423 f.; alla breve in D, I. 422 f.; early fugue in E minor, I. 219 f.; in G (unpublished), I. 322; in G minor, I. 400 f.; three fugues on themes by Corelli and Albinoni, I. 424-429. (Passacaglio)—I. 587-589; (Pastorale), III. 213 note; (Preludes) in A minor, C and G, I. 398; in C minor, III. 208; (Preludes and Fugues), eight short (C, D minor, E minor, F, G, G minor, A minor, B flat), I. 399 f.; in A, I. 589 f.; in A minor (early), I. 319; in A minor, III. 209; in B minor, III. 209 f.; in C, I. 401 f.; in C, III. 208 f.; in C (grand), III. 209 f.; in C minor (early), I. 251, 318; two in C minor, I. 590-592; in C minor, III. 208; in D, I. 405 f.; in D minor, arranged from violin sonata in G minor, II. 81; in E minor, I. 402-404; in E minor (grand), III. 209 f.; in E flat (early), I. 323; in E flat III. 209 ff.; in F, I. 590 f.; in F, III. 208; in F minor, I. 590 f.; in G, I. 405; in G, III. 208 f.; in G minor, I. 406 f.; in G minor, II. 23-25; in G minor (" Reinken "), II. 25; (Sonatas, early), in D, I. 243-249; six sonatas, III. 211-213; (Toccatas), in C, with fugue, II. 417 f.; in C, I. 627 f.; in D minor, I. 404; in D minor, with fugue, III. 208 f.; in F, I. 111; III. 208; variations on " Vom Himmel hoch," III. 294. (Orgelbüchlein, das (little organ-book), I. 597-603, 611-615, 647, 650, II. 48, III. 217, 220; latest organ works, III. 207-222.
TRUMPET—concerto for trumpet, flute, oboe, and violin, II. 132 f.
VIOL DA GAMBA—three sonatas with clavier obbligato, II. 117-121.

VIOLA POMPOSA—suite, II. 100.
VIOLIN ALONE—three suites (partien), II. 93-99; the chaconne, II. 95-98; three sonatas, II. 77-84; style of the violin works, II. 70.
VIOLIN WITH INSTRUMENTS—(see under Clavier), concertos, II. 126-128.
VIOLONCELLO ALONE—six suites, II. 99-101; dates and character of the string solos, II. 71, 652; summary of the later instrumental works, III. 135.

BACH, JOH. PHILIPP, of Meiningen, I. 10.
BACH, JOH. VALENTIN, I. 156; III. 298.
BACH, LIPS, I. 9.
BACH, MARIA BARBARA, Sebastian's first wife, I. 328 f., 339 f.; II. 11 f.
BACH, NIK. EPHRAIM, the painter, I. 11.
BACH, SAMUEL ANTON, I. 10 f.; III. 239.
BACH, STEPHAN, of Brunswick, I. 12, 39.
BACH, VEIT, I. 4-6.
BACH, WENDEL, I. 11.
BACH, WILHELM FRIEDEMANN, II. 8, 239, 331 f., III. 29, 205, 212, 225, 227 f., 230 f., 246, 248, 268 f., 276, 292, 294, 356, 359.
BACH, use of the name in music, III. 202, 205-207.
BÄHR (Beer), JOH., I. 218, 226; II. 325.
BALLETTS, II. 72.
BARON, E. G., III. 167, 231.
BASSO CONTINUO, origin and history of, II. 101 f.
BASSO OSTINATO, I. 208, 212.
BECCAU, JOACHIM, his Passion music, II. 496.
BEHRNDT, GOTTFRIED, III. 499.
BELLERMANN, CONSTANTIN, I. 634 f.
BENDA, FRANZ, III. 227, 231, 236.
BENDELER, J. P., III. 15.
BERLIN, band at, III. 231; operas, 232.
BERTUCH, GEORG, III. 235 f.
BESSER, JOH. FRIEDRICH, II. 18.
BIEDERMANN, J. G., III. 255-261.
BINDERSLEBEN, Bachs of, I. 4.
BIRKENSTOCK, JOH. ADAM, I. 514.
BIRNBAUM, JOH. ABRAHAM, III. 251-254.
BÖHM, GEORG, I. 194-196, 203-210; his influence on Bach, I. 210-217, 241; III. 217, 219 f.
BÖRNER, ANDR., I. 222.
BOSE, family, III. 250.
BOURRÉE, II. 75.
BRANDENBURG, I. 582; Christian Ludwig, Margrave of, II. 128 f.: concertos (see under J. S. Bach).
BRANLE, II. 72.
BROCKES, B. H., II. 496 f., 520 f.
BRÜHL, Count, III. 246.
BRUHNS, NIK., I. 260, 268, 287, 508; his violin playing, II. 70.
BRUNNER, JOH. SEB., I. 625.
BÜTTNER, G. C., I. 165 f.
BUFFARDIN, P. G., II. 155 f.
BÜNTING, III. 267.
BURCK, JOACHIM (Möller) von, I. 335; his Passion-music, II. 481.

414　INDEX.

BUTTSTEDT, J. H., I. 118, 125, 127, 480 ff.
BUXTEHUDE, DIETRICH, his life, I. 257-263; his works, I. 127, 263-310; his cantatas, I. 290-310; his organ compositions, I. 267-291, 608 f., 615; his performances at Lübeck (Abendmusiken), I. 258, 626; his style, I. 110; his position in art, I. 265 ff.; his death, I. 311; his influence on Bach, I. 276, 278, 283, 285, 296, 318-326; III. 153 note, 207, 209, 211, 218; compared with Böhm, I. 205; with Pachelbel, I. 284 f., 289.

CALDARA, III. 29, 263.
CALOVIUS, III. 266.
CANTATA, history of, I. 73, 292, 473-478, 576 ff., II. 348.
CANTOR's duties at Leipzig, II. 184 f., 190; III. 307-315.
CANZONE, I. 97, 421 ff.
CARISSIMI, G., I. 90.
CASSEL, music at, I. 513 f.
CELLE, music at, I. 201 f.
CHACONNE, I. 279-282, 447, 587 f., II. 405; distinguished from the Passacaglia, III. 170.
CHOIRS in Leipzig, II. 202-204, 207-209.
CHORALE, treatment of the, I. 97-118, 127; by Bach, I. 604-619, II. 459; his ornate accompaniments, I. 312; interludes in, I. 593 f.; four-part setting of, I. 500; in older Passion-settings, II. 488 f.; at the close of the cantatas, III. 99 f.
CHRISTIAN, Duke of Weissenfels, I. 566; III. 223.
CHRISTIANA EBERHARDINE, Princess, II. 613-615.
CHRISTMAS ceremonies in Leipzig, II. 369-373, 571 f.; Christmas in Bach's music, III. 221.
CHURCH MUSIC in the 18th century, I. 478 f., 486-488; in Leipzig, II. 192 f., 265-281.
CIACONA (see Chaconne).
CLAVIER-PLAYING, I. 126, 202; II. 34-41; Bach's treatment of the clavier, II. 43-46; III. 127.
" CLEOFIDE," Hasse's, III. 226.
COLLEGIUM MUSICUM, III. 17-19.
CONCERTO, its form, I. 408-417, 487; II. 125-129, 136-138; compared with the modern sonata-form, III. 151.
CONCERTS, public, first instituted, III. 16-21.
CONCERTS, public, in Leipzig, III. 18 f.
CONDUCTING, II. 325-332.
CORELLI, ARCANGELO, I. 424 f.; his sonatas, II. 76, 78, 84.
CORNO da tirarsi, II. 428 note, 449.
CORRENTE (see Courante).
CÖTHEN, music at, II. 2 f.; royal family of, III. 224 f.; Princess Charlotte of, II. 157, III. 224; Princess Friederike of, II. 147, 157; Prince Emanuel Leberecht of, II. 2; Prince Leopold of, II. 1-8, 147, 157, II. 618 f.

COUPERIN's style of fingering, II. 37; his suites, II. 86 f., 95; his allemande for two claviers, III. 144.
COURANTE, II. 72 f., 85 f., 88 f., III. 158 f.
CRASSELIUS, III. 112.
CRÜGER, JOH., III. 123 f.
CUNCIUS, CHR., organ builder, I. 515-520.

DANCE-FORMS in the suite, II. 72 f.
DEMANTIUS, CHR., II. 482.
DEYLING, Dr. S., II. 227-230, III. 6 ff.
DIESKAU, C. H. von, III. 176.
DIEUPART, CH., I. 202, II. 91, III. 263.
DOLES, J. F., III. 20, 244, 255 f., 260.
DORN, J. C., III. 249.
DRESDEN, music at, I. 583, II. 337, 367.
DRESE, ADAM, I. 164-170, 367, 369, 519.
DRESE, JOH. SAM., I. 391 f., III. 300.
DRESE, JOH. WILH., I. 392, III. 300.
DRESE, WILH. FR., I. 167.

EASTER dramas or oratorios, II. 588-590.
EBERLIN, DAN., I. 129.
ECHO SONGS, II. 586-588.
EFFLER, JOH., I. 39, 107, 221.
EILMAR, Archdeacon, I. 359-361, 363.
EINICKE, G. F., III. 243, 257 f.
EISENACH, organ of St. George's church at, I. 38; residence of the Bach family, I. 24, 38, 173 f.; music at, I. 183 f., 486.
EMMERLING, Kammermusicus at Brandenburg, II. 129.
ENDERLEIN, poet, III. 256.
ERDMANN, G., I. Preface v., 189 ff., 525 f.; II. 253 f.
ERFURT, residence of Bach family, I. 14 ff., 153 f.; its condition during the thirty years' war, I. 14 ff.
ERICH, J. C., II. 260.
ERLEBACH, P. H., I. 124 note, 153, 357, 474 f.; his influence on Bach, I. 351.
ERNESTI, J. A., II. 261, III. 1-16, 230, 280, 287, 307-315.
ERNESTI, J. H., II. 200 f., 242, 602, 613.

" FABER " Canon, III. 237.
FALSETTO singing, II. 310.
FASCH, C. F., III. 128.
FASCH, J. F., II. 181, 349.
FEDELI, RUGGIERO, I. 513.
FEMALE voices in Protestant church music, I. 326 f., 569.
FINGERING, methods of, II. 34-43.
FLEISCHER, J. C., I. 140.
FLEMMING, Count, II. 584.
FRANCISCI, J., III. 236.
FRANCK, SALOMO, I. 526-530, III. 267.
FREDERICK the Great and Bach, III. 192-197, 231-234.
FREISLICH, J. C., I. 545.
FRENCH music, I. 201 ff., II. 72 ff., 85 ff., 139 f.
FRESCOBALDI, G., I. 108-116, 206, III. 263; his influence on Bach, I. 420, 422

INDEX. 415

FREYLINGHAUSEN, J. A., his hymn book, I. 366-370.
FROBERGER, J. J., I. 110, 236; his influence on Bach, I. 323 f., 627, III. 187 note.
FROHNE, J. A., I. 358-361.
FUGUE, I. 79 f., 99 f., 116, 341, 422 f.
FUNCKE, Cantor in Luneburg, II. 487.
FUX, J., III. 123, 125 f.

GABRIELI, G., I. 82, 84, 97, 124, II. 76.
GAGLIARDA, II. 72 f.
GALLUS, author of a Passion-setting, II. 482 f.
GAMBA, II. 117 f.
GANDERSHEIM, Bachs of, I. 11 f., 575.
GARTNER, K. C., III. 260 note.
GASPARD DE ROUX, II. 86.
GASSMANN, III. 27.
GAUDLITZ, G., II. 231 f.
GAVOTTE, II. 75.
GERBER, CHR., I. 480.
GERBER, H. N., II. 293 f., III. 126 f., 240 f.; his accompaniments to Albinoni's sonata, III. 388-398.
GERHARD, J., III. 266.
GERHARDT, hymn-writer, III. 113.
GERLACH, C. G., II. 225, 673 f.; III. 20, 248.
GERMAN courts, their musical culture, I. 10 f., 157 f., 164 f., 168-171, 200 f., 377-382, II. 1-6, 128 f., 613-621, 626, 630, 632, III. 222-225, 231-234.
GERSTENBÜTTEL, J., II. 19.
GESE, B., II. 483.
GESNER, J. M., I. 390, II. 242, 256-263, 678.
GEWANDHAUS concerts instituted, III. 18.
GHRO, J., II. 73.
GIGUE and Giga, III. 159 f.
GIOVANNINI, II. 661-663.
GLEITSMANN, P., I. 169 f., 172, 226.
GLUCK's " Iphigenia," III. 161 note.
GOLDBERG, J. G., II. 660, III. 246.
GÖRNER, J. G., II. 210-212, 214, 215, 223, 274 f., 282, 290, 674, III. 20, 276, 357, 360.
GOTTSCHED, J. C., III. 241, 250 f.
GRÄBNER, CHR., II. 646.
GRÄBNER, CHR., junr., III. 249.
GRÄFE, III. 251.
GRÄFENRODE, Bachs of, I. 1 f.
GRAFF, J., I. 118.
GRAUN, J. G., II. 499 f., III. 27, 227 f., 231, 243.
GRAUPNER, C., II. 183.
GRIGNY, N., I. 202, II. 91, III. 263.
GUNTHER, Count, his letter, III. 296.

HALLE, organ at, I. 515, 521; organist's post at, I. 517.
HAMBURG, Election of an organist at, II. 18-21.
HAMMERSCHMIDT, ANDR., I. 37, 43 f., 49, 55 f., 58, 60, 69, 124, 302, II. 589 f.
HANDEL, G.F., his friendship with Mattheson, I. 261; at Lübeck, I. 261 f.; his attempts to meet Bach, II. 9; his style of fingering, II. 37; his suites, II. 88; his concerti grossi,
II. 137 f.; his orchestra, II. 305 f.; his influence on Bach, II. 426; his Passion music, II. 494, 497 f.; his secular cantatas compared with Bach's, II. 623 f.; his suites, III. 157, 163 f.; his development, III. 222; connection with Agricola, III. 243; vocal works transcribed by Bach, III. 263.
HANKE, G. B., II. 179.
HARPSICHORD, its character, II. 31; attempts to use it in church music, II. 107; its technique (see Clavier).
HARRER, cantor after Bach, III. 248.
HARTWIG, CARL, III. 249.
HASSE, J. A., II. 327, III. 27, 226, 228.
HASSE, FAUSTINA, III. 226, 228.
HAUPTMANN, M., his opinion of "Gottes Zeit," I. 466.
HAYDN, J., II. 330, III. 27.
HEBESTREIT, P., I. 583.
HEINICHEN, J. D.. II. 1, 37, 43, 183.
HEITMANN, J. J., II. 19 f.
HELBIG, J. F., II. 12, 206.
HENNICKE, J. C., II. 635.
HENRICI (Picander), II. 340-347; 505-508, III. 43, 94 f.
HERDA, E., I. 188 ff.
HERTEL, J. C., III. 236.
HERTHUM, C., I. 222, 226.
HILDEBRAND, Z., II. 46 ff., 282 f.
HILLER, J. A., II. 454 f., III. 20, 27.
HOFFMANN, CHR., I. 19.
HOFFMANN, J. C., III. 250, 270.
HOFFMANN, MELCHIOR, II. 205.
HOMILIUS, G. A., III. 244 f.
HUDEMANN, Dr. L. F., II. 647, III. 229 note, 250 note.
HUNOLD, C. F., II. 492 f.
HURLEBUSCH, III. 263 f.
HYMN-BOOKS in Leipzig, II. 278.

INTRADA, II. 72.
IPHIGENIA, Gluck's, III. 161 note.
ITALIAN oratorios, II. 527.

JENA, residence of the Bach family, I. 131 ff.
JENA, organ at, I. 137 f.

KAUFFMANN, G. F., I. 118, II. 183.
KAYSERLING, Baron von, III. 8, 226 f., 246.
KEISER, R., II. 493, 497, 708 f., III. 228, 263.
KELLNER, CHRISTIANE PAULINE, I. 569.
KELLNER, J. P., II. 632, 653, III. 238, 248.
KESSLER, G., I. 11.
KIESEWETTER, J. C., I. 186, 391.
KIRCHBACH, H. C. von, II. 616.
KIRCHHOFF, G., I. 118, III. 191 note.
KIRNBERGER, J. P., II. 294, 296, 299, 307, 712, III. 245, 264; his accompaniments to Bach, II. 106; his theoretical works, III. 116 f., 121; his opinion of Bach, III. 116 f., 119 f., 122, 133, 348-350.
KITTEL, J. C., I. Preface, xi., II. 104, 301, 658, 678, III. 246 f.

KNÜPFFER, III. 28.
KOCH, J. S., I. 343.
KÖHLER, J. F., III. 11.
KÖNIG, J. U., II. 494 f.
KORTTE, G., II. 627.
KRAMER, cantor of Dondorf, II. 479, III. 235.
KRAUSE, G. T., III. 4, 309 f.
KRAUSE, J. G., III. 3-9; his Coffee cantata, II. 641.
KREBS, J. C., III. 241, 250.
KREBS, J. L., III. 7, 241 f., 269, 287-289, 309.
KREBS, J. T., I. 523, III. 41.
KREBS, J. T., junr., III. 241.
KRIEGEL, ABRAHAM, III. 5, 276.
KRIEGER, J., suites, II. 91 f.
KUHNAU, J., I. 237-240, II. 181, 332-336; connection with Leipzig opera, II. 208 f.; with the Collegium musicum, III. 17; his memorial to the town of Leipzig, III. 302-305; his Christmas cantata, II. 676 f.; his Whitsunday cantata, II. 677; his Passion music, II. 491 f., 499; his short mass, III. 28; his hymn for Whitsuntide, III. 34; his Bible sonatas, I. 236, 243 ff.; his influence on Bach, I. 243-245, 320, II. 371, III. 155.
KÜHNEL, A., I. 514.
KUNSTPFEIFFER, see Stadtpfeiffer.
KÜTTLER (Kittler), S., II. 240, III. 6 f.

LÄMMERHIRT, T., I. 339, II. 153.
LANGULA, state of music at, I. 343.
LAUTENCLAVIERE, I. 140.
LAUTENCLAVICYMBEL, II. 46 f.
LEGRENZI, G., I. 423-425.
LEIDING, G. D., I. 260.
LEIPZIG, schools, II. 189, 193, 200, 204; music at, II. 673; cantors at, II. 181-185; musical societies at, III. 276; opera at, II. 203 ff.; organs at, II. 675 f.; III. 305-307; hymn books at, II. 278; concerts at, III. 18 f.
LEO, LEONARDO, II. 639.
LIBRETTISTS for Cantatas, II. 339.
LINIKE, C. B., II. 4.
LINKE, of Schneeberg, III. 242.
LIPPE, Count von, III. 270.
LORBEER, J. C., I. 392.
LOTTI, A., I. 585 f., III. 28, 236, 263; his influence on Bach, II. 638, III. 28.
LÖW, J. J., I. 192, 194.
LÜBECK, VINCENTIUS, I. 200, 260, III. 155; his son, II. 19.
LÜBECK, organ in Marienkirche at, I. 257-259, III. 298 f.
LULLY, J. B., I. 124.
LÜNEBURG, music at, in 17th and 18th centuries, I. 189 ff.; organ at, I. 217; St. Michael's schools, I. 217.
LUSTIG, J. W., III. 228 f.
LUTHER, III. 266 note.
LUTHERAN SERVICE in Leipzig, II. 263-281.
LUTHERANISM, attitude towards Catholicism,

MACHOLD, J., II. 481.
MADRIGAL, verse-form, I. 469-479.
MANCINUS, T., II. 478.
MANIEREN, II. 297 f., 312, 317, 320-323.
MARCHAND, J. L., I. 583-585, 644-647; his suites, II. 86, 95.
MARCUS, M. F., II. 4.
MARPURG, F. W., III. 117, 126, 159 note, 204.
MASS, the, in Protestant service, III. 26-29.
MATTHESON, J., I. 261, 481-483; II. 16 f., 21, 23, 27, 79 f., 82-84, III. 204, 228 f., 234, 255, 257.
MEININGEN, residence of the Bach family, I. 10; state of music at, I. 581 f.
MERULO, C., I. 97.
MEYER, J., I. 480.
MINOR SCALE, the modern, III. 130 f.
MINUET, II. 75.
MIZLER, L. C., II. 104; III. 22-25, 125.
MODES, the ecclesiastical, I. 80, 84; III. 129-134.
MÖLLER, G., I. 378.
MÜLLER, A. F., 624-626.
MÜLLER, K., I. 514.
MONJOU, Demoiselles de, II. 4.
MONOCHORD, Neidhardt's, I. 137 f.
MOTETT, history of, I. 53-59, 73 f.; Schütz's, I. 54; anon., I. 55-58; Gabrieli's, I. 82; Michael Bach's, I. 59-73; the form as treated by Bach, II. 604-612; accompaniment of, II. 607-612; its use in the German service, II. 596 f.
MOTZ, G., I. 480.
MUFFAT, G., I. 109, 117, II. 95.
MÜHLHAUSEN, state of music at, I. 340 ff.; the constitution, I. 344 ff., 357, restoration of the Blasiuskirche organ, I. 354-358, 370.
MÜLLER, A. F., II. 624 ff.
MÜLLER, H., III. 266.
MÜLLER, J. J., II. 3.
MUSICAL societies in Leipzig, III. 16-25, 219, 258, 276.
MUSICIANS' GUILD, I. 143-153.
MUSICIANS' SALARIES, I. 20, 24, 29 f., 33, 39, 108, 185, 188, 191 f., 223, 337, 380, 516, 518, III. 19, 299.
MUSIKVEREINE, students', III. 17, 19.
MÜTHEL, Bach's pupil, III. 246-248, 274.
MYSTERIES, or Miracle-plays, II. 500-503, 567, 570-572.

NÄGELI, editor of the Wohlt. Clav., II. 668.
NEIDHARDT, J. G., I. 137; II. 41.
NEUKIRCH, B., II. 494.
NEUMEISTER, E., I. 470-478, 630 f., II. 339, 493, III. 267.
NICHELMANN, Chr., III. 231, 248.
NIEDT, F. E., III. 120.
NIKOLAIKIRCHE, organ, II. 286.

OBOE d'amore, II. 682.
OLEARIUS, J. C., I. 166.
OLEARIUS, J. G., I. 30, 166, 313 f.

INDEX. 417

OPEN-AIR performances, II. 621 f.
OPERA, in Germany, I. 199, 466-469, II. 338; in Leipzig, II. 202-206.
OPUS-NUMBERS, III. 156.
ORATORIO, history of, I. 43, II. 306, 499, 570 f.
ORCHESTRA, organisation of, II. 303 ff.
ORCHESTRAL SUITES (Orchesterpartien), II. 139-146.
ORGAN-PLAYING, history of, I. 96 f., 102, 126, II. 35 ff.; organ employed in services, II. 278-281; its use in the chorale cantatas, III. 104 f.; organs in Leipzig (specification), II. 281-290, 675 f., III. 305-307.
OVERTURES, I. 124.

PACHELBEL, HIERONYMUS, his toccata for two claviers, III. 144.
PACHELBEL, J., I. 38, 94, 107-109; his works, I. 110-125, 127; his organ chorales, I. 247, III. 100 f., 213 f., 217, 220; his influence on Bach, I. 111-123, 255 f.; compared with Böhm, I. 205-209; with Buxtehude, I. 284 f., 289, 297.
PADUANA, II. 72.
PALESTRINA, III. 29.
PARTIE, partita, I. 127, II. 74 f., III. 219 f.
PASSACAGLIO, I. 279-282, II. 405, III. 170.
PASSEPIED, II. 75, 91.
PASSION MUSIC, its origin and history, II. 477 ff.; at Meissen, II. 478; passion plays, II. 500-503, 567; settings before Bach's, II. 538-539.
PAURBACH, composer of a Passion setting, II. 501.
PERFORMANCES, sacred, at Christmas, Easter, &c., II. 570 f.
PESTEL, J. E., II. 94.
PETRI, SAMUEL, II. 301-309, III. 234.
PETZOLD, C., I. 583.
PFEIFFER, A., III. 266.
PIANOFORTE, suited to Bach's works, II. 45.
PICANDER (Henrici), II. 340-347, 505-508, III. 43, 94 f.
PIETISM, I. 30 f., 166-168; at Mühlhausen, I. 358-370; its influence on music, I. 362 f., 479-486.
PISENDEL, J. G., I. 583, II. 205 f., III. 227.
PITCH (see Tuning) of the Leipzig organs, II. 676-678.
PITSCHEL, T. L., III. 262 f.
PONICKAU, J. C., von, II. 411 f.
PRAETORIUS, M., II. 34.
PREAMBLE, the, I. 105.
PRELUDES, for organ, II. 291.
"PROGRAMME MUSIC," I. 236-245, 246 f., 494, 549.
PROTESTANTISM, its effect on music, I. 112 f.
PROTESTANT element in German masses, III. 26-29, 34 f.

QUANTZ, flute player, III. 231.
QUODLIBETS, I. 154, III. 172, 174-176.

RAMBACH, J. J., II. 346, III. 267.
RAMEAU, J. P., III. 124 ff.
REALISM in music, I. 553.
RECITATIVE, I. 494-498, II. 311-320.
REFORMATION FESTIVAL, II. 271.
REICHE, G., II. 248.
REIMANN, J. B., III. 114, 236.
REINECCIUS, G. T., I. 389 f.
REINKEN, J. A., I. 195-200, 429-431, II. 15, 18; his suites, II. 87, f.; his sonata, III. 366-384.
RESE, J. L., II. 4.
RICHTER, J. C., II. 52.
RIGAUDON, II. 75.
RINGK, J., II. 633.
RITTER, C., II. 93.
RIVINUS, J. F., II. 621.
ROCHLITZ, III. 277.
ROCKHAUSEN, residence of the Bachs, I. 2 f.
ROLLE, C. C., II. 299, 307.
ROLLE, C. E., II. 2.
ROLLE, C. F. (of Quedlinburg), I. 520, II. 181 f.
ROMANESCA, II. 72.
RÖMHILD, J. T., I. 10, III. 235.
ROTHE, J. C., II. 491.
RUHLA, residence of the Bachs, I. 10 f.

SALARIES of musicians, I. 20, 24, 29 f., 33, 39, 108, 185, 188, 191 f., 223, 337, 380, 516, 518, III. 19, 299.
SARABANDE, II. 75, 91.
SAX-MERSEBURG, DUKE OF, III. 228, 235.
SCANDELLI, author of a Passion setting, II. 588.
SCARLATTI, DOM., II. 77, III. 160 note, 182.
SCHEIBE, J., organ builder, II. 282 f., 290, 292, 298 f., 317, 334.
SCHEIBE, J. A., II. 645-647, 675, III. 21, 252-255, 305-307.
SCHEIDEMANN, H., II. 16.
SCHEIDT, S., I. 97-99, 126.
SCHEMELLI, G. C. and C. F., I. 307-370, III. 109-114, 242.
SCHIEFERDECKER, J. C., I. 311; II. 346.
SCHIFF, C., I. 480 note, 179.
SCHMIDT, BALTHASAR, III. 219, 238, 250, 269, 290, 294.
SCHMIDT, JOH., III. 238, 248.
SCHMIDT, J. C., of Zella, I. 434 note, 135, II. 633.
SCHMIDT, J. M., III. 258 f., 275.
SCHNEIDER, J., II. 5, 262, III. 242.
SCHNITKER, ARP., II. 18.
SCHOLARS' CHOIR, at Eeisenach, I. 183.
SCHOTT, G. B., II. 183, 206, 673.
SCHRÖTER, C. G., II. 301, III. 257 f.
SCHUBART, J. M., I. 343, 522.
SCHÜBLER, J. G., III. 219, 238.
SCHULER, HANS, I. 2.
SCHULTZ, CHR., II. 478-481.
SCHULZ, A. P., III. 129.

SCHUMANN, R., I. 618, II. 98, III. 202.
SCHÜTZ, H., I. 43 f., 54, 63, 79, 469, II. 479, 483-486, 588.
SCHWANENBERGER, G. H. L., III. 237.
SCHWARZBURG, LUDWIG GÜNTHER, Count of, I. 33, 157, 163 f.; ANTON GÜNTHER, Count of, I. 164 f., 168 f.
SCHWEINFURT, residence of the Bach family, I. 156.
SCHWENKE, C. F. G., II. 668.
SCHWERIN, Duke of, III. 247.
SEBASTIANI, J., II. 487, 490, 515.
SEEBACH, J. G., II. 490, 496, 510.
SERENATA, I. 125.
SERVICES, Order of, in Leipzig, II. 266-281.
SERVICES, for saint's days, III. 42.
SILBERMANN, G., II. 46, III. 226, 232.
SILESIAN WAR, III. 64 f.
SINFONIE, instrumental movement in a vocal work, I. 296.
SONATA, I. 124 f., 293, 408, II. 75-77, III. 21 f.
SORGE, G. A., III. 255.
SPEE, FR., II. 587.
SPENER, the pietist, III. 266 f.
SPETH, J., I. 109.
SPIESS, JOS., II. 4.
SPORCK, Count von, III. 42 f., 179 f., 280.
STADT-PFEIFFER, or KUNST-PFEIFFER (town musicians), I. 15, 18, 143-153, 183, 191, 225, II. 140 f.
STAUBER, J. L., I. 339.
STÖLZEL, G. H., II. 53, 206, 497, III. 235.
STOPPE, D. III. 237.
STÖRMTHAL, organ dedication at, II. 365.
STRATTNER, G. C., I. 391, 541.
STRAUBE, RUDOLPH, III. 245.
STRICKER, A. R., II. 3.
STRUNCK, DELPHIN, I. 98.
STRUNGK, N. A., I. 201, II. 70, 203 f.
SUITE, history of, II. 72-74, 84-93; compared with the sonata, II. 92 f.
SWEELINCK, J. P., I. 97, 195, II. 73.
SYMPHONY, see Sinfonie.

TAULER, the mystic, III. 266.
TELEMANN, G. P., I. 50, 410, 474 f., 486 ff., II. 181-183, 204 f., 311, III. 34, 228 f., 234 f., 263, 276; his works, "Gleichwie der Regen," I. 491-501; his Passion music, II. 497; "Uns ist ein Kind," I. 487-491; "Wer sich selbst erhöhet," II. 15; his influence on Bach, II. 437, 694 f.; his connection with the Musikverein, III. 18, 20.
TEMPERAMENT, I. 137-139, II. 41 f.
THEILE, J., I. 99.
THIRTY YEARS' WAR, I. 3, 13 f., 15 ff., 30 f.; its influence on German mnsic, I. 40 f., II. 73 f.
THOMASKIRCHE ORGANS, II. 282-284; accounts of, III. 305-307.
THOMASSCHULE, II. 189-203.
THOROUGH-BASS, II. 102-110, 293 ff.; Bach on thorough-bass, III. 315-347, 348.

THURINGIA, original home of the Bach family, I. 1 f.; music in, I. 57 note, 96 f., 108 f, 190; state during the thirty years' war, I. 13.
TILGNER, GOTTFRIED, I. 474, 480, 482, II. 339.
TOCCATA, I., 97, 109.
TORLÉE J. F., II. 4.
TOSI, III. 244.
"TOWER MUSIC," III. 21.
TOWN MUSICIANS, see Stadt-pfeiffer.
TRANSCHEL, CHR., III. 245.
TREIBER, J. P., I. 226-228.
TRIER, organist at Zittau, II. 674, III. 20, 248.
TRIO, in instrumental chamber music, II. 141 f.
TROMBA DA TIRARSI, II. 428 note, 449.
TRUMPETS, tuning of, I., 346 note.
TUNING, I. 317-319, II. 41 f, 323-325.

UNIVERSITY CHURCH ORGANS, II. 287 f.
UTHE, J. C., I. 326 f.

VARIATIONS, history and development of the form, I. 126-130, II. 88 f., III. 169 f., 17 f., 219 f.
VETTER, D., II. 209, 432.
VETTER, NIK., I. 118.
VIADANA, LUD., II. 101.
VIOLA POMPOSA, II. 69, 100, III. 227, 250.
VIVALDI, J., I. 411-416, III. 239.
VOGLER, J. C., I. 522, II. 262 f., III. 242.
VOGLER, J. G., II. 206.
VOGLER, ABT., on Bach's chorales, III. 134 f. note.
VOIGT, of Erfurt, III. 234.
VOLTA, II. 72.
VOLUMIER, J. B., I. 583.
VOLUNTARIES FOR ORGAN, II. 291.
VOPELIUS, author of a Passion setting, II. 478-481.
VULPIUS, MELCHIOR, I. 69, II. 478.

WAGNER, G. G., II. 716.
WAGNER, PAUL, II. 278; III. 267.
WALTHER, JOH., author of a Passion setting, II. 478.
WALTHER, JOH. GOTTFRIED, I. 118, 123, 336, 381-389, II. 37.
WALTHER, J. J., II. 71, 94.
WARSAW, peace of, III. 64 f.
WATERMARKS, used by Bach, II. 661, 679-683, 686, 689, 692, 694-697, 699-703, 713 f., III. 281, 284 ff.
WECHMAR, residence of the Bach family, I. 4 ff.
WEIMAR, music at, I. 220 ff., 377 ff.; specification of organ at, I. 380; Reformation Jubilee at, I. 586; orchestra at, III. 299 f.; Duke Wilhelm Ernst of, I. 375, 379, 566 ff.; Prince J. Ernst of, I. 409 f., 630; Duke Ernst August of, III. 229 f., 236.

INDEX.

WEISS, SYLVIUS LEOPOLD, III. 167, 227, 251.
WEISSENFELS, music at, I. 472, 566-569, 594; III. 223; Dukes of, I. 566, III. 223.
WELTIG, A. E., III. 224.
WENDER, J. F., organ builder, I. 222, 336, 354 ff.
WERKMEISTER, ANDR., I. 260; II. 41.
WESTHOFF, J. P., I. 221.
WIEDEBURG, Capellmeister at Gera, II. 19.
WILDE, of Leipzig, III. 240.
WILDERER, J. H., III. 29.
"WILLST du dein Herz mir schenken," II. 661.

WOLFF, J. H., music for his marriage, by Bach, II. 634.
WÜLCKEN, J. C., III. 224, 237.
WÜRBEN, COUNT VON, III. 236.

ZACHAU, F. W., I. 121, 515, III. 34.
ZEHMISCH, III. 18.
ZELENKA, DISMAS, I. 583, II. 138, III. 29, 228.
ZIEGLER, CASPAR, I. 469 f., III. 280.
ZIEGLER, J. G., I. 524.
ZSCHORTAU, organ at, III. 21.

ERRATA.

VOL. I.

Page 35, line 6 from bottom, for "Cembalo" read "Harpsichord."
,, 43, line 19, for "Gospel" read "Gospels."
,, 49, bottom line, for "Gospel" read "Gospels."
,, 164, line 14, omit comma.
,, 165, for three bottom lines, read "Of his sacred melodies, which were composed partly . . . . . and partly to his own verses, that."
,, 166, line 22, for "Cantata" read "Cantate."
,, 186, line 4 of note, for "Lüneberg" read "Lüneburg."
,, 190, bottom line of note 15, for "that year" read "1870."
,, 208, line 5, for "Manier" read "mannerism."
,, 249, note 89, at end add "6."
,, 251, line 8, for "*manieren*" read "mannerisms."
,, 296, note 125, for "Bach Soc." read "B. G."
,, 328, line 12, for "turned" read "turned up."
,, 351, note 24, for "B. S." read "B. G."
,, 399, note 81, after "Cah. 8." insert "(247)."
,, 515, line 19, for "Zachaus" read "Zachau."
,, 521, line 2 from bottom, for "Zachaus" read "Zachau."
,, 526, note 235, for "App. B. V." read "App. B. VI. (Vol. III., p. 300)."
,, 537, note 248, for "Brocke" read "Brockes."
,, 640, for lines 6—7, read "has only a figured-bass accompaniment, one has a single instrument concertante, and one is accompanied," &c.

VOL. II.

Page 9, note 17, for "Siculs" read "Sicul."
,, 11, line 1, for "Brock's" read "Brockes'."
,, 80, line 4, for "C minor" read "C major."
,, 82, line 8 from bottom, for "in augmentation" read "as a double fugue."
,, 100, note 142, last line, for "D major" read "G major."
,, 161, line 18, for "minor" read "major."
,, 172, at end of note 247, add "(200)."
,, 186, note 13, for "App. B. VI." read "App. B. II. (Vol. III., p. 301)."
,, 200, omit note 40.
,, 203, for note 43 read "App. B. IV. D."
,, 205, for note 45 read "App. B. IV. A."
,, 208, for note 50 read "App. B. IV. E."

ERRATA.

Page 214, for note 59 read " App. B. IV. C."
,, 232, note 86, for " see App. B. VII." read " given, in an abridged form, in App. B. IV. A."
,, 280, for note 159 read " App. B. IV. B."
,, 293, note 195, last line but two, for " supplement I." read " supplement VI."
,, 344, line 26, for " Brocke " read " Brockes."
,, 345, line 19, for " Brocke's " read " Brockes'."
,, 348, line 10, for " Brocke " read " Brockes."
,, 405, line 21, for " der " read " dir."
,, 415, line 9, for the " rhythm only " read " the rhythm alone being identical with that of the theme."
,, 418, line 14, for " *Schalmei* " read " *schalmeien*."
,, 426, line 19, for " Brocke's " read " Brockes' "
,, 468, for note 527 read " ante, p. 357."
,, 536, line 7, for " Enter not into judgment " read " Herr, gehe nicht ins Gericht."
,, 551, note 634, for " No. 3 " read " No. VII."
,, 633, note 785, for " note " read " ante."
,, 671, in the second musical example (line 6) the first of the group of four semiquavers in the lower part should be an F sharp, not a tied E natural.
,, 698, line 18, for " note 3 of Vol. III." read " note 3 of App. A. to Vol. III."
,, 702, line 6, for " sie " read " sei."
,, 702, line 8, for " Sündenkneckt " read " Sündenknecht."
,, 715, line 32, for " sie " read " sei."
,, 717, line 7, for " Academy for Singing " read " Singakademie."

VOL. III.

Page 11, after note 7 insert " and will be found in an abridged form in App. B. (to Vol. III.), No. I."
,, 109, note 151, line 4, for (4 A.) read (VIII. A.)
,, 113, note 161, line 2, for (4 B.) read (VIII. B.)
,, 133, note 214, for " App. B. XII." read " App. B. XIV."
,, 239, note 459, last line, for " No. 4 " read " No. IX."